Proceedings of
ASIAN MATHEMATICAL CONFERENCE 1990

Proceedings of ASIAN MATHEMATICAL CONFERENCE 1990

University of Hong Kong — August 14 – 18, 1990

Editors

- **Li Zhong**
  Peking University
- **K. P. Shum**
  The Chinese University of Hong Kong
- **C. C. Yang**
  The University of Science and Technology, Hong Kong
- **Yang Le**
  Academia Sinica, Beijing

World Scientific
Singapore • New Jersey • London • Hong Kong

Published by

World Scientific Publishing Co. Pte. Ltd.
5 Toh Tuck Link, Singapore 596224
USA office: 27 Warren Street, Suite 401-402, Hackensack, NJ 07601
UK office: 57 Shelton Street, Covent Garden, London WC2H 9HE

British Library Cataloguing-in-Publication Data
A catalogue record for this book is available from the British Library.

ASIAN MATHEMATICAL CONFERENCE, 1990
Proceedings of the Conference

ISBN-13 978-981-02-0749-6
ISBN-10 981-02-0749-2

Preface

This Proceedings collects 94 papers which were presented in the first Asian Mathematical Conference held in Hong Kong 8–13, August 1990.

The First Asian Mathematical Conference was jointly organized by the Southeast Asian Mathematical Society and Hong Kong Mathematical Society in collaboration with many Asian mathematical sociaties from China, Japan, Taiwan, Korea, India, Pakistan, Mongolia, Russia, and Bangladesh. The conference was a success. There were over 400 participants from all over the world and most of the invited addresses were delivered by the mathematicians from Asia. The atmosphere of the conference was good, it provides an excellent oppportunity for Asian mathematicians to get acquainted with each other and to discuss mathematical problems of a wide range of interest. Ideas and new directions always originated from exchange and discussions of this sort.

Because of the success of this conference, the next Asian Mathematical Conference has been scheduled to be held in Bangkok, Thailand in 1995. In order to keep up the momentum, it has also been suggested by many participants that the Asian Mathematical Conference should be held once every five years. This implies that the mathematicians in Asia will have a regular opportunity to meet, to discuss and to enhance their mathematical knowledge. This series of conferences will no doubt promote mathematical research activites throughout Asia.

SEAMS (that is, the Southeast Asian Mathematical Society) was first established under the auspices and with the prompting of ASIHL — The Association of Southeast Asian Institute of Higher Learning. Members of SEAMS now include Hong Kong, Singapore, The Philippines, Thailand, Indonesia, Vietnam, Brunei, and Malaysia. It is forseeable that the membership of SEAMS will be increased soon. The aim of the Society is to promote cross cultural and international links for mathematicians in the region.

SEAMS has surprised many of our friends with its strength and vigor since it was founded in 1972. In fact, its impact is no longer limited to the Southeast Asian region, it now extends over a much wider area, encompassing the whole of Asia. The successfulness of the First Asian Mathematical Conference is a concrete evidence of the influence which SEAMS has been quietly exercising on mathematics for many years. In future, we would like not only to host conferences but also encourage colleagues to take part in international academic activities, to speak and present papers, and to develop research as a consequence.

On behalf of SEAMS, I would like to thank Hong Kong Mathematical Society for their hard work in organizing AMC 90. I would also like to express out thanks and appreciation to the following organizations and societies for giving us financial support: UNESCO Offices at Jakarta and Paris; Wei Lun Foundation; K. C. Wong Educational Foundation; Croucher Foundation; Pei Wah Education Foundation; Hang Seng Bank Ltd; Wing Lung Bank Ltd; The Chinese Mathematical Society located at Beijing; The Mathematical Society of Japan; the Mathematical Society located at Taipei; and the Hong Kong Mathematical Society. Without their generous support, the First Asian Mathematical Conference will never be so successful.

K. P. Shum
President of SEAMS (90–91)

Contents

Proceedings of

ASIAN MATHEMATICAL CONFERENCE 1990

Proceedings of the Asian Mathematical
Conference, 1990

Equations-solving and Theorems-proving Zero-set Formulation and Ideal Formulation*

WU WEN-TSUN
Institute of Systems Science
Chinese Academy of Sciences
Beijing 100080, China

§1. Two Kinds of Problems

There are two main kinds of problems in mathematics: viz. equations-solving and theorems-proving. While the latter was of Greek origin, the former was of no doubt originated and well-developed in ancient China. In appearance these two kinds of problems are little connected to each other. This is however not so, at least in some important cases. Let us restrict ourselves for the two kinds of problems to the simplest, nevertheless the most fundamental, case where the equations or the hypothesis and conclusions of theorems are all expressible in terms of polynomials (abbr. pols) in a ring $K[X_1, \cdots, X_n] = K[X]$ with K a field of characteristic 0. For a polynomial set (abbr. polset) PS and a polynomial (abbr. pol) G both in $K[X]$ let us denote by Zero(PS/G) the totality of zeros of the equations $PS = 0$ for which $G \neq 0$ in arbitrary extension field of K. It will be denoted simply by Zero(PS) if G is lacking. The set Zero(PS) is just the usual algebraic variety over the basic field K associated to the ideal Ideal(PS) with PS as basis. The problem of solving the equations $PS = 0$ is then equivalent to the explicit determination of Zero(PS) as a point set or the precise structure of the corresponding algebraic variety. Suppose now a theorem $T = \{HYP,\ \text{CONC}\}$ is given with hypothesis $HYP = 0$ and conclusion CONC$= 0$, where HYP is a polset and CONC is a pol. Then a Zero in Zero(HYP) is just a geometrical configuration verifying the hypothesis of the theorem T in question. To prove the theorem T consists of then verifying whether CONC$= 0$ on Zero(HYP) or more generally of determining on what part of Zero(HYP) one would have CONC$= 0$. If one knows the precise structure of Zero(HYP), then this would be relatively easily done. In this sense theorems-proving may be thus considered merely as a particular application of the method of equations-solving.

For both equations-solving and theorems-proving, there are now two ways of formulation and two ways of approach through the notions of *zero-sets* and *ideals* resp. We shall discuss these two ways in successive paragraphs.

---

* The present paper is partially supported by NSFC Grant JI85312

§2. Zero-set Formulation

For the determination of the precise structure of the zero-set Zero(PS) of a polset PS in $K[X]$ let us consider first the following particular case. First of all let us arrange the variables $X_1, \cdots, X_n$ in the above natural way. For a non-zero non-constant pol P in $K[X]$ with leading variable X_c the subscript c will be called the class of P, and the the degree of P, considered as a pol in X_c alone, will then be simply called the degree of P. For a non-zero constant pol the class and degree will then simply be set to be both 0. For a pol P with class $c > 0$ and degree d we may then write it in the following standard form:

$$P = I * X_c^d + \text{lower degree terms in} \quad X_c. \tag{2.1}$$

In the above form (2.1) all the coefficients of X_c-powers will be themselves pols in the ring $K[X_1, \cdots, X_{c-1}]$. In particular, the leading coefficient I, which will play an important role in our method, is called the initial of pol P. For a non-zero constant pol the initial will be left undefined.

We now introduce a partial ordering among the non-zero pols in $K[X]$ according to their leading variables, or more precisely first according to their classes and then according to their degrees. In particular any non-zero constant pol will be one of lowest order.

With these notions a polset which may be arranged in a sequence of of pols in steadily ascending order will be called an ascending set (abbr. *asc-set*). The asc-set consisting of a single non-zero constant pol will be said to be *trivial* . Consider now a *non-trivial* asc-set ASC with pols arranged in ascending order, say

$$(ASC)\ F_1,\ F_2,\ \cdots,\ F_r. \tag{2.2}$$

Let the classes of F_i be c_i and the initials be I_i. Then we have

$$0 < c_1 < c_2 < \cdots, < c_r.$$

Sometimes we exige that for any pair $j < i$, the degree of X_j in I_i is less than the degree of F_i. The asc-set will then be said to be in *strong form* or of *type 1* , to distinguish from that with this condition neglected which will be said to be in *weak form* or of *type 0* .

For an asc-set ASC and a pol P let us divide P successively by the pols in ASC in reverse order to remove the leading variables as much as possible. The final result thus obtained will be called the *remainder* of P w.r.t. the asc-set ASC, to be denoted as *Remdr*(P/ASC). It is clear that for a trivial ASC the remainder is equal to 0, while for a nontrivial asc-set ASC as given by (2.2) there will be a formula of the following form:

$$\text{Remdr}(P/ASC) = \text{PROD}_i\ I_i^{s_i} * P \mod ASC \tag{2.3}$$

In (2.3) the product $PROD$ runs over $i = 1, \cdots, r$, and s_i are some non-negative integers which may be chosen as small as possible to make the notion of remainder to be unique, though not an indispensable one for the theory. Moreover, let us write for simplicity Y_i for X_{c_i} and the pols F_i in the standard forms

$$F_i = I_i * Y_i^{d_i} + \text{ lower degree terms in } Y_i, i = 1, \cdots, r. \tag{2.4}$$

We see from the very form that for the non-trivial asc-set ASC the zero-set Zero(ASC) may be considered as well-determined, while for a trivial asc-set ASC it is clear that Zero(ASC) $=$ *empty*. In view of this our method of solving $PS = 0$ for an arbitrary polset PS consists in reducing the problem to the above asc-set case, cf. e.g. [WU3,6]. The result may be formulated in the form of following

Well-Ordering Principle. There is an algorithm which permits to determine for any polset PS an asc-set CS in a finite number of steps such that

$$\text{Zero}(PS) \subset \text{Zero}(CS), \text{ and} \tag{2.5}$$

$$\text{Remdr}(P/PS) = 0, \text{ for all } P \text{ in } PS. \tag{2.6}$$

Moreover, let I_i be the initials of pols in CS, and J be the product of all such initials, then we have

$$\text{Zero}(CS/J) \subset \text{Zero}(PS) \subset \text{Zero}(CS), \tag{2.7}$$

$$\text{Zero}(PS) = \text{Zero}(CS/J) + SUM_i \text{ Zero}(PS[+CS] + \{I_i\}). \tag{I}$$

In (I) the square bracket [] means that we may add CS or not subject to our choice.

The asc-set CS of particular interest asoociated to the given polset is characterized by the two properties (2.5) and (2.6). It is called a *characteristic set* (abbr. *char-set*) of PS. As already indicated the zero-set Zero(CS/J) is well-determined. If we proceed further for each summand in (I) then we will arrive finally to the following

Zero-Decomposition Theorem (Special Form). There is an algorithm which permits to determine for any polset PS a finite number of asc-sets ASC_j in a finite number of steps such that

$$\text{Zero}(PS) = SUM_j \text{ Zero}(ASC_j/J_j), \tag{II}$$

in which each J_j is the product of all initials of pols in the corresponding asc-set ASC_j.

Consider a non-trivial asc-set ASC as given by (2.1) with pols F_i in the form (2.4). We shall say that ASC is *irreducible* if the pols F_i are irreducible in successive extended fields obtained from the preceding one by adjoining by means of the equations $F_i = 0$. The successive solutions of $F_i = 0$ will give rise to a zero GZ of Zero(ASC/J) $\subset$ Zero(ASC) which will be called a *generic zero* of the irreducible asc-set ASC. Let us recall that a point $x' = (x'_1, \cdots, x'_n)$ with all x'_i in some extension field K' of K is a *specialization* of a point $x'' = (x''_1, \cdots, x''_n)$ with all x''_i in some extension field K'' of K if for any pol P in $K[X]$, the equation $P(x') = 0$ would imply $P(x'') = 0$. The totality of all specializations of the generic point GZ is then an irreducible algebraic variety with GZ as a generic point in the ordinary sense which will be denoted by $Spec(GZ)$, or Var[ASC] since GZ is determined by ASC. Note that here square bracket has been used. In fact, Var[ASC] is a subvariety of the algebraic variety Zero(ASC) but usually only a proper one. Moreover this variety is usually entirely different from the one associated to the ideal Ideal(ASC) with ASC as a basis. This latter one may well be denoted by Var(ASC) which is just the variety Zero(ASC) or Zero (Ideal(ASC)).

With these terminologies and notations we have now the following

Zero-Decomposition Theorem (General Form). There is an algorithm which permits to determine for any polset PS a finite number of irreducible asc-sets IRR_k in a finite number of steps such that

$$\text{Zero}(PS) = \text{SUM}_k \ \text{Var}[IRR_k], \tag{III}$$

the summation being uncontractible and unique up to an order of summands. Moreover, finite basis of the ideal associated to each irreducible variety Var$[IRR_k]$ may be determined by computations if required.

The above formulas (I)–(III) give the precise structure of Zero(PS) and form the underlying principles of our method to solve arbitrary systems of multivariate polynomial equations $PS = 0$. It is in essence a method of elimination which may be resorted to the creations of the Chinese scholars in our ancient time. However, for the above presentation in modern and systematic form we owe in fact in the main to the american mathematician J.F.Ritt. Most of the terminologies, not to mention the techniques, may be found in his two classic monographs [R1] and [R2].

Applications of our method have been made on various sorts of problems in pure and applied mathematics as well as other related disciplines beyond mathematics. The most gigantic scale application is on mechanical theorem proving (abbr. MTP) of elementary geometries. In fact, let PS be the hypothesis polset HYP of a theorem $T = \{HYP, \text{CONC}\}$ in question and let us decompose Zero(HYP) as in (III), for example. Then we have the following

General MTP-Principle. A theorem $T = \{HYP, \text{CONC}\}$ is true on the whole component Var$[IRR_k]$ if and only if

$$\text{Remdr}(\text{CONC}/IRR_k) = 0. \tag{2.8}$$

There are also MTP-Principles corresponding to (I) and (II), which, though less general and not so precise, are yet easier to apply. In fact, actually up to now only such special forms of principles have been used in MTP which are already so powerful that hundreds of non-trivial theorems have already been proved or even discovered in this way. However, there are already survey papers and even books devoted to MTP, cf.e.g. [CH], [W-G], and [WU1,2]. We shall therefore leave this aside and satisfy ourselves in giving only descriptions of other applications, see (a)–(i) below.

(a) Factorization of Multivariate Pol in a Finite Extension Field ([H-W]).

Factorization plays an important role in computer algebra. Powerful methods of factorization as developed by S.H. Wang and others have even been implemented in various softwares. However, these implemented ones deal principally with the factorization of a multivariate pol in the rational field Q. On the other hand Hu and Wang, during a course on mathematics mechanization given by the present author in the Graduate School belonging to our Academy in year 1984, have discovered a factorization method which is valid for pols in an arbitrary finite extension field of Q. This factorization method is based on our polynomial equations-solving method and is quite different from the known ones. It has been tested for examples given in papers of S.H.Wang by hand calculations alone and seems to be quite efficient. Unfortunately it has not yet been turned into programs to be run on the computers. For more details cf. their papers as well as some mimeographed lecture notes of the present author (unpublished).

(b) Discriminant Systems ([Gao], [WDWu]).

The Petrov classification of space-time universe in relativity theory raised the problem, as a generalization of discriminant, of finding the criterion for a quartic equation to have roots with partition into sets of equal ones of type (1,1,1,1), (1,1,2), (1,3), or (2,2). Formerly it was solved in making use of explicit form of solutions of a quartic equation which cannot be generalized to equations of higher degree n. Gao solved this problem for general n and gave an algorithm for getting general criteria in the form of *discriminant systems* . Explicit forms of such discriminant systems were also given by Gao for $n \leq 5$. Some complements have also been given by W.D.Wu.

(c) A Problem of Ulam ([Liu]).

The Ulam problem in question is to find in the complex number field monic pols $f(x)$ in a single variable x such that the roots of $f(x) = 0$ are just the non-leading-coefficients of $f(x)$. In other words, the problem is to find such $f(x)$ in complex coefficients with

$$f(x) = x^n + x_1 * x^{n-1} + \cdots + x_n = \mathrm{PROD}_i\ (x - x_i).$$

Liu proved that for each n there are just $n!$ such pols (counting with multiplicities) and have determined such concrete pols up to $n = 4$. For example, for $n = 3$ the only such pols are 6 in number, as listed below:

$$x^3, x^3 + x^2 - x - 1 = (x-1) * (x+1)^2, x^3 + x^2 - 2 * x = x * (x-1) * (x+2),$$

and

$$x^3 + r_i * x^2 - x/r_i + (1/r_i - 2 * r_i) = (x - r_i) * (x + 1/r_i) * (x - 1/r_i + 2 * r_i), i = 1, 2, 3.$$

In the last three pols r_i are the three different roots of the equation $2 * r^3 + 2 * r^2 - 1 = 0$.

(d) Geometric Loci Equations ([WU10], [Li]).

Let us consider the geometric locus of a point $X = (X_1, \cdots, X_n)$ verifying some geometrical conditions $h_i(X,p) = 0, i = 1, \cdots, r$, with $p = (p_1, \cdots, p_t)$ the parameter set. The determination of the locus reduces to the problem of eliminating p from the equations $h_i = 0$. Let us denote by Proj the projection of the space $K^{n+r}(X_1, \cdots, X_n, p_1, \cdots, p_t)$ to the space $K^n(X_1, \cdots, X_n)$, and by HS the polset $\{h_1, \cdots, h_r\} \subset K[X,p]$. Then the problem amounts to the determination of the set $PZ = \text{Proj Zero}(HS)$. In contrast to what one unconciously believes in the usual treament of geometrical loci, PZ is not an algebraic variety in the ordinary sense and cannot be expressed in the form of Zero(PS) for some polset PS in $K[X]$. On the other hand we had pointed out that PZ can always be expressed in the form

$$PZ = \text{SUM}_j \ \text{Zero}(ASC_j/G_j),$$

for some asc-sets ASC_j and pols G_j all in $K[X]$ which are determined by some algorithm. Simplifications and concrete examples were given in the paper of Li.

(e) Implicitization of Parametric Forms ([LI], [G-C]).

An algebraic variety over basic field K is by definition expressed in implicit form $PS = 0$ with $PS \subset K[X]$. In circumstances it may also be expressed in parametric form

$$X_i = P_i/Q_i, i = 1, \cdots, n$$

with $P_i, Q_i \subset K[p_1, \cdots, p_t]$, where $p = (p_1, \cdots, p_t)$ is the parameter set. The geometric modelling and other related problems raised the question of turning an algebraic variety in parametric form to one in implicit form. This is a special case of (d) with $h_i(X,p) = Q_i * X_i - P_i$ and is thus in general impossible. However, the problem is of great interest in itself, besides its applications in geometric modelling et al. In this respect Gao and Chou had made a detailed and deep study of various theoretical questions, especially in the case $n = 2$ or 3, cf. their forthcoming paper in MM-Preprints, No.6.

(f) A Pyramid Problem — Beijing Theorem ([DKWa])

Consider a regular pyramid P of square base. J.Z.Zhang suggested the problem about the possibility of cutting the pyramid into a regular pentagon planar section. D.K.Wang solved the problem by means of our general method with the result that this is possible when and only when the height of the pyramid is equal to half of the diagonal of the base square. Moreover, in that case there will also planar sections of the form of a regular star outside the pyramid. Wang had also extended his results to non-regular pyramids of rhombic base. We would like to mention that in the regular square-base case Deakin of Australia had given a simple direct proof and had called the above result a *Beijing Theorem* .

(g) Automatic Discovering of Unknown Relations ([WU4,5]).

By deliberately arranging the variables involved in some definite order, our general method will furnish us a means of discovering the explicit form of unknown relations without *a priori* knpwledge of what the actual form may be. As a typical example we may mention the automatic discovery of Heron-Qin formula of the area of a triangle in terms of its 3 sides. We may also mention the extensions to the volume of a tetrahedron in terms of its 6 edges, and the area of a hyperbolic triangle in terms of its 3 sides. The above principle of automatic discovering has naturally a wide range of possible applications. It would be interesting to see whether the method may also be applied to the determination of the volume a hyperbolic tetrahedron in terms of its 6 edges. We remark that such a determination would involve highly transcendental functions like the Lobachevski ones. However, it seems that this is not an unsurmountable barricade to the application of our method.

(h) Inequalities Proving ([WU7]).

The proving of an inequality may be turned into the determination of the minimax value or the least-greatest value of a function under some restricted equality conditions. The latter one may in turn be turned, by means of Lagrangian multipliers, into the problem of equations-solving. It follows that our method will be applicable so far the final equations turn to be all polynomial ones. This principle has been applied to prove various kinds of inequalities, algebraic, trigonometric, or geometrical. It may also be applied to mechanical proving of geometry theorems involving geometrically order relations or algebraically inequalities which is yet to be done.

(i) Non-linear Programming ([TJWu], [WU9]).

The method described in (h) may well be applied to non-linear programming. Some test problems for non-linear programming occuring in the literature have been solved in this way entirely different from the usual known numerical methods. A representative problem of chemical equilibrium has also been treated in some detail which throws some light about the feasibility, efficiency, etc. of our general method. Further tests and concrete problems are in course of studying.

As problems in physical sciences and technology lead quite often to the problems of polynomial equations-solving, it is natural that our general method will have a wide range of applications far beyond mathematics. Thus our method has been applied to physics, to inverse kinematic equations of robotics, to Steward platforms, to automic control, to limit cycles of planar differential systems, and many others to come. However, we shall leave aside all these for later occasions.

§3. Ideal Formulation

For polsets PS the modern algebraic geometry pays much attention to the properties of the algebraic variety Zero(PS) via the study of the ideal Ideal(PS), rather than the actual determination of Zero(PS). The computer scientists are however much interested on the computational aspects of Ideal(PS). Such computational problems were studied by Hermann early in 1926, cf. [H]. Among such problems we may cite in particular the following one of fundamental importance:

Membership Problem. Given an ideal Ideal(PS) of basis $PS \subset K[X]$, to decide whether a given pol $P \in K[X]$ belongs to Ideal(PS) or not.

Though the above problem, along with many others, had been solved in Hermann's paper, it was settled in a much more elegant way by the introduction by Buchberger in 1965 of some other basis of Ideal(PS), the so-called *Groebner Basis* (abbr. GB) which reflects the intrinsic properties of the ideal in a simple manner and enjoys a lot of beautiful properties. The membership problem was settled by the following

Buchberger Criterion. A pol P will belong to an ideal Ideal(PS) if and only if P will be reduced to 0 by simple divisions by pols in the Groebner basis of Ideal(PS).

Buchberger and his followers had also proved the following more general

Representation Theorem. Let GB be the Groebner basis of an ideal Ideal(PS). Then w.r.t. GB any pol $P \in K[X]$ may be represented, unique up to syzygies of pols G_i in GB, in the form

$$P = SUM_i\ A_i * G_i + R, \tag{3.1}$$

In (3.1) the pol R, called the rest of P w.r.t. GB, has no terms divisible by the leading terms of any G_i and the vanishing of R is necessary and sufficient for P to belong to Ideal(PS).

Besides the algorithm of Buchberger for the introduction of Groebner basis of an ideal, there are others to introduce the same basis in different ways and under different names. One was due to the French school of informaticians in seventies under the name of *standard basis* (abbr. *SB*) which may be dated back to the well-known papers of Hironaka on desingularity of algebraic varieties in 1964, cf.e.g.[Br],[Br-G],[Gal],[Giu]. (The author would like to take this opportunity to express his gratitudes to mathematicians in Honkong and Taiwan in proccuring for him some zeroxed papers of the French mathematicians inaccessible to him in China.) Another one was due to the present author. During the study of differential ideals along the line of development of Riquier, Janet, E.Cartan, Ritt, and Thomas on partial differential equations, the author discovered that some basis enjoying some well-behaved properties may be associated to differential ideals of certain special type. In further specializing the differential ideals we get some basis of polynomial ideals which turn out to be just the Groebner basis of the ideal, cf.[WU11]. Our algorithm seems to be closely related to that of the French school but the exact relations are yet not quite clear. In any way Wu Wenda has shown that our method of construction of the Groebner basis seems to have some advantages in actual computations, cf. his forthcoming paper in MM-Preprints, No.6. Besides, our method permits to give a complement to the above Representation Theorem in that the representation may be so determined to be completely unique without intervention of syzygies.

The Groebner basis of an ideal has grat importance in theoretical studies of computer algebra and related subjects. Besides, the basis has also a wide range of applications. In particular in [BU] of Buchberger it was shown how the *GB* of an ideal will furnish a method of polynomial equations-solving which is particularly successful in the so-called 0-dimensional case. In the eightieth Kutzler-Stifter, Kapur, Chou, et al have also shown how to apply the *GB*-method to *MTP*, cf.e,g.[K-S],[K],[CH]. The literature about *GB*, both for theoretical studies and for practical applications, is too vast to discuss here. Instead of this a comparison of the ideal-theoretic approach here with the zero-set approach as described in Sect 2 may be of some interest.

Monomials in variables $X = (X_1, \cdots, X_n)$ may naturally be arranged in some definite order, say the lexicographic one. The pols in $K[X]$ may then be partially ordered according to their leading monomials, in contrast to the partial ordering according to leading variables in the case of zero-set approach. While for the zero-set formulation we have the notion of *asc-set* consisting of of pols well-arranged in ascending order, for the ideal formulation we have the notion of *autoreduced set* consisting of pols no one of which will have its leading monomial a multiple of that one of any other pol in the set. Now w.r.t. an asc-set any pol will have a *remainder* obtained by successive divisions by pols in the asc-set in order to remove as much as possible the leading variables. On the other hand w.r.t an autoreduced set any pol will have a *rest* obtained by successive divisions by pols in the autoreduced set in order to remove as much as possible the leading monomials. Suppose given an arbitrary polset *PS*. In incessantly choosing from the polsets asc-sets and adjoining remainders of remaining pols w.r.t. the asc-set chosen we will get ultimately some asc-set of particular interest, viz. the *char-set* of the given polset *PS*. On the other hand, in incessantly choosing from the polsets autoreduced sets and adjoining rests of remaining pols w.r.t. the autoreduced set chosen we will get ultimately some autoreduced set of particular interest, viz. the *Groebner basis* of the ideal associated to the given polset PS. The final char-set of a polset is connected to the polset via the zero-sets by some formulas as exhibited in the Well-Ordering Principle of Sect 2. On the other hand the final Groebner basis GB of a given polset PS is connected to PS in having the same generated ideal, viz.

$$\text{Ideal}(GB) = \text{Ideal}(PS).$$

In conclusion, there are two different ways of approach to equations-solving and theorems-proving in the polynomial case both of which have a wide range of applications, theoretical as well as practical. It is difficult to say which way is the better in each separate case. What we can say is that each way has its own merits and its own insufficiencies and the two ways should be considered as complementary to each other.

References

[Br] J.Briancon, *Weierstrass prepare a la Hironaka*, Asterisque, 7–8 (1973), 67–76.

[Br-G] J.Briancon & A.Galligo, *Deformations de points de R^2 ou C^2*, Asterisque, No.7–8 (1973), 129–138.

[BU] B.Buchberger, *Ein algorithmitisches Kriterium fur die Losbarkeit eines algebraishes Gleichungssystems*, Aeq.Math, 4 (1970), 374–383.

[CH] —, *Mechanical geometry theorem-proving*, Reidel, (1988).

[Gal] A.Galligo, *Algorithmes de calcul de base standards*, Calsif 3 (1982-3), 43–69.

[Gao] Gao Xiao-shan, *The discriminant systems of unary equations and their computation*, MM-Res.Preprints, No.1 (1987), 13-32.

[Giu] M.Giusti, *Bases standard et profondeur*, Calsif 3 (1982-3), 71–74.

[H] G.Hermann, *Die Frage der endlich vielen Schritte in der Theorie der Polynomideale*, Math.Annalen, 95 (1925), 736-788.

[H-W] Hu Sen & Wang Dongming, *Fast factorization of polynomials over rational number field or its extension fields*, Kuxue Tongbao, 31 (1986), 150–156.

[K] D.Kapur, *Using Groebner bases to reason about geometry problems*, J. Symb. Comp., 2 (1986), 399–408.

[K-S] B.Kutzler & S.Stifter, *On the application of Buchberger's algorithm to automated geometry theorem proving*, J.Symb.Comp., 2 (1986), 389–398.

[Li] Z.M.Li, *Automatic implicitization of parametric objects*, MM-Res. Preprints, No.4 (1989), 54–62.

[Liu] Z.J.Liu, *On one Ulam's fixed point problem*, MM-Res.Preprints, No.5 (1990), 39–46.

[R1] J.F.Ritt, *Differential equations from the algebraic standpoint*, Amer. Math. Soc., 1932.

[R2] —, *Differential algebra*, Amer. Math. Soc., 1950.

[DKWa] D.K.Wang, *A mechanization proving of a group of space geometry problems*, MM-Res.Preprints, No.5 (1990), 66–81.

[W-G] D.M.Wang & X.S.Gao, *Geometry theorems proved mechanically using Wu's method — Part on Euclidean geometry*, in MM-Res. preprints, No.2 (1987), 75–106.

[TJWu] T.J.Wu, *Some test problems on applications of Wu's method in nonlinear programming problems*, MM-Res.Preprints, No.6 (1991).

[WDWu] Wu Wen-da, *A note to the discriminant systems for the unary equations*, MM-Res. preprints, No.3 (1989), 33–38.

[WU1] Wu Wen-tsun, *Basic Principles of Mechanical Theorem Proving in Geometries (Part on Elementary Geometries)*, (in Chinese), Science Press, (1984).

[WU2] —, *Basic principles of mechanical theorem-proving in elementary geometries*, J. Sys. Sci. & Math. Scis., 4 (1984), 207–235. Re-published in J. of Automated Reasoning, 2 (1986), 221–252.

[WU3] —, *On zeros of algebraic equations — an application of Ritt principle*, Kexue Tongbao, 31 (1986), 1–5.

[WU4] —, *A mechanization method of geometry and its applications*, I. Distances, areas, and volumes, J. Sys. Sci. & Math. Scis.,6 (1986), 204-216

[WU5] —, *A mechanization method of geometry and its applications, I. Distances, areas, and volumes in euclidean and non-euclidean geometries*, Kuxue Tongbao, 32 (1986), 436–440

[WU6] —, *A zero structure theorem for polynomial-equations-solving and its applications*, MM-Res. Preprints, No.1 (1987), 2–12.

[WU7] —, *A mechanization method of geometry and its applications, 3. Mechanical proving of polynomial inequalities and equations-solving*, MM-Res. preprints, No.2, (1987), 1–17.

[WU8] —, *On the foundation of algebraic differential geometry*, MM-Res. preprints, No.3 (1989), 1–26. Also in Sys. Sc.& Math. Scis, 2(1989), 289–312.

[WU9] —, *On the chemical equilibrium problem and equations-solving*, MM-Res. preprints, 4 (1989), 22–39. Also in Acta Math.Sci., 10 (1990), 361–374.

[WU10] —, *On a projection theorem of quasi-varieties in elimination theory*, MM-Res. preprints, No.4 (1989), 40–48. Also in Chin. Ann. of Math., 11B(1990), 220–226.

[WU11] —, *On the construction of Groebner basis of a polynomial ideal based on Riquier-Janet theory*, MM-Res.Preprints, No.5 (1990) 5–22. Also in Sys. Sci.& Math.Scis., 4 (1991), 193–207.

Proceedings of the Asian Mathematical
Conference, 1990

Uniquely Path Colorable Graphs

ESPERANZA B. ARUGAY
College of Science & Mathematics
MSU-Iligan Institute of Technology
Iligan City, Philippines 9200

§1. Introduction

The graph under consideration throughout our discussion is a simple undirected graph, i.e., a finite graph without multiple edges or loops. Let G be a simple graph. The path chromatic number of G, denoted by $\chi_\infty(G)$ is the minimum number of colors the vertices of G may be given so that each monochromatic color class induces a path or paths. Let $\chi_\infty(G) = t$. G is said to be uniquely path colorable if there is only one partition of its vertices into t sets such that each set induces a path or paths. In short, we shall call the partitioning of the vertices as path partition.

One example is the one shown in Fig. 1.

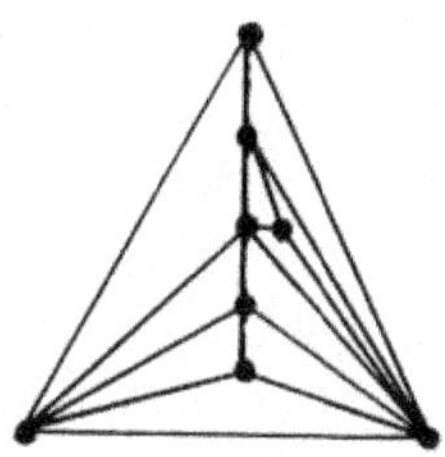

Fig. 1. A uniquely path colorable graph with path chromatic number equal to 2

In verifying that a graph is uniquely path colorable, it is sufficient to show that vertices of G has a unique path partition.

The next theorem resembles that of John Mitchem [5].

Theorem 1. *Let G be a graph of order p. G is uniquely path colorable and $\chi_\infty(G) = t$, where $1 \leq t < p$ if and only if there is a unque partition of its vertices into t sets such that each set induces a path or linear forest.*

Proof. $\Longrightarrow$: This follows from the definition of uniquely path colorable graphs.

$\Longleftarrow$: Suppose there is a unique path partition of $V(G)$ into t sets and that $\chi_\infty(G) \neq t$. We now assume that $\chi_\infty(G) = m < t$ and let $V_1, V_2, \cdots, V_m$ be a path partition of $V(G)$. Note that at most one of the V_i, say V_m, has less than two points. From the $V_i, i =$

$1, 2, \cdots, m-1$, form a partition of $V(G)$ into t sets such that each set induces a path or linear forest. WLOG, let $|V_1| \leq |V_2| \leq \cdots \leq |V_{m-1}|$.

Case 1. $|V_1| = t - m + 1$.

Form singleton subsets of V_1. The sets $V_i, i > 1$, together with the sets in the partition of V_1 form exactly t sets. Thus we obtain a partition of $V(G)$ into sets $W_1, W_2, \cdots, W_t$ such that each set induces a path or linear forest. Now, $t < p$ implies that at least one set, say W_j, contains at least two elements. From the partition $W_i, 1 \leq i \leq t$, we form another path partition of $V(G)$ by adding $v \in W_j$ to a W_i with $|W_i| = 1$. Thus we have two distinct path partitions of $V(G)$ into t sets, a contradiction to our hypotheses. This contradiction implies that $\chi_\infty(G) = t$.

Case 2. $|V_1| > t - m + 1$.

Partition V_1 into $t-m+1$ sets such that each set induces a path or linear forest. Clearly this is possible since the components of V_1 are paths. This partition together with the $V_i, i > 1$ form a partition of $V(G)$ into sets $W_i, 1 \leq i \leq t$, such that each set induces a path or linear forest. A similar reasoning as in Case 1 follows. Therefore $\chi_\infty(G) = t$.

Case 3. $|V_1| < t - m + 1$.

Partition V_1 into singleton sets and V_2 into sets such that the partitions of V_1 and V_2 together with the sets $V_i, 3 \leq i \leq m$, form a partition of $V(G)$ into t sets such that each set induces a path or linear forest. If this doesn't occur, continue the process by partitioning V_3 until we obtain a partition of $V(G)$ into sets $W_i, i = 1, 2, \cdots, t$ such that each set induces a path or linear forest. The same reasoning follows as in Case 1. Therefore, $\chi_\infty(G) = t$

Hence, in all cases, we have shown that $\chi_\infty(G) = t$.

Lemma 1. *Let G be uniquely path colorable graph with $\chi_\infty(G) = t$. For $m \leq t$, the union of any m of the t sets induces a uniquely path colorable subgraph G^* of G with $\chi_\infty(G^*) = m$.*

Proof. Suppose there exists an $m, m < t$ such that the union of these m color classes induces a graph G^* which is not uniquely path colorable. WLOG, let $V_1, V_2, \cdots, V_m$ be such sets where $V(G^*) = \cup V_i, i = 1, 2, \cdots, m$. Since G^* is not uniquely path colorable, there exists another partition $V_1^*, V_2^*, \cdots, V_m^*$ of $V(G^*)$ in which each subset induces a path or linear forest. Thus $V_1^*, V_2^*, \cdots, V_m^*, V_{m+1}, V_{m+2}, \cdots, V_t$ is another path partition of $V(G)$ contradicting uniqueness. Therefore, G^* must be uniquely path colorable and from Theorem 1, $\chi_\infty(G^*) = m$.

The next theorem gives us a necessary condition for a graph to be uniquely path colorable. This significant result will be used later to formulate sufficient conditions for a graph to be uniquely path colorable.

Theorem 2. *If G is uniquely path colorable with $\chi_\infty(G) = t, t \geq 2$, and $V_1, V_2, \cdots, V_t$ is a path partition of $V(G)$ such that $|V_1| \leq |V_2| \leq \cdots \leq |V_t|$, then $|V_1| \geq 2$ and $|V_1| + |V_2| \geq 7$.*

Proof. Suppose V_1 contains only one point v. We can color at least one point, say w, in V_2 with the same color as v such that $V_1 \cup \{w\}, V_2 - \{w\}, V_3, \cdots, V_t$ partitions $V(G)$ into subsets each of which induces a graph whose components are paths or linear forest contradicting the uniqueness of G. Thus $|V_1| \geq 2$. To complete the proof, we take up several cases.

Case 1. $|V_1| = 2$. Let $V_1 = \{v_1, v_2\}$.

Subcase 1.1. Assume $|V_2| = 4$. Since $\langle V_2 \rangle$ is a linear forest, V_2 can be partitioned into two independent sets C_1 and C_2 with $|C_1| = |C_2| = 2$. Now, $C_1 \cup \{v_1\}$ and $C_2 \cup \{v_2\}$ both induce graphs whose components are paths only. This contradicts the fact that $\langle V_1 \cup V_2 \rangle$ is uniquely path colorable from Lemma 1. Hence $|V_2| \neq 4$ when $|V_1| = 2$.

Subcase 1.2. Assume $|V_2| = 3$. Again V_2 can be partitioned into two sets C_1 and C_2 such that C_1 contains two independent points and C_2 is a singleton. Now, $C_1 \cup \{v_1\}$ and $C_2 \cup \{v_2\}$ induce graphs whose components are paths only. A similar reasoning follows as in Subcase 1.1. Hence $|V_2| \neq 3$ when $|V_1| = 2$.

Subcase 1.3. Assume $|V_2| = 2$. Let $V_2 = \{w_1, w_2\}$. Then $\{v_1, w_1\}$ and $\{v_2, w_2\}$ both induce graphs whose components are paths only. Using the same argument as in Subcase 1.1, we arrive at $|V_2| \neq 2$ when $|V_1| = 2$.

Thus, the subcases show that when $|V_1| = 2$, then $|V_2| \geq 5$. This implies that $|V_1| + |V_2| \geq 7$.

Case 2. $|V_1| = 3$. Let $V_1 = \{v_1, v_2, v_3\}$.

Subcase 2.1. Assume $|V_2| = 3$. Let $V_2 = \{w_1, w_2, w_3\}$. Then two points of V_1, say v_1, v_2 are non-adjacent as are two points, say w_1 and w_2 of V_2. The sets $\{v_1, v_2, w_3\}$ and $\{w_1, w_2, v_3\}$ induce graphs whose components are paths only. This contradicts again the uniqueness of $\langle V_1 \cup V_2 \rangle$. Hence $|V_2| \neq 3$ when $|V_1| = 3$. Thus $|V_2| \geq 4$ when $|V_1| = 3$. Therefore, $|V_1| + |V_2| \geq 7$.

Case 3. $|V_1| = 4$. Since $|V_1| \leq |V_2|$, then $|V_1| + |V_2| \geq 8 > 7$.

Hence in all cases, if G is uniquely path colorable, $|V_1| \geq 2$ and $|V_1| + |V_2| \geq 7$.

The next corollary is a consequence of the theorem.

Corollary 2.1. *If G is a uniquely path colorable graph with $\chi_\infty(G) \geq 2$ then $|G| \geq 7$.*

We shall now present some methods of constructing uniquely path colorable graphs. The following proposition is analoguous to a result of Harary et al. [4].

Proposition 1. *Let G be a uniquely path colorable graph such that $\chi_\infty(G) = t, t \geq 2$ and $V_1, V_2, \cdots, V_t$ be a unique path partition of $V(G)$. If a new point v is added and joined to all points in $V_i, i = 2, 3, \cdots, t$, then the resulting graph G^* is also uniquely path colorable and $\chi_\infty(G^*) = t$.*

Proof. WLOG, let $V_1, V_2, \cdots, V_t$ be such that $|V_1| \leq |V_2| \leq \cdots \leq |V_t|$. From a previous result, $|V_1| \geq 2$ and V_2 has at least four points. Let v be a new point and joined to all points of $V_i, i = 2, 3, \cdots, t$. Clearly, $\langle V_i \cup \{v\} \rangle$ is not a path or a linear forest and $V_1 \cup \{v\}, V_2, V_3, \cdots, V_t$ partitions $V(G^*)$ into t sets such that each set induces a path or a linear forest. Moreover, any partition of $V(G^*)$ must partition $V(G)$ according to its unique coloring.

In the next theorem, we will see a class of uniquely path colorable graphs G such that $\chi_\infty(G) = 2$. This is analogous to the work of J. Mitchem [5]. First we recall some definitions.

Let G_1 and G_2, be mutually disjoint graphs. The join of G_1 and G_2, denoted by $G_1 + G_2 = \sum_{i=1}^{2} G_{i_1}$ is the graph which consists of $G_1 \cup G_2$ together with all edges joining $V(G_1)$ with $V(G_2)$.

Theorem 3. *Let G_1 and G_2 be disjoint graphs which are paths or linear forests with $|V(G_1)| \geq 2, |V(G_1)| + |V(G_2)| \geq 7$ and $|E(G_i)| \geq 1, i = 1, 2$. Then $G = G_1 + G_2$ is uniquely path colorable and $\chi_\infty(G) = 2$.*

Proof. Clearly, $\chi_\infty(G) = 2$ and $V(G_1) = V_1, V(G_2) = V_2$ is a path partition of $V(G)$.

We now claim that this is unique. Suppose C_1 and C_2 is another path partition of $V(G)$. For every pair V_i, $C_j, i = 1,2, j = 1,2$, neither one is a subset of the other. To see this, WLOG, suppose $V_1 \subset C_1$, then there exists a vertex $c \in C_1 - V_1$. Since V_1 has an edge, say $[v,w], \langle\{v,w,c\}\rangle$ form a cycle in $\langle C_1\rangle$. A contradiction to the fact that C_1 induces paths only. Hence no V_i is a subset of C_j. Furthermore, if $C_j \subset V_i$, say, $C_1 \subset V_1$, then $V_2 \subset C_2$ which is impossible as shown above. Therefore, no C_j is a subset of V_i. Next, we explore other possibilities.

Case 1. One C_j contains two or more points from each of V_i. This case results to a cycle in $\langle C_j\rangle$. Hence no C_j contains two or more points from each of V_i.

Case 2. One C_j contains a point from some V_i. We will show that this case is impossible. WLOG, suppose C_1 contains one point from V_1, i.e., $|C_1 \cap V_1| = 1$.

Subcase 2.1. $|V_1| = 2$ and $|V_2| \geq 5$.

Since for some i, $|C_i \cap V_2| \geq 3$ then $\langle C_i\rangle$ contains a $K_{1,3}$, a contradiction.

Thus, in this subcase, no C_j contains one point from some V_i when $|V_1| = 2$ and $|V_2| \geq 5$.

Subcase 2.2. $|V_1| \geq 3$ and $|V_2| \geq 4$.

Again if $|C_1 \cap V_2| \leq 2$ then $|C_2 \cap V_2| \geq 2$ and $\langle C_2\rangle$ contains a cycle otherwise $\langle C_1\rangle$ contains a $K_{1,3}$, a contradiction.

Therefore, in all cases, G is uniquely path colorable.

The above results were done in the light of following suggestions of Frank Harary [2].

References

[1] Chartrand, G. and Geller, D., *On Uniquely Colorable Planar Graphs*, J. Combin. Theory, 6(1969), 271–278.

[2] Harary, F., *Conditional Colorability in Graphs*, Graphs and Applications., Proc. Ist. Col. Symp. Graph Theory. Wiley, New York (1985), 127–136.

[3] Harary, F. *Graph Theory*, Addison-Wesley, Reading, MA (1969).

[4] Harary, F., Hedetniemi, S. and Robinson, R., *Uniquely Colorable Graphs*, J. Combin. Theory, 6(1969), 264–270; 9 (1970), 221.

[5] Mitchem, John, *Uniquely k-Arborable Graphs*, Israel J. Math., 10 (1971), 17–25.

Proceedings of the Asian Mathematical
Conference, 1990

Groups and Invariant Rings Associated with Valuation Rings

H.H. BRUNGS
University of Alberta
Edmonton, Alberta
Canada T6G 2G1

1. The notion of a valuation on a field as a mapping into an ordered group (Krull [5]) can directly be extended to a skew field F (Schilling [6]): A valuation on F is a mapping v_{in} into a totally ordered set $\Gamma \cup \{\infty\}$, where Γ is an ordered group and $\infty\infty = \infty g = g\infty = \infty, \infty > g$ for all g in Γ such that the following conditions hold:

1) $v_{in}(a) = \infty$ if and only if $a = 0$;

2) $v_{in}(x + y) \geq \min\{v_{in}(x), v_{in}(y)\}$ for all x, y in F;

3) $v_{in}(xy) = v_{in}(x)v_{in}(y)$ for all x, y in F.

The subset $R = \{a \in F \mid v(a) \geq v(1)\}$ of F is a subring of F that satisfies the following two conditions:

(T) R is total, i.e. $x \in F\backslash R$ implies $x^{-1} \in R$;

(I) $dRd^{-1} = R$ for all $0 \neq d \in F$, i.e. R is invariant in F.

Conversely, given a total and invariant subring R of the skew field F, then $H(R) = \{aR \mid 0 \neq a \in F\}$ is an ordered group with $aRbR = abR$ as operation and $aR \leq bR$, if and only if $aR \supseteq bR$, defining the order. The corresponding valuation maps 0 to $\infty, 0 \neq a$ in F to aR.

However, there can exist total subrings B of F which are not necessarily invariant. For a division algebra D finite dimensional over its center K, $[D : K] = n^2$, we say that B, a total subring of D, is an extension of V, a total subring (i.e. a valuation subring) of K, if $B \cap K = V$.

In [1] the following result was proved:

Theorem 1. a) *The valuation ring V of K has an extension B in D if and only if the set S of elements in D integral over V is a subring.*

b) *Any two extensions of V in D are conjugate in D and the number m of such extensions is $\leq n$.*

It was also proved that in the case where S is a subring and $B = B_1, \cdots, B_m$ are the extensions of V one has $S = \bigcap_{i=1}^m B_i = \bigcap dBd^{-1}, 0 \neq d \in D$. Even though B is not necessarily invariant one can associate with B the group $\Gamma = \{aB \mid 0 \neq a \in F \ni aB = Ba\}$ which can be totally ordered as before and which can be considered as a group containing the value group Γ_0 of V.

As a special case of a result proved for more general valuation rings in central simple algebras in [7] (see [3]) one obtains the following equation:

$$[D : K] = [\Gamma : \Gamma_0][B/J(B) : V/J(V)]m^2\delta(B),$$

where $J(R)$ denotes the Jacobson radical of a ring R and $\delta(B)$ the defect of B which is equal to 1 if $\mathrm{char}(V/J(V)) = 0$ and to p^a for some integer $a \geq 0$ if $\mathrm{char}(V/J(V)) = p$.

Let V, K, B and D be as in Theorem 1 with $M = \{B_1 = B, \cdots, B_m\}$ the set of extensions of V in D. Then one can consider the group $P(B)$ of all permutations induced on M by inner automorphism of D. This group is investigated in [4] where it is shown that $P(B)$ is solvable and a bound for the order of $(P(B))$ and various examples are given.

2. We consider the general situation where B is a total subring of a skew field F. Associated with B is a group G and an invariant subring R of D. In [2] the relationship between properties of R and the group G is investigated.

Let $W = \{aB \mid 0 \neq a \in F\} \cup \{\infty\}$ with ∞ as its largest element and $aB \leq bB$ if and only if $aB \supseteq bB$. Then W is a totally ordered set, but a multiplicative structure can no longer be defined. The mapping v from F to W satisfies the conditions 1) and 2) in section 1 and condition 3'): $v(x) \leq v(y)$ implies $v(zx) \leq v(zy)$ for x, y, z in F.

If on the other hand a mapping v exists from F into a totally ordered set W such that 1), 2) and 3') hold, then $B_v = \{a \in F \mid v(a) \geq v(1)\}$ is a total subring of F, since $v(x) < v(1)$ implies $x \neq 0$ and $v(1) = v(x^{-1}x) \leq v(x^{-1})$. We define the group

$$G = \{\widetilde{x} \mid o \neq x \in F\}$$

of order preserving bijections of W with $\widetilde{x}(w) = \widetilde{x}(v(a)) = v(xa)$ for $w \in W, a \in F$ as the group associated with B. The operation in G is given by $\widetilde{x}\widetilde{y} = \widetilde{xy}$ and $\widetilde{x} \leq \widetilde{y}$ if and only if $\widetilde{x}(w) \leq \widetilde{y}(w)$ for all w in W defines a partial order on W. We say that G is lattice ordered if the infimum and the supremum exists for any two elements $\widetilde{x}, \widetilde{y}$ in G.

Associated with B is the invariant subring

$$R = \bigcap dBd^{-1}, \quad 0 \neq d \in F$$

of F and we denote with $H(R) = \{aR \mid 0 \neq a \in F\}$ the partially ordered group of non-zero cyclic R-submodules of F with $aRbR = abR$ as operation and $aR \leq bR$ if and only if $aR \supseteq bR$ defining a partial order. The mapping φ from G to $H(R)$ with $\varphi(\widetilde{x}) = xR$ defines an order preserving isomorphism between G and $H(R)$. It follows that the definition of G is left-right symmetric and that G is also isomorphic to F^*/R^* where F^* and R^* are the groups of units of F and R, respectively.

We say that R is a distributive ring if $A \cap (B + C) = (A \cap B) + (A \cap C)$ holds for any ideals A, B, C of R and say R is F-distributive if the same equation holds for R-submodules A, B, C of F. It follows that R is F-distributive if $aR \cap (bR + cR) = (aR \cap bR) + (aR \cap cR)$ holds for any a, b, c in F and distributivity implies F-distributivity if F is the skew field of quotients of R. The ring R is called F-Bezout if all finitely generated R-submodules of F are cyclic R-submodules and R is called Bezout if all finitely generated ideals of R are principal. The commutative distributive domains are exactly the Prüfer domains.

If G and hence $H(R)$ is lattice ordered, then there exists for aR, bR in $H(R)$, a cR in $H(R)$ with $cR = \sup\{aR, bR\}$. Hence $cR \subseteq aR \cap bR$, and for every $d \in aR \cap bR$ we have $dR \geq cR, dR \subseteq cR$, and $cR = aR \cap bR$ follows. The next result states that the converse is also true.

Lemma 2. *$H(R)$ is lattice ordered if and only if $aR \cap bR \in H(R)$ for a, b in F^*.*

One obtains the following interesting corollary:

Corollary. *If G is lattice ordered, then F is the skew field $Q(R)$ of quotients of R.*

Proof. For any a in F^* there exists c in F^* with $aR \cap R = cR$ and $c = ar$ for c, r in R follows, i.e. $a = cr^{-1} \in Q(R)$. □

The next result shows that there is a linkage between some properties of R and G.

Theorem 3. *Let R be a distributive ring with $Q(R)$=F. Then R is a Bezout ring if and only if G is lattice ordered.*

If R is invariant, then G is totally ordered. In this case R is distributive and G is lattice ordered. If F is finite dimensional over its center, then $R = \cap B_i$ is distributive and Bezout and G is lattice ordered. In the next section we discuss the construction of a total subring B of a skew field F such that G is not lattice ordered and R is distributive in one case, not distributive in the other. Whether the fourth possibility, G lattice ordered, R not distributive, can occur is left unanswered.

3. Let K be a commutative field, A a group of automorphisms of K which can be considered as the homomorphic image $\varphi(\Gamma)$ of the free group Γ which can be ordered. The skew field $F = K((\Gamma, \varphi))$ of skew power series $d = \Sigma\gamma k_\gamma, \gamma \in \Gamma$ with well ordered support$\{\gamma \mid k_\gamma \neq 0\}, k_\gamma \in K$, exists with multiplication defined by $k_\gamma = \gamma k^{\varphi(\gamma)}$. The subring $B = \{\Sigma_\gamma k_\gamma \in F \mid \gamma \geq \rho\ k_l \in B_0\}$ is a total subring of F if B_0 is a total subring of K. If we write $R_0 = \cap\sigma(B_0), \sigma \in A$, we obtain $R = \cap dBd^{-1}, d \in F^*$, and $R = \{\Sigma_\gamma k_\gamma \in B \mid k_l \in R_0\}$.

It is possible to check that properties like F-distributivity, F-Bezout or $H(R)$ lattice ordered hold for R with respect to F if and only if the corresponding properties hold for R_0 with respect to K.

To construct the first example let $K = L(x)$ for L an algebraically closed field and let B_0 be the valuation ring of the x-adic valuation of K. Let A be the subgroup of the automorphism group of K generated by the L-automorphisms σ, σ_l with $\sigma_l(x) = x - l, l \in L$, and $\sigma(x) = \frac{1}{x}$ and $R_0 = L$. Hence $1 \cdot R_0 \cap xR_0 = \{0\}$, the condition of Lemma 2 is not satisfied for R and G is not lattice ordered. Similarly, L is not K-distributive, and hence R is not F-distributive.

For the second example let L be algebraically closed as before and let $K = L(x)(\{\sqrt{(x-l)^3+1} \mid l \in L\})$. Let A be the subgroup of $\mathrm{Aut}(K)$ generated by the Galois group of K over $L(x)$ and the set of automorphisms $\{\sigma_l \mid l \in L, \sigma_l(x) = x - l\}$ i.e. σ_l is an extension to K of the L-automorphism that maps x to $x - l$. It turns out that $R_0 = \bigcap\sigma(B_0), \sigma \in A, B_0$ an extension in K of the x-adic valuation ring in $L(x)$, is the integral closure of $L[x]$ in K and is therefore distributive. By considering the order of ideals in the class group of some finite extensions of $L(x)$ contained in K one can show that R_0 is not Bezout. The properties of R_0 translate to R in F and one obtains by Theorem 2 an example for B in F such that R is distributive, $Q(R) = F, R$ not Bezout and hence G not lattice ordered.

References

[1] H.H. Brungs and J. Gräter, *Valuation rings in finite dimensional division algebras*, J. Algebra, 120(1989), 90–99.

[2] H.H. Brungs and J. Gräter, *Value groups and distributivity*, to appear in Canad. J. Math.

[3] J. Gräter, *The "Defektsatz" for central simple algebras*, to appear in Trans. Amer. Math. Soc.

[4] J. Gräter, *A note on valued division algebras*, preprint.

[5] W. Krull, *Allgemeine Bewertungstheorie*, J. reine u. angew. Math., 167(1932), 160–196.

[6] O.F.G. Schilling, *Non commutative valuations*, Bull. Amer. Math. Soc., 51(1945), 297–304.

[7] A.R. Wadsworth, *Dubrovin valuation rings and Henselization*, Math. Ann., 283(1989), 301–328.

Proceedings of the Asian Mathematical
Conference, 1990

A Stochastic Model for Distributed Control Processes' Behaviour

MARC BUI
INRIA
Domaine de Voluceau, Rocquencourt
BP 105 78153 LE CHESNAY Cedex,
FRANCE

This paper presents a new model which describes phenomena of distributed computing through a behavioural study of distributed algorithms. This model is based on the interconnexion of N Markov chains, each representing a distributed process. Our model differs from the various and usual ones (see [3]–[7]) since it handles formal specification of distributed systems only through local considerations.

Let be N finite homogeneous Markov chains which states spaces are respectively ${}^k\mathcal{X} = \{1, \cdots, \nu_k\} (k = 1, \cdots, N)$ and transition matrices are respectively ${}^kM (k = 1, \cdots, N)$.

The notation ${}^kM$ expresses the fact that each transition matrix depends on a multi-dimensional parameter ρ_k characterizing it, for example $\rho_k = ({}^kp_{11}, \cdots, {}^kp_{ij}, \cdots, {}^kp_{\nu_k\nu_k})$. Thus, these ${}^kM$ are matrices M_{ρ_k}.

The distributed system is made of a network of processes logically represented by the interconnexion of N Markov chains. Then there exists a set of relations between the parameters $\rho_1, \cdots, \rho_N$ which defines and characterizes the network.

$$\mathcal{R}_j(\rho_1, \cdots, \rho_N) = 0, \quad j = 1, \cdots, N. \tag{1}$$

These relations are constraints (which can be linear). We denote $\rho = (\rho_1, \cdots, \rho_N) \in \mathcal{R}$ iff the $\rho_1, \cdots, \rho_N$ satisfy (1).

For every $k \in \{1, \cdots, N\}$, suppose that there exists only one acyclic ergodic class ${}^k\mathcal{E}$ and at most one transient class ${}^k\mathcal{P}$. It is the condition (C).

For the chain number k, if $i \in {}^k\mathcal{E}$, let ${}^kT_i$ be the mean recurrence time of the state i. Recall that ${}^kT_i = 1/\{\lim_{n\to\infty} {}^km_{ji}^{(n)}\}$, this latter expression is independent of j, ${}^km_{ji}^{(n)}$ being the term (j, i) of the matrix $({}^kM)^n$.

If j and $j' \in {}^k\mathcal{P}$, let us denote by ${}^kS_{jj'}$ the total mean sojourn time in j' starting from j. These ${}^kS_{jj'}$ are given by matrix $(I - {}^kW)^{-1}$ where ${}^kW$ is the submatrix restriction of ${}^kM$ to the transient states (states of ${}^k\mathcal{P}$).

A decision function is a real function f of variables ${}^kT_i$ and ${}^kS_{jj'}, i \in {}^k\mathcal{E}, j$ and $j' \in {}^k\mathcal{P}$, and $k \in \{1, \cdots, N\}$. As these ${}^kT_i$ and ${}^kS_{jj'}$ are expressed with terms of matrices $({}^kM)^n$, i.e. in terms of ρ_k, f is function of variables ρ_k. The definition of a decision function f necessitates that the network verifies condition (C).

For each problem, following its context, a decision function will be defined and its role will be to rule the functioning of the system: so the optimum policies will be decided. An optimality criterion is based on the optimization (maximization or minimization) of a decision function f, under some constraints.

Our model $(S, S^0, \mathcal{T}, \mathcal{F})$ is based upon the interconnection of N Markov chains:

• S, the set of system-states, is here the set $\prod_{k=1}^{N}({}^k\mathcal{X})$

•S^0, the set of initial system-states is a subset of S

• $\mathcal{T}$ is the set of functioning rules. Each functioning rule, denoted here by M_ρ, is a N-tuple of transition matrices $(M_{\rho_1}, \cdots, M_{\rho_N})$, where $\rho \in \mathcal{R}$.

We are only interested by functioning rules to which we can associate a decision function f (that is to say only by the rules associated to a Markovian network verifying condition (C)).

A functioning rule is said to be optimal if and only if its ρ maximizes (resp. minimizes) the decision function f when the imposed optimality criterion is the maximization (resp. minimization) of f.

A functioning rule M_ρ is said to be bad if and only if its ρ maximizes (resp. minimizes) the decision function f when the imposed optimality criterion is the minimization (resp. maximization) of f.

Every functioning rule which is not bad is said to be advisable. Optimal and advisable functioning rules are good functioning rules.

• For each problem, it is question of finding one or several optimal functioning rules, or if not possible to find advisable ones. We denote by $\mathcal{Q} = \{M_\rho, \rho \in \mathcal{R}_0, \mathcal{R}_0 \subset \mathcal{R}\}$; this subset of $\mathcal{T}$, it is the set of good functioning rules of the problem.

Now, we are going to study some applications of our model.

§1. The Dining Philosophers' Problem

The problem considers a network of N finite homogeneous Markov chains with the same state space. $\forall k \in \{1, \cdots, N\}, {}^k\mathcal{X} = \mathcal{X} = \{1, \cdots, 4\}$, and with the following transition matrice:

$$M_{\rho_k} = \begin{pmatrix} 1-\alpha_k & \alpha_k & 0 & 0 \\ 0 & 1-\beta_k & \beta_k & 0 \\ 0 & 0 & 1-\gamma_k & \gamma_k \\ 1 & 0 & 0 & 0, \end{pmatrix},$$

where $\rho_k = (\alpha_k, \beta_k, \gamma_k) \in]0,1[^3$, for $k \in \{1, \cdots, N\}$.

The interconnection into network is expressed here by the relations:

$$L_k(\alpha_k, \beta_{k+1}) \equiv \alpha_k + \beta_{k+1} - 1 = 0 \quad (k = 1, \cdots, N-1),$$

$$L_N(\alpha_N, \beta_1) \equiv \alpha_N + \beta_1 - 1 = 0.$$

The set of $\rho_k = (\alpha_k, \beta_k, \gamma_k)$ verifying these constraints is denote by $\mathcal{R}$.

The set S of system-states is here $\mathcal{X}^N$.

The set S^0 of initial system-states is here the set $\mathcal{X}^N$ itself.

The set $\mathcal{T}$ of functioning rules is here $(M_\rho)_{\rho\in\mathcal{R}}$.

The decision function is here:

$$f = \sum_{k=1}^{N} T_3 = \sum_{k=1}^{N} \frac{\alpha_k\beta_k + \beta_k\gamma_k + \gamma_k\alpha_k + \alpha_k\beta_k\gamma_k}{\alpha_k\beta_k}.$$

To minimize it corresponds to obtain the desired functioning rule.

$\forall(\gamma_1,\cdots,\gamma_N) \in (]\,0,1\,[)^N$, the unique optimal functioning rule is $M_\rho = (M_{\rho_1},\cdots,M_{\rho_N})$ where

$$\rho_k = \left(\frac{1}{1+\sqrt{\dfrac{\gamma_{k+1}}{\gamma_k}}},\quad \frac{\sqrt{\dfrac{\gamma_{k+1}}{\gamma_k}}}{1+\sqrt{\dfrac{\gamma_{k+1}}{\gamma_k}}},\quad \gamma_k \right), \qquad k \in \{1,\cdots,N\}$$

and

$$\rho_N = \left(\frac{1}{1+\sqrt{\dfrac{\gamma_1}{\gamma_N}}},\quad \frac{\sqrt{\dfrac{\gamma_1}{\gamma_N}}}{1+\sqrt{\dfrac{\gamma_1}{\gamma_N}}},\quad \gamma_N \right).$$

If the $\gamma_1,\cdots,\gamma_N$ are bounded by below, that is to say $(\gamma_1,\cdots,\gamma_N) \in \prod_{k=1}^{N}[c_k,1[$, then the set $\mathcal{Q}$ of good functioning rule is the singleton $\{M_\rho\}$, where

$$\rho_k = \left(\frac{1}{1+\sqrt{\dfrac{c_{k+1}}{c_k}}},\quad \frac{\sqrt{\dfrac{c_{k+1}}{c_k}}}{1+\sqrt{\dfrac{c_{k+1}}{c_k}}},\quad c_k \right), \quad k \in \{1,\cdots,N-1\}$$

and

$$\rho_N = \left(\frac{1}{1+\sqrt{\dfrac{c_1}{c_N}}},\quad \frac{\sqrt{\dfrac{c_1}{c_N}}}{1+\sqrt{\dfrac{c_1}{c_N}}},\quad c_N \right).$$

Details are given in [1], [2].

§2. The Mutual Exclusion Problem

The problem considers a network of N homogeneous and finite Markov chains with the same state space $\forall k \in \{1,\cdots,N\}, {}^k\mathcal{X} = \mathcal{X} = \{1,2,3,4\}$ and with the following transition matrices:

$$M_{\rho_k} = \begin{pmatrix} \alpha_k & \beta_k & 1-\alpha_k-\beta_k & 0 \\ 1-\gamma_k & \gamma_k & 0 & 0 \\ 0 & 0 & \delta_k & 1-\delta_k \\ 1-\theta_k & 0 & 0 & \theta_k \end{pmatrix}$$

and where $\rho_k = (\alpha_k,\beta_k,\gamma_k,\delta_k,\theta_k) \in]0,1[^5$ and $1-\alpha_k-\beta_k > 0$.

The interconnection into network is expressed by the relations:

$$L_1 \equiv \sum_{k=1}^{N} \alpha_k - 1 = 0$$

and

$$L_2 \equiv N - 2 - \sum_{k=1}^{N} \beta_k = 0,$$

which shows that $N > 3$.

The set of $\rho_k = (\alpha_k, \beta_k, \gamma_k, \delta_k, \theta_k)$ verifying these constraints is denoted by $\mathcal{R}$.

The set S of system states is here $\mathcal{X}^N$.

The set S^0 of initial system states is here the set $\mathcal{X}^N$ itself.

The set $\mathcal{T}$ of functioning rules is here $(M_\rho)_{\rho \in \mathcal{R}}$.

The decision function is here

$$f = \sum_{k=1}^{N} \frac{{}^kT_2}{{}^kT_3} = \sum_{k=1}^{N} \frac{(1-\gamma_k)}{(1-\delta_k)} \frac{(1-\alpha_k-\beta_k)}{\beta_k}.$$

To minimize it corresponds to obtain the desired functioning rule.

If $\forall k \in \{1, \cdots, N\}, (\gamma_k, \delta_k) = (\gamma, \delta)$ and $\forall (\theta_1, \cdots, \theta_N) \in]0,1[^N$, the unique optimal functioning rule is $M_\rho = (M_{\rho_1}, \cdots, M_{\rho_N})$ where

$$\rho_k = \left(\frac{1}{N}, \frac{N-2}{N}, \gamma, \delta, \theta_k \right).$$

If γ is bounded by above and δ is bounded by below that is to say if $\gamma \in]\,0, c\,[$ and $\delta \in [\,d, 1\,[$, the set $\mathcal{Q}$ of good functioning rules is the set of M_ρ where

$$\rho_k = \left(\frac{1}{N}, \frac{N-2}{N}, c, d, \theta_k \right)$$

when $(\theta_1, \cdots, \theta_N)$ vary $]0,1[^N$. The fairness solution belongs to this set $\mathcal{G}$.

Details are given in [2].

§3. The Deadlock Problem

The problem considers a network of N finite homogeneous Markov chains, with the same state space $\forall k \in \{1, \cdots, N\}, {}^k\mathcal{X} = \mathcal{X} = \{1, \cdots, 4\}$ with the following transition matrice:

$$M_{\rho_k} = \begin{pmatrix} \alpha_k & 1-\alpha_k & 0 & 0 \\ \beta_k & \gamma_k & \delta_k & 1-\beta_k-\gamma_k-\delta_k \\ 1 & 0 & 0 & 0 \\ 0 & 0 & 0 & 1 \end{pmatrix}$$

and where $\rho_k = (\alpha_k, \beta_k, \gamma_k, \delta_k) \in]0,1[^4$ and where $\beta_k + \gamma_k + \delta_k < 1$.

The interconnection into network is expressed here by the relations:

$$L_1 \equiv \sum_{k=1}^{N} \alpha_k - 1 = 0,$$
$$L_2 \equiv \sum_{k=1}^{N} \beta_k - 1 = 0,$$
$$L_3 \equiv \sum_{k=1}^{N} \gamma_k - 1 = 0,$$
$$L_4 \equiv \sum_{k=1}^{N} \delta_k - 1 = 0.$$

The set of $\rho_k = (\alpha_k, \beta_k, \gamma_k, \delta_k)$ verifying these constraints is denoted by $\mathcal{R}$.
The set S of system states is here $\mathcal{X}^N$.
The set S^0 of system initial states is here the set $\mathcal{X}^N$ itself.
The set $\mathcal{T}$ of functioning rules is here $(M_\rho)_{\rho \in \mathcal{R}}$.
The decision function is here:

$$f = \sum_{k=1}^{N} \sum_{i=1}^{3} {}^k S_{2i} \equiv \sum_{k=1}^{N} \frac{(\beta_k + \delta_k) + (1 - \alpha_k)(1 + \delta_k)}{(1 - \alpha_k)(1 - \beta_k - \gamma_k - \delta_k)}$$

from which we get the expression of f.

We remark also that this study have a meaning only for $N \geq 4$ because of condition $\beta_k + \gamma_k + \delta_k < 1$.

The set $\overline{\mathcal{Q}}$ of bad functioning rules is the set of $M_\rho = (M_{\rho_1}, \cdots, M_{\rho_M})$ where

$$\rho_k = \left(\frac{1}{N}, \quad \beta_k, \quad \frac{1}{N}, \quad -\frac{3N+1}{3N-2}\beta_k + \frac{5N-1}{N(3N-2)}\right)$$

when $(\beta_1, \cdots, \beta_N) \in (\,]0, \frac{5N-1}{N(2N+1)}[\,)^N$. The fairness belongs to this set. Its complementary set $\mathcal{Q}$ is the set of advisable functioning rules.

Details are given in [2].

A fourth example, the multi-rendezvous problem, is under study, and shall be published nextly.

§4. Conclusion

We have introduced a modelisation which helps to handle with distributed control problems. We have shown with this approach that one can formally find good functioning properties (safety and liveness properties) in a systematic way for the considered distributed control problem. Our study brings out that dispositions for fairness between processors are good for the optimization of some distributed control algorithms such as mutual exclusion protocols; but concerning deadlock problems, fairness must be avoided, no hierarchy produces more opportunities for being deadlocked. Thus, we have an efficient theoretical tool for reasoning about distributed algorithms allowing to find optimal tuning for distributed system and with the particularity that global behaviour are obtained only from local behaviour's study.

References

[1] Bui M., *Réglage pour un fonctionnement optimal d'un réseau de processeurs en algorithmique distribuée*, Revue R. de Math. Pures et Appl., No. 3(1990), 197–202.

[2] Bui M., *On optimal management of resources in distributed networks*, in Optimization, 21(5)(1990), 785–796.

[3] Brand D., Zafiropulo P., *On communicating finite-state machines*, Journal of the ACM, 30(2)(1983) 323–342, April.

[4] Chandy K.M., Misra J., *Parallel Program Design: A Foundation*, Addison-Wesley, 1988.

[5] Iosifescu M., *Finite Markov processes and their applications*, Wiley, 1980.

[6] Lavallée I., *Eléments d'algorithmique parallèle et distribuée, Hermès*, Paris, 1990.

[7] Le Lann G., *Distributed System: towards a formal approach*, IFIP Congress, (1977), 155–160.

Proceedings of the Asian Mathematical
Conference, 1990

Saddle Values and Limit Cycles Generated by Separatrix of Quadratic Systems

CAI SUILIN, GUO GUANGYUAN
Department of Mathematics
Zhejiang University
Hangzhou 310027, China

In the recent study of quadratic systems with saddle points, Cai [1] introduced the following method which is similar to the one used in computing the P–L constants of a weak focus, and defined the P–L constants of a weak saddle point.

A simple saddle point of a system is called a weak saddle point if the divergence of the system vanishes at that point.

Consider a system with a weak saddle point which, without loss of generality, is supposed to be the origin o. If the system is analytic in some neighborhood of the origin, then it can be written as

$$\frac{dx}{dt} = x + \sum_{i=2}^{\infty} P_i(x,y), \quad \frac{dy}{dt} = -y + \sum_{i=2}^{\infty} Q_i(x,y), \tag{1}$$

where P_i and Q_i are homogeneous polynomials in x and y, of degree i for each $i \geq 2$. Finally, we set

$$F(x,y) = xy + \sum_{j=3}^{\infty} F_j(x,y), \tag{2}$$

such that

$$\left.\frac{dF}{dt}\right|_{(1)} = \sum_{m=1}^{\infty} U_m(x,y)^{m+1}, \tag{3}$$

in which F_j is a homogeneous polynomial of degree j in x and y whose coefficients are unknown except for F_2, which is xy, and U_m are constants to be determined.

Definition 1. The U_m in (3) is called the P–L constant of order m of the weak saddle point o of system (1), $m = 1, 2, \cdots$.

Now let us calculate the $U_m (m = 1, 2, \cdots)$. By (1), (2), and (3), we have

$$\sum_{j=3}^{\infty}[D(F_j) + G_{j-1}(F_2, F_3, \cdots, F_{j-1})] = \sum_{m=1}^{\infty} U_m(xy)^{m+1}, \tag{4}$$

where

$$D \equiv x\frac{\partial}{\partial x} - y\frac{\partial}{\partial y}, \quad G_j(F_2, F_3, \cdots, F_j) \equiv \sum_{i=2}^{j} E_i(F_{j-i+2}),$$

$$E_i \equiv P_i(x,y)\frac{\partial}{\partial x} - Q_i(x,y)\frac{\partial}{\partial y}.$$

The terms of the same degree on each side of (4) are equal, that is,

$$D(F_j) + G_{j-1}(F_2, F_3, \cdots, F_{j-1}) = U_{j/2-1}(xy)^{j/2}, \quad j = 3, 4, \cdots, \tag{5}$$

where $U_{j/2} = 0$ if j is odd. Assume inductively that $F_2, F_3, \cdots, F_{j-1}$ are known, then so is $G_{j-1}(F_2, F_3, \cdots, F_{j-1})$ which clearly is homogeneous and of degree j. We write it as

$$G_{j-1}(F_2, F_3, \cdots, F_{j-1}) = \sum_{i=0}^{j} b_{i,j-i} x^i y^{j-i},$$

and suppose

$$F_j(x, y) = \sum_{i=0}^{j} a_{i,j-i} x^i y^{j-i},$$

where $a_{i,j-i}$ are to be determined. Then (5) reduces to

$$\sum_{i=0}^{j} [(2i - j)a_{i,j-i} + b_{i,j-i}] x^i y^{j-i} = U_{j/2-1}(xy)^{j/2}. \tag{5'}$$

Notice that $U_{j/2-1} = 0$ and $2i - j \neq 0$ if j is odd, so F_j can be uniquely determined by

$$a_{i,j-i} = \frac{b_{i,j-i}}{j - 2i}, \quad (i = 0, 1, \cdots, j). \tag{6}$$

When j is even, say $j = 2m$, then (6) still holds for $i \neq m$; thus (5') is solvable iff $U_{j/2-1} = b_{i/2,j/2}$. When this condition holds the P–L constants are

$$U_{j/2-1} = b_{j/2,j/2}, \tag{7}$$

and one can choose any number for $a_{j/2,j/2}$. By induction, F_i and U_k can be determined for $i = 3, 4, \cdots$, and $k = 1, 2, \cdots$. So we have the following

Lemma 1. *For j odd, there exists a unique F_j satisfying (5) where "unique" means that F_j is unique for the $F_2, F_3, \cdots, F_{j-1}$ determined, and when j is even, there exist more than one F_j satisfying (5), but all of them can be written as $F_j^{(1)} = F_j + AF_2^{j/2}$, where A is a constant and F_j is any one of them.*

From the above discussion, we see that the value of $U_{j/2-1}$ depends on $F_2, F_3, \cdots, F_{j-1}$. Generally, different $F_2, F_3, \cdots, F_{j-1}$ determine different values of $U_{j/2-1}$. But this does not happen to the first nonzero P–L constant as the following theorem indicates.

Theorem 1. *The value of the first nonzero P–L constant of the saddle point o as well as its order does not depend on the choice of $F_j(x, y) (j \geq 3)$ in (2).*

To prove this theorem, we first prove the following two lemmas.

Lemma 2. *Suppose that $F_2 = xy$, $F_3, \cdots, F_{2k+1}$ and $U_{j/2-1} = 0$ $(j = 4, 5, \cdots, 2k+1)$ and $U_k \neq 0$ satisfying (5). Then for any constant A_1,*

$$F_j^{(1)} = \begin{cases} F_j & j = 2, 3, \\ F_j + A_1 \sum\limits_{\substack{s+r=j \\ s,r>1}} F_s F_r, & j = 4, 5, \cdots, 2k+2, \end{cases}$$

and $U_{j/2-1}$ *also satisfy (5) for* $j = 4, 5, \cdots, 2k+2$.

Proof. For $j = 4$, we have by Lemma 1 that $D(F_4^{(1)}) = D(F_4 + A_1F_2^2) = D(F_4) + 2A_1F_2D(F_2) = D(F_4)$; hence $F_4^{(1)}$ satisfies (5). Suppose inductively that $F_2^{(1)}, F_3^{(1)}, \cdots, F_j^{(1)}$ satisfy (5):

$$D(F_j^{(1)}) + G_{j-1}(F_2^{(1)}, F_3^{(1)}, \cdots, F_{j-1}^{(1)}) = U_{j/2-1}F_2^{j/2},$$

$$j = 4, 5, \cdots, 2k+1, \text{for some } k.$$

Then for $j = 4, 5, \cdots, 2k+1$,

$$\begin{aligned} & D(F_{j+1}^{(1)}) + G_j(F_2^{(1)}, F_3^{(1)}, \cdots, F_j^{(1)}) \\ = \; & D(F_{j+1}) + G_j(F_2, \cdots, F_j) \\ & + 2A_1 \sum_{\substack{s+r=j+1 \\ s,r>1}} F_rD(F_s) + \sum_{i=2}^{j-2} E_j\left(A_1 \sum_{\substack{s+r=j-i+2 \\ s,r>1}} F_sF_r\right) \\ = \; & U_{(j-1)/2}F_2^{(j+1)/2}. \end{aligned}$$

Therefore (5) is satisfied for $j = 4, 5, \cdots, 2k+2$. This proves the lemma.

In the same way we can prove

Lemma 3. *Under the conditions of Lemma 2, we have for any constant* $A_{m-1}, m = 3, 4, \cdots, k+1$,

$$F_j^{(m-1)} = \begin{cases} F_j, & j = 2, \cdots, 2m-1; \\ F_j + A_{m-1} \sum\limits_{\substack{\lambda_1+\cdots+\lambda_m=j \\ \lambda_i>1}} F_{\lambda_1}\cdots F_{\lambda_m}, & j = 2m, \cdots, 2k+2, \end{cases}$$

and $U_{j/2-1}(j = 4, 5, \cdots, 2k+2)$ *satisfy (5)* .

Proof of Theorem 1. Suppose $F_2 = xy, F_3, \cdots, F_{2k+2}$ and $U_{j/2-1}(j = 4, 5, \cdots, 2k+2)$ satisfy (5) :

$$D(F_j) + G_{j-1}(F_2, F_3, \cdots, F_{j-1}) = U_{j/2-1}F_2^{j/2}, \quad j = 4, 5, \cdots, 2k+2,$$

where $U_{j/2-1} = 0$ for $j = 4, 5, \cdots, 2k+1$, and $U_k \neq 0$.

If $\widetilde{F}_2 = xy, \widetilde{F}_3, \cdots, \widetilde{F}_{2k+2}$ and $\widetilde{U}_1, \widetilde{U}_{3/2}, \widetilde{U}_2, \cdots, \widetilde{U}_{\widetilde{k}}$ are other choices which satisfy (5), then it suffices to prove that $\widetilde{U}_{j/2-1} = U_{j/2-1} = 0, (j = 4, 5, \cdots, 2k+1); \widetilde{U}_k = U_k \neq 0$. To this end, note that $\widetilde{F}_2 = F_2$ implies that $\widetilde{F}_3 = F_3$ by Lemma 1. Since U_1 is uniquely determined by F_2 and F_3, we have $\widetilde{U}_1 = U_1 = 0$. Again by Lemma 1, there exists some constant A_1 such that $\widetilde{F}_4 = F_4 + A_1F_2^2$. Applying Lemma 2 to this A_1, we get some $F_i^{(1)}$ which together with $U_{i/2-1}$ satisfy (5) for $i = 3, 4, \cdots, 2k+2$. Since $F_2^{(1)} = \widetilde{F}_2, F_3^{(1)} = \widetilde{F}_3, F_4^{(1)} = F_4 + A_1F_2^2 = \widetilde{F}_4$, we have by Lemma 1 $F_5^{(1)} = \widetilde{F}_5$ which in turn gives $\widetilde{U}_2 = U_2 = 0$. Also , Lemma 1 implies that there exists some constant A_2 such that $\widetilde{F}_6 = F_6^{(1)} + A_1F_2^3$. Let $F_i^{(1)(2)} = F_i^{(1)}(i = 2, 3, 4, 5)$ and $F_6^{(1)(2)} = F_6^{(1)} + A_2F_2^3$. From Lemma 3 we know that $F_i^{(1)(2)}$ and $Ui/2-1$ $(i = 4, 5, \cdots, 2k+2)$ satisfy (5). Then $\widetilde{F}_7 = F_7^{(1)(2)}$ by Lemma 1; therefore $\widetilde{U}_3 = U_3 = 0$.

Following this deduction, we have that $F_i^{(1)\cdots(j)}$ and $U_{i/2-1}$ satisfy (5) for $j = 1, 2, \cdots, k; i = 3, 4, \cdots, 2k+2$, that is , $D(F_i^{(1)\cdots(k)}) + G_{i-1}(F_2^{(1)\cdots(k)}, \cdots, F_{i-1}^{(1)\cdots(k)}) = U_{i/2-1}F_2^{i/2}$, and $F^{(1)\cdots(k)} = \widetilde{F}_i, \quad i = 2, 3, \cdots, 2k+2$. So we have

$$D(\widetilde{F}_i) + G_{i-1}(\widetilde{F}_2, \widetilde{F}_3, \cdots, \widetilde{F}_{i-1}) = U_{i/2-1}F_2^{i/2}.$$

On the other hand, by assumption we have

$$D(\widetilde{F}_i) + G_{i-1}(\widetilde{F}_2, \widetilde{F}_3, \cdots, \widetilde{F}_{i-1}) = \widetilde{U}_{i/2-1}F_2^{i/2}.$$

Thus for $i = 4, 5, \cdots, 2k+2$, we have $U_{i/2-1} = \widetilde{U}_{i/2-1}$. Therefore $\widetilde{U}_i = U_i = 0, \quad i = 1, 2, \cdots, k-1; \quad \widetilde{U}_k = U_k \neq 0$. The Proof is completed.

The following corollary is obvious.

Corollary. *If there exists a sequence of polynomials $F_2 = xy, F_3, \cdots$ with all $U_m = 0$ for $m = 1, 2, \cdots$ satisfying (5), then for every sequence of polynomials $\widetilde{F}_2 = xy, \widetilde{F}_3, \cdots$ with $\widetilde{U}_1, \widetilde{U}_2, \cdots$ satisfying (5), we have $\widetilde{U}_1 = \widetilde{U}_2 = \cdots = 0$.*

When all the P–L constants are zero, we get a series,

$$F = xy + \sum_{j=3}^{\infty} F_j(x, y).$$

Does this series converge? This is answered in the following. First, from [2] §23.B, we know that there exists an analytic transformation

$$x = u + \cdots, \quad y = v + \cdots, \tag{8}$$

which transforms (1) into the follwoing normal form

$$\begin{aligned} \frac{du}{dt} &= u + u(a_1uv + a_2(uv)^2 + \cdots + a_i(uv)^i + \cdots), \\ \frac{dv}{dt} &= -v + v(b_1uv + b_2(uv)^2 + \cdots + b_i(uv)^i + \cdots), \end{aligned} \tag{9}$$

in some nieghborhood of the origin o. Based on this fact, C. Rousseau [3] gave

Definition 2. The value $a_i + b_i (i = 1, 2, \cdots)$ in (9) is called the i–th saddle value of the weak saddle point o of system (1), and the order of the first nonzero saddle value is called the order of this weak saddle point.

According to this definition, direct calculation of the saddle values involves the calculation of the transformation (8) and the normal form (9), which is surprisingly difficult even for a quadratic system. The following theorem reveals the relation between the saddle values and the P–L constants and thus provides a much simpler method for calculating the first nonzero saddle value and the order of the weak saddle point.

Theorem 2. *The first nonzero saddle value and the order of a weak saddle point are the same as the first nonzero P–L constant of the saddle and its order respectively. Moreover, all the saddle values are zero iff all the P–L constants are zero.*

Proof. Suppose first that o is a weak saddle point of order m. By definition, (9) gives

$$\frac{d}{dt}(uv)\,|_{(9)} = (a_m + b_m)(uv)^{m+1} + \cdots, \tag{10}$$

where $a_m + b_m \neq 0$ is the m-th saddle value of o. From (8) , we get two analytic functions in some neighborhood of the origin o,

$$u = x + \cdots, \quad v = y + \cdots. \tag{11}$$

Set

$$F = uv = xy + \sum_{j=3}^{\infty} F_j(x, y). \tag{12}$$

In view of (10), we have

$$\begin{aligned} \left.\frac{dF}{dt}\right|_{(1)} &= \left.\frac{d}{dt}(uv)\right|_{(9)} = (a_m + b_m)(uv)^{m+1} + \cdots \\ &= (a_m + b_m)(xy)^{m+1} + \cdots. \end{aligned}$$

From this follows the first part of the theorem. Now suppose all the saddle values of o are zero. Then (9) gives $\frac{d}{dt}(uv)\mid_{(9)}= 0$, which implies $\frac{dF}{dt}\mid_{(1)}\equiv 0$, where F is defined by (12). Then the conclusion of the second part of the theorem follows. The proof is completed.

Corollary. *If all the P–L constants of the saddle point o are zero, then there exists in some neighborhood of o an analytic function $F(x, y)$ such that*

$$\left.\frac{dF}{dt}\right|_{(1)} = 0. \tag{13}$$

Moreover, we can choose xy and some $F_3, F_4, \cdots$ with $U_i = 0 \quad (i = 1, 2, \cdots)$ satisfying (5) such that $F(x, y)$ is of the form

$$F = xy + \sum_{j=3}^{\infty} F_j(x, y); \tag{14}$$

and $F(x, y) = c$ is a first integral of the system in some neighborhood of o for any real number c.

Proof. By Theorem 2, $a_m + b_m = 0$ for all positive integers m. We choose F to be the one defined by (12). Clearly (13) holds; thus $F(x, y) = c$ is a first integral of the system (1) for any real number c. Since every $F = xy + \sum_{j=3}^{\infty} F_j(x, y)$ satisfying (13) can be got formally by the procedure described before Lemma 1, so is the F defined by (12). This implies that we can choose a sequence of $F_i(x, y)(i = 3, 4, \cdots)$ satisfying (5) such taht series (14) converges in some neighborhood of the origin. The proof is completed.

Theorem 2 provides a method for computing the first nonzero saddle value of the weak saddle point and its order. This was shown in [1], which proves to be much easier than calculating them directly.

Early in 1951, E.Leontovich already had defined saddle values of a weak saddle point. Suppose $P(x, y)$ and $Q(x, y)$ in system (1) are analytic (in paper [4], it is assumed that $P(x, y)$ and $Q(x, y)$ have continuous derivatives of sufficiently high order in some neighborhood of the saddle point, say o. Then (1) can be transformed into the following form by an analytic transformation,

$$\frac{d\eta}{d\xi} = -\frac{\eta}{\xi}(1 + c_1\xi\eta + c_2\xi^2\eta^2 + \cdots + c_i\xi^i\eta^i + \cdots). \tag{15}$$

Using this form, Leontovich gave the following definition of saddle values of the saddle point o:

Definition 3. The i-th saddle value of o is defined to be c_i in (15). The order of o as a weak saddle point is by definition equal to the subscript of the first nonzero saddle value.

The following theorem assures that the two definitions define the same thing.

Theorem 3. *The two definitions above coincide, and in case all of the saddle values of one definiton are zero, then so are the ones of the other definition.*

Proof. Suppose $a_i + b_i = 0, i = 1, 2, \cdots, m-1$ and $a_m + b_m \neq 0$ in (9) . Then we have

$$\frac{dv}{dt} = -\frac{v}{u}[1 - (a_m + b_m)(uv)^m + \cdots],$$

where "$\cdots$" stands for a power series in uv which begins with the term of degree $m+1$. Comparing with (15), we have $c_i = 0 (i = 1, 2, \cdots, m-1); c_m = -(a_m + b_m) \neq 0$. Note that $a_i + b_i = 0$ iff $c_i = 0$ for all i. This completes the proof.

By Theorem 3, we can interpret the order of a saddle point in two ways.

In paper [5], we proved

Theorem A. *A quadratic system which has a weak saddle point of order two or three does not have any cycle or separatrix loop.*

From the proof of this theorem, we can draw a more strong conclusion.

Theorem B. *Suppose that a quadratic system has a weak saddle point such that the first saddle value of it is zero. Then either there does not exist any cycle or separatrix loop in the plane or there exist some separatrix loops (may be an infinite separatrix loop), each of which encloses a centre.*

The author of paper [5] concluded without proof that small perturbation [4]on a quadratic system never generates four limit cycles in the neighborhood of the separatrix loop. Now we prove

Theorem 4. *Consider a quadratic system with a separatrix cycle such that the unique singular point on it is a simple saddle point. If, moreover, the singular point enclosed by the separatrix cycle is not a centre, then the number of limit cycles generated in the neighborhood of the separatrix cycle by small perturbation on the sytem can not exceed 3.*

Proof. By Theorem B, it suffices to consider the case where the unique singular point on the separatrix cycle, say $S^{(1)}$ is a rough saddle point or a weak one of order one. If the unique singular point on the separatrix cycle $S^{(1)}$, is a rough one, that is, the value of the divergence of the system is nonzero at that point, then following a theroem in [6], we know that there can exist at most one limit cycle in the neighborhood of $S^{(1)}$. If the unique singular point on $S^{(1)}$ is a weak saddle point of order one, then $c_1 \neq 0$. If the value of separatrix [4] $a_1 \neq 0$, then according to Theorem 3 in [4], the number of limit cycles generated in the neighborhood of $S^{(1)}$ by small perturbation on the system can not exceed two. Finally, if $a_1 = 0$, the same theorem assures that the number of limit cycles generated in the neighborhood of $S^{(1)}$ by small perturbation on the system can never exceed three. In both cases, Theorem 4 holds.

References

[1] Cai Suilin, *The Weak Saddle and Separatrix of Quadratic System,* Acta Math. Sinica 30 (4) (1987), 553–559.

[2] V. I. Arnold, *Geometric Methods in the Theory of Ordinary Differential Equations,* Springer–Verlag, 1983.

[3] C. Rousseau, *Example of a Quadratic System with Two Limit Cycles Appearing in a Homoclinic Loop Bifurcation,* J. of Diff. Eqs., 66 (1) (1987), 140–150.

[4] E.A.Leontovich, *The Creation of Limit Cycles from a Separatrix,* Doklady AN SSSR Vol. 78, No.4 (1951).

[5] Zhang Pingguang and Cai Suilin, *Quadratic Systems with Second–order or Third–order,* Acta Math. Sinica, 30 (4) (1987), 560–565.

[6] A.A.Andronov, E.A.Leontovich, I.I. Gordon, and A.G.Maier, *Theory of Bifurcation of Dynamical System on a Plane,* Wiley, New York , 1973.

Proceedings of the Asian Mathematical
Conference, 1990

Variational-iterative Schemes as Applied to the Non-linear Burgers' Equation

JAMES CALDWELL
City Polytechnic of Hong Kong
Kowloon, Hong Kong

§1. Introduction

In this paper we consider a technique for solving nonlinear differential equations such as those which govern viscous fluid flow. The aim is to isolate a linear self-adjoint operator (e.g. the Laplaciam) from the rest of the equation. Then we construct a variationally equivalent formulation which can be solved interatively.

As a test example we use the steady-state version of the one dimensional Burgers' equation, namely

$$u_t + uu_x = \nu u_{xx}, \tag{1}$$

where $u = u(x, t)$ and ν is a parameter. We shall examine the results and obtain measures of convergence of the method.

§2. Complementary Variational Principles

Complementary variational principles have been introduced by Kato [8] and others up to Walpole [9] for the linear equation

$$T\phi = f, \tag{2}$$

in a suitably chosen real Hilbert space H where T is a completely continuous self-adjoint operator and f is some function belonging to H. Burrows and Perks [5, 6] have carried out further work which is an application of these earlier ideas. These principles provide upper and lower bounds for $\langle \phi | f \rangle$ where $\langle \quad \rangle$ denotes the inner product of H. Arthurs et al. [1–4] have also introduced complementary variational principles for the solution of equations of the form

$$T\phi = f(\phi), \tag{3}$$

where f is a non-linear function of ϕ. These provide upper and lower bounds to a functional under certain conditions. In fact, under sertain boundary conditions complementary extremum principles can be found when

$$T - \frac{df}{d\phi} > 0 \quad \text{or} \quad T - \frac{df}{d\phi} < 0. \tag{4}$$

In later work Burrows and Perks [7] have extended their linear theory to deal with the non-linear equation (3). A variational-iterative scheme is used so that the simple linear theory is applied to a sequence of equations, the solutions of which converge to the solutions of (3).

By considering the hyleras functional

$$J(\Delta, \Phi) = \langle \Phi | T\Phi \rangle - 2\langle \Phi | f \rangle + \Delta \langle T\Phi - f | T\Phi - f \rangle, \tag{5}$$

where Δ is a real constant, Burrows and Perks show that the real quantity

$$S = \min_{\psi \in H} \{ J(\Delta_1, w_p) - J(\Delta_2, \psi) \}, \tag{6}$$

provides a measure of the convergence criteria. Here ψ denotes the exact solution, Δ_1 and Δ_2 minimum and maximum principles, and w_p the limit of the sequence of trial functions $\{w_{n,p}\}$ containing p variational parameters for each iterate.

§3. Application to Non-linear Equations

The basic idea is to rearrange the non-linear equation

$$A\phi = f(\phi), \tag{7}$$

into the form (3) where T is self-adjoint and has a discrete spectrum. Then we iterate with the sequence of equations

$$T_{\Psi_{n+1}} = f(\Phi_{n+1}), \tag{8}$$

to give Φ_{n+1} as a variational approximation to ψ_{n+1}. Under certain conditions the sequence $\{\Phi_{n+1}\}$ converges to ϕ.

§4. Acceleration of Convergence

A discussion and choice of starting Φ_0 is given by Burrows and Perks. Equation (2) is often rewritten as

$$T\phi = bT\phi + (1-b)T\phi, \tag{9}$$

where b is a relaxation parameter chosen to produce rapid convergence. Equation (8) then becomes

$$T\psi_{n+1} = T\phi_n + b\{f(\phi_n) - T\phi_n\}. \tag{10}$$

In the next section we consider the application to Burgers' equation.

§5. Application to Burgers' Equation

For simplicity we consider how these variational principles can be applied to the steady state version of equation (1), namely

$$uu_x = \nu u_{xx}, \quad x \geq 0, \tag{11}$$

under the boundary conditions

$$u(o) = 0, \quad u(\infty) = -2\nu. \tag{12}$$

For checking purposes the exact solution is

$$u = -2\nu \tan hx. \tag{13}$$

We transform from the infinite domain $[0, \infty]$ to the finite domain $[0, 1]$ using the substitution

$$y = 1 - e^{-x}.$$

Then the boundary conditions become

$$\phi(o) = 0, \quad \phi(1) = -2 \quad (0 \leq y \leq 1) \tag{14}$$

on making the substitution $\phi = u/\nu$. Equation (11) then becomes

$$-\frac{d^2\phi}{dy^2} = -y\frac{d^2\phi}{dy^2} - (1+\phi)\frac{d\phi}{dy}, \tag{15}$$

with exact solution

$$u = \frac{2\{(1-y)^2 - 1\}}{(1-y)^2 + 1}. \tag{16}$$

It should be noted that this solution also satisfies $\phi'(1) = 0$. We now try to find the variational solution of (15) over $0 \leq y \leq 1$ and compare the results with the exact solution. Consider the Hilbert space of functions satisfying the boundary conditions $\phi(o) = 0, \phi'(1) = 0$ and let the inner product be defined by

$$\langle \phi_1 | \phi_2 \rangle = \int_0^1 \phi_1(y)\phi_2(y)dy.$$

Then (15) has the form of equation (2) with

$$f(\phi) = -y\frac{d^2\phi}{dy^2} - (1+\phi)\frac{d\phi}{dy}. \tag{17}$$

Note that the operator $T = -\frac{d^2}{dy^2}$ is symmetric since

$$\langle \phi_2 | T\phi_1 \rangle = \langle T\phi_2 | \phi_1 \rangle.$$

The eigenvalues of T are defined by

$$-\frac{d^2\phi_i}{dy^2} = \lambda_i\phi_i. \tag{18}$$

On imposing the boundary conditions, this leads to

$$\lambda_i = \{\frac{(2i+1)\pi}{2}\}^2 \quad (i = 0, 1, 2, \cdots) \tag{19}$$

and

$$\phi_i(y) = a_i \sin\frac{(2i+1)\pi y}{2}. \tag{20}$$

This suggests that we take

$$\phi_0(y) = a_0 \sin\frac{\pi y}{2}$$

as a first approximation in the variational approach. The eigenvalues λ_i of T are discrete and $\lambda_i \geq (\pi/2)^2$, and so we take

$$\Delta = \Delta_1 = 0$$

and

$$\Delta = \Delta_2 = -(2/\pi)^2$$

in $J(\Delta, \phi_{n+1})$ to give minimum and maximum principles respectively at the unique stationary point $\phi_{n+1} = \psi_{n+1}$. We use

$$J(\Delta_2, \psi) \leq -\langle\phi|f\rangle \leq J(\Delta_1, \psi), \tag{21}$$

where $J(\Delta_2, \Phi)$ represents a minimum principle at the unique stationary point $\psi = \phi$ with

$$\Delta = \Delta_2 \leq -\frac{1}{\lambda_i}$$

for all i. This gives

$$J(\Delta_2, \psi) \leq -\langle\phi|f\rangle \leq J(\Delta_1, \Phi). \tag{22}$$

In out particular case

$$J(-4/\pi^2, \Phi_{n+1}) \leq -\langle\psi_{n+1}|f(\phi_n)\rangle \leq J(0, \Phi_{n+1}). \tag{23}$$

Defining the functional by

$$G(\Phi_{n+1}) = J(0, \Phi_{n+1}) \tag{24}$$

we first apply this variational-iterative approach using the one-parameter trial function

$$w_{n,1} = a_{n,1} \sin\frac{\pi y}{2}$$

to equation (15) subject to the boundary conditions in (14).

§6. Discussion and Results

The results obtained from the 1-parameter trial function suggest that we require more complex trial functions. Hence the calculations were repeated with trial functions of the form

$$w_{n,m} = \sum_{j=1}^{m} a_{n,m}^{(j)} \sin \frac{(2j-1)\pi y}{2}, \tag{25}$$

where the form of $w_{n,m}$ is suggested by the eigenfunctions in equation (19). In each case the measure of convergence is provided by

(a) $\{G(\phi) - G(\phi_p)\}$,

(b) $S_p = -\frac{4}{\pi^2}\langle Tw_p - f(w_p) \mid Tw_p - f(w_p)\rangle$,

where ϕ_p is the best p-parameter variational approximation obtained.

To find ϕ_p we use the approach based on the functional

$$G(w_{n,p}) = \langle w_{n,p}|Tw_{n,p}\rangle - 2\langle w_{n,p}|\bar{f}(w_{n-1,p})\rangle, \tag{26}$$

where

$$\bar{f}(w_{n-1,p}) = (1-b)Tw_{n-1,p} + bf(w_{n-1,p}) \tag{27}$$

and proceed in the usual way with a suitable choice of the initial values for the iterative scheme. This ensures rapid convergence of the variational scheme.

The method has been used to produce variational approximations Φ_p of increasing accuracy (as measured by S_p) for $p = 3$ up to $p = 7$. Table 1 shows the variations of the measures of convergence S_p and $\{G(\phi) - G(\phi_p)\}$ with p. The method is clearly converging. This is confirmed by Table 2 which shows the variation of the error in the p-parameter approximation $\{\phi(y) - \phi_p(y)\}$ with $y = 0.1(0.1)1.0$.

§7. Full Burgers' Equation

The encouraging results obtained for the steady-state case of Burgers' equation gives confidence in the application of complementary variational principles to Burgers' equation itself and this work is in progress.

We illustrate by considering Burgers' equation (1) under the boundary and initial conditions,

$$\begin{cases} u(o,t) = u(1,t) = 0, & t > 0, \\ u(x,0) = \sin \pi x, & 0 \leq x \leq 1. \end{cases} \tag{28}$$

In this case, $u(x,t) \to 0$ as $t \to \infty$. Therefore we assume

$$u(x,t) \; t \rightleftarrows \infty \; e^{-at}, \quad \alpha > 0.$$

It is best to substitute $T = 1 - e^{-t}$ and then equation (1) gives

$$(1-T)u_T + uu_x = \nu u_{xx}, \tag{29}$$

inside the square $0 \leq x \leq 1, 0 \leq T \leq 1$. Equation (29) can be rewritten as

$$u_{xx} + u_{TT} = (1+\nu)u_{xx} + u_{TT} - uu_x - (1-T)u_T, \tag{30}$$

which has the form

$$\nabla^2 u = F(u). \tag{31}$$

We write

$$u = f(x,T) + \phi,$$

where $f(x,T)$ obeys the boundary conditions of the problem and hence $\phi = 0$ on the boundary δV of the square V. Equation (31) then gives

$$\nabla^2 \phi = F(u) - \nabla^2 f = G(\phi). \tag{32}$$

Hence we can take $T = \nabla^2$ i.e. a completely continuous self-adjoint operator.

For the conditions in (28) we have

$$F(u) = (1+\nu)u_{xx} + u_{TT} - uu_x - (1-T)u_T,$$

and the boundary condition is $f(x,T) = (1-T)\sin \pi x$, on the boundary of the square V. Therefore

$$-\nabla^2 f = \pi^2(1-T)\sin \pi x.$$

This means that we are solving

$$\nabla^2 \phi = F(u) + \pi^2(1-T)\sin \pi x,$$

where ∇^2 is now symmetric on ϕ. The eigenvalues of ∇^2 are defined by

$$\phi_{xx} + \phi_{TT} = -\lambda\phi$$

and this gives

$$\phi(x,T) = P(x)Q(T)$$

with

$$P(x) = A\sin n\pi x \text{ and } Q(T) = C\sin N\pi T.$$

Therefore the eigenvalues are

$$\lambda = \pi^2(n^2 + N^2), \quad n, N = 1, 2, 3, \cdots$$

and the eigenfunctions are

$$V_{n,N} = a^{(n,N)} \sin n\pi x \ \sin N\pi T.$$

We are now looking for a solution u of the form

$$u = (1-T)\sin n\pi x + \sum_n \sum_N a^{(n,N)} \sin n\pi x \ \sin N\pi T.$$

For the 1-parameter approximation we try

$$\Phi_{(1,1)} = (1-T)\sin \pi x + a^{(1,1)} \sin \pi x \ \sin \pi T.$$

Them we apply the same techniques as before to maximise the functional, obtain bounds and hence find a measure of convergence for each of the trial functions used.

References

[1] A.M. Arthurs, *Complementary Variational Principles*, Oxford Univ. Press (Clarendon), Oxford, 1970.

[2] A.M. Arthurs and N. Anderson, *Dual extremum principles for hear transfer problem with variable thermal properties*, Quart. J. Mech. Appl. Math., 35(3)(1977), 337–343.

[3] A.M. Arthurs and C.W. Coles, *Error bounds and variational methods for non-linear differential and integral equations*, J. Inst. Math. Appl., 7(1977), 324–330

[4] A.M. Arthurs and P.D. Robinson, *Complementary variational principles for DEL2PHI =F(PHI) with applications to Thomas-Fermi and Liouville Equations*, Proc. Cambridge Philos. Soc., 65(1969), 535.

[5] B.L. Burrows and A.J. Perks, *Variational calculations for scattering lengths*, J.Phys. B, 10(1977), 1591–1600.

[6] B.L. Burrows and A.J. Perks, *Variational calculations for phase shifts*, J. Phys. B, 12(1979), 2687–2695.

[7] B.L. Burrows and A.J. Perks, *Complementary variational principles and variational iterative principles*, J. Phys. A, 14(1981), 797–808.

[8] T. Kato, Math. Ann., 126(1953), 253.

[9] L.J. Walpole, J. Inst. Math. Appl., 14(1974), 113.

Table 1. Variation of the measures of convergence $\{G(\phi) - G(\phi_p)\}$ and S_p with p.

P	$\{G(\phi) - G(\phi_p)\}$	S_p
1	+3.8681	-1.1296
2	-0.1185	-0.1059
3	-0.0416	-0.0725
4	-0.0160	-0.0576
5	-0.0045	-0.0488
6	-0.0016	-0.0429
7	-0.0004	-0.0394

Table 2 Variation of the error $\varepsilon_p(y) = \{\phi(y) - \phi_p(y)\} \times 10^4$ of ϕ_p for $y = 0.1(0.1)1.0$.

y	ε_1	ε_2	ε_3	ε_4	ε_7
0.1	+1969	+118	+25	+27	+34
0.2	+3646	-67	-9	-20	-36
0.3	+4960	-194	-53	-20	-5
0.4	+5874	-250	-115	-84	-58
0.5	+6389	-281	-164	-113	-87
0.6	+6556	-276	-149	-98	-69
0.7	+6474	-257	-86	-43	-9
0.8	+6271	-242	-52	+8	+6
0.9	+6081	-234	-74	-15	-12
1.0	+6006	-232	-95	-50	-23

Proceedings of the Asian Mathematical
Conference, 1990

On the Fourier-Stieltjes Transform of Cantor Functions

HARRY M. CARPIO
Msu-Iligan Institute of Technology
9200 Iligan City, The Philippines

§1. Introduction

A mapping w from $\mathbb{R}^n$ into $\mathbb{R}^n$ is called a similitude if $|w(x) - w(y)| = r|x - y|$, for all x, y in $\mathbb{R}^n$ and for some $r > 0$. The number r is called the Lipschitz constant of w and is denoted by Lip w. If $r < 1$, then w is a contracting similitude. Throughout this note we shall consider a fixed family $\Omega = \{w_1, w_2, \cdots, w_m\}$ of $m \geq 2$ contracting similitudes with Lipschitz constants $r_1, r_2, \cdots, r_m$, respectively. The positive real number denoted by s throughout satisfying $\sum_{j=1}^{m} r_j^s = 1$ is called the similarity dimension of E.

Theorem 1.1 (Hutchinson [2]). *The family $\Omega = \{w_1, \cdots, w_m\}$ of m contracting similitudes on $\mathbb{R}^n$ determines a unique non-empty compact set E for which*

$$E = \bigcup_{j=1}^{m} w_j(E). \tag{1.1}$$

Furthermore, there exists a unique Borel probability measure μ on $\mathbb{R}^n$ with compact support such that

$$\mu = \sum_{j=1}^{m} r_j^s w_j(\mu), \tag{1.2}$$

where $w_j(\mu)$ is the image measure of μ under the mapping w_j. The support of the measure μ is E.

The measure μ is called a Cantor measure and the unique distribution function c corresponding to μ is called the Cantor function belonging to Ω. This is an n-dimensional generalization of the classical Cantor-Lebesgue function, which is a distribution function on $\mathbb{R}$. The main problem we consider here is to characterize the Fourier-Stieltjes transform of this type of Cantor functions. Recall that if ν is a finite Borel measure on $\mathbb{R}^n$, the Fourier-Stieltjes transform of ν is the complex-valued function on $\mathbb{R}^n$ given by

$$\hat{\nu}(t) := \int \exp i\langle t, x\rangle d\nu(x), \quad t \in \mathbb{R}^n.$$

The function $\hat{\nu}$ is uniformly continuous on $\mathbb{R}^n$. If f is a distribution function on $\mathbb{R}^n$, the Lebesgue-Stieltjes measure determined by f is denoted by μ_f. The Fourier-Stieltjes transform of f is defined by $\widehat{f}(t) := \widehat{\mu}_f(t)$. For the small amount of Fourier Analysis needed here, see Bauer [1, Ch. VIII].

§2. Preliminaries

We shall establish notations, state definitions and collect facts that are needed later. It was shown in [2, p. 717] that a similitude w on $\mathbb{R}^n$ can be written as $w(x) = r\Phi(x) + a$, where r =Lip w and Φ is an orthogonal transformation on $\mathbb{R}^n$. The adjoint of Φ is denoted by Φ^*. we recall that $\langle \Phi^*(x), y\rangle = \langle x, \Phi(y)\rangle$, for all x and y in $\mathbb{R}^n$, where $\langle -,-\rangle$ is the ordinary dot product in $\mathbb{R}^n$. For any positive integer k, the set of k-tuples of numbers $1, 2, \cdots, m$ will be denoted by $C_k^{(m)}$. Each element of $C_k^{(m)}$ is a multi-index and is written $j_1 \cdots j_k$. For each $\alpha \in C_k^{(m)}$, the following notations will be used:

$$w_\alpha = w_{j_1\cdots j_k} := w_{j_1} \circ \cdots \circ w_{j_k}, \tag{2.1}$$

$$\Phi_\alpha = \Phi_{j_1\cdots j_k} := \Phi_{j_1} \circ \cdots \circ \Phi_{j_k}, \tag{2.2}$$

$$r_\alpha = r_{j_1\cdots j_k} := r_{j_1} \cdots r_{j_k}, \tag{2.3}$$

$$X_\alpha = X_{j_1\cdots j_k} := w_{j_1} \dots w_{j_k}(x), \quad X \subseteq \mathbb{R}^n, \tag{2.4}$$

$$x_\alpha = x_{j_1\cdots j_k} := w_{j_1} \dots w_{j_k}(x), \quad x \in \mathbb{R}^n. \tag{2.5}$$

If $w_{j_1}, \cdots, w_{j_k}$ are similitudes with Lipschitz constants $r_{j_1}, \cdots, r_{j_k}$, respectively, their composition $w_{j_1\cdots j_k}$ is also a similitude with Lipschitz constant $r_{j_1\cdots j_k}$. If 0 is the zero element of $\mathbb{R}^n$, then for each integer t, with $1 \leq t < k$,

$$0_{j_1\cdots j_k} = 0_{j_1\cdots j_t} + r_{j_1\cdots j_t}\Phi_{j_1,\cdots,j_t}(0_{j_{t+1}\cdots j_k}). \tag{2.6}$$

For each positive integer k, we write

$$\sum_{\alpha\in C_k^{(m)}} A_\alpha := \sum_{j_1=1}^{m} \cdots \sum_{j_k=1}^{m} A_{j_1\cdots j_k}.$$

§3. Transforms of Cantor Functions

In the proof of Theorem 1.1, Hutchinson considered an operator H on the set $\mathcal{M}^c$ of all finite Borel probability measures on $\mathbb{R}^n$ with compact support. This operator is given by

$$H(\mu) := \sum_{j=1}^{m} r_j^s w_j(\mu), \quad \mu \in \mathcal{M}^c.$$

The second part of Theorem 1.1 asserts the existence of a unique probability measure μ in $\mathcal{M}^c$ such that $H(\mu) = \mu$. By straightforward calculation we have, for each $\mu \in \mathcal{M}^c$,

$$\widehat{H(\mu)}(t) = \sum_{j=1}^{m} r_j^s \exp i\langle t, 0_j\rangle \widehat{\mu}(r_j\Phi_j^*(t)), \tag{3.1}$$

where 0 is the zero element of $\mathbb{R}^n, r_j = \text{Lip}\,w_j$, and Φ_j^* is the adjoint of the orthogonal transformation Φ_j. As mentioned earlier, the unique Borel probability measure μ for which $H(\mu) = \mu$ is called a Cantor measure. The distribution function c associated with μ is called the Cantor function belonging to Ω.

Definition 3.1. Let $C(\mathbb{R}^n, \mathbb{C})$ be the class of all continuous, complex-valued functions f on $\mathbb{R}^n$. We define the operator $T : C(\mathbb{R}^n, \mathbb{C}) \rightarrowtail C(\mathbb{R}^n, \mathbb{C})$ by

$$T(f)(t) := \sum_{j=1}^{m} r_j^s \exp i\langle t, 0_j\rangle f(r_j \Phi_j^*(t)). \tag{3.2}$$

Moreover, we put $T^1 := T$ and $T^{k+1} := T^k \circ T = T \circ T^k$, for $k \geq 1$. The reader will note that equation (3.2) is analogous to equation (3.1).

Proposition 3.2. *Let μ be a Borel probability measure on $\mathbb{R}^n$ with compact support. Then*

(i) $T(\widehat{\mu}) = \widehat{H(\mu)}$;

(ii) $T(\widehat{\mu}) = \widehat{\mu}$ *if and only if* $\mu = \mu_c$;

(ii) $T(\widehat{c}) = \widehat{c}$.

Proof. Part (i) obtains easily by comparing (3.1) and (3.2). The rest follows from Theorem 1.1 and the Uniqueness Theorem of Fourier-Stieltjes transforms.

Remark. We want to prove that if $T(f) = f$, then $f = \widehat{c}$. If $f = \widehat{\mu}$ for some Borel probability measure μ with compact support, then this assertion follows from Proposition 3.2 (ii). The general case is a consequence of the following theorem.

Theorem 3.3. *For every function $f \in C(\mathbb{R}^n, \mathbb{C})$ with $f(0) = 1, T^k(f)$ converges uniformly to $\widehat{c}$ on every bounded subset of $\mathbb{R}^n$.*

Proof. By (2.6), it follows by induction that

$$T^k(f)(t) := \sum_{\alpha \in C_k^{(m)}} r_\alpha^s \exp i\langle t, 0_\alpha\rangle f(r_\alpha \Phi_\alpha^*(t)). \tag{3.3}$$

We fix $\rho > 0$ and let $\varepsilon > 0$ be given. Since f and $\widehat{c}$ are continuous and since $f(0) = 1 = \widehat{c}(0)$, there exists a $\delta > 0$ such that $|f(z) - \widehat{c}(z)| < \varepsilon$ whenever $|z| < \delta$. For this δ, there exists an integer $N > 0$ such that if $k \geq N$, then $r_\alpha \rho < \delta$, for each multi-index α in $C_k^{(m)}$. Suppose now that $|t| < \rho$ and let α be a multi-index in $C_k^{(m)}$. Since Φ_α is orthogonal, so is Φ_α^*, and it follows that if $k \geq N$, then

$$|r_\alpha \Phi_\alpha^*(t)| \leq r_\alpha \rho < \delta. \tag{3.4}$$

Thus if $k \geq N$, then

$$|f(r_\alpha \Phi_\alpha^*(t)) - \widehat{c}(r_\alpha \Phi_\alpha^*(t))| < \varepsilon. \tag{3.5}$$

Hence, if $k \geq N$, we have, by (3.5) and Proposition 3.2, that

$$\begin{aligned} |T^k(f)(t) - \widehat{c}(t)| = {} & |T^k(f)(t) - T^k(\widehat{c})(t)| \\ < {} & \sum_{\alpha \in C_k^{(m)}} r_\alpha^s |\exp i \leq t, 0_\alpha\rangle|\varepsilon = \varepsilon \sum_{\alpha \in C_k^{(m)}} r_\alpha^s = \varepsilon. \end{aligned}$$

Thus $T^k(f)$ converges to $\hat{c}$ uniformly on each ball of radius ρ about the origin and hence on each bounded subset of $\mathbb{R}^n$.

Corollary 3.4. *Let $f \in \mathcal{C}(\mathbb{R}^n, \mathbb{C})$ and suppose $f(0) = 1$. Then*
(i) $T^k(f)$ converges pointwise to $\hat{c}$;
(ii) If $T(f) = f$, then $f = \hat{c}$.

Corollary 3.5. *Let $a \in \mathbb{R}^n$. Then for each t in $\mathbb{R}^n$*

$$\hat{c}(t) = \lim_k \sum_{\alpha \in C_k^{(m)}} r_\alpha^s \exp i\langle t, a_\alpha \rangle. \tag{3.6}$$

Proof. Let $f(x) = 1$, for all x. It follows from Corollary 3.4 (i) and equation (3.3) that

$$\hat{c}(t) = \lim_k \sum_{\alpha \in C_k^{(m)}} r_\alpha^s \exp i\langle t, 0_\alpha \rangle.$$

The argument used in Theorem 3.3 is used here again to show that $|\exp i\langle t, 0_\alpha\rangle - \exp i\langle t, a_\alpha\rangle|$ tends to 0 as k tends to ∞. This yields the conclusion.

Definition 3.6. Let $a \in \mathbb{R}^n$ and let $\mathcal{E}_a$ be the Dirac distribution at a. The real-valued function defined by

$$\mathcal{B}_k^a(x) := \sum_{\alpha \in C_k^{(m)}} r_\alpha^s \mathcal{E}_{a_\alpha}(x), \quad x \in \mathbb{R}^n, \tag{3.7}$$

is a generalized Bernoulli distribution on $\mathbb{R}^n$.

Theorem 3.7. *Let $a \in \mathbb{R}^n$. Then*

$$\hat{c}(t) = \lim_k \widehat{\mathcal{B}_k^a}(t).$$

Hence $\mathcal{B}_k^a$ converges weakly to $\hat{c}$, for each a in $\mathbb{R}^n$.

Proof. Straightforward calculation using Corollary 3.5.

Theorem 3.8. *Suppose that $w_j(x) = r\Phi(x) + 0_j$ for each $j = 1, 2, \cdots, m$. Then*

$$\hat{c}(t) = \prod_{p=1}^{\infty} \hat{\mathcal{B}}\left[r^{p-1}[\Phi^*]^{p-1}(t)\right], \tag{3.8}$$

where $\mathcal{B}$ is the generalized Bernoulli distribution given by

$$\mathcal{B}(x) := \sum_{j=1}^{m} r^s \mathcal{E}_{0_j}(x). \tag{3.9}$$

In particular, if $\Phi(x) = x$, then

$$c(x) = *_{p=1}^{\infty} \mathcal{B}(r^{1-p}x). \tag{3.10}$$

Proof. Let α be a multi-index in $C_k^{(m)}$. Then it follows from equation (2.6) that

$$0_\alpha = \sum_{p=1}^{k} r^{p-1}\Phi^{p-1}(0_{j_p}).$$

Therefore,

$$\begin{aligned}\sum_{\alpha\in C_k^{(m)}} r_\alpha^s \exp i\langle t, 0_\alpha\rangle &= \sum_{\alpha\in C_k^{(m)}} r^{ks} \exp i\langle t, \sum_{p=1}^{k} r^{p-1}\Phi^{p-1}(0_{jp})\rangle \\ &= \sum_{\alpha\in C^{(m)}} \left[\prod_{p=1}^{k} r^s \exp i r^{p-1}\langle t, \Phi^{p-1}(0_{jp})\rangle\right] \\ &= \prod_{p=1}^{k}\left[\sum_{j=1}^{m} r^s \exp i\langle r^{p-1}[\Phi^*]^{p-1}(t), 0_j\rangle\right] \\ &= \prod_{p=1}^{k}\left[\int \exp i\langle r^{p-1}[\Phi^*]^{p-1}(t), x\rangle d\mathcal{B}(x)\right] \\ &= \prod_{p=1}^{k}\widehat{\mathcal{B}}\left[r^{p-1}[\Phi^*]^{p-1}(t)\right].\end{aligned}$$

By Corollary 3.5, we have

$$\widehat{c}(t) = \lim_k \prod_{p=1}^{k}\widehat{\mathcal{B}}[r^{p-1}[\Phi^*]^{p-1}(t)].$$

This proves equation (3.8). To prove equation (3.10), we note that for each integer p we have

$$\mathcal{B}(\widehat{r^{1-p}x})(t) = \widehat{\mathcal{B}}(r^{p-1}t).$$

It follows from the Continuity Theorem of Fourier-Stieltjes transforms that equation (3.10) holds.

I wish to thank Prof. D. Kolzov, who initiated the study of Cantor functions on R^n, for his advice.

References

[1] Bauer, Heinz, *Probability Theory and Elements of Measure Theory*, Academic Press Inc., London, 1981.

[2] Hutchinson, J., *Fractals and self-similarity*, Indiana J. Math., 30(1981), 713–747.

[3] Hille, E. and J.D. Tamarkin, *Remarks on a known example of a monotone continuous function*, Amer. Math. Monthly, 36(1929), 255–264.

Proceedings of the Asian Mathematical
Conference, 1990

Skew-Circulant Preconditioners for Skew-Hermitian Type Toeplitz Systems

RAYMOND H. CHAN & XIAO-QING JIN
Department of Mathematics
University of Hong Kong
Hong Kong

§1. Introduction

Let A_n be a Toeplitz matrix of order n having the following form:

$$A_n = \begin{bmatrix} a_0 & a_1 & & a_{n-2} & a_{n-1} \\ -\bar{a}_1 & a_0 & a_1 & \cdot & a_{n-2} \\ -\bar{a}_2 & \cdot & \cdot & \cdot & \cdot \\ \cdot & \cdot & \cdot & \cdot & a_1 \\ -\bar{a}_{n-1} & \cdot & -\bar{a}_2 & -\bar{a}_1 & a_0 \end{bmatrix}, \tag{1}$$

where a_0 is a real number. Obviously, $A_n = a_0 I + H_n$, where I is the identity matrix and H_n is a skew-Hermitian Toeplitz matrix. We call A_n the skew-Hermitian type Toeplitz matrix. We are interested in solving the system $A_n x = b$. This kind of problems often appear in solving discretized hyperbolic differential equations.

The idea of using preconditioned conjugate gradient method with circulant preconditioner for solving symmetric Toeplitz systems was first proposed by Strang [4]. Chan and Strang [1] then proved that if A_n is a symmetric Toeplitz matrix with entries which are Fourier coefficients of a positive function in the Wiener class, then the eigenvalues of $\tilde{C}_n^{-1} A_n$ are clustered around 1. Here $\tilde{C}_n$ is the symmetric circulant matrix which copies the central diagonals of A_n. These results are generalized to Hermitian positive definite Toeplitz systems in Chan [2]. However, only positive definite matrices are discussed in these papers. In this paper, we use the preconditioned conjugate gradient method with skew-circulant preconditioner S_n for solving the skew-Hermitian type Toeplitz systems. We show in §2 that if the generating function $f(\theta)$ is of the form $a_0 + ig(\theta)$, where $g(\theta)$ is a real-valued function in the Wiener class, then the spectrum of $(S_n^{-1}A_n)^*(S_n^{-1}A_n)$ clusters around 1. The numerical result is given in §3.

§2. Spectrum of the Normalized Preconditioned System

Let us begin by supposing that the entries $a_{jl} = a_{l-j}$ of A_n are Fourier coefficients of the complex generating function

$$f(\theta) = \sum_{-\infty}^{\infty} a_k e^{-ik\theta}$$

defined on $[0, 2\pi)$. Since A_n is of the form given by (1), we see that $a_{-k} = -\bar{a}_k$. It is clear that $f(\theta)$ also can be written as

$$f(\theta) = a_0 + ig(\theta),$$

where $g(\theta)$ is a real-valued function. In fact,

$$\begin{aligned} g(\theta) = & \sum_{k=-\infty, k\neq 0}^{\infty} (-ia_k)e^{-ik\theta} = \sum_{k=1}^{\infty}(-i)(a_k e^{-ik\theta} + a_{-k}e^{ik\theta}) \\ = & \sum_{k=1}^{\infty}(-i)(a_k e^{-ik\theta} - \overline{a_k e^{-ik\theta}}) = 2\sum_{k=1}^{\infty} \mathrm{Im}(a_k e^{-ik\theta}), \end{aligned}$$

where $\mathrm{Im}(x)$ denotes the imaginary part of the complex number x. We assume that $g(\theta)$ is a function in the Wiener class, i.e., its Fourier coefficients $\{-ia_k\}$ is in l_1. We then have

$$\sum_{-\infty}^{\infty} |a_k| = |a_0| + \sum_{\substack{k=-\infty, \\ k\neq 0}}^{\infty} |-ia_k| \equiv c_0 < \infty. \tag{2}$$

It is clear that the solution of the system $A_n x = b$ is the same as the solution of the normalized preconditioned system $(P_n^{-1}A_n)^*(P_n^{-1}A_n)x = (P_n^{-1}A_n)^* P_n^{-1}b$ for any preconditioner P_n. When the conjugate gradient method is applied to solve the normalized preconditioned system, the convergence rate of the method depends on the spectrum of $(P_n^{-1}A_n)^*(P_n^{-1}A_n)$. The more clustered the eigenvalues are, the faster the convergence.

For skew-Hermitian type Toeplitz matrix A_n given by (1), we note that it can always be partitioned as

$$A_n = \frac{1}{2}C_n + \frac{1}{2}S_n,$$

where

$$C_n = \begin{bmatrix} a_0 & a_1 - \bar{a}_{n-1} & \cdot & a_{n-2} - \bar{a}_2 & a_{n-1} - \bar{a}_1 \\ -\bar{a}_1 + a_{n-1} & a_0 & \cdot & \cdot & \cdot \\ \cdot & \cdot & \cdot & \cdot & \cdot \\ -\bar{a}_{n-2} + a_2 & \cdot & \cdot & a_0 & a_1 - \bar{a}_{n-1} \\ -\bar{a}_{n-1} + a_1 & \cdot & \cdot & -\bar{a}_1 + a_{n-1} & a_0 \end{bmatrix},$$

and

$$S_n = \begin{bmatrix} a_0 & a_1 + \bar{a}_{n-1} & \cdot & a_{n-2} + \bar{a}_2 & a_{n-1} + \bar{a}_1 \\ -(a_{n-1} + \bar{a}_1) & a_0 & \cdot & \cdot & \cdot \\ \cdot & \cdot & \cdot & \cdot & \cdot \\ -(a_2 + \bar{a}_{n-2}) & \cdot & \cdot & a_0 & a_1 + \bar{a}_{n-1} \\ -(a_1 + \bar{a}_{n-1}) & \cdot & \cdot & -(a_{n-1} + \bar{a}_1) & a_0 \end{bmatrix}.$$

Clearly C_n is circulant and S_n is skew-circulant. We will use S_n as preconditioner for A_n. We show in the following that if the generating function $f(\theta) = a_0 + ig(\theta)$, where $g(\theta)$ is a real-valued function in the Wiener class, then

I. S_n and its inverse are uniformly bounded in the l_2-norm;

II. $S_n^{-1}A_n$ can be partitioned into the sum of $I + L_n + M_n$, where L_n is a low rank matrix and M_n is a matrix with small l_2-norm;

III. the spectrum of $(S_n^{-1}A_n)^*(S_n^{-1}A_n)$ clusters around one.

For claim I, we have

Theorem 1. *Let $f(\theta) = a_0 + ig(\theta)$, where $g(\theta)$ is a real-valued function in the Wiener class and a_0 is a nonzero real number. Then for all n, the skew-circulant matrix S_n and its inverse are uniformly bounded in the l_2-norm.*

Proof. We note that $S_n = a_0 I + \hat{H}_n$, where $\hat{H}_n$ is a skew-Hermitian Toeplitz matrix. Therefore, the eigenvalues of S_n can be expressed as $\lambda_j(S_n) = a_0 + b_j i$, where b_j are real. Hence

$$|\lambda_j(S_n)| \geq |a_0| > 0, \quad for \quad j = 0, \cdots, n-1. \tag{3}$$

In particular, we see that S_n is invertible. We also notice that the eigenvalues $\lambda_j(S_n)$ of S_n also can be expressed as

$$\lambda_j(S_n) = a_0 + \sum_{p=1}^{n-1} (a_p + \bar{a}_{n-p}) e^{\frac{(2j+1)\pi i p}{n}}, \quad \text{for} \quad j = 0, \cdots, n-1,$$

see Davis [3]. Hence we have by (2)

$$|\lambda_j(S_n)| \leq |a_0| + \sum_{p=1}^{n-1} |a_p + \bar{a}_{n-p}| \leq 2 \sum_{p=-\infty}^{\infty} |a_p| = 2c_0, \quad \text{for} \quad j = 0, \cdots, n-1.$$

Since S_n is a normal matrix, we have

$$\|S_n\|_2 = |\lambda_{\max}(S_n)| \leq 2c_0, \quad \|S_n^{-1}\|_2 = |\lambda_{\min}^{-1}(S_n)| \leq \frac{1}{|a_0|}.$$

Here $\lambda_{\max}(S_n)$ and $\lambda_{\min}(S_n)$ denote respectively the largest and the smallest eigenvalues in absolute value for S_n. □

We remark that when $a_0 = 0$, the matrix A_n is skew-Hermitian and the generating function $f(\theta) = ig(\theta)$. Then if we assume that $g(\theta) \geq g_{\min} > 0$ for all $\theta \in [0, 2\pi)$, then instead of getting (3), we can get $|\lambda_j(S_n)| \geq g_{\min} > 0$, for $j = 0, \cdots, n-1$, see Chan and Strang [1]. Thus we have

Theorem 1'. *Let $f(\theta) = ig(\theta)$, where $g(\theta)$ is a real positive function in the Wiener class. Then for all n sufficiently large, the skew-circulant matrix S_n and its inverse are uniformly bounded in the l_2-norm.*

For claim II, we first show that the matrix $C_n - A_n$ can be expressed as the sum of a low rank matrix and a matrix with small l_2-norm.

Lemma 1. *Let $f(\theta) = a_0 + ig(\theta)$, where $g(\theta)$ is a real-valued function in the Wiener class. Then for all $\varepsilon > 0$, there exists an $N > 0$, such that for all $n > N, C_n - A_n = W_n^{(N)} + U_n^{(N)}$, where $\|W_n^{(N)}\|_2 \le \varepsilon$ and $\text{rank}(U_n^{(N)}) \le 2N$.*

Proof. Define $B_n = C_n - A_n$. It can be easily checked that B_n is skew-Hermitian Toeplitz matrix with entries $b_{ij} = b_{j-i}$ given by

$$b_l = \begin{cases} 0 & l = 0, \\ a_{l-n} & 0 < l < n, \\ -\bar{b}_{-l} & 0 < -l < n. \end{cases}$$

Since g is in the Wiener class, by (2) for all given $\varepsilon > 0$, there exists an $N > 0$, such that

$$\sum_{l=N+1}^{\infty} |a_l| < \varepsilon.$$

Let $U_n^{(N)}$ be the n-by-n matrix obtained from B_n by replacing the $(n-N)$-by-$(n-N)$ leading principal submatrix of B_n by the zero matrix. Then rank $(U_n^{(N)}) \le 2N$. Let $W_n^{(N)} \equiv B_n - U_n^{(N)}$. The leading $(n-N)$-by-$(n-N)$ block of $W_n^{(N)}$ is the leading $(n-N)$-by-$(n-N)$ principal submatrix of B_n, hence this block is a Toeplitz matrix. Thus

$$\begin{aligned} \|W_n^{(N)}\|_1 \le \; & 2\sum_{l=1}^{n-N-1} |b_l| = 2\sum_{l=1}^{n-N-1} |a_{l-n}| = 2\sum_{l=N+1}^{n-1} |a_{-l}| \\ = \; & 2\sum_{l=N+1}^{n-1} |a_l| \le 2\sum_{l=N+1}^{\infty} |a_l| < 2\varepsilon. \end{aligned}$$

Since $W_n^{(N)}$ is skew-Hermitian, we have $\|W_n^{(N)}\|_\infty = \|W_n^{(N)}\|_1$. Thus

$$\|W_n^{(N)}\|_2 \le (\|W_n^{(N)}\|_1 \cdot \|W_n^{(N)}\|_\infty)^{\frac{1}{2}} < \varepsilon.$$

Hence the spectrum of $W_n^{(N)}$ lies in $(-\varepsilon, \varepsilon)$. □

Since $C_n - A_n = A_n - S_n$, we have by Lemma 1,

$$\begin{aligned} S_n^{-1}A_n = \; & I + S_n^{-1}(A_n - S_n) = I + S_n^{-1}(C_n - A_n) \\ = \; & I + S_n^{-1}(W_n^{(N)} + U_n^{(N)}) = I + M_n + L_n, \end{aligned}$$

where $M_n = S_n^{-1}W_n^{(N)}$ and $L_n = S_n^{-1}U_n^{(N)}$. By Theorem 1, we thus have our main result.

Theorem 2. *Let $f(\theta) = a_0 + ig(\theta)$, where $g(\theta)$ is a real-valued function in the Wiener class and a_0 is a nonzero real number. Then for all $\varepsilon > 0$, there exists an $N > 0$, such that for all $n > N, S_n^{-1}A_n$ can be written in the form $I + M_n + L_n$, where $\|M_n\|_2 \le \varepsilon$ and* $\text{rank}(L_n) \le 2N$.

Claim III now is an immediate Corollary of Theorem 2.

Corollary 1. *Let $f(\theta) = a_0 + ig(\theta)$, where $g(\theta)$ is a real-valued function in the Wiener class and a_0 is a nonzero real number. Then for all n sufficiently large, the spectrum of $(S_n^{-1}A_n)^*(S_n^{-1}A_n)$ clusters around one. More precisely, for all $\varepsilon > 0$, there exists an $N > 0$*

such that for all $n > N$, at most 4N eigenvalues of $(S_n^{-1}A_n)^*(S_n^{-1}A_n) - I$ can have absolute values exceeding ε.

Proof. By Theorem 2, we have for all $\varepsilon > 0$, there exists an $N > 0$, such that for all $n > N, S_n^{-1}A_n = I + M_n + L_n$ where $\|M_n\|_2 \le \varepsilon$ and $\text{rank}(L_n) \le 2N$. Hence

$$\begin{aligned}(S_n^{-1}A_n)^*(S_n^{-1}A_n) - I = & \ (I + M_n^* + L_n^*)(I + M_n + L_n) - I \\ = & \ M_n + L_n + M_n^* + M_n^*M_n + M_n^*L_n + L_n^* + L_n^*M_n + L_n^*L_n \\ = & \ \hat{M}_n + \hat{L}_n,\end{aligned}$$

where $\hat{M}_n = M_n + M_n^* + M_n^*M_n$ and $\hat{L}_n = L_n^*(I + L_n + M_n) + (I + M_n^*)L_n$.

Thus $\|\hat{M}_n\|_2 \le 2\varepsilon + \varepsilon^2$ and $\text{rank}(\hat{L}_n) \le 4N$. Since both $\hat{M}_n$ and $\hat{L}_n$ are Hermitian, by Cauchy Interlace Theorem, see Wilkinson [5], at most $4N$ eigenvalues of $(S_n^{-1}A_n)^*(S_n^{-1}A_n) - I$ have absolute values exceeding $2\varepsilon + \varepsilon^2$. □

Similar to Theorem 1', when $a_0 = 0$ but $g(\theta)$ is positive, we have

Theorem 2' *Let $f(\theta) = ig(\theta)$, where $g(\theta)$ is a real positive function in the Wiener class. Then for all $\varepsilon > 0$, there exists an $N > 0$, such that for all $n > N, S_n^{-1}A_n$ can be written in the form $I + M_n + L_n$, where $\|M_n\|_2 \le \varepsilon$ and $\text{rank}(L_n) \le 2N$.*

Corollary 1'. *Let $f(\theta) = ig(\theta)$, where $g(\theta)$ is a real positive function in the Wiener class. Then for all n sufficiently large, the spectrum of the normalized preconditioned matrix $(S_n^{-1}A_n)^*(S_n^{-1}A_n)$ clusters around one.*

Hence if the conjugate gradient method is applied to solve the normalized preconditioned system

$$(S_n^{-1}A_n)^*(S_n^{-1}A_n)x = (S_n^{-1}A_n)^*S_n^{-1}b,$$

we can expect a fast convergence rate.

§3. Numerical Result

We consider the following problem in this section.

Let A_n be a Toeplitz matrix with diagonals given by

$$a_k = \begin{cases} 1 & k = 0, \\ (1+k)^{-1.1} & k > 0, \\ -a_{-k} & k < 0. \end{cases}$$

Obviously the generating function of A_n is in the Wiener class. Table 1 shows the number of iterations required to get $\frac{\|r_k\|_2}{\|r_0\|_2} < 10^{-7}$. Here r_k is the residual vector at the k-th iteration. We use the vector of all ones for the right hand side vector b and the zero vector as our initial guess. We note that as n increases, the number of iterations increases for the normalized system, while it remains a constant for normalized preconditioned system.

Table 1. Number of Iterations for Different Systems

n	$A_n^* A_n$	$(S_n^{-1} A_n)^*(S_n^{-1} A_n)$
16	8	8
32	12	8
64	16	8
128	18	8
256	19	8

References

[1] R. Chan and G. Strang, *Toeplitz Equations by Conjugate Gradients with Circulant Preconditioner*, SIAM J. Sci. Statist. Comput., 10(1989), 104–119.

[2] R. Chan, *Circulant Preconditioners for Hermitian Toeplitz Systems*, SIAM J. Matrix Anal. Appl., 10(1989), 542–550.

[3] P. Davis, *Circulant Matrices*, John Wiley & Sons, Inc., New York, 1979.

[4] G. Strang, *A Proposal for Toeplitz Matrix Calculations*, Stud. Appl. Math., 74(1986), 171–176.

[5] J. Wilkinson, *The Algebraic Eigenvalue Problem*, Clarendon Press, Oxford, 1965.

Proceedings of the Asian Mathematical
Conference, 1990

Regularity of Weak Solutions for a Class of Degenerate Elliptic Systems in Diagonal Form

CHEN BAOYAO
Department of Mathematics
Zhongshan University
Guanzhou 510275, China

§ 1. Weighted Sobolev Space $H^{1,p}(\Omega, \mathbb{R}^N, \lambda)$

We say that $\lambda(x)$ belongs to Muckenhoupt class A_p, if $\lambda(x) \geq 0$ and satisfies

$$L_p = \sup_{B_R} \left\{ \frac{1}{|B_R|} \int_{B_R} \lambda(x) dx \right\} \left\{ \frac{1}{|B_R|} \int_{B_R} \lambda(x)^{\frac{-1}{p-1}} dx \right\}^{p-1} < +\infty, \quad 1 < p < +\infty$$

for all balls $B_R \subset \mathbb{R}^n$ with center x_0 and radius R, where $|B_R| = \text{meas } B_R$.

We shall use the following notations:

$$\lambda^{\pm 1}(B_R) = \int_{B_R} \lambda^{\pm 1}(x) dx, \quad u_R = \frac{1}{\lambda(B_R)} \int_{B_R} u \lambda dx,$$

$$⨍_{B_R} u dx = \lambda^{-1}(B_R) \int_{B_R} u \lambda dx.$$

Suppose $\lambda(x) \in A_2$ and denote $D = \{x \in \Omega, \lambda(x) = 0\}$, where $\lambda(x) = |x_n|^\alpha \in A_2$, $(0 < \alpha < 1)$. We know that diam D can attain $n-1$, although meas $D = 0$.

Let $\Omega \subset \mathbb{R}^n$ be a bounded domain, and $H^{1,p}(\Omega, \lambda)$ denote the weighted Sobolev space with norm:

$$\|u\|_{H^{1,p}(\Omega,\lambda)} = \|u\|_{L^p(\Omega,\lambda)} + |\nabla u|_{L^p(\Omega,\lambda)},$$

$$\|u\|_{L^p(\Omega,\lambda)} = \left(\int_\Omega |u|^p \lambda dx \right)^{1/p},$$

$$H^{1,p}(\Omega, \mathbb{R}^N, \lambda) = [H^{1,p}(\Omega, \lambda)]^N.$$

Denote the critical imbedding exponent of $H^{1,p}(B_R, \lambda)$ by r.

According to [C] and [F], we have the following lemmas:

Lemma 1. *Suppose that $\lambda(x) \in A_2$ is bounded. Then there exist positive constants R_0, C_1 and C_2, such that*

$$C_1 R^{\frac{2r}{r-2}} \leq \lambda(B_R) \leq C_2 R^n, \quad \forall R \leq R_0, \quad \frac{2n}{n-1} < r \leq \frac{2n}{n-2}. \tag{1.1}$$

Lemma 2. *Suppose $\lambda(x) \in A_p$. There exists a positive constant C such that for all $u \in H^{1,p}(B_R, I\!R^N, \lambda)$ and all $k \in [1, r/p]$, we have*

$$\left\{\frac{1}{\lambda(B_R)}\int_{B_R}|u-u_R|^{kp}\lambda(x)dx\right\}^{1/kp} \leq CR\left\{\frac{1}{\lambda(B_R)}\int_{B_R}|\nabla u|^p\lambda(x)dx\right\}^{1/p}. \tag{1.2}$$

By these lemmas we can obtain the weighted Sobolev–Poincaré inequality,

$$\left\{\int_{B_R}|u-u_R|^r\lambda dx\right\}^{1/r} \leq C\left\{\int_{B_R}|\nabla u|^2\lambda dx\right\}^{1/2}. \tag{1.3}$$

By (1.1) –(1.3) , for all $\rho \in (0, R)$ and $2(1-\frac{2}{r}) < s < 2$ we will have

$$R^2 ⨍_{B_R}|u|^r\lambda dx \leq c\left[\int_{B_R}(|\nabla u|^2+|u|^r)\lambda dx\right]^{1-\frac{2}{r}} ⨍_{B_R}(R^2|\nabla u|^2+|u|^2)\lambda dx, \tag{1.4}$$

$$\int_{B_R}|u|^{\frac{2}{2-s}}\lambda dx \leq C\lambda(B_R)^{\frac{2}{r}-1}[\int_{B_R}|u|^{\frac{2r}{r-2}\cdot\frac{s-1}{2-s}}\lambda dx]^{1-\frac{2}{r}}\int_{B_R}(R^2|\nabla u|^2+|u|^2)\lambda dx, \tag{1.5}$$

$$\begin{gathered}\int_{B_\rho}|u|^2\lambda dx \leq C\left(\frac{\lambda(B_\rho)}{\lambda(B_R)}\right)^{1-\frac{2}{r}}\int_{B_R}(R^2|\nabla u|^2+|u|^2)\lambda dx,\\ ⨍_{B_\rho}|u|^2\lambda dx \leq c(\tfrac{\rho}{R})^2 ⨍_{B_R}(R^2|\nabla u|^2+|u|^2)\lambda dx.\end{gathered} \tag{1.6}$$

§ 2. $C^{0,\alpha}_{loc}(\Omega, I\!R^N)$ Regularity of Weak Solutions

Consider the quasilinear elliptic systems in diagonal form

$$-D_\alpha[a^{\alpha\beta}(x,u)D_\beta u^i + a_i^\alpha(x,u)] = B_i(x,u,\nabla u), \quad i=1,2,\cdots,N, \tag{2.1}$$

satisfying the degenerate ellipticity condition

$$\lambda(x)|\xi|^2 \leq a^{\alpha\beta}(x,u)\xi_\alpha,\xi_\beta \leq \Lambda\lambda(x)|\xi|^2, \quad \forall\xi\in I\!R^n, \tag{2.2}$$

where $\lambda(x) \in A_2$ is bounded.

We first state three well-known conditions:

I.

$$|a_i^\alpha(x,u)| \leq \lambda(x)(|u|^{r/2}+|f(x)|), \tag{2.3}$$

$$\begin{aligned}|B_i(x,u,p)| &\leq \lambda(x)(|p|^{2(1-\frac{1}{r})}+|u|^{r-1}+|g(x)|,\\ f(x),g(x) &\in L^p(\Omega,\lambda), \quad p\geq 2n,\end{aligned} \tag{2.4}$$

(called the controllable growth condition);

II.

$$|a^{\alpha\beta}(x,u)| \leq \lambda(x)(|u|^{\frac{1}{2-s}}+|f|), \quad 2(1-\frac{1}{r}) < s < 2, \tag{2.5}$$

$$|B_i(x,u,p)| \le \mu\lambda(x)(|p|^s + |u|^{\frac{s}{2-s}} + |g(x)|), \tag{2.6}$$

(called the middle growth condition);

III.

$$|B_i(x,u,p)| \le \mu\lambda(x)(|p|^2 + |g(x)|), \tag{2.7}$$

(called the natural growth condition.)

Let $u(x)$ be a weak solution to (2.1) , $a^{\alpha\beta}(x) = a^{\alpha\beta}(x, u(x))$. Let us consider the boundary problem:

$$\begin{aligned} -D_\alpha(a^{\alpha\beta}(x)D_\beta v^i) = 0, \qquad x \in B_R, \quad i = 1, 2, \cdots, N \\ v^i - u^i \in H_0^1(B_R, \lambda). \end{aligned} \tag{2.8}$$

According to $[Y]$ and using (1.1) we have the following lemma.

Lemma 3. *Suppose that $0 < \rho < R/32$, $R <$ dist $(x, \partial\Omega)$. Then the boundary problem (2.8) has a unique solution . Moreover, we have the estimate:*

$$\fint_{B_\rho} \rho^2 |\nabla u|^2 \lambda dx = C\left(\frac{\rho}{R}\right)^{2n+2\delta} \fint_{B_R} R^2 |\nabla u|^2 \lambda dx, \tag{2.9}$$

where $\delta \in (0,1)$ is some constant.

Theorem 1. *Suppose the conditions (2.2)–(2.4) hold. Then all $H^1(\Omega, I\!R^N, \lambda)$ weak solutions of (2.1) belong to $C^{0,\alpha}_{loc}(\Omega, I\!R^N)$.*

Theorem 2. *Suppose the conditions (2.2) (2.5) and (2.6) hold. Then all $H^1 \cap L^{\frac{2r}{r-2}\cdot\frac{s-1}{2-s}}(\Omega, I\!R^N, \lambda)$ weak solutions of (2.1) belong to $C^{0,\alpha}_{loc}(\Omega, I\!R^N)$.*

Theorem 3. *Suppose the conditions (2.2) (2.3) and (2.7) hold, and $2\mu M < 1$, $M =$* $\sup_\Omega |u|$ *. Then all $H^1(\Omega, I\!R^N, \lambda) \cap L^\infty(\Omega, I\!R^N)$ weak solutions of (2.1) belong to $C^{0,\alpha}_{loc}(\Omega, I\!R^N)$.*

Proof of Theorme 2. Let $u \in H^1 \cap K^{\frac{2r}{r-2}\cdot\frac{s-1}{2-s}}(\Omega, I\!R^N, \lambda)$ be a weak solution of (2.1) . Spliting $u = v + w$, w^1 obviously satisfies the equation,

$$-D_\alpha[a^{\alpha\beta}(x,u)D_\beta w^i + a_i^\alpha(x,u)] = B_i(x,u,\nabla u).$$

Set the test function $\phi^i = w^i$. By (1.4) we get

$$\begin{aligned} \int_{B_R} |\nabla w|^2 \lambda dx \le \quad & C\int_{B_R} (|\nabla u|^s + |u|^{s/(2-s)} + |g|)\,|w|\,\lambda dx \\ & + C\int_{B_R} (|u|^{2/(2-s)} + |f|^2)\lambda dx, \end{aligned}$$

$$\begin{aligned} \int_{B_R} |\nabla u|^s\,|w|\,\lambda dx \le \quad & \varepsilon\int_{B_R} |\nabla w|^2 \lambda dx + C\int_{B_R} |\nabla u|^2 \lambda dx \\ & \cdot\left(\int_{B_R} |w|^{\frac{2r}{r-2}\cdot\frac{s-1}{2-s}} \lambda dx\right)^{(\frac{2}{s}-1)(1-\frac{2}{r})}, \end{aligned}$$

$$\begin{aligned} \int_{B_R} (|u|^{s/(2-s)} + |g|)\,|w|\,\lambda dx \le \quad & \varepsilon\int_{B_R} |\nabla w|^2 \lambda dx \left(\int_{B_R} |w|^{\frac{2r}{r-2}\cdot\frac{s-1}{2-s}} \lambda dx\right)^{1-2/r} \\ & + C\int_{B_R} |u|^{2/(2-s)} \lambda dx + C\lambda(B_R)^{1-2/ps}. \end{aligned}$$

Noting that $w \in H^1 \cap L^{\frac{2r}{r-2}\cdot\frac{s-1}{2-s}}(\Omega, \mathbb{R}^N, \lambda)$, where ε and R are small enough by (1.5) , we get

$$\fint_{B_\rho} \rho^2 |\nabla w|^2 \lambda dx \leq C(\tfrac{\rho}{R})^2 \left[\left(\int_{B_R} (|u| + |w|)^{\frac{2r}{r-2}\cdot\frac{s-1}{2-s}} \lambda dx \right)^{(-\frac{2}{r})(\frac{2}{s}-1)} + \left(\int_{B_R} |\nabla u|^2 \lambda dx \right)^{s-1} \fint_{B_R} (R^2 |\nabla u|^2 + |u|^2) \lambda dx \right] + CR^{2n+2\alpha}.$$

Denote

$$\phi(R) = \fint_{B_R} (R^2 |\nabla u|^2 + |u|^2) \lambda dx.$$

By Lemma 2.1 we get for $\rho < R$,

$$\phi(\rho) \leq C \left[\left(\frac{\rho}{R} \right)^{2n+2\delta} + \varepsilon_1 \right] \phi(R) + CR^{2n+2\alpha}. \tag{2.10}$$

When $\rho^2 > R^2/2^k$, (2.10) is trivial.

For simplicity, we assume $\alpha \leq \delta$. According to Lemma 3.1 in [G] Chapter III, we have

$$\int_{R_\rho} |\nabla u| dx \leq (\fint_{B_\rho} |\nabla u|^2 \lambda dx)^{1/2} \leq C\rho^{n-1+\alpha}.$$

Applying Morrey's theorem , we conclude Theorem 2. Theorems 1 and 3 , can be proved similarly.

References

[C] Chen Baoyao, Lin Changhao, *L^p–Estimates for the Gradient of weak solutions to a class of degenerate nonlinear elliptic systems.* Advance in Math., Vol. 19. No.1, 1990.

[G] Giaquinta, M., *Multiple intergrals in the calculus of variations and nonlinear elliptic systems.* Princeton University Press, 1983.

[F] Fabes, E.B., C.E. Kenig, R.P. Serapioni, Comm. in P.D.E. 1982, 7.

[Y] Yang Shixin, Huang Qinbo, Jour . of Xiaman Univ., 1988; 27 (1).

Proceedings of the Asian Mathematical
Conference, 1990

Interaction of Shock and Sound Wave for System of Gas Dynamics

CHEN SHUXING
Institute of Mathematics
Fudan University
Shanghai 200433, China

Since the eighties the study of discontinuous solution for multidimensional hyperbolic system of conservation laws has been considerable developed. For instance, the local existence of multidimensional shock, rarefaction wave or sound wave, results on reflection of shock and interaction of two shocks or two progressing waves with weak singuarity were obtained (see [1–6]). In this paper we are going to discuss the problem on interaction of shock and sound wave for steady flow in gas dynamics. The main conclusion is: when two such waves meet together, the shock will continue propagating, while the sound wave may split into two waves. The conclusion is natural from the viewpoint of physics, eventhough its rigorous proof is not as easy.

We consider the solution with shocks for the steady Euler system. The shock is always assumed to be weak, hence the flow can be approximately regarded as isentropic. In this case the system

$$\frac{\partial}{\partial x}\begin{pmatrix} \rho u \\ p+\rho u^2 \\ \rho uv \\ \rho uw \end{pmatrix} + \frac{\partial}{\partial y}\begin{pmatrix} \rho v \\ \rho uv \\ p+\rho v^2 \\ \rho vw \end{pmatrix} + \frac{\partial}{\partial z}\begin{pmatrix} \rho w \\ \rho uw \\ \rho vw \\ p+\rho w^2 \end{pmatrix} = 0 \tag{1}$$

is a good description of the flow. In (1) u, v, w, p, ρ represent the components of velocity, pressure, and density respectively. The unknown functions also satisfy Bernoulli's relation

$$\frac{1}{2}(u^2+v^2+w^2) + \frac{\gamma p}{(\gamma-1)\rho} = \text{const.},$$

where $p = p(\rho)$ is a given positive increasing function. In what follows we will only consider supersonic flow, and assume that $v > a(= (p'(\rho))^{1/2})$ without loss of generality. Thus the system is a y-hyperbolic one.

If $S : x = \psi(y, z)$ is a shock in the flow field, then the Rankine-Hugoniot conditions on S are

$$-\begin{pmatrix} [\rho u] \\ [p+\rho u^2] \\ [\rho uv] \\ [\rho uw] \end{pmatrix} + \begin{pmatrix} [\rho v] \\ [\rho uv] \\ [p+\rho v^2] \\ [\rho vw] \end{pmatrix}\psi_y + \begin{pmatrix} [\rho w] \\ [\rho uw] \\ [\rho vw] \\ [p+\rho w^z] \end{pmatrix}\psi_z = 0, \quad \text{on } x = \psi(y,z). \tag{2}$$

If $R: x = \varphi(y,z)$ is a sound wave, then all parameters $U = {}^t(u,v,w,p)$ of the flow on R are continuous, but the gradients of them may have jumps. In this case, R must be a characteristic surface of the system, and it satisfies one of the following equations:

$$\begin{aligned} &-u + v\frac{\partial\varphi}{\partial y} + w\frac{\partial\varphi}{\partial z} + a\left(1 + \left(\frac{\partial\varphi}{\partial y}\right)^2 + \left(\frac{\partial\varphi}{\partial z}\right)^2\right)^{1/2} = 0 \\ &-u + v\frac{\partial\varphi}{\partial y} + w\frac{\partial\varphi}{\partial z} = 0 \\ &-u + v\frac{\partial\varphi}{\partial y} + w\frac{\partial\varphi}{\partial z} - a\left(1 + \left(\frac{\partial\varphi}{\partial y}\right)^2 + \left(\frac{\partial\varphi}{\partial z}\right)^2\right)^{1/2} = 0. \end{aligned} \tag{3}$$

Assume that in the region $y < 0$ the function U, the shock S and the sound wave R are given, S and R are away from each other, they meet at the origin and satisfy

$$\psi(0,0) = \varphi(0,0) = \frac{\partial\psi}{\partial z}(0,0) = \frac{\partial\varphi}{\partial z}(0,0), \quad \frac{\partial\psi}{\partial y}(0,0) > \frac{\partial\varphi}{\partial y}(0,0). \tag{4}$$

Our goal is to determine the solution U of (1) as well as the location of possible singularities of the solution, so that U satisfies (1) on $y > 0$ except some surfaces, on which U satisfies (2) in case S or one of (3) in case R.

We assume that the initial data have some smoothness away from S and R. More precisely, let $s > 6, \lambda$ be a sufficiently large number greater than s, and U be H^λ smooth on $y < 0$ except on S and $R, \varphi \in H^\lambda(y < 0), \psi \in H^{\lambda+1}(y < 0)$, then the following conclusion holds.

Theorem 1.1. *Under the above assumptions we have:*

1) *The intersection of S and R:*

$$y = \eta(z), \quad x = \xi(z)$$

can be uniquely determined for small z, and there exist functions $\tilde\psi(y,z), \tilde\varphi(y,z)$ and $\tilde U(x, y, z)$ *for* $0 < y < \eta(z)$ *and small z, such that* $\{\tilde U, \tilde\varphi, \tilde\psi\}$ *satisfies (1)–(3) and*

$$\tilde U(x,0,z) = U(x,0,z), \quad \tilde\psi(0,z) = \psi(0,z), \quad \tilde\varphi(0,z) = \varphi(0,z);$$

moreover, $\tilde\psi(y,z) \in H^{\lambda+1}, \tilde\varphi(y,z) \in H^\lambda, \tilde U(x,y,z) \in H^\lambda$ *in* $0 < y < \eta(z)$ *except* $S: x = \tilde\psi(y,z)$ and $R: x = \tilde\psi(y,z)$.

2) *There exist functions* $\tilde{\tilde\psi}(y,z), \tilde{\tilde\varphi}_i(y,z) (i = 2,3)$ *and* $\tilde{\tilde U}(x,y,z)$ *for* $y > \eta(z)$ *and small z, such that* $\{\tilde{\tilde U}, \tilde{\tilde\varphi}_1, \tilde{\tilde\varphi}_2, \tilde{\tilde\psi}\}$ *satisfies (1)–(3), and*

$$\tilde{\tilde U}(x,\eta(z),z) = U(x,\eta(z),z),$$

$$\tilde{\tilde\psi}(\eta(z),z) = \tilde{\tilde\varphi}_i(\eta(z),z) = \xi(z),$$

moreover, $\tilde{\tilde\psi}(y,z) \in H^{s+1}, \tilde{\tilde\varphi}_i(y,z) \in H^s, \tilde{\tilde U}(x,y,z) \in H^s$ *in* $y > \eta(z)$ *away from* $S: x = \tilde{\tilde\psi}(y,z)$ and $R_i: x = \tilde{\tilde\varphi}_i(y,z)$.

The domain $y < 0$ is divided by S and R into three subdomains $D_I : x < \psi(y,z), D_{II} : \psi(y,z) < x < \varphi(y,z)$ and $D_{III} : x > \varphi(y,z)$. Correspondingly, the solution U in these subdomains is denoted by U_I, U_{II} and U_{III}. In order to determine the intersection of S and R, we extend U_{II} to D_I as an H^λ function. By using the result in [3] we can find a local solution in $y > 0$ with a sound wave near the origin. The sound wave and the solution on the two sides of the sound wave are denoted by $x = \varphi(y,z), U_1^-(x,y,z)$ and $U_2^-(x,y,z)$. After that, we extend U_{II} to D_{III} as an H^λ function. Using the result in [1] we can find a solution in $y > 0$ with a shock near the origin. The shock and the solution on the two sides of the shock are denoted by $x = \psi(y,z), U_2^+$ and U_3^+. Since

$$\frac{\partial\psi}{\partial y}(0,0) > \frac{\partial\varphi}{\partial y}(0,0) \tag{5}$$

by the implicit function theorem we can determine a function $y = \eta(z)$ for small z, such that

$$\psi(\eta(z),z) = \varphi(\eta(z),z)(= \xi(z)). \tag{6}$$

The curve $y = \eta(z), x = \xi(z)$ is the expression of the intersection Γ_0. By using the property of finite propagation speed for hyperbolic system we may succesively prove $U_2^-(x,y,z) = U_2^+(x,y,z)$ in the whole domain $0 < y < \eta(z), \psi(y,z) < x < \varphi(x,z)$, and $\eta(z), U_2^\pm$ are independent of the way of the extension of U. Hence Theorem 1.1 holds.

Consider the solution in the domain $y > \eta(z)$. Starting from Γ_0 there are two upward characteristic surfaces — $R_2 : x\varphi_2(y,z)$ and $R_3 : x = \varphi_3(y,z)$ — on the left side of S, and no upward characteristic surface on the right side of S (see [7]). The leftest characteristic surface R_3 and the solution U^- of the system (1.1) in the domain left to R_3 are known. On the other hand, the solution in the domain right to S can also be known, but the place of S is to be determined.

In order to determine the solution in between R_3 and S, we reduce our problem to a Goursat problem in a wedged domain by a series of coordinate transformations. The final form of the equation becomes

$$LU = A_1\frac{\partial U}{\partial x'} + A_2\frac{\partial U}{\partial y'} + A_3\frac{\partial U}{\partial z'} = 0 \quad \text{in} \quad \Omega : 0 \le x' \le y' \tag{7}$$

where $U = {}^t(U_{(1)}, U_{(2)})$ is a vector with double components of ${}^t(u,v,w,p)$ at two points with some symmetry, the cofficients metrices A_i depend on the unknown functions U, Φ and ψ. The function Φ describes the characteristic surfaces and satisfies

$$d_1\Phi_{y'} + d_2\Phi_{z'} + d_3 = 0 \quad \text{in} \quad \Omega. \tag{8}$$

Meanwhile, the boundary conditions are transformed into

$$\begin{aligned} U_{(2)} &= U^-, && \text{on} \quad \Sigma_2 : x' = y', \\ F(U,\psi) &= h_1\frac{\partial\psi}{\partial y_4} + h_2\frac{\partial\psi}{\partial z_4} + h_3 = 0 && \text{on} \quad \Sigma_2 : x' = y', \\ \psi &= 0 && \text{on} \quad \Gamma : x' = y' = 0, \end{aligned}$$

$$U_{(1)} = U_{(2)} \quad \text{on} \quad \Sigma_1 : x' = 0,$$

$$\Phi(x', x', z') = x' \quad \text{on} \quad \Sigma_2 : x' = y'.$$

This problem is called (G). Now in order to prove the second part of Theorem 1.1 we only need to prove the existence of a smooth solution of (G) in $0 < y' < \delta$ for small δ.

In order to solve the problem (G) we introduce the following iterative scheme:

$$\begin{cases} L_\nu U_{\nu+1} \equiv L(U_\nu, \Phi_\nu, \psi_\nu) U_{\nu+1} = 0 & \text{in } \Omega, \\ U_{(2),\nu+1} = U^- & \text{on } \Sigma_2, \\ f_\nu(U_{\nu+1}, \psi_{\nu+1}) = f_\nu(U_\nu, \psi_\nu) - F(U_\nu, \psi_\nu) & \text{on } \Sigma_2, \\ \psi_{\nu+1} = 0 & \text{on } \Gamma, \\ U_{(1),\nu+1} = U_{(2),\nu+1} & \text{on } \Sigma_1, \end{cases} \tag{9}$$

$$\begin{cases} d_{1\nu} \dfrac{\partial \Phi_\nu}{\partial y} + d_{2\nu} \dfrac{\partial \Phi_\nu}{\partial z} + d_{3\nu} = 0 & \text{in } \Omega, \\ \Phi_\nu = x' & \text{on } \Sigma_2, \end{cases} \tag{10}$$

where f_ν is the Frechet derivative of F_ν, and $f_\nu(\delta V, \delta\theta) = h_1 \frac{\partial \delta\theta}{\partial y} + h_2 \frac{\partial \delta\theta}{\partial z} + M\delta V$. For $\nu = 0$, (U_0, Φ_0, ψ_0) is an asymptotic solution, which satisfies

$$\begin{aligned} L_0 U_0 &= O(y'^\lambda) && \text{in } \Omega, \\ U_{(2),0} &= U^- && \text{on } \Sigma_2, \\ F(U_0, \psi_0) &= O(y'^{\lambda+1}) && \text{on } \Sigma_2, \\ \psi_0 &= 0 && \text{on } \Gamma, \\ U_{(1),0} &= U_{(2),0} && \text{on } \Sigma_1. \end{aligned} \tag{11}$$

$$\begin{aligned} d_{10} \frac{\partial \Phi_0}{\partial y'} + d_{20} \frac{\partial \Phi_0}{\partial z'} + d_{30} &= 0 && \text{in } \Omega, \\ \Phi_0 &= x' && \text{on } \Sigma_2. \end{aligned} \tag{12}$$

Such an asymptotic solution can be constructed mainly by determining the coefficients of its expansion in series successively.

Introducing the weighted L^2 norm $\| \quad \|_{L^2_\lambda}$ with weight $y^{-2\lambda}$, weighted Sobolev's norm

$$\|u\|_{k,\lambda,\delta} = \left\{ \sum_{|\alpha| \le k} \lambda^{2(k-\alpha')} \|\partial^\alpha u\|^2_{L^2_{\lambda-\alpha'}(\Omega_\delta)} \right\}^{1/2}, \quad \alpha' = \alpha_x + \alpha_y,$$

$$|f|_{k,\lambda,\delta} = \left\{ \sum_{|\alpha| \le k} \lambda^{2(k-\alpha_y)} \|\partial^\alpha f\|^2_{L^2_{\lambda-\alpha_y}(\Sigma_{2,\delta})} \right\}^{1/2},$$

and tangential weight Sobolev's norm

$$\|u\|'_{k,\lambda,\delta} = \left\{ \sum_{|\alpha| \le k} \lambda^{2(k-\alpha')} \|D^\alpha u\|^2_{L^2_{\lambda-\alpha'}(\Omega_\delta)} \right\}^{1/2},$$

where D represents tangential operators as $D_1 = (x-y)\frac{\partial}{\partial y}, D_2 = y\frac{\partial}{\partial y}, D_3 = \frac{\partial}{\partial z}$ and denoting

$$|||(U,\Phi,\psi)|||_{s,\lambda,\delta} = \lambda\|U\|_{s,\lambda+\frac{1}{2},\delta} + \lambda\|\Phi\|_{s,\lambda+\frac{1}{2},\delta} + |U|_{s,\lambda,\delta} + |\psi|_{s+1,\lambda+1,\delta}$$

we may establish energe estimates for the problem (11) and (12). For sufficiently large λ and small $\delta U_\nu = U_\nu - U_0, \delta\psi_\nu = \psi_\nu - \varphi_0$, the main estimates for $\delta\Phi_\nu = \Phi_\nu - \Phi_\nu, \delta U_{\nu+1}$ and $\delta\psi_{\nu+1}$ are

$$\lambda^2\|\delta\Phi_\nu\|^2_{s'\lambda,\delta} \leq C(\|\delta U_\nu\|^2_{s',\lambda-1,\delta} + |\delta\psi_\nu|^2_{s'+1,\lambda-\frac{1}{2},\delta}), \quad s' \leq s, \tag{13}$$

$$\lambda\|\delta U_{\nu+1}\|^2_{s',\lambda+\frac{1}{2},\delta} + |\delta U_{(1),\nu+1}|^2_{s',\lambda,\delta} + |\delta\psi_{\nu+1}|^2_{s'+1,\lambda+1,\delta} \leq C\delta, \quad s' \leq s. \tag{14}$$

To establish these estimates we will meet two difficulties, one is caused by the appearance of gradient of unknown function in the coefficients of (7), another one is caused by the fact that the boundary $y = 0$ is characteristic. To overcome these difficulties we should take advantage of the special form of the system and proceed with delicate analysis. The details can be referred to [8].

(13) and (14) indicate that the map from $(\delta U_\nu, \delta\Phi_\nu, \delta\psi_\nu)$ to $(\delta U_{\nu+1}, \delta\Phi_{\nu+1}, \delta\psi_{\nu+1})$ is an inner map in the set E_δ, where

$$\begin{aligned}
E_\delta = \{&(\delta U, \delta\Phi, \delta\psi) \in H^s_{\lambda+1}(\Omega_\delta) \times H^s_{\lambda+\frac{1}{2}}(\Omega_\delta) \times H^{s+1}_{\lambda+1}(\Sigma_{2,\delta});\\
&\delta U|_{\Sigma_{2,\delta}} \in H^s_\lambda, \delta U_{(2)} = 0 \quad \text{on} \quad \Sigma_{2,\delta}, \delta\Phi = 0 \quad \text{on} \quad \Sigma_{2,\delta}, \delta U_{(1)} = \delta U_{(2)} \quad \text{on} \quad \Sigma_{1,\delta}\\
&|||(\delta U, \delta\Phi, \delta\psi)||| \leq K\}.
\end{aligned} \tag{15}$$

On the other hand, we can also establish the estimate as

$$\begin{aligned}
&|||(\delta U_{\nu+2} - \delta U_{\nu+1}, \delta\Phi_{\nu+2} - \delta\Phi_{\nu+1}, \delta\psi_{\nu+2} - \delta\psi_{\nu+1})|||_{0,\lambda,\delta}\\
\leq\ & C\lambda^{-1}|||(\delta U_{\nu+1} - \delta U_\nu, \delta\Phi_{\nu+1} - \delta\Phi_\nu, \delta\psi_{\nu+1} - \delta\psi_\nu)|||_{0,\lambda,\delta}
\end{aligned}$$

which leads to the convergence of the sequence $\{(\delta U_\nu, \delta\Phi_\nu, \delta\psi_\nu)\}$ in the space $L^2_{\lambda+\frac{1}{2}}(\Omega_\delta)\times L^2_{\lambda+\frac{1}{2}}(\Omega_\delta) \times H^1_{\lambda+1}(\Sigma_{2,\delta})$ when $\lambda > 2C$. The limit $(\delta U, \delta\Phi, \delta\psi)$ also belongs to E_δ by Banach-Saks theorem, because of $|||(\delta U_\nu, \delta\Phi_\nu, \delta\psi_\nu)|||_{s,\lambda,\delta}$ is uniformly bounded. Therefore, we confirm that the solution $(U_0 + \delta U, \Phi_0 + \delta\Phi, \psi_0 + \delta\psi)$ is the solution to the problem (G). Correspondingly, the second part of Theorem 1.1 is thus proved.

References

[1] A. Majda, *The stability of multi-dimensional shock fronts, The existence of multi - dimensional shock fronts*, Memoirs of A. M. S. #275, # 281 (1983).

[2] S. Alinhac, *Existence d'ondes de rarefaction por des systemes quasi-lineaires hyperboliques multidimensonnels*, Comm. in PDEs, 14(1989), 173–230.

[3] G. Metivier, *Ondes sonique*, Sem. EDP, Ecole Polytechnique, Paris (1987–88).

[4] Chen Shuxing, *On reflecton of multidimensional shock front*, Jour. Diff. Eqs., 80(1989), 199–236.

[5] G. Metivier, *Interaction de deux chocs pour un systeme de deux lois de conservation, en dimension deux d'espace*, Trans, Amer. Math Soc., 296(1986), 431–479.

[6] S. Alinhac, *Interaction d'ondes simples pour des equations completement non lineaires*, Ann. Sc. de 1'E'N'S (to appear).

[7] R. Courant & K. O. Friedrichs, *Supersonic flow and shock waves*, Wiley-Interscience New York (1948).

[8] Chen Shuxing, *On interaction of shock and sound wave* (to appear).

[9] K. O. Friedrichs, *Symmetric positive linear differential equations*, Comm. Pure Appl. Math., 11(1958), 333–418.

Proceedings of the Asian Mathematical
Conference, 1990

Solving Assignment Problem Through Data Processing

M.C. Cheng
Department of Mathematics
University of Malaya
59100 Kuala Lumpur, Malaysia

§ 1. Introduction

A standard linear programming problem (LP) can be solved in a finite number of iterations by the simplex method with anti-cycling (see Dantzig [3]). With respect to an optimal basic solution, the variables of a (LP) can be classified into basic and non basic. Various methods of characterizing and identifying basic and non basic variables of an optimal basic solution are readily available, see Cheng [1], Karwan etc. [5]. However, in a general (LP) even though a basic variable of an optimal basic solution can be identified, its final value can only be determined after the last basic variable in a set of variables which constitute an optimal basic solution has been identified.

The transportation problem (TP) is a special case of the (LP). A much faster method [4] than the simplex method [3] of obtaining an initial basic solution for (TP) is to allow a variable to satisfy a row constraint or a column constraint at a time, until all the constraints are satisfied. Such an initial basic solution is usually primal feasible but not dual feasible, and often requires many iterations of improvement. Recently Cheng [2] has proposed a new algorithm of constructing an initial basic solution which is near optimal or even optimal by processing the given data of (TP).

The assignment problem (AP) is a special case of the (TP) . There are two general [6] analytical methods of solving an (AP), namely the Hungarian method developed by Kuhn and the Bradford method developed by Mack [8]. In both methods, an initial solution which is dual feasible but not primal feasible is constructed and is iterated upon until it is both primal and dual feasible. Both methods exploit the property that the positions of the basic variables of an optimal assignment remain unchanged when the same quantity is added to or subtracted from every element of a row or a column of the given cost matrix of (AP). However, there is no guarantee that the initial solution used is anywhere close to an optimal assignment. This is because a large number of variables of the initial solution are not zero reduced cost variables of an optimal assignment, and consequently require many iterations before an optimal assignment is derived.

In this paper, we extend our method [2] of solving the (TP) through data processing to the (AP). We are happy to present an algorithm for solving the (AP). This algorithm attempts to identify zero reduced cost variables, non basic variables and strictly positive reduced cost variables (whose cost can consequently be reduced by certain amount and yet enable these variables to remain non basic variables of the original problem) of an optimal basic solution through merely processing the given data.

§2 presents the various characterizations of non basic variables of an optimal basic solution of (AP). For strictly positive reduced cost variables it is shown that the cost of these variables may be reduced by a certain amount and yet enable these variables to remain as non basic variables with respect to the same optimal basic solution. §3 gives various sufficient conditions for zero reduced cost variables of an optimal basic solution for (AP).

Basing upon the above characterizations, a new algorithm for solving the (AP) through processing the given data is presented in §4. The algorithm attempts to construct an optimal or near optimal initial basic solution by assigning value to as many variables as possible which can be identified as zero reduced cost variables of an optimal basic solution. An example is presented to illustrate the usefulness and versatility of the new algorithm. This example also establishes the existence of a subclass of the class of (AP) in which it is possible to construct an optimal initial basic solution.

§5 discusses some of the pertinent points related to our algorithm.

§ 2. Characterizations of non Basic Variables of an Optimal Basic Solution of Assignment Problem

Consider the assignment problem:

$$(AP): \quad \text{Minimize} \quad \sum_{i=1}^{n}\sum_{j=1}^{n} c_{ij}x_{ij}$$

$$\text{subject to} \quad \sum_{j=1}^{n} x_{ij} = 1, \quad i = 1,2,\cdots,n,$$

$$\sum_{i=1}^{n} x_{ij} = 1, \quad j = 1,2,\cdots,n,$$

$$x_{ij} = 0 \text{ or } 1, \; i = 1,2,\cdots,n, \; j = 1,2,\cdots,n.$$

The dual to the above problem is:

$$\text{Maximize} \quad \sum_{i=1}^{n} u_i + \sum_{j=1}^{n} v_j,$$

$$\text{subject to} \quad u_i + v_j \leq c_{ij}, \; i = 1,2,\cdots,n, \; j = 1,2,\cdots,n.$$

For $i = 1,2,\cdots,n, j = 1,2,\cdots,n$, the reduced cost of x_{ij} with respect to an optimal basic solution is

$$w_{ij} = c_{ij} - u_i - v_j.$$

In an (AP), there are a total of n! possible assignments or basic solutions which increases like an exponential function as n increases. An optimal basic solution is a basic solution which minimizes the objective function.

Definition 1. x_{ij} is a basic variable of an optimal basic solution of (AP) if $x_{ij} = 1$ and $w_{ij} = 0$.

Definition 2. x_{ij} is a non basic variable of an optimal basic solution of (AP) if $x_{ij} = 0$ and $w_{ij} \geq 0$.

Observe that $w_{ij} > 0$ is a sufficient condition for x_{ij} to be a non basic variable of an optimal basic solution, and such a variable is called a strictly positive reduced cost variable.

Observe also that every optimal basic solution of an (AP) has exactly n basic variables with one such basic variable in each row and each column. Thus if $x_{p_iq_i}$, where p_i and q_i, $i = 1, 2, \cdots, n$ are two permutations of the sequence $(1, 2, \cdots, n)$, are basic variables of an optimal basic solution of (AP), then $w_{p_iq_i} + w_{p_jq_j} \leq w_{p_iq_j} + w_{p_jq_i}, i \neq j, i = 1, 2, \cdots, n, j = 1, 2, \cdots, n$, which can be transformed to

$$c_{p_iq_i} + c_{p_jq_j} \leq c_{p_iq_j} + c_{p_jq_i},$$
$$i \neq j, \quad i = 1, 2, \cdots, n, \quad j = 1, 2, \cdots, n.$$

Denote

$$N = \{ \quad i; \quad i = 1, 2, \cdots, n, \quad n \text{ a positive integer}\}$$

$$R_i = \{ \quad k; \quad k \in N \text{with} x_{ik} \text{not an identified non basic variable of}$$
$$\text{an optimal basic solution}\},$$

$$L_j = \{ \quad l; \quad l \in N \text{with} x_{lj} \text{ not an identified non basic variable}$$
$$\text{of an optimal basic solution}\}.$$

Thus if x_{rs} is a basic variable of an optimal basic solution of (AP) with $n > 1$, then there exists another basic variable $x_{ij}, i \neq r, j \neq s$ such that

$$c_{rs} + c_{ij} \leq c_{rj} + c_{is}.$$

Conversely, if x_{rs} is such that the above inequality is false, then x_{rs} is a non basic variable of an optimal basic solution. Hence we have proved the following theorem:

Theorem 1. *A variable x_{rs} of (AP) is a non basic variable of an optimal basic solution if it satisfies any one of the following conditions:*

(i) $\exists$ *a zero reduced cost varialble x_{ij} such that*

$$c_{rs} + c_{ij} > c_{is} + c_{rj}, \quad \text{for} \quad i \neq r, j \neq s;$$

(ii) $\exists$ *a row $i, i \neq r$ such that*

$$c_{rs} + c_{ij} > c_{is} + c_{rj}, \quad j \in R_i - \{s\};$$

(iii) $\exists$ *a column $j, j \neq s$ such that*

$$c_{rs} + c_{ij} > c_{is} + c_{rj}, \quad i \in L_j - \{r\}.$$

If x_{rs} is identified as a non basic variable of an optimal basic solution by Theorem 1, then a strictly positive value δ_{rs} may be associated to each of the cases of Theorem 1 as

(i) $c_{rs} + c_{ij} - (c_{is} + c_{rj})$;

(ii) (a) Minimum $\{c_{rs} + c_{ij} - (c_{is} + c_{rj}), j \in R_i - \{s\}, i \neq r, x_{is}$ is a non basic variable. $\}$
or (b) Minimum $\{c_{rs} + c_{i_k j} - (c_{i_k s} + c_{rj}), j \in R_{i_k} - \{s\}, k = 1,2, r \neq i_1 \neq i_2, x_{i_1 s}$ and $x_{i_2 s}$ are not detected non basic variables.$\}$;

(iii) (a) Minimum $\{c_{rs} + c_{ij} - (c_{is} + c_{rj}), i \in L_j - \{r\}, j \neq s, x_{rj}$ is a non basic variable.$\}$
or (b) Minimum $\{c_{rs} + c_{ij_k} - (c_{is} + c_{rj_k}), i \in L_{j_k} - \{r\}, k = 1,2, s \neq j_1 \neq j_2, x_{rj_1}$ and x_{rj_2} are not detected non basic variables.$\}$.

It is readily proved that

Theorem 2. *Let x_{rs} be identified by Theorem 1 as a non basic variable of an optimal basic solution for (AP) with associated value δ_{rs} . If c_{rs} is changed to $c_{rs} - \delta_{rs}$, then x_{rs} remains a non basic variable with respect to the same optimal basic solution.*

§ 3. Characterizations of Zero Reduced Cost Variables of an Optimal Basic Solution for (AP)

In this section, several sufficient conditions for a variable to have zero reduced cost with respect to an optimal basic solution are derived. This is easily done by the method of contradiction.

Theorem 3. *A variable x_{rs} of (AP) is a zero reduced cost variable of an optimal basic solution if it satisfies any of the following conditions:*

(a) *There exist non basic variables x_{rk}, for all $k \in N - \{s\}$.*

(b) *There exist non basic variables $x_{\ell s}$, for all $\ell \in N - \{r\}$.*

(c) *For $j \in R_r - \{s\}, i \in L_s - \{r\}$, every (i,j) satisfies one of the following conditions:*

(i) $c_{rs} + c_{ij} \leq c_{is} + c_{rj}$;

(ii) $c_{rs} + c_{ij} - \delta_{ij} \leq c_{is} + c_{rj}$ *for a strictly positive reduced cost variable x_{ij};*

(iii) *there exists a zero reduced cost variable $x_{pq}, p \neq i, p \neq r, q \neq j, q \neq s$ such that*

$$c_{rs} + c_{pj} + c_{iq} \leq c_{pq} + c_{is} + c_{rj},$$

whenever $c_{rs} + c_{ij} > c_{is} + c_{rj}$;

(iv) *there exists either R_p or $L_q, p \neq i, p \neq r, q \neq j, q \neq s$ such that either*

$$c_{rs} + c_{pj} + c_{ik} \leq c_{pk} + c_{is} + c_{rj} \quad \text{for} \quad k \in R_p - \{j, s\},$$

$$\text{or } c_{rs} + c_{lj} + c_{iq} \leq c_{lq} + c_{is} + c_{rj} \quad \text{for} \quad l \in L_q - \{i, r\},$$

whenever $c_{rs} + c_{ij} > c_{is} + c_{rj}$.

Let $AP(N,N)$ denotes an optimal basic solution for an (AP) with n row constraints and n column constraints in the set N. If x_{ij} is a basic variable of $AP(N,N)$, then the optimum value is

$$V(AP(N,N)) = c_{ij} + V(AP(N - \{i\}, N - \{j\})).$$

This shows that an optimal initial basic solution for a given (AP) can be constructed if basic variable(s) can be identified at every stage so that the size of the given problem can be progressively reduced by satisfying a row constraint and a column constraint simultaneously at a time.

§ 4. An Algorithm for Constructing an Optimal or Near Optimal Initial Basic Solution

In general, there are two known analytical methods for solving an (AP), namely the Hungarian method proposed by Kuhn [6] and the Bradford method proposed by Mack [8]. Both methods start with an initial basic solution which is dual but not primal feasible, and utilize the property that the positions of the basic variables of an optimal assignment remain unchanged if the same quantity is added to or subtracted from every element of a row or a column of the given cost matrix. However there is no guarantee that the initial solution so used is anywhere close to an optimal basic solution. The main reason is because a large number of variables of the initial solution are not zero reduced cost variables of an optimal basic solution. Hence both existing methods require many iterations in general before an optimal solution is obtained.

We are happy to present an alternative algorithm which constructs an initial basic solution which is optimal or near optimal through processing the data of the cost matrix of a given (AP) as follows:

Step (1). Use Theorem (3) to identify zero reduced cost variables of an optimal basic solution. If none can be identified, go to Step (2), otherwise go the Step (5).

Step (2). Use Theorem (1) to detect non basic variables of an optimal basic solution. If at least one new non basic variable has been identified, go to Step (1), otherwise go to Step (3).

Step (3). Use Theorem (2) to reduce the cost of those strictly positive reduced cost variables x_{rs} by δ_{rs}. If at least one strictly positive reduced cost variable has been reduced go to Step (1) , otherwise go to Step (4).

Step (4). Assign value to a variable which has enabled the most number of variables to be identified as non basic variable of an optimal basic solution. Designate other variables in the corresponding row and column as non basic variables. If an initial basic solution has been constructed, go to Step (6), otherwise go to Step (1).

Step (5). Assign value to a zero reduced cost variable of an optimal basic solution, one at a time, and designate other variables in the corresponding row and column as non basic variables. If an initial basic solution has been constructed, go to Step (6), otherwise go to Step (1).

Step (6). Test the initial basic solution for optimality. If optimal, go to Step (8), otherwise go to Step (7).

Step (7). Improve until optimality conditions are satisfied. Go to Step (8).

Step (8). Print an optimal basic solution . End.

We illustrate the usefulness and the versatility of the above algorithm by the following example.

Example. Consider an (AP) with designated assignments (i, i) $i = 1, 2, \cdots, 7$ not permitted and whose cost matrix is:

-	10	10	10	10	10	9
15	-	13	9	17	7	3
12	10	-	12	11	6	4
8	16	17	-	12	13	6
6	11	14	23	-	12	2
5	13	16	10	24	-	5
4	17	7	13	11	13	-

Since certain specified assignments are not permitted, we assign a large value to each of the corresponding cost so as to enable them to be identified as non basic variables. In the above example, it is sufficient to assign $c_{ii} = 30, i = 1, 2, \cdots, 7$.

Step (1) fails to identify any zero reduced cost variable. Step (2) and Step (3) identify the following non basic variables and calculate and reduced the corresponding δ_{ij} as follows:

r, s	1, 1	2,2	3,3	4,4	5,4	5,5	6,5	6,6	7,7	4,2	7,
Pivot	R_7, R_6	L_6, L_7	L_6, L_2	R_6, R_2	R_6, R_2	R_4, R_7	L_4	L_4, L_2	L_1, L_3	R_5	L_5
$\delta_{r,s}$	21	17	17	15	1	9		16	24		

Step (1) fails to identify any zero reduced cost variable and go to Step (2).

Step (2) uses Theorem (1) to identify the following as non basic variables and return to Step (1).

r, s	1,7	2,1	2,5	3,1	3,4	4,3	4,6	5,3	5,6	6,3	6,7	7,2
Pivot	R_2	R_4	R_1	R_4	R_2	R_1	R_3	L_7	L_7	L_1	R_5	R_1

By Step (1), Theorem 3(c)(i), (ii) and (iv) identify x_{73} (use R_6 for both $(i, j) = (1, 1)$ and $(i, j) = (2, 1)$ when applying Theorem 3(c) (iv)) as zero reduced cost variable. Step (5) assigns $x_{73} = 1$, designates other variables in row seven and column three as non basic variables and return to Step (1).

By Step (1), Theorem 3(c)(i), (ii) and (iv) together identify x_{45} (for $(i, j) = (1, 1), (i, j) = (3, 1)$ use R_6 , and for $(i, j) = (1, 7)$ use L_2 when applying Theorem 3 (c)(iv)) as zero reduced cost variable. Step (5) assigns $x_{45} = 1$, designates other variables in row four and column five as non basic variables and return to Step (1).

By Step (1), Theorem 3(c)(i) , (ii) and (iv) together identify x_{61} (for $(i, j) = (5, 4)$ use L_7 when applying Theorem 3(c)(iv)) as zero reduced cost variable. Step (5) assigns $x_{61} = 1$, designates other variables in row six and column one as non basic variables and return to Step (1).

By Step (1) , Theorem 3(c)(i), (ii) and (iv) together identify x_{57} (for $(i, j) = (2, 2)$ use R_3 when applying Theorem 3(c)(iv)) as zero reduced cost variable. Step (5) assigns $x_{57} = 1$, designates other variables in row five and column seven as non basic variables and return to Step (1).

By Step (1), Theorem 3(c)(i), (ii) and (iv) together identify x_{12} (for $(i,j) = (3,4)$ use R_2 when applying Theorem 3(c)(iv)), x_{24} (for $(i,j) = (1,6)$ use R_3 when applying Theorem 3(c)(iv)) and x_{36} (for $(i,j) = (2,2)$ use R_1 when applying Theorem 3(c) (iv)) as zero reduced cost variables. Step (5) assigns $x_{12} = 1, x_{24} = 1, x_{36} = 1$, designates other variables in rows one, two and three and columns two , four and six as non basic variables. Since an initial basic solution has been constructed, it proceeds to Step (6). Step (6) verifies that the initial basic solution is optimal with one set of u_i, v_j as given below:

i	1	2	3	4	5	6	7
u_i	2	2	1	5	2	3	0
v_i	2	8	7	7	7	5	0

Step (8) prints out an optimal assignment as $x_{12} = 1, x_{24} = 1, x_{36} = 1, x_{45} = 1, x_{57} = 1, x_{61} = 1$ and $x_{73} = 1$.

§ 5. Discussion

The basic computation in our data processing involves $c_{rs} + c_{ij}$ and whether or not $c_{rs} + c_{ij} > c_{is} + c_{rj}$ for $i \neq r, j \neq s$. There are altogether $n^2(n-1)^2$ such computations. Through processing the informations contained in such computations, non basic variable, strictly positive reduced cost variable and zero reduced cost variable of an optimal basic solution are identified. Additional computations are necessary only after determining those strictly positive reduced cost variables x_{rs} and after computing the amount δ_{rs} that c_{rs} can be reduced in order that x_{rs} will remain a non basic variable of the original problem. Hence the total amount of computation involved in our algorithm is at most polynomial.

It is hoped that our algorithm can be extended to solve the combinatorially hard travelling salesman problem [7].

References

[1] Cheng, M.C., *Generalized theorems for permanent basic and non basic variables*, Mathematical Programming, 31 (1985), 229–234.

[2] Cheng, M.C., *A new algorithm for transportation problem* , SEA Bull. Maths., Vol. 13, No. 2(1989), 103–114.

[3] Dantzig, G.B., *Maximization of a linear function of variables subject to linear inequalities*, in : T.C. Koopmans (ed.), Activity Analysis of Production and Allocation, John Wiley & Sons, Inc. New York, 1951, 339–347.

[4] Dantzig, G.B., *Application of the simplex method to a transportation problem*, in: T.C. Koopmans (ed.), Activity Analysis of Production and Allocation, John Wiley & Sons, Inc., New York, 1951, 359–373.

[5] Karwan, M.H., Lotfi, V., Telgen, J. and Zionts, S., *Redundancy in Mathematical Programming: A state-of-the-art survey*, Lecture notes in Economics and Mathematical Systems, Vol. 206, Springer-verlag, Berlin, Heidelberg, New York, Tokyo, 1983.

[6] Kuhn, H.W., *The Hungarian method for the assignment problem*, Nav. Res. Logistics Quart., 2(1955), 83–97.

[7] Lawler, E.L., Lenstra, J.K., Rinnooy Kan, A.H.G. and Shmoys, D.B. (eds.), The Travelling Salesman Problem –A Guided Tour of Combinatorial Optimization, John Wiley, New York, 1985.

[8] Mack, C. *The Bradford method for the assignment problem*, The New J. of Stats. and Op. Res., 1(1969), 17–29.

Proceedings of the Asian Mathematical
Conference, 1990

Value Distribution of Analytic Functions with Slow Growth the Open Unit Disk

T.Y. PETER CHERN
Department of Applied Mathematics
Kaohsiung Polytechnic Institute
Kaohsiung County, Taiwan, 80048, China

§1. Introduction

A meromorphic function f in $|z| < 1$ is admissible if $\limsup_{r\to 1^-} T(r,f)/\log(1/(1-r)) = +\infty$, where $T(r,f)$ is Nevanlinna's characteristic function. The Second Fundamental Theorem of Nevanlinna [5] provides an inequality to evaluate the lower bound of the counting function $N(r,f)$, but practically this theorem can only be applied when f is admissible [4, p.5].

In this paper concerning non-admissible functions analytic in the open unit disk, Anderson, Clunie and Pommerenke [1, p.20] obtained $(1/2)\log\log(1-r)^{-1}$ as an upper bound for counting functions of Bloch functions which are necessary non-admissible. It is natural to ask what the lower bounds for counting functions of non-admissible function classes are. A method working for some of them shall be developed and examples involving lacunary series showing the estimates being sharp up to a multiplicative constant shall be given.

The following notation shall be used throughout. Let D denote the unit disk $\{z : |z| < 1\}$, $\mathbb{C}$ the complex plane, and $\overline{\mathbb{C}} = \mathbb{C} \cup \{\infty\}$. A function f in D is called an α-Bloch function $(\alpha > 0)$ [10, p.31] if f is analytic in D and satisfies

$$\sup\{(1-|z|)^\alpha |f'(z)| : z \in D\} < +\infty. \tag{1.1}$$

The set of all α-Bloch function is denoted by B^α. A Bloch function is precisely a 1-Bloch function. Let $B_0^\alpha(\alpha > 0)$ be the set of functions f analytic in D such that

$$(1-|z|)^\alpha |f'(z)| \to 0 \quad \text{as} \quad |z| \to 1^-. \tag{1.2}$$

B_0^α is called the space of little α-Bloch functions. A little Bloch function is precisely a little 1-Bloch function. Let f be a meromorphic function in $\mathbb{C}$. If $a \in \overline{\mathbb{C}}$, we denote by $n(r,a)$ the number of solutions $f(z) = a$ in $|z| \le r$. For $a \in \overline{\mathbb{C}}$, the Nevanlinna counting function [1, (3.2), p.20] of the value a is defined by

$$N(r,a) = \int_0^r \frac{n(t,a) - n(0,a)}{t}\,dt + n(0,a)\log r. \tag{1.3}$$

A function f in D is said to be in the Shapiro-Shields class S^{-k} $(0 < k < +\infty)$ [8, (16), p.224] if f is analytic in D and satisfies

$$\sup\{(1-|z|)^k\,|f(z)| : z \in D\} < +\infty. \tag{1.4}$$

Let $L_a^P(D)(1 \leq p < +\infty)$ be the Bergman space of all analytic functions f such that $|f|^P$ is integrable on D.

§2. The Value Distribution of Bloch Functions

In this section, we develop a method to evaluate the lower bound of $N(r,f)$ as $r \to 1^-$ for analytic functions in D with slow growth, especially for Bloth functions. A lacunary series with Hadamard gaps means a function f analytic in D which can be expressed as a power series in the form

$$f(z) = \sum_{k=0}^{\infty} a_k z^{n_k}, \text{ with } a_k \neq 0,\ n_{k+1}/n_k \geq q > 1. \tag{2.1}$$

Proposition 2.1. *If $f(z)$ is a lacunary series with Hadamard gaps in D, and if there is a positive constant K such that*

$$\liminf_{r \to 1^-} (1-r)^2 \int_0^{2\pi} |f'(re^{i\theta})|^2\, d\theta \geq K \tag{2.2}$$

then $\liminf_{r\to 1^-} \dfrac{T(r,f)}{\log\log 1/(1-r)} \geq C_q$, *where C_q is a positive constant depending only on q.*

This result is quite useful and could be used to estimat a lower bound of $T(r,f)$ for some f in B^α $(\alpha \geq 1)$, $L_a^p(D)$ $(p \geq 1)$, S^{-k} $(k > 0)$, and some classes of lacunary series with Hadamard gaps.

Proposition 2.2. *There exists a Bloch function in D which is not little Bloch such that*

$$\liminf_{r\to 1^-} \frac{N(r,f,a)}{\log\log 1/(1-r)} \geq K, \text{ for each } a \in \mathbb{C}, \text{ where } K \text{ is a positive constant.} \tag{2.3}$$

For example, $f(z) = \sum_{k=0}^{\infty} z^{n_k}$ with $n_k = 2^k$ will be a Bloch function which meets the requirement of Proposition 2.2 (see [4]).

§3. The Value Distribution of Little Bloch Functions

Let B_0 denote the set of all little Bloch functions. We first show that $\frac{1}{2}\log^+(o(\log\frac{1}{1-r}))$ is an upper bound of $T(r,f)$ as $r \to 1^-$, for $f \in B_0$ (Proposition 3.1). We next obtain a result about the lower bound of $T(r,f)$ for some little Bloch functions in D (Proposition 3.2). The result of Proposition 3.1 is sharp in the sense of Proposition 3.2. Furthermore, in Proposition 3.3, we obtain a result about the lower bound of the counting function of some little Bloch functions. By applying Proposition 3.3, we show that there exists a little Bloch function f in D such that for each $a \in \mathbb{C}$, (2.3) holds (Proposition 3.4). The main results in this section are as follows.

Proposition 3.1. *If $f \in B_0$, then $T(r,f) \leq \frac{1}{2}\log^+(o(\log\frac{1}{1-r}))$ as $r \to 1^-$.*

Corollary 3.1. *If $f \in B_0$, then $N(r,f,a) \leq \frac{1}{2}\log^+(o(\log\frac{1}{1-r})) + O(1)$ as $r \to 1^-$ for each complex number a.*

Proposition 3.2. *If $\varphi(r)$ is a non-negative strictly decreasing continuous function in the open interval $(0,1)$ with $\varphi(1^-) = 0$, then there exists a little Bloch function f in D such that $T(r,f) \geq K\log^+[\varphi(r)\log 1/(1-r)]$ as $r \to 1^-$, where K is a positive constant.*

Proposition 3.3. *If $\varphi(r)$ is a non-negative strictly decreasing continuous function in the open interval $(0,1)$ with $\varphi(1^-) = 0$ and $\sum_{k=2}^{\infty}\varphi[1-\frac{1}{2^{k-1}}] = +\infty$, then there exists a little Bloch function g in D such that $N(r,g,a) \geq K\log^+[\varphi(r)\log\frac{1}{1-r}]$ as $r \to 1^-$, for each $a \in \mathbb{C}$, where K is a positive constant.*

Proposition 3.4. *There exists a little Bloch function g in D such that (2.3) holds.*

Proposition 3.1 provides an upper bound of $T(r,f)$ for little Bloch functions. The proof, similar to that of [1, theorem 3.1, p.20], is omitted here. Corollary 3.1 is an immediate consequence of the First Fundamental Theorem of Nevanlinna and Proposition 3.1. For example, $f(z) = \sum_{k=2}^{\infty} a_k z^{2^k}$, with $a_k = \sqrt{\varphi(1-1/2^{k-1})}$, will be a good example for Proposition 3.2.

For a detailed proof of Proposition 3.3, see [4]. Proposition 3.4 is an easy consequence of Proposition 3.3 by putting $\varphi(r) = (\log\frac{1}{1-r})^{-1/2}$.

§4. Counting Functions of Analytic Functions with Slow Growth

Theorem 4.1. *If $f \in S^{-k}$, $k > 0$, then $T(r,f) = O(\log\frac{1}{1-r})$ as $r \to 1^-$ and for each $a \in \mathbb{C}$, $N(r,f,a) = O(\log\frac{1}{1-r})$ as $r \to 1^-$.*

The proof of Theorem 4.1, similar to that of Proposition 3.1, is omitted here. The following Corollary 4.1 and Corollary 4.2 are immediate consequences of theorem 4.1, and the facts that $B^\alpha \subset S^{-2(\alpha-1)}$ [3], and $L_a^p \subset S^{-2/p}$ [3].

Corollary 4.1. *If $f \in B^\alpha$ ($\alpha > 1$), then $T(r,f) = O(\log\frac{1}{1-r})$ and for each $a \in \mathbb{C}$, $N(r,f,a) = O(\log\frac{1}{1-r})$ as $r \to 1^-$.*

Corollary 4.2. *If $f \in L_a^p(D)$, then $T(r,f) = O(\log\frac{1}{1-r})$ and for each $a \in \mathbb{C}$, $N(r,f,a) = O(\log\frac{1}{1-r})$ as $r \to 1^-$.*

Theorem 4.2. *If $\alpha > 1$, then there exists a function $f \in B^\alpha$ such that for each $a \in \mathbb{C}$, $\liminf_{r\to 1^-} N(r,f,a)/\log(1/(1-r)) \geq K > 0$, where K is a constant.*

For a detailed proof of theorem 4.2, see [4]. The following Corollary 4.3 is an immediate consquence of Theorem 4.2, and the facts that $B^{k/2+1} \subset S^{-k}$ [3], and $S^{-1} \subset L_a^p(D)$ [3].

Corollary 4.3. *If $k > 0$, then there exists a $f \in S^{-k}$ such that for each $a \in \mathbb{C}$, $\liminf_{r\to 1^-} N(r,f,a)/\log(1/(1-r)) \geq L$, where L is a positive constant.*

Remark. Combining above results in this section, we obtain that the least upper bound and the greatest lower bound of $N(r,a)$, $a \in \mathbb{C}$, for functions in B^α ($\alpha > 1$), $L_a^p(D)$ ($1 <$

$p < +\infty$), S^{-k} $(0 < k < +\infty)$ all have the same magnitude $O(\log \frac{1}{(1-r)})$.

References

[1] J.M. Anderson, J. Clunie, Ch. Pommerenke, *On Bloch function and Normal functions*, J. Reine Angew Math., 270 (1974), 12–37.

[2] E. Beller, *Zeros of A^p functions and related classes of Analytic functions*, Israel J. Math., 22(1)(1975), 68–80.

[3] E. Beller, *Factorization for Non-Nevanlinna classes of Analytic functions*, Israel J. Math., 27(1977), 320–330.

[4] T.Y.P. Chern, *A study of value distribution, zero sets, inequalities, and coefficient conditions for analytic functions with slow growth and Lacunary series with Hadamard gaps*, PH.D. Dissertation, Michigan State University, 1988.

[5] W.K. Hayman, *Meromorphic functions*, Oxford, 1964.

[6] T. Murai, *Une remarqe sur la distribution des valeus des series de Taylor aleatoies*, C. R. Acad. Sci. Paris, t. 287(20 nobembre 1978), 931–934.

[7] T. Murai, *Sur la distribution des valeurs des series Lacunaires*, J. London Math. Soc., 21 (1980), 93–110.

[8] H. S. Shapiro and A. L. Shields, *On the zeros of functions with finite Dirichlet integral and some related function spaces*, Math. Z., 80 (1962), 217–229.

[9] S. Yamaschita, *Gap series and α-Bloch functions*, Yokohama Math. J., 28(1980), 31-36.

[10] A. Zygmund, *Trigonometric series, I, II*, Camgridge Press, 1968.

Proceedings of the Asian Mathematical
Conference, 1990

Theory of Computation and Complex Analytic Dynamics

C.T. CHONG*
National University of Singapore

The classical theory of computation, dating back to the works of Gödel, Church, Kleene, Turing and Post in the 1930's and 40's, was developed on the discrete model of the integers **Z**. In this model, expressed in the form of a Turing machine, basic arithmetic operations can be simulated. Fundamental among the considerations of pioneers of the classical theory was the pre-eminent position held by the notion of an effective procedure. Their work provided a precise mathematical formulation for the intuitive idea of effectivity on the integers (and on objects coded by them), hitherto only hinted at by number theorists and algebraists. This formulation also gave a thoretical basis for the basic principles of operation of the digital computer that was commercially introduced more than a decade later. Computation theory (otherwise known as recursion theory) on **Z** has since flourished and is now a very active area of research in the field of mathematical logic (see the book by Soare [So] for a sophisticated treatment of the subject of recursively enumerable sets, and the paper by Slaman [S1] for some of the latest developments in the theory of degrees of unsolvability).

Over the years logicians have considered various domains for a possible (satisfactory) theory of computation (cf. for example Friedman [Fr], Fenstad [Fe]), but until recently there has not been a conscious effort to apply the results to concrete situations like the real or complex number fields where classical mathematical techniques and theorems could come into play. One should mention, as an exception to this general statement, the contributions of Friedman and Simpson (see for example Simpson [Si]) on reverse mathematics, in which the second order theory of arithmetic plays a crucial role in the investigations of the complexity of statements of classical theorems of mathematics. However, the emphasis there is not quite the same in spirit as that of Gödel et al. The work of Blum, Shub and Smale in [BSS] provides, among other things, the first instance where a theory of computation is consciously applied to domains of classical mathematics such as $\mathbf{R}^n$ or **C** (indeed their work applies as

* Part of the work was done while the author held a membership at the Mathematical Sciences Research Institute, Berkeley, during the period January-June 1990. Grateful thanks are due to Stephen Smale, Lenore Blum, Bodil Branner and Stuart Cowan for many stimulating discussions and helpful introduction to the subjects mentioned in the title. Blum in particular was instrumental in initiating the author into this exciting new area of mathematics/theoretical computer science. We wish to record here the financial support received from the National University of Singapore during this period.

well to any algebraic structure like a ring or a field. Here we will focus our attention on the real and complex domains). It paves way for investigations into effectivity questions in the areas of analysis and geometry, subjects which did not show much connection with ideas of computation in the past. This is significant since one may now consider computational questions analysis and geometry on the same footing as those in number theory, using a different (continuous) model of computation.

In this paper we discuss the key ideas of the BSS theory, introduce an associated computation machine with oracles, and report on the work done by us relating these machines to the classification problem (in terms of the degrees of unsolvability) of Julia sets of quadratic polynomials in complex dynamical systems. The full paper will appear in Chong [Ch].

Computation Machines

We begin with an example that illustrates the basic idea of the BSS theory (cf. Blum [B1] for an extended discussion). Given a polynomial f over the complex field, define the Newton function as

$$N_f(z) = z - (f(z)/f'(z)).$$

$N_f(z)$ is rational, and it is straightforward to see that $f(z) = 0$ if and only if $N_f(z) = z$. The advantage of transforming the problem "look for a root of $f(z)$" into "look for a fixed point of $N_f(z)$" (a fixed point of $N_f(z)$ is a point of the type $N_f(z) = z$) is that the latter allows one to perform a series of iterations in an effective manner to approximate to the root z. Specifically, if z_0 is a root of f, then there is a neighborhood U of z_0 such that for any $z \in U, |z_0 - N_f(z_0)| \leq |z_0 - z|$, and strict inequality holds if $z \neq z_0$. This means that one may begin with a point $z_1 = z \in U$, and form the orbit $z_1, z_2, \cdots, z_{n+1}, \cdots$, where $z_{n+1} = N_f(z_n)$. The sequence $\{z_n\}$ converges to z_0 . Assuming that arithmetic operations on the complex numbers are basic (commonly accepted in scientific computation), one then obtains an algorithm with which to compute z_0 to any degree of accuracy (by computing the difference $|z_n - z_m|$). Let $\varepsilon > 0$ be small, so that the disk $D_\varepsilon(z_0)$ centered at z_0 with radius ε lies completely in U. The following diagram gives a picture of how a typical machine works:

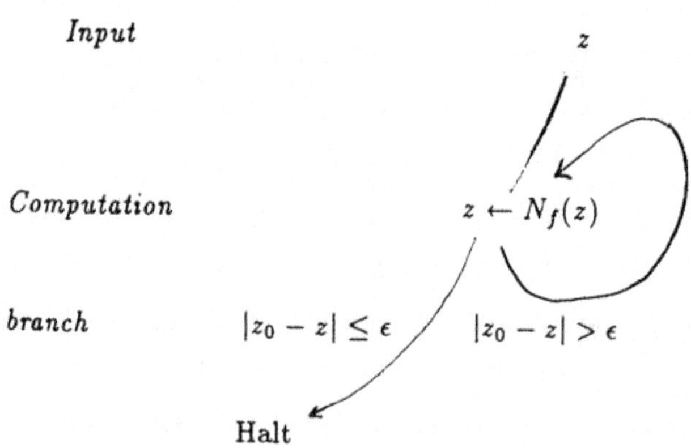

The symbol $z \longleftarrow N_f(z)$ means "replace z by $N_f(z)$". What the above flow chart shows is that there is a 'machine' which on input z halts eventually if some iteration of z under the map N_f takes z to within ε-distance of the root z_0. On the other hand, if no iteration of N_f does it, then the machine never halts on input z (goes into an infinite loop).

In general, a machine M consists of an input node, a finite number of computation nodes (such as N_f in the above illustration), a finite number of branch nodes, and a finite number of output nodes. The machine configuration can best be described in terms of a flow chart, as was done above. We consider only finite dimensional machines (in [BSS], infinite dimensional machines were treated to handle universal machines). the input space is $\mathbf{R}^n$ for some n, and the functions associated with computation nodes are rational functions from $\mathbf{R}^n$ to $\mathbf{R}^m$, where n and m have to be suitably prescribed. At each branch node, there is associated a rational function $h : \mathbf{R}^n \to \mathbf{R}$, where again n has to be of an appropriate value. In the example above, the function is $h(z) = \varepsilon - |z_0 - z|$, so that the machine halts if $h(z) \geq 0$, and goes through a round of iteration otherwise.

A set $A \subset \mathbf{R}^n$ is recursively enumerable (abbreviated r.e.) if there is a machine M for which A is its halting set (i.e. the set of z for which M eventually halts). A is recursive or decidable if both A and its complement in $\mathbf{R}^n$ are r.e. Clearly a set is recursive if and only if there is an algorithm to decide its membership. The next result gives a topological description of r.e. sets.

Theorem 1[BSS]. *An r.e. set is a countable union of semi-algebraic sets.*

A basic semi-algebraic set is a set that satisfies a finite set of algebraic equalities and /or inequalities. A semi-algebraic set is a finite union of basic semi-algebraic sets. It follows from Theorem 1 that the set of irrational numbers is not r.e., although the set of rationals is. Thus we have an example of an r.e. non-recursive set. On the other hand, the unit circle is a recursive set (just verify if (x, y) satisfies $x^2 + y^2 = 1$), as are all solution sets of algebraic equations.

Our next set of examples concern complex analytic dynamics. Let us briefly review some of the basic notions (cf. Blanchard [B] for a more comprehensive discussion of the topic). The symbol $\overline{\mathbf{C}}$ denotes the complex Riemann sphere, and we consider rational functions defined on $\overline{\mathbf{C}}$. Let g be a rational function and let g^n denote the n-fold iteration of g. The Fatou set of g, denoted $F(g)$, is defined to be the set of all points $z \in \overline{\mathbf{C}}$ where the sequence $\{g^n\}$ forms a normal family locally. The Julia set of g, denoted $J(g)$, is its complement. The Fatou-Julia theory says that the Julia set of g is also the closure of repelling periodic points: z is a periodic point if there is an n such that $g^n(z) = z$ (the least such n is called the period of z). Attracting periodic points are those for which $|(g^n)'(z)| < 1$, and repelling periodic points are those where $|(g^n)'(z)| > 1$. If $u = |(g^n)'(z)| = 1$, then the point z is an indifferent rational point if u is a root of unity, and an indifferent irrational point otherwise.

The Julia set of the map $p(z) = z^2$ is the unit circle, while the Julia set of $p_{-2}(z) = z^2 - 2$ is the interval $[-2, 2]$. On the other hand, the Julia set of $p_c(z) = z^2 + c$, where c is a non-trivial solution of the equation $(c^2 + 2)^2 + c = 0$, looks like the following:

Douady's rabbit

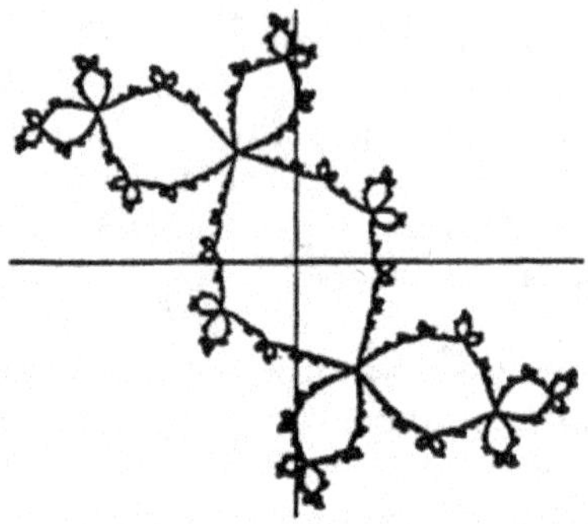

By identifying the complex numbers with $\mathbf{R}^2$, computation machine can be introduced to study the Fatou and Julia sets of rational functions. While the first two examples give rise to r.e. (indeed recursive) Julia sets, the third one does not. In fact, one has

Theorem 2[BSS]. *Most Julia sets are not r.e. If g is rational and J(g) is r.e., then either J(g) is empty, all of* $\overline{\mathbf{C}}$*, one point, an analytic arc, or an analytic Jordan curve.*

It follows from Theorem 2 that an r.e. Julia set has an integral Hausdorff dimension. Let $p_c = z^2 + c$, and denote by $HD(J(g))$ the Hausdorff dimension of $J(g)$. David Ruelle [R] has proved that for $|c|$ small, the map $|c| \longmapsto HD(J(p_c))$ is real analytic. This shows that for $|c| > 0$ and small, $J(p_c)$ is not r.e. Topologically it is a Jordan curve which is nowhere differentiable. In fact for such a c, 0 is an attracting fixed point and $F(p_c)$ is an r.e. set . $J(p_c)$ is then a co-r.e. set (complement of an r.e. set) which is not r.e.

Julia sets of rational functions g share several common properties:

(a) $J(g)$ is compact;

(b) For every $z \in J(g)$, every neighborhood U of z, there is an n such that $g^n(U) \supset J(g)$;

(c) $J(g)$ is chaotic, in the technical sense of the word;

(d) The backward orbit of every $z \in J(g)$ is dense in $J(g)$;

(e) $J(g)$ is completely invariant under $g : g(J(g)) = J$ and $g(F(g)) = F(g)$.

In the case when $g = p_c$ for some c, we have

(f) $J(g)$ is perfect, and furthermore it is either connected or totally disconnected.

The next result was observed by several people (it was hinted at in [BSS], and also noted by Qi Feng, among others):

Fact. Any open set is r.e.

Thus every Fatou set is r.e., and one has at once a collection of (natural) examples of r.e. non-recursive sets in $\mathbf{R}^2$. There is associated with r.e. sets the notion of the topological complexity of the machine used to generate these sets. By definition this is the least number of branch nodes needed to define a machine whose halting set is the r.e. set in question. This subject has been studied by Smale [Sm] and Hirsch [H] for various sets. It is an important notion in complexity theory.

Sullivan [Su] has obtained a complete classification of periodic domains of $F(g)$ (i.e. domains D of $F(g)$ such that $g^n(D) = D$ for some n). These are either attracting, superattracting, parabolic, Siegel, or Herman (cf. [B] for details). The next result shows that these domains have different topological complexities:

Theorem 3[C]. *The topological complexity is equal to 1 for attracting, superattracting, Siegel and Herman domains. It is greater than 1 for parabolic domains.*

Machines with Oracles

Confronted with the knowledge that most Julia sets are not r.e., one now attempts to introduce some order into this collection of non-computable sets. Hence, is there a method of comparing the relative levels of non-computability among Julia sets? In other words, if one assumes that there is complete membership information about $J(g)$, can one, or what additional conditions are needed to, answer membership questions about $J(f)$? To begin with, $J(p-2)$ and $J(p_0)$ are certainly equally complicated (or easy) to compute. Since both are recursive, each can be trivially computed from the other, Non-r.e. Julia sets, however, are much more difficult to handle.

A finite-dimensional computation machine M with oracles has one input node, a finite number of computation nodes, a finite number of branch nodes, a finite number of output nodes, and in addition, a finite number of query nodes. The query nodes are used to answer questions like "Is z in B?". Based on such answers, the program may direct the machine to go through a particular computation. A simple example of a machine with oracles is the following. This is one that uses knowledge about B to compute A:

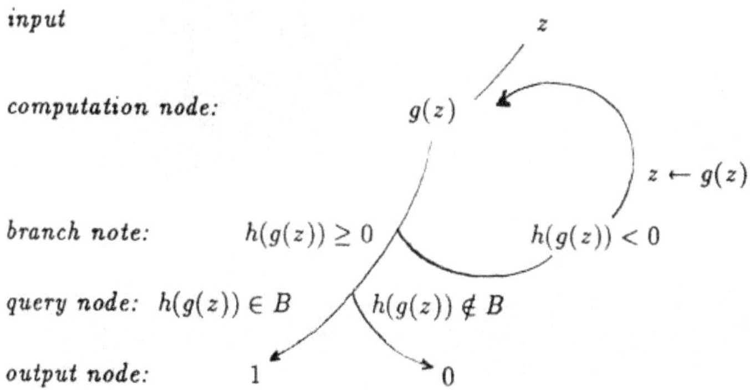

We say that A is reducible to B, written $A \leq^* B$, if there is a computation machine M with oracle B such that for any z, M halts with the answer '1' if $z \in A$, and M halts with the answer '0' if $z \notin A$. It is not difficult to verify that $\leq^*$ is reflexive and transitive. We say that A is equivalent to B, written $A \equiv^* B$, if $A \leq^* B$ and $B \leq^* A$. Subsets of $\overline{\mathbf{C}}$ are divided into equivalence classes under $\equiv^*$, called the degrees of unsolvability $\mathcal{D}$. The degree of a recursive set is the simplest among all the degrees, since any set in that degree is reducible to any other set. Denote the degree of A by $\deg(A)$.

Fact. There is a natural binary operation $\vee$ which makes $\mathcal{D}$ an upper-semilattice. Under $\vee$, $\deg(A)\vee\deg(B)$ is the least upper bound of $\deg(A)$ and $\deg(B)$.

The next result provides a topological interpretation for the computational relation $\leq^*$.

Theorem 4 [C]. *If $J(p_c) \leq^* J(p_d)$, and $J(p_c)$ is non-recursive, then the set of local homeomorphisms from $J(p_c)$ to $J(p_d)$ is relatively open and dense in $J(p_c)$.*

The best way to describe the degrees of unsolvability of Julia sets of quadratic polynomials is in terms of the Mandelbrot set. This is defined as

$$\mathcal{M} = \{c \mid p_c^n(0) \not\to \infty\}.$$

The topological structure of $\mathcal{M}$ is still not completely understood. Douady and Hubbard [DH] showed that $\mathcal{M}$ is connected. There remains a number of fundamental questions on $\mathcal{M}$ which remains unanswered, including several of computational nature. It is not difficult to show that the complement of $\mathcal{M}$ is an r.e. set. But Penrose [P] and others have asked: Is $\mathcal{M}$ decidable? This question is equivalent to asking whether $\mathcal{M}$ is r.e., and the main difficulty appears to be due to the lack of understanding of the boundary of $\mathcal{M}$. The next figure (shaded) shows the Mandelbrot set obtained by computer simulation.

The Mandelbrot set

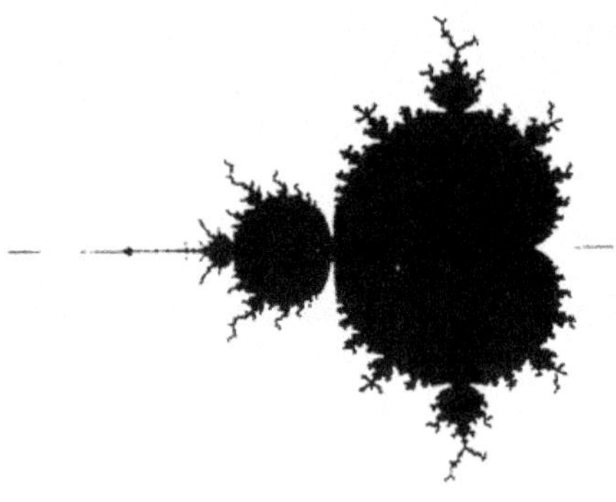

The Friedberg-Muchnik Theorem in recursion theory (cf. [So]) states the existence of two r.e. sets neither of which is reducible to the other. The proof uses an intricate finite-injury priority argument and by induction constructs two r.e. sets with the desired properties. Here we consider $\mathcal{M}$ as the parameter space, and apply Theorem 4 to obtain a 'natural' Friedberg-Muchnik pair of r.e. sets.

Theorem 5 [C]. *If $c \in \mathcal{M}$ and $d \in \overline{\mathbb{C}} \backslash \mathcal{M}$, and $J(p_c)$ is not recursive, then $J(p_c)$ and $J(p_d)$ have incomparable degrees, i.e. neither is reducible to the other.*

Now within $\mathcal{M}$ itself, incomparability is also not an unusual phenomenon. We introduce a new notion. A hyperbolic omponent of the interior of $\mathcal{M}$ is a component H such that p_c contains an attracting periodic point for each $c \in \mathcal{M}$. It is known that for each $p/q \in (0,1)$, there is a hyperbolic component $W_{p/q}$ contained in the interior of $\mathcal{M}$. Let

$$\alpha_c \longmapsto \text{multiplier of } \alpha_c,$$

where α_c is the attracting periodic point of c, and multiplier of α_c is the value $(p_c^n)'(\alpha_c)$.

Douady and Hubbard [DH] have proved that this map is a conformal map of $W_{p/q}$ to the open unit disk. Let the centre of $W_{p/q}$ be the preimage of 0.

Theorem 6 [C]. *If c and d are centres of different hyperbolic components as above, then $J(p_c)$ and $J(p_d)$ have incomparable degrees.*

The next step is to investigate the degrees of Julia sets $J(p_c)$ with c's lying on the same hyperbolic component. Define the main cardiod of $\mathcal{M}$ to be to points c which have an attractive fixed point.

Theorem 7 [C]. *If $0 < |c| < |d|$ are on the main cardiod, then $J(p_c)$ and $J(p_d)$ have incomparable degrees.*

Thus it appears that for Julia sets of quadratic polynomials, incomparable degrees are more common than comparable ones. The full picture of $\mathcal{D}$ is however still only vaguely understood at this point.

References

[B] P. Blanchard, *Complex analytic dynamics on the Riemann sphere*, Bulletin Amer. Math. Soc., 11(1984), 85–141

[Bl] L. Blum, *Lectures on a theory of computation and complexity over the reals (or an arbitrary ring)*, preprint.

[BSS] L. Blum, M. Shub and S. Smale, *On a theory of computation and complexity over the real numbers: NP completeness, recursive functions, and universal machines*, Bulletin Amer. Math. Soc., 21(1989), 1–46.

[C] C.T. Chong, *Reducibility of Julia sets in complex analytic dynamics*, in preparation

[DH] A. Douady and H. Hubbard, *Itération des polynomés quadratiques complexes*, C.R. Acad. Paris, 294(1982), 123–126.

[Fe] J.E. Fenstad, *General Recursion Theory*, Springer Verlag, 1980.

[Fr] H. Friedman, *Algorithmic procedures, generalized Turing algorithms, and elementary recursion theory, in: Logic Colloquium '69*, North-Holland, 1971, 361–390.

[H] M. Hirsh, *Applications of Topology to Lower Bound Estimates in Computer Science*, Ph. D. Thesis, Univ. California at Berkeley, 1990.

[P] R. Penrose, *The Emperor's New Mind*, Oxford Univ. Press, 1989.

[R] D. Ruelle, *Repellers for real analytic maps*, Ergodic Theory and Dynamical Systems 2(1982), 99–105.

[Si] S. Simpson, *Reverse mathematics, in: Recursion Theory*, Proc. Symposia in Pure Math., Amer. Math. Soc. 1985.

[Sl] T.A. Slaman, *Degree structures*, Proc. International Congress of Mathematicians, Kyoto 1990, to appear.

[Sm] S. Smale, *On the topology of algorithms*, I., J. Complexity, 3(1987), 81–89.

[So] R. Soare, *Recursively Enumerable Sets*, Springer Verlag, 1987.

[Su] D. Sullivan, *Quasiconformal homeomorphisms and dynamics, III*, preprint.

Proceedings of the Asian Mathematical
Conference, 1990

A Fundamental Integral on the Real Line

SOEPARNA DARMAWIJAYA
Department of Mathematics
Gadjah Mada University
Yogyakarta, INDONESIA

§0. Introduction

On the real line, there are at least three kinds of integral of Riemann type, that is, ordinary Riemann integral, completely continuous Riemann integral and approximately continuous integral of Burkill. Each of the three integrals depends on a calculus involving a collection of point-interval pairs.

After observing and compiling the fundamental properties of the collections, we shall have an "axiomatic" collection of sets, from which we develop the concepts in calculus Consequently, we define an integral which is called a fundamental integral.

§1. A Basic Structure

If $x \in [a,b]$ then a family of measurable sets $\mathcal{D}_x$ is called a system of fundamental sets at x if the following axioms are satisfied:

(A1) For every $D_x \in \mathcal{D}_x$ and $u < x < v$ we have $D'_x = D_x \cap (u,v) \in \mathcal{D}_x$;

(A2) If D'_x, $D''_x \in \mathcal{D}_x$ then $D_x = D'_x \cap D''_x \in \mathcal{D}_x$.

(A3) Let A be an index set. If $D^\sigma_x \in \mathcal{D}_x$ for every $\sigma \in A$ then

$$D_x = \bigcup_{\sigma \in A} D^\sigma_x \in \mathcal{D}_x.$$

(A4) For every $D_x \in \mathcal{D}_x$, D_x contains x and there exist $u, v \in D_x$ with $u < x < v$.

(A5) If $D_x \in \mathcal{D}_x, u \in D_x$ and $u < x$ then there is $u_1 \in D_x$ such that $u < u_1 < x$; also, if $D_x \in \mathcal{D}_x, v \in D_x$ and $x < v$ then there is $v_1 \in D_x$ such that $x < v_1 < v$.

Any member D_x of a system $\mathcal{D}_x$ of fundamental sets at x is called a fundamental set of x. If $\mathcal{D}_x$ is a system of fundamental sets at x, then for a fixed $D_x \in \mathcal{D}_x$

$$J_x = \{(u,v); u, v \in D_x \text{ and } u \le x \le v\}$$

is called a system of fundamental open intervals at x. Then (A1), (A4) and (A5) imply

$$\bigcap_{(u,v) \in J_x} [u,v] = \{x\}.$$

It is clear that

$$\triangle = \{[u,v]; u,v \in D_x, u < v, u \le x \le v, \text{ and } x \in [a,b]\}$$

is a closed cover of the interval $[a,b]$ and it is called a fundamental full cover or, in short, a FFC of $[a,b]$; $\{D_x\}$, the family of fundamental sets associated with $\triangle$, is called the generator of $\triangle$. A partition

$$\mathcal{D} = \{a = a_0, a_1, a_2, \cdots, a_n = b; x_1, x_2, \cdots, x_n\} = ([u,v]; x)$$

with $a_{i-1}, a_i \in D_{x_i} \in \{D_x\}$ and $a_{i-1} \le x_i \le a_i$, is called a fundamental partition associated with FFC $\triangle$ or, in short, $\triangle$-partition on $[a,b]$ and a_i, x_i are called points of the $\triangle$-partition.

In what follows, we shall always assume that for every $x \in [a,b]$ there exists $\mathcal{D}_x$, a system of fundamental sets at x.

(A6) For any FFC $\triangle$ of $[a,b]$ and $x \in [a,b]$ there exist u, v with $u < x < v$ such that for any $s,t \in (u,v)$ we have $D_s \cap D_t \ne \phi$.

Theorem 1.1. *If (A6) is satisfied and $\triangle$ is a FFC of [a, b] then there exists a $\triangle$-partition on every closed sub-interval of [a, b].*

Proof. This proof follows that of Lemma 1.1. of Thomson [7]. Let $\{D_x\}$ be the generator of our FFC $\triangle$ of $[a,b]$. If $\triangle$ fails to contain a $\triangle$-partition on some closed sub-interval I of $[a,b]$, by repeated bisection of I there must be a sequence of closed sub-intervals $\{I_n\}$ where $I_{n+1} \subset I_n \subset I$, $\lim_{n\to\infty} \mu(I_n) = 0$, and there is no $\triangle$-partition on I_n. So, we see that there is a unique $x_0 \in I_n$ for every n. By (A6), there exists u, v with $u < x_0 < v$ such that for any $s,t \in (u,v)$ implies $D_s \cap D_t \ne \phi$. Further, there is a positive integer n_0 such that $I_n = [x,y] \subset (u,v)$ if $u \ge n_0$. It is clear that $D_x \cap D_y \ne \phi$ and it means that there is a $\triangle$-partition on I_n for $n \ge n_0$. It contradicts our assumption.

It might be that $(A6)$ is a stronger condition than necessary. However, in what follows we shall always assume that for every FFC $\triangle$ of $[a,b]$ there is a $\triangle$-partition on every closed sub-interval of $[a,b]$.

Examples.

(1) Let $\delta : [a,b] \to I\!R$ be a positive valued function on $[a,b]$. Then we have:

A system of fundamental sets at $x \in [a,b]$ generated by δ is the family of all fundamental sets (intervals) $D_x = (s,t)$ with $x - \delta(x) < s < x < t < x + \delta(x)$. It is not difficult to show that $(A1)$, (A2), (A3), (A4) and (A5) hold true.

For a fixed $D_x \in \mathcal{D}_x$

$$J_x = \{(u,v); u,v \in D_x = (s,t) \text{ and } u \le x \le v\}$$

is a system of fundamental open intervals at x. And,

$$\begin{aligned}\triangle = \;& \{[u,v]; u,v \in D_x, u \le x \le v \text{ and } x \in [a,b]\} \\ = \;& \{[u,v]; x - \delta(x) < s < u \le x \le v < t < x + \delta(x) \text{ and } x \in [a,b]\}\end{aligned}$$

is a fundamental full cover of $[a,b]$ associated with δ; usually it is called a completely full cover of $[a,b]$.

(2) If $x \in [a,b]$ we define D_x, a fundamental set of x, to be a set of density 1 at x and $x \in D_x$. Then $\mathcal{D}_x$, a system of fundamental sets at x, is the collection of all D_x. For a fixed $D_x \in \mathcal{D}_x$,

$$J_x = \{(u,v); u, v \in D_x \text{ and } u \le x \le v\}$$

is a system of fundamental open intervals at x and

$$\Delta = \{[u,v]; u, v \in D_x, u \le x \le v \text{ and } x \in [a,b]\}$$

is a fundamental full cover of $[a,b]$ and usually it is called an approximately full cover of $[a,b]$.

§2. Some Concepts in Calculus

If $g : [a,b] \to \overline{I\!R}$ then a number l is called a fundamental limit or, in short, f-limit of $g(x)$ as x approaches $x_0 \in [a,b]$, written by

$$\text{f-}\lim_{x \to x_0} g(x) = l,$$

if for every $\varepsilon > 0$ there is a fundamental set $D_{x_0} \in \mathcal{D}_{x_0}$ such that for every $u \in D_{x_0}$ and $u \ne x_0$ implies

$$|g(u) - l| < \varepsilon.$$

Clearly, the number l is unique. If

$$\text{f-}\lim_{x \to x_0} g(x) = g(x_0)$$

we say: the function g is fundamentally continuous or, in short, f-continuous at x_0. If g is f-continuous at every point of $E \subset [a,b]$, we say the function g is f-continuous on E and in this case we write $g \in C_f(E)$.

$$\overline{\text{f-}\lim_{x \to x_0}} g(x) = \inf_{D_{x_0} \in \mathcal{D}_{x_0}} \sup\{g(x); x \in D_{x_0}\},$$

if it exists, is called the upper f-limit of $g(x)$ as $x \to x_0$; also,

$$\text{f-}\underline{\lim_{x \to x_0}} g(x) = \sup_{D_{x_0} \in \mathcal{D}_{x_0}} \inf\{g(x); x \in D_{x_0}\},$$

if it exists, is called the lower f-limit of $g(x)$ as $x \to x_0$. We obtain that

$$\text{f-}\lim_{x \to x_0} g(x) = l,$$

(the limit exists) if and only if $\overline{\text{f-}\lim_{x \to x_0}} g(x) = \text{f-}\underline{\lim_{x \to x_0}} g(x) = l$.

If there exists a number $g_f(x_0) = D_f g(x_0)$ such that

$$\text{f-}\lim_{x \to x_0} \frac{g(x) - g(x_0)}{x - x_0} = D_f g(x_0),$$

we say that the function g is f-derivable at x_0 and the number $D_f g(x_0)$ is called the f-derivative of g at x_0. The upper and the lower f-derivative of g at x_0, $\overline{D}_f g(x_0)$ and $\underline{D}_f g(x_0)$, are defined in the usual way; and we have

(i) $\underline{D}_f g(x_0) \leq \bar{D}_f g(x_0)$

(ii) $\underline{D}_f[-g(x_0)] = -\bar{D}_f g(x_0)$

(iii) $\bar{D}_f\{g(x_0) + h(x_0)\} \leq \bar{D}_f g(x_0) + \bar{D}_f h(x_0)$

(iv) $\underline{D}_f\{g(x_0) + h(x_0)\} \geq \underline{D}_f g(x_0) + \underline{D}_f h(x_0)$.

Let $g_n, g : [a,b] \to \overline{I\!R}$ for every n. We say the sequence $\{g_n\}$ is locally fundamentally convergent to g at $x \in [a,b]$ or, in short, locally f-convergent to g at $x \in [a,b]$ if for every $\varepsilon > 0$ there exist a positive integer n_x and a fundamental set $D_x \in \mathcal{D}_x$ such that for every $n \geq n_x$ and $u \in D_x$ imply

$$|g_n(u) - g(u)| < \varepsilon.$$

If $\{g_n\}$ is locally f-convergent to g at every point of $E \subset [a,b]$ we say $\{g_n\}$ is locally f-convergent on E.

Theorem 2.1. *If g_n is f-continuous at $x \in [a,b]$ for every n and $\{g_n\}$ is locally f-convergent to g at x then g is f-continuous at x.*

Proof. Given $\varepsilon > 0, g_n$ is f–continuous at x if and only if there is $D_x^n \in \mathcal{D}_x$ such that for every $u \in D_x^n$ it implies

$$|g_n(u) - g_n(x)| < \varepsilon/3.$$

Since $\{g_n\}$ is locally f-convergent to g at x, there is $D_x \in \mathcal{D}_x$ and a positive integer m_x such that for any $u \in D_x$ and $n \geq m_x$ we have

$$|g_n(u) - g(u)| < \varepsilon/3.$$

By (A1), we have $D_x^0 = D_x^n \cap D_x \in \mathcal{D}_x$ with $n = m_x$. Therefore, for any $u \in D_x^0$ and $n = m_x$ we have

$$\begin{aligned} |g(u) - g(x)| \leq & \ |g(u) - g_n(u)| + |g_n(u) - g_n(x)| + |g_n(x) - g(x)| \\ < & \ \varepsilon/3 + \varepsilon/3 + \varepsilon/3 = \varepsilon, \end{aligned}$$

that is, g is f-continuous at x.

Let E be a closed sub-set of $[a,b]$ and $\{[a_n, b_n]\}$ the sequence of all closed continuous intervals of E with respect to $[a,b]$. We say a function $F : [a,b] \to I\!R$ is absolutely fundamentally continuous on E or, in short, $F \in AC_f^*(E)$ if for every number $\varepsilon > 0$ there is a number $\eta > 0$ such that for every $x_k \in E(k = 1,2,\cdots,p)$ with $x_{k-1} < x_k$ and non overlapping $D_{x_k}(k = 1,2,\cdots,p)$ with $D_{x_k} \in \mathcal{D}_{x_k}$ and if $u_k \in D_{x_k}$ with $\sum_{k=1}^{p} |x_k - u_k| < \eta$ then

$$\sum_{k=1}^{p} |F(x_k) - F(u_k)| < \varepsilon.$$

It is easy to see that if $F \in AC_f^*(E)$ then it is f-continuous on E. Let $\{F_n\}$ be a sequence of functions defined on $[a,b]$ and $E \subset [a,b]$ a closed subset. We say $F_n \in AC_f^*(E)$ uniformly in n, if for any number $\varepsilon > 0$ there is a number $\eta > 0$ such that for any $x_k \in E(k = 1, 2, \cdots, p)$ with $x_{k-1} < x_k$ and non overlapping $D_{x_k}(k = 1, 2, \cdots, p)$ with $D_{x_k} \in \mathcal{D}_{x_k}$ and if $u_k \in D_{x_k}$ with $\sum_{k=1}^{p} |x_k - u_k| < \eta$ then

$$\sum_{k=1}^{p} |F_n(x_k) - F_n(u_k)| < \varepsilon$$

for every n. If there is a sequence of closed sets $\{E_n\}$ such that $[a,b] = \cup_n E_n$ and $F \in AC_f^*(E_n)$ for every n, we say F is generalized fundamentally absolutely continuous on $[a,b]$ or, in short, $F \in ACG_f^*[a,b]$.

Theorem 2.2. *Let $E \subset [a,b]$ be a closed subset. If $\{F_n\}$ is a sequence of functions defined on $[a,b]$ such that $F_n \in AC_f^*(E)$ uniformly in n and $\{F_n\}$ is locally f-convergent to F on $[a,b]$, then $F \in AC_f^*(E)$.*

Proof. Since $F_n \in AC_f^*(E)$ uniformly in n, then for any number $\varepsilon > 0$ there is a number $\eta > 0$ such that for every $x_k \in E(k = 1, 2, \cdots, p)$ with $x_{k-1} < x_k$ and non overlapping $D_{x_k}(k = 1, 2, \cdots, p)$ with $D_{x_k} \in \mathcal{D}_{x_k}$ and if $u_k \in D_{x_k}$ with $\sum_{k=1}^{p} |x_k - u_k| < \eta$ then

$$\sum_{k=1}^{p} |F_n(x_k) - F_n(u_k)| < \varepsilon/3$$

for every n. Also $\{F_n\}$ is locally f-convergent to F on $[a,b]$ then for the number ε above and $x_k \in E$ there is a fundamental set $D''_{x_k} \in \mathcal{D}_{x_k}$ and a positive integer n_k such that if $u \in D''_{x_k}$ and $n \geq n_k$ then

$$|F_n(u) - F(u)| < \frac{\varepsilon}{3}p\,.$$

Therefore, if $u_k, x_k \in D_{x_k}$ with $D_{x_k} = D'_{x_k} \cap D''_{x_k}$ we have

$$\begin{aligned}\sum_{k=1}^{p} |F(x_k) - F(u_k)| \leq\ & \sum_{k=1}^{p} |F(x_k) - F_n(x_k)| + \sum_{k=1}^{p} |F_n(x_k) - F_n(u_k)| \\ & + \sum_{k=1}^{p} |F_n(u_k) - F(u_k)| < \frac{\varepsilon}{3} + \frac{\varepsilon}{3} + \frac{\varepsilon}{3} = \varepsilon\end{aligned}$$

and the proof is complete.

Let F be a function defined on $[a,b]$ and $E \subset [a,b]$ a closed set. We call F is a fundamentally bounded variation function on E or, in short, $F \in BV_f^*(E)$ if there is a number $M \geq 0$ such that for any $x_k \in E(k = 1, 2, 3, \cdots, p)$, with $x_{k-1} < x_k$ and non overlapping $D_{x_k}(k = 1, 2, \cdots, p)$ with $D_{x_k} \in \mathcal{D}_{x_k}$ and if $u_k \in D_{x_k}$ we have

$$\sum_{k=1}^{p} |F(x_k) - F(u_k)| \leq M.$$

It is clear that $AC_f^*(E) \subset BV_f^*(E)$. If there is a sequence of closed sets $\{E_n\}$ and $[a,b] = \cup_n E_n$ such that $F \in BV_f^*(E_n)$ for every n we say F is a generalised fundamentally bounded variation function on $[a,b]$ or, in short, $F \in BVG_f^*[a,b]$.

§3. A Fundamental Integral

Definition 3.1 (Variational integral). A function g: $[a,b] \to \overline{\mathbb{R}}$ is said to be fundamentally variationally integrable on $[a,b]$ or, in short, $g \in VR_f^*[a,b]$ if there exists $G : [a,b] \to \mathbb{R}$ such that for any $\varepsilon > 0$ there exist a non-decreasing function $\phi : [a,b] \to \mathbb{R}$ with $\phi(b) - \phi(a) < \varepsilon$ and a fundamental set $D_x \in \mathcal{D}_x$ for every $x \in [a,b]$ such that

$$|G(v) - G(u) - g(x)(v-u)| \leq \phi(v) - \phi(u)$$

for every $u, v \in D_x$ with $u \leq x \leq v$.
The function G in Definition 3.1 is called VR_f^*-primitive of g on $[a,b]$, and it is unique in the sense if there is H, another VR_f^*-primitive of g, then $H(x) = G(x) + k$ for every $x \in [a,b]$ where k is a constant.

The number

$$(VR_f^*) \int_a^b g = G(b) - G(a)$$

is called the value of VR_f^*-integral of g on $[a,b]$. For simplicity, we may take $G(a) = 0$. It is easy to see that if $g \in VR_f^*[a,b]$ then $g \in VR_f^*[a,t]$ for every $t \in [a,b]$ and

$$(VR_f^*) \int_a^t g = G(t).$$

Further, $VR_f^*[a,b]$ is a linear space. Let $\triangle'$ be a FFC of $[a,b]$ generated by $\{D'_x\}$ and $\triangle''$ a FFC of $[b,c]$ generated by $\{D''_x\}$. Then $\{D_x\}$, with $D_x = D'_x$ if $x \in [a,b)$, $D_x = D'_x \cap D''_x$ if $x = b$, and $D_x = D''_x$ if $x \in (b,c]$, generates a FFC $\triangle$ of $[a,c]$. It implies that if $g \in VR_f^*[a,b]$ and $g \in VR_f^*[b,c]$ then $g \in VR_f^*[a,c]$ with VR_f^*-primitive on $[a,c]$ being G:

$$G(x) = \begin{cases} G_1(x) & \text{if } x \in [a,b], \\ G_1(b) + G_2(x) & \text{if } x \in [b,c], \end{cases}$$

where G_1 is a VR_f^*-primitive of g on $[a,b]$ with $G_1(a) = 0$ and G_2 is a VR_f^*-primitive of g on $[b,c]$ with $G_2(b) = 0$.

Theorem 3.2. *If $g \in VR_f^*[a,b]$ with VR_f^*-primitive G, then G is f-continuous on $[a,b]$ and $D_f G(x) = f(x)$ a.e. on $[a,b]$.*

Proof. Since $g \in VR_f^*[a.b]$ with VR_f^*-primitive G, for any $\varepsilon > 0$ there exist a fundamental set $D_x \in \mathcal{D}_x$ for $x \in [a,b]$ and a non-decreasing function $\phi : [a,b] \to \mathbb{R}$ with $\phi(0) - \phi(a) < \varepsilon$ such that, for $u, v \in D_x$ and $u \leq x \leq v$,

$$|G(v) - G(u) - g(x)(v-u)| \leq \phi(v) - \phi(u).$$

It implies

$$|G(v) - G(u)| \leq |g(x)|(v-u) + \phi(v) - \phi(u).$$

For a fixed $x \in [a,b]$ and if $|g(x)| \neq 0$, by (A1), we may take D_x such that $|g(x)|(v-u) < \varepsilon$ for $u, v \in D_x$ with $u \leq x \leq v$. Therefore, we have

$$|G(v) - G(u)| < 2\varepsilon$$

a.e., G is f-continuous at $x \in [a,b]$. If $g(x) = 0$, it is clear also that g is f-continuous there.

Since $D_i\phi(x) = D\phi(x) = \lim\limits_{\substack{u\to x\\ v\to x}} \dfrac{\phi(v)-\phi(u)}{v-u}$ exists a.e. on $[a,b]$,

$$|G(v) - G(u) - g(x)(v-u)| \le \phi(v) - \phi(u)$$

implies

$$\lim_{v\to x}\left|\frac{G(v)-G(x)}{v-x} - g(x)\right| \le \phi'(x)$$

almost everywhere on $[a,b]$; and, since ϕ' is Lebesque integrable on $[a,b]$,

$$(L)\int_a^b \lim_{v\to x}\left|\frac{G(v)-G(x)}{v-x} - g(x)\right| \le \phi(b) - \phi(a) < \varepsilon.$$

Hence, we obtain $D_f G(x^+) = g(x)$ almost everywhere on $[a,b]$. Similarly, we obtain $D_f G(x^-) = g(x)$ almost everywhere on $[a,b]$. Hence, $D_f G(x) = g(x)$ almost everywhere on $[a,b]$.

Given a function $g : [a,b] \to \overline{I\!R}$, a FFC $\triangle$ of $[a,b]$, and a $\triangle$-partition

$$\begin{aligned} \mathcal{D} &= \{a = a_0, a_1, a_2, \cdots, a_n = b; x_1, x_2, \cdots, x_n\} \\ &= ([u,v], x) \qquad \text{(in short)}. \end{aligned}$$

The associated Riemann sum of g over the $\triangle$-partition $\mathcal{D}$ is

$$S(g; \triangle - \mathcal{D}) = \sum_{k=1}^{n} g(x_k)(a_k - a_{k-1}) = \mathcal{D}\sum g(x)(v-u).$$

Definition 3.3. A function $g : [a,b] \to \overline{I\!R}$ is said to be fundamentally Riemann integrable on $[a,b]$ or, in short, $g \in R_f^*[a,b]$ if there is a number A such that for every $\varepsilon > 0$ there exists a FFC $\triangle$ of $[a,b]$ such that for every $\triangle$-partition $\mathcal{D} = ([u,v]; x)$ on $[a,b]$ it implies

$$|S(g; \triangle - \mathcal{D}) - A| = |\mathcal{D}\sum g(x)(v-u) - A| < \varepsilon.$$

The number A is unique, it will be written by

$$A = (R_f^*)\int_a^b g = (R_f^*)\int_a^b g(t)dt$$

and it is called the value of R_f^*-integral g on $[a,b]$.

Theorem 3.4. *$g \in R_f^*[a,b]$ if and only if for every $\varepsilon > 0$ there exist a FFC $\triangle$ of $[a,b]$ such that for any two $\triangle$-partitions $\mathcal{D}_1$ and $\mathcal{D}_2$ it implies*

$$|S(g; \triangle - \mathcal{D}_1) - S(g; \triangle - \mathcal{D}_2)| < \varepsilon.$$

Theorem 3.5. *If $g \in R_f^*[a,b]$ then $g \in R_f^*[a,x]$ for every $x \in [a,b]$. By Theorem 3.5., we have a function $G : [a,b] \to I\!R$:*

$$G(x) = (R_f^*)\int_a^x g = (R_f^*)\int_a^x g(t)dt$$

with $G(a) = 0$. *Therefore, we may say:* $g \in R_f^*[a,b]$ *if there is a function* $G : [a,b] \to \mathbb{R}$ *such that for every* $\varepsilon > 0$ *there is a FFC* $\triangle$ *of* $[a,b]$ *such that for any* $\triangle$*-partition* $\mathcal{D} = ([u,v]; x)$ *it implies*

$$|\mathcal{D}\Sigma\{G(v) - G(u) - g(x)(v-u)\}| < \varepsilon.$$

In terms of FFC, we also have Henstock's Lemma.

Theorem 3.6(Henstock's Lemma). *Let* $g \in R_f^*[a,b]$, *i.e., there is a function* $G : [a,b] \to \mathbb{R}$ *such that for every* $\varepsilon > 0$ *there exists a FFC* $\triangle$ *of* $[a,b]$ *such that for any* $\mathcal{D} = ([u,v]; x)$ *it implies*

$$|\mathcal{D}\Sigma\{G(v) - G(u) - g(x)(v-u)\}| < \varepsilon.$$

Then for every partial sum Σ_1 *of* $\mathcal{D}\Sigma$ *we have*

$$|\Sigma_1\{G(v) - G(u) - g(x)(v-u)\}| < 2\varepsilon.$$

Using Henstock's Lemma we shall prove the following important theorem.

Theorem 3.7. *The* R_f^**-integral and the* VR_f^**-integral are equivalent, i.e.,* $g \in R_f^*[a,b]$ *if and only if* $g \in VR_f^*[a,b]$.

Proof. Sufficiency: $g \in VR_f^*[a,b]$ if and only if there is a function $G : [a,b] \to \mathbb{R}$ such that for every $\varepsilon > 0$ there exist a fundamental set $D_x \in \mathcal{D}_x$ for every $x \in [a,b]$ and a non–decreasing function $\phi : [a,b] \to \mathbb{R}$ with $\phi(b) - \phi(a) < \varepsilon$ such that

$$|G(v) - G(u) - f(x)(v-u)| \leq \phi(v) - \phi(u)$$

for every $u, v \in D_x$ with $u \leq x \leq v$. The collection $\{D_x\}$ generates an FFC $\triangle$ of $[a,b]$. It implies that for every $\triangle$-partition $\mathcal{D}D$ on $[a,b]$ we have

$$\begin{aligned} & |\mathcal{D}\Sigma\{G(v) - G(u) - g(x)(v-u)\}| \\ \leq\ & \mathcal{D}\Sigma|G(v) - G((u) - g(x)(v-u)| \\ \leq\ & \mathcal{D}\Sigma\{\phi(v) - \phi(u)\} \\ =\ & \phi(b) - \phi(a) < \varepsilon, \end{aligned}$$

i.e., $g \in R_f^*[a,b]$.

Necessity: $g \in R_f^*[a,b]$ if and only if there is a function $G : [a,b] \to \mathbb{R}$ such that for every $\varepsilon > 0$ there exists a FFC $\triangle$ of $[a,b]$ such that for any $\triangle$-partition $\mathcal{D}$ on $[a,b]$ it implies

$$|\mathcal{D}\Sigma\{G(v) - G(u) - g(x)(v-u)\}| < \varepsilon.$$

We define an interval function σ as follows. If $[c,d] \subset [a,b]$,

$$\sigma(c,d) = \sup_P |P\Sigma\{G(v) - G(u) - g(x)(v-u)\}|,$$

where P is a $\triangle$-partition on $[c,d]$. By Henstock's Lemma we have $\sigma(c,d) < 2\varepsilon$. Now, we define a function ϕ on $[a,b]$: $\phi(x) = \sigma(a,x)$ for every $x \in [a,b]$, then $\phi(a) = 0, \phi(b) =$

$\phi(b) - \phi(a) < 2\varepsilon$ and ϕ is a non-decreasing function on $[a, b]$. Hence, for any $D_x \in \{D_x\}$, where $\{D_x\}$ generates $\triangle$, we have

$$|G(v) - G(u) - g(x)(v - u)| \leq \phi(v) - \phi(u)$$

for every $u, v \in D_x$ with $u \leq x \leq v$, that is, g is VR_f^*-integrable on $[a, b]$.

Now, it is easy to prove the following theorem.

Theorem 3.8. *$g \in R_f^*[a, b]$ if and only if there is a function $G:[a, b] \to \mathbb{R}$ such that for every $\varepsilon > 0$ there exists a FFC $\triangle$ of $[a, b]$ such that for any $\triangle$-partition $\mathcal{D} = ([u, v], x)$ it implies*

$$\mathcal{D}\Sigma|G(v) - G(u) - g(x)(v - u)| < \varepsilon.$$

The following Cauchy property and its proof follow those of Bullen [1, Theorem 15 page 147–148]:

Theorem 3.9 (Cauchy property). *If $g \in R_f^*[\alpha, \beta]$ for every $[\alpha, \beta] \subset [a, b]$ with $a < \alpha < \beta < b$ and*

$$\text{f-}\lim_{\alpha \to a^+, \beta \to b^-} (R_f^*) \int_\alpha^\beta g = I,$$

exists, then $g \in R_f^*[a, b]$ and $(R_f^*) \int_a^b g = I$.

Proof. It is sufficient to prove that if $g \in R_f^*[a, \beta]$ for every $a < \beta < b$ and $\text{f-}\lim_{\beta \to b} (R_f^*) \int_a^\beta g = I$ then $g \in R_f^*[a, b]$ and $(R_f^*) \int_a^b g = I$.

Let $a = \beta_0 < \beta_1 < \beta_2 < \cdots < b$ and $\lim_{n \to \infty} \beta_n = b$.

Since $g \in R_f^*[\beta_{k-1}, \beta_k]$ then there is a number I_k such that for every $\varepsilon > 0$ there is a FFC $\triangle_k$ of $[\beta_{k-1}, \beta_k]$ such that for any $\triangle_k$-partition $\mathcal{D}_k$ on $[\beta_{k-1}, \beta_k]$ we have

$$|I_k - \mathcal{D}_k\Sigma g(x)(v - u)| < \frac{\varepsilon}{2^k},$$

where $I_k = (R_f^*) \int_{\beta_{k-1}}^{\beta_k} g$. Since $\text{f-}\lim_{\beta \to b} (R_f^*) \int_a^\beta g = I$ then for $\varepsilon > 0$ there is $\sigma > 0$ and $D_b \in \mathcal{D}_b$ such that

$$\left|I - (R_f^*) \int_a^x g\right| < \varepsilon \quad \text{and} \quad |(b - x)g(x)| < \varepsilon$$

for every $x \in A = D_b \cap (b - \delta, b]$. It is clear that $\triangle = [\cup\triangle_k] \cup \{[x, b]; x \in A\}$ is a FFC on $[a, b]$.

Hence, if $\mathcal{D}_k = \{a = a_0, a_1, a_2, \cdots, a_n = b; x_1, x_2, \cdots, x_n\}$ is a Δ-partition on $[a, b]$, then

$$\begin{aligned} \left| I - \sum_{k=1}^{n} g(x_k)(a_i - a_{i-1}) \right| &\leq \left| (R_f^*) \int_a^{a_{n-1}} g - \sum_{i=1}^{n-1} g(x_k)(a_i - a_{i-1}) \right| \\ &\quad + \left| I - (R_f^*) \int_a^{a_{n-1}} \right| + |g(b)(b - a_{i-1})| \\ &< \sum_{k=1}^{n-1} \frac{\varepsilon}{2^k} + \varepsilon + \varepsilon < 3\varepsilon, \end{aligned}$$

that is, $g \in R_f^*[a, b]$ and $(R_f^*) \int_a^b g = I$.

Theorem 3.10 (Harnack property). *Let E be a perfect subset of $[a, b]$ such that $a, b \in E$ and $\{[a_n, b_n]\}$ the sequence of closed continuous intervals of E with respect to $[a, b]$. If I_E is the characteristic function on $E, gI_E \in R_f^*[a_n, b_n]$, and $g \in R_f[a_n, b_n]$, and $g \in R_f[a_n, b_n]$ for every n, such that*

$$\sum_{n=1}^{\infty} \sup \left\{ \left| (R_f^*) \int_\alpha^\beta g \right| ; [\alpha, \beta] \subset [a_n, b_n] \right\} < \infty$$

then $g \in R_f^*[a, b]$ and

$$(R_f^*) \int_a^b g = (R_f^*) \int_a^b gI_E + \sum_{n=1}^{\infty} (R_f^*) \int_{a_n}^{b_n} g.$$

Proof. By our hyphotesis, for every number $\varepsilon > 0$ there are:
(i) a FFC Δ' of $[a, b]$ such that for any Δ'-partition $\mathcal{P}' = ([u, v]; x)$ on $[a, b]$ implies

$$\left| (R_f^*) \int_a^b gI_E - \mathcal{P}'\Sigma(gI_E)(x)(v - u) \right| < \frac{\varepsilon}{4},$$

(ii) a FFC Δ_n of $[a_n, b_n]$, for every n, such that for any Δ_n-partition $\mathcal{P}_n' = ([u, v]; x)$ on $[a_n, b_n]$ implies

$$\left| (R_f^*) \int_{a_n}^{b_n} g - \mathcal{P}_n'\Sigma g(x)(v - u) \right| < \frac{\varepsilon}{2^n},$$

(iii) and a positive integer n_0 such that

$$\sum_{n > n_0} \left| (R_f^*) \int_{a_n}^{b_n} g \right| < \frac{\varepsilon}{4}.$$

It is sufficient to prove that $g(I - I_E) = g - gI_E \in R_f^*[a, b]$ and

$$(R_f^*) \int_a^b g(I - I_E) = \sum_{n=1}^{\infty} (R_f^*) \int_{a_n}^{b_n} g.$$

Let $\{D'_x\}$ and $\{D^n_x\}$ be the generators of $\triangle'$ and $\triangle_n$ respectively. We define $D_x = D'_x$ if $x \in E$ and $D_x = D'_x \cap D^n_x$ if $x \in [a_n, b_n]$. If $\triangle$ is the FFC of $[a,b]$ generated by $\{D_x\}$, $\mathcal{P} = ([u,v];x)$ is an arbitrary $\triangle$-partition on $[a,b]$, and $\mathcal{P}_n = ([u,v];x) = \{a^n_0, a^n_1, \cdots, a^n_p; x^n_1, x^n_2, \cdots, x^n_p\}$ is an arbitrary $\triangle$-partition on $[a_n, b_n]$ for every n, then

$$\begin{aligned}
&\left|\sum_{n=1}^{\infty}(R^*_f)\int_{a_n}^{b_n} g - \mathcal{P}\Sigma\, g(I - I_E)(x)(v-u)\right| \\
\leq\ &\left|\sum_{n>n_0}(R^*_f)\int_{a_n}^{b_n} g\right| + \sum_{n\leq n_0}\left|(R^*_f)\int_{a_n}^{b_n} g - \sum_{k=1}^{p} g(x^n_k)(a^n_k - a^n_{k-1})\right| \\
<\ &\frac{\varepsilon}{4} + \sum_{n>n_0}\frac{\varepsilon}{2^n} < 2\varepsilon
\end{aligned}$$

and the proof is complete.

§4. Further Extensions

In a usual way we may define fundamental Perron integral, fundamentally special Denjoy integral and find their relationship with R^*_f-integral. Also, we may develope the corresponding convergence theorems.

References

[1] Bullen, P.S., *The Burkill approximately continuous integral*, J. Austral. Math. Soc. (Series A), 35(1983), 236–253.

[2] Burkill, J.C., *The approximately continuous Perron integral*, Math. Z., 34 (1931), 270–278.

[3] Kubota, Y., *An elementary theory of the special Denjoy integral*, Math. Japonica 37, no. 5(1980), 507–220.

[4] Lee Peng Yee, *Lanzhou lectures on Henstock integration*, Series in Real Analysis Volume II (1989), World Scientific Publ. Co., Singapore—New Jersey—Hong Kong.

[5] Muldowney, P., *A general theory of integration in function spaces*, Pitman Research Notes in Mathematics Series, 135(1987), Longman Sci. & Technical.

[6] Ostaszewski, K.M., *Henstock integration in the plane*, Memoirs American Math. Soc., 63 no. 353(1986).

[7] Thomson, B.S., *On full covering*, Real Anal. Exchange, 6(1980–1981), 77–93.

[8] Thomson, B.S., *Real functions*, Springer-Verlag, LN 1170(1985).

Proceedings of the Asian Mathematical
Conference, 1990

Manifolds of Constant Width

B.V. DEKSTER*)

Bodies of constant width in spherical and hyperbolic spaces were considered in [1], [8], [9], [10], [11]. A short review of these results can be found in the second half of §11 in [3]. The notion of a body of constant width was extended to Riemannian manifolds in [4]. Here, we review some results from [4] and some other recent results in this area.

Let M^n be a regular Riemannian manifold, not necessarily complete. A set $C \subset M^n$ will be called definitely convex if

(i) C is compact and each two points of C can be connected in C by a rectifiable curve;

(ii) any shortest path in C between two points of C is a geodesic (the path exists due to (i));

(iii) any geodesic segment in C is the unique shortest path in C between its ends;

(iv) each geodesic segment in C contains no pair of conjugate points along this segment.

For instance, let M^2 be a unit sphere in R^3 and C be a geodesic of length $\frac{3}{2}\pi$, then C is not definitely convex because (iv) fails even though (i), (ii) and (iii) hold. Let M^2 be a cylinder of unit radius in R^3 and $C \subset M^2$ be a circular arc of length $\frac{3}{2}\pi$, then C is definitely convex. A circumference on the last M^2 is not definitely convex since (iii) fails while (i), (ii) and (iv) hold. The domain $y \leq x^2$ of the xy-plane is not definitely convex since (i) and (ii) fail while the other two conditions hold.

By a shortest path, we will mean a shortest path within the definitely convex set C under consideration; by the diameter of C, we mean its diameter with respect to the metric $\rho(x, y)$ being length of a shortest path in C between x and $y \in C$. Thus the diameter of the arc of length $\frac{3}{2}\pi$ in the first two examples above is $\frac{3}{2}\pi$.

If a definitely convex set C is homeomorphic to a ball, it will be called a definitely convex body. Let $K \subset M^n$ be a definitely convex body. Its boundary and diameter will be denoted by ∂K and diam K. We will call K a body of constant width $W > 0$, it for any $p \in \partial K$ and any exterior unit normal n of K at p, there exists a geodesic pq of length W having direction $-n$, such that $K \supset pq$ but K contains no longer geodesic $pq' \supset pq$ (then obviously $q \in \partial K$ and by (iii), pq is a shortest path in K); thus $D \overset{\text{def}}{=} \text{diam } K \geq W$. For the definition of exterior normal see [2, §4.5]). We use K, W and D as defined above throughout the rest of the paper. The segment pq will be called geodesic normal.

Theorem 1 [4, Remark 3]. $D = W$. *The geodesic normal pq above is a double geodesic normal (i.e. the direction of pq at q is an exterior normal of K).*

*) Supported by a Canadian NSERC Grant.

Theorem 2 [7, Theorem 1]. *A definitely convex body in M^n is of constant width if and only if each of its geodesic normals is a double geodesic normal.*

Theorem 3 [4, Remark 4]. *Let K be the maximum sectional curvature in K and let W satisfy $W < \pi/\sqrt{k}$ if $k > 0$. Denote by $\kappa = \kappa(k, W)$ the curvature of a circumference of radius W in Euclidean, hyperbolic or spherical 2-space of curvature k. Fix a $\lambda < \kappa$. Then, at each point $p \in \delta K$, the body K can be locally supported by a regular hypersurface whose normal curvatures on the side of K all equal λ. (See [4] for more exact presentation.)*

The quantity $r_i = \max_{x\in K} \min_{y\in\partial K} xy$ will be called inradius of K and each point in K at the distance r_i from ∂K will be called incentre of K. The quantity $r_c = \min_{x\in K} \max_{y\in K} xy$ will be called circumradius of K. Each point $x \in K$ for which $\max_{y\in K} xy = r_c$ will be called circumcenter of K.

Theorem 4 [5, Theorem].

(A) $r_i \le r_c$.

(B) *Each incenter of K is its circumcenter; each circumcenter of K is its incenter.*

(C) $r_i + r_c = W$.

(D) *Denote by k the maximum sectional curvature in K. The body K has a unique center (i.e., in-and circumcenter at the same time, see (B)) when $k \le 0$ and also when $k > 0$ and $r_c < \frac{\pi}{2\sqrt{k}}$. (This condition holds e.g. when $W \le \pi/2\sqrt{k}$.)*

Let $C \subset M^n$ be a definitely convex body, $a, b \in \partial c$ and n_a, n_b be exterior unit normals at a and b respectively. Suppose n_a turns into $-n_b$ after its parallel translation along the chord ab. Then the normals n_a, n_b are said to be antipodal. (Symmetry of this relation is obvious.)

For a pair of antinodal normals n_a, n_b we define a breadth $B(n_a, n_b)$ putting

$$B(n_a, n_b) = ab \cos\alpha,$$

where $a \le \pi/2$ is the angle between n_a and the direction of the chord ba at a (or between n_b and the direction of ab at b which is the same). Note that $B(n_a, n_b)$ in R^n is the standard width of C in the direction $n_a = -n_b$. To further justify the definition of $B(n_a, n_b)$ in M^n, consider an osculating Euclidean space along ab. The image of the body C in this space (within a neighbourhood of ab) has clearly the standard width $B(n_a, n_b)$ in the direction $n_a = -nb$.

Note that n_a can have a few antipodal normals. The following Theorem guarantees at least one.

Theorem 5 [6, Theorem 1]. *Let $c \subset M^n$ be a definitely convex body. Then each of its exterior unit normals has an antipodal one.*

Let n_a be a normal of C. Denote by N the set of all its natipodal normals. Put

$$W(n_a) = \max_{n\in N} B(n_a, N).$$

By compactness, the maximum exists and

$$W(n_a) = B(n_a, n_x) \quad \text{for some} \quad n_x \in N.$$

The quantity $W(n_a)$ will be called the width of C for normal n_a. Since

$$B(n_a, n_x) = ax\cos\alpha \le ax \le \operatorname{diam} C,$$

one has

$$W(n_a) \leq \text{diam } C.$$

Obviously, in R^n, the width $W(n_a) = B(n_a, n)$ for any $n \in N$. It is the standard width in the direction n_a.

Theorem 6 [6, Theorem 2]. *A definitely convex body C in M^n is of constant width W if and only if its width $W(m) = W$ for each exterior unit normal m of C.*

We will denote by M_k^n a spherical, hyperbolic or Euclidean n-dimensional space of curvature k. Consider in M_k^2 a quadrangle $abcd$ such that

(i) a is symmetric to b and c is symmetric to d with respect to a geodesic in M_k^2;

(ii) a and d are on the same side of the geodesic;

(iii) $0 < cd = ac = bd = W$ where $W < \frac{\pi}{\sqrt{k}}$ if $k > 0$;

(iv) $0 \leq ab = \Delta \leq W$. In addition,

$$\Delta \geq 3W - 2\pi/\sqrt{k} \quad \text{when} \quad k > 0.$$

Denote by $\gamma = \gamma(\Delta, W, k)$ the angle $cab = dba$ or its limiting value if $\Delta = 0$. One can easily check that the quadrangle exists, and that $\gamma < \frac{\pi}{2}$. (The last line in (iv) guarantees the existence for $W > \frac{2\pi}{3\sqrt{k}}$.) Trigonometry shows that

$$\cos\gamma = \tan\frac{\Delta\sqrt{k}}{2}\cot W\sqrt{k} + \tfrac{1}{2}\sec\frac{\Delta\sqrt{k}}{2}\sec\frac{W\sqrt{k}}{2} \quad \text{if} \quad k > 0;$$

$$\cos\gamma = \frac{W+\Delta}{2W} \quad \text{if} \quad k = 0;$$

$$\cos\gamma = \tanh\frac{\Delta\sqrt{-k}}{2}\coth W\sqrt{-k} + \tfrac{1}{2}\operatorname{sech}\frac{\Delta\sqrt{-k}}{2}\operatorname{sech}\frac{W\sqrt{-k}}{2} \quad \text{if} \quad k < 0.$$

Theorem 7 [4, Theorem]. *Let K b a body of constant width W in $M_k^n, n \geq 2$, satisfying:*

$$W \leq \frac{\pi}{2\sqrt{k}} \quad \text{when} \quad k > 0$$

let $a, b \in \partial k$, $ab = \Delta \geq 0$ and n_a, n_b be exterior normals at a and b respectively, then the angle ξ between n_a and n_b after their parallel translation along the chord ab to the same point satisfies

$$\xi \leq \pi - 2\gamma(\Delta, W, k).$$

The equality here can hold for any triple (Δ, W, k).

References

[1] Blaschke, W., *Einige Bemerkungen über Kurven und Flächen von konstanter Breite.*, Ber. d. Verh. d. Sächs. Akad. Leipzig 67(1915), 290–297.

[2] Burago, Yu. D. and Zalgaller, V.A., *Convex sets in Riemannian spaces of non-negative curvature*, Russian Math. Surveys, 32(3)(1977), 1–57.

[3] Chakerian, G.D. and Groemer, H., *Convex bodies of constant width, Convexity and its application*, Birkhäuser, Basel-Boston, Mass., (1983), 49–96.

[4] Dekster, B.V., *Bodies of constant width in Riemannian manifolds*. To appear in: "*Victor Klee Festschrift*", *Applied Geometry and Discrete Mathematics*, the DIMACS series on Discrete Mathematics and Computer Science, copublished by AMS and ACM.

[5] Dekster, B.V., *In- and circumcenters of manifolds of constant width*, *Geometriae Dedicata*, to appear.

[6] Dekster, B.V., *Width of a manifold and the constant width*, preprint.

[7] Dekster, B.V., *Double normals charakterise manifolds of constant width*, preprint.

Proceedings of the Asian Mathematical
Conference, 1990

On Amenable Banach Algebras

DENG HONGJUN, HU YAOHUA
Department of Mathematics
Hubei University
Wuhan 430062, China.

Introduction

Amenable Banach algebras were introduced and studied by B.E. Johnson in his definitive monograph [1]. This class of Banach algebras arises out of the cohomology theory for Banach algebras, the algebraic version of which was developed by Hochschild [2,3]. Other relevent papers are Kamowitz [4], Guichardet [5,6], Johnson [7–11], Helemskii [12, 13, 14], Helemskii and Sheinberg [15], Lau [16] and Racher [17].

In this paper, we extend the concept of amenable algebra to n–amenable algebra, and prove several properties which are extension of 1–amenable algebras appeared in [1] and [2], and give a sufficient condition for a Banach algebra being 2–amenable.

§ 1. Notations

Let A be a Banach algebra over complex field C. Then a Banach space X over C is a Banach A-module if it is a two–sided A–module and there is a positive real number k such that $\|ax\| \leq k\|a\|\|x\|$, and $\|xa\| \leq k\|x\|\,\|a\|$, for all $a \in A, x \in X$. Let A be a Banach algebra with unit. Consider the category $\mathcal{A}$ whose objects are A–modules and whose morphisms are continuous homomorphic mappings called below A–module morphisms. For A–modules X and Y, the set of morphisms of X into Y is denoted by $\mathrm{Hom}(X, Y)$, while the (generally larger) set of all linear operators from X into Y is denoted by $B(X, Y)$. A morhpism $\phi \in \mathrm{Hom}(X, Y)$ is admissible if its kernel has a Banach complement in X and its image is closed and has a Banach complement in Y.

Recall the definition ([1]) of cohmology groups of a Banach algebra. Let A be a Banach algebra, and let Y be a normed bimodule over A. An n–cochain ($n = 0, 1, \cdots$) is a continuous n–linear operator $f : A \times A \times \cdots \times A_n \longrightarrow Y$. The n–cochains form an abelian group under addition, denoted by $L^n(A, Y)$. The boundary operator $\delta^n : L^n(A, Y) \longrightarrow L^{n+1}(A, Y)$ is defined by

$$\begin{aligned}\delta T(a_1, \cdots, a_{n+1}) = \;& a_1 T(a_2, \cdots, a_{n+1}) - T(a_1 a_2, \cdots, a_{n+1}) + T(a_1, a_2 a_3, \cdots, a_{n+1}) \\ & + \cdots + (-1)^n T(a_1, \cdots, a_{n-1}, a_n a_{n+1}) + (-1)^{n+1} T(a_1, \cdots, a_n) a_{n+1}.\end{aligned}$$

The equality $\delta^{n+1}\delta^n = 0$ is easily verified, hence the cochains form a complex

$$0 \longrightarrow L^0(A, Y) \longrightarrow L^1(A, Y) \longrightarrow L^2(A, Y) \longrightarrow \cdots \qquad (1.1)$$

The nth cohomology group $H^n(A,Y)$ is called the nth cohomoloty group of the Banach algebra A with coefficients in the (normed) bimodule Y. We will denote Ker δ^{n+1} by $Z^n(A,Y)$, and Imδ^n by $B^n(A,Y)$. By ([12] Lemma 6.1) we have:

Lemma 1.1. *For any left normed A-module X and Y we have, up to an isomorphism,*

$$Ext^n(X,Y) = H^n(A, B(X,Y)).$$

Let X be a Banach A-module, X^* be the dual space of X if we define,

$$(ya, x) = (y, ax),$$

$$(ay, x) = (y, xa), \quad a \in A, x \in X, y \in X^*$$

then the dual X^* of X becomes a Banach A-module. The tensor product $A\hat{\otimes}A$ is a Banach A-module if we define

$$a(b\otimes c) = ab \otimes c, \quad (b \otimes c)a = b\otimes ca, \quad a,b,c \in A.$$

π will denote the continuous linear map $A\hat{\otimes}A \longrightarrow A$ defined by $\pi(a\otimes b) = ab(a,b \in A)$.

The complex 1.1 can be defined for any Banach A-module Y. In this paper we will restrict us to the case in which the Banach A-module Y is the dual of an A-module X.

Definition 1.2. Let A be a Banach algebra, we say A is n-amenable, if $H^n(A,X^*) = 0$, for any A-module X.

Standard algebraic formula shows ([1], 1.a) how, given A, X, n, p, we can find a Banach A-module X_1 , with $H^{n+1}(A,X^*) \cong H^n(A,X_1^*)$. Thus we have the following

Corollary 1.3. *If a Banach algebra A is n-amenable then A is $n+p$-amenable for all $n \geq 1, p \geq 0$.*

§ 2. Fundamental Properties of n-amenable Algebras

Let A be a Banach algebra, X a Banach A-module and $\widetilde{A}$ the algebra, obtained by adjoining an identity e to A. Then X becomes a unital $\widetilde{A}$-module by defining $ex = x = xe$, for all $x \in X$. We denote the $\widetilde{A}$-module by $\widetilde{X}$, then we have ([1], 1.d)

$$H^1(\widetilde{A},\widetilde{X}) \cong H^1(A,X).$$

More general we can rather less easily show that the same ismorphism between the higher cohomology groups ([3], Theorem 2) holds that is

(a) $H^n(\widetilde{A},\widetilde{X}) \cong H^n(A,X)$ for all n.

Next we have ([1], 1.c).

(b) Let A be a Banach algebra with identity element e. If X is a Banach A-module then

$$H^n(A,X) \cong H^n(A,eXe).$$

Lemma 2.1. *Let A be any Banach algebra, B be a Banach algebra with an identity element e, hence $A\oplus B$ is a Banach algebra. Let X be an $A\oplus B$-module. If $T \in Z^m(A\oplus$*

$B, X)$, then there exists an $S \in L^{m-1}(A \oplus B, X)$ such that $(T - \delta S)(c_1, \cdots, c_m) = 0$ whenever one of the $c_i (\in A \oplus B)$ is equal to e.

Proof. We prove this by induction on m.

If $m = 1$, set $S = eT(e) - T(e)e$. Then

$$(T - \delta S)(c) = T(c) - cS + Sc = T(c) - ceT(e) + cT(e)e + eT(e)c - T(e)ec.$$

Since $T(e) = eT(e) + T(e)e$ (by $\delta T(e.e) = 0$), we have $eT(e)e = 0$ and so $(T - \delta S)(c) = 0$ when c is equal to e.

Now suppose that the lemma has been proved for all $m \leq n - 1$ and consider $T \in Z^n(A \oplus B, X)$. If $\widetilde{T}$ is the corrosponding element of $Z^{n-1}(A \oplus B, L^1(A \oplus B, X))$ ([20] p.261) there exists, by our hypothesis, an element $\widetilde{S} \in L^{n-2}(A \oplus B, L^1(A \oplus B, X))$ such that $(T - \delta S)(c_1, \cdots, c_{n-1}) = 0$, whenever one of the c_i is equal to e. Define $S \in L^{n-1}(A \oplus B, X)$ by setting

$$S(c_1, \cdots, c_{n-1}) = (\widetilde{S}(c_1, \cdots, c_{n-2}))(c_{n-1}).$$

Then $(T - \delta S)(c_1, \cdots, c_n) = 0$ whenever one element of the $c_i (i = 1, \cdots, n - 1)$ is equal to e. Now write $T' = T - \delta S$ and set $S'(c_1, \cdots, c_{n-1}) = T'(c_1, \cdots, c_{n-1}, e) - 2T'(c_1, \cdots, c_{n-1}, e)e$. Using the fact that T' is a cocycle (i.e. $\delta T' = 0$) we see that $\delta S'(c_1, \cdots, c_n) = \delta T'(c_1, \cdots, c_n, e) + (-1)^{n+1} T'(c_1, \cdots, c_{n-1}, c_n) + (-1)^n T'(c_1, \cdots, c_n)e + (-1)^{n+1} T'(c_1, \cdots, c_{n-1}, e)c_n$.

It follows that the element $S + (-1)^{n+1} S'$ satisfies the requirement of the Lemma for $m = n$.

Lemma 2.2. *If A, B, X are as in Lemma 2.1, and if $X' = \{x \in X \mid xb = 0 = bx$ for all $b \in B\}$ then $H^n(A \oplus B, X) \cong H^n(A, X') \oplus H^n(B, X)$.*

Proof. Denote by I the identity operator on X. If T is any element of $L^n(A \oplus B, X)$, we define, for $a_1, \cdots, a_n \in A$, and $b_1, \cdots, b_n \in B$

$$T_1(a_1, \cdots, a_n) = (I - e)T(a_1, \cdots, a_n)(I - e)$$

and

$$T_2(b_1, \cdots, b_n) = eT(b_1, \cdots, b_n)e.$$

It is easily verified that these mappings are such that $(\delta T)_1 = \delta T_1$ and $(\delta T)_2 = \delta T_2$.

It follows that the mapping $T \to T_1 + T_2$ induces a homomorphism of $H^n(A \oplus B, X)$ into $H^n(A, X') \oplus H^n(B, eXe)$.

Furthermore, if $T_1' \in Z^n(A, X')$ and $T_2' \in Z^n(B, eXe)$ we define $T(a_1 + b_1, \cdots, a_n + b_n) = T_1'(a_1, \cdots, a_n) + T_2'(b_1, \cdots, b_n)$. Then $T \in Z^n(A \oplus B, X), T_1 = T_1'$(since $T_1'(a_1, \cdots, a_n) \in X'$) and $T_2 = T_2'$ (since $T_2'(b_1, \cdots, b_n) \in eXe$). Hence our mapping is a homomorphism onto.

Now consider a $T \in Z^n(A \oplus B, X)$ such that T_1 and T_2 are coboundaries. Then there exists $S_1' \in L^{n-1}(A, X')$ and $S_2' \in L^{n-1}(B, eXe)$ such that $T_1 = \delta S_1'$ and $T_2 = \delta S_2'$. On the other hand, applying Lemma 2.1 to T we obtain an $S \in L^{n-1}(A \oplus B, X)$ such that $(T - \delta S)(c_1, \cdots, c_n) = 0$ whenever one of the element C_i is equal to e. It is easily seen that this implies that :

$$(T-\delta S)(a_1+b_1,\cdots,a_n+b_n)=(T-\delta S)_1(a_1,\cdots,a_n)+(T-\delta S)_2(b_1,\cdots,b_n).$$

That is

$$(T-\delta S)(a_1+b_1,\cdots,a_n+b_n)=(\delta S_1',-\delta S_1)(a_1,\cdots,a_n)(\delta S_2'-\delta S_2)(b_1,\cdots,b_n).$$

Now define

$$S'^*(a_1+b_1,\cdots,a_{n-1}+b_{n-1})=S_1'(a_1,\cdots,a_{n-1})+S_2'(b_1,\cdots,b_{n-1})$$

and

$$S^*(a_1+b_1,\cdots,a_{n-1}+b_{n-1})=S_1(a_1,\cdots,a_{n-1})+S_2(b_1,\cdots,b_{n-1}).$$

Then

$$T-\delta S=(\delta(S'^*-S^*))_1+(\delta(S'^*-S^*))_2=\delta(S'^*-S^*)$$

or

$$T=\delta(S-S^*+S'^*).$$

Thus our homomorphism is an isomorphism .

Finally by (b), $H^n(B,eXe)\cong H^n(B,X)$. We have completed the proof of Lemma 2.2.

Corollary 2.3. *Let A and B be Banach algebras over complex number field C. If X is a Banach $(A\oplus B)$-module, then $H^n(A\oplus B,X)\cong H^n(A,X)\oplus H^n(B,X)$.*

Proof. If A has an identity element e_1, we have $e_1X'e_1=e_1Xe_1$, where $H^n(A,X')\cong H^n(A,X)$. By (a) $H^n(\widetilde{A},\widetilde{X})=H^n(A,X)$ for all n. Thus we have

$$H^n(A\oplus B,X)\cong H^n(\widetilde{A\oplus B},\widetilde{X})\cong H^n(\widetilde{A},\widetilde{H})\oplus H^n(\widetilde{B},\widetilde{X})\cong H^n(A,X)\oplus H^n(B,X).$$

Proposition 2.4. *Let A be a Banach algebra. Suppose J is a complementable closed two-sided ideal of A, J and A/J are n-amenable algebras. Then A is also n-amenable.*

Proof. By our hypothesis, we have $A\cong J\oplus A/J$. Thus

$$H^n(A,X^*)\cong H^n(A/J,X^*)\oplus H^n(J,X^*)=0.$$

When $n=1$, we obtain the result of ([1] Proposition 5.1)

Lemma 2.5. *Let A be a Banach algebra, X a neo-unital A-module and J a closed two-sided ideal of A with a bounded approximate identity. Then , for any $T\in Z^n(A,X^*), n\in N$, and $S\in L^{n-1}(J,X^*)$ such that $\delta S=T\mid_J$, there is a $\widetilde{S}\in L^{n-1}(A,X^*)$ with $\delta\widetilde{S}=T,\widetilde{S}\mid_J=S$.*

Proof. We show by induction that for $i=0,1,\cdots,n-1$, there is $S_i\in L^{n-1}(A,X^*)$ with

(a) $\delta S_i(a_1,\cdots,a_n)=T(a_1,\cdots,a_n)\quad a_1,\cdots,a_{n-i}\in J, a_{n-i+1},\cdots,a_n\in A.$

(b) $S_i\mid_J=S.$

Denoting $S_\alpha(a_1,\cdots,a_{n-1}) = S(a_1e_\alpha,\cdots,a_{n-1}e_\alpha)$, we see that S_α is a bounded net in $L^{n-1}(A,X^*)$ and so has a subset S_β which is convergent to a limit S_0 in the W^*-topology induced by identifying $L^{n-1}(A,X^*)$ with $L_{n-1}(A,X^*)$ (see [1], p.5). Thus

$$\lim(X,S(a_1e_\beta,\cdots,a_{n-1}e_\beta)) = (X,S_0(a_1,\cdots,a_{n-1}))$$

for all $a_1,\cdots,a_{n-1}\in A, x\in X$. Since $S(a_1e_\alpha,\cdots,a_{n-1}e_\beta)\to S(a_1,\cdots,a_{n-1}))$ in norm. Thus $S_0\mid_J = S$. Also $\delta S_0\mid_J = \delta S = T\mid_J$. So S_0 satisfies (a) and (b).

If S_i satisfies (a) and (b) we put $T' = T-\delta S_i$ and define

$$S'_\alpha(a_1,\cdots,a_{n-1}) = T'(a_1e_\alpha,\cdots,a_{n-i-2}e_\alpha,e_\alpha,a_{n-i-1},\cdots,a_{n-1}).$$

As before we find a subsequence S'_γ, W^*-convergent to $S'\in L^{n-1}(A,X^*)$. Note that $S'(b_1,\cdots,b_{n-1}) = 0$ for $b_1,\cdots,b_{n-1}\in J$. Since $T'(b_1e_\alpha,\cdots,b_{n-i-2}e_\alpha,e_\alpha,b_{n-i-1},\cdots,b_{n-1}))-T'(b_1,\cdots,b_{n-i-2},e_\alpha,b_{n-i-1},\cdots,b_{n-1})\longrightarrow 0$ in norm as runs along the net, we see that

$$\begin{aligned}&(X,S'(b_1,\cdots,b_{n-i-2},b_{n-i-1},\cdots,b_{n-1})\\ =\ &\lim(X,T'(b_1e_\gamma,\cdots,b_{n-i-2}e_\gamma,e_\gamma,b_{n-i-1},\cdots,b_{n-1}))\\ =\ &\lim(X,T'(b_1,\cdots,b_{n-i-2},e_\gamma,b_{n-i-1},\cdots,b_{n-1})).\end{aligned}$$

Thus

$$\begin{aligned}&(X,\delta S'(a_1,\cdots,a_{n-i-1},a_{n-i},\cdots,a_{n-1})\\ =\ &\lim(X,a_1T'(a_2,\cdots,a_{n-i-1},e_\alpha,a_{n-i},\cdots,a_n))\\ &+\sum_{j=1}^{n-j-2}(-1)^jT'(a_1,\cdots a_ja_{j+1},\cdots,e_\alpha,a_{n-i},\cdots a_n)),\end{aligned}$$

where $a_1,\cdots,a_{n-i-1}\in J$. $a_{n-i-1},\cdots,a_n\in A$.

And the remaining terms in $\delta S'$ are zero by (a).

Comparing this with the equation

$$\delta T'(a_1,\cdots,a_{n-i-1},e_\alpha,a_{n-i},\cdots,a_n) = 0$$

we see that

$$\begin{aligned}&(x,\delta S'(a_1,\cdots,a_{n-i-1},a_{n-i},\cdots,a_n)\\ =\ &\lim_\alpha(x,(-1)^{n-i}T'(a_1,\cdots,a_{n-i-2},a_{n-i-1}e_\alpha,a_{n-i},\cdots,a_n)\\ =\ &(x,(-1)^{n-i}T'(a_1,\cdots,a_{n-i-1},a_{n-i},\cdots,a_n).\end{aligned}$$

Thus $S_{i+1} = S_i + (-1)^{n-i}S'$ has the required properties. The lemma then follows by taking $\widetilde{S} = S_{n-1}$ because

$$(T-\delta\widetilde{S})(a_1,\cdots,a_n) = 0 \quad \text{if} \quad a_1\in J \quad \text{and as}$$

$$\delta(T-\delta\widetilde{S})(a_1,\cdots,a_{n+1}) = 0 \quad \text{taking} \quad a_1\in J \quad \text{we get}$$

$$a_1(T-\delta\widetilde{S})(a_2,\cdots,a_{n+1}) = 0.$$

Now if $y \in X^*$ and $a_1 y = 0$ for all $a_1 \in J$ then $0 = (x, a_1 y) = (xa_1, y)$ for all $x \in X, a_1 \in J$. So $y = 0$ as X is neo-unital thus we see $T = \delta\widetilde{S}$.

Proposition 2.6. *Let A be a Banach algebra, J a closed two-sided ideal of A with a bounded approximate identity, and J n-amenable, then A is also n-amenable.*

Proof. By ([1] Proposition 1.8), it is enough to show that $H^n(A, X^*) = 0$ for any neo-unital A-module X. Given $T \in Z^n(A, X^*)$, then there is a $S \in L^{n-1}(J, X^*)$ such that $\delta S = T|_J$, since J is n-amenable. So we have , by Lemma 2.5, $\widetilde{S} \in L^{n-1}(A, X^*)$, with $\delta\widetilde{S} = T, \widetilde{S}|_J = S$. Thus $H^n(A, X^*) = 0$.

Moreover, for any Banach algebra A, let $\widetilde{A}$ be A if A has a unit and the unitarization of A otherwise. Then $\widetilde{A}$ is n-amenable if A is n-amenable.

Proposition 2.7. *Let A be a 2-amenable algebra and J a closed amenable two-sided ideal of A. Then the quotient algebra A/J is also 2-amenable.*

Proof. Suppose X is a A/J-module, hence an A-module, and $Q^{(2)} : A \times A \longrightarrow A/J \times A/J$ is the product of the classical quotient of A into A/J. For any $T \in Z^2(A/J, X^*), TQ^{(2)} \in Z^2(A, X^*)$, there is a $S \in L^1(A, X^*)$ such that $TQ^{(2)} = \delta^2 S$ and $TQ^{(2)}|_J = 0$, hence $\delta S(b_1, b_2) = b_1 S(b_2) - S(b_1, b_2) + S(b_1)b_2 = 0, b_1, b_2 \in J$, so $S|_J \in Z^1(J, X^*) = B^1(J, X^*)$. This implies $S|_J = \delta y$ for some $y \in X^*$, and $\delta y(a) = ay - ya$, but the action of J on X^* is trivial , so $S|_J = 0$. There is a $\widetilde{S} : A/J \longrightarrow X^*$ such that $\widetilde{S} = SQ^{(2)}$. Thus $T = \delta\widetilde{S}$.

Proposition 2.8. *If A is 2-amenable, and J is a closed two-sided ideal with an approximate identity, then J is 2-amenable.*

Proof. It is enough to show $H^2(J, X^*) = 0$ for all unital J-module X- ([1] Propositon 1.8) . By ([19]. Proposition 3.7) there is an algebraic homomorphism $Q : A \longrightarrow \Delta(J)$ such that X becomes an A-module. By ([1] Proposition 1.9), for any $D \in Z^2(J, X^*)$ there is a $\widetilde{D} \in Z^2(\Delta(J), X^*)$, such that $\widetilde{D}|_J = D$ then $\widetilde{D}Q^{(2)} \in Z^2(A, X^*)$, hence $DQ^{(2)} = \delta^2 T$ for some $T \in L^1(A, X^*)$, so $D = \delta(T|_J)$ since the restriction of Q to J is the identity map.

Let A be a Banach algebra, J a closed two-sided ideal of A with an approximate identity by Proposition 2.6 and Prposition 2.8, A is 2-amenable if and only if J is 2-amenable. If A is a C^*-algebra then every (two-sided) closed ideal of A has a bounded approximate identity ([21] Lemma 4.2.11). Hence A is 2-amenable if and only if one of two-sided ideal of A is two-amenable.

Proposition 2.9. *Let A, B be two Banach algebras, and A is 2-amenable with an approximate, then $A\hat{\otimes}B$ is 2-amenable.*

Proof. It is trivial by Proposition 2.6.

§ 3. Criterion of Amenable Algebras

Theorem 3.1. *Any commutative Banach algebra with an approximate identity is 2-amenable.*

Proof. Suppose A is a commutative Banach algebra with an apporoximate identity, and $\widetilde{A}$ is the unitarization of A. We will show that $H^2(A, X^*) = 0$, for any neo-unital A-module. Given $T \in Z^2(\widetilde{A}, X^*)$, then $0 = \delta T(a, b, c) = aT(b, c) - T(ab, c) + T(a, bc) - T(a, b)c$. Take $c = e$, then $aT(b, e) = T(ab, e)$, take $a = e$, then $T(e, bc) = T(e, b)c$.

Setting $S(a) = T(a, e) = aT(e, e)$, then we have $(x, \delta S(a, b)) = (x, aS(b) - S(ab) +$

$S(a)b) = (x, aS(e)b - aS(e)b + aS(e)b) = (bxa, S(e)) = (x, aT(e,e)b) = (x, T(a,b))$ for all $a, b \in A$. Hence $T(a,b) = \delta S(a,b)$.

Corollary 3.2. *All commutative C^*-algebras are 2-amenable .*

Proof. Because every C^*-algebra has an approximate identity, so every commutative C^*-algebra is 2-amenable by Theorem 3.1

References

[1] B.E. Johnson, *Cohomology in Banach Algebras,* Memoirs of the American Math. Soc., 127 (1972).

[2] G. Hochschild, *On the cohomology groups of associative algebra,* Ann. of Math., 46 (1945), 58–67.

[3] G. Hochschild, *On the cohomology theory for associative algebra,* Ann. of Math., 47 (1946), 568–579.

[4] H.Kamowits, *Cohomology groups of commutative Banach algebras,* Trans. Amers. Math., Soc., 102 (1962), 352–372.

[5] A. Guichardet, *Sur lhomologie et la cohomology des algebras de Banach,* C.R. Acad. Sci. Paris ser., A262 (1966), 38–41.

[6] A. Guichardet, *Sur lhomologie et la cohomologie des groups locolement compacts* C.R. Acad. Sci. Paris ser., A262 (1966), 118–120.

[7] B.E. Johnson, *Approximate diagonals and cohomology of certain annihilatror Banach algebras,* American J. Math., 94 (1972), 685-698.

[8] B. E. Johnson, *Low dimensional cohomology of Banach algebras,* Pro. of Symposis in pure Math., 38 (1982), part 2.

[9] B.E. Johnson, *Perturbations of Banach Algebras,* Proc. London Math. Soc., 34 (1977), 439–458.

[10] B. E. Johnson, *The wedderburn decomposition of Banach algebras with finite-dimensional radical,* Amer. J. Math., 90 (1968), 866–876.

[11] B.E. Johnson, *Introduction to cohomology in Banach algebras,* Algebras in Analysis. Academic Press. New York, 1075. pp. 84–100.

[12] A. Ja. Helemskiy, *On the homological demension of normed modules over Banach algebras* (in Russian), Mat. Skornik, 81(123) (1970), 430–444.

[13] Helemskii. A.Y., *On a method for calculating and estimating the globle homological dimension of Banach algebras,* Math. USSR-Sb. 16 (1972), 125–138.

[14] Helemskii. A.Y. Flat , *Banach modules and amenable algebras,* Trudy Moskow. Math. Soc., 70 (1978), 156-160.

[15] Helemskii. A. Y. and Sheinberg, M.V., *Amenable Baach algebras,* Functional Anal. Appl., 13 (1979), 32–37.

[16] Lau. A.T., *Characterizations of amenable Banach algebras,* Proc. Amer. Math. Soc., 70 (1978), 156–160.

[17] Racher. G. *On amenable and compact groups Monatsh,* Math., 92 (1981), 305–311.

[18] Jean–Paul, *Pier Amenabl locally compact groups,* New York Chieherter Brisbane Toronto Singapore, 1984.

[19] Robert C. Busby, *Double centralisers and extension of C^*-algebras.*

[20] J.R. Ringrose, *Cohomology of Banach algebras Lectures on Operator algebras,* 359–434 (Springer lectures notes in Math., No. 247).

[21] Richard V. Kadison, John R. Ringrose, *Fundamentals of the theory of operator algebras,* Academic Press, INC. 1983.

[22] J.W. Bunce. *Finite operators and amenable C^*-algebras,* Pro. American Math. Soc., 56 (1976), 145–151.

Proceedings of the Asian Mathematical
Conference, 1990

Continuity of Set Valued Metric Projection in Normed Almost Linear Spaces

K.S. Dersanambika, P. Govindarajulu
Department of Mathematics
SRI Venkateswara University
Tirupati India

V. Subba Rao
Department of Mathematics
Indian Institute of Technology
Madras-600 036 India

§1. Introduction

G. Godini [1] introduced normed almost linear spaces (nals) over the real field R. She has shown that the problem of best simultaneous approximation in normed linear space becomes a particular case of the problem of best approximation in nals.

In the framework of a normed almost linear spaces we consider the problem of continuity, semi-continuity and radial continuity of metric projection which has been studied extensively in normed linear spaces by several authors.

§2. Preliminaries

We recall some definitions, notations from [1].

Definition 2.1. An almost linear space (als) is a set X together with two mappings $s : X \times X \to X$ and $m : R \times X \to X$ satisfying the conditions L_1)–L_8) given below. For $x, y \in X$ and $\lambda \in R$ we denote $s(x, y)$ by $x + y$ and $m(\lambda, x)$ by λx. Let $x, y, z \in X$ and $\lambda, \mu \in R$. $L_1)(x+y)+z = x+(y+z)$; $L_2)x+y = y+x$; L_3) There exists an element $0 \in X$ such that $x + 0 = x$ for each $x \in X$; L_4) $1x = x$; L_5) $\lambda(x+y) = \lambda x + \lambda y$; $L_6)0x = 0$; L_7) $\lambda = (\mu x) = (\lambda\mu)x$; $L_8)(\lambda+\mu)x = \lambda x + \mu x$ for $\lambda \geq 0, \mu \geq 0$.

$-1x$ is denoted by $-x$ and in the sequel $x - y$ means $x + (-y)$. Let $V_X = \{x \in X : x - x = 0\}$. By L_1–L_8 it follows that V_X is a linear space.

Definition 2.2. A norm on the als X is a functional $\|\cdot\| : X \to R$ satisfying N_1–N_3 below. Let $x, y, z \in X$ and $\lambda \in R$. $N_1)\|x - z\| \leq \|x - y\| + \|y - z\|$; $N_2)\|\lambda x\| = |\lambda|\,\|x\|$; $N_3)\|x\| = 0$ iff $x = 0$.

Definition 2.3. An als X together with $\|\cdot\| : X \to R$ satisfying N_1–N_3 is called a normed almost linear space (nals).

The norm of a nals X generate a semi metric ρ on X with the following properties (for details refer [3]). Let $x, y, z \in X, v \in V_X$ and $\lambda \in R$. (1) $|\,\|x\| - \|y\|\,| \leq \rho(x, y)$; (2) $\rho(x, v) = \|x - v\|$; (3) $\rho(x+z, y+z) = \rho(x, y)$; (4) $\rho(\lambda o x, \lambda o y) = |\lambda|\rho(x, y)$; (5) $\lim_{\lambda \to \lambda_0} \rho(\lambda o x, x) = \rho(\lambda_o o x, x), \lambda_0 > 0$.

Definition 2.4. Let X be a nals and let G be a non-empty subset of V_X. We define $R_X(G)$ in the following way: $x \in R_X(G)$ if for each $g \in G$, there exists $v_g \in V_X$ such that $\|x-g\| = \|v_g - g\|$ and $\|x-v\| \geq \|v_g - v\|$ for all $v \in V_X$.

Definition 2.5. Let X be a nals and let G be a non-empty subset of V_X. We define $R_X^1(G)$ in the following way. $x \in R_X^1(G)$ if for each $g \in P_G(x)$ there exists $v_g \in V_X$ such that

(1) $\|x-g\| = \|v_g - g\|$, (2) $g \in P_G(v_g)$, (3) $\|x-g\| \geq \|v_g - v\|$ for all $v \in G$.

Example 2.1. Let $E = R^2$ with the norm $\|x\|_E = \sqrt{x_1^2 + x_2^2}$ where $x = (x_1, x_2)$. Let X be the collection of all bounded, convex non-empty subsets A of E. For A, A_1, A_2 of X and $\lambda \in R$, define

$$s(A_1, A_2) = A_1 + A_2 = \{a_1 + a_2 : a_1 \in A_1, a_2 \in A_2\},$$

$$m(\lambda, A) = \lambda A = \{\lambda a : a \in A\}.$$

The element 0 in X is the set $\{0\}$. Then X is an als and $V_X \equiv E$. For $A \in X$, let $\|A\| = \text{Sp}_{a\in A}\|a\|_E$.

It is easy to see that $\|\cdot\|$ satisfies $N_1 - N_3$. Hence X is a nals. Let

$$A = \{\lambda z_1 + (1-\lambda)z_2 : 0 \leq \lambda \leq 1, z_1 = (-1,1), z_2 = (0,1)\}.$$

Then $A \in X$. Let $G = \{(x_1, x_2) : x_1 = x_2, x_1, x_2 \in R\}$. We observe that $d(x, G) = \sqrt{2}$ and $g_0 = (0,0) \in P_G(A)$. Further we have $v_0 = (-1,1) \in V_X$ such that

(1) $d(x, G) = \sqrt{2} = d(v_0, G)$, identifying x as the element $A \in X$

(2) $\|x - g_0\| = \|v_0 - g_0\|$

(3) $\|x - g\| \geq \|v_0 - g\|$ for all $g \in G$.

Thus $A \in R_{X^1}(G)$. Furthermore $V_X \subset R_{X^1}(G)$.

§3. Approximative Compactness and Continuity

Let X be a nals. If G is a subset of X and $x \in X$ let (i) $d(x,G) = \inf_{g\in G}\|x-g\|$ and (ii) $P_G(x) = \{g_0 \in G : d(x,G) = \|x - g_0\|\}$.

Throughout the section we assume that G is a subset of V_X.

Definition 3.1. A sequence $\{y_n\}$ in G is said to be a minimising sequence for $x \in X$, if $\lim_n \|x - y_n\| = d(x, G)$.

Definition 3.2. A set $G \subset V_X$ is said to be approximatively compact (approximatively weakly compact) if for any $x \in V_X$, each minimising sequence $\{y_n\} \subset G$ has a subsequence converging (converging weakly) to an element of G.

Proposition 3.1. *Let G be a subspace of V_X and $x \in X$. If $g_0 \in P_G(x)$, then $g_0 \in P_G(z)$ for every $z = \lambda g_0 + (1-\lambda)x, \lambda \in R$.*

Definition 3.2. Let X be a nals. The metric projection P_G is said to be upper semi continuous (usc) (resp. lower semi continuous (lsc)) if $\{x \in X : P_G(x) \cap A \neq \emptyset\}$ is closed (resp. open) for each closed [resp. open] for each closed [resp. open] set A in X. Further P_G is said to be continuous if P_G is both usc and lsc.

Proposition 3.2. *Let X be a nals and let $G \subset V_X$ is proximinal in X. If G is approximatively compact in V_X and $R_{X^1}(G)$ closed then P_G is upper semi continuous on $R_{X^1}(G)$.*

Proof. It is sufficient to prove that $F = \{x \in R_{X^1}(G) : P_G(x) \cap A \neq \emptyset\}$ is closed in X, whenever A is closed.

For, let $\{x_n \subset F\}$ such that $\rho(x_n, x) \to 0, x \in X$. Let $g_n \in P_G(x_n) \cap A$, for each n. Then it follows

$$\lim_n d(x_n, G) = d(x, G), \tag{1}$$

since $d(x_n, G) = \inf_{g \in G} \rho(x_n, g) \leq \rho(x_n, x) + d(x, G)$. Further

$$\begin{aligned} \big| \|x - g_n\| - d(x, G) \big| &= \big| \|x - g_n\| - \|x - g_0\| \big| \\ &\leq \big| \|x - g_n\| - \|g_n - x_n\| \big| + \big| \|g_n - x_n\| - \|g_0 - x\| \big| \\ &\leq \rho(x - g_n, x_n - g_n) + |d(x_n, G) - d(x, G)| \\ &\leq \rho(x_n, x) + |d(x_n, G) - d(x, G)|. \end{aligned}$$

In view of equation (1) and $\rho(x_n, x) \to 0$ as $n \to \infty$, we get $\lim_n \|x - g_n\| = d(x, G)$.

Since $P_G(x)$ is non-empty and $R_{X^1}(G)$ is closed, $x \in R_{X^1}(G)$ and $g_0 \in P_G(x)$, then there exists $v_0 \in V_X$ satisfying (1), (2), (3) in Definition 2.5.

$$d(v_0, G) \leq \lim_n \|v_0 - g_n\| \leq \lim_n \|x - g_n\| = d(x, G) = d(v_0, G).$$

Thus $\lim_n \|v_0 - g_n| = d(v_0, G)$.

Since G is approximatively compact, there exists a subsequence $\{g_n^1\}$ of $\{g_n\}$ such that

$$\rho(g_n^1, g^1) \to 0 \quad \text{as} \quad n \to \infty, \quad \text{for some} \quad g^1 \in G.$$

$$\big| \|g^1 - v_0\| - \|g_n^1 - v_0\| \big| \leq \rho(g^1 - v_0, g_n^1 - v_0) \leq \rho(g^1, g_n^1) \to 0 \quad \text{as} \quad n \to \infty.$$

Hence $\|g^1 - v_0\| = \lim_n \|g_n^1 - v_0\| = d(v_0, G)$. Thus $g^1 \in P_G(v_0)$

$$\big| \|g^1 - x\| - \|g_n^1 - x\| \big| \leq \rho(g^1 - x, g_n^1 - x) \leq \rho(g^1, g_n^1).$$

From this it follows that

$$d(x, G) \leq \|g^1 - x\| = \lim_n \|g_n^1 - x\| = d(x, G).$$

Thus $g^1 \in P_G(x)$. Since A is closed and $\{g_n^1\} \subset A$ and $\rho(g_n^1, g^1) \to 0$ it follows that $g^1 \in A$. Thus $g^1 \in P_G(x) \cap A$, hence $x \in F$. Thus P_G is usc on $R_{X^1}(G)$.

Theorem 3.1. *Let X be a nals G a convex subset of V_X. Let G be proximinal in X. If V_X is strictly convex and G is approximatively compact in V_X, then P_G is both usc and lsc at any point of $R_{X^1}(G)$.*

The proof runs on the same line as in Proposition 3.2, hence omitted.

Theorem 3.2. *Let X be a nals and G a convex subset of V_X such that G is proximinal in X. Suppose V_X is strictly convex and has the property* (H) *(that is $\|x_n\| \to \|x\|$ and*

$x_n \xrightarrow{w} x$ imply that $\rho(x_n, x) \to 0$) and if G is weakly approximatively compact set in V_X, then P_G is both usc and lsc at any point of $R_{X^1}(G)$.

Proof. Proof runs on the same lines as in the Proposition 3.2, hence it is omitted.

Theorem 3.3. *Let X be a nals and G a convex subset of V_X. Suppose V_X is strictly convex and has the property (H), then if G is boundedly weakly compact, P_G is both lsc and usc at any point of $R_X(G)$.*

Proof. By Proposition 4.2 of Godini [4], it follows that G is proximinal in X. Let $x \in R_X G$. Since V_X is strictly convex, we get $P_G(x)$ is singleton. Let $P_G(x) = \{g_0\}$. For $d(x, G\} = 0$ the result follows easily. Let $x \in X$ such that $d(x, G) > 0$. Let $\{x_n\}$ in X such that $\rho(x_n, x) \to 0$ and $g_n \in P_G(x_n).\{g_n\}$ is bounded in V_X, since

$$\|g_n\| \leq 2\|x_n\| = 2\rho(x_n, 0) \leq 2(\rho(x_n, x) + \rho(x, 0)).$$

Since G is boundedly weakly compact, we get $g_n \to g^1$. Since

$$d(x, G) \leq |x - g^1\| \leq \liminf_n \|x - g_n\| = d(x, G).$$

Thus $g^1 \in P_G(x) = \{g_0\}$. Hence $g_n \xrightarrow{w} g_0$. Now $x \in R_x(G), g_0 \in G$, there exists $v_0 \in V_X$ such that

(1) $\|x - g_0\| = \|v_0 - g_0\|$ and (2) $\|x - g_n\| \leq \|v_0 - g_n\|$, for all n.

Choose $\{g_n^1\}$ of $\{g_n\}$ such that $\limsup_n \|v_0 - g_n\| = \lim_n \|v_0 - g_n^1\|$. Since $v_0 - g_n^1 \xrightarrow{w} v_0 - g_0$, we have

$$\begin{aligned} d(x, G) = \|x - g_0\| = &\ \|v_0 - g_0\| \leq \liminf_n \|v_0 - g_n\| \leq \limsup_n \|v_0 - g_n\| \\ \leq &\ \lim_n \|v_0 - g_n^1\| \leq \liminf_n \|x - g_n^1\| = d(x, G). \end{aligned}$$

Thus $\|v_0 - g_n^1\| \to \|v_0 - g_0\|$. Now $v_0 - g_n^1 \xrightarrow{w} v_0 - g_0, \|v_0 - g_n^1\| \to \|v_0 - g_0\|$ and since V_X has the property (H), we get $\rho(g_n^1, g_0) \to 0$. Since $\{x_n\}$ is arbitrary it follows that P_G is both lsc and usc on $R_X(G)$.

Corollary 3.1. *Let X be a nals and suppose that V_X is an E-space, Let G a convex subset of V_X. Then P_G is both lsc and usc at any $x \in R_X(G)$.*

The above result extends the Theorem 5.11 of Godini [4].

Corollary 3.2. *Let X be a nals and G a convex subset of V_X. Suppose V_X is strictly convex and has property (H). Then if G is weakly compact, P_G is both usc and lsc on $R_{X^1}(G)$.*

§4. Radial Continuity

Definition 4.1. Let $G \subset X$ and $x_0 \in X$. We say that P_G is outer radially lower continuous (ORL-continuous) at x_0 if, for each $g_0 \in P_G(x_0)$ and each open set W such that $P_G(x_0) \cap W \neq \emptyset$, there exists a neighbourhood U of x_0 such that $P_G(x) \cap W \neq \emptyset$ for all $x \in U \cap \{g_0 + \lambda(x_0 - g_0) : \lambda \geq 1\}$.$P_G$ is called ORL-continuous if it is ORL-continuous at each point of X.

Clearly every lsc metric projection is ORL-continuous. But the converse need not be true.

Lemma 4.1. *Let $G \subset V_X$ and $x_0 \in X$. Then the following statements are equivalent.*

1. *P_G is ORL-continuous at x_0.*
2. *For each $g_0, g_1 \in P_G(x_0)$ and $\varepsilon > 0$, there exists $\delta > 0$ such that $P_G(x) \cap B(g_1, \varepsilon) \neq \emptyset$ for each x in $\{g_0 + \lambda(x_0 - g_0) : 1 \leq \lambda < 1 + \delta\}$*
3. *For each g_0, g_1 in $P_G(x_0)$ and every sequence $\{x_n\}$ in $\{g_0 + \lambda(x_0 - g_0) : \lambda \geq 1\}$ with $\rho(x_n, x_0) \to 0$ there exists $g_n \in P_G(x_n)$ such that $\rho(g_n, g_1) \to 0$.*

Proof. (1)$\Longrightarrow$(2) is clear. To prove (2) $\Longrightarrow$ (3). Let $g_0, g_1 \in P_G(x_0)$. Let $x_n = g_0 + \lambda_n(x_0 - g_0), \lambda_n > 1$. Given $\rho(x_n, x) \to 0$. It follows that $\lambda_n \to 1$. For $\varepsilon = 1$, there exists $\delta > 0$ such that $P_G(x) \cap B(g_1, 1) \neq \emptyset$ for each $x \in G_{\delta_1}$ where $G_{\delta_1} = \{g_0 + \lambda(x_0 - g_0) : 1 < \lambda < 1 + \delta_1\}$. Choose n_1 such that $x_n \in G_{\delta_1}$ for every $n \geq n_1$ and $y_1 \in P_G(x_n) \cap B(g_1, 1)$. For $\varepsilon = 1/2$, there exists $\delta_2, 0 < \delta_2 < \delta_1$ such that $x_n \in G_{\delta_1}$ for all $n \geq n_2$. Choose $y_2 \in P_G(x_{n_2}) \cap B(g_1, 1/2)$. Continuing in this way, we obtain a sequence of integers $\{n_k\}$, a decreasing sequence of positive integers $\{\delta_k\}$ and a sequence $\{y_k\}$ such that $\delta_k \to 0, x_n \in G_{\delta_k}$ for all $n \geq n_k, y_k \in P_G(x_{n_k}) \cap B(g_1, 1/k)$. We define a sequence $\{g_n\}$ by taking $g_n \in P_G(x_n)$ for $n = 1, 2, \cdots, n_1 - 1$ and $g_{n_k} = y_k$ for all k and $g_n \in P_G(x_n) \cap B(g_1, 1/k)$ for all $n_k < n < n_{k+1}$. Then it follows that $\rho(g_n, g_1) \to 0$ as $n \to \infty$.

To prove (3) $\Longrightarrow$ (1). Suppose (3) holds but (1) fails. Then there exists $g_0 \in P_G(x_0)$ and an open set W with $P_G(x_0) \cap W \neq \emptyset$ such that for every neighbourhood U of x_0, there exists an x in $U \cap \{g_0 + \lambda(x_0 - g_0) : \lambda > 1\}$ such that $P_G(x) \cap W = \emptyset$. Choose $g_1 \in P_G(x_0) \cap W$. Then for every n, there exists $x_n = g_0 + \lambda_n(x_0 - g_0)$ with $1 < \lambda_n < 1 + 1/n$ such that $P_G(x) \cap W = \emptyset$. Here $\lim_n \lambda_n = 1$ and $\rho(x_n - x_0) \to 0$. But if $g_n \in P_G(x_n)$ then $g_n \notin W$ and $\rho(g_n, g_1) \to 0$ which is a contradiction to (3). Therefore (1) is true.

Definition 4.2. Let $G \subset V_X$ an $x_0 \in X. P_G$ is said to be inner radially lower (abbrev IRL) continuous at x_0 if for every $g_0 \in P_G(x_0)$ and each open set W with $W \cap P_G(x_0) \neq \emptyset$, there exists a neighbourhood U of x_0 such that $P_G(x) \cap W \neq \emptyset$ for every x in $U \cap \{g_0 + \lambda(x_0 - g_0) : 0 \leq \lambda \leq 1\}$. P_G is called IRL-continuous if it is IRL continuous at each point.

Clearly each lsc metric projection is IRL-continuous. The converse need not be true.

Lemma 4.2. *Let $G \subset V_X$ and $x_0 \in X$. The following statements are equivalent.*

(1) *P_G is IRL-Continuous at x_0.*

(2) *For each g_0, g_1 in $P_G(x_0)$ and each $\varepsilon > 0$, then there exists $\delta > 0$ such that $P_G(x) \cap B(g_1, \varepsilon) \neq \emptyset$ for all x in $\{g_0 + \lambda(x_0 - g_0) : 1 - \delta < \lambda \leq 1\}$.*

(3) *For each g_0, g_1 in $P_G(x_0)$ and each sequence x_n in $\{g_0 + \lambda(x_0 - g_0) : 0 \leq \lambda \leq 1\}$ with $\rho(x_n, x_0) \to 0$ $d(g_1, P_G(x_n)) \to 0$ (that is there exists $g_n \in P_G(x_n)$ such that $\rho(g_n, g_1) \to 0$).*

Proof. Proof runs on the same lines as that of Lemma 3.2 in [2], hence it is omitted.

Theorem 4.1. *Let $G \subset V_X$. If $P_G(x)$ it is convex for $x \in X$, then P_G is IRL-Continuous.*

Proof. If $P_G(x) = \emptyset$, the result is trivial. Let $g_0, g_1 \in P_G(x)$ and $x_n \in [x, g_0]$ with $\rho(x_n, x) \to 0$. Thus $x_n = \lambda_n g_0 + (1 - \lambda_n)x = (1 - \lambda_n)(x - g_0) + g_0$, where $0 \leq \lambda_n \leq 1$. Since

$$d(x_n, G) \leq \|x_n - g_0\| = \rho(x_n, g_0) \leq \rho(x_n, x) + \rho(x, g_0),$$

and

$$d(x,G) = \lim_n d(x_n, G),$$

we have

$$\rho(x,g_0) = d(x,G) = \lim_n d(x_n,G) \leq \lim_n \rho(x_n,g_0) \leq \rho(x,g_0).$$

Thus $\rho(x,g_0) = \lim_n \rho(x_n,g_0) = \lim_n (1-\lambda_n)\rho(x,g_0)$. This is true if and only if $\lambda_n \to 0$. Since $P_G(x)$ is convex and $g_0, g_1 \in P_G(x)$, we get $g_n = (1-\lambda_n)g_1 + \lambda_n,\ g_0 \in P_G(x)$. Also $\rho(g_n, g_1) \to 0$. Also $\|x_n - g_n\| = (1-\lambda_n)\|x - g_1\| = (1-\lambda_n)\|x - g_0\| = \|x_n - g_0\| = d(x_n, G)$ [By Proposition 3.1]. Thus $g_n \in P_G(x_n)$. Hence by Lemma 4.2, we get P_G is IRL-continuous.

Remark. In general, the converse of the Theorem 4.1 is false. For example let $X = R_{\geq 0}XR$ and let $z_1 = (x_1, y_1), z_2 = (x_2, y_2)$ where $x_i \in R_{\geq 0}, y_i \in R, i = 1,2$. We say that $z_1 + z_2 = (x_1 + x_2, y_1 + y_2) = s(z_1, z_2)$ and $\lambda z_1 = (|\lambda| x_1, \lambda y_1) = m(\lambda, z_1)$. Then X is an als. Let as define the norm on X as $\|z\| = \max\{|x|, |y|\}$. Let $G = \{(0,0), (0,1/2), (0,-1/2)\}, z = (1,0)$. Then clearly $P_G(z) = G$ hence $P_G(z)$ is not convex. But it is easily seen that P_G is IRL-continuous at z.

Corollary 4.1. *If G is a Chebyshev set, then P_G is IRL-continuous.*

Corollary 4.2. *If G is a convex subset of V_X, then P_G is IRL-continuous.*

We conclude that Definition 4.1, Lemma 4.2 and Lemma 4.3 of [2] may also be extended in this new framework.

References

[1] B. Brosowski and F. Deutsch, *On some geometrical properties of suns*, J. Approx. Theory, 10(1974), 295–302.

[2] B. Brosowski and F. Deutsch, *Radial continuity of set valued metric projections*, J. Approx. Theory, 11(1974), 236–253.

[3] K.S. Dersanambika, P. Govindarajulu and V. Subba Rao, *Best approximations in normed almost linear spaces*, Communicated.

[4] G. Godini, *A framework for best simultaneous approximation-normed almost linear spaces*. J. Approx. Theory, 43(1985), 338–358.

[5] G. Godini, *Best approximation in normed almost linear spaces*, Constructive Theory of Functions, 84(1984), 356–363.

Proceedings of the Asian Mathematical
Conference, 1990

Best Approximations in Normed Almost Linear Spaces

K.S. Dersanambika, P. Govindarajulu
Department of Mathematics
SRI Venkateswara University
Thirupathy, India

V. Subba Rao
Department of Mathematics
Indian Institute of Technology
Madras-600 036, India

§1. Introduction

G. Godini [1] introduced normed almost linear spaces (nals) over the real field R. She has shown that the problem of best simultaneous approximation in normed linear space becomes a particular case of the problem of best approximation in nals.

In the framework of a normed almost linear space we extend some results both from the theory of best approximation and from the theory of best simultaneous approximation in normed linear spaces.

§2. Preliminaries

We recall some definitions, notations from [1].

Definition 2.1. An almost linear space (als) is a set X together with two mappings $s : X \times X \to X$ and $m : R \times X \to X$ satisfying the conditions L_1)–L_8) given below. For $x, y \in X$ and $\lambda \in R$ we denote $s(x, y)$ by $x + y$ and $m(\lambda, x)$ by λx. Let $x, y, z \in X$ and $\lambda, \mu \in R$. L_1) $(x + y) + z = x + (y + z)$; L_2) $x + y = y + x$; L_3) There exists an element $0 \in X$ such that $x + 0 = x$ for each $x \in X$; L_4) $1x = x$; L_5)$\lambda(x + y) = \lambda x + \lambda y$; L_6) $0x = 0$; L_7) $\lambda(\mu x) = (\lambda\mu)x$; L_8)$(\lambda + \mu)x = \lambda x + \mu x$ for $\lambda \geq 0, \mu \geq 0$.

$-1x$ is denoted by $-x$ and in the sequel $x - y$ means $x + (-y)$. Let $V_X = \{x \in X : x - x = 0\}$. By L_1)–L_8) it follows that V_X is a linear space.

Definition 2.2. A norm on the als X is a functional $\|\cdot\| : X \to R$ satisfying N_1)–N_3) below. Let $x, y, z \in X$ and $\lambda \in R$. N_1)$\|x - z\| \leq \|x - y\| + \|y - z\|$; N_2) $\|\lambda x\| = |\lambda| \, \|x\|$; N_3)$\|x\| = 0$ iff $x = 0$.

Definition 2.3. An als X together with $\|\cdot\| : X \to R$ satisfying $N_1 - N_3$ is called a normed almost linear space (nals).

The norm of a nals X generate a semi metric ρ on X with the following properties [for details refer [3]]. Let $x, y, z \in X, v \in V_X$ and $\lambda \in R$. (1) $| \, \|x\| - \|y\| \, | \leq \rho(x, y)$; (2) $\rho(x, v) = \|x - v\|$; (3) $\rho(x + z, y + z) = \rho(x, y)$; (4) $\rho(\lambda o x, \lambda o y) = |\lambda| \rho(x, y)$; (5) $\lim_{\lambda \to \lambda_o} \rho(\lambda o x, x) = \rho(\lambda_o o x, x), \lambda_o > 0$.

Definition 2.4. Let X be a nals and let G be a non-empty subset of V_X. We define $R_X(G)$ in the following way: $x \in R_X(G)$ if for each $g \in G$, there exists $v_g \in V_X$ such that $\|x - g\| = \|v_g - g\|$ and $\|x - v\| \geq \|v_g - v\|$ for all $v \in V_X$.

For examples see [1].

§3. Study of Sets $R_X(G)$ and $P_G(x)$

Let X be a nals. If G is a subset of X and $x \in X$ let (i) $d(x, G) = \inf_{g \in G} \|x - g\|$ and (ii) $P_G(x) = \{g_0 \in G : \|x - g_0\| = d(x, G)\}$. G is said to be proximinal (resp. Chebyshev) with respect to A in X if $P_G(x)$ is non-empty (resp. singleton) for each $x \in A$.

Throughout the section we assume that G is a subset of V_X.

Proposition 3.1. *Let V_X be a strictly convex normed linear space and let G be proximinal with respect to $R_X(G)$. If the values of the metric projection P_G are convex, then G is Chebyshev with respect to $R_X(G)$.*

Proof. Suppose $g_1, g_2 \in P_G(x)$. Since $P_G(x)$ is convex,

$$\frac{g_1 + g_2}{2} \in P_G(x). \text{ Since } x \in R_X(G) \text{ and } \frac{g_1 + g_2}{2} \in G$$

there exists $v_0 \in V_x$ such that

$$\begin{aligned}
\|x - \tfrac{g_1 + g_2}{2}\| &= \|v_0 - \tfrac{g_1 + g_2}{2}\| \text{ and} \\
\|x - v\| &\geq \|v_0 - v\| \text{ for all } v \in V_X \\
d(x, G) &= \|x - \tfrac{g_1 + g_2}{2}\| = \|v_0 - \tfrac{g_1 + g_2}{2}\| \\
&= \|1/2(v_0 - g_1) + 1/2(v_0 - g_2)\| \\
&\leq 1/2\|v_0 - g_1\| + 1/2\|v_0 - g_2\| \\
&\leq 1/2\|x - g_1\| + 1/2\|x - g_2\| \\
&= 1/2d(x, G) + 1/2d(x, G) \\
&= d(x, G).
\end{aligned}$$

Thus $\|v_0 - g_1\| = \|v_0 - g_2\| = \|v_0 - (g_1 + g_2)/2\|$. Since V_X is strictly convex we get $g_1 = g_2$. Hence G is Chebyshev.

Lemma 3.1. *Let G be a convex subset of V_X. If g_1, g_2 are best approximations to x, then $\lambda g_1 + (1 - \lambda) g_2, 0 \leq \lambda \leq 1$ is also a best approximation to x.*

Corollary 3.1. *Let G be convex proximinal with respect to $R_X(G)$ and let V_X be a strictly convex normed linear space. Then G is Chebyshev with respect to $R_X(G)$.*

The following two propositions follow from the definitions.

Proposition 3.2. *Let G be a subset of V_X. Consider the following statements.*

(1) G is compact.

(2) G is boundedly compact.

(3) *G is proximinal.*

Then (1) $\Longrightarrow$ (2) $\Longrightarrow$ (3).

Proposition 3.3. *Let X be a nals. Let G be a subset of V_X. Consider the following statements:*

(1) *G is weakly compact.*

(2) *G is boundedly weakly compact.*

(3) *G is proximinal.*

Then (1) $\Longrightarrow$ (2) $\Longrightarrow$ (3).

§4. Applications to Best Simultaneous Approximation

Theorem 4.1. *Let X be a nals and $G \subset V_X$ be a non-empty compact convex subset of V_X and V_X be strictly convex. Then for each $x \in R_X(G)$, $P_G(x)$ is singleton.*

Proof. By Proposition 3.2, G is proximinal with respect to X. Suppose g_1, g_2 are two best approximations to $x \in R_X(G)$ in G. Therefore, $\|x - g_i\| = d(x,G), i = 1,2$. Then $\|x - (g_1 + g_2)/2\| = d(x,G)$. Since $x \in R_X(G)$, for the element $(g_1 + g_2)/2 \in G$, there exists $v_0 \in V_X$ such that

$$\|x - (g_1 + g_2)/2\| = \|v_0 - (g_1 + g_2)/2\| \text{ and } \|x - g_i\| \geq \|v_0 - g_i\|, \quad i = 1,2.$$

Then

$$\begin{aligned} d(x,G) = \;& \|x - (g_1 + g_2)/2\| = \|v_0 - (g_1 + g_2)/2\| \\ \leq \;& \frac{\|v_0 - g_1\| + \|v_0 - g_2\|}{2} \leq \frac{\|x - g_1\| + \|x - g_2\|}{2} \\ = \;& d(x,G). \end{aligned}$$

Thus $\|v_0 - g_1\| = \|v_0 - g_2\| = \|v_0 - (g_1 + g_2)/2\|$. Since V_X is strictly convex, we obtain $g_1 = g_2$.

The following corollary is an immediate application of Theorem 4.1 to best simultaneous approximation in normed linear spaces.

Corollary 4.1. *Let X be a strictly convex normed linear space, G a compact convex subset of X and F a subset of X which is remotal with respect to G. Then, there exists a unique best simultaneous approximation in G.*

This result was proved by T.D. Narang [5].

Theorem 4.2. *If G is a boundedly weakly sequentially compact and convex subset of V_X and V_X is strictly convex, then for each $x \in R_X(G)$, $P_G(x)$ is singleton.*

The proof of the Theorem follows on the same lines as in Theorem 4.1.

As an immediate application we get.

Corollary 4.2. *If X is a strictly convex normed linear space, G a boundedly weakly sequentially compact and convex subset of X and F is a farthest point set with respect to G, then there exists a unique best simultaneous approximation to F from G.*

Sastry and Naidu have given Corollary 4.2 in [7, Theorem 3].

Definition 4.1. The normed linear space E is said to be strictly convex with respect to its linear subspace G if the conditions $x, y \in E, \|x\| = \|y\| = \|(x+y)/2\|, x - y \in G$ imply that $x = y$.

Corollary 4.3. *Let X be a normed almost linear space and G a proximinal subspace with respect to $R_X(G)$. If V_X is strictly convex with respect to G then $P_G(x)$ is singleton for each $x \in R_X(G)$.*

Theorem 4.3. *If V_x is strictly convex with respect to G and G is reflexive subspace of V_X, then there exists one and only one best approximation to each $x \in R_X(G)$.*

This theorem was proved by G. Godini [2]. It is a particular case of Corollary 3, and has immediate consequences to the best simultaneous approximation, that is

Corollary 4.4. *Let X be a strictly convex normed linear space and G a reflexive subspace of X. Then there exists one and only one best simultaneous approximation from the elements of G to any set F which is remotal with respect to G.*

Definition 4.2. Let X be a nals and let G be a non empty subset of V_X. We define $R_{X^1}(G)$ in the following way: $x \in R_{X^1}(G)$ if for each $g \in P_G(x)$ there exists $v_g \in V_X$ such that

(1) $\|x - g\| = \|v_g - g\|$,

(2) $g \in P_G(v_g)$,

(3) $\|x - v\| \geq \|v_g - v\|$ for all $v \in G$.

Example 4.1. Let $E = R^2$ with the norm $\|x\|_E = \sqrt{x_1^2 + x_2^2}$ where $x = (x_1, x_2)$. Let X be the collection of all bounded, convex non-empty subsets A of E. For A, A_1, A_2 of X and $\lambda \in R$, define

$$s(A_1, A_2) = A_1 + A_2 = \{a_1 + a_2 : a_1 \in A_1, a_2 \in A_2\},$$

$$m(\lambda, A) = \lambda A = \{\lambda a : a \in A\}.$$

The element 0 in X is the set $\{0\}$. Then X is an als and $V_X \equiv E$. For $A \in X$, let $\|A\| = \sup_{a \in A} \|a\|_E$. It is easy to see that $\|\cdot\|$ satisfies $N_1 - N_3$. Hence X is a nals. Let

$$A = \{\lambda z_1 + (1-\lambda) z_2 : 0 \leq \lambda \leq 1, z_1 = (-1, 1), z_2(0, 1)\}.$$

Then $A \in X$. Let $G = \{(x_1, x_2) : x_1 = x_2, x_1, x_2 \in R\}$. We observe that $d(x, G) = \sqrt{2}$ and $g_0 = (0, 0) \in P_G(A)$. Further we have $v_0 = (-1, 1) \in V_X$ such that

(1) $d(x, G) = \sqrt{2} = d(v_0, G)$, identifying x as the element $A \in X$,

(2) $\|x - g_0\| = \|v_0 - g_0\|$,

(3) $\|x - g\| \geq \|v_0 - g\|$ for all $g \in G$.

Thus $A \in R_{X^1}(G)$. Furthermore $V_X \subset R_{X^1}(G)$.

Theorem 4.4. *Let V_X be a locally uniformly convex space with respect to the subspace G. If $P_G(x)$ is non-empty for each $x \in R_{x^1}(G)$, then $(G, R_{X^1}(G))$ has the property (P_1)). (The pair (G, A) has the property P_1 as well as P_2 refer Godini [2]).*

Proof. Suppose $(G, R_{X^1}(G))$ does not have property (P_1). Then there is an element $x \in R_{X^1}(G)$ and $\varepsilon > 0$ such that for every $n \in N$, there exists $g_n \in G$ such that $\|x - g_n\| < d(x, G) + 1/n$ and $d(g_n, P_G(x)) \geq \varepsilon$. Let $g_0 \in P_G(X)$. Then $d(g_n, g_0) \geq \varepsilon$ for every n. Since $x \in R_{X^1}(G), g_0 \in P_G(x)$, there exists $v_0 \in V_X$ such that

(1) $\|x - g_0\| = \|v_0 - g_0\|$,
(2) $g_0 \in P_G(v_0)$,
(3) $\|x - v\| \geq \|v_0 - v\|$ for all $v \in G$.

From (2) we get $g_0 \in P_G(v_0) \Longrightarrow d(x, G) = d(v_0, G)$. Let $g_n^* = (g_0 + g_n)/2$. Then it follows that $\|v_0 - g_n^*\| \geq d(v_0, G)$. Let $u_0 = v_0 - g_0$ and $u_n = v_0 - g_n$. Then $\|u_0\| = \|v_0 - g_0\| = d(v_0, G)$ and

$$\|u_n\| = \|v_0 - g_n\| \leq \|x - g_n\| \leq d(x, G) + 1/n.$$

Hence $\lim_n \|u_n\| = d(x, G) = d(v_0, G)$. Consider

$$\|(u_0 + u_n)/2\| = \|v_0 - (g_n + g_0)/2\| = \|v_0 - g_n^*\| \geq d(v_0, G).$$

Therefore $\lim_n \|(u_0 + u_n)/2\| \geq d(v_0, G)$. Now $u_0 - u_n = g_n - g_0 \in G$. But $\|u_0 - u_n\| = \|g_n - g_0\| \geq \varepsilon > 0$ for all n which is a contradiction to the local uniform convexity of the space V_X with respect to G. Hence $(G, R_{X^1}(G))$ has the property (P_1).

Corollary 4.5. *Let V_X be a locally uniformly convex space and let G be closed convex subset of V_X such that $P_G(x)$ is non-empty for each $x \in R_{X^1}(G)$. Then $(G, R_{x^1}(G))$ has the property (P_1).*

Corollary 4.6. *Let X be a locally uniformly convex normed linear space. Let G be a closed convex subsets of X and let U be a collection of all closed convex subsets of X which admits farthest points with respect to G such that $\mathrm{rad}_G(A) = d(x, G)$ for some $x \in A$. Thus, if $\mathrm{Cent}_G(A)$ (refer [1]) is non-empty for each A in U, the pair (G, U) has the property (P_1).*

From this corollary we have the following result.

Corollary 4.7. *Let X be a locally uniformly convex space. Let V be a closed convex subset of X and U the class of all compact convex non-empty subsets F of X such that $\mathrm{rad}_V(F) = d(x, V)$ for some $x \in F$. If $\mathrm{Cent}_V(F) \neq \emptyset$ for every $F \in U$, then the pair (V, U) has the property (P_1).*

The following Lemma 4.1 and Theorem 4.5 are generalizations of Godini's result in [2] and the proof will follow on the same lines of Godini's

Lemma 4.1. *Let X be a nals and $G \subset X$.*

(i) *If $x, y \in X$, then $|d(x, G) - d(y, G)| \leq \rho(x, y)$.*

(ii) *If $x \in X, y \in Dom(P_G)$ and $g \in P_G(y)$ then*

$$\|x - g\| \leq d(x, G) + 2\rho(x, y).$$

Theorem 4.5. *Let X be a nals, $G \subset X$ and $A \subset Dom(P_G)$. If (G, A) has the property (P_1) then $P_G|_{Dom(P_G)}$ is upper Hausdorff semi continuous at any $x \in A$. If (G, A) has the property (P_2) then $P_G|_A$ is Hausdorff continuous on A.*

Theorem 4.6. *Let X be a nals. Let V_X be a locally uniformly convex space with respect to its subspace G. If $P_G(x)$ is non-empty for each $x \in R_{X^1}(G)$ then $P_G|_{Dom(P_G)}$ is upper Hausdorff semi continuous at any $x \in R_{X^1}(G)$.*

This theorem follows immediately from Theorem 4.5 and 4.4.

Theorem 4.7. *Every one dimensional subspace of V_X is a Chebyshev set in $R_X(G)$ if and only if V_X is strictly convex.*

References

[1] G. Godini, *A frame work for best simultaneous approximations-Normed almost linear space*, J. Approx. Theory, 43(1985), 338–358.

[2] G. Godini, *Best approximation in normed almost linear spaces*, Constructive Theory of Functions', 84(1984), 356–363.

[3] G. Godini, *On normed almost linear spaces*, Math. Ann., 279(1988), 449–455.

[4] Dietrich Brass, *Nonlinear approximation theory*, Springer Verlog Berlin, 1986.

[5] T.D. Narang, *On best simultaneous approximation*, J. Approx. Theory, 39(1983), 93–96.

[6] J. Mach, *Continuity properties of Chebyshev centers*, J. Approx. Theory, 29(1980), 223–230.

[7] K.P.R. Sastry and S.V.R. Naidu, *On best simultaneous approximation in normed linear spaces*, Proc. Nat. Acad. Sciences, India, Sect. A, 48(1978), 249–250.

[8] B.N. Sahney and S.P. Singh, *On best simultaneous approximation in Banach spaces*, J. Approx. Theory, 35(1982), 222–224.

Proceedings of the Asian Mathematical
Conference, 1990

On a Mathematical Theory of Logics with Uncertainty

PHAN DINH DIEU
NCSR of Vietnam, Hanoi

§ 1. Introduction

In mathematical theories the used logic is mainly the classical formal logic with two truth values: *true* and *false*; every mathematical sentence is either true or false, not otherwise *(law of excluded middle)*. However, in the practice of life most of humain knowledges are somewhat uncertain, and we have to reason in approximate rather that precise terms upon these uncertain knowledges. Some formalisms to dealing with this uncertainty were developed, among them the most important was the probabilistic method of inference with the Bayes' rule. In recent years , with the development of Computer Science, the problem of ≪ approximate reasoning ≫ is investigated intensively and becomes one focus of research in Artificial Intelligence. A variety of formalisms and techniques have been proposed from various approaches. A review of these theories can be found, for example in [1,2], or in [3] with a comparative study. This situation motivates some authors to find a common *paradigm* capable of representing most of the improtant different approaches (see, e.g. [4]). A such attempt is made by the author in [5]. Following this way, in this paper we continue to develop the mathematical theory presented in [5] to include more capacity of reasoning on uncertain knowledges of other types.

The paper is structured as follows: the section 2 presents the various methods of representation of the truth value with uncertainty of a sentence (by a real number, by an interval, and by a fuzzy number), and the method of defining the truth value of a sentence based upon the ≪ possible words analysis ≫ and considered relatively with respect to a given knowledge base. The theory of interval-valued probabilistic logic is presented in the section 3, and the logic with fuzzy number valued uncertainty –in the section 4. In the section 5 is considered the possibility of developing a theory of single valued probabilistic logic by introducing an additional assumption such as the assumption of conditional independence or the maximum entropy hypothesis.

§ 2. Possible Worlds Analysis and Definition of Truth Values

In the classical logic (without uncertainty) we have two truth values: false (0) and true (1). For representing the uncertainty we consider the following methods of giving the truth value of a sentence:

a) by a real number α in the interval [0,1]: α is the truth probability of the considered sentence;

b) by an interval $[\alpha, \beta] \subseteq [0,1]$: we consider that the truth probability of the considered sentence is some number in $[\alpha, \beta]$;

c) by a fuzzy number A characterized by its membership function $\mu_A : [0,1] \to [0,1]$. For every $\alpha \in [0,1]$, $\mu_A(\alpha)$ is the measure of certainty of the fact that the truth probability of the considered sentence is α.

The single valued estimation is used in most of the theories of probabilistic logic. The interval valued one is used extensively in recent years, it gives us a flexible method of representation of the certainty measure, and at the same time it provides a common framework to compare and combine various theories in probabilistic reasoning (see, e.g., [4,6,7,8,5]). The fuzzy number valued approach is used in the fuzzy logic in relation with the representation of linguistic truth values (see, e.g., [9,10,11]). The theory of fuzzy numbers is presented, e.g., in [12, 13, 14]. For a given fuzzy number A and a real $\alpha (0 \le \alpha \le 1)$ we put

$$A^\alpha = \{x \mid \mu_A(x) \ge \alpha\},$$

and call it the α-cut of A. If A is the truth value of the sentence S, then $x \in A^\alpha$ means that there is a chance $\ge \alpha$ for that the truth probability of S is x.

The truth value of a sentence is defined through a possible worlds analysis and in relation to a given knowledge base as follows :

Following Nilsson in [15], we start with the (two valued) ordinary first-order predicate logic. We denote by W the set of all interpretations of this logic. We could imagine that each interpretation is a ≪possible world ≫ for this logic as a whole, as well as for any set of its formulas.

Let $\Gamma = \{S_1, \cdots, S_L\}$ be a finite set of sentences, and $S_N \notin \Gamma (N = L+1)$. We denote $\Sigma = \{S_1, \cdots, S_L, S_N\}$. A subset $W \subseteq W$ is said to be a Σ-equivalence class of possible worlds corresponding to the valuation vector $U = (u_1, \cdots, u_N)$ if it contains all interpretations in which the sentence S_i gets the truth value $u_i (i = 1, \cdots, N, u_i = 0$ or $1)$.

Suppose that there are K consistent valuation vectors $U_1, \cdots, U_k$ for Σ, and the Σ-equivalence classes of possible worlds corresponding to them are $W_1, \cdots, W_K$. The partition $W = W_1 \cup \cdots \cup W_K$ is called the Σ-standard partition of W. The classes $W_1, \cdots, W_K$ comprise a sample space over which we can define a probability distribution $\mathbf{P} = (p_1, \cdots, p_K)$; every p_j can be considered as the probability that our ≪ actual world ≫ is in the class W_j. Therefore, if π_i denotes the truth probability of the sentence $S_i (i = 1, \cdots, N)$, then we have the equation:

$$\Pi = M.P \tag{1}$$

where $\Pi = (\pi_1, \cdots, \pi_N)$, M is the $N \times K$-matrix, whose j-th column is the vector U_j (all the vectors P, Π, and U_j are understood to be expressed in the column form).

As a probability distribution, the vector P is assumed to satisfy the condition:

$$\sum_{j=1}^{K} p_j = 1, \quad p_j \ge 0 \quad (j = 1, \cdots, K). \tag{2}$$

We consider first the case of interval-valued uncertainty. Let the knowledge base $\boldsymbol{B}$ be given by the set of sentences $\Gamma = \{S_1, \cdots, S_L\}$ together with their truth values $[v_{i1}, v_{i2}] (i = 1, \cdots, L)$:

$$\boldsymbol{B} = \{\langle S_i, [v_{i1}, v_{i2}]\rangle \mid i = 1, \cdots, L\}. \tag{3}$$

Let S_N be a new sentence, and $\Sigma = \Gamma \cup \{S_N\}$ as above. A value $\pi_N \in [0,1]$ is said to be feasible for the truth probability of S_N with respect to $\boldsymbol{B}$ if there exist a probability distribution P and L values $\pi_i (i = 1, \cdots, L)$ such that $\pi_i \in [v_{i1}, v_{i2}]$ and the equation (1) holds. The set of all feasible values such defined is denoted by $F(S_N, \boldsymbol{B})$.

It is not difficult to show that $F(S_N, \boldsymbol{B})$ is also an interval (possible empty), $F(S_N, \boldsymbol{B}) = [v_{N1}, v_{N2}]$, we define this interval to be the truth value of S_N w.r.t. the knowledge base $\boldsymbol{B}$. We say also that $< S_N, [v_{N1}, v_{N2}] >$ is a knowledge derived from the knowledge base $\boldsymbol{B}$, and we write:

$$\boldsymbol{B} \quad \vdash \quad \langle S_N, [v_{N1}, v_{N2}]\rangle.$$

If every interval $[v_{i1}, v_{i2}] (i = 1, \cdots, L)$ is reduced to a single number π_i , then we obtain a knowledge base with single-valued truth probabilities. But , in this case $F(S_N, \boldsymbol{B})$ is not necessarily reduced to a real number, it is still generally an interval.

We now consider the case of fuzzy number-valued uncertainty. Let us give a knowledge base

$$\boldsymbol{B} = \{\langle S_i, A_i\rangle \mid i = 1, \cdots, L\}, \tag{4}$$

where A_i is a fuzzy number-the truth value of the sentence S_i . We may observe that there is a chance $\geq \alpha$ for the truth probability of S_N to be π_N if and only if there is a chance $\geq \alpha$ for the truth probability of every S_i to be equal to $\pi_i (i = 1, \cdots, L)$ and a probability distribution P such that the equation (1) holds. Thus, we can define for every $\alpha \in [0,1]$ an interval

$$A_N^\alpha = F(S_N, \{\langle S_i, A_i^\alpha\rangle \mid i = 1, \cdots, L\}).$$

It is easy to see that if $\alpha \leq \beta$ then $A_N^\beta \subseteq A_N^\alpha$, therefore, we can define a fuzzy number A_N such that for every α $(0 \leq \alpha \leq 1)$ the interval A_N^α is its α-cut. This fuzzy number is defined to be the truth value of S_N w.r.t. the knowledge base $\boldsymbol{B}$, and denoted by $F(S_N, \boldsymbol{B})$.

Thus , the approach of possible worlds analysis and of defining the truth value of a sentence relatively w.r.t. a given knowledge base may provide us with an unified framework to develop a common mathematical theory for logics with uncertainty, regardless of that the considered truth values are given by a point, by an interval, or by a fuzzy number.

§ 3. Probabilistic Logic with Interval-valued Uncertainty

A general method of computing the interval $F(S_N, \boldsymbol{B})$ can be given as follows: We denote by D the K-dimensional space of probabilistic distributions $P = (p_1, \cdots, p_K)$, and by E the N-dimensional linear space of vectors $\Pi = (\pi_1, \cdots, \pi_N)$ of truth probabilities of sentences $S_1, \cdots, S_N$. The simplex (2) in the space D is mapped by the transformation (1) into a set C in the space E, which is the convex hull of K points $U_1, \cdots, U_N$ in E. The set

$$\{(\pi_1, \cdots, \pi_N) \mid v_{i1} \leq \pi_i \leq v_{i2} \quad (i = 1, \cdots, L)\} \tag{5}$$

is a cylinder D in the space E. If $C \cap D = \emptyset$ then $F(S_N, \boldsymbol{B}) = \emptyset$; if $C \cap D \neq \emptyset$ then it is a convex set in E, and hence there are two functions $h_1(\pi_1, \cdots, \pi_L)$ and $h_2(\pi_1, \cdots, \pi_L)$ such that it can be expressed in the form

$$h_1 \quad (\pi_1, \cdots, \pi_L) \le \pi_N \le h_2(\pi_1, \cdots, \pi_L),$$
$$(\pi_1, \cdots, \pi_L) \in C' \cap D',$$

where C' and D' are the projections of C and D, respectively, on the subspace of coordinates $(\pi_1, \cdots, \pi_L)$. Then the interval $F(S_N, \boldsymbol{B}) = [v_{N1}, v_{N2}]$ is defined by :

$$v_{N1} = \min\{h_1(\pi_1, \cdots, \pi_L) \mid (\pi_1, \cdots, \pi_L) \in C' \cap D'\},$$
$$v_{N2} = \max\{h_2(\pi_1, \cdots, \pi_L) \mid (\pi_1, \cdots, \pi_L) \in C' \cap D'\}.$$

Thus, in order to get the values v_{N1} and v_{N2} we have to solve two optimization problems, whose objective functions are ususally piecewise linear, and the constraints are given by linear inequalities.

By the general method presented above we can find systematically the formulas giving the (interval-valued) truth values of a given sentence w.r.t. a given knowledge base. Most of formulas already obtained by other methods can be found also by this method. In the following we shall give a number of formulas related to the basic logical connectives and inference rules. In these formulas, $v_1(S)$ and $v_2(S)$ denote the lower bound and the upper bound of the interval -truth value of S; the truth value of the sentence on the left side is defined relatively w.r.t. to the knowledge base consisting of the sentences occuring on the right side together with their truth values.

Basic logical connectives:

$$\begin{cases} v_1(A \wedge B) = \max(0, v_1(A) + v_1(B) - 1), \\ v_2(A \wedge B) = \min(v_2(A), v_2(B)), \end{cases}$$

$$\begin{cases} v_1(A \wedge B) = \max(0, v_1(A) + v_1(B) - v_2(A \vee B)), \\ v_2(A \wedge B) = \min(1, v_2(A) + v_2(B) - v_1(A \vee B)), \end{cases}$$

$$\begin{cases} v_1(A \wedge B) = \max(0, v_1(A) + v_1(A \rightarrow B) - 1), \\ v_2(A \wedge B) = v_2(A) + v_2(A \rightarrow B) - 1, \end{cases}$$

$$\begin{cases} v_1(A \vee B) = \max(v_1(A), v_1(B)), \\ v_2(A \vee B) = \min(1, v_2(A) + v_2(B)), \end{cases}$$

$$\begin{cases} v_1(A \vee B) = \max(0, v_1(A) + v_1(B) - v_2(A \wedge B)), \\ v_2(A \vee B) = \min(1, v_2(A) + v_2(B) - v_1(A \wedge B)), \end{cases}$$

$$\begin{cases} v_1(A \vee B) = 1 + v_1(B) - v_2(A \rightarrow B), \\ v_2(A \vee B) = \min(1, v_2(B) - v_1(A \rightarrow B) + 1, \end{cases}$$

$$\begin{cases} v_1(A \to B) = \max(1 - v_2(A), v_1(B)), \\ v_2(A \to B) = \min(1, 1 - v_1(A) + v_2(B)), \end{cases}$$

$$\begin{cases} v_1(A \to B) = 1 - v_2(A) + v_1(A \wedge B), \\ v_2(A \to B) = \min(1, 1 - v_1(A) + v_2(A \wedge B)), \end{cases}$$

$$\begin{cases} v_1(A \to B) = 1 + v_1(B) - v_2(A \vee B), \\ v_2(A \to B) = \min(1, v_2(B) - v_1(A \vee B) + 1), \end{cases}$$

$$\begin{cases} v_1(\neg A) = 1 - v_2(A), \\ v_2(\neg A) = 1 - v_1(A), \end{cases}$$

$$\begin{cases} v_1(B \mid A) = \max(0, (v_2(A) + v_1(B) - 1)/v_2(A)), \\ v_2(B \mid A) = \min(1, v_2(B)/v_1(A)) \quad \text{(assume that} \quad v_1(A) \neq 0), \end{cases}$$

$$\begin{cases} v_1(A \wedge B \mid C) = \max(0, v_1(A \mid C) + v_1(B \mid C) - 1)), \\ v_2(A \wedge B \mid C) = \min(v_2(A \mid C), v_2(B \mid C)), \end{cases}$$

$$\begin{cases} v_1(A \vee B \mid C) = \max(v_1(A \mid C), v_1(B \mid C)), \\ v_2(A \vee B \mid C) = \min(1, v_2(A \mid C) + v_2(B \mid C)). \end{cases}$$

Inference rules:

Modus ponens: $\begin{cases} v_1(B) = \max(0, v_1(A) + v_1(A \to B) - 1)), \\ v_2(B) = v_2(A \to B). \end{cases}$

Modus tollens: $\begin{cases} v_1(A) = 1 - v_2(A \to B), \\ v_2(A) = \min(1, \ 1 - v_1(A \to B) + v_2(B)). \end{cases}$

Resolution principle: $\begin{cases} v_1(A \vee B) = \max(0, v_1(A \vee C) + v_1(B \vee \neg C) - 1), \\ v_2(A \vee B) = 1. \end{cases}$

Probabilistic modus ponens: $\begin{cases} v_1(B) = v_1(A).v_1(B \mid A), \\ v_2(B) = 1 - v_1(A) + v_1(A).v_2(B \mid A). \end{cases}$

We now consider the particular case when the truth values of the sentences in the given knowledge base $\boldsymbol{B}$ are reduced to single real numbers $\pi(S_i)(i = 1, \cdots, L)$. Then we write:

$$\boldsymbol{B} = \{< S_i, \pi(S_i) > \mid i = 1, \cdots, L\}. \tag{6}$$

If for any given L-tuple of real numbers $(\pi(S_1), \cdots, \pi(S_L))(0 \leq \pi(S_i) \leq 1)$ the interval $F(S_N, \boldsymbol{B})$ is reduced to a single real number $\pi(S_N)$, then we say that the set $\Gamma = \{S_1, \cdots, S_L\}$ is a defining set of S_N, and we have a function:

$$\pi(S_N) = f(\pi(S_1), \cdots, \pi(S_L)). \tag{7}$$

It is easy to see that , if Γ is a defining set of S_N and the equation (7) holds, then when the knowledge base $\boldsymbol{B}$ is given by (3) , $F(S_N, \boldsymbol{B})$ will be an interval $[v_{N1}, v_{N2}]$ where v_{N1} and v_{N2} are defined by :

$$v_{N1} = \min\{f(\alpha_1, \cdots, \alpha_L) \mid v_{i1} \leq \alpha_i \leq v_{i2}, i = 1, \cdots, L\},$$

$$v_{N2} = \max\{f(\alpha_1, \cdots, \alpha_L) \mid v_{i1} \leq \alpha_i \leq v_{i2}, i = 1, \cdots, L\}.$$

The following theorems could be useful for the design of heuristic algorithms of approximate reasoning when the logic developed here is used:

Theorem 1. *Let $\boldsymbol{B}$ and $\boldsymbol{B}'$ be two knowledge bases, and let S be a sentence different from sentences in $\boldsymbol{B}$ and $\boldsymbol{B}'$. Suppoe that*

$$\boldsymbol{B} \quad \vdash \quad \langle S, [u_1, u_2]\rangle \quad \text{and} \quad \boldsymbol{B}' \quad \vdash \quad \langle S, [v_1, v_2]\rangle.$$

Then , if $\boldsymbol{B}' \subseteq \boldsymbol{B}$ then $[u_1, u_2] \subseteq [v_1, v_2]$.

Theorem 2. *Let $\boldsymbol{B}$ be a knowledge base, and let T and S be two distinct sentences not contained in $\boldsymbol{B}$. Suppose that*

$$B \vdash \langle S, [u_1, u_2]\rangle \quad \text{and} \quad \boldsymbol{B} \vdash \langle T, [t_1, t_2]\rangle$$

$$\boldsymbol{B} \cup \{\langle T, [t_1, t_2]\rangle\} \vdash \langle S, [v_1, v_2]\rangle.$$

Then, $[u_1, u_2] = [v_1, v_2]$.

Theorem 3. *Let $\boldsymbol{B}_1$ and $\boldsymbol{B}_2$ be two knowledge bases, and let S be a sentence not occuring in $\boldsymbol{B}_1$ and $\boldsymbol{B}_2$. If*

$$\boldsymbol{B}_1 \cup B_2 \quad \vdash \quad \langle S, [v_1, v_2]\rangle,$$

and

$$\boldsymbol{B}_1 \quad \vdash \quad \langle S, [u_1, u_2]\rangle, \quad \boldsymbol{B}_2 \quad \vdash \quad \langle S, [t_1, t_2]\rangle,$$

then$[v_1, v_2] \subseteq [u_1, u_2] \cap [t_1, t_2]$.

Theorem 4. *Let $\boldsymbol{B}$ be a knowledge base containing the knowledge $\langle T, [t_1, t_2]\rangle$. Suppose that*

$$\boldsymbol{B} \vdash \langle S, [u_1, u_2]\rangle,$$

and

$$\boldsymbol{B}\setminus \quad \{\langle T, [t_1, t_2]\rangle\} \quad \vdash \quad \langle Z, [z_1, z_2]\rangle,$$
$$\{\langle T, [t_1, t_2]\rangle, \langle Z, [z_1, z_2]\rangle\} \quad \vdash \quad \langle S, [v_1, v_2]\rangle.$$

Then we have $[u_1, u_2] \subseteq [v_1, v_2]$.

Theorem 5. *Let us give a knowledge base $\boldsymbol{B}$:*

$$\boldsymbol{B} = \{\langle S_i, [v_{i1}, v_{i2}]\rangle \mid i = 1, \cdots, L\},$$

where $S_1 = A_1, S_i = A_{i-1} \to A_i (i = 2, \cdots, L)$. Suppose that

$$\boldsymbol{B} \quad \vdash \quad \langle A_L, [u_1, u_2]\rangle,$$

and

$$\{\langle S_i, [v_{11}, v_{12}]\rangle, \langle S_2, [v_{21}, v_{22}]\rangle\} \quad \vdash \quad \langle A_2, [t_{21}, t_{22}]\rangle,$$

$$\{\langle A_{i-1}, [t_{i-1,1}, t_{i-1,2}]\rangle, \langle S_i, [v_{i1}, v_{i2}]\rangle\} \quad \vdash \quad \langle A_i, [t_{i1}, t_{i2}]\rangle \quad (i = 3, \cdots, L).$$

Then we have $[u_1, u_2] = [t_{L1}, t_{L2}]$.

The Theorem 1 asserts the monotonicity of the deductions, the Theorem 2 states that the derived knowledge does not increase the ≪ power of deduction ≫ of a given knowledge base. In the practice of approximate reasoning, it is not easy at every stage to handle a large knowledge base as a whole; therefore, at every stage of reasoning we use often a part of our knowledge base or some knowledge derived from this base. The Theorems 3 and 4 state that the result obtained in such manner is always more rude then, but not contray to , the result obtained by deduction from the knowledge base as a whole. The Theorem 5 gives us the continuous connecting rule of deduction by modus ponens in our formalism (this rule was considered, e.g., in [16]).

The relation between the definition of defining sets in this formalism and the axioms of the probability theory is established by the following theorem:

Theorem 6. *The set $\Gamma = \{S_1, \cdots, S_L\}$ is a defining set of the sentence S, and*

$$\pi(S) = f(\pi(S_1), \cdots, \pi(S_L))$$

if and only if this equation is deduced from the following standard axioms of the Probability theory:

R1. If A is a tautology, then $\pi(A) = 1$,

R2. If $\rceil(A \wedge B)$ is a tautology, then $\pi(A \vee B) = \pi(A) + \pi(B)$.

By using this last theorem we obtain the following formulas, in each of which the sentences occuring in the right side form a defining set of the sentence on the left side:

$$\pi(A \wedge B) = \pi(A) + \pi(B) - \pi(A \vee B)$$

$$\pi(A \wedge B) = \pi(A) + \pi(A \to B) - 1$$

$$\pi(A \vee B) = \pi(A) + \pi(B) - \pi(A \wedge B)$$

$$\pi(A \vee B) = 1 + \pi(B) - \pi(A \to B)$$

$$\pi(A \to B) = 1 - \pi(A) + \pi(A \wedge B)$$

$$\pi(A \to B) = 1 + \pi(B) - \pi(A \vee B)$$

$$\pi(B) = \pi(A) + \pi(A \to B) - \pi(B \to A).$$

§ 4. Logic with Fuzzy Number-valued Uncertainty

In the framework of the formalism presented in this paper we would try to find a mathematical foundation for the fuzzy logic, and first of all to find a method of computing the fuzzy truth value of a sentence with respect to a given knowledge base consisting of a set of sentences given together with their fuzzy truth values. In most of recent works on the fuzzy logic the fuzzy numbers used as the truth values of sentences are usually assumed to be convex (and sometimes even normal). With the definition of fuzzy truth values given in this paper (see Section 2) we can prove:

Theorem 7. *Let us give a knowledge base*

$$\boldsymbol{B} = \{(S_i, A_i) \mid i = 1, \cdots, L\},$$

where $A_i(i = 1, \cdots, L)$ is a convex fuzzy number having support in [0,1], and let S be a sentence different from sentences in $\boldsymbol{B}$. Then $F(S, \boldsymbol{B})$ is also a convex fuzzy number having support in [0,1].

By this theorem we can assume that the fuzzy numbers used as truth values of sentences are always convex ones having support in the unity interval (they may be not normal).

Let A be a fuzzy number characterized by its membership function $\mu_A : R \to [0, 1]$. We denote by A^* the fuzzy number whose membership function is defined by :

$$\mu_{A^*}(x) = \begin{cases} \mu_A(x), & \text{if } 0 \leq x \leq 1, \\ 0, & \text{otherwise.} \end{cases}$$

Theorem 8. *If $\Gamma = \{S_1, \cdots, S_L\}$ is a defining set of the sentence S, and*

$$\pi(S) = f(\pi(S_1), \cdots, \pi(S_L)),$$

then the fuzzy truth value A of the sentence S with respect to the knowledge base

$$\boldsymbol{B} = \{\langle S_i, A_i\rangle \mid i = 1, \cdots, L\}$$

is given by

$$A = f(A_1, \cdots, A_L)^*.$$

The last expression $f(A_1, \cdots, A_L)$ is understood in the sense of the arithmetic of fuzzy numbers [12, 13].

The development of fuzzy logic in the formalism presented here, and a comparative study of this logic to other theories of fuzzy logic will be the subject of a further work.

§ 5. Probabilistic Logic with Single-valued Uncertainty

As has been well known, in difference from the case of the classical logic, in the logics with uncertainty the truth probability of a compound sentence is not, in general, expressed as a function of truth probabilities of its components. As has been shown in [17], the logical connectives cannot be the truth functional w.r.t. probabilities, or any other purely numeric uncertainty measure. In our formalism we see easily that, even when the interval-truth values of all the sentences in the given knowledge base $\boldsymbol{B}$ are reduced to single real numbers, the truth value $F(S, \boldsymbol{B})$ is not necessarily reduced to a single value. This fact holds even when the set of sentences in $\boldsymbol{B}$ is that of all components of S.

Thus , in order to get a single value for $F(S, \boldsymbol{B})$ we have to add some assumptions to the given knowledge base $\boldsymbol{B}$. The most important kinds of such assumptions considered in the litterature are those of conditional independence and of maximum entropy.

In the framework of the formalism developed in this paper the assumption of maximum entropy can be formulated in clear manner, hence it can provide a general method to compute the truth probability , and therefore, to develop a theory of probabilistic logic with single-valued truth probabilities.

Let us give a knowledge base

$$\boldsymbol{B} = \{\langle S_i, \alpha_i\rangle \mid i = 1, \cdots, L\},$$

where α_i is the single valued truth probability of the sentence S_i . Suppose that $W = W_1 \cup \cdots \cup W_K$ is the Σ-standard partition of W into Σ–equivalence classes of possible worlds. Let M be the $N \times K$-matrix defined as in the section 2, and let M^* denote the $L \times K$-matrix obtained from M by removing its last row. It is evident that the truth probability of S_N depends on the probability distribution $P = (p_1, \cdots, p_K)$ over the classes $W_1, \cdots, W_K$. This distribution must satisfy the conditions:

$$\sum_{j=1}^{K} p_j = 1, \quad p_j \geq 0, \tag{8}$$

$$M^*.P = (\alpha_1, \cdots, \alpha_L). \tag{9}$$

According to the assumption of maximum entropy we select among feasible solutions of (8)–(9) for P that P with maximum entropy (see [16]). This vector P is the solution of the following problem:

$$\begin{cases} H = \sum_{j=1}^{K} p_j \log p_j \rightarrow \max \\ \text{subject to constraints (8)–(9) .} \end{cases}$$

After having the distribution P with the maximum entropy we obtain the value α_N for the sentence S_N by the equation

$$\alpha_N = r_N \cdot P,$$

where r_N is the last row of the matrix M.

We denoter $\alpha_N = p(S_N, \boldsymbol{B})$ and write also $\boldsymbol{B} \quad \vdash \quad \langle S_N, \alpha_N\rangle$.

By this method a theory of probabilistic logic with single-valued truth probabilities can be developed. In this theory we have, for example, the following formulas (in these formulas the related knowledge base is understood to comprise the sentences occurring in the right side together their truth probabilities):

$$\begin{aligned}
p(A \wedge B) &= p(A).p(B), \\
p(A \wedge B) &= p(A) + p(B) - p(A \vee B), \\
p(A \wedge B) &= p(A) + p(A \rightarrow B) - 1, \\
p(A \vee B) &= p(A) + p(B) - p(A) \cdot p(B), \\
p(A \vee B) &= p(A) + p(B) - p(A \wedge B), \\
p(A \vee B) &= 1 + p(B) - p(A \rightarrow B), \\
p(A \rightarrow B) &= 1 - p(A) + p(A) \cdot p(B),
\end{aligned}$$

$$\begin{aligned}
p(A \to B) &= 1 - p(A) + p(A \wedge B),\\
p(A \to B) &= 1 + p(B) - p(A \vee B),\\
p(B) &= p(A)/2 + p(A \to B) - 1/2,\\
p(A) &= 1 - p(A \to B) + p(B)/2,\\
p(A \vee B) &= (p(A \vee C) + p(B \vee \neg C))/2,\\
p(B) &= p(A) \cdot p(B \mid A) - p(A)/2 + 1/2.
\end{aligned}$$

In the application of this theory in the practice of approximate reasoning one of our concerns is about the consistency of deductions. This property of consistency can be expressed as follows: Let us give a knowledge base

$$\boldsymbol{B} = \{\langle S_i, \alpha_i \rangle \mid i = 1, \cdots, L\},$$

and let A and S be two distinct sentences different from the sentences in $\boldsymbol{B}$. We denote

$$\Gamma = \{S_1, \cdots, S_L, A\} \quad \text{and} \quad \Sigma = \{S_1, \cdots, S_L, S\}.$$

Suppose that $\boldsymbol{B} \vdash \langle A, \alpha \rangle$. We form a new knowledge base $\boldsymbol{B}^* = \boldsymbol{B} \cup \{\langle A, \alpha \rangle\}$. A question arises: whether $p(S, \boldsymbol{B}) = p(S, \boldsymbol{B}^*)$ is true or not? We have:

Theorem 9. *1) If the standard partitions of the set W of possible worlds are the same w.r.t. Γ and Σ then $p(S, \boldsymbol{B}) = p(S, \boldsymbol{B}^*)$.*

2) If these standard partitions are different, then it is not necessarily true that $p(S, \boldsymbol{B}) = p(S, \boldsymbol{B}^*)$.

By this theorem we see that after adding a derived knowledge to the given knowledge base its capacity of deduction may be changed. With the assumption of maximum entropy we hope that no any additional information is introduced in the process of deduction except that contained in the initial knowledge base. But , may be , this is not true because when choosing the standard partition for the set of all the possible worlds we have accepted some type of additional information.

Finally we notice that there is a number of formulas obtained by using the assumption of conditional independence can be found also by the maximum entropy hypothesis in the framework of our formalism. For example, in [18] by using the assumption of conditional independence it was shown that if $A = D \wedge F$ and $B = E \wedge F$ then

$$p(C \mid A \wedge B) = \left(1 + \frac{1 - p(C \mid A)}{p(C \mid A)} \frac{1 - p(C \mid B)}{p(C \mid B)} \frac{p(C \mid F)}{1 - p(C \mid F)}\right)^{-1}.$$

This is a formula used rather often in the processes of knowledge acquisition, it can be proved also by the method of entropy maximization in the formalism presented above.

§ 6. Conclusion

In the paper is presented the foundation for a mathematical theory of logics with uncertainty, in which the truth value of a sentence is defined through a possible worlds analysis and in relation with a given knowledge base. The different methods of representation of the truth values with uncertainty are considered: this value may be a real number, a subinterval in the unity interval, or a convex fuzzy number . In the framework of the presented formalism one can find on an unified basis many results already obatained in various theories of logics with uncertainty. At the sam etime in the paper is given a number of theorems which could be useful for the design of heuristic algorithms of approximate reasoning in the Artificial Intelligence.

References

[1] H. Prade, *A synthetic review of approximate reasoning techniques.* Proc. 8th Intern. Joint Conference on Artificial Intelligence, 1983, v. 1, 130–136.

[2] H. Prade, *A computational approach to approximate and plausible reasoning with application to expert systems.* IEEE Trans. on pattern analysis and machine intelligence, 7, 1985, 260–279.

[3] A. Goicoechea, *Expert system models for inference with imperfect knowledge: a comparative study.* J. Statistical planning and inference, 20, 1988, 245–277.

[4] B.N. Grosof, *An inequality paradigm for probablistic knowledge. The logic of conditional probability intervals,* in L.N.Kanal and J.F.Lemmer (eds.) Uncertainty in Artificial Intelligence, Elsevier Sci. Publ., 1986, 259–275.

[5] D.D. Phan, *Probabilistic logic for approximate reasoning,* in I. Plander (ed.) Artificial Intelligence and Information–Control Systems of Robots–89, Elsevier Sci. Publ., 1989, 107–112.

[6] L. A. Appelbaum and E.H. Rupini., *ARIES: A tool for inference under conditions of imprecisin and uncertainty.* Proc., SPIE vol.458 Applications of Artificial Intelligence II, Arlington, Va. 1985.

[7] J. Baldwin, *Support logic programming,* in A.I.Jones et al.(eds.) Fuzzy sets theory and applications. D.Reidel Publ. Co., 1986.

[8] R.M. Tong and L.A. Appelbaum, *Experiments with interval-valued uncertaity,* in J.F. Lemmer and L.N. Kandal (eds.) *Uncertainty in Artificial Intelligence, II,* Elsevier Sci. Publ., 1988, 63–75.

[9] R.E. Bellman and L.A. Zadeh, *Local and fuzzy logics,* in J.M. Dunn and J.Epstein (eds.) Modern use of multiple-valued logic, D.Reidel Publ. Co., 1977, 103–165.

[10] P.B. Bonissone, *A fuzzy sets based linguistic approach: theory and applications.* in M: M. Gupta and E.Sanchez (eds.), Approximate Reasoning in decision analysis., North Holland Publ., 1982, 329–339.

[11] P.B. Bonissone et al., *RUM: A layered architecture for reasoning with uncertainty,* Proc. 10th Intern. Joint Conference on Artificial Intelligence, 1987, v.2, 891–898.

[12] M. Mizumoto and K.Tanaka, *Some properties of fuzzy numbers.* in M.Gupta et al. (eds.) Advances in fuzzy set theory and applications, North holland Publ. 1979, 153–164.

[13] P. Dubois and H. Prade, *Possibility theory. An approach to computerized processing of uncertainty.* Plenum press, New York and London, 1988.

[14] A. Kandel, *Fuzzy number techniques with applications,* Addison Wesley Publ., 1986.

[15] N. Nilsson, *Probabilistic logic*, Artificial Intelligence, 28, 1986, 71–87.

[16] S. Lee and K.G. Shin, *Uncertain inference using belief functions*, Proc. 3th Conf. on Artificial Intelligence Applications, Hyatt Orlando, 1987, 238–249.

[17] A. Bundy, *Incidence calculus: a mechanism for probabilistic reasoning*, Journal of Automated Reasoning, 1, 1985, 263–283.

[18] T. Havranék and Z. Kouba, *On the expert system EPROB*, Journal of New generation Computer systems, Berlin, 1, 1988, 87–96.

Proceedings of the Asian Mathematical
Conference, 1990

Affine Geometry in Complex Function Spaces and Algebras

A.J. Ellis
University of Hong Kong

§1. Compact Convex Sets

Let K be a complex convex subset of a locally convex Hausdorff space E. Denote by $A(K)$ (respectively $A_{\mathbb{C}}(K)$) the Banach space of all continuous real-valued (respectively complex-valued) affine functions on K, with the supremum norm. The set of extreme points of K will be denoted by ∂K. The centre of $A(K)$ consists of those functions $f \in A(K)$ such that for each $g \in A(K)$ there corresponds an $h \in A(K)$ with $h(x) = f(x)g(x)$ for all $x \in \partial K$.

A closed face F of K is said to be split if there exist a disjoint face F' of K such that each $x \in K$ has a unique representation $x = \lambda y + (1 - \lambda)z$, with $y \in F$, $z \in F'$ and $0 \leq \lambda \leq 1$. The family $\{\partial F : F \text{ is a closed split face of } K\}$ forms the closed sets for the facial topology of ∂K. A results of Alfsen and Andersen shows that the family of real-valued functions on ∂K which are continuous for the facial topology coincides with the restriction to ∂K of the centre of $A(K)$. The sets of constancy in ∂K for the centre of $A(K)$ form a family of sets of the form $\{\partial F_\alpha\}$, where $\{F_\alpha\}$ is a family of closed split faces of K called the Šilov decomposition for K. If the Šilov decomposition is trivial, that is if the centre of $A(K)$ consists only of the constant functions, then $A(K)$ is said to be antisymmetric.

The compact convex set K is called a simplex (or Choquet simplex) if it satisfies the following geometric condition: whenever $x \in E$ and $\lambda > 0$ the set $K \cap (\lambda K + x)$ is either empty or equal to $\mu K + y$ for some $\mu \geq 0$ and $y \in E$. In finite dimensions this definition may be shown to coincide with the usual one for an n-simplex. The property that K is a simplex is equivalent to the Banach dual space $A(K)^*$ being a vector lattice for the natural partial ordering. The property that $A(K)$ is a vector lattice is equivalent to the condition that K is a simplex with ∂K closed, called a Bauer simplex, and this in turn is equivalent to $A(K)$ coinciding with its centre.

Results of Alfsen and of Ellis show that K is a simplex if and only if every closed face of K is split.

General references for the results mentioned above are Alfsen [1] and Asimow and Ellis [3].

For two compact convex sets K_1 and K_2 we will write $K_1 \cong K_2$ there exists an affine homeomorphism mapping K_1 onto K_2.

§2. Complex Function Spaces

Let X be a compact Hausdorff space and let $C_{\mathbb{C}}(X)$ denote the Banach space of all continuous complex-valued functions on X with the supremum norm. By a complex function space M on X we will mean a closed linear subspace of $C_{\mathbb{C}}(X)$ which contains constants and separates points of X. We say that M is self-adjoint if $f \in M$ implies that $\bar{f} \in M$. The space $A_{\mathbb{C}}(K)$ is an example of a self-adjoint complex function space on K.

Let Σ denote the closed unit ball of M^* endowed with the w^*-topology, and define the state space S of M to be $S = \{\phi \in \Sigma : \phi(1) = 1\}$ and the complex state space Z of M to be $Z = \text{co}(S \cup -iS) \subset M^*$. The sets Σ, S and Z are all w^*-compact convex sets. The space X is naturally embedded in S by the natural evaluation map, and the sets ∂S and $\overline{\partial S}$ are then subsets of X, referred to as the Choquet boundary and Šilov boundary of M respectively. It is often convenient to restrict M to its Šilov boundary.

In the case where $M = A_{\mathbb{C}}(K)$ it is known that the embedding of K coincides with S, and that S is a closed split face of Z. Conversely, if S is a closed split face of Z then M is self-adjoint and M may be represented as $A_{\mathbb{C}}(S)$.

We may define a real-linear mapping $\theta : M \to A\lambda(Z)$ by $(\theta f)(z) = re(z(f))$, for $f \in M$ and $z \in Z$. For each $x, y \in X$ we have $(\theta(u + iv))(\lambda x - i(1 - \lambda)y) = \lambda u(x) + (1 - \lambda)v(y)$. Consequently we have $\|\theta(u + iv)\| = \max\{\|u\|, \|v\|\}$, and it may be shown that θ is a real-linear homeomorphism of M onto $A(Z)$.

It is known that $\theta(u + iv)$ belongs to the centre of $A(Z)$ if and only if $(u + v)M$ and $(u - v)\overline{M}$ are both contained in M (cf. [6, p.623]). If M is self-adjoint then $A(Z)$ cannot be antisymmetric. In general $A(Z)$ will be antisymmetric if and only if $\{h \in C_{\mathbb{R}}(X) : hM \subseteq M\} = \{\text{constants}\}$ and $\{h \in C_{\mathbb{R}}(X) : h\overline{M} \subseteq M\} = \{0\}$.

For complex function spaces M_1, M_2 we will write $M_1 \cong_{\mathbb{C}} M_2$ if there exists a complex-linear isometric isomorphism between M_1 and M_2. If M_1 and M_2 are isometric spaces we will write $M_1 \cong M_2$; in this case we note that the Mazur-Ulam theorem gives a real-linear isometric isomorphism between M_1 and M_2.

§3. Self-adjoint Complex Function Spaces

If M is a self-adjoint complex function space then $M \cong_{\mathbb{C}} A_{\mathbb{C}}(S)$. If, for two such spaces M_1 and M_2, we have $M_1 \cong_{\mathbb{C}} M_2$ we may ask whether $A(S_1) \cong A(S_2)$ or $S_1 \cong S_2$. If $S_1 \cong S_2$ then it is straightforward to prove that $A(S_1) \cong A(S_2)$ and $A_{\mathbb{C}}(S_1) \cong A_{\mathbb{C}}(S_2)$. However, even for finite-dimensional sets, we may have $A(S_1) \cong A(S_2)$ (and consequently $A_{\mathbb{C}}(S_1) \cong_{\mathbb{C}} A_{\mathbb{C}}(S_2)$) while $S_1 \ncong S_2$. It is also possible to have $A_{\mathbb{C}}(S_1) \cong A_{\mathbb{C}}(S_2)$ while $A(S_1) \ncong A(S_2)$. These results are discussed in [5] where the positive result is proved that if the Šilov decomposition for S_1 is finite then $A_{\mathbb{C}}(S_1) \cong_{\mathbb{C}} A_{\mathbb{C}}(S_2)$ implies that $A(S_1) \cong A(S_2)$. In particular this conclusion holds if S_1 is finite-dimensional.

For self-adjoint complex function spaces M_1 and M_2 we get corresponding results that $A(S_1) \cong A(S_2)$ implies that $A(Z_1) \cong A(Z_2)$, but not conversely in general; if the Šilov decomposition for S_1 is finite then the converse implication holds.

It may be shown that $A(S_1) \cong A(S_2)$ implies that $S_1 \cong S_2$ for certain compact convex sets; for example if both S_1 and S_2 are simplexes, or if both are state spaces of uniform algebras, or if both are state spaces of unital C^*-algebras.

If M is a complex function space such that Z is a simplex then, since S is necessarily a split face of Z, it follows that M is self-adjoint. Further, since $A(Z)^*$ is a vector lattice Kakutani's theorem shows that $A(Z)^*$ is an AL-space. Hirsberg and Lazar [12] showed in face that M^* is a complex $L^1(\mu)$ space in this situation, and conversely that $M^* \cong_{\mathbb{C}} L^1(\mu)$ implies that Z is a simplex. Such spaces M are known as complex Lindenstrauss spaces.

A similar characterisation of complex Lindenstrauss spaces M amongst complex function spaces is that M is self-adjoint and S is a simplex; that is S is a simplex and is a split face of Z.

There are other geometrical characterisations of complex Lindenstauss spaces in terms of L-ideals. A closed face F of a compact convex set K is split if and only if the real-linear span J of F is an L-ideal in $A(K)^*$, that is J is complemented in $A(K)^*$, $A(K)^* = J \oplus J'$, where $\|\phi + \phi'\| = \|\phi\| + \|\phi'\|$ for $\phi \in J$, $\phi' \in J'$. It was proved in [9] that a complex Banach space M is a complex Lindenstrauss space if and only if for each closed face F of Σ the complex-linear span of F is an L-ideal in M^*.

Hirsberg and Lazar's results also show that a self-adjoint complex function space M is a complex Lindenstrauss space if and only if Σ has the property that each $\phi \in \Sigma$ can be represented by a unique maximal probability measure on Σ. This latter property of Σ was shown by Fuhr and Phelps [11] to be equivalent to each proper face of Σ being a (possibly non-compact) simplex; such sets Σ are called simplexoids.

We should also mention here, without detailed discussion, that there is a large amount of literature concerning the characterisation of Lindenstrauss spaces in terms of intersection properties of closed balls, due to Lindenstrauss [15], Hustad [13], Lima [14] and others. We just mention the result that M is a complex Lindenstrauss space if and only if it is an $E(3)$-space, that is whenever $\{B(m_j, r_j)\}$, $j = 1, 2, 3$, are closed balls in M such that $\cap_{j=1}^3 B(\phi(m_j), r_j) \neq \phi$ for each $\phi \in \Sigma$, then $\cap_{j=1}^3 B(m_j, r_j) \neq \phi$

§4. Uniform Algebras

A complex function space on X which is a pointwise algebra on X is called a uniform algebra on X. For a uniform algebra M on X much is known concerning the relationship between geometric properties of Z and algebraic and analytic properties of M (cf. [3, Section 4.7]). We recall that an intersection of peak sets for M is called a generalised peak set, and that a subset E of X such that $M|E = C_{\mathbb{C}}(E)$ is called an interpolation set for M.

It is known that if F is a closed face of S, for a uniform algebra M, then $F \cap X$ is a generalised peak set for M if and only if $\mathrm{co}(F \cup -iF)$ is a split face of Z, and $F \cap X$ is a generalised peak interpolation set for M if and only if F is a split face of Z.

For a uniform algebra M the space $A(Z)$ is antisymmetric if and only if $M \cap C_{\mathbb{R}}(X)$ equals the real constants, that is M is antisymmetric in the standard sense. The Šilov decomposition $\{E_\alpha\}$ for M consists of the sets of constancy for $M \cap C_{\mathbb{R}}(X)$, and coincides with $\{F_\alpha \cap X\}$, where $\{F_\alpha\}$ is the Šilov decomposition for Z. The Bishop decomposition $\{E_\beta\}$

for M consists of the maximal sets of antisymmetry for M, and coinicides with $\{E_\beta \cap X\}$, where $\{F_\beta\}$ is the family of maximal split faces of Z such that the centre of $A(F_\beta)$ is trivial; the family $\{F_\beta\}$ is called the Bishop decomposition for Z.

The essential set for a uniform algebra M on X is the annihilator in X of the largest algebra ideal of $C_{\mathbb{C}}(X)$ which is contained in M. It may be shown that the essential set for M has the form $F \cap X$, where F is the smallest closed face of S such that $\mathrm{co}(F \cup -iS)$ is a split face of Z.

If M_1 and M_2 are uniform algebras then $Z_1 \cong Z_2$ if and only if $A(Z_1) \cong A(Z_2)$ (cf. [10]). The equivalence $M_1 \cong_{\mathbb{C}} M_2$ implies that M_1 and M_2 are algebraically isomorphic (cf. Nagasawa [16]) and that $Z_1 \cong Z_2$. In general the equivalence $Z_1 \cong Z_2$ does not imply that $M_1 \cong_{\mathbb{C}} M_2$. Indeed there exist compact Hausdorff spaces X_1, X_2 such that $X_1 \cup X_1'$ is homeomorphic to $X_2 \cup X_2'$, where X_1', X_2' are disjoint copies of X_1, X_2 respectively, while X_1 and X_2 are not homeomorphic. For such spaces we may take $M_j = C_{\mathbb{C}}(X_j)$, $j = 1, 2$, and we have $M_1 \not\cong_{\mathbb{C}} M_2$ while $A(Z_1) \cong C_{\mathbb{R}}(X_1 \cup X_1') \cong C_{\mathbb{R}}(X_2 \cup X_2') \cong A(Z_2)$ (cf. [4]).

We say that Z_1 and Z_2 are real-equivalent if there exists an affine homeomorphism $\psi : Z_1 \to Z_2$ such that $\psi(S_1) = S_2$. It may be shown that $M_1 \cong_{\mathbb{C}} M_2$ if and only if Z_1 and Z_2 are real-equivalent. As a consequence of this result it follows that if M_1 has only finitely many distinct idempotents then $Z_1 \cong Z_2$ if and only if $M_1|Y_1 \cong_{\mathbb{C}} M_2|Y_2$ and $M_1|(X_1 \backslash Y_1) \cong_{\mathbb{C}} \overline{M}_2|(X_2 \backslash Y_2)$, where Y_1, Y_2 are open and closed subsets of the Šilov boundaries X_1, X_2 of M_1, M_2 respectively, and where $\overline{M}_2$ denotes the complex conjugate algebra of M_2. It is also known that if M_1, M_2 are uniform algebras on their essential sets X_1, X_2 then $Z_1 \cong Z_2$ if and only if $M_1 \cong (M_2|E) \oplus (\overline{M}_2|(X_2 \backslash E))$, for some open and closed subset E of X_2 (cf. [6]).

In the context of the preceding results we can not replace $\overline{M}_2$ by M_2 since it is possible to have $M \not\cong_{\mathbb{C}} \overline{M}$ for a uniform algebra M (see [8]). The equivalence $M \cong_{\mathbb{C}} \overline{M}$ for a uniform algebra is itself equivalent to the geometric property of Z that there exists an affine homeomorphism $\psi : Z \to Z$ such that $\psi(S) = -iS$ and $\psi(-iS) = S$.

It would be of interest to have geometric characterisations of the state space and complex state space of a uniform algebra, analogous to the characterisation of the state space of a unital C^*-algebra due to Alfsen, Hanche-Olsen and Shultz [2]. In the absence of such a characterisation we may study the relationship between the convex sets S, Z and Σ for complex function spaces and the corresponding sets for uniform algebras.

Simple examples show that it is possible to have $S_1 \cong S_2$ and $\Sigma_1 \cong \Sigma_2$ while $Z_1 \not\cong Z_2$, with M_1 a uniform algebra and M_2 a complex function space which is not an algebra. On the other hand we can have $S_1 \cong S_2$ and $Z_1 \cong Z_2$ while $\Sigma_1 \not\cong \Sigma_2$, with M_1 a uniform algebra and M_2 a complex function space which is not an algebra. It is perhaps surprising that if M_1 and M_2 are uniform algebras and M is a complex function space such that $\Sigma \cong \Sigma_1$ and $Z \cong Z_2$ then M is also uniform algebra (cf. [6, 7]).

This result depends partly upon the following fact: If M is a complex function space and if M_1 is either a uniform algebra or a self-adjoint complex function space on its Šilov boundary X_1 then $\Sigma \cong \Sigma_1$ implies that $M \cong_{\mathbb{C}} (M_1|E) \oplus (\overline{M}_1|(X_1 \backslash E))$, where E is an open and closed subset of X_1. An analogue of this result holds for any complex Banach space M. A sufficient condition that is required of M_1 in the proof of this result is the geometric condition that if $y \in \partial S_1$ then $\text{co}\{y, -iy\}$ is a face of Z_1. The geometric condition holds for a uniform algebra because each $y \in \partial S_1$ is a split face of Z_1, and for a self-adjoint complex function space since S_1 is a split face of Z_1.

The results just mentioned also have an application to complex Lindenstrauss spaces. In fact if M is a complex Banach space such that $M \cong M_1$ for a complex Lindenstrauss space M_1 then $M \cong_{\mathbb{C}} M_1$.

References

[1] E.M. Alfsen, *Compact Convex Sets and Boundary Integrals*, Springer-Verlag, Berlin, 1971.

[2] E.M. Alfsen, H. Hanche-Olsen and F.W. Shultz, *State spaces of C^*-algebras*, Acta Math., 144 (1980), 267–305.

[3] L. Asimow and A.J. Ellis, *Convexity Theory and its Applications in Functional Analysis*, Academic Press, London, 1980.

[4] A.J. Ellis, *Equivalence for complex state spaces of function spaces*, Bull. London Math. Soc., 19(1987), 359–362.

[5] A.J. Ellis, *Isometries of self-adjoint complex function spaces*, Math. Proc. Camb. Phil. Soc., 105 (1989), 133–138.

[6] A.J. Ellis, *Algebraic structure in complex function spaces*, Proc. Amer. Math. Soc., 107 (1989), 621–626.

[7] A.J. Ellis, *Real characterizations of function algebras amongst function spaces*, Bull. London Math. Soc. (in press).

[8] A.J. Ellis, *Real linear isometries of complex function spaces*, Proc. Function Spaces Conference, Edwardsville, April 1990 (to appear).

[9] A.J. Ellis, T.S.S.R.K. Rao, A.K. Roy and U. Uttersrud, *Facial characterizations of complex Lindenstrauss spaces*, Trans. Amer. Math. Soc., 268 (1981), 173–186.

[10] A.J. Ellis and W.S. So, *Isometries and the complex state spaces of uniform algebras*, Math. Zeitschrift, 195 (1987), 119–125.

[11] R. Fuhr and R.R. Phelps, *Uniqueness of complex representing measures on the Choquet boundary*, J. Funct. Anal., 14 (1973), 1–27.

[12] B. Hirsberg and A.J. Lazar, *Complex Lindenstrauss spaces with extreme points*, Trans, Amer. Math. Soc., 186 (1973), 141–150.

[13] O. Hustad, *Intersection properties of balls in complex Banach spaces whose duals are L^1-spaces*, Acta Math., 132 (1974), 283–313.

[14] A. Lima, *Intersection properties of balls and subspaces in Banach spaces*, Trans. Amer. Math. Soc., 227 (1977), 1–62.

[15] J. Lindenstrauss, *Extensions of compact operators*, Memoirs Amer. Math. Soc. No. 48 (1964).

[16] M. Nagasawa, *Isomorphisms between commutative Banach algebras with an application to rings of analytic functions*, Kōdai Math. Sem. Rep., 11 (1959), 182–188.

Proceedings of the Asian Mathematical
Conference, 1990

Non-periodic Ground and Gibbs States

A.C.D. VAN ENTER
Institute for Theoretical Physics
P.O. Box 800, 9700 AV Groningen
The Netherlands

Since the experimental discovery of quasicrystals[1] there has been a surge of interest in the possibility of non-crystalline (=non-periodic) long-range order. One of the as yet unsolved problems is how to derive such a non-periodic long-range order property from a statistical mechanical model. In fact[2], even the more fundamental problem of how to prove *crystalline* long-range order for a reasonably realistic interaction is essentially unsolved. In this sorry state of affairs one often resorts to the study of toy models in order to obtain any detailed results at all. This is what we will also do in this paper. To be more specific, we will consider one-dimensional lattice models. At each point of the lattice $\mathbb{Z}$, there is a two valued "spin" variable ω_i. Thus the configuration space $\Omega = \{-1,+1\}^{\mathbb{Z}}$ and

$$\omega_i(\omega) = +1$$

means that site i is occupied in configuration ω and

$$\omega_i(\omega) = -1$$

means that site i is empty in configuration ω.

Thus every configuration is a doubly infinite sequence of plus and minus ones. Although the lattice structure by its nature already seems to impose some periodicity, this turns out to be the case.

A (translation invariant) interaction is defined to be a set of functions $\Phi_X : \{-1,+1\}^X \to \mathbb{R}$ labeled by the finite subsets X of $\mathbb{Z}$.

By taking $\Phi_X(\{\omega_i(\omega)_{i\in X}\})$ the Φ_X become functions on Ω. They are translation-invariant if $\Phi_{\tau_i X} = \Phi_X$ where τ_i denotes the shift over a distance i.

A state of the system is a probability measure on Ω.

A *ground* state (zero temperature state) for Φ is a measure μ such that

$$\sum_{X\cap\Lambda\neq\emptyset} \{\mu'(\Phi_X) - \mu(\Phi_X)\} > 0$$

for all μ' which differ from μ only $\{\omega_i\}_{i\in\Lambda}$, the configurations in Λ, and all finite Λ. A Gibbs state (positive temperature state) for Φ is a state such that

$$\frac{\mu(\omega^{1,\Lambda}|\omega^{\Lambda^c})}{\mu(\omega^{2,\Lambda}|\omega^{\Lambda^c})} = \exp - \sum_{X\cap\Lambda\neq\emptyset} \{\Phi_X(\omega^{1,\Lambda}\omega^{\Lambda^c}) - \Phi_X(\omega^{2,\Lambda}\omega^{\Lambda^c})\}$$

for μ-almost all configurations $\omega^{1,\Lambda}\omega^{\Lambda^c}$ and $\omega^{2,\Lambda}\omega^{\Lambda^c}$ which differ only on the sites in Λ, for every Λ [2–4]. For each shift invariant Φ there is at least one shift invariant ground state and at least one shift invariant Gibbs state.

We will say that a shift invariant ground state has non-periodic long-range order if it has no periodic configuration in its support. We will say that a shift invariant Gibbs state has non-periodic long-range order if it is not an extreme point in the set of Gibbs states and cannot be decomposed (in its Choquet decomposition) into periodic extremal Gibbs states (Gibbs measures, in contrast to ground states, always form a Choquet simplex [3,4]).

We have constructed a 4-body interaction [5] (that is, $\Phi_X = 0$ for $|X| > 4$) which as its ground state configurations (point measures on Ω) has the Thue-Morse sequences. Thue-Morse sequences are constructed as follows: We start by taking the configuration which has $\omega_i = +1$ for all $i \in \mathbb{Z}$. At the first step we flip every second spin. At the nth step we flip all blocks of 2^n spins within the previous $n-1$st configuration from site $(2k+1)2^n+1$ to $(2k+2)2^n$ for every k. The limit of this sequence of periodic configurations of period 2.2^n is a non-periodic Thue-Morse sequence. The closure of its orbit under translations supports only one shift invariant measure μ_{TM} (it is a "strictly ergodic" system) [6]. We have shown that the Thue-Morse (the configurations in this support) sequences are characterized by the following property: Let i, j, k, l be four points such that $j-i = l-k = 2^r$ and $l-i = (2p+1)2^r$ for some p and r. Then the Thue-Morse sequences are precisely those sequences in which it is never true that $\omega_i(\omega) = \omega_j(\omega)$ and at the same time $\omega_k(\omega) = \omega_l(\omega)$. Thus any interaction which assigns a positive energy to all these forbidden 4 point patterns and is zero elsewhere has the Thue-Morse sequences as its ground state configurations, and the shift invariant Thue-Morse state μ_{TM} as its only shift invariant ground state. (Note that as a function of p and r, the interaction can decay arbitrarily fast.)

As for Gibbs states [7], we use the characterization of shift invariant Gibbs states as tangent functionals to the pressure function [3]

$$P(\Phi) = \lim_{\Lambda\to\mathbb{Z}} \frac{1}{|\Lambda|} \ln \sum_{\{\omega_i\}_{i\in\Lambda}} \exp - \sum_{X\subset\Lambda} \Phi_X(\omega)$$

and apply the Bishop-Phelps theorem in Israel's version [3 Ch.5] which states that every P-bounded functional can be approximated by tangent functionals to P in a suitable topology. We thus show the existence of a tangent functional (shift-invariant Gibbs state) for an element in the space of 4 body interactions, which is uniformly close to the Thue-Morse measure μ_{TM} on suitable four point observables. The forbidden patterns in μ_{TM} have a uniformly low density in this Gibbs state. Its Choquet decomposition–decomposition at infinity–into extremal Gibbs states contains some non-periodic Gibbs state in its support.

A somewhat related construction for non-periodic ground states was independently found by Aubry [8] (see also [9]). His result is more general in that it applies to *all* strictly ergodic system, but weaker in the sense that he has to consider arbitrarily many body interactions. This prevents a direct generalization of his results to positive temperature.

Summarizing, we have shown the possibility of non-periodic long-range order originating from a translation invariant finite-body interaction. As for ground states, this interaction can decay arbitrarily fast. However at positive temperature we do not have control over the rate of decay of the interaction, except that it is summable, so that any finite region interacting with its environment has finite energy.

Acknowledgements. This work has been done in collaboration with C.S. Gardner, C. Radin and especially J. Miekisz whom I thank for all they taught me. Part of it was made possible by a fellowship of the Royal Netherlands Academy of Arts and Sciences.

References

[1] D. Schechtman, I. Blech, D. Gratias and J.W. Cahn, Phys.Rev.Lett., 53(1984), 1951–1954.

[2] C. Radin, Int. Journal of Modern Physics B1(1987), 1157–1191.

[3] R.B. Israel, *Convexity in the Theory of Lattice Gases*, Princeton University Press, Princeton NJ, 1979.

[4] H.O. Georgii, *Gibbs measure and phase transitions*, De Gruyter, Berlin-New York, 1988.

[5] C.S. Gardner, J. Miekisz, C. Radin and A.C.D. van Enter, J.Phys., A22(1989), L 1019–1023.

[6] M. Keane, Zeit.Wahr., 10(1968), 335–353.

[7] A.C.D. van Enter and J. Miekisz, Comm. Math. Phys., 134(1990), 647–651.

[8] S. Aubry, J.Physique Coll. C3 Supp No. 3 50(1989) 97–106.

[9] C. Radin, *Disordered ground states of classical lattice systems*, University of Texas preprint.

Proceedings of the Asian Mathematical
Conference, 1990

Applications of Quasirandom Sequence in Statistics

FANG KAI-TAI
Institute of Applied Mathematics
Chinese Academy of Sciences
Hong Kong Baptist College

WANG YUANG
Institute of Mahtematics
Chinese Academy of Sciences
Beijing 100080, China

§1. Introduction

There are increasing new problems which are arisen everyday in statistics and its applications, and it requires more results in pure mathematics for the solutions of these problems. In the past 15 years *differential geometry* and *group theory* have been applied to statistics with fruiful results, for examples, [1], [2], [4], [5] and [6]. In this paper we will give a brief survey on the applications of number-theoretic methods to statistics.

First of all we will give the definition of quasirandom F-sequence (QRFS). Let $F(\mathbf{x})$ be a continuous multivariate distribution in R^s and n be a given integer. We are required to find n points, $\mathbf{x}_1, \cdots, \mathbf{x}_n$, in R^s such that they have a good representation for $F(\mathbf{x})$.

Definition 1. Let $\mathcal{P} = \{\mathbf{x}_k, k = 1, \cdots, n\}$ be a set of points of R^s and $F_n(\mathbf{x})$ be its empirical distribution. Then

$$D_F(n, \mathcal{P}) = \sup_{\mathbf{x} \in R^s} |F_n(\mathbf{x}) - F(\mathbf{x})|$$

is called the F-discrepancy of $\mathcal{P}$ with respect to $F(\mathbf{x})$.

Therefore, the F-discrepancy is a measure for representation of $\mathcal{P}$ with respect to $F(\mathbf{x})$ and is just the Kolmogorov-Smirnov distance.

Definition 2. Let $\mathcal{P}_n$ be a sequence of sets of points and K_n be the number of points of $\mathcal{P}_n$ satisfying $K_n \to \infty$ as $n \to \infty$. If $D_F(n, \mathcal{P}_n) \to 0$ as $n \to \infty$, $\mathcal{P}_n$ is called a quasirandom F-sequence (QRFS).

Let D be a bounded and closed domain of R^s and Denote by $F_D(\mathbf{x})$ the uniform distribution on D. The corresponding QRFS gives sets of points which are uniformly scattered on D and are extremely important in simulation, experimental design, optimization, etc.

Recently Shaw [19] gave a detailed survey on the applications of QRFS with respect to $F_D(\mathbf{x})$ where $D = C^s$ is a unit cube in Bayesian statistics. But he considered only the applications of QRFS to numerical integrations. In this paper, we shall point out that QRFS have wider applications in statistics. In particular it can give the solutions for many open problems where they were difficult for obtaining satisfactory solutions in a comparatively long time.

§2. Quasirandom Sequences of a Multivariate Distribution

There are a number of methods for generating QRFS with respect to $F_D(\mathbf{x})$ and $D = C^s$ (cf. [18]).

Wang and Fang [21] and [22] proposed an approach to generate QRFS with respect to $F_D(\mathbf{x})$ where D, for examples, can be taken the following domains

$$A_s = \{(x_1, \cdots, x_s) : 0 \le x_1 \le x_2 \le \cdots \le \mathbf{x}_s \le 1\},$$

$$B_s = \{(x_1, \cdots, x_s) : x_1^2 + \cdots + x_s^2 \le 1\},$$

$$U_s = \{(x_1, \cdots, x_s) : x_1^2 + \cdots + x_s^2 = 1\},$$

$$V_s = \{(x_1, \cdots, x_s) : x_i \ge 0, i = 1, \cdots, s, x_1 + \cdots + x_s \le 1\},$$

$$T_s = \{(x_1, \cdots, x_s) : x_i \ge 0, i = 1, \cdots, s, x_1 + \cdots + x_s = 1\}.$$

With their method, Fang, Yuan and Bentler [14] suggested a method to obtain QRFS for many useful classes of distributions, such as the elliptical distributions, the multivariate, Liouville distributions and the multivariate l_1-norm symmetric distributions.

§3. Evaluation of Probabilities and Moments of a Multivariate Distribution

For most practical problems the evaluation of probabilities and moments of a multivariate distribution can be transferred to the canonical form

$$I = \int_D f(\mathbf{x}) dv,$$

where D is a bounded domain and dv is the volume element of D. We may use the Monte Carlo method to find approximation of I, but its efficiency is low, and the rate of convergence is $O_p(1/\sqrt{n})$. When $D = C^s$ many authors suggested to use QRFS instead of random numbers, the corresponding rate of convergence can be reached to $O(n^{-1}(\log n)^{s-1})$. Wang and Fang [21] generalized the method for D being not necessary to be C^s and proposed the following three methods to get approximation of I: a) the characteristic function method; b) transformation method; c) direct method. These methods have been used by Zhang [24] for evaluating mixture moments of order statistics.

§4. Optimization and Its Applications in Statistics

Let D be a closed and bounded domain of R^s and f be a continuous and real function on D. We are required to find a point $\mathbf{x}^* \in D$ such that the f arrives its global maximum M over D at $\mathbf{x} = \mathbf{x}^*$. For most well-known methods, f is required to be unimodal/convex and differentiable in D. Fang and Wang [9] proposed a sequential number-theoretic method for optimization (SNTO) which requives only the continuity of f. This is a great advantage in many cases. With SNTO Fang and Yuan [12] gave a method to maximum likelihood estimation and likelihood ratio critieria. Some applications of SNTO to nonlinear regression models (Fang and Zhang [15]), system of nonlinear equations (Fang and Wang [9]), and vector quantizers (Fang, Yuan and Bentler [14]) were obtained also. A detailed review and discussion are given in Wang and Fang [23].

§5. Experimental Design

Consider an experimental design in which each factor has s levels. When s is large (for example $s > 11$), the usual designs require $O(s^2)$ experiments which are too much for most laboratories. How to reduce the number of experiments? As usual an experimental design should satisfy: uniformity, harmonic and reqularity. Wang and Fang [20], [7] suggested a new design based on uniformity only, the uniform design, which requires only $O(s)$ experiments. Since 1980 a lot of nice results were obtained by uniform design in many fields such as watch industry, textile industry, military sciences, computer software and agriculture experiments in China. The uniform design can also apply to experiments with mixtures (cf. Wang and Fang[21]).

§6. Geometric Probability, Simulation and Statistical Inference

Let D be a domain where a simulation is required to work on. In many cases it is difficult to do simulation when D is not a rectangle, for example, D is the unit sphere U_s. With QRFS with respect to $F_D(\mathbf{x})$ Fang and Wei [10] [11] gave the solutions for two open geometric probability problems which can not be solved for a long time.

Fang, Yuan and Bentler [13] applied QRFS to testing multinormality and robust estimation. Their work shows that a wider applications of quasirandom F-sequence to the projection pursuit will be possible.

References

[1] Amari, S. -I., *Differential-Geometrical Methods in Statistics*, Lecture Notes in statistics, 28, Spring-Verlag, New York (1985).

[2] Amari, S.-I. et al., *Differential Geometry in Statistical Inference*, IMS, Hayward, California (1987).

[3] Anderson, T. W., *An Introduction to Multivariate Statistical Analysis*, 2nd edition, Wiley, New York (1984).

[4] Eaton, M. L., *Group Invariance Applications in Statistics*, CBMS Regional Conference Series in Probability and Statistics, Vol. 1, IMS, Hayward, California (1989).

[5] Efron, B., *Defining the curvature of a statistical problem with applications to second-order efficiency (with discussion)*, Ann. Statist. 3(1975), 1189–1242.

[6] Diaconis, P., *Group Representations in Probability and Statistics*, IMS Lecture Notes-Monograph Series; 11(1989) Hayward, California.

[7] Fang, K. T., *Experimental design by uniform distribution*, Acta Math. Appl. Sinica, 3(1980), 363–372.

[8] Fang, K. T. and Wang, Y., *A sequential algorithm for optimization and its applications to regression analysis*, in Lecture Notes in Contemporary Mathematics, eds by Yang, L. and Wang, Y., Science Press, (1990a), 17–28.

[9] Fang, K. T. and Wang, Y., *A sequential algorithm for solving a system of nonlinear equations*, to appear in Math. Numerical Sinica (English Ser.) (1990b).

[10] Fang, K. T. and Wei, G., *The distributions of a class of the first hitting time*, to appear in Acta Math. Appl. Sinica, (1989a).

[11] Fang, K. T. and Wei, G., *The applications of number theoretic method to some geometric probability problems in statistics*, The Third Japan-China Symposium on Statistics, Contributed Papers, Tokyo, (1989b) 42–45.

[12] Fang, K. T. and Yuan, K. H., *A unified approach to maximum likelihood estimation*, Chinese J. Appl. Prob. and Statist., 6(1990), 412–418.

[13] Fang, K. T. Yuan, K. H. and Bentler, P. M., *Applications of sets of points uniformly distributed on a sphere to testing multinormality and robust estimation*, Technical Report 044, Institute of Appl. Math., Academic Sinica, (1989).

[14] Fang, K. T., Yuan, K. H. and Bentler, P. M., *Quantizer and representative points of elliptically contoured distributions*, invited talk in " The First Conference on Recent Developments in Statistical Research", Hong Kong, (1990).

[15] Fang, K. T. and Zhang, J. T., *A new algorithm for calculation of estimates of parameters of nonlinear regression modellings*, to appear in Acta Math. Appl. Sinica, (1990).

[16] Fang, K. T. and Zhen, H. N., *A maximum symmetric differences principle and its applications in uniform design*, to appear in Chinese J. Appl. Prob. Statist., 6(1990).

[17] Fang, K. T., Kotz, S., and Ng, K. W. (1990), *Symmetric Multivariate and Related Distributions*, Chapman and Hall, London.

[18] Hua, L. K. and Wang, Y. (1981), *Applications of Number Theory to Numerical Analysis*, Springer-Verlag and Science Press, Berlin and Beijing.

[19] Shaw, J. E. H., *A quasirandom approach to intergration in Bayesian statistics*, Ann.

Statist., 16(1988), 859–914.

[20] Wang, Y. and Fang, K. T., *A note on uniform distribution and experimental design*, Chinese Science Bulletin (English Ser), 26(1981), 485–489.

[21] Wang, Y. and Fang, K. T., *Number theoretic method in applied statistics*, Chinese Ann. Math., Ser. B, 11(1990a), 41–55.

[22] Wang, Y. and Fang, K. T., *Number theoretic method in applied statistics (II)*, Chinese Ann. Math. Ser. B, 11(1990b), 76–86.

[23] Wang, Y. and Fang, K. T., *A sequential number-theoretic method for optimization and its applications in statistics*, in Statistics in China: Achievement and Progress, Longman Academic Scientific and Technical, (1991) 139–156.

[24] Zhang, J. T., *Computation of moments of order statistics*, J. Graduate School, 7(1990), 1–29.

Proceedings of the Asian Mathematical
Conference, 1990

Observations on the Construction of a Foundation for the Handling of Uncertainty

DON FAUST
Department of Mathematics and Computer Science
Northern Michigan University
Marquette, Michigan, United States

§1. Motivation

With developments in computing which include larger knowledge storage capacities and faster inference engines, and the prospect that further enhancements in these areas will continue, the automation of reasoning about and control of more and more complex domains becomes possible. This phenomenon, of course, is precisely what is witnessed by the recent and projected future developments in the area of knowledge-based systems.

In many of these domains recently being considered for automation, some of the knowledge to be represented and processed has degrees of certainty other than 0 (false) and 1 (true). For example, in a process control environment it may well be the case that a fact may have a certainty based upon many sensors; faults intrinsic to the sensors themselves, or intrinsic to the environment, may affect the certainty level to which the fact is known to be the case; for instance, the momentary absence of a reading from one of the sensors upon which the certainty level of the fact is based may cause a decrease in the certainty level attached to the fact during that moment.

More generally, information is sometimes unclear, imprecise, uncertain; yet we must be able to process it. Conclusions to be drawn in such domains of uncertain knowledge, as well as in domains where knowledge is clear, or "crisp", will sometimes have a degree of uncertainty to them. This may be by necessity, because of uncertain knowledge. Or it may be desired, for example because crisp conclusions are not required for the triggering of appropriate action, or because a real time processing environment precludes the luxury of complete argumentation to a crisp conclusion.

§2. Logical Foundations for Handling Uncertainty

It would seem most productive to proceed by extending the foundation upon which the incisive and fruitful crisp logic and model theory, which has developed over the past sixty years, was based. This foundation was built upon Tarski's mathematical realization of the correspondence theory of truth [Tarski]. There it was clarified that the predicates 'true' and 'false' applied to syntactic objects, and that a sentence (a syntactic object) was true or false in a given model (a semantic object) according as what the sentence described was or was not the case in the model. To use Tarski's initial and informal example.

'it is snowing' is true if it is snowing. (0)

Here the quoted occurrence of " it is snowing" denotes the sentence itself, while the second unquoted occurrence denotes a state of affairs (using a naive semantics). This clarification which Tarski made is certainly simple enough. But we are deeply indebted to him, because he clearly delineated this notion of satisfaction, which formed a cornerstone for the very fruitful development of classical logic and model theory: of crisp languages, crisp models and a crisp satisfaction relation.

It is quite natural, then, that we return to this cornerstone of crisp logic in attempting to lay a foundation for a logic for handling uncertainty. Continuing informally, with Tarski's example, and choosing the set of real numbers between 0 and 1 inclusive, denoted $[0, 1]$, as our truth space: let us say that "s-snowing denotes "snowing with certainty s", and that "s-true" denotes "true with certainty s", for $s \in [0,1]$. The following possibilities immediately present themselves:

it is s-snowing' is true if it is s-snowing (1)

and

'it is snowing' is s-true if it is s-snowing. (2)

For completeness, let us list exhaustively the eight possibilities:

	Language	**Satisfaction**	**Models**	
(0')	crisp	crisp	crisp	←—classical logic
	crisp	crisp	fuzzy	←— *a
	crisp	fuzzy	crisp	←— *b
(2')	crisp	fuzzy	fuzzy	←— as in (2) above
	fuzzy	crisp	crisp	←— *c
(1')	fuzzy	crisp	fuzzy	←— as in (1) above
	fuzzy	fuzzy	crisp	←— *d
	fuzzy	fuzzy	fuzzy	←— *e

If crisp modeling of an application domain is possible, then certainly that should be our choice. In such case there is no need to introduce any fuzziness whatsoever, and we should have crisp language associated by crisp satisfaction to crisp models: that is, classical logic (0' above). From this point of view, then, *b, *c, and *d offer inferior solutions and thus are not of interest.

On the other hand, if the domain at hand demands fuzzy modeling, then there would seem to be exactly the two possibilities indicated above. (1') is the possibility of using fuzzy language, connected to the fuzzy models by way of a crisp satisfaction relation. (2'), on the other hand, is the possibility of using crisp language, connected to the fuzzy models by way of a fuzzy satisfaction relation. In this case, involving fuzzy modeling, *a and *e are inferior and not of interest here.

§3. A Prototype Implementation: Certainty Prolog

Classical Prolog implements only a proper subset of first-order logic, namely Horn clause logic. Some of the extensions others have studied and implemented include the addition of a useful equality predicate, the addition of a weak form of negation, and the recent extension to constraint logic programming. What we have done, guided especially by the needs of expert system development environments, is to implement an extension which seeks to provide machinery for the handling of uncertainty. Predicates may have a certainty attached to them (a real between 0 and 1), and Prolog rules of the following form are supported, where the c_i and $f(c_1, \cdots, c_n)$ are certainties:

$$p^{f(c_1,\cdots,c_n)}(\bar{x}) \longleftarrow P_1^{c_1}(\bar{x}_1), p_2^{c_2}(\bar{x}_2), \cdots, p_n^{c_n}(\bar{x}_n).$$

Here $f(c_1, \cdots, c_n)$ expresses the certainty of the rule head p as a function of the certainties c_i of the predicates in the rule body. Retracing certainty attachments, the inference engine can provide the certainty level at which a given goal has been proven, or alternatively the maximum certainty level at which a given goal can be proven. In a number of ways this implemented prototype, Certainty Prolog, is providing us with a testbed for further development of a general foundation for logics adequate for the representation and processing of uncertainty.

§4 Further Considerations: Questions; Prospectus

There are certainly important philosophical questions related to any framework departing from classical logic. For example, as regards the area of the "methodology of the exact sciences" Quine and others have argued that classical logic should suffice because one should analyze and precisify a domain, to the point where classical logic suffices for the formal codification and processing of the domain's knowledge. While acknowledging that this is an admirable ideal in the methodology of science, out position here is based on an examination of the types of domains which artificial intelligence workers are now able, through advances in technology, to begin automating (e.g. where domain knowledge is based on multiple sensors of time-varying reliability and domains themselves are often real-time embedded and involve resource-limited reasoning). Whether at some future time science may advance to the point where we come to know the nature of our universe intimately, and it is found that indeed classical logic codification is both appropriate and possible, is an interesting question to ponder. However, at present we are nowhere near that point. In fact, at present, we are trying to codify and automate domains which involve a lot of vagueness and uncertainty. Seen from this point of view, research on nonclassical logics and their applications to current technology should not be seen as making a statement about the yet-to-be-found essential character of scientific knowledge and the methodology of the exact sciences. Rather, such research is simply attempting to provide frameworks which are helpful in addressing current knowledge representation and knowledge processing problems. See also [Faust] in this regard.

There is of course much to be done in investigation of the truth spaces which may be most appropriate. It may be ultimately determined that we should want to have the truth space be a meta-variable of the logic, determined according to the nature of the application domain. Susan Haack's paper [Haack] gives a lucid analysis of degrees of fuzzification and their appropriateness. We have worked here with the truth space [0, 1] of real numbers, but that was done just to be able to set aside, for future consideration, the large and important problem of truth spaces. For example, using a hyperreal truth space would provide machinery for expressing and manipulating infinitesimal changes in certainty and may be very productive. At the foundational level we should stipulate only minimal conditions on the truth space (for example that it be a lattice or have some minimal topological structure [see e.g. Chang]): the goal is to have enough truth space structure to knit effectively the truth space with the rest of the foundational framework, and yet keep the framework general enough to be broadly applicable.

The difficult problem of "what to do about negation" is a good example showing that we must, in our departure from classical logic, be (1) bold enough so that there is some hope of providing machinery adequate to the tasks ahead, (2) conservative enough so that important properties such as consistency and completeness persist, and (3) delicate enough so that the machinery is broadly applicable. The saying 'absence of evidence is not evidence of absence' indicates the deep problems surrounding negation. Just as humans do routinely, our extended machinery should provide for the processing of a sentence φ by considering the evidence in faver of φ and the evidence against φ. Semantically, $\mathcal{O} \models \varphi$ should hold when the structure $\mathcal{O}$ confirms φ at level s and refutes φ (confirms $\neg\varphi$) at level t. Syntactically, $\Sigma \frac{s}{t} \varphi$ should hold when there are two proofs from the axioms Σ: one which proves φ at certainty level s, and one which refutes φ (proves $\neg\varphi$) at certainty level t.

Surely much remains to be done. However, if we build upon the firm foundation of classical logic which has been laid for handling that which is certain, we may be able to develop a similarly firm foundation for handling that which is uncertain.

References

[1] Chang, C.C. and H.J. Keisler, *Continuous Model Theory*, Princeton University Press, 1966.

[2] Faust, Don, *The Representation and Processing of Uncertain Information: Status Report and Prospectus for Information Technology Applications*, Proc. of Internat. Conference on Information Technology (ICIT 1990-Malaysia), E 275–E 295.

[3] Haack, Susan, *Do we need fuzzy logic?*, Int. J. of Man-Machine Studies, 11(1979), 437–445.

[4] Tarski, Alfred, *Logic, Semantics, Metamathematics: Papers from 1923 to 1938* (tran. Woodger), Oxford University Press, 1956.

Proceedings of the Asian Mathematical
Conference, 1990

A Resolution to the Inconclusive Portion of Kneser's Test of Oscillation via Asymptotics*

H. GINGOLD
Department of Mathematics
218 Eiesland Hall
West Virginia University
Morgantown, Wv 26506

§1. Introduction

The purpose of this article is to sketch a resolution to an inconclusive portion of a test of Kneser in oscillation theory and to enhance "complex oscillation theory". Consider a second order linear ordinary differential equation

$$y'' = \phi(x)y \tag{1.1}$$

on an interval $[a, b)$ where b is allowed to be unbounded. To better understand the nature of our result we adopt the following convention.

Convention 1.1. We say that the equation (1.1) is nonoscillatory at b^- if no nontrivial solution of (1.1) on $[a, b)$ possesses an infinite number of zeros on $[a, b)$ which have an accumulation point at b. A nontrivial solution of (1.1) is said to be oscillatory at b^- if it possesses an infinite number of zeros on $[a, b)$ with accumulation point at b^-.

A simplified version of an oscillation test which goes back to Kneser can be formulated as follows. Let $\phi(x)$ be a continuous real valued mapping on $[a, b)$. Assume that

$$\lim_{x\to\infty} (x \ln x)^2[\phi(x) + \frac{1}{4x^2}] = \gamma. \tag{1.2}$$

Then $\gamma > -\frac{1}{4}$ implies that (1.1) is nonoscillatory at $b = \infty, \gamma < -\frac{1}{4}$ implies that (1.1) is oscillatory at $b = \infty$. If $\gamma = -\frac{1}{4}$ the test is inconclusive. See [5] p. 462.

* This work was supported in part by grants from the National Aeronautics and Space Administration, NAG-1-741, and the Office of Naval Research, N001488K0319.

§2. Preliminaries

In this section we introduce a few conventions and mappings which will be needed in the sequel.

It is convenient to consider the companion system of (1.1) namely,

$$Y' = \begin{bmatrix} 0 & 1 \\ \phi(x) & 0 \end{bmatrix} Y, \qquad Y = \begin{bmatrix} y_1 & y_2 \\ y_1^1 & y_1^2 \end{bmatrix}, \tag{2.1}$$

Our asymptotic analysis will be applied directly to the system (2.1).

Capital letters like Y stand for two by two matrix functions.

Throughout this work, we employ a variety of expressions which involve the logarithms and roots of complex valued mappings as follows.

Convention 2.1. Given a mapping $f(x)$ such that $f(x) \not\equiv 0$ and $f(x)^{-1}$ belong to $C([a,b])$, the k-th root of $f(x)$ will be defined by

$$\sqrt[k]{f(x)} := \exp k^{-1} \ln f(x). \tag{2.2}$$

The mapping

$$\ln f(x) = \ln |f(x)| + i \arg f(x), \quad i = \sqrt{-1}, \tag{2.3}$$

is determined as follows. We choose a point $x_0 \in [a,b)$ and we select a convenient branch of $\ln f(x_0)$. Unless otherwise stated, this branch will be given by the principal value of the logarithm. We then define $\ln f(x)$ for all $x \in [a,b)$ by the analytic continuation of the branch chosen above. We then define $\arg f(x)$ by

$$\arg f(x) := \operatorname{Im} \ln f(x), \quad x \in [a,b). \tag{2.4}$$

Next we need

Assumption 2.2. The mapping $\phi(x)$ has the following properties.
(i) $\phi(x) \in C^3([a,b))$. (ii) $\arg \phi(x) \in C([a,b])$. iii) $\phi(x) \neq 0$ for $x \in [a,b)$.

Next, we turn to the determination of certain roots and quantities.

Notation 2.3. We define on $[a,b)$ the following mappings;

$$d_1 = \sqrt{\phi}, \quad c_1 = \frac{1}{4}\frac{\phi'}{\phi}, \quad l_1 = \frac{c_1}{d_1} \tag{2.5}$$

$$d_{k+1} = \sqrt{d_k^2 + c_k^2}, \quad c_{k+1} = \frac{i}{2}\left(\arctan \frac{c_k}{d_k}\right), \quad l_{k+1} = \frac{c_{k+1}}{d_{k+1}}, \quad k = 1, 2. \tag{2.6}$$

The definition of $\arctan z$ is given via the logarithmic function,

$$\arctan z - \arctan z_0 = \int_{z_0}^{z} \frac{ds}{1+s^2} = \frac{1}{2i} \ln \left[\frac{z-i}{z+i}\frac{z_0+i}{z_0-i}\right]. \tag{2.7}$$

§3. A Theorem of Asymptotic Decomposition

The subject of this section is

Theorem 3.1. *The linear differential system (2.1) possesses a fundamental matrix solution*

$$Y = W_1 W_2 U(I+Q)T \tag{3.1}$$

on $[a,b)$ under the conditions specified below.

$$W_1 = \begin{bmatrix} q^{-1} & q^{-1} \\ q & -q \end{bmatrix}, \quad q = \sqrt[4]{\phi(x)}, \tag{3.2}$$

$$W_2 = \begin{bmatrix} s+s^{-1} & -s+s^{-1} \\ i(s-s^{-1}) & -i(s+s^{-1}) \end{bmatrix}, \quad s = \begin{bmatrix} \dfrac{1-il_1(x)}{1+il_1(x)} \end{bmatrix}^{1/4}, \tag{3.3}$$

$$U = m^{-1}\begin{bmatrix} 1 & -g \\ \bar{g} & 1 \end{bmatrix}, \quad g = \sqrt{1+l_2^{-2}(x)} - l_2^{-1}(x), \quad m = (1+g\bar{g})^{1/2}. \tag{3.4}$$

The matrix $Q = (q_{jk})$, $j,k = 1,2$, is such that $(I + Q(X))$ together with its inverse is continuous on $[a,b]$. Moreover

$$Q(b^-) = 0. \tag{3.5}$$

The matrix T is a lower triangular matrix which is a fundamental matrix solution of the matrix initial value problem

$$T' = \begin{bmatrix} \lambda_1 & 0 \\ r_{21} & \lambda_2 \end{bmatrix} T, \quad T(b^-) = I. \tag{3.6}$$

The entities in (3.6) are given by

$$\lambda_1 = d_3 - (g\bar{g}' - mm')m^{-2}, \quad \lambda_2 = d_3 - (\bar{g}g' - mm')m^{-2}, \tag{3.7}$$

$$\tilde{b} = m^{-2}[-2d_2\bar{g} + (1-g^{-2})c_2], \quad r_{21} = \tilde{b} - m^{-2}\bar{g}. \tag{3.8}$$

The matrix T is explicitly given by

$$T = \begin{bmatrix} e_1 & 0 \\ t_{21} & e_2 \end{bmatrix}, \tag{3.9}$$

where

$$e_1 = \exp\int_b^x \lambda_1(t)dt, \quad e_2 = \exp\int_{b_2}^x \lambda_2(t)dt, \quad t_{21} = \delta(x,b,x)e_1, \tag{3.10}$$

$$\delta(x,t_1,t_0) = \int_{t_1}^x r_{21}(s)\exp\left(\int_{t_0}^s (\lambda_1(u) - \lambda_2(u)du)\right) ds, \quad r_{21} = \tilde{b} - m^{-2}\bar{g}'. \tag{3.11}$$

Formula (3.1) is valid under the following conditions

$$l_1^2(x) \not\equiv -1 \quad \text{but} \quad l_1^2(b^-) = -1, \tag{3.12}$$

$$\sup\left|\operatorname{Re} 2\int_t^x d_3(s)ds\right| \leq \ln\theta < \infty \tag{3.13}$$

for some positive number θ. The mapping $h(x,b)$ given by

$$h(x,b) = \left| \int_x^b m^{-2} |g'| [(1+\theta) + (2+\theta)\theta c(x,t) + \theta^2 c^2(x,t)] \right| dt \tag{3.14}$$

is continuous on $[a,b]$ where

$$c(x,t) = \theta \left| \int_t^x [\tilde{b} + m^{-2} \tilde{g}'] ds \right| . \tag{3.15}$$

Proof. The proof requires familiarity with [2]. There are two main steps in our proof. One consists of appropriate linear transformations. The second step consists of a final step of asymptotic decomposition.

It can be verified by a straightforward calculation that the transformations

$$Y = W_1 W_2 Y_2, \quad Y_2 = UV \tag{3.16}$$

take the differential system (2.1) into

$$Y_2' = \begin{bmatrix} d_2 & c_2 \\ c_2 & -d_2 \end{bmatrix} Y_2, \quad V' = \begin{bmatrix} \lambda_1 & r_{12} \\ r_{21} & \tilde{\lambda}_2 \end{bmatrix} V \tag{3.17}$$

with

$$r_{12} = m^{-2} g' . \tag{3.18}$$

(Elaboration on the construction of the matrix W_2 could be found also in [3] formula (4.43). If one normalizes there in W_2 by

$$\mu^{-1} = [1 + l_1^2(x_1)]^{1/4} [\sqrt{1 + i l_1(x_1)} + \sqrt{1 - i l_1(x_1)}] \tag{3.19}$$

we obtain the matrix given in (3.3).)

We turn to the question of nonoscillation in the next section.

§4. Nonoscillation

We intend to show that under the conditions assumed (1.1) is nonscillatory at b^-. This is formulated in the theorem below.

Theorem 4.1. *Let conditions of Theorem 3.1 hold. Assume that*

$$l_1^2(x) \not\equiv 1, \quad l_1^2(b^-) = -1, \quad l_2^2(b^-) = -1, \quad |t_{21}(b^-)| = \infty.$$

Then, equation (1.1) is nonoscillatory at b^-.

Proof. Define and denote by y_1, y_2 and y_1', y_2' two linearly independent solutions of (1.1) and their derivatives via the relation

$$Y = W_1 W_2 U (I + Q) T = \begin{bmatrix} y_1 & y_2 \\ y_1' & y_2' \end{bmatrix} . \tag{4.1}$$

A straightforward but lengthy calculation reveals that two linearly independent solutions of (1.1), namely $y_1(x)$ and $y_2(x)$ are given by

$$y_1 = (qm)^{-1}h_1 t_{21} k_2[1+\tilde{q}_{11}], \tag{4.2}$$

$$y_2 = (qm)^{-1}h_1 k_2 e_2[1+\tilde{q}_{12}], \tag{4.3}$$

where

$$\tilde{s}(x) = \left[\frac{1-il_2(x)}{1+il_2(x)}\right]^{1/4}, \quad h_1 = \tilde{s}+\tilde{s}^{-1}, \quad h_2 = -\tilde{s}+\tilde{s}^{-1}, \quad k_2(x) = (-g+h_2h_1^{-1}), \tag{4.4}$$

$$\lim_{x\to b^-} h_2(x)h_1^{-1}(x) = +1 \text{ or } -1, \quad g(b^-) = \pm i, \quad \tilde{q}_{11}(b^-) = \tilde{q}_{12}(b^-) = 0. \tag{4.5}$$

Therefore, neither $y_1(x)$ not $y_2(x)$ is oscillatory at b^-. A general solution of (1.1) is give by

$$y(x) = \alpha y_1(x) + \beta y_2(x), \tag{4.6}$$

with α, β complex value constants. Let $\alpha \neq 0$. Then

$$y(x) = \alpha y_1(x)\left[1+\frac{\beta}{\alpha}\Delta(x)\right], \quad (x) = \frac{e_2}{t_{21}(x)}\cdot\frac{1+\tilde{q}_{12}(x)}{1+\tilde{q}_{11}(x)}. \tag{4.7}$$

Since $|t_{21}(b^-)| = \infty$ we have $\Delta(b^-) = 0$. Therefore, no nontrivial solution of (1.1) is oscillatory at b^-. A straightforward calculation now reveals that for families of differential equations (1.1) with $\phi(x)$ complex valued and $\gamma = -\frac{1}{4}$ in (1.2), Theorem 4.1 guarantees nonoscillation.

References

[1] H. Gingold, *On the location of zeros of oscillatory solutions*, Trans. Amer. Math. Soc., 279(1983), 471–496.

[2] H. Gingold, *On a family of nonoscillatory equations* $y'' = \phi(x,\varepsilon)y$, J. Math. Anal. Applic., 135(1988), 309–325.

[3] H. Gingold, *An invariant asymptotic formula for solutions of second-order linear ODE's.* J. Asym. Anal., 1, (1988), 317–350.

[4] P. Hartman, *Ordinary differential equations*, Wiley, New York, 1964.

[5] E. Hille, *Lectures on differential equations*, Addison-Wesley, Reading Mass., 1969.

[6] K. Kreith, *Oscillation theory*, Lecture Notes in Math., Vol. 324, Springer-Verlag, Berlin, Heidelberg and New York, 1973.

[7] C.A. Swanson, *Comparison and oscillation theory of linear differential equations*, Academic Press, New York, 1968.

Proceedings of the Asian Mathematical
Conference, 1990

Globally Analytic Triangularization of a Matrix Function – an Algorithm

HARRY GINGOLD†
Department of Mathematics
West Virginia University
Morgantown, WV 26506 USA

PO-FANG HSIEH
Department of Mathematics and Statistics
Western Michigan University
Kalamazoo, MI 49008 USA.

§1. Introduction

Similarity transformations of a matrix play very important roles in various disciplines of mathematical sciences. There is a well-developed theory for canonical transformations of matrices with constant entries. The same cannot be said if the entries of the matrix are functions of one or several variables. Given an n by n matrix function A with entries belonging to $C^k[a,b]$, where $[a,b]$ is a real interval, the following problem is of prime importance: Determine a nonsingular matrix T such that

$$T^{-1}AT = B, \quad \text{or} \quad AT - TB = 0, \tag{1.1}$$

where the entries of B and T also belong to $C^k[a,b]$. Moreover, we require that B be "as simple as possible". If A is analytic in a domain D, we require B and T to be analytic. This problem is related to a "simplification" problem in the theory of ordinary differential equations (e.g., see [GH1] and [GH2]). The main result of this paper is the following theorem and a constructive algorithm associated with its proof.

Theorem 1.1. *Let $A(t)$ be an n by n matrix function with analytic entries on $[a,b]$, where $-\infty \le a < b \le \infty$. Assume that every eigenvalue of $A(t)$ is real on $[a,b]$. Then, there exists a unitary matrix $U(t)$ analytic on $[a,b]$ such that*

$$Q(t) = U^{-1}(t)A(t)U(t), \tag{1.2}$$

where $Q(t)$ is an upper-triangular matrix whose entries are analytic functions of t on $[a,b]$.

An algorithm is to be provided to construct the matrix $U(t)$ in the spirit of Schur's theory depending on the analyticity of all the eigenvalues on $[a,b]$. The labeling of the eigenvalues is important to assure the analyticity of these eigenvalues. For instance, for

$$F(t) = \begin{bmatrix} 0 & t \\ t & 0 \end{bmatrix}, \qquad t \in (-\infty,\infty), \tag{1.3}$$

† The work of this author is partially supported by grants from NASA, NAG-1-741 and the Office of Naval Research, N0001488K0319.

we take $\pm t$ as eigenvalues, rather than $\pm|t|$.

Theorem 1.1 is valid also when the interval $[a,b]$ is replaced by (a,b), ($[a,b)$ or $(a,b]$). Namely, we have

Corollary 1.1. *Let $A(t)$ be an n by n matrix function with analytic entries on (a,b), ($[a,b)$ of $(a,b]$), where $-\infty \leq a < b \leq \infty$. Assume that every eigenvalue of $A(t)$ is real on (a,b), ($[a,b)$ or $(a,b]$). Then, there exists a unitary matrix $U(t)$ analytic in (a,b), ($[a,b)$ or $(a,b]$), such that*

$$Q(t) = U^{-1}(t)A(t)U(t), \tag{1.4}$$

where $Q(t)$ is an upper-triangular matrix whose entries are analytic functions of t on (a,b), ($[a,b)$ or $(a,b]$).

In fact, by applying Theorem 1.1 to a closed subinterval in (a,b), and let the closed subinterval approach (a,b), we have Corollary 1.1. In the same manner, we can have the same result for a half-open interval.

Such analytic reduction of a matrix function attracted the attention of many experts. In a series of papers related to the spectral theory, Rellich [R1, R2] proved a special case of this theorem by an entirely different method applying a perturbation of the result for a constant matrix. Leavitt [L] proved that if all the entries and the eigenvalues of $A(x)$ are analytic in a compact domain in the complex x-plane, then there exists a matrix $T(x)$ analytic in that domain such that $B(x)$ is upper-triangular. Similar problems are investigated by many experts such as in [S], [HS], [Br], [Gi1] and [Gi2]. Wasow [W] investigated the conditions that, when $A(x)$ and $B(x)$ are analytic functions in a domain, the pointwise existence of $T(x)$ assures the existence of an analytic matrix $T(x)$ in a subdomain. Wasow's results were studied and extended by many experts, for example, [OS], [M], [F] and [Gu].

Even though the global existence of an analytic and unitary matrix $U(t)$ for Rellich' s diagonalization [R1, R2] for a Hermitian matrix $U(t)$ was claimed in the literature(e.g., [Ba]), its proof has not been seen anywhere. Because of many important results, such as those pertaining to quantum mechanics (cf. [Gi3], [GH1] and [GH2]) and conservation laws of a hyperbolic system (cf. [GT1] and [GT2]), hinge on the global existence of such a matrix $U(t)$, thus we find it compelling to give a detailed proof of a global version of Rellich's Theorem. Since the diagonalization could be obtained as a by-product of triangularization, in this paper we will focus on a constructive method of triangularization, which will be useful as a numerical algorithm as well. Therefore in this paper, we study the global triangularization of a matrix $A(t)$, which is analytic on a finite or infinite real interval and all of its eigenvalues are real valued in this real interval, by a unitary matrix analytic on the same interval. This is an extension of Schur's decomposition (see [Sc] or, e.g, [Be], [St]) of a constant matrix to one with analytic entries. If $A(t)$ is Hermitian, then this result actually gives the diagonalization of $A(t)$.

The main result of this paper is strong in the sense that if a matrix $A(t)$ has $C^\infty[0,1]$ entries, there is no guarantee that it has even continuous eigenvectors. For instance, see [Gi2], [K p.111] [R1] or [R2].

Moreover, the triangular form of a matrix with analytic entries is the best simplification we can hope for, since it is nor always possible to find an analytic nonsingular matrix to reduce it to a Jordan canonical form. For instance, let

$$A(t) = \begin{bmatrix} 0 & 0 \\ t & 0 \end{bmatrix}. \tag{1.5}$$

If it is reducible to a Jordan canonical form by $T(t)$, then

$$\begin{aligned} T^{-1}(t)A(t)T(t) &= \begin{bmatrix} 0 & 0 \\ 1 & 0 \end{bmatrix}, \quad (t \neq 0), \\ I^{-1}A(0)I &= \begin{bmatrix} 0 & 0 \\ 0 & 0 \end{bmatrix}, \quad (t = 0). \end{aligned} \tag{1.6}$$

Thus $T(t)$ must be singular at $t = 0$.

It is also noteworthy that the results in this paper cannot be extended to a matrix of several variables. For instance (see [K p.116] or [RSz p. 379]) , let

$$A(t_1, t_2) = \begin{bmatrix} t_1 & t_2 \\ t_2 & -t_1 \end{bmatrix}. \tag{1.7}$$

Then the eigenvalues of $A(t_1, t_2)$ are

$$\lambda_\pm(t_1, t_2) = \pm\sqrt{t_1^2 + t_2^2}, \tag{1.8}$$

which are real when both t_1 and t_2 are real, analytic at (t_1, t_2) where $t_1^2 + t_2^2 \neq 0$ but not analytic where $t_1^2 + t_2^2 = 0$.

Since the proof of Theorem 1.1 is elaborated in several steps and each step is significant in its own right, each of these steps is stated as a proposition in §2. The details of these proofs are given in [GH3] and [GH4].

The process of proving Theorem 1.1 itself provides an algorithm for computing the desired $U(t)$.

If $A(t)$ is a Hermitian matrix, we have the following theorem

Theorem 1.2. *In addition to the assumptions of Theorem 1.1, if $A(t)$ is assumed to be Hermitian on $[a, b]$, then there exists a unitary matrix $U(t)$ analytic on $[a, b]$ such that*

$$Q(t) = U^{-1}(t)A(t)U(t) \tag{1.9}$$

is a diagonal matrix whose elements are analytic on $[a, b]$.

This is the global version of Rellich's Theorem [R1, R2]. In fact, since $A(t)$ is Hermitian, it is easy to see that $Q(t)$ given by (1.2) satisfies the relation $Q^*(t) = Q(t)$ in addition to being upper-triangular. Thus $Q(t)$ is diagonal.

Furthermore if $A(t)$ is a periodic matrix function with all the eigenvalues periodic and having the same period, then $U(t)$ in (1.2) and (1.9) can be obtained to be periodic with the same period. The conditions that all the eigenvalues of a periodic analytic matrix to be periodic, along with the method of determining their common period, will be investigated in a forthcoming paper.

§2. Steps of Proof

The global analyticity of $U(t)$ depends on the global analyticity of all eigenvalues of $A(t)$. The local analyticity of each of the eigenvalues to be derived from the assumption that they are all real valued for $t \in [a,b]$ depends on the absolute convergence of the Puiseux expansion of each eigenvalue as a root of the charactersitic equation. Thus, it is necessary for us to extend the discussion of the concern of this paper to a matrix function $\tilde{A}(x)$ of a complex variable x which is analytic in a domain containing $[a,b]$ and satisfies $\tilde{A}(t) = A(t)$ for $t \in [a,b]$ (this is possible by the analytic function theory). We will have to establish first.

Proposition 2.1. *Let $\tilde{A}(x)$ be an n by n matrix analytic in a simply connected domain $\mathcal{D}$ containing $[a,b]$. Assume that every eigenvalue of $\tilde{A}(x)$ is real for $x \in [a,b]$. Then there exists a simply connected subdomain $\mathcal{D}_1$ of $\mathcal{D}$ containing $[a,b]$ such that all the eigenvalues of $\tilde{A}(x)$ are analytic in $\mathcal{D}_1$.*

Since an Hermitian matrix has only real eigenvalues, we have

Corollary 2.1. *Let $\tilde{A}(x)$ be an n by n matrix analytic in a simply connected domain $\mathcal{D}$ containing $[a,b]$. If $\tilde{A}(x)$ is Hermitian for real $x \in [a,b]$, then there exists a simply connected subdomain $\mathcal{D}_1$ of $\mathcal{D}$ containing $[a,b]$ such that all the eigenvalues of $\tilde{A}(x)$ are analytic in $\mathcal{D}_1$.*

In order to construct the analytic unitary matrix $U(t)$ for Theorem 1.1, we have to construct an analytic eigenvector with unit length on $[a.b]$ for each of the analytic eigenvalues. Using Proposition 2.1, we can establish:

Proposition 2.2. *Let $\tilde{A}(x)$ be an n by n matrix analytic in a simply connected domain $\mathcal{D}$ containing $[a,b]$. Let $\tilde{\lambda}(x)$ be an eigenvalue of $\tilde{A}(x)$ analytic in $\mathcal{D}$. Then, there exists a simply connected subdomain $\mathcal{D}_2$ of $\mathcal{D}$ containing $[a,b]$ and a corresponding eigenvector $\tilde{u}(x)$ analytic in $\mathcal{D}_2$ with unit length on $[a,b]$.*

A constructive algorithm for proving Proposition 2.2 is developed in [GH3, GH4]. The proof of Proposition 2.2 itself is more involved and needs a reduction process of a singular matrix function with analytic entries to a Smith form (see [Ga] or [GLR]) by Matrices of meromorphic entries.

By restricting the result of Proposition 2.2 to $[a,b]$, we can establish:

Proposition 2.3. *Let $A(t)$ be an n by n matrix analytic on $[a,b]$ and $\lambda(t)$ be an eigenvalue of $A(t)$ analytic on $[a,b]$. Then there exists a unitary matrix $U(t)$ analytic on $[a,b]$ such that*

$$U^{-1}(t)A(t)U(t) = \begin{bmatrix} \lambda(t) & a_2(t) & \cdots & a_n(t) \\ 0 & & & \\ \vdots & & A_1(t) & \\ 0 & & & \end{bmatrix}, \tag{2.1}$$

where $A_1(t)$ is an $(n-1)$ by $(n-1)$ matrix whose entries and the function $a_2(t), \cdots, a_n(t)$ are analytic on $[a,b]$.

By the principle of mathematical induction and using Proposition 2.3, Theorem 1.1 can be proved.

Remark 1. It is noteworthy that $F(t)$ in (1.3) is Hermitian when t is real but is not when t is complex. Thus, it is crucial to assume in Corollary 2.1 that $\tilde{A}(x)$ is Hermitian when $x \in [a,b]$.

Remark 2. Propositions 2.1 is not true for a matrix with entries in several variables. For example, the eigenvalues of $A(x_1, x_2)$ given by (1.7) are real when both x_1 and x_2 are real but they are not analytic at (0, 0).

§3. Algorithm

For a matrix $A(t)$ satisfying Theorem 1.1, by the process established in its proof, we can have the following algorithm it find $U(t)$ satisfying (1.2).

I. Compute an analytic eigenvalue $\lambda_1(t)$.

II. Compute two matrices $L_1(t)$ and $R_1(t)$ such that

$$L_1(t)[A(t) - \lambda_1(t)I_n]R_1(t) = \mathrm{diag}\{d_1(t), d_2(t), \cdots, d_m(t), 0, \cdots, 0\} \quad (1 \leq m < n).$$

III. Let $u_j(t) = R_1(t)e_j, (j = 1, 2, \cdots, n)$, to get a set of linearly independent vectors $\{u_1(t), u_2(t), \cdots, u_n(t)\}$.

IV. Apply Gram-Schmidt process to $\{u_1(t), u_2(t), \cdots, u_n(t)\}$ and obtain a set of orthogonal vectors $\{v_1(t), v_2(t), \cdots, v_n(t)\}$.

V. Normalize this set by $\widehat{v}_j(t) = v_j(t)/\|v_j(t)\|, (j = 1, 2, \cdots, n)$ to obtain an orthonormal set of vectors $\{\widehat{v}_1(t), \widehat{v}_2(t), \cdots, \widehat{v}_n(t)\}$, analytic on $[a, b]$. In particular, $\widehat{v}_1(t)$ is an eigenvector of $A(t)$ corresponding to $\lambda_1(t)$.

VI. Let $U_1(t) = [\widehat{v}_1(t), \widehat{v}_2(t), \cdots, \widehat{v}_n(t)]$. Then $U_1(t)$ reduces $A(t)$ to a matrix in the form of (2.1).

VII. If $\dim A_1(t) \geq 2$ go to Step I for $A_1(t)$. Continue this process to get $U_2(t), \cdots, U_{n-1}(t)$.

VIII. Put $U(t) = U_{n-1}(t)U_{n-2}(t)\cdots U_1(t)$, then $U(t)$ satisfies (1.2).

References

[Ba] H. Baumgärtel, *Analytic Perturbation Theory for Matrices and Operators*, Birkhäuser Verlag, Basel, 1985.

[Be] R. Bellman, *Introduction to Matrix Analysis*, 2nd ed., McGraw Hill, New York, 1970.

[Br] B.L.J. Braaksma, *Global reduction of linear differential systems involving a small singular parameter*, SIAM J. Math. Anal., 2(1971), 149–165.

[F] S. Friedland, *On pointwise analytic similarity of matrices*, Israel J. Math., 35(1980), 89–108.

[Ga] F.R. Gantmacher, *The Theory of Matrices*, Vol. 1, Chelsea Publ. Co, New York, 1959.

[Gi1] H. Gingold, *On the existence of a global simplifying transforming matrices*, SIAM J. Math. Anal., 9(1978), 1076–1082.

[Gi2] H. Gingold, *On continuous triangularization of matrix functions*, SIAM J. Math. Anal., 10(1979), 709–720.

[Gi3] H. Gingold, *In general, the less degeneracy the less transition. A principle for time dependent Hamiltonian systems in quantum mechanics.* J. Math. Phys., 28(1987), 2400–2406.

[GH1] H. Gingold and P.F. Hsieh, *Global approximations of perturbed Hamiltonian differential equations with several turning points*, SIAM J. Math. Anal., 18(1987), 1275–1293.

[GH2] H. Gingold and P.F. Hsieh, *Analytic solution of a Hamiltonian system in intervals with several turning points*, SIAM J. Math. Anal., 19(1988), 1142–1150.

[GH3] H. Gingold and P.F. Hsieh, *An algorithm for Globally Analytic Triangularization of a Matrix Function*, Mathe. & Stat. Report, No. 81, Western Michigan Univ., 1989, 35 Pp.

[GH4] H. Gingold and P.F. Hsieh, , *Globally Analytic Triangularization of a Matrix Function*, to appear.

[GT1] H. Gingold and V. Trutzer, *A uniform representation for parametrized degenerate hyperbolic system*, preprint.

[GT2] H. Gingold and V. Trutzer, *On long time behavior of time dependent degenerate hyperbolic system*, preprint.

[GLR] I. Gohberg, P. Lancaster and L. Rodman, *Matrix Polynomials*, Academic Press, New York, 1982.

[Gu] R.M. Guralnick, *Similarity of matrices over local rings*, Lin. Alg. and Appl., 41(1981), 161–174.

[HS] P.F. Hsieh and Y. Sibuya, *A global analysis of matrices of functions of several variables*, J. Math. Anal., 14(1966), 332–340.

[K] T. Kato, *Perturbation Theory for Linear Operators*, Springer-Verlag, New York, 1965.

[L] W.G. Leavitt, *A normal form for matrices whose elements are holomorphic functions*, Duke Math. J., 15(1948), 463–472.

[M] B. McDonald, *Similarity of matrices over artinian principal ideal rings*, Lin. Alg. and Appl., 21(1978), 153–162.

[OS] J. Ohm and H. Schneider, *Matrices similar on a Zariski-open set*, Math. Z., 85(1964), 373–381.

[R1] F. Rellich, *Störungstheorie der Spektralzerlegung*, I. Math. Ann., 113(1936), 600–619; II, ibid., 113(1937), 677–685; III, ibid., 116(1939), 555–570; IV, ibid., 117(1940), 356–382; V. ibid., 118(1942), 462–484.

[R2] F. Rellich, *Perturbation Theorey of Eigenvalue Problems*, Gordon and Breach, New York, 1969.

[RSz] F. Riesz and B. Sz-Nagy, *Functional Analysis*, Frederick Ungar Publ. Co., New York, 1955.

[Sc] I. Schur, *Über die characteristichen Wurzeln einer linearen Substitution mit einer Anwendung auf die Theorie der Integral Gleichungen*, Math. Ann., 66(1909), 488–510.

[Si] Y. Sibuya, *Some global properties of matrices of functions of one variable*, Math. Ann., 161(1965), 67–77.

[St] G. Strang, *Linear Algebra and Its Applications*, 3rd ed., Harcourt Brace Jovanovich, San Diego, 1988.

[W] W. Wasow, *On holomorphically similar matrices*, J. Math. Anal. Appl., 4(1962), 202–206.

Proceedings of the Asian Mathematical
Conference, 1990

Sheaf Representations of Rings and Their Rings of Quotients

Normal g.p.p. rings with Köthe radical

R. GONCHIGDORZH
Institute of Mathematics
Mongolian Academy of Sciences
Ulan-Bator 51, P.O.Box 112
Mongolian People's Republic

The recent papers [8], [10] and [6] have been devoted to the study of noncommutative and commutative generalized p.p. rings with identity. In the noncommutative case the existence of classical right rings of quotients was assumed. In the present paper we continue these investigations for noncommutative normal generalized p.p. rings not necessarily having an identity element. In this case we can not assume existence of the classical ring of quotients, because it may happen that such a ring has no cancellative element. So, it is a natural question to find a construction of ring-extensions for generalized p.p. normal rings not necessarily having identity which can be considered as a substitution of the construction of the classical rings of quotients. We shall present a construction of a ring of right π-regular quotients of a normal generalized p.p. ring. This is a generalization of the rings of right regular quotients of reduced rings introduced by the author in [4]. The latter one was used in [5] for characterizing semihereditary and hereditary reduced rings without assuming the existence of an identity element.

The first section is concerned with preliminaries and there we recall some definitions and results needed later.

In Section 2 we shall define rings of right π-regular-quotients of normal generalized p.p. rings and we shall give some sufficient and necessary conditions for the existence of rings of right π-regular-quotients.

In the third section we shall consider normal generalized p.p. rings with Köthe radical and there a characterization will be obtained for a normal generalized p.p. ring having a ring of right π-regular-quotients with Pierce stalks which are local rings.

§1. Preliminaries

Throughout this paper all rings considered are associative and without assuming the existence of identity elements, unless otherwise indicated. Let R be a ring and S be a subset of R. Then $r_R(S)$ and $l_R(S)$ denote the right and left annihilator of S in R, respectively, and $\text{ann}_R(S)$ will stand for $r_R(S)$ when $r_R(S) = l_R(S)$. $E(R)$ and $B(R)$ denote the set of all idempotents and the set of all central idempotents of the ring R, respectively. A ring R is said to be a normal ring if $E(R) = B(R)$ i.e. if every idempotent of R is central.

Definition 1.1. A ring R is said to be a *right generalized p.p. ring* if $E(R)R = R$ and for any $a \in R$ there exist a positive integer n and an idempotent $e \in E(R)$ with $r_R(a^n) = r_R(e)$. A *left generalized p.p. ring* is defined analogously. A ring R is a *generalized p.p. ring* (abbreviated g.p.p. ring) if it is both right and left generalized p.p. ring.

It is clear that if the ring R has an identity this definition coincides with the usual one given in [8], [10]. An extension of the concept of g.p.p. ring to rings without identity was made in [7]. There it was not assumed that $E(R)R = R$ and of course, in such a ring some nil direct summand may occur. Nil rings, however, can be considered as a trivial case of g.p.p. rings. Therefore, in our definition of a g.p.p. ring R we have supposed the condition $E(R)R = R$ for excluding such trivial cases.

Lemma 1.2 (cf [8, Corollary 4]). *Let R be a normal ring. Then R is g.p.p. ring if and only if $R \cdot E(R) = R$ and for any $a \in R$ there exist a positive integer n and an idempotent $e \in E(R)$ such that for every $K \geq n$ $\text{ann}_R(a^k) = \text{ann}_R(e)$ and $a^K e = a^K$.*

Proof. The necessity is obvious.

Let R be a g.p.p. ring and a be an arbitrary element of R. Then there are integers $m, n > 0$ and idempotents $e, f \in E(R)$ such that $r_R(a^n) = r_R(e), l_R(a^m) = r_R(f)$. Then for all $K \geq \max\{n, m\}$ we have $r_R(a^k) = ann_R(e)$ and $l_R(a^k) = ann_R(f)$ (cf. [8, Lemma 3]). For, let $a^{n+1}b = 0, b \in R$. Then $0 = eab = aeb$ and hence $a^n eb = 0, r_R(a^{n+1}) = r_R(a^n)$. In a similar way we have $l_R(a^{m+1}) = l_R(a^m)$.

Now, let $r_R(a^k) = ann_R(e), l_R(a^k) = ann_R(f)$. Then $0 = a^K(f - ef) = (f - ef)a^K$, and hence $f(f - ef) = 0, f = ef$. Similarly we have $e = ef$ and therefore $e = f$ and $r_R(a^k) = ann_R(e) = l_R(a^k)$ i.e. $ann_R(a^K) = ann_R(e)$. Because $E(R) \cdot R = R$, we have an idempotent $e' \in E(R)$ such that $a^K - ea^K = e'(a^K - ea^K) = (e' - e'e)a^K$. Since $e(e' - ee') = 0$, we have $0 = (e' - ee')a^K = a^K - ea^K$ and so, $a^K = ea^K$.

Definition 1.3. An element $r \in R$ is called *reduced* if there is a central idempotent $e \in B(R)$ such that $ann_R(r) = ann_R(e)$ and $r = er$. In this case it is easy to verify that the central idempotent e is unique and we call it associated idempotent of the reduced element r.

Corollary 1.4 (cf. [9, Theorem 1]). *Let R be a normal ring. Then the following are equivalent:*

1) R is g.p.p. ring,

2) a) $E(R) \cdot R = R$,

b) every $a \in R$ can be written in the form $a = r + n$ where r is a reduced element and n is a nilpotent element, and the sum is orthogonal, i.e. $rRn = nRr = (0)$.

Proof. 1)$\Longrightarrow$ 2): Let R be a g.p.p. ring. Then by Lemma 1.2 for some integer $K > 0$ a^K is reduced element with an associated idempotent $e \in B(R)$. Putting $r = ea$ and $n = a - ea$ we have $rRn = nRr = 0$ and $(ea)^K = ea^K = a^K = (ea + a - ea)^K = (ea)^K + n^k$. Hence n is nilpotent and, of course, r is reduced.

The implication 2)$\Longrightarrow$ 1) is obvious, because if r is a reduced element, then for every integer $K > 0$ r^K is reduced.

Now, let R be a ring and $B(R)$ be the set of all central idempotents of R. Then $B(R)$ forms a Boolean ring under the following operations

$$e \oplus f = e + f - 2ef$$

$$e \cdot f = ef.$$

The Boolean spectrum spec $B(R)$ of the ring R consisting of all prime ideals of $B(R)$ with the Zarisky topology is a totally disconnected, local compact Hausdorff space (i.e. Stone space) and we denote it by $X(R)$. An important property of the space $X(R)$, the *partition property* of $X(R)$ can be formulated as follows.

Lemma 1.5. *Let U be an open compact subset of $X(R)$ and $\{U_\alpha, \alpha \in I\}$ be an open cover of U. Then there exists a finite set $\{V_i, i = 1, 2, \cdots, n\}$ of open compact subsets in $X(R)$ such that*

i) *for every $i \le n$ there is an index $\alpha \in I$ with $V_i \subseteq U_\alpha$,*

ii) $V_i \cap V_j = \emptyset$ *if $i \neq j$,*

iii) $U = \bigcup_{i=1}^n V_i$.

We define a presheaf on $X(R)$ as follows. Let U be an open subset in $X(R)$. Then we write $P(U) = R/\cap_{x \in U} xR$ (we note that x is a prime ideal in $B(R)$ and xR is an ideal in R) and we define

$$\rho_V^U : R/\bigcap_{x \in U} xR \to R/\bigcap_{y \in V} yR, \quad \text{where } V \subseteq U,$$

to be the canonical projection. It is easy to get that, in fact, this presheaf is a sheaf and a calculation gives that the stalks $P_x = R/xR$ for all $x \in X(R)$. We shall write R_x instead of P_x. This sheaf P is called a *Pierce sheaf* of the ring R and the stalks are called the *Pierce stalks* of R. An adaptation of [11, Proposition 1.1] gives

Lemma 1.6. *Let R be a ring with $B(R) \cdot R = R$. Then every Pierce stalks R_x of R is a ring with identity. Moreover, each Pierce stalk R_x is indecomposable (i.e. without non-trivial central idempotents) if and only if R is a normal ring.*

In the rest, we suppose that R is a normal ring with $B(R) \cdot R = R$. Let $S(R) = \cup_{x \in X(R)} R_x$ be the display space and $\hat{r} : X(R) \to S(R)$ be a map defined by an element

$r \in R$ by $\hat{r}(x) = r_x$ for all $x \in X(R)$. Then $\hat{r}$ is a continuous global section of the Pierce sheat of R and the Pierce sheaf $P(R) = (S(R), X(R))$ is a *sheaf of rings* $R_x, x \in X(R)$ *with identity* in the sense of [3, Definition 2.2]. Moreover, the set of all global sections $\Gamma = \Gamma(X(R), S(R))$ forms a ring under naturally defined ring operations and a map $i : R \to \Gamma$ with $i(r) = \hat{r}$ is a monomorphism of rings. Hence we can identity the ring R with the subring $i(R) = \widehat{R} = \{\hat{r} | r \in R\}$. It is well known that $\widehat{R} = \Gamma$ if and only if R has an identity element. Let σ be a global (continuous) section in Γ. Then supp σ denotes the *support* of σ, i.e. supp $\sigma = \{x \in X(R) | \sigma(x) \neq 0_x\}$. We shall end this section with a useful remark: a subset $U \subset X(R)$ is open and compact if and only if U =supp$\hat{e}$ for an idempotent $e \in B(R)$.

More details one can find in [1], [3] or in [11], [12].

§2. The Ring of π-Regular-quotients of Normal g.p.p. Rings

In order to define the ring of π-regular quotients of normal g.p.p. rings, at first we shall establish some facts about Pierce stalks of normal g.p.p. rings, about reduced elements and about some extensions of normal g.p.p. rings.

Lemma 2.1. *Let R be a normal g.p.p. ring. Then for each $x \in X(R)$ every zero-divisor of the pierce stalk R_x is nilpotent.*

Proof. Let $r_x \in R_x$. By Lemma 1.2 for a positive integer m and an idempotent $e \in B(R)$ we have $r^m e = r^m, ann_R(r^m) = ann_R(e)$. So if $e_x = 0_x$, then $r_x^m = r_x^m e_x = 0_x$ and r_x is nilpotent. Let $e_x \neq 0_x$ and $r_x \cdot s_x = 0_x$ for an element $s_x \in R_x$. Then $rsf = 0$ for some idempotent $f \in B(R)$ with $f_x \neq 0_x$. Moreover, also $ef = f$ holds. Hence $0 = (sf)e = sf$ and, in particular, $0_x = s_x f_x = s_x$. Therefore, r_x is right non-zero-divisor in R_x. Similarly we can show that r_x is a left non-zero-divisor in R_x.

Corollary 2.2. *Let R be a normal g.p.p. ring. Then an element $r \in R$ is reduced if and only if for every $x \in$ supp rr_x is a non-zero-divisor in R_x and in this case supp r is open compact in $X(R)$. Moreover, $ann_R(r) = ann_R(e$ if and only if supp r =supp e.*

Proof. Obvious.

Definition 2.3. Let R and $\widetilde{R}$ be rings with $R \subset \widetilde{R}$. Then $\widetilde{R}$ is called a unital extension of R, if for every element $\tilde{r} \in \widetilde{R}$ we have $E(R)\tilde{r} \neq 0, \tilde{r}E(R) \neq 0$ whenever $\tilde{r} \neq 0$.

Lemma 2.4. *Let R be a normal g.p.p. ring with a unital extension $\widetilde{R}$. Then for every element r or R there exists at most one element $r^* \in \widetilde{R}$ such that*

$$r^* r^{m+1} = r^m, \quad ann_{\widetilde{R}}(r^*) = ann_{\widetilde{R}}(r^m) = ann_{\widetilde{R}}(e) \tag{$*$}$$

for some integer $m > 0$ and idempotent $e \in B(R)$.

If there exists such an element $r^* \in R$ then we call it the *π-regular-inverse of the element* r and throughout this paper the notation r^* will be used only in this sense.

Proof. Let $r^*, r^! \in R$ satisfy condition $(*)$ for some element $r \in R$. In view of Lemma 1.2 we may suppose that the integer and the idempotent appearing in $(*)$ are the same. So we have $(r^* - r^!)r^{m+1} = 0$ and hence $(r^* - r^!)e = 0$ (Lemma 1.2). Moreover, if f is an arbitrary idempotent of R then $(r^* - r^!)(r - ef) = 0$, since $e(f - ef) = 0$. Therefore, we have

$$(r^* - r^!)f = (r^* - r^!)(f - ef + ef) = 0$$

and hence $(r^* - r^!)B(R) = 0$. Because $\widetilde{R}$ is a unital extension of R the last equality yields $r^* = r^!$.

It is clear that in any unital extension every nilpotent element has a π-regular-inverse which is zero. We know that every element of a normal g.p.p. ring is an orthogonal sum of a reduced element and a nilpotent element (Corollary 1.4). Moreover, Lemma 2.1 and Corollary 2.2 imply that these reduced and nilpotent parts are uniquely determined by the element . So we have

Lemma 2.5. *Let R be a normal g.p.p. ring and r be an element of R with orthogonal decomposition $r = a + b$, where a is a reduced, b is a nilpotent element. Then the element r has a π-regular inverse r^* in unital extension if and only if the element a has a^* in this extension and $a^* = r^*$.*

Definition 2.6. Let R be a normal g.p.p. ring and Q be a unital extension of R. Then Q is said to be the *ring of right π-regular-quotients* of R if the following conditions are satisfied:

(i) every element $r \in R$ has π-regular inverse r^* in Q,

(ii) every element $q \in Q$ has a form $q = rs^*$ for some elements $r, s \in R$.

The ring of right π-regular quotients Q of the ring R (for brevity we denote it by $Q^r_\pi(R)$) has the following universal property, if it exists: Let P be a unital extension of R such that every element of R has π-regular-inverse in P. Then there exists a unique homomorphism.

$$\varphi : Q^r_\pi(R) \to P$$

such that $\varphi(r) = r$ for all $r \in R$. Therefore, $Q^r_\pi(R)$ is unique up to isomorphism. Moreover, we can analogously define the ring of left π-regular-quotients $Q^l_\pi(R)$ of R. The above universal property says that if both $Q^r_\pi(R)$ and $Q^l_\pi(R)$ exist, then they are naturally isomorphic.

Before formulating the main result of this section in view of Lemma 2.5 we notice that conditions (i) and (ii) in Definition 2.6 can be changed by the following conditions:

(i)$'$ every reduced element of R has a π-regular-inverse in Q,

(ii)$'$ every element $q \in Q$ has the form $q = rs^*$ for some $r, s \in R$ where s is reduced.

Theorem 2.7. *Let R be a normal ring. Then the following are equivalent:*

(1) R is a g.p.p. ring having a ring of right π-regular-quotients,

(2) a) every Pierce stalk R_x of R is a right Ore ring with identity and each zero-divisor in R_x is nilpotent,

b) for every element $r \in R$ the set

$$\{x \in X(R), r_x \text{ is non-zero-divisor}\}$$

is open compact in $X(R)$ and supp $r \subseteq$supp f for some $f \in B(R)$.

(3) R is g.p.p. ring with right π-Ore-condition: that is for every pair of elements $a, b \in R$ where b is reduced, there exist $c, d \in R$ such that

$$ad = bc, \quad ann_R(d) = ann_R(b).$$

Proof. (1)$\Longrightarrow$ (2): a) By Lemma 2.1 it is sufficient to prove that R_x is right Ore ring for all $x \in X(R)$. For, let $a_x, b_x \in R_x$ be non-zero elements and let b_x be a non-zero-divisor in R_x. Then we have $b^*a = cd^*$ for some $c, d \in R$ and by Lemma 2.5 we conclude that both b and d are reduced. Hence $bb^*ad = bcd^*d$ and noting that $b_xb_x^* = 1_x, d_x^*d_x = 1_x$ (Corollary 2.2) we have the right Ore condition: $a_xd_x = b_xc_x$, where d_x is non-zero-divisor in R_x.

Condition (2): b) follows immediately from Definition 1.1 and Corollaries 1.4 and 2.2.

(2)$\Longrightarrow$ (3). At first we prove that R is g.p.p. ring. Let r be an arbitrary element of R. By the assumption there are idempotents e, f such that

$$\text{supp } e = \{x \in X(R), r_x \text{ is non-zero-divisor}\}$$

and supp$r \subseteq$ supp$f$. Since $ann_R(re) = ann_R(e)$ and $rf = r, r = re + r(f-e)$, we see that re is a reduced element and $reRr(f-e) = 0$. Now we prove that $r' = r(f-e)$ is nilpotent. Since supp$(f-e)$ is open compact and r'_x is nilpotent for every $x \in$ supp$(f-e)$ by the standard Pierce sheave argument (cf. Lemma 1.5) we get an orthogonal set of idempotents $\{e_1, \cdots, e_n\}$ with $e_1 + e_2 + \cdots + e_n = f - e$ and integers $K_i, i = 1, 2, \ldots, n$ such that $(r'e_i)^{K_i} = 0$ for all i. Then taking $K = \max\{K_i \mid i = 1, 2, \cdots, n\}$ we have $(r')^K$=0, as desired.

Now we will test the π-Ore-condition. We take elements $a, b \in R$, where b is reduced. By Corollary 2.2 for every $x \in$ supp bb_x is non-zero-divisor in R_x, and hence by our assumption there are elements $c_x, d_x \in R_x$, where d_x is a non-zero-divisor in R_x, such that $a_x, d_x = b_xc_x$. Noting that if r_x is non-zero-divisor in R_x we can assume that r is reduced (see Corollary 1.4) and using again the standard Pierce sheave argument we get an orthogonal set $\{e_1, \cdots, e_n\}$ of idempotents and a set $\{c_1, \cdots, c_n, d_1, \cdots, d_n\}$ of elements of R such that each d_ie_i is reduced and $ad_ie_i = bc_ie_i, e_1 + e_2 + \cdots + e_n = e$, where supp$e$ = supp$b$. Hence, putting $d = \sum_{i=1}^{n} d_ie_i, c = \sum_{i=1}^{n} c_ie_i$ we get the desired equality $ad = bc$ with $ann_R(d) = ann_R(b)$, where d is reduced (see Corollary 2.2).

The implication (3) $\Longrightarrow$ (2) is obvious (see the proof of (1)$\Longrightarrow$ (2)).

Now we prove that condition (1) follows from conditions (2) and (3). Assume conditions (2) and (3). It is well known that R is a subdirect product of its Pierce stalks $R_x, x \in X(R)$. So, for every reduced element $r \in R$ we can construct a unique element $r^! \in \prod_{x \in X(R)} Q_{cl}^r(R_x)$, where $Q_{cl}^r(R_x)$ is the classical right ring of quotients of the ring R_x, as follows:

$$r_x^! = \begin{cases} r_x^{-1}, & \text{if } x \in \text{supp } r, \\ 0_x, & \text{otherwise.} \end{cases}$$

Let Q be a subring of $\prod_{x \in X(R)} Q^r_{cl}(R_x)$ generated by the set

$$R \cup \{r^!, r\text{-reduced element of } R\}.$$

We shall prove $Q = Q^r_\pi(R)$. It is clear that Q is a unital extension of R. As we have noted, we have to test conditions (i)$'$ and (ii)$'$.

Condition (i)$'$ is obvious, because $r^!$ is the π-regular-inverse of the reduced element $r \in R$. Further we shall write r^* instead of $r^!$.

Testing of (ii)$'$: It is enough to prove the following two conditions:

(ii)$'_1$: for every pair of elements $a, b \in R$, where b is reduced, there exist elements $c, d \in R$ such that d is reduced and $b^* a = cd^*$.

(ii)$'_2$: for every pair of reduced elements $a, b \in R$ there exist elements $c, d \in R$ such that d is reduced and $a^* + b^* = cd^*$.

The first one follows immediatley from the π-Ore-condition in (3).

For the second one let e and f be the associated idempotents of a and b, respectively (see Definition 1.3). We orthogonalize the sum $a^* + b^*$ by

$$a^* + b^* = a^*(e - ef) + b^*(f - ef) + (a^* + b^*)ef$$

where the summands are mutually orthogonal (we recall, two elements $u, v \in R$ are orthogonal, if $uRv = vRu = (0)$). Hence, if we can find $c, d \in R$ with $(a^* + b^*)ef = cd^*$ we will have

$$a^* + b^* = (e \oplus f + c)[a(e - ef) + b(f - ef) + d]^*$$

(we could suppose, without loss of generality, that $c = cef, d = def$). Therefore our problem is reduced to the case $e = f$. Let $e = f$. By the π-Ore-condition we have $as = bt$ with $ann_R(s) = ann_R(b)$ for some elements s and t of R, where s is reduced. Moreover, since a, b and s are reduced elements with the same associated idempotent e and for every $x \in \operatorname{supp} et_x$ is not a right zero-divisor. Hence t_x is nor nilpotent. Therefore t_x is a non-zero-divisor in R_x and we can assume that t is a reduced element with associated idempotent e. So we have $a^* = s(bt)^*$ and

$$a^* + b^* = s(bt)^* + eb^* = s(bt)^* + tt^*b^* = (s + t)(bt)^*$$

as desired. Thus the theorem is proved.

§3. Normal g.p.p. Rings with Köthe Radical

We recall that a ring R is said to be a *ring with Köthe radical* $K(R)$ if $K(R)$ is a nil ideal of R containing all one-sided nil ideals of R. It is a well known open question whether every ring has Köthe radical. We also recall that a ring R is called a *local ring* if the ring $R/J(R)$ is a division ring, where $J(R)$ denotes the Jacobson radical of R.

From Lemma 2.1 and from the proof of the implication (2) $\Longrightarrow$ (3) in Theorem 2.7 we have

Proposition 3.1. *Let R be a normal ring. The R is a g.p.p. ring if and only if the following conditions are satisfied:*

1) in each Pierce stalk R_x of R every zero-divisor is nilpotent,

2) for every element $r \in R$ the set

$$\{x \in X(R), r_x \text{ is non-zero-divisor in } R_x\}$$

is open compact in $X(R)$ and supp $r \subseteq$ supp f for an idempotent $f \in B(R)$.

An addition which is needed for the main result of this section and also has its own interest is the following.

Proposition 3.2. *Let R be a right Ore ring. Then the following are equivalent:*

1) for an arbitrary element $r \in R$ either r is non-zero-divisor, or $r+s$ is non-zero-divisor for all non-zero-divisor $s \in R$,

2) $Q^r_{cl}(R)$ is a local ring.

Proof. 1)$\Longrightarrow$2). Let $q = rs^{-1} \in Q$ be a non-invertable element of Q, where $Q = Q^r_{cl}(R)$ and $r, s \in R$. Then r is zero-divisor, and hence by the assumption $r+s$ is a non-zero-divisor and $1 + rs^{-1} = (r+s)s^{-1}$ is invertable. Now we are to verify that

$$J(Q) = \{q \in Q, q \text{ is not invertable}\}.$$

The inclusion $\subseteq$ is obvious. Let q be a non-invertable element. Now q is a zero-divisor in Q. Let p be an arbitrary element of Q. If q is a right zero-divisor (the left case is similar) then qp is a right zero-divisor, and hence, as we have proved, $1+qp$ is invertable. Thus qQ as a quasi-regular right ideal must be contained in $J(Q)$, in particular, $q \in J(Q)$. Therefore $Q/J(Q)$ is a division ring and Q is a local ring.

2)$\Longrightarrow$1) Let r be a right zero-divisor in R, and s be a non-zero-divisor in R. We have to verify that $r+s$ is a non-zero-divisor. Since $q = rs^{-1}$ is a non-invertable element of Q by assumption $q \in J(Q)$ and $1 + rs^{-1}$ is invertable. Therefore $r + s = (1 + rs^{-1})s$ is invertable in Q and $r+s$ is a non-zero-divisor in R.

Theorem 3.3. *Let R be a normal ring.*

(A) The following conditions are equivalent:

a) R is a g.p.p. ring with Köthe radical,

b) in each Pierce stalk R_x every zero-divisor is nilpotent, and the sum of nilpotent elements in R_x is nilpotent and for every element $r \in R$ the set

$$\{x \in X(R), r_x \text{ is non-zero-divisor}\}$$

is open and compact and supp $r \subseteq$ supp f for some $f \in B(R)$

(B) Moreover, if R is g.p.p. ring with $Q^r_\pi(R)$ the above conditions are equivalent with the following condition:

C) Every Pierce stalk Q_x of the ring $Q^r_\pi(R)$ on $x \in X(R)$ is a local ring.

Proof. At first we prove the following

Lemma 3.4. *Let R be a normal g.p.p. ring. Then R has Köthe radical if and only if the set of all nilpotent elements of R is an ideal of R.*

Proof of the Lemma. The necessity is obvious. Let R be normal g.p.p. ring with Köthe radial K and let a be a nilpotent element of R. Then for every element $b \in R$ ab is nilpotent, because $(ab)_x = a_x b_x$ is a zero-divisor, hence the reduced part of the element ab is zero. Therefore aR is a nil right ideal. In particular, $a \in K$ and K is an ideal of all nilpotent elements of R.

Now we continue the proof of Theorem 3.3.

a)$\Longrightarrow$ b): Let a_x, b_x be nilpotent elements of R_x. Then for some $e \in B(R)$ with $e_x = 1_x$, the elements ae and be are nilpotent. So, by the assumption and by Lemma 3.4, $ae + be$ is nilpotent, and hence $a_x + b_x$ is nilpotent. The rest of condition b) follows from Proposition 3.1.

b) $\Longrightarrow$ a). It is enough to verify that the sum of nilpotent elements of R is nilpotent and this can be done by the standard Pierce sheave argument (see the proof of Theorem 2.7 (2) $\Longrightarrow$ (3)).

B) b)$\Longrightarrow$c). Let R be a ring with $Q^r_\pi(R)$ and let condition b) in (A) be satisfied. By Corollary 2.8 $Q_x = Q^r_{cl}(R_x)$ for all $x \in X(R)$. Moreover, R_x satisfies condition 1) of Proposition 3.2, because if $r_x + s_x$ and r_x are nilpotent, then s_x is nilpotent, too. Therefore by this proposition Q_x is a local ring.

c) $\Longrightarrow$ b): Let $Q_x = Q^r_{cl}(R_x)$ be a local ring, and $a_x, b_x \in R_x$ be nilpotent. If $a_x + b_x$ is not nilpotent, then it is non-zero-divisor and hence by Proposition 3.3 $a_x = (a_x + b_x) - b_x$ is non-zero-divisor. This is, however, impossible and so $a_x + b_x$ is nilpotent.

Corollary 3.4. *Let R be a normal g.p.p. ring with classical right ring of quotients Q. Then every Pierce stalk of the ring Q is a local ring if and only if R has Köthe radical.*

For the proof we note that $B(Q) = B(R)$.

With respect to Corollary 3.4 we have the following

Conjecture. Let Q be a classical right ring of quotients of a normal g.p.p. ring with identity. Then every Pierce stalk of the ring Q is a local ring.

Acknowledgement. The author wishes to express his indebtedness and gratitude to Prof. L.A. Bokut, for his permanent attention.

References

[1] W.D. Burgess, W. Stephenson, *Pierce sheaves of non-commutative rings*, Comm. Algebra, 4(1976), 51–75.

[2] W.D. Burgess, W. Stephenson, *Rings all of whose Pierce stalks are local*, Canad. Math. Bull., 22(1979), 159–164.

[3] J. Dauns, K.H. Hofmann, *The representation of biregular rings by sheaves*, Math. Zeitschr., 91(1966), 103–123.

[4] R. Gonchigdorzh, *Rings of regular-quotients of reduced rings (in Russian)*, Algebra i Logika, 26(1987), N2.

[5] R. Gonchigdorzh, *Reduced p.p rings and semihereditary and hereditary rings (in Russian)*, Algebra i Logika (to appear).

[6] Y. Hirano, *On generalized p.p. rings*, Math. Journal Okayama Univ., 25(1983), 7–12.

[7] Y. Hirano, H. Tominaga, *Rings decomposed into direct sums of nil rings and certain reduced rings*, Math. Journal Okayama Univ., 27(1985), 35–38.

[8] M. Ôhori, *On non-commutative generalized p.p. rings*, Math. Journal Okayama Univ., 26(1984), 157–168.

[9] M. Ôhori, *On strongly π-regular rings and periodic rings*, Math. Journal Okayama Univ., 27(1985), 49–52.

[10] M. Ôhori, *Some studies on generalized p.p. rings and hereditary rings*, Math. Journal Okayama Univ., 27(1985), 53–70.

[11] R.S. Pierce, *Modules over commutative regular rings*, Mem. Amer. Math. Soc., No. 70, Amer. Math. Soc., Providence, 1967.

Proceedings of the Asian Mathematical
Conference, 1990

Two-Point Homogeneous Spaces and G-Structures

P. JA. GRUSHKO
Irkutsk, USSR

§1. Introduction

A metric space (M,d) is said to be a two point homogeneous space if for each pairs $(x_1,x_2),(y_1,y_2)$ such that $d(x_1,x_2)=d(y_1,y_2)$ there is an isometry F such that $Fx_i=y_i$. All these spaces were described by Wang H. C. and Tits J. [2], [3]. It is known that a Riemannian space is a two point homogeneous spaces if and only if the group of all isometries is transitive on the bundle of unit tangent vectors. Now we say that a G-structure B is isotropic if the group of N-automorphisms of B is transitive on the bundle of rays in TM. There is a complete list of such objects [6].

Furthermore, a N-transitive G-structure is said to be an almost N-isotropic structure if the codimension of every nonzero orbit of the isotropy subgroups on tangent spaces is less than or equal to 1. There is the complete list of these G-structures. For example, the contact structure is an almost isotropic structure such that G is not an irreducible group. On the other hand , pseudoriemannian manifolds of constant curvature are also almost isotropic G-structures.

§2. G-Structures

We denote by $\mathrm{Aut}(V)$ ($\mathrm{End}(V)$) the group of all linear automorphisms of a vector space V (the Lie algebra of all endomorphisms of V respectively). Let $G\subset \mathrm{Aut}\ (V)$ be a Lie subgroup and $\widehat{G}\subset \mathrm{End}(V)$ the Lie algebra of this subgroup. Let G acts to the right without fixed points on a manifold B such that there exists the principal bundle

$$\pi : B\to M=B \pmod{G}.$$

Definition 1. A geometric structure $\mathcal{B}=\{G,V,B,\omega\}$ is defined by a canonical form $\omega : TB\to V$ such that $\omega R_g^*=g^{-1}\omega$ and there is an exact sequence of vector bundles $O\to\widehat{G}\times B\xrightarrow{j} TB\xrightarrow{\overline{\omega}} V\times B\to O$ where $\overline{\omega}(t_b)=(\omega(t_b),b)$. The map j is defined by the formula

$$j_b(\lambda)=\left.\frac{d}{dt}\right|_{t=0} R_{\exp(t\lambda)}b,\quad b\in B,\quad \lambda\in\widehat{G}.$$

In fact, B is a reduction of the bundle of frames $F(M)$ to a subgroup $G\subset\mathrm{Aut}(V)$. Let $N\subset N(G)$ be a subgroup of the normalizer of G.

Definition 2. A N-automorphism of a structure $\mathcal{B}$ is a triple $\varphi = \{k, l, f\}$ where k is an automorphism of $G, l \in N, f$ is a diffeomorphism of B such that $l(gv) = kg \cdot lv, g \in G, v \in V, f(R_g b) = R_{kg} fb, g \in G, b \in B, \omega f^* = l\omega$. We denote by $N\mathrm{Aut}\mathcal{B}$ the group of all N-automorphisms of $\mathcal{B}$.

For example, $\{K = \mathrm{Int}_g, l = g, f = R_g\}$ is a G-automorphism which induces the identity map on the base M, thus, if $G \subset N \subset N(G)$, then G is a normal subgroup in $\mathcal{D} = N$ Aut $\mathcal{B}$.

Definition 3. A structure $\mathcal{B}$ is said to be N-transitive if it is transitive with respect to N-automorphisms.

Thus, for each $b_1, b_2 \in B$ there is $\{k, l, f\} \in N$ Aut $\mathcal{B}$ such that $fb_1 = b_2$.

Let L_0 be a Lie subgroup in L. We denote $\theta(x, dx) = L_x^* - 1dx$ the Maurer-Cartan form of L. Let $\chi : \widehat{L} \to V = \widehat{L}/\widehat{L}_0$ be the natural projection. Then the isotropy representation $\varepsilon : L_0 \to$ Aut (V) is defined by $\varepsilon(x)\chi y = \chi(Ad_x y), x \in L_0, y \in \widehat{L}$. If Ker $\varepsilon = \{e\}$ then $\{G = L_0, B = L, V = \widehat{L}/L_0, \omega = \chi\theta\}$ is the homogeneous structure corresponding to (L, L_0). If there exists a commutative ideal $J \subset \widehat{L}$ such that $\widehat{L} = \widehat{L}_0 \oplus J$ then we have the flat structure.

If L_1 =Ker $\varepsilon \neq \{e\}$ then $\{G = L_0/L_1, B = L(\mathrm{mod}\ L_1), V = \widehat{L}/\widehat{L}_0, \omega\}$ is said to be a quasi-homogeneous structure corresponding to (L, L_0) where $\omega q^* = \chi\theta, q : L \to B$ is the natural projection. All e-transitive G-structures of finite type are quasihomogeneous.

Now we can extend the class of quasi-homogeneous structures as follows. Let D be a Lie group, $\mathcal{D}_1 \subset \mathcal{D}_0 \subset \mathcal{D}$ be Lie subgroups, and $G \subset \mathcal{D}$ a normal subgroup such that $D_0 = D_1 \cdot G, \mathcal{D}_1 \cap G = \{e\}$. If $0 \to \widehat{\mathcal{D}}_0 \to \widehat{\mathcal{D}} \xrightarrow{\mu} V \to 0$ is an exact sequence of linear spaces then the isotropy representation $\varepsilon : \mathcal{D}_0 \to$ Aut (V) is defined by

$$\varepsilon(x)\mu g = \mu(Ad_x y), \quad x \in \mathcal{D}_0, y \in \widehat{\mathcal{D}}.$$

Clearly $\varepsilon(G) = e$, thus, $\varepsilon(\mathcal{D}_0) = \varepsilon(\mathcal{D}_1)$.

Definition 4. We say that $\mathcal{D} = \{\mathcal{D}, \mathcal{D}_0, \mathcal{D}_1, G, \mu, V, \Lambda\}$ is a N-agregate if $\Lambda : \mathcal{D} \to$ Aut V is a representation such that $\Lambda|_{\mathcal{D}_1} = \varepsilon|_{\mathcal{D}_1}. \Lambda|_G$ is a monomorphism, $\Lambda\mathcal{D} = N$ and D_1 does not contain any normal subgroup of D.

Now we shall construct a N-transitive G-structure corresponding to a N-agregate as follows. Put $V = \widehat{\mathcal{D}}/\widehat{\mathcal{D}}_0, B = \mathcal{D}(\mathrm{mod}\ \mathcal{D}_1), M = \mathcal{D}(\mathrm{mod}\ \mathcal{D}_0)$. Let $q : \mathcal{D} \to B$ be the natural projection. We put $R_g(qx) = q(g^{-1}x), \omega q^x(x, dx) = \Lambda(x)\mu\theta(x, dx), dx \in T_x$. We call $\{G, B, V, \omega\}$ the nearly homogeneous structure, corresponding to $\mathcal{D}$. Each $a \in \mathcal{D}$ induces a N-automorphism $\{k, l, f\}$ such that $l = \Lambda\alpha, k =$Int $e, fqx = qax, x \in \mathcal{D}$.

Theorem [4]. *If there is a transitive Lie group D of N-automorphisms of $\mathcal{B}$ then $\mathcal{B}$ is a nearly homogeneous structure.*

Here D_1 is the isotropy subgroup of a fixed point $b_0 \in B, \Lambda\{k, l, f\} = l, \mu = \omega\widetilde{q}^* e$, where $\widetilde{q} : \mathcal{D} \to B$ is defined by the formula $\widetilde{q}(f) = fb_0$.

Example. Let D =Spin (7), $G = \{e\}, D_0 = D_1 = G_2, V = R^7, \Lambda$ is induced by the standard representation of $SO(7)$ =Spin $(7)/Z_2$. Then we get the absolute parallelism of S^7 =Spin$(7)/G_2$ which is not e-transitive.

§3. The Structure Function and a Fan

Let $s : V \to T_b$ be a linear map such that $\omega s = id$. Define $Q \in \Lambda^2 V^* \otimes V$ by $Q(v_1, v_2) = d\omega(sv_1, sv_2)$. The equivalence class of Q in the Spencer cohomology $H^{2,-1}(V, \widehat{G}) = (\Lambda^2 V^* \otimes V)/\partial(V^* \otimes \widehat{G})$ defines the structure function $C : B \to H^{2,-1}(V, \widehat{G})$, [1]. The action of N on $H^{2,-1}(V, \widehat{G})$ is defined by $Q^l(v_1, v_2) = Q(l^{-1}v_1, l^{-1}v_2)$. If $\{k, l, f\} \in N\mathrm{Aut}\mathcal{B}$ then $C(fb) = c(b)^l$. If $\mathcal{B}$ is flat (e-transitive) then its structure function is zero (constant, respectively). If $\mathcal{B}$ is N-transitive then $\{c(b)/l \in B\}$ is contained in an orbit of N. As a rule, the structure function is a good tool only if it is constant. Now we shall construct a more delicate invariant of N-transitive structures.

Let V be a linear space, $N \subset \mathrm{Aut}V$, $H \subset N$ a Lie subgroups, $G \subset N$ a normal subgroup. We put $Z = G \cap H, W = G \cdot H$. Let $\rho : \widehat{W} \to \widehat{G}$ be a linear map such that $\rho|_{\widehat{G}} = id, \rho(\widehat{Z}) \subset \widehat{H}$. If $\Gamma i \in V^* \otimes \widehat{N}, Q_i \in H^{2,-1}(V, \widehat{Z})$ then we put $(\Gamma_1, Q_1) \sim (\Gamma_2, Q_2)$ if there exists $\gamma \in V^* \otimes \widehat{N}$ such that

$$\Gamma_2 = \Gamma_1 + \gamma, \quad Q_2 = \{Q_1 - \partial(\rho - \gamma)\}(\mathrm{mod}\ \partial(V^*(\otimes \hat{z})).$$

The equivalences classes are called fans, [5].

Example. Let $\mathcal{D} = \{\mathcal{D}; \mathcal{D}_0, \mathcal{D}_1, G, V, \mu, \Lambda\}$ be a N-agregate. Let $\Sigma : V \to \widehat{\mathcal{D}}$ be a linear map such that $\mu \cdot \Sigma = id$. We denote $\widehat{\Lambda} = \Lambda_e^* : \widehat{\mathcal{D}} \to \widehat{N}$. Put $Q(v_1, v_2) = \mu[\Sigma v_1, \Sigma v_2], \Gamma v = \widehat{\Lambda} \cdot \Sigma v, H = \Lambda \mathcal{D}_1$. The equivalence class of $\{Q, \Gamma\}$ is called the fan of $\mathcal{D}$.

Now denote

$$\begin{aligned} T^h(v_1, v_2) = \;& hQ(v_1, v_2) - Q(hv_1, v_2) - Q(v_1, hv_2) \\ & + \rho\{[h, \Gamma v_1] - \Gamma(hv_1)\} \cdot v_2 - \rho\{[h, \Gamma v_2] - \Gamma(hv_2)\} \cdot v_1, \\ K(v_1, v_2, v_3) = \;& \sigma\{Q(Q(v_1, v_2), v_3) - Q^{\Gamma v_3}(v_2, v_2) - \rho([\Gamma v_1, \Gamma v_2] \\ & - \Gamma(\Gamma v_1 \cdot v_2 - \Gamma v_2 \cdot v_1 - Q(v_1, v_2))) \cdot v_3\}. \end{aligned}$$

Then the structure equations holds

$$[h, \Gamma v] - \Gamma(hv) \in \widehat{W}, v \in V, h \in \widehat{H},$$

$$[\Gamma v_1, \Gamma v_2] - \Gamma(\Gamma v_1 \cdot v_2 - \Gamma v_2 \cdot v_1 - Q(v_1, v_2)) \in \widehat{W},$$

$$T^h = 0 \quad (\mathrm{mod}\ \partial V^* \otimes \widehat{Z}), \quad K = 0 \quad (\mathrm{mod}\ \partial(\Lambda^2 V^* \otimes Z)).$$

§4. Almost N-Isotropic Structures

Definition 5. A group $H \subset \mathrm{Aut}\, V$ is said to be perfect if the codimension of all nonzero orbits is equal to 0 or 1.

Clearly, a group H is perfect if and only if $\dim \widehat{H} v \geq \dim V - 1$ for each $v \neq 0$. There are two cases:

1) The group H is irreducible on V (i.e. the identity representation of H on V is irreducible),

2) The group H is reducible. The there is an H-invariant hyperplane $V_0 \subset V$ such that H acts transitively on $V_0 \backslash \{0\}$ and for each $e_0 \in V/V_0$ we have $He_0 + Re_0 = V$.

Thus, there is a group $\overline{H} \subset \mathrm{Aut}\ V_0$ transitive on $V_0 \backslash \{0\}$ such that

$$H = \left\{ \begin{pmatrix} \lambda & 0 \\ d & A \end{pmatrix} / a \in V_0, A \in \hat{H}, \lambda > 0 \right\}.$$

Here either λ is independent or $\lambda = \psi(A)$ where $\psi : \overline{H} \to R^+$ is a homomorphism.

Definition 6. A structure $\mathcal{B}$ is said to be an almost N-isotropic structure if there is a finite dimensional Lie group $\mathcal{D} \subset N$ Aut $\mathcal{B}$ transitive on B such that the codimension of each nonzero orbit on TM is 0 or 1.

Let D_m be the isotropy subgroup of $m \in M$ and $H_m \subset \mathrm{Aut}\ T_m$ the group of linear transformations of the tangent space T_m induced by D_m. It is clear that our structure is almost N-isotropic if and only if the group H_m is perfect. Thus, we shall study aggregates $\{\mathcal{D}, \mathcal{D}_0, \mathcal{D}_1, G, \mu, V, \Lambda\}$ such that the group $H = \Lambda \mathcal{D}_1$ is perfect.

Clearly, the group N is perfect. Since G is a normal subgroup in N it isn't hard to enumerate these groups. We shall not present the complete list of almost N-isotropic structures because it is too extensive. Most of these structures are quasi-homogeneous.

Put $L = \ker\Lambda$, $L_0 = L \cap \mathcal{D}_0$, $b_0 = qe \in B$.

Lemma 1. *If $\Lambda\mathcal{D} = \Lambda\mathcal{D}_0$. Then the group L is transitive on the base M and the orbit $\overline{B} = L \cdot b_0 \subset B$ is a reduction of the structure group G to the subgroup $\varepsilon(L_0) \subset G$. It is the quasi-homogeneous structure corresponding to the pair (L, L_0).*

In fact, if $d \in \mathcal{D}$ then there exists $d_0 \in \mathcal{D}$ such that $\Lambda d = \Lambda d_0$. Thus, $x = d \cdot d_0^{-1} \in L$ and $d(\mathrm{mod}\ D_0) = x(\mathrm{mod}\ D_0)$.

Lemma 2. *The group $\varepsilon(L_0)$ coincides with $Z = H \cap G$.*

Proof. If $x \in L_0$ then $\varepsilon x \in \varepsilon\mathcal{D}_0 = \varepsilon\mathcal{D}_1 = \Lambda\mathcal{D}_1 = H$ on the other hand, $x = d_1 \cdot g$ where $d_1 \in \mathcal{D}_1, g \in G$ and $\Lambda d_1 \cdot \Lambda g = 1$. Thus, $\varepsilon x = \varepsilon d_1 = \Lambda d_1 = \Lambda g^{-1} \in G$.

Conversely, if $Z \in Z = H \cap G$ then $Z = \Lambda d_1 = \Lambda g$. Thus, $\Lambda(d_1 g^{-1}) = 1$ and $x = d_1 \cdot g^{-1} \in L_0$. Now $\varepsilon x = \varepsilon(d_1 g^{-1}) := \varepsilon d_1 = \Lambda d_1 = z$.

Corollary. *If N=W and the group Z is perfect then the quasi-homogeneous structure corresponding to (L, L_0) is almost N-isotropic.*

Since $[h, \Gamma v] = \Gamma(hv)\ (\mathrm{mod}\ \widehat{W})$, $h \in \hat{H}$ we see that $\Gamma(\hat{H}V) = 0\ (\mathrm{mod}\ \widehat{W})$ if the action of $\hat{H}$ on $\hat{N}/\widehat{W}$ is trivial. Thus, if $\hat{H}.(\hat{N}/\widehat{W}) = 0$ then $\Gamma(v_0) = 0\ (\mathrm{mod}\ \widehat{W})$. If $\hat{H} \cdot V = V$ then $\Gamma = 0$ and $W = N$.

Corollary. *If the group Z is perfect, $\hat{H} \cdot V = V$ and $\hat{H}$ acts trivially on $\hat{N}/\widehat{W}$ then our G-structure is an extension of the structure group of a quasi-homogeneous almost N-isotropic structure.*

In conclusion, we shall consider the almost N-isotropic e-structure (absolute parallelisms).

§5. N-Isotropic e-Structures

Example. Let $G = \{e\}, B = M = R^+ \times R^{n-1} = \{(t, x)/t > 0, x \in R^{n-1}\}, \omega(t, x, dt, dx) = (tpdt, tqdx)$ where p, q are constants. If $q = 0$ then $dw = 0$ and our e–structure is flat. Suppose $q \neq 0$. Let N be a set of matrices $\begin{pmatrix} l_{11} & 0 \\ l_{21} & l_{22} \end{pmatrix}$ where $l_{11} > 0, l_{21} \in R^{n-1}, l_{22} \in \tilde{N} \subset$

$GL(n-1)$. An arbitrary N-automorphism is given by the formula

$$f(t,x) = (\alpha(t,x), \beta(t,x)), \quad \alpha(t,x) \in R^{+}, \beta(t,x) \in R^{n-1},$$

where

$$\alpha^{p} d\alpha = l_{11} t^{p} dt, \quad \alpha^{q} d\beta = l_{21} t^{p} dt + l_{22} t^{q} dx.$$

If $p \neq -1$ then $\alpha(t,x) = t l_{11}^{1/(p+1)}$,

$$\beta(t,x) = \frac{t^{p-q+1}}{p-q+1} l_{11}^{-q/(p+1)} l_{21} + l_{11}^{-q/(p+1)} l_{22} x + c.$$

Since e-automorphisms are $f(t,x) = (t, x+c)$ the group of e-automorphisms is not transitive. The isotropy subgroup of the point $m = (t=1, x=0)$ is defined by $l_{11} = 1, c + l_{21}/(p-q+1) = 0$. Now

$$df(1,0,dt,dx) = (dt, dt \cdot l_{21} + l_{22} dx).$$

Thus, $H = \{l \in N / l_{11} = 1\}$.

It $p = -1$ then our e-structurec is transitive. An arbitrary N-automorphism is defined by $f(t,x) = (c_0 t, c_0^{-q}(l_{22}x - \frac{l_{21}}{t} + c))$. An arbitrary e-automorphism is defined by $f(t,x) = (c_0 t, c_0^{-q} x + c)$. The group H is the same.

In both cases the structure is almost N-isotropic if and only if $\widehat{N} l_0 = V_0 + (p+1) R \cdot l_0$ and the group $\widetilde{N}$ is transitive on $V_0 \backslash \{0\}$.

Theorem. *There are three classes of almost N-isotropic structures:*

1) *the flat structure (the absolute parallelism of a vector space),*

2) *the absolute parallelism of* S^7,

3) *the e-structure defined by* $\omega = (t, x, dt, dx) = (t^p dt, t^q dx)$.

Proof. Suppose H is irreducible. Choose a P_H-invariant subspace $\nabla \subset \widehat{N}$ such that $\widehat{N} = \widehat{H} \oplus \nabla$ where P_H is the semisimple component of $\widehat{H}$. If $\Gamma : V \to \widehat{N}$ then $[h, \Gamma v] = \Gamma(hv), h \in P_H, v \in V$. Thus, $\Gamma = 0$ or $P_H = G_2, V = R^7$. Since $hQ(v_1, v_2) = Q(hv_1, v_2) + Q(v_1, hv_2)$ we have $Q = 0$ except the case when $P_H = G_2, V = R^7$.

Suppose H is reductive. Then $V = R \cdot e_0 \oplus V_0, \widehat{H} \cdot V_0 = V_0, \widehat{H} \cdot e_0 \supset V_0$. Using the structure equation we have $\Gamma V_0 = 0, \Gamma e_0 = \begin{pmatrix} \lambda & 0 \\ 0 & \mu \cdot 1 \end{pmatrix}$, where $\lambda, \mu \in R, Q(V_0, V_0) = 0, Q(e_0, u) = cu, u \in V_0, c \in R$. If $H \cdot e_0 = V$ then $\Gamma = 0, Q = 0$. This fan corresponds to the structure which is described in the example where $q = -c, p = \lambda - 1$.

References

[1] Singer I. M., Sternberg S. *On the infinite groups of Lie and Cartan*, Ann. Inst. Fourier (Grenoble), 1965, 15, 1–114.

[2] Tits J. *Sur certaines classes d'espaces homogenes de groupes de Lie*, Acad. Roy. Belg. Cl. SciMem. Coll., 1955, 29.

[3] Wang. H. C. *Two point homogeneous spaces*, Ann. Math., 1962, 55, 177–192.

[4] Grushko P. Ja., *Conjugate transitive structure of finite type*, Izv. VUZ. Math., 1981, 2, 24–29.

[5] Grushko P. Ja., *On a problem of Cartan's equivalence* , Sib. Math. J., 1981, 22, I, 68–80.

[6] Grushko P. Ja., *Conjugate isotopic space*, Izv. VUZ. Math., 1982, 15–19.

Proceedings of the Asian Mathematical
Conference, 1990

Some Notes on Eigenvalue Problems of Elliptic Equations of High Order

GU YONGGENG
Institute of Systems Science
Chinese Academy of Sciences
Beijing 100080, China

§1. Introduction

Let Ω be a smooth bounded domain in R^n with $n \geq 2$. We are concerned with the eigenvalue problem

$$\begin{cases} (-\Delta)^t u = \Lambda u, & \text{in } \Omega, t = 1, 2, \cdots \\ \frac{\partial^j u}{\partial v^j}\Big|_{\partial\Omega} = 0, & j = 0, 1, 2, \cdots, t-1, \end{cases} \tag{1.1}$$

where v is the unit outward normal to $\partial\Omega$.

According to the theory of eigenvalue, problem (1.1) has discrete real eigenvalues. They can be lined up as

$$0 < \Lambda_1 \leq \Lambda_2 \leq \cdots \leq \Lambda_k \cdots; \quad \Lambda_k \to \infty \quad \text{as} \quad k \to \infty.$$

Let $u_k(x)$ be the eigenfunction corresponding to $\Lambda_k, k = 1, 2, \cdots$. Suppose that they are orthonormal, i.e.

$$\int_\Omega u_j(x) u_k(x) dx = \delta_{jk} = \begin{cases} 1, & \text{when } j = k, \\ 0, & \text{when } j \neq k. \end{cases}$$

When $t = 1$, (1.1) is the usual eigenvalue problem of Laplace equation with Dirichlet boundary condition. For the convenience of the following discussion, we denote it by

$$\begin{cases} -\Delta\varphi = \lambda\varphi, & \text{on } \Omega, \\ \varphi|_{\partial\Omega} = 0. \end{cases} \tag{1.2}$$

The eigenvalues and corresponding eigenfunctions of problem (1.2) are denoted by $\{\lambda_k\}$ and $\{\varphi_k\}$ respectively.

As you know, the eigenvalue problem of elliptic equations is a very important and basic problem in mathematics and physics. It not only has very definite mechanical and physical backgrounds, but also possesses an outstanding position in the study of modern theory of partial differential equations.

Early in 1912, H. Weyl proved the asymptotic formula for the eigenvalues of problem (1.2),

$$\lambda_k \sim C_n \left(\frac{k}{|\Omega|}\right)^{2/n}, \quad k \to \infty, \tag{1.3}$$

where

$$C_n = (2\pi)^2 \left(\frac{n}{\omega_{n-1}}\right)^{2/n}, \tag{1.4}$$

$|\Omega|$ is the volume of the domain Ω, and ω_{n-1} is the area of the $n-1$ dimensional unit sphere S^{n-1}. In 1960, Polya posed the famous "Polya conjecture": [4]

$$\lambda_k \geq C_n \left(\frac{k}{|\Omega|}\right)^{2/n}, \quad k = 1, 2, \cdots. \tag{1.5}$$

When $n = 2$, Polya himself proved that the conjecture is right for some special planar domain. Until now, the best result in this direction is obtained by Li and Yau[1]:

$$\sum_{j=1}^{k} \lambda_j \geq \frac{nkC_n}{n+2} \left(\frac{k}{|\Omega|}\right)^{2/n}, \quad k = 1, 2, \cdots \tag{1.6}$$

and

$$\lambda_k \geq \frac{n}{n+2} C_n \left(\frac{k}{|\Omega|}\right)^{2/n}, \quad k = 1, 2, \cdots. \tag{1.7}$$

In [1], Yau pointed out that formula (1.6) showed that Polya conjecture is in an average sense.

In this paper we mainly discuss the lower estimates of the high order eigenvalue problem (1.1). We obtain the following estimates in a form similar to (1.6) and (1.7),

$$\sum_{j}^{k} \Lambda_j \geq \frac{nkC_n^t}{n+2t} \left(\frac{k}{|\Omega|}\right)^{2t/n}, \quad k = 1, 2, \cdots, \tag{1.8}$$

and

$$\Lambda_k \geq \frac{n}{n+2t} C_n^t \left(\frac{k}{|\Omega|}\right)^{2t/n}, \quad k = 1, 2, \cdots, \tag{1.9}$$

where C_n is a constant given by (1.4), depending only on n.

§2. Estimation of the First Eigenvalues

For $l = 1, 2, \cdots$, let

$$\dot{C}^l(\Omega) = \{u(x) \in C^l(\overline{\Omega}) \cap C^{2l}(\Omega), \frac{\partial^j u}{\partial v^j}|_{\partial\Omega} = 0, j = 1, 2, \cdots, l-1\},$$

$$\dot{C}^l(R^n) = \left\{ \begin{array}{c} u(x) \in C^l_{loc}(R^n), \\ \dfrac{\partial^j u}{\partial x_1^{j_1} \cdots \partial x_n^{j_n}} \to 0, \quad |x| \to \infty, \quad j = \sum_{i=1}^n j_i, \quad j = 1, 2, \cdots, l-1. \end{array} \right\}$$

Lemma 2.1. *For all $u \in \dot{C}^l(\Omega)$, we have the integration equality*

$$\sum_{i_1, \cdots, i_l = 1}^n \int_\Omega \left| \frac{\partial^l u}{\partial x_{i_1}, \cdots, \partial x_{i_l}} \right|^2 dx = \begin{cases} \int_\Omega |\Delta^h u|^2 dx, & \text{when } l = 2h, \\ \int_\Omega |\nabla(\Delta^{k-1} u)|^2 dx, & \text{when } l = 2h-1, \end{cases} \quad h = 1, 2, \cdots. \tag{2.1}$$

Now let us establish a relationship between λ_1 and Λ_1.

Theorem 2.1. *Suppose λ_1 and Λ_1 are the first eigenvalues of problem (1.2) and (1.1) respectively. Then we have $\Lambda_1 \geq \lambda_1^t$.*

Proof. Let $u_1(x)$ be the eigenfunction of problem (1.1) corresponding to the first eigenvalue Λ_1 with $\int_\Omega u_1^2 dx = 1$. Then $u_1(x) \in \dot{C}^t(\Omega) \subset \dot{C}^m(\Omega), \forall m \leq l$. According to the minimum principle of eigenvalues, Poincare inequality and equality (2.1), we have

$$\lambda_1 \leq \begin{cases} \dfrac{1}{\lambda_1^{t-1}} \int_\Omega |\Delta^h u_1|^2 dx, & \text{when } t = 2h, \\ \dfrac{1}{\lambda_1^{t-1}} \int_\Omega |\nabla(\Delta^{h-1} u_1)|^2 dx, & \text{when } t = 2h-1. \end{cases} \quad h = 1, 2, \cdots.$$

According to the minimum principle of eigenvalues, we have $\Lambda_1 = \int_\Omega |\Delta^h u_1|^2 dx$ (or $\int_\Omega |\nabla(\Delta^{h-1} u)|^2 dx$). Then we obtain $\lambda_1 \leq \frac{\Lambda_1}{\lambda_1^{t-1}}$, that is, (2.2), thus completing the proof.

§3. Estimation of High Order Eigenvalues

In this section, we mainly discuss the estimation of the kth eigenvalue $\Lambda_k (k = 1, 2, \cdots)$ of problem (1.1). First, we introduce two lemmas.

Lemma 3.1. *Suppose $\{u_k(x)\}$ is a set of eigenfunctions of problem (1.1). Then $\{u_k(x)\}$ forms a base of $L^2(\Omega)$.*

Lemma 3.2. *Suppose $f(z) : R^n \to R^1$ is a bounded function, and $\|f(z)\|_{L^\infty(R^n)} = K_1 < \infty$, $\||z|^{2t} f(z)\|_{L^1(R^n)} = K_2 < \infty$. Then*

$$\|f(z)\|_{L^1(R^n)} \leq \left(K_1 \frac{\omega_{n-1}}{n}\right)^{\frac{t}{n+t}} \left(K_2 \frac{n+2t}{n}\right)^{\frac{n}{n-2t}}, \tag{3.1}$$

where the constant $t > 0$ is given.

Theorem 3.1. *Let $\{\Lambda_j\}$ be the eigenvalue series of problem (1.1). Then we have*

$$\sum_{j=1}^{k} \Lambda_j \geq \frac{nk}{n+2t} C_n^t \left(\frac{k}{|\Omega|}\right)^{\frac{2t}{n}}, \quad k = 1, 2, \cdots,$$

and

$$\Lambda_k \geq \frac{n}{n+2t} C_n^t \left(\frac{k}{|\Omega|}\right)^{\frac{2t}{n}}, \quad k = 1, 2, \cdots,$$

where constant C_n is given by (1.4).

Proof. Suppose $\{u_j(x)\}$ are the eigenfunctions corresponding to eigenvalues $\{\Lambda_j\}$. Without loss of generality, we assume that $\{u_j(x)\}$ have been ortho-normalized. We see that $u_j(x) \in \dot{C}^t(\Omega) \cap C^{t-1}(\overline{\Omega}), j = 1, 2, \cdots$. Let $u_j(x) = 0$ on $R^n \backslash \Omega$. We still call these functions $u_j(x), j = 1, 2, 3, \cdots$, after such an extension. It is clear that $u_j(x) \in \dot{C}^t(R^n)$. Let

$$U(x, y) = \sum_{j=1}^{k} u_j(x) u_j(y), \quad (x, y) \in R^n \times R^n, \quad k = 1, 2, \cdots.$$

So supp $U(x, y) = \Omega \times \Omega$. Let $\widehat{U}(z, y)$ be the Fourier transformation of $U(x, y)$ about x, that is ,

$$\begin{aligned} \widehat{U}(z, y) = & \ (2\pi)^{-n/2} \int_{R^n} U(x, y) e^{izx} dx = \sum_{j=1}^{k} \widehat{u}_j(z) u_j(y) \\ = & \ (2\pi)^{-n/2} \sum_{j=1}^{k} u_j(y) \int_{R^n} u_j(x) e^{izx} dx, \end{aligned}$$

where $\widehat{u}(z)$ is the Fourier transformation of $u_j(x)$. Suppose

$$f(z) = \int_\Omega |\widehat{U}(z, y)|^2 dy.$$

Then

$$\begin{aligned} f(z) = & \int_\Omega |\widehat{U}(z, y)|^2 dy = (2\pi)^{-n} \int_\Omega \left[\int_\Omega U(x, y) e^{izx} dx\right]^2 dy \\ = & \ (2\pi)^{-n} \int_\Omega |T(z, y)|^2 dy, \end{aligned} \tag{3.2}$$

where

$$T(z, y) = \sum_{j=1}^{k} \int_\Omega u_j(x) e^{izx} dx u_j(y).$$

According to Lemma 3.1, $\{u_j(x)\}$ form the base of $L^2(\Omega)$. So $\{u_1(x), \cdots, u_k(x)\}$ expands into a subspace in $L^2(\Omega)$. Clearly $T(z, y) \in \{u_1(y), \cdots, u_k(y)\}$ is the projection of e^{izx} on $\{u_1(x), \cdots, u_k(x)\}$. Hence

$$\int_\Omega |T(z, y)|^2 dy \leq \int_\Omega |e^{izx}|^2 dx, \quad \forall z \in R^n. \tag{3.3}$$

Combining (3.2) and (3.3), we get

$$f(z) \le (2\pi)^{-n}|\Omega|. \tag{3.4}$$

Now we turn to estimating $\|f(z)\|_{L^1(R^n)}$. Using Planchel's formula, we have

$$\begin{aligned}\|f(z)\|_{L^1(R^n)} &= \int_{R^n}\int_{\Omega}|\widehat{U}(z,y)^2 dydz| \\ &= \sum_{j=1}^{k}\int_{R^n}|\widehat{u}_j(z)|^2 dz = \sum_{j=1}^{k}\int_{\Omega}|u_j(x)|^2 dx = k.\end{aligned} \tag{3.5}$$

Applying the formula $i\widehat{\frac{\partial}{\partial x_l}u_j}(z) = z_l\widehat{u}_j(z)$ $l = 1,2,\cdots,n$, we can get easily that $-\widehat{\Delta u_j}(z) = |z|^2\widehat{u}_j(z)$ $j = 1,2,\cdots,k$, and

$$\widehat{(-\Delta)^h u_j}(z) = |z|^{2h}\widehat{u}_j(z), \quad h = 1,2,\cdots.$$

Now let us estimate

$$\||z|^{2t}f(z)\|_{L^1(R^n)}.$$

Let $t = 2h$. We have

$$\begin{aligned}\||z|^{2t}f(z)\|_{L^1(R^n)} &= \int_{R^n}\int_{\Omega}|z|^{2t}|\widehat{U}(z,y)|^2 dydz \\ &= \sum_{j,l=1}^{k}\int_{R^n}|z|^{2t}\widehat{u}_j(z)\widehat{u}_l(z)dz\int_{\Omega}u_j(y)u_l(y)dy \\ &= \sum_{j=1}^{k}\int_{R^n}|\widehat{\Delta^h u_j}(z)|^2 dz \\ &= \sum_{j=1}^{k}\int_{\Omega}|\Delta^h u_j(x)|^2 dx = \sum_{j=1}^{k}\Lambda_j.\end{aligned} \tag{3.6}$$

For $t = 2h-1, h = 1,2,\cdots$, following the same method, we can get the estimate of (3.6). Combining (3.4), (3.5) and (3.6), for $K_1 = (2\pi)^{-n}|\Omega|, K_2 = \sum_{j=1}^{k}\Lambda_j$, we apply Lemma 3.2 to get

$$\begin{aligned}k \le\ & (K_1\frac{\omega_{n-1}}{n})^{\frac{2t}{n-2t}}(K_2\cdot\frac{n+2t}{n})^{\frac{n}{n+2t}} \\ =\ & (2\pi)^{\frac{-2nt}{n-2t}}\left[\frac{|\Omega|\omega_{n-1}}{n}\right]^{\frac{2t}{n-2t}}\left[\left(\sum_{j=1}^{k}\Lambda_j\right)\frac{n+2t}{n}\right]^{\frac{n}{n-2t}}.\end{aligned} \tag{3.7}$$

From (3.7) we get the estimate (1.8). Substituting $\Lambda_1,\cdots,\Lambda_{k-1}$ with Λ_k in the left side of (1.9), we obtain (1.9). Thus the proof is complete.

References

[1] Shing-Tung Yau and Richard Schoen, *Differential Geometry*, Science Press, 1988.

[2] G.N. Hile and M.H. Protter, *Inequalities for eigenvalues of the Laplacian*, Indiana Univ. Math. J., 29(1980), 523–538.

[3] R. Courant and D. Hilbert, *Methods of Mathematical Physics*, Vol. I. II. New York Interscience, 1953.

[4] Polya, *On the eigenvalues of vibrating membrane*, London Math. Soc. Ser. 3, 11(1961), 414–433.

Proceedings of the Asian Mathematical
Conference, 1990

Certain Hyperelliptic Curves of Genus Two

SANG G. HAHN
Department of Mathematics, KAIST
Yusung-gu, Taejon 305–701
South Korea

Leonard Euler studied rational solutions of several Diophantine equations of the form $y^2 = f(x)$ where $f(x)$ is a certain polynomial of degree three or four [2]. As one can easily show, his method was essentially doubling or tripling an obvious rational point of certain elliptic curves he considered. Let us explain actually how did he managed to do so. Consider the equation

$$E : y^2 = x^3 + ax + b^2,$$

where $a, b \in \mathbf{Q}, b \neq 0$. We see that E denotes an elliptic curve over $\mathbf{Q}$ if $4a^3 + 27b^4 \neq 0$. Then there is one obvious rational point $x = 0, y = b$.

Now assume that there is another rational solution $x = t, y = b + pt$ where p is rational number to be determined. Then

$$(b + pt)^2 = t^3 + at + b^2.$$

By letting $p = \frac{a}{2b}$, we get $p^2t^2 = t^3$.

So $x = p^2, y = b + p^3$ is another rational solution of E if $a \neq 0$. One can easily check that, if we let P denote the rational point $(0, b)$ of E, then the new point $(p^2, b + p^3)$ is actually $-2P$ [5].

Theorem 1. *Consider the elliptic curve*

$$E : y^2 = x^3 + ax + b^2, \qquad a, b \in \mathbf{Z}.$$

If $2b \nmid a$, then rank $E(\mathbf{Q}) > 0$.

Proof. Immediate from Lutz-Nagell theorem [5].

Suppose we were looking for a rational point of the form

$$x = t, \qquad y = b + pt + qt^2$$

where p and q are once again rational numbers to be determined. Then as a result we will get $-3P$ by letting

$$p = \frac{a}{2b}, \quad q = -\frac{p^2}{2b}, \quad t = \frac{1 - 2pq}{q^2}.$$

When combined with Lutz-Nagell theorem for torsion points we see that $E(\mathbf{Q})$ has non-zero rank if a and b are positive integers and a is not divisible by $2b$ [5].

If the degree of $f(x)$ is greater than four, then Euler's method does not work in general because the genus of the curve $E: y^2 = f(x)$ will be usually greater than one [5]. Thereby we can not assume any group structure on the set of rational points on E and by a theorem of Mordell-Faltings there are only finitely many rational points on E [3]. Similar technique applies when $f(x)$ is a quartic polynomial and $f(0)$ is a square of a non-zero rational number.

In this paper we will show that for certain quintic polynomials $f(x)$, Euler's method finds *non-obvious* rational point.

Let $f(x) = x^5 + ax^3 + bx^2 + cx + d^2, a, b, c, d \in \mathbf{Q}, d \neq o$. Let $E: y^2 = f(x)$. We have a trivial solution

$$x = 0, \qquad y = d.$$

Now assume there is another rational solution

$$x = t, \quad y = d + pt + qt^2.$$

Then

$$(d + pt + qt^2)^2 = t^5 + at^3 + bt^2 + ct + d^2.$$

We want to make this into a linear equation in t. Let

$$2dp = c, \quad p^2 + 2dp = b, \quad 2pq = a.$$

This implies

$$p = \frac{c}{2d}, \quad q = \frac{4bd^2 - c^2}{8d^3}, \quad a = \frac{c(4bd^2 - c^2)}{8d^4}.$$

We see that there is a restriction on a, i.e. $f(x)$ is not the most general quintic polynomial.

Theorem 2. *Consider*

$$y^2 = x^5 + \frac{c(4bd^2 - c^2)}{8d^4}x^3 + bx^2 + cx + d^2$$

where $b, c, d \in \mathbf{Q}, d \neq 0$. Then there is a rational solution

$$x = q^2, \qquad y = d + \frac{c}{2d}q^2 + q^5,$$

where $q = \dfrac{4bd^2 - c^2}{8d^3}$.

Example 3. Consider

$$E: y^2 = x^5 - 60x^3 + x^2 + 8x + 1.$$

Then $x = 225/4, y = 752143/32$ is a rational solution. Also $f(x) = x^5 - 60x^3 + x^2 + 8x + 1$ is irreducible over the field of rational numbers. Hence $f(x)$ has no multiple roots, the genus of E is two [4].

In case $c = 0$, we are looking at equations $y^2 = x^5 + bx^2 + d^2, b, d \in \mathbf{Q}, d \neq 0$. The discriminant of the above equation is $D = 3125d^8 + 108b^5d^2$ [3]. And $D \neq 0$ if $d \neq 2^2 3^3 r^5$ for any rational number r. Thereby we get an infinite family of genus two curves with non-obvious rational point.

We could also look for rational point of the form

$$x = t, \qquad y = d + pt + qt^2 + rt^3.$$

By letting

$$p = \frac{c}{2d}, \quad q = \frac{b - p^2}{2d}, \quad r = \frac{a - 2pq}{2d},$$

we get the equation

$$r^2t^6 + 2qrt^5 + (q^2 + 2pr)t^4 = t^5.$$

Once again, *if we assume* $q^2 + 2pr = 0$, we get

$$t = \frac{1 - 2qr}{r^2}.$$

Theorem 4. *Consider*

$$y^2 = x^5 + \left(\frac{24bc^2d^2 - 5c^4 - 16b^2d^4}{32cd^4}\right)x^3 + bx^2 + cx + d^2,$$

where $b, c, d \in \mathbf{Q}, c, d \neq 0$.

If $c^2 \neq 4bd^2$, then there is a rational solution

$$x = t, \quad y = d + pt + qt^2 + rt^3$$

where

$$p = \frac{c}{2d}, \quad q = \frac{b - p^2}{2d}, \quad r = \frac{a - 2pq}{2d}, \quad t = \frac{1 - 2qr}{r^2}.$$

Proof. Clear from the conditions $q^2 + 2pr = 0, r \neq 0$.

Remark. If $c^2 = 4bd^2$ in the above theorem, then it reduces to the case of Theorem 2.

Example 4. Consider $y^2 = x^5 + \frac{3}{32}x^3 + x^2 + x + 1$. Then $x = 4528/81, y = 1379665567/59049$ is a rational solution.

Considering that Euler was able to get $-2P$ or $-3P$, one may ask whether the non-obvious points are actually $\pm 2P$ or $\pm 3P$ in the Jacobian of the corresponding hyperelliptic curve. We will check that by David Cantor's explicit addition law [1].

In Example 2, trivial rational point was $P = (0, 1)$ and non-obvious rational point was $Q = (225/4, 752143/32)$. Let J denote the Jacobian of the curve $E : y^2 = x^5 - 60x^3 + x^2 + 8x + 1$. Let ∞ denote the point of (the model of) E at infinity. Then by explicit law, one can see that

$$Q - \infty \neq k(P - \infty) \quad \text{in} \quad J$$

for $k = 2, 3$.

We get some natural questions after this.

-How far can Euler's 'elementary' method can be extended for elliptic curves? For example, can $4P$ or $5P$ be found by a slight change of his method?

-Can this method be used to find curves with genus ≥ 2 which have 'several' non-obvious rational points?

-Under what extra condition, rank $J(\mathbf{Q}) > 0$?

References

[1] David Cantor, *Computing in the Jacobian of a Hyperelliptic curve*, Math. Comp., Vol. 48 No. 177(1987), 95–101.

[2] Leonard Euler, *Elements of Algebra*, English edition, Springer-Verlag, New York, 1972.

[3] Barry Mazur, *On the Arithmetic of Curves*, AMS Colloquium Lecture, 1984.

[4] T. Sasaki, Y. Kanada, and S.Watanabe, *Calculation of Discriminant of High Degree Equations*, Tokyo J. Math., Vol. 4 No. 2 (1981), 493–500.

[5] Joseph Silverman, *The Arithmetic of Elliptic Curves*, Springer-Verlag, New York, 1986.

Proceedings of the Asian Mathematical
Conference, 1990

Extensions of MV and BCK-algebras*

C.S. Hoo
Department of Mathematics
University of Alberta
Edmonton, Alberta, Canada

We recall (see [12]) that a BCK-algebra is a non-empty set X with a constant 0 and a binary operation $*$ satisfying the axioms:
(1) $\{(x*y)*(x*z)\}*(z*y)=0$ (2) $\{x*(x*y)\}*y=0$ (3) $x*x=0$ (4) $x*y=0$ and $y*x=0$ imply that $x=y$ (5) $0*x=0$.

We can define a partial ordering $\leq$ on X by $x\leq y$ if and only if $x*y=0$. Then 0 is the smallest element of X. We say that X is bounded if there exists a largest element 1, and that X is commutative if $x*(x*y)=y*(y*x)$ for all $x,y\in X$. In this case, $x*(x*y)$ is the greatest lower bound $x\wedge y$ of x and y.

An MV-algebra (see [1], [3], [13], [14]) is an algebra $X=(X,+,\cdot,-,0,1)$ of type (2, 2, 1, 0, 0) satisfying the identities:
(1) $(x+y)+z=x+(y+z)$ (2) $x+0=x$ (3) $x+y=y+x$ (4) $x+1=1$ (5) $\overline{\overline{x}}=x$ (6) $\overline{0}=1$ (7) $x+\overline{x}=1$ (8) $(\overline{x}+y)^{-}+y=(x+\overline{y})^{-}+x$. (9) $x\cdot y=(\overline{x}+\overline{y})^{-}$.
Then the least upper bound $x\vee y=x+\overline{x}y$ and the greatest lower bound $x\wedge y=(x+\overline{y})y$ make X into a distributive lattice with 0 and 1 as the bounds, and $x\leq y$ if and only if $x\wedge y=x$.

Given a bounded commutative BCK-algebra $X\neq\{0\}$, we can define $+,\cdot,-$ by $\overline{x}=1*x, x\cdot y=x*\overline{y}=x*(1*y)$ and $x+y=(\overline{x}\cdot\overline{y})^{-}=1*\{(1*x)*y\}$. Then we obtain an MV-algebra. Conversely, given an MV-algebra X, we can define $*$ by $x*y=x\overline{y}$. Then we obtain a bounded commutative BCK-algebra. In fact, this sets up a natural equivalence between the categories of bounded commutative BCK-algebras and MV-algebras, where the morphisms in these categories are the maps which preserve the operations and the constants (see [13]).

We say that a commutative BCK-algebra X is unitarily extendible if and only if for each $e\notin X$ we can extend the commutative BCK-algebra structure on X to $X\cup\{e\}$ with e as the largest element in $X\cup\{e\}$. An MV-algebra is unitarily extendible if and only if it is unitarily extendible as a commutative BCK-algebra. In these cases, X is a BCK-subalgebra of $X\cup\{e\}$, but observe that in the case of MV-algebras, we do not have X as an MV-subalgebra of $X\cup\{e\}$.

* This work was supported by NSERC Grant A3026

Given a BCK-algebra X and an element $e \notin X$, we can always make $X \cup \{e\}$ into a BCK-algebra by defining $e * e = 0, e * x = e$ and $x * e = 0$ for all $x \in X$. Then e is the largest element of $X \cup \{e\}$ and X is a BCK-subalgebra of $X \cup \{e\}$. This construction is due to Iséki. However, it is easily checked that $X \cup \{e\}$ is not commutative in general even if X is commutative. Further, as Cornish [6] pointed out, there is no special way to adjoin e to X so that it is the largest element of $X \cup \{e\}$. We consider here the question of how to adjoin e to a commutative BCK-algebra X so that $X \cup \{e\}$ is a commutative BCK-algebra with e as the largest element, and also the question of when this can be done. We shall show that for MV-algebras, there is a unique way, provided it can be done at all.

Theorem 1. *A commutative BCK-algebra X is unitarily extendible* $\Longleftrightarrow$ *there exists a function* $f : X - \{0\} \to X - \{0\}$ *such that* $f(f(x)) = x$ *and* $x * f(y) = y * f(x)$ *for all* $x, y \in X - \{0\}$.

Proof: $\Longrightarrow$ Suppose that the commutative BCK-algebra X is unitarily extendible to a bounded commutative BCK-algebra $X \cup \{1\}$ for $1 \notin X$. We shall consider X as a BCK-subalgebra of $X \cup \{1\}$ with the same symbol for the operation in both. Here 1 is the largest element of the algebra $X \cup \{1\}$. Define $f : X - \{0\} \to X - \{0\}$ by $f(x) = 1 * x$. We observe that if $x \in X - \{0\}$ then $1 * x \in X - \{0\} \subset X \cup \{1\}$. In fact, working in $X \cup \{1\}$, if $1 * x = 0$ then $1 \leq x$, a contradiction, while if $1 * x = 1$, then $x = 1 \wedge x = 1 * (1 * x) = 0$, again a contradiction. We check that $f(f(x)) = 1 * (1 * x) = 1 \wedge x = x$, and $x * f(y) = x * (1 * y) = \{1 * (1 * x)\} * (1 * y) = \{1 * (1 * y)\} * (1 * x) = (1 \wedge y) * (1 * x) = y * (1 * x) = y * f(x)$ for all $x, y \in X - \{0\}$.

$\Longleftarrow$ Suppose that $f : X - \{0\} \to X - \{0\}$ exists with $f(f(x)) = x$ and $x * f(y) = y * f(x)$ for all $x, y \in X - \{0\}$. Let $1 \notin X$. We can extend the BCK-algebra operation $*$ on X to $X \cup \{1\}$ by: $1 * x = f(x)$ for all $x \in X - \{0\}, 1 * 0 = 1, 1 * 1 = 0$, and $x * 1 = 0$ for all $x \in X$. We now verify the BCK-algebra axioms for $X \cup \{1\}$. Clearly, 0 and 1 are the bounds for $X \cup \{1\}$ assuming that it is a BCK-algebra. Axioms (3), (4) and (5) are obvious. We first verify that $*$ satisfies $x * (x * y) = y * (y * x)$ for all $x, y \in X \cup \{1\}$. This is clearly true if $x, y \in X$ or if $x = y = 1$. We need only check the case $x \in X$ and $y = 1$. In this case we have $x * (x * y) = x * (x * 1) = x * 0 = x$, while $y * (y * x) = 1 * (1 * x)$. This is $f(f(x)) = x$ if $x \neq 0$ and is $1 * (1 * 0) = 1 * 1 = 0 = x$ if $x = 0$. Thus $x * (x * y) = y * (y * x)$ for all $x, y \in X \cup \{1\}$. It remains for us to verify Axioms (1) and (2). Axiom (2) is easily verified. In order to verify Axiom (1), let $P = \{(x * y) * (x * z)\} * (z * y)$. If $x, y, z \in X$ or if $x = y = z = 1$, then clearly $P = 0$. If $x, y \in X$ and $z = 1$ then $P = \{(x * y) * 0\} * (1 * y) = (x * y) * (1 * y)$. If $y = 0$, this gives $P = (x * 0) * (1 * 0) = x * 1 = 0$. If $y \neq 0$, we have $P = (x * y) * f(y) = \{x * f(f(y))\} * f(y)$. We have two subcases here: $x = 0$ and $x \neq 0$. If $x = 0$ we have $P = (0 * y) * f(y) = 0$. If $x \neq 0$, we have $P = \{f(y) * f(x)\} * f(y) = \{f(y) * f(y)\} * f(x) = 0 * f(x) = 0$. If $x, z \in X$ and $y = 1$, we have $P = \{0 * (x * z)\} * 0 = 0$. If $y, z \in X$ and $x = 1$, we have $P = \{(1 * y) * (1 * z)\} * (z * y)$. If $y = 0$, this gives $P = \{1 * (1 * z)\} * z$. Now if $z = 0$ also, we have $P = \{1 * (1 * 0)\} * 0$ while if $z \neq 0$, we have $P = \{f(f(z))\} * z = z * z = 0$. We now consider the case $y \neq 0$. In this case $P = \{f(y) * (1 * z)\} * (z * y)$. If $z \neq 0$, this is $P = \{f(y) * f(z)\} * (z * y) = \{z * f(f(y))\} * (z * y) = (z * y) * (z * y) = 0$ while if $z = 0$, this is $P = \{f(y) * 1\} * (0 * y) = 0 * (0 * y) = 0$. We now dispose off the remaining cases. If $x \in X$ and $y = z = 1$, then $P = (0 * 0) * 0 = 0$. If $y \in X$ and $x = z = 1$, then

$P = \{(1 * y) * 0\} * (1 * y)$. If $y = 0$ this is $P = \{(1 * 0) * 0\} * (1 * 0) = 0$ while if $y \neq 0$ this is $P\{f(y) * 0\} * f(y) = f(y) * f(y) = 0$. If $z \in X$ and $x = y = 1$, then $P = \{0 * (1 * z)\} * 0 = 0$. Thus $X \cup \{1\}$ is a commutative BCK-algebra.

Remark. Thus if a commutative BCK-algebra X admits a function $f : X - \{0\} \to X - \{0\}$ satisfying $f(f(x)) = x$ and $x * f(y) = y * f(x)$ for all $x, y \in X - \{0\}$, then for each $e \notin X$ we can make $X \cup \{e\}$ into a bounded commutative BCK-algebra, that is, an MV-algebra, by $x * e = 0$ for all $x \in X, e * e = 0, e * 0 = e$ and $e * x = f(x)$ for all $x \in X - \{0\}$.

Theorem 2. *An MV-algebra X is unitarily extendible $\iff$ X has a unique atom which is also the smallest element of $X - \{0\}$.*

Proof. $\Longrightarrow$ If X is unitarily extendible, then it is unitarily extendible as a commutative BCK-algebra. Hence by Theorem 1, there exists a function $f : X - \{0\} \to X - \{0\}$ such that $f(f(x)) = x$ and $x * f(y) = y * f(x)$ for all $x, y \in X - \{0\}$. Let $1 \in X$ be the largest element of X and let $x_0 = f(1)$. Then for all $x \in X - \{0\}$ we have $x_0 * x = x_0 * f(f(x)) = f(x) * f(x_0) = f(x) * f(f(1)) = f(x) * 1 = 0$. Thus $x_0 \leq x$ for all $x \in X - \{0\}$. Clearly x_0 is the unique atom of X.

$\Longleftarrow$ Suppose that $x_0 \in X$ is the unique atom of X and is the smallest element of $X - \{0\}$. Define $f : X - \{0\} \to X - \{0\}$ by $f(x) = 1 * (x * x_0)$ for $x \in X - \{0\}$. Observe that if $x \neq 0$ then $1 * (x * x_0) \neq 0$ because if $1 * (x * x_0) = 0$ then $x * x_0 = 1$. This means that $x = 1$ and hence $x_0 = 1 * (1 * x_0) = 1 * (x * x_0) = 0$, a contradiction. Using the MV-algebra operations on X, we have for $x \neq 0, f(x) = (x\overline{x}_0)^- = \overline{x} + x_0$. Hence $f(f(x)) = \overline{f(x)} + x_0 = (\overline{x} + x_0)^- + x_0 = x\overline{x}_0 + x_0 = x_0 \vee x = x$. Also, if $x, y \in X - \{0\}$, then $x * f(y) = x * (\overline{y} + x_0) = x * (\overline{y} + x_0)^- = xy\overline{x}_0$ while $y * f(x) = y * (\overline{x} + x_0) = y(\overline{x} + x_0)^- = yx\overline{x}_0$. Thus $x * f(y) = y * f(x)$. Since X is a commutative BCK-algebra, it follows by Theorem 1 that X is unitarily extendible as a commutative BCK-algebra and hence also as an MV-algebra.

Remark. Observe that by the above if an MV-algebra X has a unique atom x_0 which is also the smallest element of $X - \{0\}$, then for each $e \notin X$, we can extend the BCK-algebra operation $*$ on X to $X \cup \{e\}$ so that e is the largest element of the MV-algebra $X \cup \{e\}$ as follows: $x * e = 0$ for all $x \in X, e * e = 0, e * 0 = e$ and $e * x = 1 * (x * x_0)$ for all $x \in X - \{0\}$. In this case X is a subalgebra of $X \cup \{e\}$ as commutative BCK-algebras, but X is not a subalgebra of $X \cup \{e\}$ as MV-algebras since $1 \in X$ is the largest element while $e \in X \cup \{e\}$ is the largest element.

Theorem 3. *If X is a unitarily extendible MV-algebra, then there exists a unique function $f : X - \{0\} \to X - \{0\}$ such that $f(f(x)) = x$ and $x * f(y) = y * f(x)$ for all $x, y \in X - \{0\}$. Hence for each $e \notin X$, there is unique way of extending the BCK-algebra operation on X to $X \cup \{e\}$ so that e is the largest element of the MV-algebra $X \cup \{e\}$.*

Proof. Since X is a unitarily extendible MV-algebra, there exists a function $f : X - \{0\} \to X - \{0\}$ such that $f(f(x) = x)$ and $x * f(y) = y * f(x)$ for all $x, y \in X - \{0\}$. Let $x_0 = f(1)$ where 1 is the largest element of X. Then x_0 is the unique atom of X which is also the smallest element of $X - \{0\}$. For each $x \in X - \{0\}$, we have $1 * f(x) = x * f(1) = x * x_0$. This is an equation in X and is true for any such function f. Thus if $g : X - \{0\} \to X - \{0\}$ is also a function such that $g(g(x)) = x$ and $x * g(y) = y * g(x)$ for all $x, y \in X - \{0\}$, then $1 * f(x) = x * x_0 = 1 * g(x)$. Hence $1 * \{1 * f(x)\} = 1 * \{1 * g(x)\}$, that is $f(x) = g(x)$ for all

$x \in X - \{0\}$. Thus there is a unique such function. Since $e \notin X$ is adjoined to X to make an MV-algebra $X \cup \{e\}$ with largest element e by $x * e = 0$ for all $x \in X, e * e = 0, e * 0 = e$ and $e * x = 1 * (x * x_0)$ for $x \in X - \{0\}$, where $x_0 = f(1)$ for such a function f, it follows that the extension is unique.

Corollary 4. *Let X be an MV-algebra and let $e \notin X$. If X is a unitarily extendible MV-algebra, then $X \cup \{e\}$ is also a unitarily extendible MV-algebra where e is the largest element of $X \cup \{e\}$.*

Let G be a commutative l-group. Recall that an element u in the positive cone G_+ is an order unit for G if for each $x \in G$ we have $x \leq nu$ for some integer n. Let $G[u] = \{x \in G \mid 0 \leq x \leq u\}$. If $x, y \in G[u]$, we can define $x * y = (x - y)_+ = (x - y) \vee 0$. Then $x * y \in G[u]$ and this makes $G[u]$ into a bounded commutative BCK-algebra (see [3], [5], [14]). The MV-algebra operations $\oplus, \cdot, -$ are given by $\bar{x} = u - x, x \oplus y = (x + y) \wedge u$, and $x \cdot y = (\bar{x} \oplus \bar{y})^- = (x + y - u) \vee 0 = (x + y - u)_+$. The lattice operations in $G[u]$ coincide with those induced from G. It is shown in [14] that given an MV-algebra A, we can find a commutative l-group G and an order unit u such that $A \cong G[u]$. Further, (G, u) is uniquely determined by A, up to isomorphism. Let $G(u) = \{x \in G \mid 0 \leq x < u\}$. Observe that if $x, y \in G(u)$ then $x * y \in G(u)$, that is, $G(u)$ is a commutative BCK-subalgebra of $G[u]$. Clearly $G(u)$ is a unitarily extendible commutative BCK-algebra.

Theorem 5. *A commutative BCK-algebra X is unitarily extendible $\Longleftrightarrow$ there exists a commutative l-group G and an order unit u such that $X \cong G(u)$.*

Proof: $\Longleftarrow$ This is obvious.

$\Longrightarrow$ Let X be a unitarily extendible commutative BCK-algebra and let $e \notin X$. Then we can extend the BCK-algebra operation on X to $X \cup \{e\}$ so that $X \cup \{e\}$ is a bounded commutative BCK-algebra with e as its largest element, that is, an MV-algebra. Then by [14] we can find a commutative l-group G and an order unit u so that $X \cup \{e\} \cong G[u]$ as MV-algebras, and hence also as BCK-algebras. Clearly, this isomorphism restricts to an isomorphism $X \cong G(u)$ of commutative BCK-algebras.

In the case of MV-algebras, we can determine, up to a certain indeterminacy, all the unitarily extendible algebras. They include the MV-algebras of C.C. Chang [3]. For each integer $m \geq 1$, let $S(m) = \{0, \frac{1}{m}, \frac{2}{m}, \cdots, \frac{m-1}{m}, 1\}$. This is an MV-algebra with operations given by

$$\begin{aligned}
\left(\frac{i}{m}\right) + \left(\frac{j}{m}\right) &= \min\left\{1, \frac{i+j}{m}\right\}, \\
\left(\frac{i}{m}\right) \cdot \left(\frac{j}{m}\right) &= \max\left\{0, \frac{i+j-m}{m}\right\}, \\
\left(\frac{i}{m}\right)^- &= \frac{m-i}{m}.
\end{aligned}$$

The BCK-algebra operation $*$ is given by

$$\left(\frac{i}{m}\right) * \left(\frac{j}{m}\right) = \begin{cases} \frac{i-j}{m} & \text{if } i > j, \\ 0 & \text{if } i \leq j. \end{cases}$$

We also have the algebra $C = \{0, c, c + c, \cdots, 1 - c - c, 1 - c, 1\}$. If we write $oc = 0, nc = c + \cdots + c$ (n terms), $1 - oc = 1, 1 - nc = 1 - c - \cdots - c$ (n terms in c), then the MV-algebra operations are given by

$$\begin{aligned}(nc)+(mc) &= (m+n)c,\\ (nc)+(1-mc) &= \begin{cases}1 & \text{if } m\le n,\\ 1-(m-n)c & \text{if } m>n,\end{cases}\\ (1-nc)+(1-mc) &= 1,\\ \overline{(nc)}=1-nc,\quad \overline{(1-nc)} &= nc \quad\text{and}\quad x\cdot y=(\overline{x}+\overline{y})^{-}\end{aligned}$$

for all x, y.

The corresponding BCK-algebra operation is given by

$$\begin{aligned}(mc)*(nc) &= \begin{cases}0 & \text{if } m\le n,\\ (m-n)c & \text{if } m>n,\end{cases}\\ (mc)*(1-nc) &= 0,\\ (1-mc)*(nc) &= 1-(m+n)c,\\ (1-mc)*(1-nc) &= \begin{cases}0 & \text{if } n\le m,\\ (n-m)c & \text{if } n>m.\end{cases}\end{aligned}$$

We first make some preliminary observations. Suppose that X is an MV-algebra with a unique atom x_0 which is also the smallest element of $X-\{0\}$. Then $\{0\}$ is a prime ideal. In fact, it can be checked that $\{x\in X \mid x\wedge x_0=0\}$ is an ideal and that it is $\{0\}$. By Theorem 2.9 of [7] it is a prime ideal. It follows that X is linearly ordered. Indeed, for any $x, y\in X$, we have $(x*y)\wedge(y*x)=0$, a result true in any MV-algebra. Since $\{0\}$ is a prime ideal, it follows that either $x*y=0$ or $y*x=0$, that is, either $x\le y$ or $y\le x$. Also we can check that $x_0^2=x_0 \Longleftrightarrow X\cong\{0,1\}=S(1)$. The following results can be found in [10].

Lemma 6. *Suppose that $nx_0\le\overline{x}_0$ for all integers $n\ge 0$. Then $(\overline{x}_0)^m+(\overline{x}_0)^n=1$ for all integers $m\ge 0, n\ge 0$.*

Lemma 7. *Suppose that $nx_0\le\overline{x}_0$ for all integers $n\ge 0$. Let $\varphi: C\to X$ be given by $\varphi(nc)=nx_0$ and $\varphi(1-nc)=(\overline{x}_0)^n$ for all integers $n\ge 0$. Then φ is a monomorphism of MV-algebras.*

We now state and prove our characterization theorem for unitarily extendible MV-algebras.

Theorem 8. *An MV-algebra X is unitarily extendible $\Longleftrightarrow$ either $X\cong S(m)$ for some integer $m\ge 1$, or X is linearly ordered and there exists a monomorphism $\varphi: C\to X$ such that $\{nx_0\}_{n=0}^{\infty}$ is an ideal of X, where $x_0=\varphi(c)$.*

Proof. $\Longleftarrow$ If $X\cong S(m)$ for some integer $m\ge 1$, then X has a unique atom which is the smallest element of $X-\{0\}$ and hence by Theorem 2, X is unitarily extendible. If X is linearly ordered and there exists a monomorphism $\varphi: C\to X$ such that $\{nx_0\}_{n=0}^{\infty}$ is an ideal of X where $x_0=\varphi(c)$, then we observe that x_0 is an atom. In fact, suppose that $x\le x_0$. Then $x\le x_0\in\{nx_0\}_{n=0}^{\infty}$ and $\{nx_0\}_{n=0}^{\infty}$ is an ideal of X. Hence $x=nx_0$ for some integer $n\ge 0$. Thus $x=n\varphi(c)=\varphi(nc)$. This means that $\varphi(nc)=x\le x_0=\varphi(c)$. Since

φ is a monomorphism, we have $nc \leq c$ and hence $n = 0$ or $n = 1$. Thus either $x = x_0$ or $x = 0$, proving that x_0 is an atom. Now suppose that x is an element of $X - \{0\}$. Since X is linearly ordered and x_0 is an atom, we have $x_0 \leq x$. Thus x_0 is the smallest element of $X - \{0\}$, and hence is the unique atom of X which is also the smallest element of $X - \{0\}$. By Theorem 2, it follows that X is unitarily extendible.

$\Longrightarrow$ Suppose that X is unitarily extendible. Then by Theorem 2, X has a unique atom x_0 which is the smallest element of $X - \{0\}$. Then either $x_0^2 = 0$ or $x_0^2 = x_0$. If $x_0^2 = x_0$ then by our comments above, $X \cong S(1)$. If $x_0^2 = 0$, that is, $x_0 * \overline{x}_0 = 0$, then $x_0 \leq \overline{x}_0$. It follows that either (i) there is an integer $n > 0$ such that $nx_0 \leq \overline{x}_0$ but $(n+1)x_0 \not\leq \overline{x}_0$ or (ii) $nx_0 < \overline{x}_0$ for all integers $n \geq 0$. In case (i) we have $x_0 \not\leq (\overline{x}_0)^{n+1}$. Since x_0 is the smallest element of $X - \{0\}$, it follows that $(\overline{x}_0)^{n+1} = 0$, that is, $(n+1)x_0 = 1$. On the other hand, $nx_0 \leq \overline{x}_0$ and hence $nx_0 \neq 1$. Thus x_0 is an atom of order $n+1$. By [8] Theorem 4.8, it follows that $X \cong S(n+1)$. It remains for us to consider case (ii). By our comments above, X is linearly ordered, and by Lemma 7, we have a monomorphism $\varphi : C \to X$ given by $\varphi(nc) = nx_0$ and $\varphi(1 - nc) = (nx_0)^- = (\overline{x}_0)^n$ for all integers $n \geq 0$. Since x_0 is an atom, it follows that $\{nx_0\}_{n=0}^{\infty}$ is an ideal, in fact, the ideal generated by the atom $x_0 = \varphi(c)$. This proves the theorem.

References

[1] L.P. Belluce, *Semisimple algebras of infinite valued logic and Bold fuzzy set theory*, Can J. Math., 38(1986), 1356–1379.

[2] A. Bigard, K. Keimel and S. Wolfenstein, *Groupes et anneaux réticulés*, Lecture Notes in Mathematics (1977), 608 Springer-Verlag, New York.

[3] C.C. Chang, *Algebraic analysis of many valued logics*, Trans. Amer. Math. Soc., 88(1958), 467–490.

[4] C.C. Chang, *A new proof of the completeness of the Lukasiewicz axioms*. Trans. Amer. Math. Soc., 93(1959), 74–80.

[5] W.H. Cornish, *Lattice ordered groups and BCK-algebras*, Math. Japon, 25(1980), 471–476.

[6] W.H. Cornish, *A large variety of BCK-algebras*, Math. Japon, 26(1981), 339–344.

[7] C.S. Hoo, *Bounded commutative BCK-algebra satisfying d.c.c.*, Math. Japon, 32(1987), 217–225.

[8] C.S. Hoo, *MV-algebras, ideals and semisimplicity*, Math. Japon, 34(1989), 563–583.

[9] C.S. Hoo, *Unitary extensions of MV and BCK-algebras*, (Submitted).

[10] C.S. Hoo, *Unitary extensions of MV and BCK-algebras (II)*, Math. Japon (to appear).

[11] K. Iséki and S. Tanaka, *Ideal theory of BCK-algebras*, Math. Japon, 21(1976), 351–366.

[12] K. Iséki and S. Tanaka, *An introduction to the theory of BCK-algebras*, Math. Japon, 23(1978), 1–26.

[13] D. Mundici, *MV-algebra are categorically equivalent to bounded commutative BCK-algebras*, Math. Japon, 31(1986), 889–894.

[14] D. Mundici, *Interpretation of AF C^*-algebras in Lukasiewicz sentential calculus*, J. Func. Analy., 65(1986), 15–63.

Proceedings of the Asian Mathematical
Conference, 1990

On a Milin's Conjecture and Its Applications

HU KE
Department of Mathematics
Jiangxi Normal University
Nanchang 330027, China

Let $f(z)$ be analytic in an open set D and $n(w) = n(w, D, f)$ the number of roots of the equation $f(z) = w$ in D. We write

$$P(R) = \frac{1}{2\pi}\int_0^{2\pi} n(Re^{i\theta})\,d\theta.$$

If $P(R) \leq P$, we call this function to be mean P-valent. Let $M = \{f(z) = z+\sum_{n=2}^{\infty} a_n z^n : f$ is analytic and mean 1-valent in $|z| < 1\}$. If $f \in M$, let $f^{\lambda}(z) = z^{\lambda}(1+\sum_{n=1}^{\infty} D_n(\lambda)z^n), \lambda > 0$. Let $M(\alpha)$ be the class of functions in M with $\lim(1-r)^2 M(r,f) = \alpha > 0$, where $M(r,f) = \max_{|z|=r}|f(z)|$. In this paper, we have

Theorem 1(I.M. Milin Conjecture[1]). *Let $f \in M$ and $\log f(z)/z = \sum_{k=1}^{\infty}\gamma_k z_0^{*}$. Then*

$$\sum_{k=1}^{\infty} k|\gamma_k|^2 r^{2k} \leq \frac{1}{2} \log M(r,f)/r. \tag{0.1}$$

In [2], this theorem has been proved for $f \in S$.

Theorem 2. *If $f \in M$, then*

$$\Big||D_n^2(1/2)| - |D_{n-1}^2(1/2)|\Big| \leq 4e^{1-c_n-1/2}, \quad n = 2,3,\cdots, \tag{0.2}$$

$$\Big||a_n| - |a_{n-1}|\Big| < 2^{3/2}e^{1-c/2}, \quad n = 2,3,\cdots,^{[3]} \tag{0.3}$$

where c is the Euler constant.

Theorem 3. *Let $f \in M(\alpha)$ and $\log f(z)/z = \sum_{k=1}^{\infty}\gamma_k z^k$. Then*

$$\sum_{k=1}^{\infty} k\left|\gamma_k - \frac{1}{k}\right|^2 < \log\frac{2}{\sqrt{\alpha}},$$

where $\theta = 0$ is a direction of maximum growth for $f(z)$.

If $f \in S$, the theorem is due to Basilevic[2].

Theorem 4. *let $f \in M(\alpha), W(z) = \log\phi^{\lambda}(z) = \sum_{k=1}^{\infty} A_k z^k, \phi(z) = (1-z)^2 f(z)/z$. Then*

$$\{\phi^{\lambda}(z)(1-z)^{-h}\}_n/d_n(h) \sim \phi^{\lambda}(r) \sim \exp\left\{\sum_{k=1}^{n} A_k\right\} \quad (h > \frac{1}{2}), \tag{0.5}$$

as $n \to \infty$, where $\theta = 0$ is a direction of maximum growth for $f(z)$ and $r = 1 - t/n, m < t < M, d_n(h) = h(h+1)(h+2)\cdots(h-n+1)/n!$

If $f(z) \in S(\alpha)$, the equation (0.5) is due to I.M. Milin.

§1. Proof of Theorem 1

In order to prove Theorem 1, we need some lemmas.

Lemma 1[3]. *If $f(z) \in M$, then*

$$1 \le \frac{(1+|z|)}{|z|}|f(z)| \le \frac{(1+r)}{|z|}|f(z)|, \quad |z| \le r < 1. \tag{1.1}$$

Lemma 2[3]. *Let $\psi(z) \in M$. Then*

$$A(r,\psi) = \frac{1}{\pi}\int_0^r t\,dt \int_0^{2\pi} |\psi'(te^{i\theta})|^2\,d\theta \le M^2(r,\psi), \quad 0 < r < 1. \tag{1.2}$$

Lemma 3. *Let $f \in M$ and $\{f(z)/z\}^{1/p} = 1 + \sum_{k=1}^{\infty} D_k(1/p)z^k$, where p is a positive integer. Then*

$$1 + \sum_{k=1}^{\infty}(kp+1)\,|D_k(1/p)|^2 r^{2k} \le \{M(r,f)/r\}^{2/p}, \quad 0 < r < 1. \tag{1.3}$$

Proof. Since $f \in M$, then $\psi_p(z) = \{f(z^p)\}^{1/p} = z + \sum_{k=1}^{\infty} D_k(1/p)z^{kp+1} \in M$[3] for $p = 1, 2, \cdots$. Using Lemma 2 and taking $r \in (0,1)$, we have

$$\begin{aligned} A(r^{1/p},\psi_p)/r^{-2/p} &= \frac{1}{\pi r^{2/p}}\int_0^{r^{\frac{1}{p}}} t\,dt \int_0^{2\pi} |\psi_p'(te^{i\theta})|^2\,d\theta \\ &\le M^2(r^{1/p},\psi_p)/r^{2/p} = \{M(r,f)/r\}^{2/p}. \end{aligned}$$

But

$$\frac{1}{\pi}\int_0^{r^{1/p}} t\,dt \int_0^{2\pi} |\psi'(te^{i\theta})|^2\,d\theta = r^{2/p}\{1 + \sum_{k=1}^{\infty}(kp+1)\,|D_k(1/p)|^2 r^{2k}.$$

(1.3) follows at once.

Lemma 4. *Let $f \in M$ and $F(z) = \{f(z)/z\}^{1/p} = 1 + \sum_{k=1}^{\infty} D_k(1/p)z^k$ where p is a positive integer. Then*

$$p^2/\pi \int_0^r t\,dt \int_0^{2\pi} |F'(te^{i\theta})|^2\,d\theta < p\{(M(r,f)/r)^{2/p} - 1\}. \tag{1.4}$$

Proof. Evidently we have

$$\begin{aligned} & p^2/\pi \int_0^r t\,dt \int_0^{2\pi} |F'(te^{i\theta})|^2\,d\theta = p^2 \sum_{k=1}^{\infty} k|D_k(1/p)|^2 r^{2k} \\ < \;& p^2 \sum_{k=1}^{\infty} k|D(1/p)|^2 r^{2k} + p \sum_{k=1}^{\infty} |D_k(1/p)|^2 r^{2k} \\ = \;& p \sum_{k=1}^{\infty} (kp+1)|D_k(1/p)|^2 r^{2k}. \end{aligned} \tag{1.5}$$

From the inequality (1.3) we see

$$1 + \sum_{k=1}^{\infty} (kp+1)|D_k(1/p)|^2 r^{2k} \le \{M(r,f)/r\}^{2/p}, \quad 0 \le r < 1.$$

Therefore

$$p^2/\pi \int_0^r t\,dt \int_0^{2\pi} |F'(te^{i\theta})|^2\,d\theta < p\{(M(r,f)/r)^{2/p} - 1\}.$$

Hence the lemma is proved.

Lemma 5. *Let $f \in M$, $w(z) = \log\{f(z)/z\}$ and $F(z) = \{f(z)/z\}^{1/p}$. Then*

$$\int_0^r t\,dt \int_0^{2\pi} |w'(te^{i\theta})|^2\,d\theta < p^2(1+r)^{4/p} \int_0^r t\,dt \int_0^{2\pi} |F'(te^{i\theta})|^2 d\theta \tag{1.6}$$

for $r \in (0,1)$.

Proof. It is obvious that

$$\begin{aligned} F'(z) = \;& \frac{d}{dz}(f(z)/z)^{1/p} = \frac{1}{p}\left\{\frac{d}{dz}[f(z)/z]\{(f(z)/z)^{1/p-1}\}\right\} \\ = \;& \frac{1}{p}(f(z)/z)'(f(z)/z)^{1/p-1} = \frac{1}{p}(\log f(z)/z)'(f(z)/z)^{1/p} \\ = \;& \frac{1}{p}w'(z)F(z). \end{aligned}$$

Applying (1.1) to the identity, we have

$$\begin{aligned} & |w'(z)| \le (1+r)^{2/p}|f(z)|/z|^{1/p}|w'(z)| \\ = \;& (1+r)^{2/p}|F(z)|\,|w'(z)| = p(1+r)^{2/p}|F'(z)| \end{aligned} \tag{1.7}$$

for $|z| \le r < 1$. Hence (1.6) follows from (1.7) immedately.

Now we are going to the proof of Theorem 1. In virture of (1.4) and (1.6), we get

$$\begin{aligned} & \frac{1}{\pi}\int_0^r t\,dt \int_0^{2\pi} |w'(te^{i\theta})|^2\,d\theta \\ < \;& p^2(1+r)^{4/p}/\pi \int_0^r t\,dt \int_0^{2\pi} |F'(te^{i\theta})|^2\,d\theta \\ \le \;& p(1+r)^{4/p}\{(M(r,f)/z)^{2/p} - 1\}. \end{aligned} \tag{1.8}$$

Since $w(z) = \log(f(z)/z) = 2\sum_{k=1}^{\infty} \gamma_k z^k$, and

$$4\sum_{k=1}^{\infty} k|\gamma_k|^2 r^{2k} = \frac{1}{\pi}\int_0^r t\,dt \int_0^{2\pi} |w'(te^{i\theta})|^2\,d\theta.$$

Now, using the last two relations, we find that

$$\begin{aligned} 4\sum_{k=1}^{\infty} K|\gamma_k|^2 r^{2k} &= \frac{1}{\pi}\int_0^{2\pi} tdt \int_0^{2\pi} |w'(te^{i\theta})|^2 d\theta \\ &< p^2(1+r)^{4/p}/\pi \int_0^r tdt \int_0^{2\pi} |F'(te^{i\theta})|^2 d\theta \\ &\le p(1+r)^{4/p}\{(M(r,f)/r)^{2/p} - 1\} \\ &= p(1+r)^{4/p}\{\exp[2/p\log(M(r,f)/r)] - 1\}. \end{aligned} \tag{1.9}$$

Put $p \to \infty$ in (1.9). Thus we get the result.

§2. Proof of Theorem 2

The proof of Theorem 2 is based on Theorem 1 and the following lemma.

Lemma 6[3]. *If $f \in M$, then*

$$(1-r_2)^2 M(r_2,f)/r_2 \le (1-r_1)^2 M(r_1,f)/r_1, \quad 0 < r_1 \le r_2 < 1. \tag{2.1}$$

Since $\psi_{1/2}(z) = (1-z)(f(z)/z)^{1/2} = 1 + \sum_{k=1}^{\infty}\{D_k(1/2) - D_{k-1}(1/2)\}z^k, D_0 = 1$. Without loss any generality, we may assume that $M(r^2,f) = |f(r^2)|$ and $r = e^{-1/2n}$. By Lebejev-Milin inequality , we have

$$\begin{aligned} \big||D_n(1/2)| - |D_{n-1}(1/2)|\big| r^n &\le \exp 1/2\left\{\sum_{k=1}^{n} k|\gamma_k - 1/k|^2 r^{2k} - \sum_{k=1}^{n} 1/k\right\} \\ &\le \exp 1/2\left\{\sum_{k=1}^{\infty} K|\gamma_k - 1/k|^2 r^{2k} - \sum_{k=1}^{n} 1/k\right\}, \end{aligned} \tag{2.2}$$

and

$$\begin{aligned} |D_n(1/2)|r^n &< \exp 1/2\left\{\sum_{k=1}^{\infty} k|\gamma_k|^2 r^{2k} - \sum_{k=1}^{n} 1/k\right\}, \\ |D_{n-1}(1/2)|r^n &< \exp 1/2\left\{\sum_{k=1}^{\infty} k|\gamma_k|^2 r^{2k} - \sum_{k=1}^{n} 1/k\right\}. \end{aligned} \tag{2.3}$$

From (2.2) and (2.3) we obtain

$$\begin{aligned}
&\left||D(1/2)|^2 - |D(1/2)|^2\right| \\
< \;& 2\exp\left\{\sum_{k=1}^{\infty} k|\gamma_k|^2 r^{2k} - \mathrm{Re}\sum_{k=1}^{\infty} \gamma_k r^{2k} - 1/2\log 1/(1-r^2) - \sum_{k=1}^{n} 1/k\right\} \\
= \;& 2\exp\left\{\sum_{k=1}^{\infty} k|\gamma_k|^2 r^{2k} - 1/2\log M(r^2,f)(1-r^2)/r^2 - \sum_{k=1}^{n} 1/k\right\} \\
\leq \;& 2\exp 1/2\left\{\log M(r,f)r/M(r^2,f)(1-r^2) - \sum_{k=1}^{n} 1/k\right\} \\
\leq \;& 2(1+r/1-r)^{1/2}\exp\left\{-\sum_{k=1}^{n} 1/k\right\}.
\end{aligned}$$

Here the inequality (2.1) and Theorem 1 have been used. Nothing that $r = e^{-1/2n}$ and $\log(n+1/2) < \sum_{k=1}^{n} 1/k - c$ (0.2) follows from (2.4).

The proof of (0.3) is analogous to those on univalent functions.[4] We omit it here.

§3. Proof of Theorem 3

Using Theorem 1 and taking $M(r^2, f) = |f(r^2)|$ and $r \in (0,1)$, we have

$$\begin{aligned}
\sum_{k=1}^{\infty} K|\gamma_k - 1/k|^2 r^{2k} &= \sum_{k=1}^{\infty} k|\gamma_k|^2 r^{2k} - 2\mathrm{Re}\sum_{k=1}^{\infty} \gamma_k r^{2k} + \log 1/(1-r^2) \\
&= \sum_{k=1}^{\infty} k|\gamma_k|^2 r2k - \log M(r^2,f)(1-r^2)/r^2 \\
&\leq \log(M(r,f)/r)^{1/2}/M(r^2,f)r^{-2}(1-r^2) \to \log 2/\sqrt{2} \qquad (r\to 1).
\end{aligned}$$

The proof is complete.

The proof of Theorem 4 is similar to the proof of Milin's Theorem[2] for $f \in S_0$. We omit it in this paper.

References

[1] M. Milin, *On a conjecture for the logarithmic coefficients of univalent functions* (in Russion) Zap. Nauchy. Smin. Leningrad Ottdd. Inst. Stekloua (LOMI), 125(1983), 135–143.

[2] M. Milin, *Univalent functions and orthonomal system*, 32–36.

[3] W.K. Hayman, *Multivalent functions*, Cambridge Univ. Press. Cambridge, 1958, 17–100.

[4] Hu Ke, *A short proof of Haymam's Theorem*, Kexue Tongbao, 14(1984), 894.

Proceedings of the Asian Mathematical
Conference, 1990

Limit Cycle for Cubic System of Kolmogorov Type with a Conic Algebraic Trajectory

HUANG QIYU
Department of Mathematics
Fujian Normal University
Fuzhou 350007, China

In this paper we shall deal with the Kolmogorov cubic system appearing in an ecosystem:

$$\begin{aligned} \frac{dx}{dt} &= xf_2(x,y) \\ \frac{dy}{dt} &= yg_2(x,y) \end{aligned} \quad \text{or} \quad \frac{dy}{dx} = \frac{yg_2(x,y)}{xf_2(x,y)}, \tag{$\overline{\text{E}}$)$_3$}$$

where $f_2(x,y)$ and $g_2(x,y)$ are quadratic polynomials of x and y. We study the existence of limit cycles for the cubic system $(\bar{E})_3$ with a conic algebraic trajectory,

$$F(x,y) = ax^2 + 2bxy + cy^2 + dx + ey + f = 0.$$

§1. $b^2 - ac \neq 0$

The system $(\overline{E})_3$ has a solution of hyperbolic or elliptic curve $F(x,y) = 0$. The general form of the cubic differential equation is

$$\frac{dy}{dx} = \frac{\sum\limits_{0 \le i+j \le 3} a_{ij} x^i y^j}{\sum\limits_{0 \le i+j \le 3} b_{ij} x^i y^j}. \tag{E)$_3$}$$

Lemma 1. *The necessary and sufficient condition that the differential equation* (E)$_3$ *has the hyperbolic or elliptic curve* $F(x,y) = 0$ *as its solution is that* (E)$_3$ *may be reduced to the following form:*

$$\frac{dy}{dx} = \frac{F_x(x,y)P_2(x,y) + H_1(x,y)F(x,y)}{-F_y(x,y)P_2(x,y) + \overline{H}_1(x,y)F(x,y)}, \tag{1}$$

where $P_2(x,y)$ *is a second-degree polynomial of* x *and* y, $H_1(x,y)$ *and* $\overline{H}_1(x,y)$ *are the linear polynomials of* x *and* y, *and*

$$F(x,y) = ax^2 + 2bxy + cy^2 + dx + ey + f \quad (b^2 - ac \neq 0).$$

Theorem 1. *The necessary and sufficient condition that the cubic system* $(\overline{\mathrm{E}})_3$ *of Kolmogorov type has the central conic* $F(x, y) = 0$ (i.e. $F(x, y) = 0$ *is a hyperbola or ellipse*) *as its particular integral curve is that* $(\overline{\mathrm{E}})_3$ *can be reduced to the following form:*

$$\frac{dy}{dx} = -\frac{y[k_1 x F_x + k_2 F + k_3 F]}{x[k_1 y F_y + k_2 F - k_3 F]}, \tag{2}$$

where $k_i (i = 1, 2, 3)$ *is any constant with* $k_1 \neq 0$ *and* $k_2^2 + k_3^2 \neq 0$.

Proof. The sufficiency is obvious. The necessity is proved as follows. For any hyperbola or ellipse $F(x, y) = 0$ in a plane, there exists a non-singular linear transformation:

$$\begin{cases} x = \alpha x' + \beta y', \\ y = \gamma x' + \delta y', \end{cases} \quad \text{or} \quad \begin{cases} \mu_1 x = y' - K_1 x', \\ \mu_2 y = y' - K_2 x'. \end{cases} \tag{3}$$

Reduce $F(x, y) = 0$ to the form:

$$F_2(x', y') = (x' - x_0)^2 + \lambda(y' - y_0)^2 - 1 = 0 \quad (\lambda = -1 \quad \text{or} + 1).$$

The two straight lines $x = 0$ and $y = 0$ are respectively reduced to

$$y' = K_1 x' \quad \text{and} \quad y' = K_2 x' \quad (K_1 \neq K_2).$$

Therefore, if $(\overline{\mathrm{E}})_3$ has a trajectory of hyperbola or ellipse $F(x, y) = 0$, then, by using the transformation (3), the equation $(\overline{\mathrm{E}})_3$ is reduced to the cubic differential system $(\overline{\mathrm{E}})_3$ which is the system of integral curves consisting of two straight lines: $y' = K_i x' (i = 1, 2; K_1 \neq K_2)$ and a hyperbola or ellipse $F_2(x', y') = 0$. By Lemma 1 the necessary and sufficient condition for $(\overline{\mathrm{E}})'_3$ to have a solution of $F_2(x', y') = 0$ is:

$$\frac{dy'}{dx'} = \frac{F_{2x'} P_2(x', y') + H_1(x', y') F_2(x', y')}{-F_{2y'} P_2(x', y') + \overline{H}_1(x', y') F_2(x', y')}, \tag{4}$$

where

$$P_2(x', y') = a_1 x'^2 + a_2 x' y' + a_3 y'^2 + a_4 x' + a_5 y' + a_6,$$

$$H_1(x', y') = A_1 x' + B_1 y' + C_1,$$

$$\overline{H}_1(x', y') = A_2 x' + B_2 y' + C_2,$$

$$F_{2x'} = 2(x' - x_0),$$

$$F_{2y'} = 2\lambda(y' - y_0).$$

Because $y' = K_1 x'$ and $y' = K_2 x'$ are also the solutions of $(\mathrm{E})'_3$, they satisfy the differential equation (4). Substituting them in (4), we get

$$\{2[(x' - x_0) + \lambda K_i (y' - y_0)] P_2\}_{y' = K_i x'} \equiv \{[K_i \overline{H}_1 - H_1] F_2\}_{y' = K_i x'} \quad (i = 1, 2). \tag{5}$$

Because the curve $F_2(x', y') = 0$ does not intersect $y' = K_i x' \quad (i = 1, 2)$, then $F_2(x', K_i x') = 0 \quad (i = 1, 2)$ does not contain the real factor of degree one. And from the identity (5), we get

$$\begin{cases} [P_2(x', y')]_{y' = K_i x'} \equiv [F_2(x', y')]_{y' = K_i x'}, \\ [K_i \overline{H}_1 - H_1]_{y' = K_i x'} \equiv 2[(x' - x_0) + \lambda K_i (y' - y_0)]_{y' = K_i x'} \quad (i = 1, 2). \end{cases} \tag{6}$$

Comparing the coefficients of the terms containing x of the same degree in (6), we have

$$\begin{cases} a_1 + K_i a_2 + K_i^2 a_3 = (1+\lambda K_i^2), \\ a_4 + K_i a_5 = -2x_0 - 2\lambda K_i y_0, \\ a_6 = x_0^2 + \lambda y_0^2 - 1 \quad (i=1,2) \end{cases} \tag{7}$$

and

$$\begin{cases} -A_1 + K_i A_2 - K_i B_1 + K_i^2 B_2 = 2(1+\lambda K_i^2), \\ -C_1 + K_i C_2 = -2x_0 - 2\lambda K_i y_0 \quad (i=1,2). \end{cases} \tag{8}$$

Solving Eq. (7), we get

$$\begin{cases} a_1 = 1 - K_1 K_2 h_1, & a_4 = -2x_0, \\ a_2 = (K_1+K_2)h_1, & a_5 = -2\lambda y_0, \\ a_3 = \lambda - h_1, & a_6 = x_0^2 + \lambda y_0^2 - 1. \end{cases} \tag{9}$$

And solving Eq. (8), we get

$$\begin{cases} A_1 = -K_1 K_2 h_2 - 2, & A_2 = h_3 - (K_1+K_2)h_2, \\ B_1 = h_3, & B_2 = h_2 + 2\lambda, \\ C_1 = 2x_0, & C_2 = -2\lambda y_0. \end{cases} \tag{10}$$

In the above equation, $h_i (i = 1, 2, 3)$ are constants. Substituting (9) and (10) in Eq. (4), we get

$$\frac{dy}{dx} = \frac{-h_1 G_2 F_{2x'} + [-K_1 K_2 h_2 x' + h_3 y'] F_2}{h_1 G_2 F_{2y'} + \{[h_3 - (K_1+K_2)h_2]x' + h_2 y'\} F_2}, \tag{11}$$

where $G_2 = (y' - K_1 x')(y' - K_2 x')$.

The inverse transformation of (3) is

$$\begin{cases} x' = \dfrac{\mu_1 x - \mu_2 y}{K_2 - K_1}, \\ y' = \dfrac{\mu_1 K_2 x - \mu_2 K_1 y}{K_2 - K_1}. \end{cases} \tag{3$'$}$$

Applying the inverse transformation (3)' to Eq. (11), we get the form (2), and then

$$\begin{aligned} \frac{dy}{dx} &= \frac{\dfrac{1}{\mu_2}\left(\dfrac{dy'}{dx'} - K_2\right)}{\dfrac{1}{\mu_1}\left(\dfrac{dy'}{dx'} - K_1\right)} \\ &= \frac{\dfrac{1}{\mu_2}(-h_1 G_2)[F_{2x'} + K_2 F_{2y'}] + \dfrac{1}{\mu_2}(y' - K_2 x')(h_3 - K_2 h_2) F_2}{\dfrac{1}{\mu_1}(-h_1 G_2)[F_{2x'} + K_1 F_{2y'}] + \dfrac{1}{\mu_1}(y' - K_1 x')(h_3 - K_1 h_2) F_2} \\ &= \frac{y[K_1 x F_x + K_2 F - K_3 F]}{-x[K_1 y F_y + K_2 F - K_3 F]}, \end{aligned}$$

where

$$
\begin{aligned}
&G_2(x', y') = \mu_1\mu_2 xy,\\
&F_x = \frac{\mu_1}{K_2 - K_1}[F_{2x'} + K_2F_{2y'}],\\
&F_y = \frac{-\mu_2}{K_2 - K_1}[F_{2x'} + K_1F_{2y'}],\\
&\mathcal{K}_1 = h_1(K_1 - K_2),\\
&\mathcal{K}_2 = \frac{1}{2}(K_1 - K_2)h_2,\\
&\mathcal{K}_3 = h_3 - \frac{(K_1 + K_2)}{2}h_2,\\
&\mathcal{K}_1 \neq 0, \quad \mathcal{K}_2^2 + \mathcal{K}_3^2 \neq 0.
\end{aligned}
$$

The proof is completed.

Theorem 2. *For the Kolmogorov type cubic system which has a solution of non-degenerate and central conic, there exists no limit cycle on the whole plane.*

Proof. From Theorem 1, $(\underline{\mathrm{E}})_3$ may be reduced to (2). Because the differential equation (2) has the integrating factor

$$\rho(x, y) = \frac{1}{xyF(x, y)},$$

the general integral curve of Eq. (2) is

$$[F(x, y)]^{\mathcal{K}_1}(xy)^{\mathcal{K}_2}\left(\frac{x}{y}\right)^{\mathcal{K}_3} = c, \tag{12}$$

when $F \neq 0, xy \neq 0$. It is analytic. Therefore, on the whole plane there exists no limit cycle for $(\overline{E})_3$; otherwise, spiral trajectories exist in the neighborhood of a limit cycle, which is incompatible with the fact that for an analytic integral curve (12) exists.

Corollary 1. *If the solution curves* $F_2(x', y') = 0$ *and* $y' = K_ix'$ $(i = 1, 2)$ *have intersecting points and the roots of* $F(x', K_ix') = 0$ $(i = 1, 2)$ *are not equal to* $\frac{x_0 + \lambda K_i y_0}{1 + \lambda K_i^2}$ $(i = 1, 2)$, *one may prove in the same way as for Theorems 1 and 2 that* $(\overline{E})_3$ *has no limit cycle in the whole plane.*

§2. $b^2 - ac = 0$

The system $(\overline{E})_3$ has a parabolic trajectory $F(x, y) = 0$.

Lemma 2. *The necessary and sufficient condition for the differential equation* $(\mathrm{E})_3$ *to have a particular integral curve in the form of a parabola* $F_2(x, y) = 0$ *is that* $(\mathrm{E})_3$ *may be reduced to the following form:*

$$\frac{dy}{dx} = \frac{2(y - y_0)R_2(x, y) + 2(x - x_0)L_2(x, y)}{(x - x_0)R_2(x, y) + L_2(x, y)}, \tag{13}$$

where $F_2(x, y) = (y - y_0) - (x - x_0)^2$; $R_2(x, y)$ and $L_2(x, y)$ *are quadratic polynomials in x and y.*

Lemma 3. *The necessary and sufficient condition for the quadratic curve $R_2(x,y)=0$ to go through three non-collinear points $P_i(i=1,2,3)$ is that $R_2(x,y)=0$ can be reduced to the expression:*

$$R_2(x,y)=\alpha_1 l_1 l_2+\alpha_2 l_2 l_3+\alpha l_3 l_1,$$

where $\alpha_i(i=1,2,3)$ are parameters, and $l_i=0,(i=1,2,3)$ are respective equations of the straight lines P_2P_3, P_3P_1 and P_1P_2.

Theorem 3. *The necessary and sufficient condition for the cubic system of Kolomogorov type $(\overline{E})_3$ to have a parabolic trajectory $F(x,y)=0$ is that $(\overline{E})_3$ may be reduced to the following form:*

$$\frac{dy}{dx}=\frac{y\{2[K_1x-K_3]F-[K_1(x-x_0)-K_2(y-y_0)+K_4]xF_x\}}{x\{2[K_2y+K_3]F+[K_1(x-x_0)-K_2(y-y_0)+K_4]yF_y\}}, \tag{14}$$

where

$$F(x,y)=ax^2+2bxy+cy^2+dx+ey+f \quad (b^2-ac=0),$$

$K_i(i=1,2,3,4)$ are constants, and (x_0,y_0) is the coordinates of the vertex of the parabola F=0.

Proof. Necessity. Evidently, there exists an affine transformation,

$$\mu_1 x=y'-K_2x';\mu_2 y=y'-K_1x', \quad (K_1\neq K_2) \tag{15}$$

with inverse

$$x'=\frac{\mu_1 x-\mu_2 y}{K_1-K_2}, \quad y'=\frac{\mu K_1 x-\mu_2 K_2 y}{K_1-K_2}, \tag{15'}$$

which can transform $F(x,y)=0$ into

$$F_2(x',y')=(y'-y'_0)-(x'-x'_0)^2=0,$$

the straight lines $x=0$ and $y=0$ into

$$y'=K_1x' \quad \text{and} \quad y'=K_2x' \quad (K_1\neq K_2),$$

and therefore, can transform equation $(\overline{E})_3$ into the cubic system $(E)'_3$. And from Lemma 2, we have obtained the necessary and sufficient condition for $F_2(x',y')=0$ to be a solution of $(\overline{E})'_3$, that is

$$\frac{dy'}{dx'}=\frac{2(y'-y'_0)R_2(x',y')+2(x'-x'_0)L_2(x',y')}{(x'-x'_0)R_2(x',y')+L_2(x',y')}. \tag{16}$$

Because $y'=K_ix'(i=1,2)$ are solutions of $(\overline{E})'_3$, substituting them in (16) we have

$$\{(K_ix'+K_ix'_0-2y'_0)R_2\}_{y'=K_ix'}\equiv\{(K_i-2x'+2x'_0)L_2\}_{y'=K_iX'}. \tag{17}$$

Because the trajectories $F_2(x',y')=0$ and $y'=K_ix'(i=1,2)$ do not intersect, therefore $\Delta=K_i^2+4K_ix'_0-4y'_0<0,(i=1,2)$. By using (17) we have the identities

$$\begin{aligned}[R_2(x',y')]_{y'=K_ix'} &\equiv (2x'-2x'_0-K_i)(A_1x'+B_1K_ix'),\\ [L_2(x',y')]_{y'=K_ix'} &\equiv (K_ix'+K_ix'_0-2y'_0)(C_1x'+D_1K_ix') \quad (i=1,2).\end{aligned}$$

This implies that the quadratic curve $R_2(x', y') = 0$ goes through the three points: $0(0,0)$,

$$P_2\{\tfrac{1}{2}(2x'_0 + K_1), \tfrac{1}{2}(2x'_0 + K_1)K_1\} \quad \text{on line} \quad y' = K_1 x', \text{ and}$$

$$P_3\{\tfrac{1}{2}(2x'_0 + K_2), \tfrac{1}{2}(2x'_0 + K_2)K_2\} \quad \text{on line} \quad y' = K_2 x'.$$

Then by Lemma 3, we have

$$R_2(x', y') = h_0(y' - K_1 x')(y' - K_2 x') + m(y' - K_1 x')l_1(x', y') + n(y' - K_2 x')l_1(x', y'), \quad (18)$$

where h_0, m and n are constants, and

$$l_1(x', y') = y' - (2x'_0 + K_1 + K_2)x' + \frac{1}{2}(2x'_0 + K_1)(2x'_0 + K_2) = 0$$

is the linear equation which, as a line, goes through points P_2 and P_3.

For the same reason, we know that the curve $L_2(x', y') = 0$ goes through the three non-collinear points: $0(0,0)$, $Q_1\{\frac{1}{K_1}(2y'_0 - K_1 x'_0), 2y'_0 - K_1 x'_0\}$ and $Q_2\{\frac{1}{K_2}(2y'_0 - K_2 x'_0), 2y'_0 - K_2 x'_0\}$, and therefore

$$L_2(x', y') = p_0(y' - K_1 x')(y' - K_2 x') + q(y' - K_1 x')\bar{l}_1(x', y') + r(y' - K_2 x')\bar{l}_1(x', y'), \quad (19)$$

where p_0, q and r are constants and

$$\bar{l}_1 = y' - \frac{K_1 K_2 x'_0}{2y'_0} x' - \frac{1}{2y_0}(2y'_0 - K_1 x'_0)(2y'_0 - K_2 x'_0).$$

Substituting (18), (19) in the identity (17) and setting $i = 1$ and $i = 2$ respectively, we obtain that .

$$n = \frac{2}{2x'_0 + K_2}, \quad r = \frac{2y'_0}{2y'_0 - K_2 x'_0} \quad \text{and} \quad m = \frac{2}{2x'_0 + K_1}, \quad q = \frac{2y'_0}{2y'_0 - K_1 x'_0}. \quad (20)$$

respectively

Now, we can get the expression of $(\bar{E})_3$ from the differential equation (16) by means of the inverse transformation (3)'. Because

$$\frac{\mu_2 dy}{\mu_1 dx} = \frac{[2(y' - y'_0) - K_1(x' - x'_0)]R_2(x', y') + [2(x' - x'_0) - K_1]L_2}{[2(y' - y'_0) - K_2(x' - x'_0)]R_2(x', y') + [2(x' - x'_0) - K_2]L_2}, \quad (21)$$

substituting (18) and (20) in the right side of fraction (21), we have, after simplification, that

$$\begin{aligned}\text{Numerator} = \ & (y' - K_1 x')\{[2(y' - y'_0) - K_1(x' - x'_0)][h(y' - K_2 x') - 2(x' - x'_0) + K_2] \\ & + [2(x' - x'_0) - K_1][p(y' - K_2 x') + K_2(x' + x'_0) - 2y'_0]\}\end{aligned} \quad (22)$$

and

$$\begin{aligned}\text{Denominator} = \ & (y' - K_2 x')\{[2(y' - y'_0) - K_2(x' - x'_0)][h(y' - K_1 x') - 2(x' - x'_0) + K_1] \\ & + [2(x' - x'_0) - K_2][p(y' - K_1 x') + K_1(x' + x'_0) - 2y'_0]\},\end{aligned} \quad (23)$$

where $h = h_0 + m + n$ and $p = p_0 + q + r + 2$. We also have the expressions

$$\left.\begin{aligned} 2(x' - x'_0) - K_1 &= \frac{K_1 - K_2}{-\mu_1}\frac{\partial F(x,y)}{\partial x}, \\ 2(x' - x'_0) - K_2 &= \frac{K_1 - K_2}{\mu_2}\frac{\partial F(x,y)}{\partial y}, \\ 2(y' - y'_0) - K_1(x' - x'_0) &= 2F(x,y) + [-(x - x_0) + \tfrac{\mu_2}{\mu_1}(y - y_0)]\frac{\partial F(x,y)}{\partial x}, \\ 2(y' - y'_0) - K_2(x' - x'_0) &= 2F(x,y) + [\tfrac{\mu_1}{\mu_2}(x - x_0) - (y - y_0)]\frac{\partial F(x,y)}{\partial y}, \end{aligned}\right\} \tag{24}$$

where (x_0, y_0) is the coordinates of the vertex of the parabola $F(x,y) = 0$. Substituting (24) in (22) and (23), the differential equation (21) is then reduced to the expression (14), where $K_1 = h_1\mu_1, K_2 = h\mu_2, K_3 = K_1 - K_2, K_4 = (p-2)(K_1 - K_2)$. Thus the theorem is proved.

Theorem 4. *For the cubic differential system of Kolmogorov type which has a parabolic trajectory, there exists no closed trajectory or limit cycle in the whole plane.*

Proof. From Theorem 3, $(\underline{E})_3$ can be reduced to

$$\begin{aligned} \frac{dx}{dt} &= x\{2[k_2 y + k_3]F + [k_1(x - x_0) - k_2(y - y_0) + k_4]yF_y\} = P_3(x,y), \\ \frac{dy}{dt} &= y\{2[k_1 x - k_3]F - [k_1(x - x_0) - k_2(y - y_0) + k_4]xF_x\} = Q_3(x,y). \end{aligned} \tag{25}$$

There exists a function $B(x,y) = 1/xyF(x,y)$ which is continuously differentiable in the region

$$G = \{(x,y); xy \neq 0, F \neq 0\}.$$

Therefore we have

$$\frac{\partial}{\partial x}(BP_3) + \frac{\partial}{\partial y}(BQ_3) = h\mu_1\mu_2\frac{1}{F(x,y)} \qquad (\mu_i \neq 0, i = 1,2).$$

When $h \neq 0$, by applying Dulac's theorem to it, we conclude that there doesn't exist any closed or singular closed trajectory for system (25) in a simply connected region of G[6]. And because the boundary line of G is a trajectory, (25) doesn't have any limit cycle or closed trajectory in the whole plane. When $h = 0, B(x,y) = 1/xyF(x,y)$ becomes the integrating factor of (25) and then there exists a regular integral curve for $(\overline{E})_3$. So there exists one limit cycle.

Corollary 2. *If the parabolic solution $F(x,y)=0$ intersects, but is not tangent to, the coordinate axis, then, there exists no limit cycle, but perhaps there exists a singular closed trajectory for $(\overline{E})_3$ in the whole plane.*

The proof is similar to that of Theorems 3 and 4 except for that here $\Delta = K_i^2 + 4K_i x'_0 - 4y'_0 > 0 \quad (i = 1,2)$.

By using the result of the above theorem we have arrived at the following theorem, which is the final one in this paper.

Theorem 5. *For the cubic system of Kolmogorov type which has a non-degenerate quadratic algebraic trajectory, there exists no limit cycle in the whole plane.*

References

[1] Chin Yuanshun, *The equation with quadratic algebras limit cycle* $dy/dx = \sum_{0\le i+j\le 2} a_{ij}x^iy^j / \sum_{0\le i+j\le 2} b_{ij}x^iy^j$, Acta. Math. Sinica, 8:1(1958), 23–35.

[2] Chelkas L.A., *Method for estimating the number of limit cycles of autonomous system*, Diff. Equ., 13:5(1977).

[3] Chen Shupin, *Limit cycle for quadratic system with the integral curve of parabolic*, Kexue Tongbao, 30:6(1985), 401–404

[4] Huang Qiyu, *The algebraic curve cycle of differential system with line solution*, Acta. Math. Applicatae Sinica, 8:2(1985), 173–181

[5] Huang Qiyu, *Limit cycle for cubic system of Kolmogorov type with centered conic solution*, Science Bulletin, 32(12)(1987), 793–797

[6] Ye Yanqian, *Theory of limit cycle* (2nd edition), Shanghai Science Press, Shanghai, China, 1984.

Proceedings of the Asian Mathematical
Conference, 1990

Limit Cycles in a General Predator–Prey System with Anorexia Predator

XUNCHENG HUANG
Department of Mathematics
New Jersey Institute of Technology
Newark, New Jersey 07102 U.S.A.

PINGTAI SUN
Department of Information Management
National Defence Management College
Taiwan, China

§1. Introduction

The problem of limit cycles has attracted the attention of many mathematicians and scientists since the famous papers of Poincare (1881–1886). Even in Hilbert's 23 problems, one can find limit cycles. Since the papers of May (1972 [7]) and Albrecht, Gatzke, and Wax (1973 [1]), ecologists and applied mathematicians have special interests in the conditions that guarantee the uniqueness of limit cycles in predator-prey systems. Recently, there have been several results concerning uniqueness of limit cycles of the predator-prey model of Kolmogorov-type. These were done by Kuang and Freedman (1988 [6]) and Huang and Merrill (1989 [5]).

The existence of limit cycles has not been proved by Kuang and Freedman (1988 [6]). For example, the system

$$\begin{aligned}\frac{dx}{dt} &= x(k-x) - x^{\frac{1}{2}}y,\\ \frac{dy}{dt} &= y(x-\gamma),\end{aligned} \tag{1.1}$$

where $k > 3\gamma$, has no limit cycle but all the assumptions (H_1)-(H_8) of Kuang and Freedman (1988 [6]) are satisfied. This can be seen by using the Dulac function $D(x,y) = x^{-\frac{1}{2}}y^{\frac{1}{2}}$. In fact,

$$\begin{aligned}&\frac{\partial}{\partial x}((x(k-x) - x^{\frac{1}{2}}y)x^{-\frac{1}{2}}y^{\frac{1}{2}}) + \frac{\partial}{\partial y}(y(x-\gamma)x^{-\frac{1}{2}}y^{\frac{1}{2}})\\ = \quad &(k-3\gamma)\tfrac{1}{2}x^{-\frac{1}{2}}y^{\frac{1}{2}} > 0, \quad \text{for} \quad k > 3\gamma.\end{aligned} \tag{1.2}$$

Huang and Merrill proved the uniqueness as well as existence theorems of limit cycles in a more general model of Kolmogorov-type (1989 [5]). But in their assumptions, the prey functional function $\psi(x)$ has only one positive root. Thus the case of anorexia phenomenon has not been considered.

In this paper, we assume that in the case of too many prey, the predator is not willing eating the prey to a certain degree. Under this assumption $\psi(x)$ has more than one positive root. The conditions obtained here for the existence and uniqueness of limit cycles are easy to test. This work can be considered as a generalization of the one in (1989 [5]).

§2. The Model

We consider the system

$$\begin{aligned}\frac{dx}{dt} &= \phi(x)(F(x) - \pi(y)),\\ \frac{dy}{dt} &= \rho(y)\psi(x),\end{aligned} \tag{2.1}$$

with the fundamental assumptions:

(H_1) $\phi, \psi, \pi, \rho \in C^1[0,\infty), \phi(0) = \pi(0) = \rho(0) = 0, F \in C^1(0,\infty), F(0) \in (0,\infty); \phi' > 0$ for $x \geq 0, \pi' > 0, \rho' > 0$ for $y \geq 0$; and there exists $0 < x_1 < x_2$ such that $\psi(x_i) = 0, i = 1,2, \psi'(x_1) > 0, \psi'(x_2) < 0, \psi(x) > 0$ for $x_1 < x < x_2$.

(H_2) The curve $F(x) - \pi(y) = 0$ is defined for all $x \geq 0$, and $\phi(x) \leq mx$ for some positive m and all $x > 0$; there exists a $K > x_1$ such that $F(K) = 0, F'(K) < 0$ and $F(x) > 0$ for $0 < x < K$, and an $x_0 \geq K$ such that $F(x_0) = 0$ and $F(x) \neq 0$ for all $x > x_0$.

Our discussion is in the region $\Omega : \{(x,y) \mid x > 0, y > 0\}$. The system (2.1) has four equilibria: $E_0(0,0), E_1(x_1,y_1), E_2(x_2,y_2)$ and $E_K(K,0)$ in $\overline{\Omega} : \{(x,y) \mid x \geq 0, y \geq 0\}$, in which E_0, E_2 are saddle points, E_K is a stable node or focus if $x_2 < K$ or saddle point if $K < x_2$, and E_1 is a stable node or focus if $F'(x_1) < 0$ or unstable if $F'(x_1) > 0$.

The assumptions $(H_1), (H_2)$ are not exactly same as those studied by Kuang and Freedman (1988, [6]), and Huang and Merrill (1989, [5]). For example, the system (2.1) with (H_1) and (H_2) consists of some anorexia predator-prey systems but [5] and [6] do not.

It is not difficult to understand that in the case of too many prey the perdator may be exhausted from eating them (see Chen [2]). Then the prey functional function $\frac{ax}{1+\omega x}$ should be replaced with $\frac{ax}{1+\omega x+\beta x^2}$, where a, ω, β are positive constants. Such a predator-prey system can be reduced to

$$\begin{aligned}\frac{dx}{dt} &= x((a - bx) - \frac{ay}{1 + \omega x + \beta x^2}),\\ \frac{dy}{dt} &= y(-c + \frac{hax}{1 + \omega x + \beta x^2}),\end{aligned} \tag{2.2}$$

which has two equilibria in Ω and two on the axes.

By the time transform $(1 + \omega x + \beta x^2)d\tau = adt$ and using $t, a, b, c, \omega, \beta, h$ instead of new parameters, we have the equivalent form of the system (2.2):

$$\begin{aligned}\frac{dx}{dt} &= x((a - bx)(1 + \omega x + \beta x^2) - y),\\ \frac{dy}{dt} &= y(-c(1 + \omega x + \beta x^2) + hx).\end{aligned} \tag{2.3}$$

Obviously, (2.3) is a special case of the system (2.1) with $\psi(x) = (1 + \omega x + \beta x^2) + hx$. In the case that $-c(1 + \omega x + \beta x^2) + hx$ has two roots $0 < x_1 < x_2$,the system (2.3) can not be included in Kuang-Freedman's model (1988, [6]) or Huang-Merrill's model (1989, [5]).

§3. Theorems and Proofs

Theorem 3.1. *If $F'(x_i) > 0, i = 1, 2$, then the system (2.1) has no limit cycle in Ω.*

Proof. E_2 is always a saddle point and $F'(x_1) > 0$ implies that E_1 is an unstable mode or focus. Let l_2 be the line $x = x_2$. We can prove the following claim:

"No limit cycle of the system (2.1) intersects l_2."

In fact, by the phase portrait analysis and the fact that indexes of the equilibria inside a limit cycle must be equal to 1, the claim is obvious (see Ye et al 1986 [8]).

Also, there is no limit cycle when $x > x_2$ and $y > 0$, because no equilibrium is in that region.

From $F'(x_i) > 0, i = 1, 2$, we have $x_2 < x_0$ and $F'(x) > 0$ for $0 < x < x_2$. The proof of Theorem 3.1 can be done by the Dulac theorem with

$$D(x, y) = \frac{1}{\phi(x)\rho(y)} \quad \text{for} \quad 0 < x < x_2,$$

since

$$\begin{aligned} &\frac{\partial}{\partial x}(\phi(x)(F(x) - \pi(y))D(x, y)) + \frac{\partial}{\partial y}(\rho(y)\psi(x)D(x, y)) \\ = \; &\frac{1}{\rho(y)}F''(x) > 0 \quad \text{for} \quad 0 < x < x_2. \end{aligned}$$

Theorem 3.2. *If $F'(x_1) > 0$ and $x_2 \le \bar{x} = \min\{x | F(x) = F(x_1), x > x_1\}$, then the system (2.1) has no limit cycle in Ω.*

Proof. By Theorem 1 (Huang, 1989 [4]), the region $A = \{(x, y) \mid x_1 \le x \le \bar{x}, \pi(y_1) \le \pi(y) \le F(x)\}$ must be inside of all the limit cycles of the system (2.1). Thus, any limit cycle would intersect the line $x = x_2$ if $x_2 \le \bar{x}$. This is a contradiction to the claim in the proof of Theorem 3.1.

Theorem 3.3. *If $F'(x_1) > 0, F'(x_2) < 0$ and $\bar{x} < x_2 < K$, then there is at least one limit cycle surrounding E_1.*

Proof. Let $l_i : x = x_i, i = 1, 2, l_3 : y = y_1$. By the phase portrait analysis, any trajectory intersecting lines $x = x_i, i = 1, 2$, above E_i will cross $x = x_i$ from right to left and below E_i left to right; and any trajectory intersecting line l_3 will cross it from above to below.

Let Γ_{E_2} be the trajectory of the system (2.1) passing the line $x = x_2$ at $D(x_2, y_D)$ with $y_D > y_2$ and approaching to E_2 from the left. This is possible because E_2 is a saddle point (see Figure 1).

Now, consider any trajectory Γ of the system (2.1) that crosses l_2 at $G(x_2, y_G)$ with $y_2 < y_G < y_D$. Since Γ can not intersect Γ_{E_2} and any trajectory crossing l_2 above E_2 must be from right to left, it will remain in the region bounded by Γ_{E_2} and l_2. Thus, the line segment $\overline{E_2 D}$ and Γ_{E_2} constitute an outer boundary of the Poincare-Bendixson region which contains E_1 as the only equilibrium. Theorem 3.3 is proved.

Remark. The condition that $F'(x_2) < 0$ is necessary for the existence of the limit cycles. Otherwise, the line $x = x_2$ will intersect the region A.

Theorem 3.4. *If*

$$\frac{F'(x)\phi(x)}{\psi(x)} \quad \text{is nonincreasing for} \quad 0 < x < x_1 \quad \text{and} \quad x_1 < x < x_2, \tag{3.1}$$

then the system (2.1) has at most one limit cycle and it is stable if it exists.

Proof. By the same argument as in the proof of Theorem 3.1, there is no limit cycle of the system (2.1) in the region $\Omega'' = \{(x,y) \mid x \geq x_2, y > 0\}$. Our discussion can be restricted in the region $\Omega' = \{(x,y) \mid 0 < x < x_2, y > 0\}$.

It is easy to check that the hypotheses of Theorem B (Huang and Merrill, 1989 [5]) are satisfied in Ω'. If we properly redefine the functions $F, , \phi, \psi$ for $x_2 \leq x < \infty$ such that the hypotheses of Theorem B (1989 [5]) are satisfied in Ω'' as well, the proof can be done with the same transforms as the one of Theorem 3.1 in Huang and Merrill (1989 [5]).

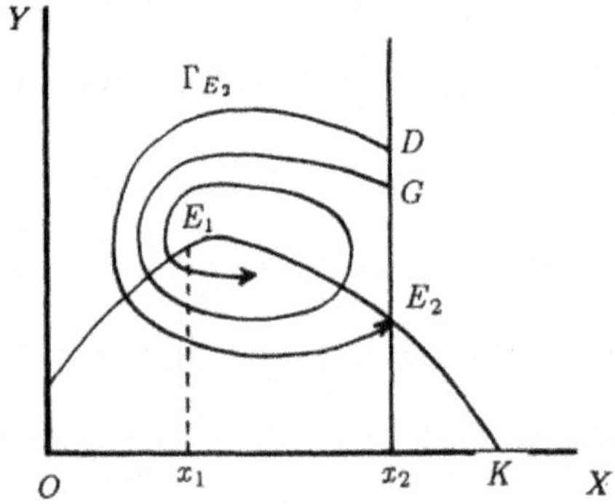

Fig. 1 $\overline{E_2D}$ and Γ_{E_2} constitute an outer boundary of the Poincare-Bendixson region

Remark. Notice that the assumption $\phi(x) \leq m_1 + m_2 x$ in [5] has been replaced by $\phi(x) \leq mx$. The latter is sufficient for that the function $\xi(u)$ is well-defined for all $u \in (-\infty, +\infty)$.

Combining Theorems 3.3 and 3.4, we have

Theorem 3.5. *If $F'(x_1) > 0, F'(x_2) < 0, \bar{x} < x_2 < K$ and $\frac{F(x)\phi(x)}{\psi(x)}$ is nonincreasing for $0 < x < x_1$ and $x_1 < x < x_2$, then the system (2.1) has a unique limit cycle which is stable.*

It is interesting to notice that the condition (3.1) is also useful in the discussion of the global stability of the equilibrium point $E_1(x_1, y_1)$.

Theorem 3.6. *If $F'(x_1) < 0$ and (3.1) holds,then E_1 is globally asymptotically stable.*

Proof. Suppose E_1 is locally but not globally asympotically stable. Then the system (2.1) has both a stable equilibrium point and a limit cycle which is stable from outside. Since (3.1) holds, by Theorem 3.4, this system has at most one limit cycle. Suppose Γ is the limit cycle and let Γ_0 be a closed curve inside Γ surrounding E_1.

Consider the annular region between Γ_0 and Γ. Since all the trajectories which intersect the boundaries $\Gamma_0 \cup \Gamma$ either cross from interior to exterior or remain on it, the Poincare-

Bendixson theorem implies that there exists at least one limit cycle Γ', which is unstable from outside (see Ye, 1987 [8]). This is a desired contradiction to the uniqueness of limit cycles. Hence, E_1 is globally asymptotically stable.

If $K < x_2$, we have the same conclusion as in Huang and Merrill (1989 [5]) .

Theorem 3.7. *Assume $K < x_2$ and (3.1) holds. If*

(i) $F(x_1) < 0$, *then E_1 is globally asymptotically stable;*

(ii) $F(x_1) > 0$, *then there exists a unique globally attracting stable limit cycle surrounding E_1.*

The proof is similar to the one in Huang and Merrill (1989 [5]). We omit it.

References

[1] Albrecht, F., H. Gatzke, and N. Wax, *Stable limit cycle in prey-predator populations*, Science 181, 1073–1074(1973).

[2] Chen, Lan-Sun, *Mathematical Ecological Models and Research Methods*, Science Press, Beijing, 1988.

[3] Freedman, H.I., *Deterministic Mathematical Models in Population Ecology*, Marcel Dekker, New York, 1980.

[4] Huang, Xun-Cheng, *Relative positions of limit cycles in a Kolmogorov-type system*, J. Phys. A: Math. Gen. 22, L317–322(1989).

[5] Huang, Xun-Cheng and Stephen Merrill, *Conditions for uniqueness of limit cycles in general predator-prey systems*, Math. Biosci., 96, 47-60(1989).

[6] Kuang, Yang and H.I. Freedman, *Uniqueness of limit cycles in Gause-type models of predator-prey systems*, Math. Biosci., 88, 67–84(1988).

[7] May, R., *Limit cycles in predator-prey communities*, Science 177, 900–902(1972).

[8] Ye, Yan-Qian et al, *Theory of Limit Cycles*, Amer. Math. Soc., Providence, 1986.

[9] Zhang, Zhi-fen, Dokl. Akad. Nauk USSR 119: 659–662 (1958).

[10] Zhang, Zhi-fen, *Proof of the uniqueness theorem of limit cycles of generalized Lienard equations*, Appl. Anal., 23: 67–76(1986).

Proceedings of the Asian Mathematical
Conference, 1990

Semi-p-Functions and p-Functions on Semigroups

HUANG ZHIRUI
Department of Mathematics
Zhongshan University
Guangzhou 510275, China

§1. Introduction

Renewal sequences and p-functions are analytic models for the study of regenerative phenomena. Their analytic properties and algebraic structure were thoroughly investigated by Kendall, Davidson [1] and Kingman[2]. Moreover, Kingman further studied generalized renewal sequences and semi-p-functions[3]. In recent years. Hou Zhengting, Liang Zhishun, the author of this paper and their graduate students have successively obtained a series of new results in this field[5,6]. Particularly, we first introduce p-functions and semi-p-functions defined on a semigroup[7,8]. This greatly expands the range of study of the theory of renewal sequences and p-functions. In this paper we further study the various properties of p-functions and semi-p-functions on a semigroup.

Throughout this paper T denotes an abelian semigroup, written additively, R, R_+ and N denote the sets of real numbers, positive real numbers and positive integers respectively. R^T, the set of all real functions defined on T, with the pointwise convergent topology, forms an abelian topological semigroup under the pointwise multiplication. For $p, q \in R^T$ and $\lambda \in R, pq = \{p(t)q(t), t \in T\}$ denotes the pointwise product of p and q, and $p^\lambda = \{(p(t))^\lambda, t \in T\}$, as long as it is well defined for λ, denotes the λ-th pointwise power of p.

For $p \in R^T, n \in N$ and $t_1, \cdots, t_n \in T$, let

$$F_n(t_1, \cdots, t_n; p) = \sum_{r=1}^{n}(-1)^{r-1} \sum_{0=k_0<k_1<\cdots<k_r=n} \prod_{j=1}^{r} p\left(\sum_{k_{j-1}+1}^{k_j} t_i\right). \tag{1}$$

We thus define a family of functions $\{F_n(\circ; p)\}$, called the F-function family of p. Some important relations about the F-function family have been obtained in [7, 8, 9].

$p \in R^T$ is called a p-function on T, if for any $n \in N$ and $t_1, \cdots, t_n \in T$, the following two conditions are satisfied:

$$F_n(t_1, \cdots, t_n; p) \geq 0, \tag{2}$$

$$\sum_{k=1}^{n} F_k(t_1, \cdots, t_k; p) \leq 1. \tag{3}$$

If only the inequality (2) is satisfied, then p is called a semi-p-function on T. We denote by $\mathcal{F}_T$ and $\mathcal{K}_T$ respectively the set of semi-p-functions on T and the set of p-functions on T. Obviously, $\mathcal{K}_T \subset \mathcal{F}_T \subset R^T$. The operation and the topology on $\mathcal{F}_T$ and $\mathcal{K}_T$ are the restrictions of those on R^T.

In [7] and [8] we showed that $p \in \mathcal{F}_N$ and $p \in \mathcal{K}_N$ are just a generalized renewal sequence and a renewal sequence respectively, and that $p \in \mathcal{F}_{R+}$ and $p \in \mathcal{K}_{R+}$ are just a semi-p-function and a p-function respectively.

§2. Fundamental Properties

Theorem 1. *Let $p, q \in \mathcal{F}_T$ (or $\mathcal{K}_T$). Then $pq \in \mathcal{F}_T$ (or $\mathcal{K}_T$).*

In this theorem, the result for $\mathcal{F}_T$ was first obtained in [7], and that for $\mathcal{K}_T$ in [8]. In [9], by using a simpler method we obtained

Lemma 1. *If $p, q \in \mathcal{F}_T$, then for any $n \in N$ and $t_1, \cdots, t_n \in T$, we have*

$$F_n(t_1, \cdots, t_n; pq) \geq F_n(t_1, \cdots, t_n; p) F_n(t_1, \cdots, t_n; q). \tag{4}$$

If $p, q \in \mathcal{K}_T$, then we also have

$$\sum_{k=1}^{n} F_k(t_1, \cdots, t_k; pq) \leq \sum_{k=1}^{n} F_k(t_1, \cdots, t_k; p) \sum_{r=1}^{n} (t_1, \cdots, t_r; q). \tag{5}$$

By this lemma we easily obtain the results in Theorem 1.

Theorem 2. *Let $\{p_\alpha\}_{\alpha \in D}$ be a net in $\mathcal{F}_T$ (or $\mathcal{K}_T$), and $\lim_\alpha p_\alpha = p \in R^T$. Then $p \in \mathcal{F}_T$ (or $\mathcal{K}_T$).*

Proof. By the given condition of convergence of $\{p_\alpha\}_{\alpha \in D}$, we have for each $t \in T$,

$$\lim_\alpha p_\alpha(t) = p(t).$$

By (1), $F_n(t_1, \cdots, t_n; p)$ is a polynomial of the values of function p, so

$$\lim_\alpha F_n(t_1, \cdots, t_n;\ p_\alpha) = F_n(t_1, \cdots, t_n;\ p). \tag{6}$$

From this the conclusion follows easily. Q. E. D.

By the above two theorems, we can see that both $\mathcal{F}_T$ and $\mathcal{K}_T$ are abelian topological semigroups and that $\mathcal{F}_T$, is a closed subsemigroup of R^T, and $\mathcal{K}_T$, included in $[0, 1]^T$, is a compact subsemigroup of R^T.

For each constant function on T, say c, we easily obtain that for any $n \in N$ and $t_1, \cdots, t_n \in T$,

$$F_n(t_1, \cdots, t_n; c) = c(1-c)^{n-1}, \tag{7}$$

$$\sum_{k=1}^{n} F_k(t_1, \cdots, t_k; c) = 1 - (1-c)^n. \tag{8}$$

Therefore, if $0 \leq c \leq 1$, then $c \in \mathcal{K}_T$; and $c \in \mathcal{F}_T$, if otherwise.

Theorem 3. *Let T be an abelian semigroup and S a subsemigroup of T. For $p \in R^T$, let p_S denote the restriction of p on S. If $p \in \mathcal{F}_T$ (or $\mathcal{K}_T$), then $p_S \in \mathcal{F}_S$ (or $\mathcal{K}_S$).*

Proof. For any $n \in N$ and $t_1, \cdots, t_n \in S \subset T$, we have

$$F_n(t_1, \cdots, t_n; p_S) = F_n(t_1, \cdots, t_n; p).$$

From this the conclusions follow easily. Q. E. D.

This theorem implies the following two well-known results[2]:

1. If $\{u_n\}$ is a renewal sequence, then for each $r \in N \{u_{nr}\}_{n \in N}$ is also a renewal sequence;
2. If p is a p-function, then for each $h \in R+$, $\{p(nh)\}_{n \in N}$ is a renewal sequence.

§3. Infinite Divisibility

In accordance with the usual definition, an element s of an abelian semigroup S is called infinitely divisible, if for each $n \in N$ there is an n-th root $s_n \in S$ of s, such that $s = s_n^n$. All infinitely divisible elements of S form a subsemigroup of S, which is denoted by $I(S)$.

In the following, w investigate the sets of positive functions in $\mathcal{F}_T$ and $\mathcal{K}_T$ denoted by $\mathcal{F}_T^+$ and $\mathcal{K}_T^+$ respectively.

Lemma 2. *$\mathcal{F}_T^+$ (or $\mathcal{K}_T^+$) is a hereditary subsemigroup of $\mathcal{F}_T$ (or $\mathcal{K}_T$).*

Lemma 3. *$p \in I(\mathcal{F}_T^+)$(or $I(\mathcal{K}_T^+)$) if and only if for each $\lambda > 0, p^\lambda \in \mathcal{F}_T^+$ (or $\mathcal{K}_T^+$).*

Proof. We only need to prove the necessity. Let $p \in I(\mathcal{F}_T^+)$. Then for any $n \in N$, its n-th root $p^{1/n} \in \mathcal{F}_T^+$. By the semigroup property of $\mathcal{F}_T^+, p^{m/n} \in \mathcal{F}_T^+$ for any $n, m \in N$, i.e., $p^r \in \mathcal{F}_T^+$ for any rational number $r > 0$. Suppose that $\lambda \in R_+$. Then there is a sequence $\{r_n\}$ of positive rational numbers, satisfying $r_n \to \lambda$, such that

$$p^{r_n}(t) \to p^\lambda(t) \quad \text{for all} \quad t \in T.$$

Obviously, $0 < p^\lambda(t) < \infty$. By Theorem 2, $p^\lambda \in \mathcal{F}_T^+$. The proof for the case $p = I(\mathcal{K}_T^+)$ is entirely similar. Q. E. D.

Let $p \in R_+^T$. For each $s \in T$, we can define a function $r_s(t; p) \in R_+^T$ by the following expression:

$$r_s(t;p) = \frac{p(t+s)}{p(t)} \quad \text{for each} \quad t \in T. \tag{9}$$

If for any $s, t_1, t_2 \in T$.

$$r_s(t_1 + t_2; p) \geq r_s(t_1, p) \geq p(s), \tag{10}$$

we call p an increasing-ratio function on T. This concept was first introduced in [10] for the positive function defined on R_+. It is easily seen that the increasing-ratio function on N is just a Kaluza sequence. We denote the set of increasing-ratio functions on T by $\mathcal{S}_T$ and its bounded subset by $\mathcal{S}_T^b$.

By using methods analogous to that in [10], we can obtain the following conclusions.

Lemma 4. *$p \in \mathcal{S}_T^b$, if and only if $p \in \mathcal{S}_T$ and*

$$p(t) \leq 1 \quad \textit{for each} \quad t \in T.$$

Lemma 5. *Let $p \in R_+^T$. Then for any $n \geq 2$ and $t_1, \cdots, t_n \in T$,*

$$F_n(t_1, \cdots, t_n; p) = \sum_{k=1}^{n-1} F_k\left(t_1, \cdots, t_{k-1}, \sum_k^{n-1} t_i; p\right)\left[r_{t_n}\left(\sum_k^{n-1} t_i; p\right) - r_{t_n}\left(\sum_{k+1}^{n-1} t_i; p\right)\right], \tag{11}$$

where $r_{t_n}(\sum_h^{n-1} t_i; p)$ is defined as $p(t_n)$.

Theorem 4. $I(\mathcal{F}_T^+) = \mathcal{S}_T$ *and* $I(\mathcal{K}_T^+) \subset \mathcal{S}_T^b$.

Proof. By Lemma 5, $\mathcal{S}_T \subset \mathcal{F}_T^+$. By the definition of r_s,

$$r_s(\circ\,;\,p)r_s(\circ\,;\,q) = r_s(\circ\,;\,pq) \quad \text{for} \quad p, q \in R_+^T, \tag{12}$$

and

$$r_s(\circ\,;\,p^\lambda) = r_s^\lambda(\circ\,;\,p) \quad \text{for} \quad \lambda > 0 \quad \text{and} \quad p \in R_+^T. \tag{13}$$

Thus $\mathcal{S}_T$ is a subsemigroup of $\mathcal{F}_T^+$, and from $p \in \mathcal{S}_T$ it follows that $p^\lambda \in \mathcal{S}_T$ for $\lambda > 0$. Therefore, $\mathcal{S}_T \subset I(\mathcal{F}_T^+)$. Conversely, if $p \in I(\mathcal{F}_T^+)$, then $p^\lambda \in \mathcal{F}_T^+$ for each $\lambda > 0$ by Lemma 3. Let

$$p(t) = e^{\varphi(t)} \quad \text{for} \quad t \in T,$$

where $\varphi(t) \in R$ for $t \in T$. Then for $t_1, t_2, t_3 \in T$,

$$\lim_{\lambda \to 0+} \frac{1}{\lambda} F_2(t_1, t_2; p^\lambda) = \quad \varphi(t_1 + t_2) - [\varphi(t_1) + \varphi(t_2)] \geq 0,$$

$$\lim_{\lambda \to 0+} \frac{1}{\lambda} F_3(t_1, t_2, t_3; p^\lambda) = \quad \varphi(t_1 + t_2 + t_3) - \varphi(t_1 + t_2) - \varphi(t_2 + t_3) + \varphi(t_2) \geq 0.$$

By the above two inequalities we obtain

$$r_{t_1}(t_2 + t_3; p) \geq r_{t_1}(t_2; p) \geq p(t_1).$$

That is, $p \in \mathcal{S}_T$. Therefore $I(\mathcal{F}_T^+) = \mathcal{S}_T$. Consequently we can easily obtain that $I(\mathcal{K}_T^+) \subset \mathcal{S}_T^b$. Q. E. D.

§4. Quasi-Exponential Functions

We call $p \in R^T$ a quasi-exponential function on T, if there is a homomorphism φ from T into the real additive group R such that

$$p(t) = e^{\varphi(t)}, \quad \text{for} \quad t \in T. \tag{14}$$

The set of all quasi-exponential functions is denoted by $\mathcal{G}_T$.

Theorem 5. *Let $p \in R_+^T$. Then the following assertions are equivalent:*

(i) $p \in \mathcal{G}_T$.

(ii) $p(t+s) = p(t)p(s)$ *for any* $t, s \in T$.

(iii) $F_n(t_1, \cdots, t_n; p) = 0$ *for any* $n \geq 2$ *and* $t_1, \cdots, t_n \in T$.

(iv) $r_s(t; p) = p(s)$ *for any* $s, t \in T$.

Proof. By the definition of quasi-exponential functions, (i)$\Longleftrightarrow$(ii). By the definition of the functions r_s, (ii) $\Longleftrightarrow$ (iv). Obviously, (iii)$\Longrightarrow$ (ii). By (3) of [9] and using induction we can deduce that (ii)$\Longrightarrow$ (iii). Q. E. D.

Lemma 6. *Let H be the set of all homomorphisms from T into the real additive group R. Then H forms a group under the pointwise additive operation and its unit is the constant function with value 0.*

Theorem 6. *$\mathcal{G}_T \subset I(\mathcal{F}_T^+)$ and it is the unit group of $\mathcal{F}_T$.*

Proof. By Theorem 5, $\mathcal{G}_T \subset I(\mathcal{F}_T^+)$. By Lemma 6, under the pointwise multiplicative operation, $\mathcal{G}_T$ is a subgroup of $\mathcal{F}_T$. Suppose that $p, q \in \mathcal{F}_T$ and $pq = 1$. By the hereditary property of $\mathcal{F}_T^+$ in $\mathcal{F}_T$, we have $p, q \in \mathcal{F}_T^+$. For any $s, t \in T$,

$$\begin{aligned} 0 = \ & F_2(s,t;1) = p(s)p(t)F_2(s,t;q) \\ & + F_2(s,t;p)q(s)q(t) + F_2(s,t;p)F_2(s,t;q). \end{aligned}$$

Then $F_2(s,t;p) = F_2(s,t;q) = 0$. By Theorem 5, $p, q \in \mathcal{G}_T$.

Corollary. *Let $p \in \mathcal{G}_T$ and $q|p$. Then $q \in \mathcal{G}_T$.*

Lemma 7. *Let $p \in R^T$ and $q \in \mathcal{G}^T$. Then for any $n \in N$ and $t_1, \cdots, t_n \in T$,*

$$F_n(t_1, \cdots, t_n; pq) = q(\sum_1^n t_i) F_n(t_1, \cdots, t_n; p). \tag{15}$$

This lemma is immediately deduced by using induction. From it we can obtain the following conclusion.

Theorem 7. *Let $p \in \mathcal{F}_T$ and $q \in \mathcal{G}_T$. Then $q \mid p$ and $pq \sim p$ (see [4] for the definition of the associate relation $\sim$).*

References

[1] D.G. Kendall and E.F. Harding, ed., *Stochastic Analysis*, John Wiley & Sons, 1973.

[2] J.F.C. Kingman, *Regenerative Phenomena*, Wiley, London, 1972.

[3] J.F.C. Kingman *Semi-p-functions*, Trans. Amer. Math. Soc., 174(1972), 257–274.

[4] I.Z. Ruzsa and G.J. Székely, *Algebraic Probability Theory*, John Wiley & Sons, 1988.

[5] Hou Zhengting, *Closure of renewal sequences for circle-operation*, Sci. Sinica ser. 25A, 8(1982), 816–824.

[6] Liang Zhishun and Huang Zhirui, *Renewal sequences, p-functions, their extensions and the related topics*, in Probability Theory and its Applications in China, AMS Cont. Math. (to appear).

[7] Huang Zhirui, *Generalized renewal sequences, semi-p-functions and their F-function clusters*, (in Chinese), Acta. Sci. Natur. Univ. Sun., 2(1982), 30–38.

[8] Huang Zhirui, *F-function clusters and their applications*, (in Chinese), Chin. Ann. Math. 5A, 3(1984), 273–286.

[9] Huang Zhirui, *Pointwise multiplication and power on the class of p-functions* (in Chinese), Chin. J. Appl. Prob. Stat., 2, 3(1986), 211–214

[10] Huang Zhirui and Liang Zhishun, *On infinitely divisible positive semi-p-functions and their I_o-class* (in Chinese), Chin. Ann. Math. 10A, 2(1989), 148–159.

Proceedings of the Asian Mathematical
Conference, 1990

A New Direction in Steady Supersonic/Hypersonic Flow Computation*

W.H. Hui
Department of Applied Mathematics
University of Waterloo
Waterloo, Ontario, Canada

Important advances have been made during the past four decades on the numerical simulations, using a shock-capturing methodology, of inviscid gas flow as governed by the Euler equations of motion. This is most evident in the recent comprehensive survey article by Yee [1].

It appears that most existing works are based on the Eulerian method of description of fluid motion. It has also become traditional to treat steady flow as an asymptotic steady state of an unsteady flow after marching a large number of time steps. In the special case where the flow Mach number is everywhere greater than unity, steady flow can be more easily computed as steady flow proper without artificially introducing the time variable. This is the way it is done in this paper. In this way, computer storage and time can be greatly reduced and accuracy increased.

Supersonic, and especially hypersonic flows are usually dominated by shock and slipline (contact line) discontinuities. Woodward and Collella [2] observed that the overall accuracy of numerical simulation of these flows is very closely related to the accuracy with thich flow discontinuities are represented. A review of current literature shows that for one-dimensional unsteady flow, the Eulerian formulation with a second order total variation diminishing (TVD) scheme or with an essentially non-oscillatory (ENO) scheme is capable of resolving a shock discontinuity in about two computational cells (see Haten [3] for ideal gas flow and Liou et al. [4] for real gas flow). However, resolution of slipline (contact line) discontinuities in the Eulerian formulation is rather poor (see Fig. 1) ; it typically takes six or more cells to resolve it [3,4] and grows worse with increasing time steps. Furthermore, Chang and Liou [5] have found, in a recent study of steady two-dimensional flow involving shock reflection, that the accuracy of the second order TVD scheme based on Eulerian formulation deteriorates with increasing Mach number M, whereas the second order ENO scheme fails to converge to a steady state for $M > 4$ and must be reduced to a first order scheme, rendering its accuracy unacceptable. Their results for $M = 10$ are reproduced here as Figs. 2 b, c.

* The author wishes to acknowledge the support of the Natural Sciences and Engineering Research Council of Canada.

These deficiencies of the Eulerian method can be avoided by using a new Lagrangian method recently developed by Loh and Hui [6] to compute steady two-dimensional supersonic/hypersonic flow. In this formulation the two independent variables are a stream function ξ and the Lagrangian time τ, the latter playing a dual role as the Lagrangian label of a fluid particle while being the time of motion. Consequently a steady flow is computed, by marching in τ, as a steady flow proper and not as the asymptotic state of an unsteady flow. In this method the "computational fluid" has the same dynamics as the (inviscid) fluid it simulates, consequently it avoids the numerical errors on crossing a shock or a slipline which are present in any Eulerian method. In fact, it turns out that the accuracy of the new Lagrangian method increases with increasing shock strength, making it the more suitable for hypersonic flow computation.

The Euler equations of motion of an inviscid, non-heat conducting perfect gas obeying the γ-law in two-dimensional steady motion are written in conservation form [6]

$$\frac{\partial E}{\partial \tau} + \frac{\partial F}{\partial \xi} = 0, \tag{1}$$

where

$$E = \begin{pmatrix} K \\ H \\ Ku + pV \\ Kv - pU \\ U \\ V \end{pmatrix}, \qquad F = \begin{pmatrix} 0 \\ 0 \\ -pv \\ pu \\ -u \\ -v \end{pmatrix}. \tag{2}$$

Here, p and ρ are the pressure and density, u and v are the x-and y-component of velocity, respectively, and

$$u = \frac{\partial x}{\partial \tau}, \qquad v = \frac{\partial y}{\partial \tau}, \tag{3}$$

$$U = \frac{\partial x}{\partial \xi}, \qquad V = \frac{\partial y}{\partial \xi}. \tag{4}$$

In (2)

$$K = \rho(uV - vU), \tag{5}$$

$$H = \frac{1}{2}(u^2 + v^2) + \frac{\gamma}{\gamma - 1}\frac{p}{\rho}. \tag{6}$$

The first two equations of (1) express conservation of mass and energy, the 3rd and 4th equations are the momentum equations, whereas the last two equations are the compatibility conditions which are a unique feature of the Lagrangian formulation, representing the geometrical deformation of the fluid particles.

In the new Lagrangian formulation the coordinate lines are the streamlines and the time lines. Consequently, the flow tangency condition on a solid boundary is satisfied exactly on a coordinate line, $\xi = \xi_0$ say. We futher remark that since sliplines are also streamlines, they must be coordinate lines.

Finding a steady flow is to solve (1) subject to the flow tangency condition on the solid boundaries present. This is done using a shock-capturing method. The first order solutions based on Godunov scheme are already given in Ref. 6. More accurate solutions have now been obtained using a high resolution total variation diminishing (TVD) scheme [7].

As a first example, the problem of shock reflection from a solid plane wall is computed for the case when the incident Mach number $M_\infty = 10$ and the shock angle =29°. The result for wall pressure distribution is plotted in Fig. 2d. It is seen to be in excellent agreement with the exact solution and is clearly superior to the corresponding Eulerian TVD scheme (Fig. 2b) or ENO scheme (Fig. 2c).

In the second example we compute a Riemann problem consisting of two vastly different uniform flows making an angle of 23° (Fig. 3). The numerical solutions for pressure and Mach number distributions show that the new Lagrangian method is capable of resolving both shock and slipline discontinuities within one cell and is in excellent agreement with the exact solutions.

In the last example, we compute a supersonic flow, $M_\infty = 1.87$, past a given shape and compare our results with Johannesen's experiments [8]. The agreement in details is again seen to be excellent.

In conclusion, the new Lagrangian method using a stream function and the Lagrangian time as independent variables for two-dimensional steady supersonic/hypersonic flow is shown to be superior to the Eulerian method in that

(a) it resolves slipline discontinuities crisply,
(b) its accuracy improves with increasing Mach number, whereas the Eulerian method deteriorates,
(c) it requires no grid generation.

References

[1] H.C. Yee, *A class of high resolution explicit and implicit shock capturing methods,* in VKI for Fluid Dynamics Lecture Series, 1(1989), 1989-04.

[2] P.Woodward and P. Collella, J. Comput. Phys., 54(1984), 14.

[3] A.Harten, in *Wave Motion: Theory, Modelling, and Computation,* eds. A.J. Chorin and A.J. Majda, 147 (Spring-Verlag Publishers, 1987).

[4] M.S. Liou, B. Van Leer and J.S. Shuen, J. Comput. Phys., 87(1990), 1.

[5] S.H. Chang and M.S. Liou, *A numerical study of ENO and TVD schemes for shock capturing,* NASA TM 101355, ICOMP-88-18(1988).

[6] C.Y. Loh and W.H. Hui, *A new Lagrangian method for steady supersonic flow computation, Part I. Godunov Scheme,* J. Comput. Phys., 89(1990), 207.

[7] W.H. Hui and C.Y. Loh, *A new Lagrangian method for steady supersonic flow computation, Part II. slip-line resolution,* submitted to J. Comput. Phys.

[8] N.H. Johannesen, Philos, Mag., 43(1952), 568.

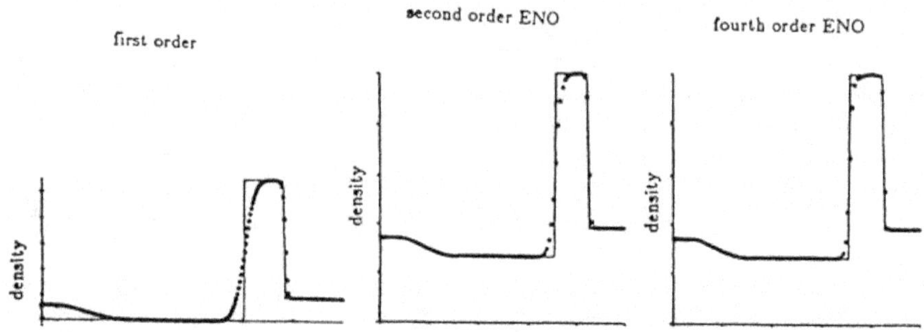

Fig 1. Flow in shock tube of perfect gas. —— exact, 0 computed by Harten [3].

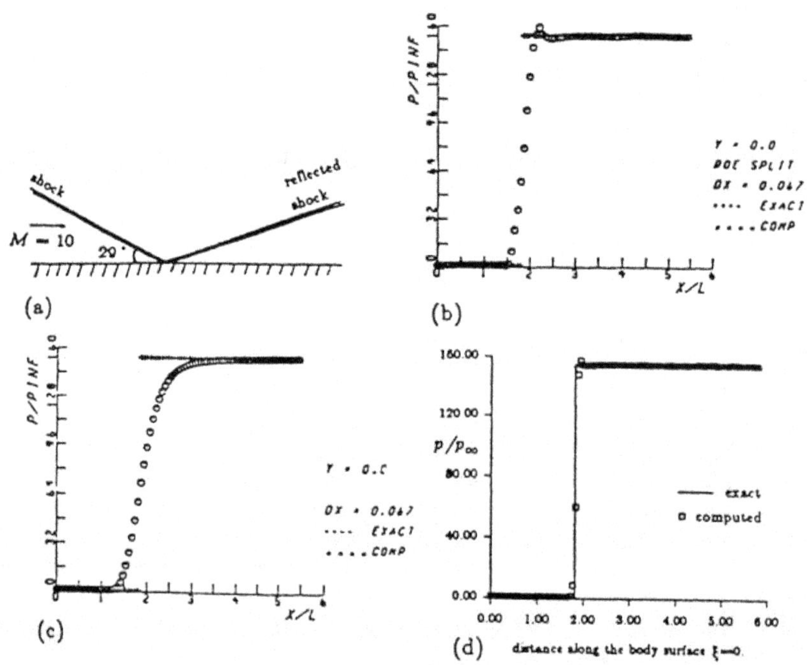

Fig 2. The shock reflection problem, $M = 10$, shock angle $= 29°$. (a) configuration, (b) computed surface pressure by TVD scheme in Eulerian formulation (courtesy of S.H. Chang & M.S. Liou [5], (c) computed surface pressure by ENO scheme (unsteady approach) in Eulerian formuation (courtesy of S.H. Chang & M.S. Liou [5], (d) computed surface pressure by TVD scheme in new Lagrangian formulation.

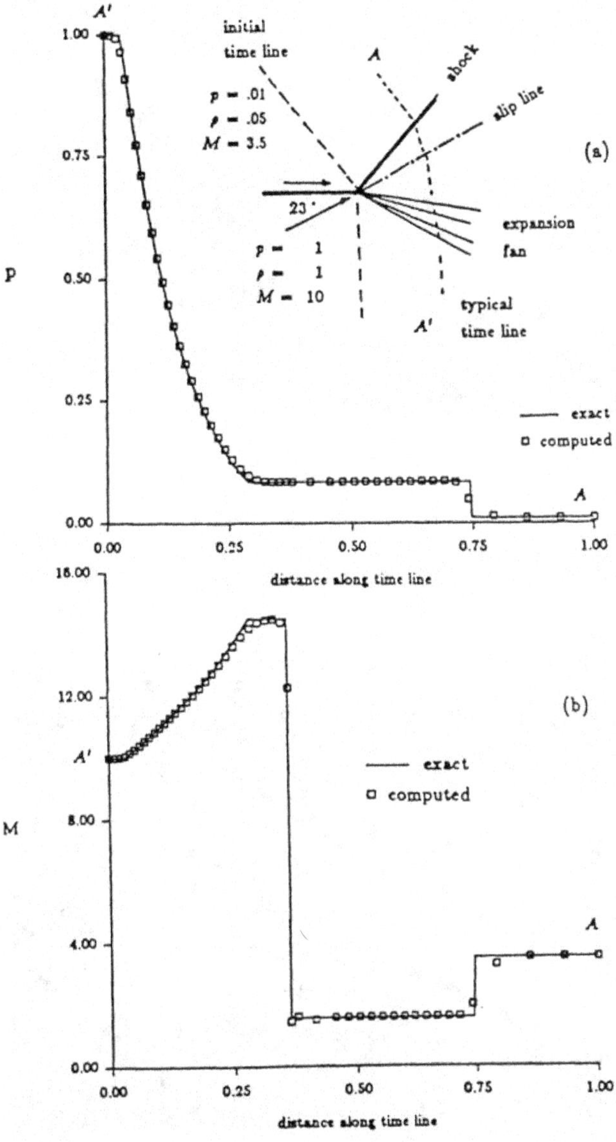

Fig 3. Riemann problem: (a) pressure, (b) Mach number, along a time line AA'

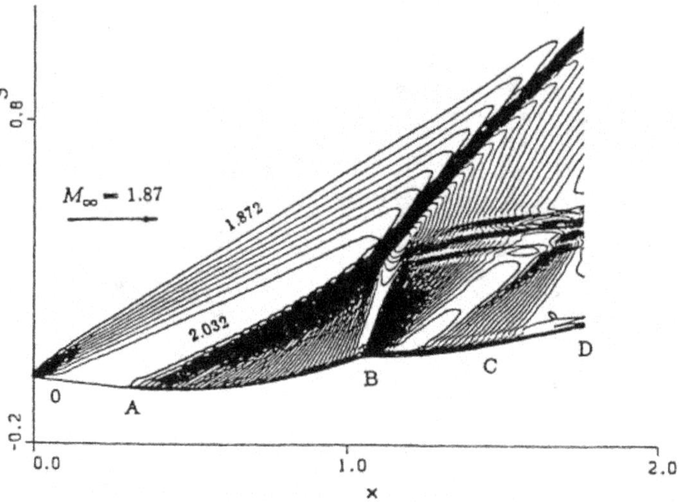

(a)

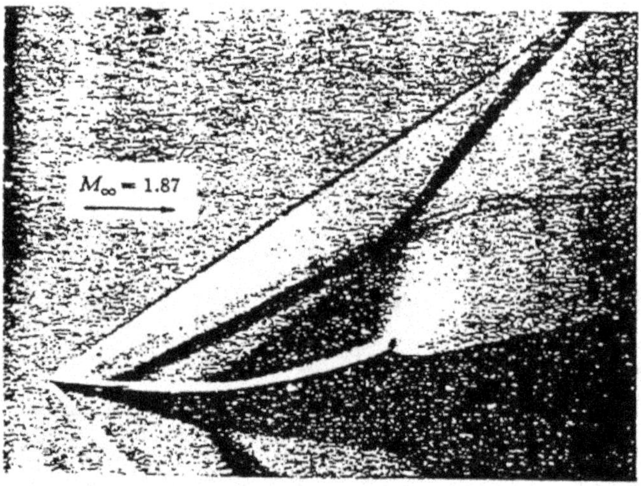

(b)

Fig 4. Computed supersonic flow field compared with experiment: (a) computed isomachs, (b) schlieren photograph of Johannesen's experiment [8].

Proceedings of the Asian Mathematical
Conference, 1990

Dense Languages and Languages Related to Primitive Words

M. Ito, & M. Katsura
Faculty of Science
Kyoto Sangyo University
Kyoto 603, Japan

Let X be a nonempty finite set and let X^* be the free monoid generated by X. Any element u of X^* is called a word over X. Let uvX^*. By $|u|$ we denote the length of u. Let $u, v \in X^*$. Then u is called a left (right) subword of v if $v = ux$ $(v = xu)$ for some $x \in X^*$. Any subset A of X^* is called a language over X. By $|A|$ we denote the cardinality of A. A language $A \subseteq X^*$ is said to be left dense (right dense, dense) if $X^*u \cap A \neq \emptyset (uX^* \cap A \neq \emptyset, X^*uX^* \cap A \neq \emptyset)$ for any $u \in X^*$. Moreover, a language $A \subseteq X^*$ is said to be two-sided dense if it is left dense and at the same time right dense. By P_A we denote the principal congruence determined by A which is defined as follows: $u \equiv v(P_A)$ if and only if $xuy \in A \iff xvy \in A$ holds for any $x, y \in X^*$. A language $A \subseteq X^*$ is said to be strongly dense if every congruence class of P_A is dense. It is easy to see that a strong dense language is dense. A language $A \subseteq X^*$ is said to be disjunctive if P_A is the identity, i.e. any congruence class of P_A is a singleton set. One of typical disjunctive languages is $Q(X)$, i.e. the set of all primitive words over X. A word $u \in X^*$ is said to be primitive if $u = v^n$ implies $n = 1$. It can easily be shown that a disjunctive language is dense. Let $Q_1(X) = \{1\} \cup Q(X)$ where 1 is the empty word and $Q_2(X) = X^* \backslash Q_1(X)$. Then $Q_1(X)$ and $Q_2(X)$ are disjunctive. We are interested in relashionships among several dense languages $A \subseteq X^*, Q_1(X)$ and $Q_2(X)$. As languages, we take recognizable languages and context-free languages.

First, we deal with dense recognizable languages. Note that a characterization of recognizable languages is the following: A language $A \subseteq X^*$ is recognizable if and only if A is accepted by an automaton.

An automaton M is a 5-tuple $M = (S, X, \delta, s_0, F)$ such that:

- S is a finite nonempty set called the set of states,
- X is a finite set called the set of inputs,
- δ is a mapping of $S \times X$ into S called the state transition function,
- s_0 is an element of S called the initial state, and
- F is a subset of S called the set of final states.

The function δ is extended to a mapping of $S \times X^*$ into S by $\delta(s, 1) = s$ and $\delta(s, xa) = \delta(\delta(s, x), a)$ for any $s \in S, x \in X^*$ and $a \in X$. In what follows, $\delta(s, x)$ is denoted by sx^M. $T(M) = \{x \in X^* \mid s_0x^M \in F\}$ is said to be the language accepted by M. A language $A \subseteq X^*$ is recognizable if and only if $A = T(M)$ holds for some automaton M.

Furthermore, let $k\min\{|Sx^M| \mid x \in X^*\}$ where $Sx^M = \{sx^M \mid s \in S\}$, and let $\mu(M) = \{w \in X^* \mid |Sw^M| = k\}$. It can easily be seen that, if $w \in \mu(M)$, then $xwy \in \mu(M)$ for any $x, y \in X^*$, i.e. $\mu(M)$ is an ideal of X^*.

Lemma 1. ([6]) *If u^i and ν^j have a common left (right) subword of length $|u| + |\nu|$, then u and ν are powers of a common word.*

Proposition 1. *Let $|X| \geq 2$ and let $A \subseteq X^*$ be a dense recognizable language. Then $A \cap Q_1(X)$ and $A \cap Q_2(X)$ are disjunctive.*

Proof. To prove that $A \cap Q_1(X)(A \cap Q_2(X))$ is disjunctive, it is enough to show that $u = \nu$ if $|u| = |v|$ and $u \equiv \nu(P_{A\cap Q_1(X)})(u \equiv v(P_{A\cap Q_2(X)}))$ for any $u, v \in X^*$ (see [8]). Suppose $u \neq v$. Let $M = (S, X, \delta, s_0, F)$ be an automaton such that $A = T(M)$. Moreover, let $w \in \mu(M)$. Since $u \equiv v(P_{A\cap Q_1(X)})(u \equiv v(P_{A\cap Q_2(X)}))$ holds and A is dense, there exist $x, y \in X^*$ such that $xuwy, xvwy \in A$. Therefore, $s_0(xuwy)^M, s_0(xvwy)^M \in F$. Note that $wy \in \mu(M)$. Since $(xuwy)^M$ is a permutation on $S(wy)^M$, there exists a sufficiently large integer $n \geq 1$ such that $((xuwy)^n)^M$ is the identity permutation on $S(wy)^M$. Then $(s_0(xuwy)^M)((xuwy)^n)^M = s_0(xuwy)^M = s_0(xuwy)^M \in F$ and $(s_0(xvwy)^M)((xuwy)^n)^M = s_0(xvwy)^M \in F$. This means that $(xuwy)^{n+1} \in A$ and $xvwy(xuwy)^n \in A$. From Lemma 1, $xvwy(xuwy)^n \in Q_1(X)$. Hence $(xuwy)(xuwy)^n \in A \cap Q_2(X)$ and $(xvwy)(xuwy)^n \in A \cap Q_1(X)$. This contradicts the fact that $u \equiv v(P_{A\cap Q_1(x)})(u \equiv v(P_{A\cap Q_2(X)}))$. Consequently, $u \neq v$, i.e. $A \cap Q_1(X)(A \cap Q_2(X))$ is disjunctive.

Regarding more detailed results, see [3].

Another characterization of recognizable languages is the following: A recognizable language is a language whose principal congruence is of finite index. We will show that there is an interesting relation between a regular language $A \subseteq X^*$ which is strongly dense and the congruence classes of P_A. Before going further, we provide some notions and lemmas.

First, we deal with the general case. An equivalence relation on X^* is called a congruence if for any equivalence classes A and B, there exists an equivalence class C such that $AB \subseteq C$ where AB means the set $\{xy \mid x \in A, y \in B\}$. Let ρ be a congruence on X^* and let $u \in X^*$. Then by $[u]_\rho$ (briefly $[u]$) we denote the equivalence class of ρ which contains u. Let ρ be a congruence on X^*. By X^*/ρ we denote the set $\{\overline{u} \mid u \in X^*\}$ with the following binary operation:

$$\overline{u}\,\overline{v} = \overline{w} \iff [u][v] \subseteq [w]$$

Remark that $\overline{u} = \overline{v}$ if and only if $[u] = [v]$. The difference between $\overline{u}$ and $[u]$ is as follows:

(1) $\overline{u}$ is an element of X^*/ρ,

(2) $[u]$ is a congruence class of ρ and a subset of X^*.

Together with the above binary operation, X^*/ρ becomes a monoid which we call the quotient monoid X^* over ρ. Here $\overline{1}$ is the identity.

Lemma 2. *Let ρ be a congruence on X^*. Then the following conditions are equivalent:*

(1) $M = X^*/\rho$ is a group.

(2) $[1]$ is left (right) dense.

(3) Every congruence class is left (right) dense.

(4) Every congruence class is dense and at least one congruence class is left (right) dense.

(5) Every congruence class is dense and at least one congruence class is two-sided dense.

(6) Every congruence class is two-sided dense.

Proof. (1) $\Longrightarrow$ (6). Let $u, v \in X^*$. Since M is a group, there exist $x, y \in X^*$ such that $\bar{u}\,\bar{x} = \bar{v}$ and $\bar{y}\,\bar{u} = \bar{v}$. Hence $ux \in [v]$ and $yu \in [v]$. This means that $uX^* \cap [v] \neq \emptyset$ and $X^*u \cap [v] \neq \emptyset$.

(6) $\Longrightarrow$ (5) $\Longrightarrow$ (4) and (6) $\Longrightarrow$ (3) $\Longrightarrow$ (2). Obvious.

(4) $\Longrightarrow$ (2). Let $u \in X^*$ and let $[u]$ be left (right) dense. Then for any $v \in X^*$ there exists $x \in X^*$ such that $xv \in [u](vx \in [u])$. Obviously, $[x][v] \subseteq [u]([v][x] \subseteq [u])$. Hence $\bar{x}\,\bar{v} = \bar{u}(\bar{v}\,\bar{x} = \bar{u})$. This means that $\bar{u}$ is a left (right) zeroid element of M (see [1] or [2]). Let K be the set of all left (right) zeroid elements of M. Then K is an ideal of M. Note that M is a simple monoid (because every congruence class of ρ is dense). Therefore, $M = K$. Hence $\bar{1}$ is a left zeroid element of M and $[1]$ is left dense.

(2) $\Longrightarrow$ (1). Since $[1]$ is left (right) dense, for any $u \in X^*$ there exists $x \in X^*$ such that $xu \in [1](ux \in [1])$, i.e. $\bar{x}\,\bar{u} = \bar{1}(\bar{u}\,\bar{x} = \bar{1})$. This means that every element of M has a left (right) inverse. Hence M is a group.

Remark 1. If M is a group, then $\rho = P_{[u]}$ for any $u \in X^*$.

Lemma 3. *Let $A \subseteq X^*$ be a recognizable language which is strongly dense. Then A is two-sided dense.*

Proof. Since X^*/P_A is a finite simple monoid, X^*/P_A is a group. Let $x \in A$. Then the congruence class $[x]$ of P_A is two-sided dense. Since $[x] \subseteq A$, A is two-sided dense, too.

Proposition 2. *Let ρ be a congruence on X^* such that every congruence class is dense. If at least one congruence class of ρ is recognizable, then the index of ρ is finite and $M = X^*/\rho$ is a finite group.*

Proof. Let $[u]$ be recognizable language where $u \in X^*$. It is almost obvious that $\rho \subseteq P_{[u]}$, i.e. any congruence class of ρ is contained in some congruence class of $P_{[u]}$. Hence $[u]$ is strongly dense. From Lemma 3, $[u]$ is two-sided dense. Therefore, from Lemma 2, M is a group. As has been seen in Remark 1, $\rho = P_{[u]}$. This means that the index of ρ is finite and M is a finite group.

Now we are ready to prove the following:

Proposition 3. *Let $A \subseteq X^*$ be a strongly dense language. Then the following assertions are equivalent:*

(1) *A is recognizable.*

(2) *At least one congruence class of P_A is recognizable.*

(3) *Every congruence class of P_A is recognizable.*

Proof. (1) $\Longrightarrow$ (3). Since A is strongly dense, every congruence class of P_A is dense. Therefore, from the proof of Lemma 3 there exists a congruence class of P_A which is two-sided dense. From Lemma 2, $M = X^*/P_A$ is a group. It follows from Remark 1 that $P_A = P_{[u]}$ for any $u \in X^*$ where $[u] = [u]_{P_A}$. This means that $[u]$ is recognizable.

(3) $\Longrightarrow$ (2). Obvious.

(2) $\Longrightarrow$ (1). Obvious from Proposition 2.

Now we will deal with a relationship between the dense property of recognizable languages and the structure of automata.

Let $M = (S, X, \delta, s_0, F)$ be an automaton and let $\emptyset \neq T \subseteq S$. If the following conditions are satisfied, then T is called a minimal stage of M:

(a) $sx^M \in T$ for any $s \in T$ and $x \in X^*$.

(b) For any $s, t \in T$, there exists $x \in X^*$ such that $sx^M = t$.

A minimal stage T is said to be accessible if there exists $x \in X^*$ such that $s_0x^M \in T$.

Lemma 4. *Let $M = (S, X, \delta, s_0, F)$ be an automaton and let $T \subseteq S$. Then T is a minimal stage of M if and only if $T = \{tx^M \mid x \in X^*\}$ for some $w \in \mu(M)$ and $t \in Sw^M$.*

Proof. ($\Longrightarrow$) Let $s \in T$ and $w \in \mu(M)$. From assertion (a), we have $sw^M \in T$. Let $t = sw^M$. Then $t \in Sw^M$ and $T = \{tx^M \mid x \in X^*\}$ from assertion (b).

($\Longleftarrow$) Assertion (a) is obviously satisfied. Let $t_1, t_2 \in T$. Then there exist $x_1, x_2 \in X^*$ such that $t_1 = tx_1^M$ and $t_2 = tx_2^M$. As has been stated in the proof of Proposition 1, $t((x_1w)^n)^M = t$ for some $n \geq 1$. Hence we have $t_1(w(x_1w)^{n-1}x_2)^M = t((x_1w)^nx_2)^M = tx_2^M = t_2$. Thus (b) is also satisfied. Hence T is a minimal stage.

Proposition 4. *Let $M = (S, X, \delta, s_0, F)$ be an automaton and let T(M) be the language accepted by M. Then T(M) is dense if and only if there exists an accessible minimal stage of M which contains a final state.*

Proof. ($\Longrightarrow$) Let $w \in \mu(M)$. There exist $x, y \in X^*$ such that $xwy \in T(M)$, i.e. $s_0(xwy)^M \in F$. since $xwy \in \mu(M)$, it follows from Lemma 4 that $s_0(xwy)^M \in T$ for some accessible minimal stage T. Thus $T \cap F \neq \emptyset$.

($\Longleftarrow$) Let $u \in X^*$ and let T be an accessible minimal stage such that $T \cap F \neq \emptyset$. There exists $x \in X^*$ such that $s_0x^M \in T$. From (a), $s_0(xu)^M \in T$. From (b), there exists $y \in X^*$ such that $s_0(xuy)^M \in F$. Hence $xuy \in T(M)$ and $T(M)$ is dense.

Finally, we provide some results on context-free languages (see [4] for details). for any $i \geq 1$. by $Q(X)^{(i)}$ we denote the set $\{q^i \mid q \in Q(X)\}$. Moreover, by $Q_3(X)$ we denote the set $\cup_{i\geq 3}Q(X)^{(i)}$. Let $u, v \in X^*$. Then u, v are said to be conjugate if $u = xy$ and $v = yx$ for some $x, y \in X^*$.

Lemma 5. ([6]) *Let $u, v \in X^*$. If $u \in Q(X)^{(i)}$ for some $i \geq 1$, then $v \in Q(X)^{(i)}$.*

Lemma 6. (Pumping lemma, [5]) *Let $A \subseteq x^*$ be a context-free language. Then there exists a positive integer p(A), depending only on A, such that every word $z \in A$ of length greater than p(A) can be written z=uvwxy with $|vwx| \leq p(A), vx \neq 1$ and $uv^nwx^ny \in A$ for any $n \geq 1$.*

Proposition 5. *Let $A \subseteq X^*$ be an infinite context-free language such that $A \subseteq Q_2(X)$. Then $A_1 = A \cap Q(X)^{(2)}$ is a context-free language and $A_2 = A \cap Q_3(X)$ is a recognizable language. More exactly, A_2 can be written as follows:*

$$A_2 = \left(\bigcup_{1\leq i\leq r} f_i^{m_i}(f_i^{k_i})^* \right) \cup F$$

where $f_i \in Q(X), m_i \geq 3, k_i \geq 1 (1 \leq i \leq r)$ and $F \subseteq Q_3(X)$ is a finite set.

Proof. Let $A \subseteq X^*$ be an infinite context-free language contained in $Q_2(X)$. Let $A_2 = A \cap Q_3(X)$. Moreover, Let $z_1 \in A_2$ and let $|z_1| > 7p(A)$. then $z_1 = uvwxy, |vwx| \leq p(A), vx \neq 1$ for some $u, v, w, x, y \in X^*$ and $z_n = uv^nwx^ny \in A$ for any $n \geq 2$. Note that $z_1 = uvwxy \in Q(X)^{(i)}, i \geq 3$ and $z_2 = uv^2wx^2y \in Q(X)^{(j)}, j \geq 2$. From Lemma 5, $z_1' = (xyuv)w \in Q(X)^{(i)}$ and $z_2' = (xyuv)vwx \in Q(X)^{(j)}$. Then we have the following computation:

$$|z_1'|/i + |z_2'|/j \leq (|xyuv| + |w|)/3 + (|xyuv| + |vwx|)/2$$
$$< 5(|xyuv|)/6 + p(A) < |xyuv|.$$

Therefore, from Lemma 1 there exist $f \in Q(X)$ such that $z_1' = f^i$ and $z_2' = f^j$. Consequently, $j > i$ and $|vx| = j - i = k|f|$ for some $k > 1$. From Lemma 5, there exists $g \in Q(X)$ such that f, g are conjugate, $z_1 = g^i$ and $z_2 = g^j$. Thus $z_1 = uvwxy = g^i$ and $z_2 = uv^2wx^2y = g^j = g^{i+k}$. From the same observation as above, $z_n = g^{i+kn}$. consequently, $z_1 \in g^i(g^k)^* \subseteq A$. Note that $|g| < p(A)/k$ and $g^n(g^k)^* \subseteq g^m(g^k)^*$ if $m \leq n$ and $m \equiv n$ (mod k). From this remark, there exist $f_1, f_2, \cdots, f_r \in Q(X), m_1, m_2, \cdots, m_r \geq 3, k_1, k_2, \cdots, k_r \geq 1$ and $F \subseteq \bigcup_{1 \leq i \leq 7p(L)} X^i$ such that $L_2 = (\bigcup_{1 \leq i \leq r} f_i^{m_i}(f_i^{k_i})^*) \cup F$. It is obvious that $F \subseteq Q_3(x)$. From the above expression of A_2, A_2 is recognizable. On the other hand, since $A_1 = A \setminus A_2$ and A_2 is recognizable, A_1 is context-free (see [5]).

Regarding context-free languages contained in $Q(X)^{(2)}$, we have the following result. To prove it, we need the following Ogden's pumping lemma.

Lemma 7. (Ogden's pumping lemma, [5]) *Let $A \subseteq X^*$ be a context-free language. Then there exists a positive integer $\overline{p}(A)$, depending only on A, such that if $z \in A$, and we mark any $\overline{p}(A)$ or more positions of z "distinguished", then we can write $z = uvwxy$, such that:*

1) *v and x together have at least one distinguished position,*
2) *vwx has at most $\overline{p}(A)$ distinguished positions, and*
3) *$uv^nwx^ny \in A$ for any $n \geq 1$.*

Proposition 6. ([4]) *Let $A \subseteq X^*$ be an infinite context-free language contained in $Q(X)^{(2)}$. Then $A = F \cup (\bigcup_{1 \leq i \leq r}\{(a_i^n b_i a_i^m)^2 \mid n, m \geq 1\}) \cup (\bigcup_{1 \leq j \leq s}\{(g_j h_j^n f_j)^2 \mid n \geq 1\})$ where $b_i a_i^2 \in Q(X)(1 \leq i \leq r), f_j g_j h_j \in Q(X)(1 \leq j \leq s)$ and $F \subseteq X^*$ is a finite set.*

As an application, we have the following:

Proposition 7. *Let $A \subseteq X^*$ be a dense context-free language. Then $A \cap Q_1(X)$ is a dense language.*

Proof. Suppose that $A \cap Q_1(X)$ is not dense. Then there exists $z \in X^+$ such that $X^*zX^* \cap (A \cap Q_1(X)) = \emptyset$. Therefore, $X^*zX^* \cap A \subseteq Q_2(X)$. Since A is dense and X^*zX^* is recognizable, $X^*zX^* \cap A$ is an infinite context-free language ([5]). Let $R = (X^*zX^* \cap A) \cap Q_3(X)$. From Proposition 5, R is recognizable. Hence $(X^*zX^* \cap A)\setminus R$ is context-free and contained in $Q(X)^{(2)}$. Let $z' \in X^+$ be a sufficiently long cube-free word (for cube-free words, see [7]). Then there exist $\alpha, \beta \in X^*$ such that $\alpha z' z \beta \in X^*zX^* \cap A$. On the other hand, since R is a context-free language (in fact, a recognizable language) contained in $Q_3(X), \alpha z' z \beta \notin R$. Hence $\alpha z' z \beta \in (X^*zX^* \cap A)\setminus R \subseteq Q(X)^{(2)}$. However, this is impossible according to Proposition 6. Consequently, $A \cap Q_1(X)$ is dense.

It is an open question whether $A \cap Q_1(X)$ is disjunctive as it is when A is a dense recognizable language.

References

[1] A.H. Clifford and D.D. Miller, *Semigroups having zeroid elements*, Amer. J. Math., 70(1948), 117–125.

[2] A.H. Clifford and G.B. Preston, *Algebraic Theory of Semigroups*, Vol. 1, Amer. Math. Soc., Providence, R.I., 1961.

[3] M. Ito and M. Katsura, *Relations between recognizable languages and disjunctive languages*, submitted.

[4] M. Ito and M. Katsura, *Context-free languages consisting of non-primitive words*, in preparation.

[5] J.E. Hopcroft and F.D. Ullman, *Introduction to automata theory, languages, and computation*, Addison-Wesley Publ. Co., Reading. Mass., 1979.

[6] M. Lothaire, *Combinatorics on Words*, Addison-Wesley Publ. Co. Reading, Mass., 1983.

[7] A. Salomaa, *Jewels of formal language theory*, Pitman Publ. Co., London, 1981.

[8] H.J. Shyr, *Free monoids and languages*, Lecture Notes, Department of Mathematics, Soochow University, Taipei, Taiwan, 1979.

Proceedings of the Asian Mathematical
Conference, 1990

Estimation of Reliability Function Normal Distribution with Known Coefficient of Variation: a Simulation Study

R.R.L. KANTAM, A. VASUDEVARAO, & V.L. NARASIMHAM
Department of Statistics
Nagarjuna University
Guntur 522 510, India

§1. Introduction

Bayesian approaches to the estimation problem are based on the principle of utilising a prior information on the distributions of the parameters to be estimated. A somewhat simple type of prior information is sometimes available to the experimenters through long association with the experimental material they may have at their disposal. In many life sciences and biological experiments it often happens that a scatter plot of cell means versus cell standard deviations in analysis of variance would yield evidence that the population cell standard deviations are proportional to population cell means where the cell observations are normally distributed. This amounts to a knowledge of population coefficient of variation (CV) in a normally distributed data. In such situations one usually comes across a sample data from $N(\theta, c\theta)$ where $c(> 0)$ is known and is equal to population CV.

Estimation of θ in $N(\theta, c\theta)$ with known c is dealt with by many authors and some of them are Searls (1964), Khan (1968), Govindarajulu and Sahai (1972), Gleser and Healy (1976), Ebrahimi (1984) and Samanta (1985).

MLE and UMVUE of reliability function $R(x; \mu, \sigma)$ in $N(\mu, \sigma)$ with no known relation between μ, σ are well-established (Sinha, 1986). Zacks and Even (1966) studied the small sample efficiencies of MLE and UMVUE of $R(x; \mu, \sigma)$ with known σ. Kantam et al. (1984) suggested the estimation of $R(x; \mu, \sigma)$ using best linear unbiased estimators (BLUEs) of μ, σ and compared the resulting estimator with MLE. Stadje (1988) described the large sample behaviours of estimators of distribution function in location-scale family obtained by substituting various estimators of location and scale parameters in the distribution function.

As not much is known about the estimation of reliability function $-R(x; c, \theta)$ in $N(\theta, c\theta)$ with known c, motivated by Stadje (1988) and in continuation to Kantam et al. (1984), three estimators of $R(x; c, \theta)$ are considered and are proved to be asymptotically as efficient as MLE of $R(x; c, \theta)$, by Kantam and Narasimham (1990).

The theoretical basis of the estimators is briefly presented in Section 2. Findings of the small sample comparison among the estimators are given in Section 3. The numerical tables obtained in this investigation are appended towards the end.

§2. Estimation of $R(x;c,\theta)$

The reliability function $R(x;c,\theta)$ in $N(\theta,c\theta)$ is given by

$$R(x;c,\theta) = 1 - F((x - c\theta)/c\theta),$$

where $F(\cdot)$ is the standard normal cumulative distribution function, $\theta > 0, c(> 0)$ is known.

Let $x_1 < x_2 < \cdots < x_n$ be a complete ordered random sample of size n drawn from $N(\theta,c\theta)$. Then we know that

$$T = \bar{x}, \tag{2.1}$$

$$T' = \frac{\sqrt{n}\Gamma((n-1)/2)S}{c\sqrt{2}\Gamma(n/2)}, \tag{2.2}$$

where

$$\bar{x} = \sum_{i=1}^{n}(x_i/n)$$

$$nS^2 = \sum_{i=1}^{n}(x_i - \bar{x})^2$$

are both unbiased estimators of θ with

$$\text{var}(T) = c^2\theta^2/n, \tag{2.3}$$

$$\text{var}(T') = \left[\frac{(n-1)\Gamma^2(\overline{n-1}/2)}{2\Gamma^2(n/2)} - 1\right]\theta^2 = d_n\theta^2 \quad \text{(say)}. \tag{2.4}$$

The MLE of θ is given by

$$T_1 = \left[\left\{4c^2S^2 + (1+4c^2)\bar{x}^2\right\}^{1/2} - \bar{x}\right] \Big/ 2c^2, \tag{2.5}$$

T_1 is best asymptotically normal (BAN) with mean θ and variance equal to Cramer-Rao lower bound (CRLB) given by

$$\text{asvar}(T_1) = c^2\theta^2/n(1+2c^2). \tag{2.6}$$

Khan(1968) proved that

$$T_2 = bT' + (1-b)T, \tag{2.7}$$

where $b = \text{var}(T)/[\text{var}(T)+\text{var}(T')]$ is the minimum variance unbiased estimator of θ in the class of unbiased estimators that are linear in T and T' and that T_2 is a BAN estimator.

BLUE of θ in $N(\theta,c\theta)$ is given by (Lloyd, 1952)

$$T_3 = \frac{(J' + cA')B^{-1}x}{(J'B^{-1}J + c^2A'B^{-1}A)}, \tag{2.8}$$

$$\operatorname{var}(T_3) = c^2\theta^2/(J'B^{-1}J + c^2A'B^{-1}A),$$

where

$$J' = (1, 1, \cdots, 1),$$

$$x' = (x_1, x_2, \cdots, x_n) - \text{the vector of ordered sample observations},$$

$$A' = (E(Z_1), E(Z_2), \cdots, E(Z_n)),$$

$$B = \operatorname{cov}(Z_i, Z_j)_{n\times n}, \quad (\text{with} \quad Z_i = \overline{x_i - \theta}/c\theta).$$

Govindarajulu and Sahai (1972) computed the coefficients to get T_3 and established that T_3 is a BAN estimator.

Gleser and Healy (1976) proposed that

$$T_4 = b_1 T + b_2 T', \tag{2.9}$$

where

$$b_1 = d_n/(d_n + n^{-1}c^2 + n^{-1}c^2 d_n)$$

$$b_2 = n^{-1}c^2 b_1/d_n$$

(with T, T', d_n given by (2.1), (2.2), (2.4) respectively) is minimum mean square error estimator (MMSEE) of θ in the class of estimators that are linear in T and T'. It can be easily seen that T_4 is also a BAN estimator (Kantam and Narasimham, 1990).

It is well-known that the MLE of $R(x, c, \theta)$ in $N(\theta, c\theta)$ is

$$R_1 = R(x; c, T_1), \tag{2.10}$$

where T_1 is given in (2.5).

Analogous to R_1, Kantam and Narasimham (1990) considered three estimators of $R(x; c, \theta)$ as

$$R_i = R(x; c, T_i), \quad i = 2, 3, 4, \tag{2.11}$$

where $T_i, i = 2, 3, 4$ are the three BAN estimators given by (2.7), (2.8), (2.9) and proved that R_i, $i = 1, 2, 3, 4$ have the same asymptotic variance

$$\operatorname{asvar}(R_i) = \frac{x^2 f^2(\overline{x - \theta}/c\theta)}{n\theta^2(1 + 2c^2)} \tag{2.12}$$

which is the CRLB for the variance of an estimator of $R(x; c, \theta), f(\cdot)$ being the standard normal density function.

In small samples R_1, R_2, R_3, R_4 may not have the same variance. At the same time the exact expressions for their variance are not mathematically tractable. Hence, the small sample behaviour of R_1, R_2, R_3, R_4 is compared by a simulation study in the next section.

§3. Small Sample Comparison of Reliability Estimates

Often when there are a number of potential estimators of a parameter or parametric function one uses variance, mean square error (MSE) as criteria to select the ' best' estimator. As mentioned in Section 2, due to the complicated distributional structures, the normal reliability estimates R_1, R_2, R_3, R_4 do not lend themselves to an analytical comparison of the variances nor their MSEs. In such situations we have to rely on empirical evidence, especially when small sample comparisons are under consideration. Thus attention will be focussed on comparing the empirical sample characteristics of the estimators calculated for simulated normal samples.

2000 random samples of size 5 each and 10 each are generated from each of the populations $N(100,1), (c = 0.01); N(10,1), (c = 0.1); N(1,1), (c = 1); N(0.1,1), (c = 10)$. The true population reliabilities 0.25, 0.5, 0.75, 0.9, 0.95 are considered for estimation. Thus in all, 20 parameter combinations (4 values of population CV, 5 values of population reliabilities) are taken in the simulation study. For each parameter combination and each size of the sample the reliability estimates R_1, R_2, R_3, R_4 are computed at values of x corresponding to the above true values. The observed mean, variance and MSE of the estimates are given in Table A.1.

For a given size of the sample, with respect to a given empirical sample characteristic, the frequency of the first preference to a particular method of estimation over all parameter combinations is presented in Table A.2. Thus the first row in Table A.2. indicates that when $n = 5$, of the 20 parameter combinations, in 10 situations R_1, in 4 situations R_2, in one situation R_3 and in 5 situations R_4 have least observed biases, so that with respect to observed bias the first preference goes R_1 in 10 times, to R_2 in 4 times, to R_3 once and to R_4 in 5 times.

Of the 20 parameter combinations (see Table A.2) irrespective of the size of the sample, the MLE$-R_1$ has the largest frequency of first preference with respect to observed MSE. On the other hand with respect to observed variance and bias the choice goes to R_1 when $n = 5$ and to R_4 when $n = 10$. R_3 is the least preferred estimator with respect to the three sample characteristics when $n = 5$ and with respect to MSE when $n = 10$.

Thus we conclude that ML method maintains its superiority over the remaining three methods to estimate normal reliability function with known CV from small and complete samples.

If the investigation extends to Type II censored samples (which is common with data of reliability studies) the estimators R_1 and R_3 only would remain in the competition as the estimators T_2, T_4 ((2.7) and (2.9)) used to get R_2, R_4 are still not known for censored samples. R_3 (based on BLUE of θ in $N(\theta, c\theta)$ from censored samples) is computationally simpler than R_1 as the MLE of θ in $N(\theta, c\theta)$ from censored samples does not have explicit expression (Ebrahimi, 1984). Hence, a comparison between R_1 and R_3 from censored samples is worth studying and is being currently considered by the authors.

Table A. 1

Obsered sample characteristics of reliability estimates

True Value of C.V.	True Value of $R(t)$	Sample Characteristic	$n = 5$				$n = 10$			
			$R1$	$R2$	$R3$	$R4$	$R1$	$R2$	$R3$	$R4$
0.01	0.25	MEAN	.262067	.262692	.264214	.262092	.256169	.256490	.249535	.256181
		VAR.	.086224	.086596	.087493	.086243	.075013	.075191	.071419	.075022
		M.S.E.	.017691	.017750	.017886	.017697	.009429	.009446	.009151	.009431
	0.5	MEAN	.492636	.493394	.495244	.492662	.495482	.495878	.487253	.495497
		VAR.	.267832	.268582	.270412	.267860	.259497	.259891	.251403	.259513
		M.S.E.	.025196	.025188	.025168	.025198	.014015	.014013	.014150	.014016
	0.75	MEAN	.726304	.726935	.728467	.726327	.736720	.737041	.729977	.736729
		VAR.	.545111	.545989	.548125	.545142	.552211	.552674	.542558	.552225
		M.S.E	.018156	.018086	.017925	.018152	.009631	.009612	.010092	.009632
	0.9	MEAN	.877238	.877624	.878553	.877253	.887627	.887811	.883681	.887632
		VAR	.776421	.777068	.778631	.776446	.791249	.791568	.784422	.791257
		M.S.E	.007393	.007345	.007235	.007390	.003520	.003508	.003796	.003520
	0.95	MEAN	.932796	.933043	.933635	.932806	.940839	.940952	.938404	.940843
		VAR.	.873064	.873509	.874577	.873083	.886510	.886719	.882017	.886517
		M.S.E	.003253	.003228	.003170	.003251	.001416	.001410	.001549	.001416
0.1	0.25	MEAN	.252825	.259119	.259046	.253008	.283236	.286590	.306163	.283143
		VAR.	.081746	.085402	.085376	.081892	.084816	.086761	.098760	.084748
		M.S.E	.017833	.018342	.018353	.017888	.005698	.005966	.008178	.005676
	0.5	MEAN	.480231	.487643	.487543	.480417	.532462	.536186	.557155	.532390
		VAR.	.255669	.262845	.262764	.255872	.289604	.293530	.316504	.289487
		M.S.E	.025438	.025202	.025221	.025455	.007141	.007344	.009348	.007097
	0.75	MEAN	.718372	.724300	.724208	.718517	.769981	.772665	.787261	.769959
		VAR.	.532990	.541168	.541048	.533182	.596285	.600344	.622923	.596216
		M.S.E	.017932	.017218	.017235	.017906	.003814	.003847	.004532	.003778
	0.9	MEAN	.875481	.878897	.878841	.875570	.908685	.910059	.917308	.908690
		VAR.	.772579	.778326	.778234	.772719	.826663	.829128	.842278	.826658
		M.S.E	.006713	.006312	.006319	.006693	.001030	.001021	.001123	.001017
	0.95	MEAN	.933328	.935407	.935373	.933386	.954414	.955189	.959206	.954422
		VAR.	.873520	.877283	.877221	.873619	.911220	.912686	.920338	.911231
		M.S.E	.002696	.002511	.002514	.002686	.000334	.000328	.000347	.000329

Table A. 1 (Continued)

True Value of C.V.	True Value of $R(t)$	Sample Characteristic	$n = 5$				$n = 10$			
			$R1$	$R2$	$R3$	$R4$	$R1$	$R2$	$R3$	$R4$
1	0.25	MEAN	.211775	.233944	.233700	.200048	.222466	.233843	.233676	.216532
		VAR.	.058728	.070812	.070818	.054422	.057539	.063383	.063390	.055218
		M.S.E	.015340	.016340	.016468	.016897	.008806	.008962	.009052	.009452
	0.5	MEAN	.446758	.463950	.463477	.432511	.468089	.476876	.476666	.462065
		VAR.	.214424	.231352	.231101	.204223	.226009	.234660	.234508	.221114
		M.S.E	.017666	.017403	.017624	.021712	.007920	.007784	.007842	.009049
	0.75	MEAN	.732484	.736656	.736448	.727083	.740716	.742972	.742917	.738881
		VAR.	.538377	.544909	.544653	.531299	.549226	.552625	.552543	.546618
		M.S.E	.002151	.002426	.002480	.003174	.000652	.000666	.000668	.000797
	0.9	MEAN	.906248	.904128	.904187	.908069	.903678	.902594	.902616	.904414
		VAR.	.821519	.817707	.817818	.824868	.816735	.814784	.814825	.818077
		M.S.E	.000272	.000278	.000281	.000343	.000115	.000115	.000115	.000132
	0.95	MEAN	.955820	.952926	.952969	.957607	.953853	.952396	.952417	.954681
		VAR.	.913866	.908383	.908468	.917308	.909980	.907213	.907255	.911567
		M.S.E	.000308	.000323	.000326	.000354	.000159	.000161	.000163	.000174
10	0.25	MEAN	.217533	.226865	.226549	.198549	.229007	.233729	.233448	.220258
		VAR.	.053895	.059347	.059317	.047153	.055687	.058256	.058181	.052203
		M.S.E	.007629	.008414	.008542	.010378	.003683	.003892	.003957	.004573
	0.5	MEAN	.491657	.492556	.492426	.486389	.495603	.496317	.496259	.493850
		VAR.	.242137	.243169	.243053	.237281	.245748	.246470	.246414	.244043
		M.S.E	.000481	.000613	.000626	.000891	.000145	.000153	.000155	.000193
	0.75	MEAN	.777280	.770706	.771010	.793116	.766744	.763294	.763519	.773619
		VAR.	.608496	.599182	.599731	.634405	.589860	.584808	.585183	.600768
		M.S.E	.005076	.005623	.005717	.007231	.002244	.002367	.002404	.002840
	0.9	MEAN	.913975	.904842	.904888	.925058	.911262	.906943	.907066	.917043
		VAR.	.838183	.822426	.822554	.858662	.832031	.824434	.824694	.842683
		M.S.E	.003028	.003711	.003757	.003559	.001759	.001936	.001974	.002005
	0.95	MEAN	.954135	.946102	.946043	.961132	.954905	.951307	.951330	.958694
		VAR.	.911793	.897154	.897070	.925159	.912668	.905982	.906049	.919918
		M.S.E	.001437	.002061	.002089	.001508	.000848	.000999	.001022	.000900

Table A.2

Observed frequemcies of the first preference

to

various reliability estimates

Sample charac-teristic	Estimator n	R_1	R_2	R_3	R_4
Mean	5	10	4	1	5
	10	2	5	5	8
Variance	5	10	4	1	5
	10	1	5	5	9
MSE	5	10	6	4	0
	10	9	6	1	4

References

[1] Ebrahimi, N., *Maximum likelihood estimation, from doubly censored samples, of the mean of a normal distribution with known coefficient of variation*, Commun. Statist. -Theor. Meth., 13:5(1984), 651–659.

[2] Gleser, L.J. and Healy, J.D., *Estimating the mean of a normal distribution with known coefficient of variation*, J. Amer. Statist. Assoc., 71(1976), 977–981.

[3] Govindarajulu, Z. and Sahai, H., *Estimation of the parameters of a normal distribution with known coefficient of variation*, Rep. Stat. Appl. Res., JUSE, 19:3(1972), 85–98.

[4] Kantam, R.R.L., Dattatreya Rao, A.V. and Narasimham, V.L. *Estimation of reliability function in normal distribution*, Proceedings of the VI Annual Conference of ISPS University of Kerala, 1984, 57–65.

[5] Kantam, R.R.L. and Narasimham, V.L., *Estimation of normal survival function with known coefficient of variation*, National Seminar on Recent Advances in Statistical Inference, 24-26 February, 1990 held at University of Rajasthan, Jaipur, India, 1990.

[6] Khan, R.A., *A note on estimating the mean of a normal distribution with known coefficient of variation*, J. Amer. Statist. Assoc., 63(1968), 1039–1041.

[7] Lloyd, E.H., *Least squares estimation of location and scale parameters using order statistics*, Biometrika, 39(1952), 88–95.

[8] Samanta, M., *Minimum MSE estimation by linear combinations of order statistics*, Austral. J. Statist., 27(1985), 189–194.

[9] Searls, D.T., *The utilisation of a known coefficient of variation in the estimation procedure*, J. Amer. Statist. Assoc., 59(1964), 1225 & 1226.

[10] Sinha, S.K., *Reliability and Life Testing*, Wiley Eastern, 1986.

[11] Stadje, W., *Estimation of distribution function of a location-scale family*, Statistics, 19(1988), 537–544.

[12] Zacks, S. and Even, M., *The efficiencies in small samples of the maximum likelihood and best unbiased estimators of reliability functions*, J. Amer. Statist. Assoc., 61(1966), 1033–1050.

Proceedings of the Asian Mathematical
Conference, 1990

Periodic Generalized Functions and Boundary Values of Holomorphic Functions

V. KARUNAKARAN, A. VIJAYAN
School of Mathematics
Madurai Kamaraj University
Madurai 625 021 India

§1. Introduction

The theory of generalized functions is an extension of the classical theory of distributions. The new theory which enables us to define arbitrary product of distributions is introduced and systematically developed in [1] and applied to problems in Physics via the differential equations (see [2]). From differentiation point of view these generalized functions behave similar to distributions in that they admit partial derivatives of any order. However from the view point of multiplications and non-linear operations these objects have properties different from those of distributions. In fact, finite products of generalized functions are closed with respect to many non-linear operations. In this paper we shall introduce the concept of periodic generalized functions which generalises the usual periodic distributions in an attempt to find the connection between these objects and the holomorphic functions in the open unit disc of the complex plane.

We assume the basic facts about the space of test functions on T^n, its dual namely the space of periodic distributions and their Fourier coefficients. In this paper, we introduce the concepts of periodic generalized functions on T, generalized holomorphic functions in U, the open unit disc and V, the exterior of the unit circle and characterise certain periodic generalized functions as having been associated to the boundary values of generalized holomorphic functions. For general information about generalized functions and their properties we refer to [1] and [2].

§2. Periodic Generalized Functions

Let $m(t)$ be the Haar measure on the unit circle T. Define $A(T) = \{\varphi \in D(T)$ such that $\int_0^{2\pi} \varphi(e^{it})dm(t) = 1\}$. For $\varphi \in A(T)$, Let us define $\varphi_j(e^{ix}) = \varphi(e^{ijx})$, $j = 1, 2, \cdots$.

Remark 2.1. $\varphi \in A(T)$ if and only if $\varphi_j \in A(T)$, $j = 1, 2, \cdots$. This follows from the definition.

Remark 2.2.

$$\hat{\varphi}_j(k) = \begin{cases} \hat{\varphi}\left[\frac{k}{j}\right] & \text{if } j \mid k \\ 0 & \text{otherwise} \end{cases} \qquad k \in Z,$$

see [4] Exercise 5 on page 7.

Let

$$D_j(e^{it}) = \sum_{0\neq s=-j+1}^{j-1} e^{ist}.$$

Define

$$\varphi_j^*(e^{ix}) = D_j(e^{ix}) + \varphi_j(e^{ix}), \quad j = 1, 2, \cdots.$$

Note that $\varphi_1^*(e^{ix}) = \varphi(e^{ix})$.

Remark 2.3. $\varphi \in A(T)$ if and only if $\varphi_j^* \in A(t)$, $j = 1, 2, \cdots$. This also follows from the definition.

Notation 2.4. Let $E(T)$ denote the set of all functions $R : A(T)XT \to \mathbb{C}; (\varphi, e^{ix}) \to R(\varphi, e^{ix})$ which are infinitely differentiable in x for each fixed $\varphi \in A(T)$.

Definition 2.5. We say that an element R of $E(T)$ is moderate if for each $\alpha = 0, 1, 2, \cdots$ and $\varphi \in A(T)$ there exist constants C (depending on α, φ) and m (depending on α) such that $|D^\alpha R(\varphi_j^*, e^{ix})| \leq C j^m$ for all $x \in R$ and for sufficiently large j.

We denote by $M(T)$ the set of all moderate elements of $E(T)$. Note that $M(T)$ is a subalgebra of $E(T)$.

The Fourier coefficients of a moderate element R are defined as

$$\widehat{R}(\varphi, k) = \int_0^{2\pi} R(\varphi, e^{ix}) e^{-ikx} dm(x), \quad k \in Z, \quad \varphi \in A(T).$$

Proposition 2.6. *$R \in M(t)$ if and only if for each $n = 0, 1, 2, \cdots$ and $\varphi \in A(T)$ there exist constants C (depending on φ, n) and m (depending on n) such that*

$$\sum_{k\in Z} (1 + k^2)^n |\widehat{R}(\varphi_j^*, k)|^2 \leq C_j^m \quad \text{for large} \quad j.$$

Definition 2.7. An elements $R \in E(T)$ is said to be null if for every $\alpha = 0, 1, 2, \cdots, \beta = 1, 2, \cdots$ and $\varphi \in A(T)$ there exist constants C (depending on φ, α, β) and L such that $|D^\alpha R(\varphi_j^*, e^{ix})| \leq C/j^\beta$ for $j \geq L$ and for all $x \in R$.

We denote this class by $N(T)$. It is easy to see that $N(T)$ is an ideal in $M(T)$.

Definition 2.8. Definie $G(T) = M(T)/N(T)$. The elements of $G(T)$ are called periodic generalized functions.

Proposition 2.9. *$R \in N(T)$ if and only if for each $n = 0, 1, 2, \cdots$, $\beta = 1, 2, \cdots$ and $\varphi \in A(T)$ there exist constants C (depending on φ, β, n) and L such that*

$$\sum_{k\in Z} (1 + k^2)^n |\widehat{R}(\varphi_j^*, k)|^2 \leq C/j^\beta \quad \text{for} \quad j \geq L.$$

Proposition 2.10. *$D'(T)$ is a subspace of $G(T)$.*

Notation 2.11. Let E denote the set of all functions $R : A(T) \to \mathbb{C}$. Then E is an algebra.

Definition 2.12. Let 'M' denote the set of all functions $R \in E$ such that for each $\varphi \in A(T)$ there exist constants C (depending on φ) and m such that $|R(\varphi_j^*)| \leq C j^m$ for large j.

It is easy to see that M is a subalgebra of E.

Definition 2.13. We denote by N the set of all functions $R \in E$ such that for each $\alpha = 1, 2, \cdots$ and $\varphi \in A(T)$ there exist constants C (depending on α and φ) and L such that $|R(\varphi_j^*)| \le C/j^\alpha$ for $j \ge L$.

Note that N is an ideal in M.

Definition 2.14. Define $\overline{G} = M/N$. The elements of $\overline{G}$ are called generalized complex numbers.

Definition 2.15. Let $G \in G(T)$ and R be a representative of G.

$$\widehat{R}(\varphi, k) = \int_0^{2\pi} R(\varphi, e^{ix}) e^{-ikx} dm(x).$$

Define $I_k(\varphi) = \widehat{R}(\varphi, k)$ for a fixed k. Then $I_k \in M$ for each $k \in Z$. Define $\widehat{G}(k) = I_k + N$. $\widehat{G}(k)'s$ are called Fourier coefficients of G.

Note that this value is independent of the representative for G. We denote this also by $\int_0^{2\pi} G(e^{ix}) dm(x)$.

Proposition 2.16. *If $S \in D'(T)$ then the new Fourier coefficients coincide with the classifical ones in the sense of generalized complex numbers.*

Definition 2.17. $G \in G(T)$ is said to be associated to a distribution $S \in D'(T)$ if for an arbitrary representative R of G.

$$\lim_{j\to\infty} \widehat{R}(\varphi_j^*, k) = \widehat{S}(k) \quad \text{for each} \quad k \in Z \quad \text{and} \quad \varphi \in A(T).$$

Definition 2.18. $G_1 \in G(T)$ is said to be associated to $G_2 \in G(T)$ if $G_1 - G_2$ is associated to the zero distribution.

Notation 2.19. Let Q denote the set of all periodic generalized functions $G \in G(T)$ such that for an arbitrary representative R of G, $\lim_{j\to\infty} \widehat{R}(\varphi_j^*, k)$ exists for every $k \in Z$ and $\varphi \in A(T)$.

§3. Generalized Holomorphic Functions

Notation 3.1. Let W be an open set in the complex plane. Let $E(W)$ denote the set of all functions

$$F : A(T)XW \to C; \quad (\varphi, z) \to F(\varphi, z)$$

which are holomorphic in z for each $\varphi \in A(T)$.

Definition 3.2. $F \in E(W)$ is said to be moderate if for each $\varphi \in A(T)$. $\alpha = 0, 1, 2, \cdots$, and compact set KCW there exist constants C (depending on K, α, φ) and m (depending on α) such that $|D^\alpha F(\varphi_j^*, z)| \le Cj^m$ for all $z \in K$ and sufficiently large j.

We denote this class by $M(W)$.

Definition 3.3. $F \in E(W)$ is said to be null if for every $\alpha = 0, 1, 2, \cdots, \beta = 1, 2, \cdots, \varphi \in A(T)$ and compact set KCW there exists constant C (depending on $\varphi, \beta, \alpha, k$) such that $|D^\alpha F(\varphi_j^*, z)| \le C/j^\beta$ for sufficiently large j and for all $z \in K$.

We denote this class by $N(W)$.

Definition 3.4. Define $G(W) = M(W)/N(W)$. The elements of $G(W)$ are called generalized holomorphic functions in W.

Definition 3.5. Let $F \in G(W)$ and H be a representative of F define $I_z(\varphi) = H(\varphi, z)$ for a fixed $z \in W$, for all $\varphi \in A(T)$. Then $I_z \in M$ for each fixed $z \in W$.

Define $F(z) = I_z + N$.

Example 1. We give an example of a generalized holomorphic function in terms of a finite set of ordinary holomorphic functions:

Let $f_1, f_2, \cdots, f_n$ be holomorphic functions in W and $C_1, C_2, \cdots, C_n \in \overline{G}$.

Define $F(\varphi, z) = \sum_{k=1}^{n} C_k(\varphi) f_k(z)$. Then $F \in M(W)$ and $H = F + N(W) \in G(W)$.

Notation 3.6. Let U be the open unit disc in the complex plane. Let $G \in G(U)$ and H be a representative of G. Then $H(\varphi, z)$ is holomorphic in U for each $\varphi \in A(T)$. Let the power series expansion of F be given by

$$H(\varphi, z) = \sum_{k=0}^{\infty} a_k(\varphi) z^k. \tag{3.1}$$

We will denote by B the set of all generalized holomorphic functions $G \in G(U)$ given by (3.1) satisfying the following

$$|a_k(\varphi_j^*)| \leq C(\varphi) j^m |k|^p \quad \text{for all} \quad k \in Z \tag{3.2}$$

for some constants $C(\varphi), m, p$ and $\lim_{j \to \infty} a_k(\varphi_j^*)$ exists for each $k \in Z$. (3.3)

Notation 3.7. Let V denote the exterior of the unit circle in the complex plane. Let $G \in G(V)$ and H be a representative of G given by

$$H(\varphi, z) = \sum_{k=0}^{\alpha} a_k(\varphi)/z^k, z \in V. \tag{3.4}$$

We will denote by P the set of all $G \in G(V)$ given by (3,4) satisfying (3.2) and (3.3).

Notation 3.8. For $G \in B$, define

$$C_H(\varphi_j^*, e^{ix}) = \lim_{r \to 1} \int_0^{2\pi} H(\varphi_j^*, re^{ix}) \varphi_j^*(e^{i(y-x)}) dm(y),$$

H is a representative of G. In this case we let $C_G = C_H + N(T)$

§4. Main Result

Theorem 4.1. *If $F \in B$, then there exists $G \in Q$ with negative Fourier coefficients as zero such that C_F is associated to G. Conversely, if $G \in Q$ with negative Fourier coefficients as zero, then there exists $F \in B$ such that C_F is associated to G.*

Corollary 4.2. *If $G \in Q$ then there exist generalized holomorphic functions $F_1 \in B$ and $F_2 \in P$ such that $C_{F_1} + C_{F_2}$ is associated to G.*

The proof of all the propositions, theorems and Lemmas sighted here will appear elsewhere.

References

[1] Colombeau, J.F., *New Generalized Functions and Multiplication of Distribution*, North Holland Mathematics Studies, 1984.

[2] Colombeau, J.F., *An Introduction to New Generalized Functions*, North Holland Mathematics Studies 113, 1985.

[3] Karunakaran, V., *Bieberbag Functions and Periodic Distributions*, Internat. J. Math. and Math. Sci., Vol. 12, No. 1(1989) 69–75.

[4] Katzelson, Y., *An introduction to Harmonic Analysis*, Dover Publications, Inc., 1976.

[5] Rudin, W., *Functional Analysis*, Mcgraw Hill, Inc. (1973).

Proceedings of the Asian Mathematical
Conference, 1990

On a Family of Hyperbolic 3-Manifolds

ANN-CHI KIM
Department of Mathematics
Pusan National University, Korea

HYUN-JONG SONG
Department of Applied Mathematics
Pusan Susan National University Korea

§0. Introduction

In HKM[2], they showed that the Fibonacci group $F(2,2n)$ is the fundamental group of closed orientable hyperbolic 3-manifolds through certain face identifications of a polyhedron Δ_n with $4n$ triangles, $6n$ edges, and $2n+2$ vertices, which can be realised in H^3 for $n \geq 4$. Similarly, we construct a family of closed orientable 3-manifolds M_n through face identifications of dual polyhedron Δ_n^* of Δ_n, where M_5 is the Seifert-Weber hyperbolic dodecahedral space. Moreover, we see that Δ_n^* can be realised in H^3 for $n \geq 4$.

§1. Construction of M_n

The polyhedron Δ_n^* consists of n-gon at the top and bottom faces, $2n$ pentagons as side faces, $6n$ edges and $4n$ vertices, as given in Figure 1.

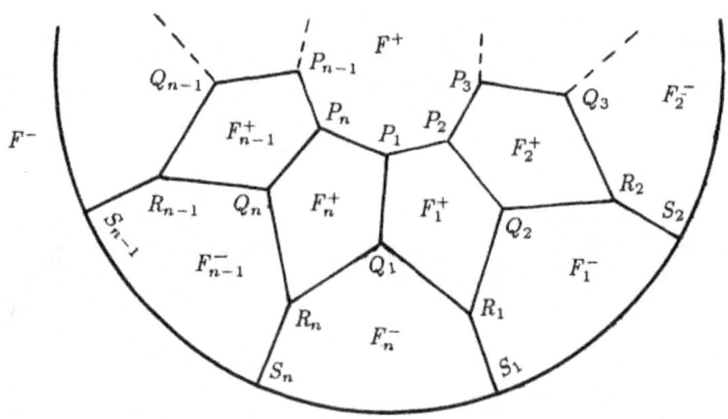

Figure 1

We may think of Δ_n^* as a tessellation of S^2 with following properties, Vertices P_i are equally spaced on the plane $z = t_1, Q_i$ on $z = t_2, R_i$ on $z = -t_2$, and S_i on $z = -t_1 (0 < t_2 < t_1)$. Moreover we may take those vertices so that if n is odd, vertices $\{P_i\}$ (resp. $\{Q_i\}$) are diagonally opposed to vertices $\{S_i\}$ (resp. $\{R_i\}$); otherwise diagonally opposed vertices of the given vertices form a new tessellation Δ_n^{*0} in such a way that each face Δ_n^* corresponds uniquely to that of Δ_n^* by diagonal opposition. For example, see Figure 2.

(i) $n = 3$

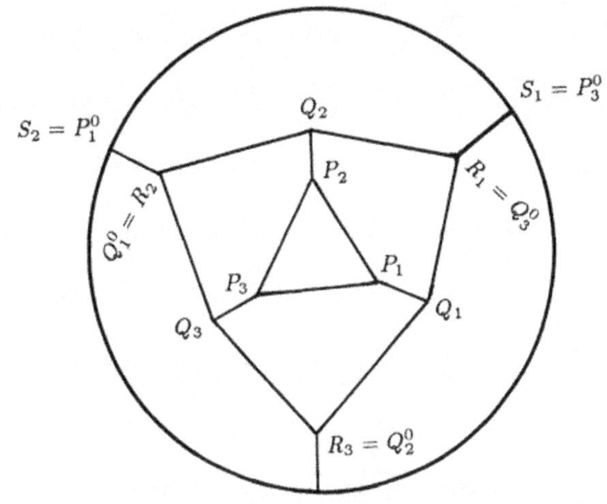

S_3 is diagonally opposed to P_2.

(ii) $n = 4$

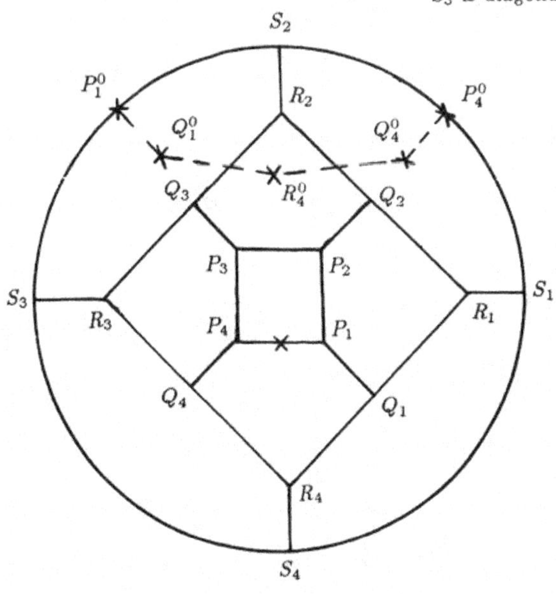

Figure 2

In order to get M_n, we identify each pair of faces of Δ_n^* in the following way.

When n is odd, each face F_i^+ is identified with diagonally opposed face $F_i^{+,0} = F_{j(i)}^-$ as follows.

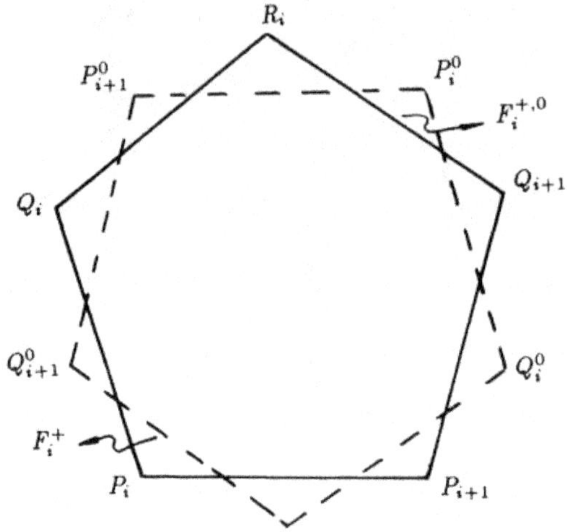

$$\overrightarrow{P_i^0 P_{i+1}^0} = \overrightarrow{Q_i P_i}, \quad \overrightarrow{P_{i+1} Q_{i+1}} = \overrightarrow{P_i^0 P_{i+1}^0}, \quad \overrightarrow{Q_{i+1} R_i} = \overrightarrow{P_{i+1}^0 Q_{i+1}^0}$$

$$\overrightarrow{R_i Q_i} = \overrightarrow{Q_{i+1}^0 R_i^0} \quad \text{and} \quad \overrightarrow{Q_i P_i R_i^0 Q_i^0}$$

Figure 3

When n is even, for each face F_i^+ we have a uniquely determined face $F_{j(i)}^-$ that can be diagonally opposed to F_i^+ after $\frac{(n-3)\pi}{2n}$ clockwise rotation, e.g., in Figure 2-(ii), for $F_i^+ = P_1Q_1R_4Q_4P_4$ we have $F_{j(i)}^- = R_3S_3S_2R_2Q_3$. Considering $F_{j(i)}^-$ as the diagonally opposed face of F_i^-, we identify them as in the odd case. Then it is easy to see that we can identify top face F^+ with bottom face F^- so that it is compatible with the above face identification. Finally, we see that M_n consists of single o-cell, $n+1$ 1-cell, where each n 1-cell comes from identification of 5 edges, the other 1-cell comes form identification of n edges, $n+1$ 2-cells and 1 3-cell.

For examples, we exhibit the case $n = 4, 5, 6$ and 7.

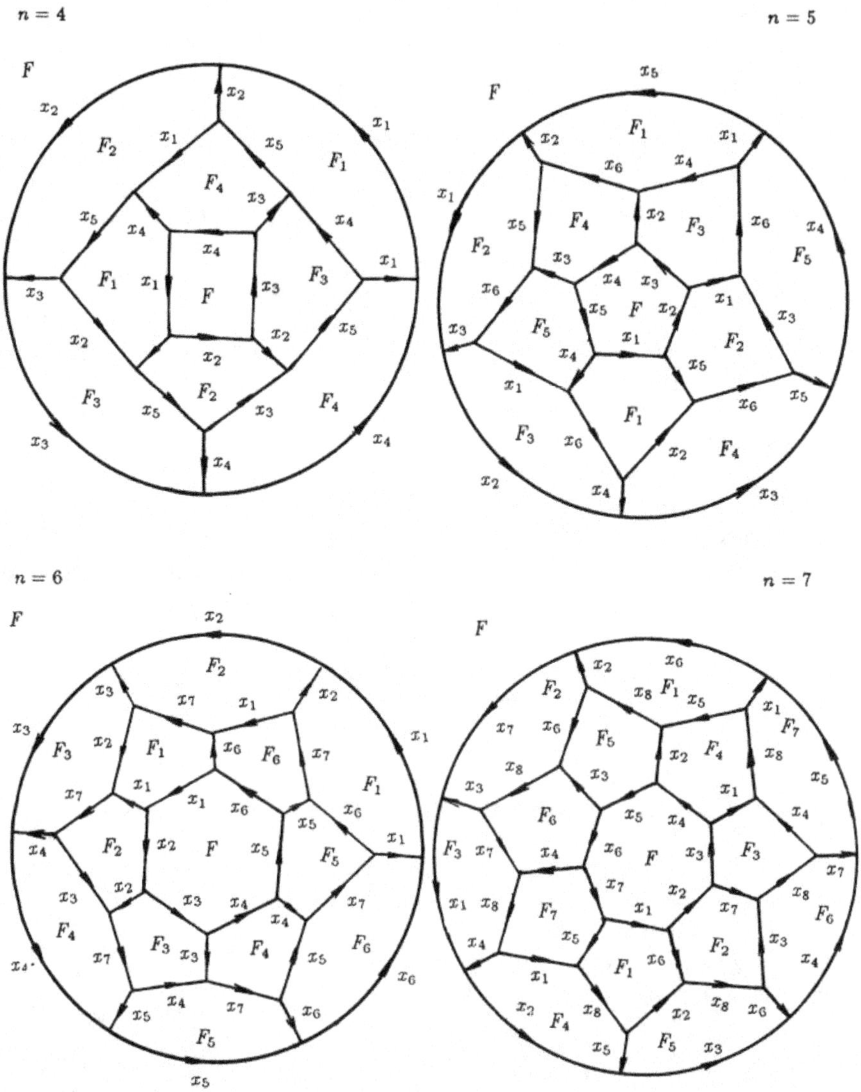

Figure 4

Suppose that

$$G_{2m} = \langle x_1, x_2, \cdots, x_{2m+1}; x_1 x_2 \cdots x_{2m} = 1, x_i x_{2m+1} x_{i+2} = x_{i+1}^2, \ i \bmod 2m\rangle$$

$$G_{2m+1} = \langle x_1, x_2, \cdots, x_{2m+2}; x_1 x_2 \cdots x_{2m+1} = 1, x_i x_{2m+1} x_{i+m+1} = x_{i+m} x_{i+1}, \ i \bmod 2m+1\rangle.$$

Then, from the above construction of M_n, we have

Theorem 1. *For $n \geq 3, G_n$ is the fundamental group of a closed orientable 3-manifolds M_n.*

§2. Realization of Δ_n^* in H^3

We want to realize Δ_n^* in H^3 for $n \geq 4$. Top and bottom faces are regular n-gons. All pentagons are congruent but not regular except $n = 5$. In this case our construction yields the regular hyperbolic dodecahedron with dihedral angle $2\pi/5$. Here we list the other properties of Δ_n^*.

P1. All edges except $Q_i R_i (1 \leq i \leq n)$ have the same length, say, x (caution: we take hyperbolic cosine of the ordinary length as our length convention.) and those exceptional edges $Q_i R_i$ have the same length, say, y.

P2. In the pentagon $P_1 Q_1 R_1 Q_2 P_2$, the length of segments $P_1 Q_2, P_2 Q_1$ and $P_2 R_1$ are the same, which is denoted by z. And so the segments $P_1 R_1$ and $Q_1 Q_2$ have the same length, say, ω.

P3. The dihedral angle at edge PQ is denoted by $\angle PQ$. Then Δ_n^* satisfies following conditions of dihedral angles.

(i) $\angle Q_i R_i = \frac{2\pi}{n}$ for $1 \leq i \leq n$.

(ii) All dihedral angles $\angle P_i P_{i+1}, \angle S_i S_{i+1}$ are the same and denoted by ε.

(iii) Dihedral angles at edges not in (i) and (ii) are all the same and denoted by μ.

(iv) $2\varepsilon + 3\mu = 2\pi$.

Thus our goal is to construct a hyperbolic polyhedron Δ_n^* satisfying properties in the above. Our construction begins with following lemma which can be easily verified through hyperbolic trigomometry.

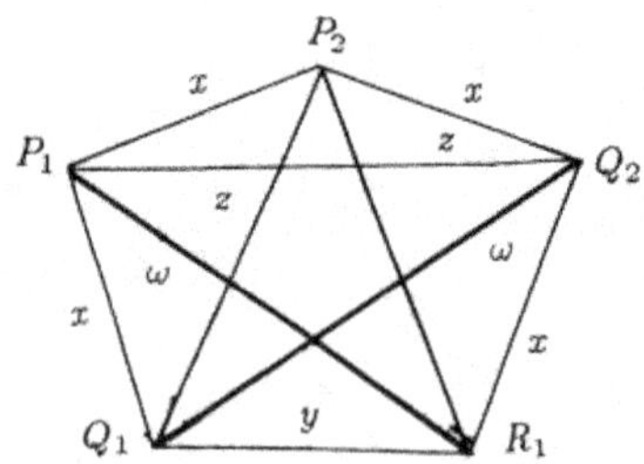

Lemma 2. *In a pentagon satisfying P_2, put $z - 1 = \beta(x - 1)$ for a positive real number β. Then we have*

(i) $w - 1 = (\beta - 1)^2 (x - 1)$,

(ii) $y - 1 = \beta(\beta - 2)^2 (x - 1)$.

Remark. The pentagon satisfying p_2 is indeed regular when $\beta^2 - 3\beta + 1 = 0$. And $\beta = 2(1 + \cos \frac{2\pi}{5})$.

Let $\beta_n = 2(1 + \cos \frac{2\pi}{n})$ and $z_n - 1 = \beta_n (x - 1)$. Consider a family of tetrahedra, e.g., $P_1 P_2 P_n Q_1$, arising from observation that each vertex belongs to exactly 3 edges in Δ_n^*. Those tetrahedra can be classified by following 2 types;

T_1:

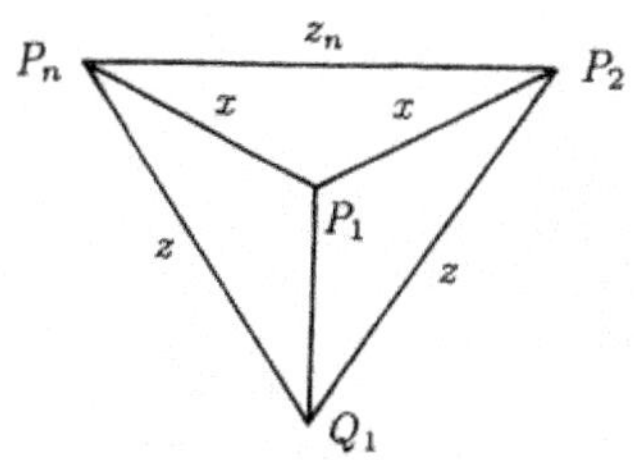

,a tetrahedron containing edge P_iP_{i+1} (or S_iS_{i+1}) of the regular n-gon.

T_2:

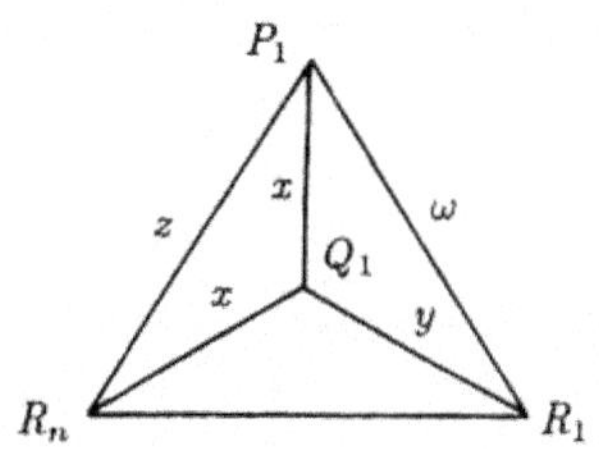

,a tetrahedron containing edge Q_iR_i.

These 2 types of tetrahedra posses edge P_iQ_i in common. Thus, given x and n, we will guarantee existence of Δ_n^* in H^3 with properties P_1 and P_2 by determining z so that dihedral angles at edge P_iQ_i measured from tetrahedra T_1 and T_2 coinside.

Hyperbolic dihedral angle is determined as follows. For proof, see [1].

Lemma 3. *Consider a tetrahedron*

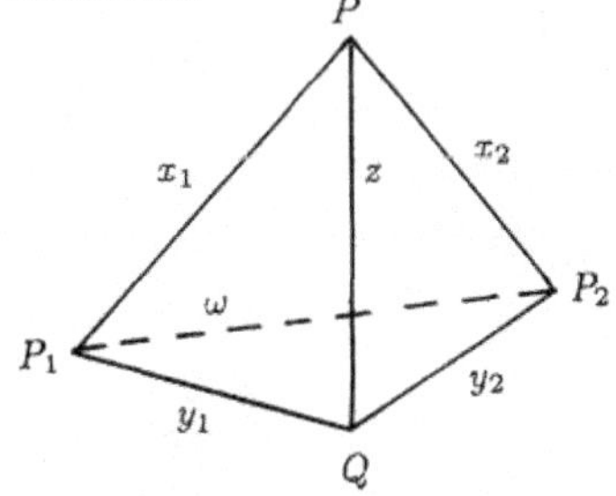

$$\cos \angle PQ = \frac{\begin{vmatrix} 1 & z & x_1 \\ z & 1 & y_1 \\ x_2 & y_2 & \omega \end{vmatrix}}{\sqrt{\begin{vmatrix} 1 & z & y_1 \\ z & 1 & x_1 \\ y_1 & x_1 & 1 \end{vmatrix}}\sqrt{\begin{vmatrix} 1 & z & y_2 \\ z & 1 & x_2 \\ y_2 & x_2 & 1 \end{vmatrix}}}. \tag{1}$$

Utilizing Lemma 3, we have

Lemma 4. *Let $\angle_{T_i}(P_1Q_1)$ be dihedral angle at edge P_1Q_1 in tetrahedron T_i for $i = 1, 2$. Then we have for $n \geq 4$.*

$$\angle_{T_1}(P_1Q_1) = \angle_{T_2}(P_1Q_1)$$

if and only if following equation has a positive real solution

$$\beta^3 - 4\beta^2 + (\beta_n + 2)\beta - \beta_n = 0. \tag{2}$$

Remark. When $n \geq 4, 2 < \beta_n = 2(1 + \cos\frac{2\pi}{n}) < 3$. Thus equation (2) has a unique positive real solution. In particular $\beta = \beta_5$ if $n = 5$.

Next Step is to determine the length x so that dihedral angle at edge Q_iR_i is $\frac{2\pi}{n}$.

Lemma 5. *In the polyhedron determined by the equation (2),*

$$\angle(Q_iR_i) = \frac{2\pi}{n}$$

if and only if

$$x = \frac{\beta^3 - 5\beta^2 + 6\beta - 4}{4 - 2\beta}. \tag{3}$$

Now let us talk about dihedral angles at rest of edges of polyhedron determined by (2) and (3). Apparently, there are 3 types of dihedral angles represented by $\angle P_1P_2, \angle P_1Q_1$ and $\angle R_1Q_2$. Put those angles ε, μ and δ respectively. Then we can compute cosine of those angles by means of (1).

Indeed we have

Lemma 6.

(i) $\cos\mu = \cos\delta = \frac{\beta - 2}{2}$,

(ii) $\cos 2\varepsilon = \frac{(\beta-2)(\beta^2-4\beta+1)}{2}$.

Finally, we can easily check that $2\varepsilon + 3\mu = 2\pi$ by utilizing following formula in Proposition 2.6, HKM [2].

Lemma 7. *Let $x_1 = \cos\alpha, x_2 = \cos\beta, x_3 = \cos\gamma$, where α, β, γ are three angles, $0 \leq \alpha_1\beta_1\gamma \leq \pi$. Then $\alpha + \beta + \gamma = 2\pi$ if and only if*

$$\begin{vmatrix} 1 & x_1 & x_2 \\ x_1 & 1 & x_3 \\ x_2 & x_3 & 1 \end{vmatrix} = 0.$$

Thus, by Maskit's theorem [3], we have

Theorem 2. *For $n \geq 4$, there is a tessellation of H^3 such that Δ_n^* with the properties P_1, P_2 and P_3 is a fundamental domain.*

Acknowledgements. The authors thank Korea Science and Engineering Foundation for their financial support for the works.

References

[1] W. Fenchel, *Elementary Geometry on Hyperbolic Space*, de Gruyter, Berlin (1989).

[2] H. Helling, A.C. Kim and J.L. Mennicke, *A Geometric Study of Fibonacci Groups*, to appear.

[3] B. Maskit, *On Poincaré's Theorem for Fundamental Polygons*, Advances in Math., 7(1971).

[4] P. Scott, *The geometry of 3-manifolds*, Bull. London Math. Soc., 15(1983), 401–487.

Proceedings of the Asian Mathematical
Conference, 1990

An Approach to Rational Diophantine Undecidability*

K.H. KIM, & F.W. ROUSH
Mathematics Research Group
Alabama State University
Montgomery, Alabama 36101, U.S.A.

§1. Introduction

In 1970 Matijasevitch completed the program of Davis, Putnam, and Robinson to prove that Diophantine equations over **Z** are undecidable. He showed [M] that one can obtain exponential functions from Fibonacci number identities and they had showed that this gives factorials and binomial coefficient [D] from which a Turing machine can be constructed.

The corresponding problem over **Q** is open. It seems reasonable to try to reduce this problem to the former result either by showing **Z** is a Diophantine set or by giving a Diophantine model of **Z**. We indicate a possible approach to the latter.

Using the theory of quadratic forms, for sets of primes like primes not of the form $(24n+1)$, we can specify that x is an integer at all primes occurring to odd powers in y; hence we have a limited form of divisibility defined.

This may allow some divisibility to be defined in abelian varieties by looking at denominators or the extent to which a point approximates an identity element. Abelian varieties [AT] of dimension 2 can be found whose isogeny ring is either the quaternions or a general quadratic form. Divisibility in the latter system was proved undecidable by Lipshitz who also showed divisibility gives a decidable Diophantine problem for imaginary quadratic rings and **Z** (see [L1] and [L2]). Some of these results were also obtained by Beltyakov. Related results on various undecidable Diophantine problems in rings are given in [D1], [D2], [KR], [P], [Sh], and [S]. Cherlin and Ershov have also obtained results of interest.

§2. An Undecidability Result for Quaternions

Let **Q** denote the integral quaternions $a+bi+cj+dk$ where $a,b,c,d \in \mathbf{Z}$,$i^2 = j^2 = k^2 = -1, ij = -ji = k, jk = -kj = i$, and $ki = -ik = j$.

* Partially supported by NSF-DMS-8820801.

Our plan is to establish a modulo m isomorphism to a matrix algebra and use divisibility of rank 1 matrices

$$\begin{bmatrix} 1 & n \\ 1 & n \end{bmatrix}, \begin{bmatrix} r & s \\ r & s \end{bmatrix}$$

to get $(1, n)$ divides (r, s), that is $s = rn$, to some large modulus. This will define multiplication.

In order to establish a modulo m isomorphism from $\mathcal{Q}$ to a matrix algebra we need to obtain that m will be a product of primes $(4i+1)$ only. Also we need that m will exceed the absolute value of a given polynomial.

Lemma 2.1. *The congruence* $n \equiv i \pmod{(u+iv)}$ *in* $\mathbf{Z}[i]$ *is solvable if and only if* u *and* v *have no common prime factors* $(4p+3)$ *given* $u \equiv v + 1 \pmod 2$. *The solution of* $n \equiv mi \pmod{(u+iv)}$ *is unique in that case.*

Proof. If u and v have common prime factors $p = 4t+3$ then $\mathbf{Z}[i]$ modulo p is a Galois field which additively is $\mathbf{Z}_p + \mathbf{Z}_p$ and we cannot solve. If $(u+iv)$ factors into primes $(p+ir)$, where $p^2 + r^2$ is a $(4t+1)$ prime p_0 then modulo each prime $(p+ir)$ the ring $\mathbf{Z}(i)$ is $\mathbf{Z}_{p_0}$ and we can solve. Moreover, the solution is unique.

Lemma 2.2. *We can define a set of* $(u+iv)$ *by Diophantine equations involving congruences over* $\mathbf{Z}[i]$ *but not variable multiplication, given* $x_1, \cdots, x_n \in \mathbf{Z}$ *for fixed* n, *such that* $(u,v) = 1$ *in at least one solution and such that in all solutions*

$$\sqrt{u^2 + v^2}$$

exceeds $2x_1 \cdots x_n$ *and is odd.*

Proof. We can take $u \equiv 0(\text{mod}(x_i + j))$ where $j = 1, \cdots, n+1$, u even, and v odd. In fact, we can take $v = 1$. The sequence $(x_i + j)$ has overlapping factors of one term with an earlier at most the product of all $(i - j)$ a fixed number n_0 which can be allowed for (taking another factor relatively prime to it). Otherwise the least common multiple will be the product.

For $i = n$, this is at least $\max\{x_i\}^{n+1}$. Unless all x_i are small $\max\{x_i\}^{n+1}(n_0)^{-1}$ is large enough. But we can allow for x_i small by taking also some large fixed modulus $m_0, u \equiv 0$ (mod m_0).

Theorem 2.3. *The system* $\langle \mathcal{Q}, + \rangle$ *with constants 0 and 1 binary relation* x *and* y *on the left, unary operations conjugacy and multiplication by* i, j, k *has undecidable Diophantine problem.*

Proof. Take a Diophantine equation $f(x_i) = 0$ over $\mathbf{Z}$. We may consider it as $f \equiv 0$(mod m) where m is as in Lemma 2.2, a least common multiple of various $(x_i + iy_i)$ and $(x_i - iy_i)$. Note that using conjugation and multiplication by i, j, k we may characterize $\mathbf{Z}$ and $\mathbf{Z}[i]$ as subgroups.

Now modulo m we may introduce congruences by the divisibility relation and we can as in Lemma 2.1 get an integer congruent to i, simultaneously requiring that the modulus be a produce of $(4i+1)$ primes. We can also identify multiples of i using i multiplication hence multiples of this integer modulo m. Now we can explicitly represent quaternions as matrices

$$\begin{bmatrix} u & v \\ u^* & v^* \end{bmatrix},$$

where "$*$" symbolically is complex conjugacy; that is in fact for a quarternion $a+bi+cj+dk$, integer u, v, u^*, and v^* are defined by linear equations. This makes explicit the modulo m isomorphism of $\mathcal{Q}$ to a general linear group.

Then we can define products modulo m, eventually monomials by

$$\begin{bmatrix} 1 & 1 \\ m & m \end{bmatrix} X = \begin{bmatrix} n & n \\ p & p \end{bmatrix}.$$

This says X induces a multiplication by n on the space of vectors (x, x) so $nm = p$. Thus, we can define the value of $f(x_i)$ modulo m and set it equal to 0.

§3. A Rational Specification of Integrality at Some Primes

It is possible by quadratic forms to specify that a variable has no odd powers of primes of certain types in its denominators. For instance

$$x^2 + y^2 - u(z^2 + w^2) = 0$$

implies u has no odd powers of primes $(4i - 1)$ in it and therefore $(u - 1)$ has none in its denominator.

The strong Hasse principle for global fields states that a quadratic form is isotropic if and only if it is isotropic locally at every prime. At odd primes p a necessary and sufficient condition for local anisotropy of a form

$$a_1x_1^2 + a_2x_2^2 + a_3x_3^2 + a_4x_4^2,$$

where a_i are nonzero locally squarefree in p, is (1) p occurs in exactly two terms, say a_3 and a_4; (2) either a_1a_2 or

$$\frac{a_3a_4}{p^2}$$

is a quadratic residue modulo p. At the reals a necessary and sufficient condition is that not all terms have the same sign.

Over the rationals, a 2-adic solution will exist if one exists modulo 8 having an odd variable.

Proposition 3.1. *For any fixed prime p we can write a Diophantine equation over* **Q** *which specifies that x is an integer locally at the prime p.*

Proof. Let p be odd, c a quadratic nonresidue modulo p. It suffices to characterize whether y has even or odd degree in p because we can let y be $(px^2 + x^{-2})$ which has even or odd degree according as x is integral or nonintegral.

We take the system of quadratic forms

$$au^2 - bv^2 + c^n y a^{n_1} b^{n_2} z^2 - pw^2,$$

where n_i is 0, 1; a, b, c are relatively prime; ab and c are nonresidues modulo p. If y has an odd power of p then such a form cannot be isotropic locally at p for both values of n for any n_1 and n_2 since ab is a nonresidue and for some n, $c^n yp$ is a nonresidue times an even power of p.

Suppose y has an even power of the prime p. Then we have local isotropy at p and at infinity for a and b positive. The only primes which could occur in exactly two terms modulo squares are those occurring in y and either a or b. But we may choose n_1 and n_2 to eliminate this possibility. (Any and-or combination of Diophantine equations can be made a single equation.) We may get local solvability at 2 by letting a and b be of the form $(8i + 1)$.

At the prime 2 a number is a sum of 3 squares if and only if it is not of the form $4^n(8k+7)$. So it has an odd power of 2 if and only if all these are sums of 3 squares: $x, 3x, 5x$, and $7x$.

Proposition 3.2. *For any given integer c, we can write a Diophantine equation specifying that a and b have no prime squarefree factor in common at which c is a nonresidue.*

Proof. We consider isotropy of a system of quadratic forms

$$d_1 x^2 - c^e d_2 y^2 + d_3 a w^2 - d_4 b c^{e_1} z^d.$$

Here e_i range over 0 and 1. If a and b have a common odd power odd prime p at which c is a nonresidue, and p does not divide d_i or c then for some choice of e_i we have local anisotropy at the prime p.

Given d_i and c the finite set of primes dividing then and 2 can be treated by Proposition 3.1.

Consider the converse, where we will further restrict the d_i but depending only on c. At the odd prime p assumed not to divide d_i or c, since it occurs in at most one term we have local isotropy. At the infinite prime we have local isotropy for d_i positive. By choosing d_1 and cd_2 congruent to 8 times the power of 2 in c we have local isotropy at 2. We assume d_i have no common factors with themselves or with c. Then no other prime occurs in two term if a and b are relatively prime to d_i and c up to squares.

We can make this happen by taking cases where we multiply a and b by product of any subset of the primes of d_i and c. Thus we have characterized existence of common prime factors relatively prime to d_i and c.

Theorem 3.3. *We can write a Diophantine equation $E(a, b, c)$ over* **Q** *which has a solution if and only if at every prime p common to the square free parts of a and b, the squarefree part of c is either a multiple of p or a residue modulo p.*

Proof. We start with the same system as above

$$d_1 u^2 - d_2 c^e v^2 + d_3 a c^{e_1} r^2 - d_4 b s^2$$

For a finite set of primes, we can use Proposition 3.1 but here we will allow the coefficients d_i to vary. We will however assume that d_i have no prime factor common to a and b. Then if the system is solvable for all choices of e and e_1 no odd prime factor common to a and b

can have c as a nonresidue (if it does not divide c). In particular, we can take care of $p = 2$ using Proposition 3.1 and henceforth assume a and b are odd.

We need to have a sufficiently wide range of choices of d_i to get a converse. For d_i, we take them of the forms

$$L = \frac{a_0^2 + zb_0^2}{a_0^2 - 2zb_0^2} \frac{2g_0^2 - zh_0^2}{g_0^2 - 2zh_0^2}$$

where z ranges over any choices from a, b, ca, and cb. Such a quotient will be relatively prime to any odd primes common to the squarefree parts of a and b. The reason is, the numerator and denominator will have the same order in this prime, because the two terms have order of different parity.

Then it only remains to show that for some such choices we get local solvability of the form. There will exist choices where d_i are relatively prime to each other and to a and b. To assure this let cab, etc. divide b_0 but not a_0. So we can only have a problem at primes of c or common to a and b. We allow d_1 to be multiplied by 2 times some modulo 8 class representatives to allow for the primes 2.

If p divides a and b but not c then c is a residue at such primes. Just let d_i be residue (we can use the Chinese remainder theorem to combine values of d_i modulo different primes including the infinite prime) . For primes p dividing c we can adjust the powers of p occurring in d_1 and d_4 and their residue classes (others be relatively prime) so that $d_1 d_4 b$ is a quadratic residue. This is done by for instance using, if p divides ac to an even power

$$\frac{a_0^2 + b_0^2 ac}{a_0^2 - 2b_0^2 ac} \quad \frac{2g_0^2 - h_0^2 ac}{g_0^2 - 2h_0^2 ac}$$

Let the two terms $a_0^2 ac, b_0^2$ have equal order in p and then we get for squared ratios $a_0^2, b_0^2, \frac{p+1}{2}$ finite residue classes in each factor. The two factors have the forms L and $L+1$ and are independent so by a counting argument some nonresidues must occur.

Theorem 3.4. *We can write a Diophantine equation over* $\mathbf{Q}$ *which specifies that x is an integer at all odd power primes of y at which c is a nonresidue, for fixed c, and another giving congruences modulo such primes.*

Proof. First we define squarefree greatest common divisor at primes where c is nonresidue (other than c). Proposition 3.2 defines that this greatest common divisor is 1. In general use $(g, xg^{-1}) = 1, (g, yg^{-1}) = 1$, and $(xg^{-1}, yg^{-1}) = 1$ to define $(x, y) = g$. The first two say g divides x and y, the last that it is the greatest common divisor.

Next we define all primes in the squarefree parts of both x and y occur in the denominator of x. This is $(x, y) = (x+1, y)$. Applied to x^{-1} this gives the same condition on the numerator.

Next we can define congruence modulo such primes. Let y_i range over numbers such that $(y, y_i) = y = y_i$ and all these primes lie in the numerator of y_i. Then x congruent to v modulo them is defined by $x - v = y_1 + y_2$ or $y_1 + y_2 + y_3$ (note that at the prime 2 we must allow these two cases, otherwise we don't need them).

Finally, it suffices to produce a full set of integral congruence classes modulo the product of these primes.

The function $x(x^2 - c)^{-1}$ always gives integers modulo these primes p since in $xy(x^2 - cy^2)^{-1}$ the denominator is not divisible by p if x and y relatively prime. By adding these we get more integral congruence classes. The values of $f(x)$ modulo different primes are independent so it suffices that for some i modulo every p the sums of i values of $f(x)$ range over all congruence classes.

The values of $f(x)$ range over at least $(p+1)2^{-1}$ congruence classes modulo p since at most pairs coincide. Suppose some z is not $f(x_i) + f(x_j)$. Note zero is, and the set is symmetric under minus. Then no two $f(x_i)$ differ by z. For every u, we eliminate one of the pair $x, z - x$. Then each pair has exactly one representative $f(x)$. But sums of two elements $f(x)$ will have at least one new element since it is not a subgroup and then the same argument means of 4 do work.

We note that if we could characterize local squares at odd power primes of x then we could characterize the fact that x has denominator a square, namely that x is a sum of terms

$$\frac{\alpha x + 1}{\alpha x - 1} + \frac{\beta(x+1)+1}{\beta(x-1)+1}$$

where α, β are such local squares. We can also using these sums characterize $y = z$ modulo these primes. If we could find representatives of integer congruence classes e.g. by an elliptic curve then we could characterize that y is an integer at these primes of x.

§4. Conclusion

This method at present provides some *heuristic support* for the thesis that Diophantine equations over **Q** are undecidable.

References

[AT] L. Auslander and R. Tolimieri, *A matrix-free treatment of the problem of Riemann matrices*, Bull. of Amer. Math. Soc., 5(1981), 263–312.

[Da] M. Davis, *Hilbert's 10th problem is unsolvable*, Amer. Math. Monthly, 80(1973), 233–267.

[D1] J. Denef, *Hilbert's 10th problem over quadratic rings*, Proc. of Amer. Math. Soc., 48(1975), 214–220.

[D2] J. Denef, *The Diophantine problem for polynomial rings and fields of rational functions*, Trans. of Amer. Math. Soc., 24(1978), 391–399.

[KR] K.H. Kim and F.W. Roush, *Problems equivalent to rational Diophantine decidability*, Jour. of Alg., 124(1989), 493–505.

[L] T.Y. Lam, *Algebraic theory of Quadratic Forms*, Benjamin, Reading , MA., 1977.

[L1] L. Lipshitz, *Undecidable existential problems for addition and division*, Proc. of Amer. Soc., 64(1977), 122–128.

[L2] L. Lipshitz, *The Diophantine problem for addition and divisibility*, Trans. of Amer. Math. Soc., 235(1978), 271–283.

[M] Y. Matijasevitch, *Enumerable sets are Diophantine*, Soviet. Math. Doklady, 11(1970), 354–357.

[P] A.T. Pheidas, *An undecidability result for power series rings of positive characteristic. II*, Proc. of Amer. Math. Soc., 100(1987), 526–530.

[Sh] A. Shlapentokh, *Extension of Hilbert's tenth problem to some algebraic number fields*, Comm. in Pure and Appl. Math., 42(1989), 936–962.

[S] B. Sturmfels, *On the decidability of Diophantine problems in combinatorial geometry*, Bull. of Amer. Math. Soc., 17(1987), 121–124.

Proceedings of the Asian Mathematical
Conference, 1990

The Asymptotic Behavior of Non-linear Dissipative Hyperbolic Equations*

WAN SE KIM

§1. Introduction

Let R be the set of all real numbers, $t_0 > 0$ in R and $\Omega = [t_0, \infty[x]0, \pi[]$. Let $C_B^2(\Omega)$ be the space of all functions $u : \Omega \to R$ which have bounded continuous partial derivatives up to order 2 with respect to both variables on Ω.

In this note, we will investigate the asymptotic stability of global solutions of non-linear dissipative hyperbolic equations of the form

$$\beta u_t + u_{tt} - u_{xx} + g(u) = 0 \quad \text{in } \Omega, \tag{1.1}$$

where $\beta(> 0) \in R, u = u(t,x)$ and $g : \Omega \to R$ is continuous.

A global solution of the problem on Ω for (1.1) will be $u \in C_B^2(\Omega)$ such that u satisfies (1.1) on Ω and satisfies Dirichlet boundary condition

$$u(t,0) = u(t,\pi) = 0, \quad t > t_0.$$

In [7], the author established the existence of a weak solution to the periodic-Dirichlet boundary value problem for the equations of the form

$$\beta u_t + u_{tt} - u_{xx} + g(u) = h(t,x)$$

with a generalized sign condition and superlinear growth in g. Our main result is related to the above problem.

For stability result, we assume $h(t,x) = 0$ and it is also another interesting question whether we could establish the asymptotic stability for the equations with non-zero forcing term $h(t,x)$ with appropriate conditions.

There have been a few authors dealing the asymptotic behavior of this type of hyperbolic equations, for example, Ficken and Fleishman [2], Haraux [3] and Rabinowitz [3]. In particular, our work is related to Rabinowitz's work in [8]. Rabinowitz treated the asymptotic stability of global solutions for the equations of the form (1.1) with $g = \varepsilon F(t,x,u)$. He discussed the asymptotic stability of any global solutions satisfying Dirichlet condition in x but his result is valied for sufficiently small ε. In our main result, our non-linear term is independent of such a parameter and we will prove that any global solution satisfying

*This research was supported by a grant from KOSEF, 1990

Dirichlet condition in x to the above equation (1.1) converges to zero exponentially. We assume only

$$ug(u) \geq 0 \quad \text{for all } u \in R \tag{H$_1$}$$

and

$$G(u) \leq ug(u) \quad \text{for all } u \in R, \tag{H$_2$}$$

where $G(u) = \int_0^u g(s)ds$.

The assumption (H_1) is nothing but a sign condition and the assumption (H_2) contains both sublinear and superlinear growth conditions in g and those two conditions (H_1) and (H_2) play important parts in driving an energy inequality. For this result, we will assume some more regularity of solutions than we do for our existence result in [7].

Our proof is based on the so called "energy method" and the main key in our proof is to set up an energy function as a function of t and an appropriate parameter ε and driving a simple differential inequality by using Dirichlet condition in x.

§2. Main Result

Theorem. *Assume g satisfies (H_1) and (H_2) also assume $g(0) = 0$. If $v \in c_B^2(\Omega)$ is any global solution to (1.1), then $v(t,x) \to 0$ uniformly as $t \to +\infty$ and in this case $v(t,x)$ decays exponentially.*

Proof. Since $g(0) = 0, u \equiv 0$ is a solution of (1.1).

Let $v(t,x)$ be any global solution and consider an energy function defined by

$$w(t) = \frac{1}{2}\int_0^\pi [v_t^2 + v_x^2 + \varepsilon\beta v^2 + 2\varepsilon v v_t]dx + \int_0^\pi G(v)dx,$$

where $\varepsilon > 0$.

Differentiate $w(t)$ with respect to t, then, we have,

$$\frac{d}{dt}w(t) = \int_0^\pi [v_t v_{tt} + v_x v_{tx} + \varepsilon\beta v v_t + \varepsilon v_t^2 + \varepsilon v v_{tt} + g(v)v_t]dx.$$

Since $v_t(t,0) = v_t(t,\pi) = 0$, we have,

$$\int_0^\pi v_x v_{tx} dx = -\int_0^\pi v_t v_{xx} dx.$$

Therefore

$$\frac{d}{dt}w(t) = \int_0^\pi v_t[v_{tt} - v_{xx} + g(v)]dx + \varepsilon\int_0^\pi v_t^2 dx + \int_0^\pi \varepsilon v[\beta v_t + v_{tt}]dx.$$

Since $v(t,x)$ satisfies equation (1.1),

$$\frac{d}{dt}w(t) = -(\beta - \varepsilon)\int_0^\pi v_t^2 dx + \varepsilon\int_0^\pi v v_{xx} dx - \varepsilon\int_0^\pi v g(v) dx.$$

Using the Dirichlet boundary condition, we get

$$\varepsilon\int_0^\pi v v_{xx} dx = -\varepsilon\int_0^\pi v_x^2 dx.$$

Hence

$$\begin{aligned}\frac{d}{dt}w(t) &= -(\beta-\varepsilon)\int_0^\pi v_t^2dx - \varepsilon\int_0^\pi v_x^2dx - \varepsilon\int_0^\pi vg(v)dx \\ &= -\left[(\beta-\varepsilon)\int_0^\pi v_t^2dx + \varepsilon\int_0^\pi v_x^2dx + \varepsilon\int_0^\pi vg(v)dx\right].\end{aligned}$$

Since $ug(u) \geq 0$ for all u,

$$\frac{d}{dt}w(t) \leq 0 \quad \text{for all} \quad 0 \leq \varepsilon \leq \beta.$$

Furthermore

$$\begin{aligned}-\frac{d}{dt}w(t) &= (\beta-\varepsilon)\int_0^\pi v_t^2dx + \varepsilon\int_0^\pi v_x^2dx + \varepsilon\int_0^\pi vg(v)dx \\ &\geq \min\{2(\beta-\varepsilon),\varepsilon\}\left[\int_0^\pi \frac{v_t^2}{2}dx + \int_0^\pi \frac{v_x^2}{2} + \int_0^\pi vg(v)dx\right] \\ &\geq \min\{2(\beta-\varepsilon),\varepsilon\}\left[\int_0^\pi \frac{v_t^2}{2}dx + \int_0^\pi \frac{v_x^2}{2}dx + \int_0^\pi G(v)dx\right]\end{aligned}$$

since $G(u) \leq ug(u)$ for all $u \in R$.

Again, by the boundary condition, we have

$$v(t,x) = \int_0^x v_x(t,x)ds.$$

Hence

$$|v(t,x)| \leq \int_0^x |v_x(t,s)|ds \leq \int_0^\pi |v_x(t,x)|dx \leq \pi^{1/2}\left[\int_0^\pi |v_x(t,x)|^2dx\right]^{1/2}.$$

Thus, we have

$$|v(t,x)|^2 \leq \pi\int_0^\pi |v_x(t,x)|^2dx.$$

Therefore, we obtain

$$\int_0^\pi |v(t,x)|^2dx \leq \pi^2\int_0^\pi |v_x(t,x)|^2dx,$$

or

$$\int_0^\pi v^2dx \leq \pi^2\int_0^\pi v_x^2dx.$$

Now

$$\begin{aligned}w(t) &= \frac{1}{2}\int_0^\pi [v_t^2 + v_x^2 + \varepsilon\beta v^2 + 2\varepsilon vv_t]dx + \int_0^\pi G(v)dv \\ &\leq \frac{1}{2}\int_0^\pi [v_t^2 + v_x^2 + \varepsilon\beta\pi^2v_x^2 + \varepsilon v^2 + \varepsilon v_t^2]dx + \int_0^\pi G(v)dx \\ &\leq \frac{1}{2}\int_0^\pi [v_t^2 + v_x^2 + \varepsilon\beta\pi^2v_x^2 + \varepsilon\pi^2v_x^2 + \varepsilon v_t^2]dx + \int_0^\pi G(v)dx \\ &= (1+\varepsilon)\int_0^\pi \frac{v_t^2}{2}dx + (1+\varepsilon\pi^2+\varepsilon\beta\pi^2)\int_0^\pi \frac{v_x^2}{2}dx + \int_0^\pi G(v)dx \\ &= (1+\varepsilon\pi^2+\varepsilon\beta\pi^2)\left[\int_0^\pi \frac{v_t^2}{2}dx + \int_0^\pi \frac{v_x^2}{2}dx + \int_0^\pi G(v)dx\right].\end{aligned}$$

Hence

$$\int_0^\pi \frac{v_t^2}{2}dx + \int_0^\pi \frac{v_x^2}{2}dx + \int_0^\pi G(v)dx \geq \frac{w(t)}{1+\varepsilon\pi^2+\varepsilon\beta\pi^2}.$$

Therefore

$$-\frac{d}{dt}w(t) \geq \frac{\min\{2(\beta-\varepsilon),\varepsilon\}}{1+\varepsilon\pi^2+\varepsilon\beta\pi^2}\cdot w(t).$$

Take $0 < \varepsilon < \beta$ such that $\varepsilon_0 = \frac{\min\{2(\beta-\varepsilon),\varepsilon\}}{1+\varepsilon\pi^2+\varepsilon\beta\pi^2} > 0$, then $-\frac{d}{dt}w(t) \geq \varepsilon_0 w(t)$ or

$$\frac{d}{dt}w(t) \leq -\varepsilon_0 w(t). \tag{1.3}$$

Multiply $e^{\varepsilon_0 t}$ on the both side of (1.3), to get

$$\frac{d}{dt}w(t)e^{\varepsilon_0 t} + \varepsilon_0 w(t)e^{\varepsilon_0 t} \leq 0.$$

Hence $\frac{d}{dt}[w(t)e^{\varepsilon_0 t}] \leq 0$ and thus $w(t)e^{\varepsilon_0 t}$ is monotone decreasing with respect to t.

On the other hand, since $G(u) = \displaystyle\int_0^u g(s)ds \geq 0$, we have

$$\begin{aligned} w(t) = {} & \frac{1}{2}\int_0^\pi [v_t^2 + v_x^2 + \varepsilon\beta v^2 + 2\varepsilon v v_t]dx + \int_0^\pi G(v)dx \\ = {} & \frac{1}{2}\int_0^\pi [(v_t+\varepsilon v)^2 + (\varepsilon\beta-\varepsilon^2)v^2 + v_x^2]dx + \int_0^\pi G(v)dx > 0. \end{aligned}$$

Therefore, $0 \leq w(t)e^{\varepsilon_0 t} \leq w(t_0)e^{\varepsilon_0 t_0}$, or $w(t) \leq w(t_0)e^{\varepsilon_0(t_0-t)}$.

Now, since

$$|v(t,x)| \leq \pi^{1/2}\left[\int_0^\pi |v_x(t,x)|^2dx\right]^{1/2},$$

$$\sup_{0\leq x\leq\pi}|v(t,x)| \leq \pi^{1/2}\left[\int_0^\pi |v_x(t,x)|^2dx\right]^{1/2}.$$

Since $w(t) \geq \frac{1}{2}\int_0^\pi v_x^2 dx$,

$$\sup_{0\leq x\leq\pi}|v(t,x)| \leq \pi^{1/2}[2w(t)]^{1/2} \leq \pi^{1/2}[2w(t_0)e^{\varepsilon_0(t_0-t)}]^{1/2}.$$

Hence, as $t \to +\infty, \sup_{0\leq x\leq\pi}|v(t,x)| \to 0$. This completes our proof.

Proceedings of the Asian Mathematical
Conference, 1990

The Schur-Weyl Duality and Its Generalizations in Group Representations

W.H. KLINK
Department of Physics and Astronomy
The University of Iowa
Iowa City, IA 52242

TUONG TON-THAT
Department of Mathematics
The University of Iowa
Iowa City, IA 52242

§1. Introduction

In 1901 I. Schur proved a theorem [Sc] that was later expanded by Weyl in a form [We] that is now called the Schur-Weyl Duality Theorem. This theorem, which is briefly reviewed in the next section, relates the representation theory of two groups, the general linear group and the symmetric or permutation group. As we shall show such a phenomenon in which representations of pairs of groups are related turns out to be quite general. It includes the complementary pairs of groups introduced in the physics literature by Moshinsky and Quesne [MQ] and the reductive dual pairs introduced in the mathematics literature by many mathematicians [Ge, GU, KV, Sa, $\cdots$], and especially Howe [Ho].

§2. Definitions and Examples of Dual Representations

Let $P^{\overbrace{(1,\cdots,1)}^{n}}(\mathbb{C}^{n\times N})$ denote the space of all polynomial functions F on $\mathbb{C}^{n\times N}$ which satisfy the covariant condition

$$F(dZ) = \det(d)F(Z), \quad \forall d = \begin{bmatrix} d_{11} & & 0 \\ & \ddots & \\ 0 & & d_{nn} \end{bmatrix}, \quad Z \in \mathbb{C}^{n\times N}.$$

Let S_n denote the symmetric group of degree n and $GL(N,\mathbb{C})$ denote the complex general linear group of order n. A joint action of $S_n \times GL(N,\mathbb{C})$ on $P^{(1,\cdots,1)}$ is defined by

$$[L(\sigma) \otimes R(g)]F(Z) = F(\sigma^{-1}Zg)$$

$$\forall F \in P^{(1,\cdots,1)}, \quad \forall \sigma \in S_n, \forall g \in GL(N,\mathbb{C}), \text{ and } \forall Z \in \mathbb{C}^{n\times N}$$

Then the Schur-Weyl Duality Theorem can be stated as follows:

The $S_n \times GL(N,\mathbb{C})$-module $P^{(1,\cdots,1)}$ is decomposed into irreducible submodules as

$$P^{(1,\cdots,1)} = \sum_{(m)} \oplus I^{(m)}$$

where (m) ranges over all r-tuples $(m_1, \cdots, m_r)$ of integers such that $m_1 \geq m_2 \geq \cdots \geq m_r \geq 0, r = \min(n, N)$, and where $I^{(m)}$ is simultaneously the isotypic component of the irreducible representation of $GL(N,\mathbb{C})$ with signature $(m) = \underbrace{(m_1, \cdots, m_r, 0, \cdots, 0)}_{N}$ and the isotypic component of the irreducible representation of S_n labeled by the Young diagram

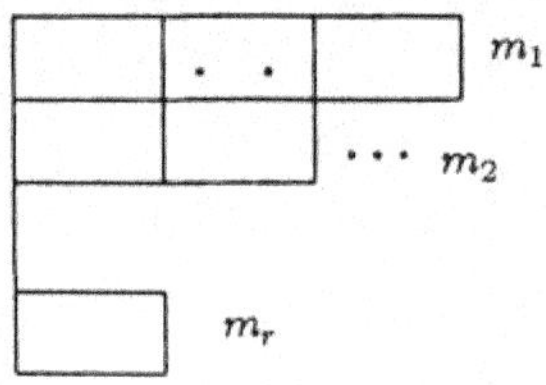

Definition 2.1. Let G and G' be two groups. Let R (resp. L) be a representation of G (resp. G') on a Hilbert space $\mathcal{H}$ s.t. the two actions commute. Assume that the representations R of G (resp. L of G') on $\mathcal{H}$ is completely reducible, i.e., is the direct sum of irreducible representations, then we say that R and L are dual if the spectral decomposition of R determines that of L completely and vice-versa.

Examples 2.2.

1) Schur–Weyl Duality Theorem.

2) Let G be any compact group, let $\mathcal{H} = L^2(G)$, and let L and R denote the left and right regular representations of G on $L^2(G)$, respectively; then the Peter-Weyl Theorem says that

$$L^2(G) = \sum_{\lambda \in \widehat{G}} (L^2(G))_\lambda,$$

where λ represents an equivalence class of an irreducible representation of G, and $L \otimes R|_{(L^2(G))_\lambda}$ is irreducible. Here $G = G'$ and $(L_2(G))_\lambda$ has dimension d_λ^2 where d_λ is the degree of λ.

3) Let $Z \in \mathbb{C}^{n \times N}$ and $d\mu(Z) = \frac{1}{\pi^{Nn}} \exp(-tr(ZZ^t))dZ$ and set

$$\mathcal{H} = \mathcal{F}(\mathbb{C}^{n\times N}) = \{f : \mathbb{C}^{n\times N} \to \mathbb{C} : f \text{ entire and } \int_{\mathbb{C}^{n\times N}} |f(z)|^2 d\mu(Z) < \infty\}.$$

Let $G = \underbrace{GL(N,\mathbb{C}) \times \cdots \times GL(N,\mathbb{C})}_{m \text{ copies}}$. let $p_1, \cdots, p_m$ be integers such that $p_1 + p_2 + \cdots + p_m = n$, and let

$$G' = \underbrace{GL(p_1,\mathbb{C}) \times \cdots \times GL(p_m,\mathbb{C})}_{m}$$

Partition $Z \in \mathbb{C}^{n\times N}$ in block form as

$$Z = \begin{bmatrix} Z_1 \\ \vdots \\ Z_m \end{bmatrix} \quad \text{with} \quad Z_i \in \mathbb{C}^{p_i\times N}, \quad 1 \le i \le m;$$

then G acts on $\mathbb{C}^{n\times N}$ via

$$Z\cdot(g_1,\cdots,g_m) = \begin{bmatrix} Z_1 g_1 \\ \vdots \\ Z_m g_m \end{bmatrix}, \quad \forall g_i \in GL(N,\mathbb{C})$$

and G' acts on $\mathbb{C}^{n\times N}$ via

$$(k_1,\cdots,k_m)\cdot Z = \begin{bmatrix} k_1^{-1} Z_1 \\ \vdots \\ k_m^{-1} Z_m \end{bmatrix}, \quad k_i \in GL(p_i).$$

These actions induce a representation of $G' \times G$ on $\mathcal{H}$ defined by

$$\begin{aligned} &[(L\otimes R)((k_1,\cdots,k_m),(g_1,\cdots,g_m))]f(Z) \\ = \;& f[(k_1,\cdots,k_m)\cdot Z\cdot(g_1,\cdots,g_m)] \quad \forall f \in H. \end{aligned}$$

Theorem 2.3. *The representation $L\otimes R$ of $G'\times G$ on H is decomposed into irreducible subrepresentations labeled by the double signature $(\mu_G, \mu_{G'})$ where $\mu_{G'}$ is an irreducible representation of G' indexed by an n-tuple of integers of the form*

$$(M_1^{(1)},\cdots,M_{p_1}^{(1)}; M_1^{(2)},\cdots,M_1^{(m)},\cdots,M_{p_m}^{(m)})$$

with $M_1^{(i)} \ge M_2^{(i)} \ge \cdots \ge M_{p_i}^{(i)} \ge 0, 1 \le i \le m$ and μ_G is a tensor product of irreducible representations of G indexed by the same n-tuple of integers. Moreover, each irreducible subspace of this orthogonal direct sum decomposition of $\mathcal{H}$ is an isotypic component of both μ_G and $\mu_{G'}$.

See [KT1] and [KT3].

§3. Applications of Dual Representations

1) The invariant theory of block diagonal subgroups of $GL(n,\mathbb{C})$. Let $G' = GL(p_1,\mathbb{C}) \times \cdots \times GL(p_m,\mathbb{C})$ and $\mathcal{H}$ be the same as in Example 4 of Section 2. Let $n = p_1 + \cdots + p_m$ and consider the algebra $S(C^{n\times n})$ of all polynomial functions on $C^{n\times n}$. Then the adjoint representation of $GL(n,\mathbb{C})$ gives rise to the coadjoint representation of $GL(n,\mathbb{C})$ on $S(C^{n\times n})$. The polynomials of $S(C^{n\times n})$ that are fixed by the restriction of this coadjoint representation to the block diagonal subgroup G' of $GL(n,\mathbb{C})$ form a subalgebra of $S(C^{n\times n})$ called the subalgebra of G'-invariants. Using the work of Procesi [Pr] on the dual G of G' and a Frobenius reciprocity theorem relating the dual action of G and G' on $\mathcal{H}$ we obtain the following.

Theorem 3.1 (See [KT4]). *If a matrix $X \in C^{n\times n}$ is partitioned in block form as*

$$X = \begin{bmatrix} X_{11} & \cdots & X_{1m} \\ \vdots & & \vdots \\ X_{m1} & \cdots & X_{mm} \end{bmatrix},$$

where each X_{ij} is a $p_i \times p_j$ matrix, $1 \le i,j \le m$, then the subalgebra of all G'-invariant polynomials is finitely generated by the constants and the functions of the form

$$\mathrm{Trace}(X_{i_1 i_2} X_{i_2 i_3} \cdots X_{i_q i_1})$$

for all $i_k = 1, \cdots, m; k = 1, \cdots, q$.

2) The resolution of the multiplicity in the decomposition of tensor products of group representations. Example 3 and Theorem 2.3 is typical of the way the " see-saw" phenomenon is used. In this example, if we consider the restriction of the representation R of $G = \underbrace{GL(N,\mathbb{C}) \times \cdots \times GL(N,\mathbb{C})}_{m}$ to the diagonal subgroup $H = \{(g, \cdots, g) : g \in GL(N,\mathbb{C})\}$ then we have the problem of decomposing an m-fold tensor product $M^{(1)} \otimes \cdots \otimes M^{(m)}$ of irreducible representations of $GL(N,\mathbb{C})$, where each $M^{(i)}, 1 \le i \le m$, is the signature of an irreducible representation of $GL(N,\mathbb{C})$. To resolve multiplicities, H-invariant differential operators (or generalized Casimir operators) are introduced. These H-invariant differential operators are related to the polynomials invariants in Application 1 (see [KT1] and [KT2]).

References

[Ge] S. Gelbart, *Examples of dual reductive pairs, Automorphic Forms, Representations and L-functions, Part I,* A Borel and W. Casselmar, eds., Proc. Sympos. Pure Math., Vol. 33, Amer. Math. Soc., Providence, RI, 1979.

[GK] K. Gross and R. Kunze, *Bessel functions and representation theory II.*, J. Funct. Anal., 25(1977), 1–49.

[Ho] R. Howe, *Dual pairs in physics: Harmonic oscillators, photons, Electrons, and singletons,* (M. Flato, P. Sally, and G. Zuckermann, Eds.) Lectures in Applied Mathematics, Vol. 21, A. M. S. Soc., Providence, RI, 1985.

[KT1] W.H. Klink and T. Ton-That, *On resolving the multiplicity of arbitrary tensor products of the $U(N)$ groups,* J. Phys. A: Math. Gen., 21(1988), 3877–3892.

[KT2] W.H. Klink and T. Ton-That, *n-Fold Tensor Products of $GL(N,\mathbb{C})$ and Decomposition of Fock spaces,* Jour. Funct. Anal., 84(1988), 1–18.

[KT3] W.H. Klink and T. Ton-That, *Calculation of Clebsch-Gordan and Racah Coefficients using symbolic Manipulation Programs,* Jour. of Comp. Physics, 84(1989), 453–471.

[KT4] W.H. Klink and T. Ton-That, *Invariant theory of the block diagonal subgroups of $GL(n,\mathbb{C})$ and generalized Casimir operators,* to appear in Jour. of Alg., 1991.

[KV] M. Kashiwara and M. Vergne, *On the Segal-Shale-Weil representations and harmonic polynomials,* Invent. Math., 44(1978), 1–48.

[MQ] M. Moshinksy and C. Quesne, *Linear canonical transformations and their unitary representations*, J. Math. Phys., 12(1971), 1772–1780.

[Pr] C. Procesi, *The invariant theory of $n \times n$ matrices*, Adv. in Math., 11(3)(1976), 306–381.

[Sa] M. Saito, *Représentations unitaires des groups symplectiques*, J. Math. Soc. Japan, 24(1972), 232–251.

[Sc] I. Schur, *Dissertation*, Berlin 1901.

[We] H. Weyl, *The classical groups. Their invariants and representations*, Princeton University Press, Princeton, N.J., 1946.

Proceedings of the Asian Mathematical
Conference, 1990

Some Results on Chromatically Unique Graphs

K.M. KOH & C.P. TEO
Department of Mathematics
National University of Singapore
Singapore

Let $P(G)$ denote the chromatic polynomial of a graph G (see [4]). Two graphs G and H are said to be chromatically equivalent, or χ-equivalent, and we write $G \sim H$, if $P(G) = P(H)$. There are non-isomorphic graphs which are χ-equivalent. A graph G is said to be chromatically unique, or χ-unique, if there does not exist any graph H such that $H \not\cong G$ (H is not isomorphic to G) and $H \sim G$ (see [2]). Assume that G contains a cycle. The girth $g(G)$ of G is the length of a shortest cycle in G. Let $\sigma(G)$ denote the number of cycles of length $g(G)$ in G. It follows from Whitney's broken-cycle theorem [8] that if G and H are graphs containing cycles such that $G \sim H$, then $g(G) = g(H)$ and also $\sigma(G) = \sigma(H)$.

The following result was established recently.

Theorem A [7]. *Let G be a 2-connected graph of order n, size m and even girth $g \geq 4$. Then*

$$\sigma(G) \leq \frac{m}{g}(m - n + 1),$$

and the equality holds iff G satisfies the condition:

(*) Every two edges of G are contained in a common cycle C_g of order g.

This result was then used in [7] to prove the conjecture [5] that all complete bipartite graphs are χ-unique (see [6] for information and a different proof). In this note we shall proceed to show that Theorem A can further be utilized to discover more families of χ-unique graphs.

We shall write $V(G), E(G), v(G), e(G)$ and diam G, respectively, for the vertex set, edge set, order, size and diameter of a graph G. While the distance between two vertices u and v in G is denoted by $d(u,v)$, the degree of u in G is denoted by deg u. For a subset A of $V(G)$, the subgraph of G induced by A is denoted by $\langle A \rangle$.

We first recall from [7] some essential features of the structures of graphs satisfying the condition (*). Let G be a 2-connected graph of order n, size m and girth $g = 2f, f \geq 2$, satisfying (*). Clearly, diam $G = f$. Let $e = ab \in E(G)$, and let

$$T_a = \{v \in V(G) \mid d(a,v) \leq f - 1 \text{ in } G - e\},$$

$$T_b = \{v \in V(G) \mid d(b,v) \leq f - 1 \text{ in } G - e\},$$

$$A = \{v \in T_a \mid d(a,v) = f - 1 \text{ in } G - e\},$$

and

$$B = \{v \in T_b \mid d(b, v) = f - 1 \text{ in } G - e\}.$$

Then we have:
(1) the subgraphs $\langle T_a \rangle$ and $\langle T_b \rangle$ of G are disjoint subtrees of G, and $V(G) = T_a \cup T_b$,
(2) $E(G) = \{e\} \cup E(\langle T_a \rangle) \cup E(\langle T_b \rangle) \cup X_e$, where X_e is the set of edges in G each joining a vertex in A to a vertex in B, and $|X_e| = m - n + 1$.

In addition, we have:

Lemma 1 [1]. *Let G be a graph with $g(G)=2f$ satisfying $(*)$.*
(i) *If $u, v \in V(G)$ such that $d(u, v) = f$, then* deg u =deg v;
(ii) *If $u, v_1, v_2 \in V(G)$ such that* deg $u \geq 3$ *and $d(v_1, u) = d(v_2, u)$* then deg v_1 = deg v_2.

Some know basic results are also needed.

Lemma 2. *Let G and H be two χ-equivalent graphs. Then $v(G) = v(H), e(G) = e(H), g(G) = g(H), \sigma(G) = \sigma(H)$, and for $k = 1, 2$, G is k-connected iff H is k-connected.*

For $f, h \geq 2$, let $\theta(h, f)$ denote the graph of order $(f-1)h + 2$ and size fh (see Figure 1) which contains two specified vertices u and v such that its edge set is decomposed into h pairwise internally-disjoint $u - v$ paths of length f each. The graph $\theta(3, f)$ is a special generalized θ-graph, which is known to be χ-uniqe [3].

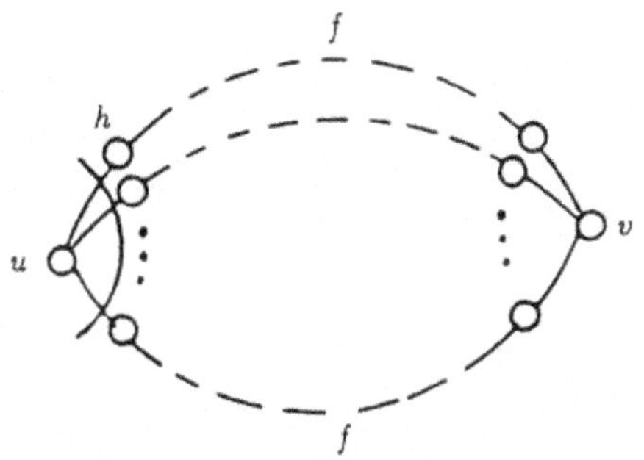

Figure 1. $\theta(h, f)$

Our first result is the following.

Theorem 1. *The graph $\theta(h, f)$ is χ-unique for $h, f \geq 2$.*

Proof. We first note that $\theta(h, 2) \cong K(h, 2)$ and $\theta(3, f)$ is a generalized θ-graph, which are all χ-unique. We may now assume that $f \geq 3$ and $h \geq 4$. Let G be a graph such that $G \sim \theta(h, f)$. Then, by Lemma 2, G is a 2-connected graph of order $n = (f-1)h + 2$ and size $m = fh$, and with $g(G) = 2f$ and $\sigma(G) = \left[\begin{smallmatrix} h \\ 2 \end{smallmatrix}\right]$. Observe that

$$\frac{m}{2f}(m - n + 1) = \frac{fh}{2f}[fh - (f-1)h - 2 + 1] = \begin{bmatrix} h \\ 2 \end{bmatrix} = \sigma(G).$$

Thus, by Theorem A, G satisfies the condition $(*)$.

Since G is not a cycle, there exists a vertex in G of degree at least three. Let $e = ab \in E(G)$ with $\deg a = l + 1 \geq 3$, and let

$$s = \min\{d(a, v) \mid \deg v \geq 3\}.$$

Since $d(u, v) \leq f$ for all $u, v \in V(G)$, we have $s \leq f$, and it follows from Lemma 1 that $s = f$ iff $g \cong \theta(h, f)$. Assume that $s < f$. We shall now refer to the structure of G and the notation introduced before. By Lemma 1, let $u_1, \cdots, u_{l+1}$ be the $l + 1$ vertices such that $d(u_i, a) = s$, $\deg u_i = t + 1$ for all $i = 1, 2, \cdots, l + 1$, and $u_1, \cdots, u_l \in T_a$ but $u_{l+1} \in T_b$. For $i = 1, 2, \cdots, l + 1$, by applying Lemma 1 again, let $v_{i_1}, v_{i_2}, \cdots, v_{i_t}$ be the t vertices such that $d(v_{ij}, u_i) = s$, $\deg v_{ij} = l + 1$ for all $i = 1, \cdots, l + 1, j = 1, 2, \cdots, t$, and $v_{ij} \in T_a$ for all $i = 1, \cdots, l, j = 1, \cdots, t$ and $v_{(l+1)j} \in T_b$ for all $j = 1, \cdots, t$. By applying Lemma 1 repeatedly, we see that $s|f$ and the structure of G can be completely determined as shown in Figure 2.

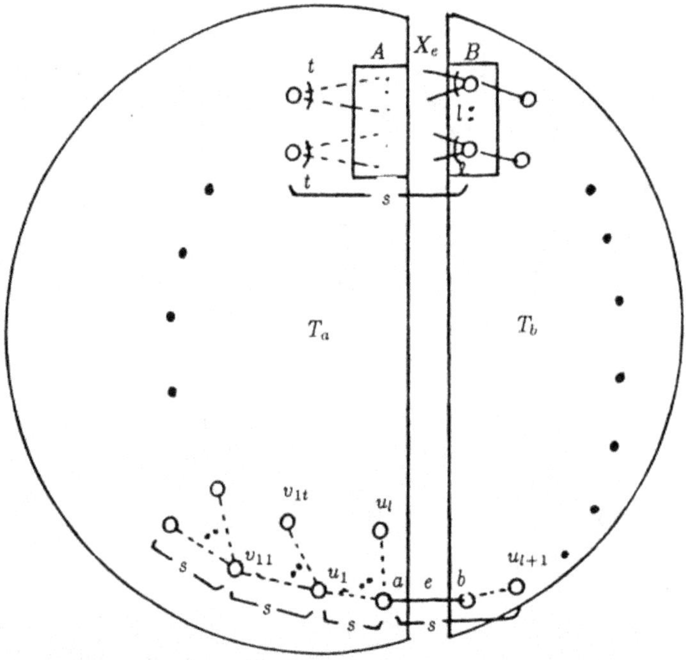

Figure 2.

Observe that for each $v \in B$, as $d(a, v) = f$, we have $\deg v = \deg a = l + 1$ by Lemma 1. Also

$$l|B| = |X_e| = m - n + 1 = h - 1.$$

If $l = h - 1$, then $|B| = 1$, and as $g(G) = 2f$, we have $G \cong \theta(h, f)$. Assume now that $l < h - 1$ and let $f = ks$ for some positive integer k. Suppose $k \geq 2$. We now count the size of G. For this purpose, let $\varepsilon_1 = l + 1$ and for each $i = 2, 3, \cdots, k$, let

$$\varepsilon_i = \sum((\deg v) - 1 \mid v \in V(G) \quad \text{and} \quad d(v,a) = (i-1)s).$$

Note that for $i = 2, 3, \cdots, k-1$, ε_i counts the number of vertices at distance ε_i from a. We have

$$fh = s \sum_{i=1}^{k} \varepsilon_i.$$

Observe that $\varepsilon_2 = (l+1)t$ and $\varepsilon_k = |X_e| + |B| = h - 1 + |B|$. Since $l|B| = h - 1$ and $l < h - 1$, we have $|B| \geq 2$. As $l \geq 2$, it follows that $l + |B| \leq l|B| = h - 1$. Thus

$$\varepsilon_i \leq \frac{\varepsilon_k}{2} = \frac{h - 1 + |B|}{2} < h - 1$$

for all $i = 1, 2, \cdots, k-1$. But then

$$\begin{aligned} fh = & \ s[(\varepsilon_1 + \varepsilon_k) + (\varepsilon_2 + \cdots + \varepsilon_{k-1})] \\ \leq & \ s[(l + 1 + h - 1 + |B|) + (h-1)(k-2)] \\ < & \ s[2h + (k-2)h] \\ = & \ (sk)h = fh, \end{aligned}$$

a contradiction. This forces that $k = 1$, i.e., $s = f$, and therefore $G \cong \theta(h, f)$. The proof is thus complete. □

For two given integers $r \geq 2$ and $g \geq 3$, let

$$n(r, g) = \min\{|V(G)| \mid G \text{ is an } r\text{-regular graph with } g(G) = g\}$$

An r-regular graph G of order n and with $g(G) = g$ is called an (r, g)-cage if $n = n(r, g)$. In th remainder of this note, we shall apply Theorem A again to show that certian cages are χ-unique.

Let G be a (2-connected) (r, g)-cage of order $n = n(r, g)$, size m and girth $g = 2f$. Let $e = ab$ be edge of G. Extend e to the subtree T of G, where $T - e = \langle T_a \rangle \cup \langle T_b \rangle$, as shown before. Observe that all vertices in $V(T) \backslash (A \cup B)$ are of degree r in T. Thus by counting the order of T, we have (see also [9]):

$$n(r, g) \geq \frac{2(r-1)^f - 2}{r - 2} \quad \text{where} \quad g = 2f.$$

Such an (r, g)-cage is called an (r, g)-Moore graph if the above equality holds. Thus, if G is an (r, g)-Moore graph, then the number $|X_e|$, when $g(G)$ is even, is equal to $|E(G) \backslash E(T)| = m - n + 1$. Hence $\sigma(G) = \frac{m}{2f}(m - n + 1)$, and we see that every Moore graph of even girth satisfies the condition (*).

It is known (see [9]) that if $r = p^c + 1$ for some prime p and positive integer c, then

$$n(r, 6) = \frac{2(r-1)^3 - 2}{r - 2},$$

and such an $(r,6)$-Moore graph is unique. We now have our second result.

Theorem 2. *All* $(r, 6)$*-cages, where* $r = p^c + 1$ *for some prime* p *and positive integer* c*, are* χ*-unique.*

Proof. Let H be such an $(r,6)$-cage. Then from the above discussion, we have

$$\sigma(H) = \frac{m}{6}(m - n + 1).$$

Suppose G is a graph such that $G \sim H$. Then G is a (2-connected) graph of order $n = n(r,6)$ and size m, and with $g(G) = 6$ and $\sigma(G) = \frac{m}{6}(m - n + 1)$. By Theorem A and Lemma 1, the structure of G is shown in Figure 2. Referring to this structure, we observe that for each $v \in B, d(a,v) = 3$ and $\deg v = l + 1$. Let k be a positive integer such that $ks = 3$.

Case (i). $k = 1$ and $s = 3$. In this case, we have $|B| = 1$ and so $G \cong \theta(3, l+1)$, which is χ-unique. But then $H \cong \theta(3, l+1)$, which is impossible.

Case (ii). $k = 3$ and $s = 1$. Since k is odd, we have $l = t$. Thus

$$l^3 = ltl = |X_e| = m - n + 1 = \frac{nr}{2} - n + 1 = \frac{n(r-2)+2}{2} = (r-1)^3,$$

which implies that $l = r - 1$. This shows that G is r-regular, i.e., an $(r,6)$-cage. Hence $G \cong H$ by uniqueness. □

References

[1] N. Homobono and C. Peyrat, *Graphs such that every two edges are contained in a shortest cycle*, Discrete Math., 76(1989), 37–44.

[2] K.M. Koh and K.L. Teo, *The search for chromatically unique graphs*, Graphs and Combinatorics, (1991), to appear.

[3] B. Loerinc, *Chromatic uniqueness of the generalized θ-graphs*, Discrete Math., 23(1978), 313–316.

[4] R.C. Read and W.T. Tutte, *Chromatic polynomials. In: Selected Topics in Graph Theory 3*, Academic Press, (1988), 15–42.

[5] P.M. Salzberg, M.A. Lopez and R.E. Giudici, *On the chromatic uniqueness of bipartite graphs*, 58(1986), 285–294.

[6] C.P. Teo and K.M. Koh, *The chromaticity of complete bipartite graphs with at most one edge deleted*, J. Graph Theory, 14(1990), 89–99.

[7] C.P. Teo and K.M. Koh, *The number of shortest cycles and the chromatic uniqueness of a graph* (submitted).

[8] H. Whitney, *A logical expansion in mathematics*, Bull. Amer. Math. Soc., 38(1932), 572–579.

[9] P.K. Wong, *Cages – A survey*, J. Graph Theory, 6(1982), 1–22.

Proceedings of the Asian Mathematical
Conference, 1990

On the Decomposition of 0-Simple Dual Semigroups

C.K. LAI & K.P. SHUM
Department of Mathematics
The Chinese University of Hong Kong
Shatin, N.T., HONG KONG

Let A be a non-empty subset of a semigroup S with zero. Then the right annihilator of A, namely $\gamma(A)$, is defined to be the set $\{x \in S|xa = 0 \text{ for all } a \in A\}$. The left annihilator of $A, l(A)$, is similarly defined. S is called dual if $\gamma(l(R)) = R$ and $l(\gamma(L)) = L$, where R and L are right ideal and left ideal of S respectfully. The study of dual semigroups was firstly initiated by St. Schwarz in 1960 [5]. Other results were obtained by A.H. Clifford and G.B. Preston[1], O. Steinfeld[8], C.K. Lai and K.P. Shum[2] and M. Satyanarayana[4]. They all have observed that the set of idempotents in a dual semigroup is non-empty, and in particular, Schwarz proved that any dual semigroup can be decomposed into 0-disjoint union of principal left (right) ideals generated by idempotents. Inspired by Schwarz[5], we establish here a decomposition theorem for 0-simple dual semigroups. We show that any 0-simple dual semigroup S can be partitioned into a set of null cells and a set of group cells. The reader is refered to Clifford-Preston [1] for all terminology not given in this paper.

§1. Preliminaries

We first quote the following results from the literature which will be used in the next section.

Theorem 1.1 (Schwarz [5]). *In a dual semigroup S, every non-zero right (left) ideal of S contains at least one 0-minimal right (left) ideal.*

Theorem 1.2 (Schwarz [5]). *Any dual semigroup can be decomposed into 0-disjoint union of principal left (right) ideals generated by idempotents, that is,*

$$S = \dot{\cup}_{e \in E} Se (= \dot{\cup}_{e \in E} eS), \text{ where } Se_i \cap Se_j = 0 (e_i S \cap e_j S = 0)$$

for $i \neq j$. Moreover, each of the summands contains a unique idempotent and a unique 0-minimal left (right) ideal of S. This form of decomposition is unique.

Theorem 1.3 (Steinfeld [8]). *The intersection of a 0-minimal left ideal L and a 0-minimal right ideal R of a semigroup S with 0 is either zero or a 0-minimal quasi-ideal of S.*

Theorem 1.4 (Steinfeld [8]). *Let e, f be non-zero idempotents of a semigroup S with 0. Suppose e, f generate the 0-minimal left (right) ideals Se and Sf (eS and fS) respectively, then Se and Sf (eS and fS) are left similar (right similar) if and only if the quasi-ideal eSf of S is non zero.*

Theorem 1.5 (Steinfeld [8]). *Let e, f be non-zero idempotents of a semigroup S with 0. Suppose e, f generate the left similar (right similar) 0-minimal left ideals Se and Sf (0-minimal right ideals eS, fS) of S respectively. Then the 0-minimal quasi-ideals eSe and fSf are isomorphic subgroups with 0.*

§2. Decomposition Theorem

Lemma 2.1. *Let S be an 0-simple dual semigroup. Then $\{eS : e \in E\}$ is the collection of all 0-minimal right ideals of S. Likewise, $\{Se : e \in E\}$ is the collection of all 0-minimal left ideals of S.*

Proof. It is well known that if I is an 0-minimal two-sided ideal containing an 0-minimal right (left) ideal of S, then I can be expressed as the union of all minimal right (left) ideals of S contained in I [1]. Since S is 0-simple, S itself can be regarded as an 0-minimal two-sided ideal of itself. Hence $S = \dot{\cup}_{\alpha\in\lambda} R_\alpha (= \dot{\cup}_{\beta\in\lambda} L_\beta)$, where $R_\alpha(L_\beta)$ is an 0-minimal right (left) ideal of S. By Theorem 1.2, we also have $S = \dot{\cup}_{e\in E} eS = \dot{\cup}_{\alpha\in\lambda} R_\alpha (*)$. Apply Theorem 1.1, there exists an 0-minimal right ideal R_α such that $R_\alpha \subset eS$ for some $\alpha \in \Lambda$. If $R \underset{\alpha\neq}{\subset} eS$, then by $(*)$, there are some R_β such that $R_\beta \cap eS \neq \{0\}$ for $\beta \neq \alpha$. Clearly, $R_\beta \cap eS$ is a non-zero right ideal of S in R_β. Therefore $R_\beta = R_\beta \cap eS \subset eS$ as R_β is 0-minimal. Thus eS contains two disjoint 0-minimal right ideals R_α and R_β. However, by the theorem of Schwarz (see Theorem 1.2), eS contains only one 0-minimal right ideal of the 0-simple dual semigroup S, this arrives a contradiction. Thus $eS = R_\alpha$ which is an 0-minimal right ideal of S. Similarly, $R_\alpha = eS$ for any $\alpha \in \lambda$. This shows that $\{eS : e \in E\}$ is the collection of all 0-minimal right ideals of S. Likewise, $\{Se; e \in E\}$ is the collection of all 0-minimal left ideals of S.

Lemma 2.2. *The intersection of a non-zero right ideals and a non-zero left ideal of an 0-simple dual semigroups S is non-zero.*

Proof. Let $e^2 = e, f^2 = f$ be non-zero idempotents of S. If $eSf = 0$, then $Sf = (SeS)f = S(eSf) = S \cdot 0 = 0$. This implies $f = 0$, which is not true. Hence $eSf \neq 0$. Now let R and L be a non-zero right and left ideal of S respectively. Since S is dual, so by Theorem 1.1, there exists an 0-minimal right ideal eS and an 0-minimal left ideal Sf such that $eS \subset R$ and $Sf \subset L$. Thus $\{0\} \neq eSf \subset eS \cap Sf \subset R \cap L$.

Lemma 2.3. *The following statements hold in any 0-simple dual semigroup.*

(i) *The elements in the set $\{eSd : e, f \in E\}$ are all distinct.*

(ii) *The elements in the set $\{eSf : e, f \in E\}$ are pairwise 0-disjoint.*

(iii) *If $e \neq f \in E$, then the elements in eSf are nilpotent elements of order 2.*

(iv) *$\{eSe : e \in E\}$ is the collection of isomorphic subgroups with 0 of S.*

Proof. (i) Assume $eSf = vSw$ for some non-zero idempotents e, f, v, w of S. Since S is 0-simple, $Sf = (SeS)f = S(eSf) = S(vSw) = (SvS)w = Sw$. Hence $Sf \cap Sw = Sf \neq \{0\}$. However, by Theorem 1.2, the union of all $Se, e \in E$ is 0-disjoint, in particular, $Sf \cap Sw = \{0\}$

for $f \neq w$. As $Sf = Sw$ so $f = w$. Similarly $e = v$. Hence, the elements in $\{eSf : e, f \in E\}$ are distinct.

(ii) By Lemma 2.1 eS is an 0-minimal right ideal of S and Sf is an 0-minimal left ideal of S respectively. In virtue of Theorem 1.3, $eSf = eS \cap Sf$ is an 0-minimal quasi-ideal of S. Consequently, all eSf must be pairwise 0-disjoint.

(iii) Since S is dual, so $ef = fe = 0$ for all $e \neq f$ in E. This implies that $(eSf)^2 = eS(fe)Sf = 0$. It is now clear that all elements of eSf are nilpotent with order 2.

(iv) By Lemma 2.2, $eSf \neq 0$ for any $e, f \in E$. So, by Theorem 1.4, Se and Sf are left similar ideals. Apply Theorem 1.5, we know immediately that eSe and fSf are isomorphic subgroups with of S.

Definition 2.4. We call $Sf(e \neq f \in E)$ be the null cell and eSe be the group cell of S.

According to Lemma 2.2 (iii) and (iv), all non-zero group cells are isomorphic and all null cells are also trivally isomorphic.

Summarizes all the above lemmas we establish the following theorem.

Theorem 2.5 (Decomposition theorem for 0-simple dual semigroups). *Any 0-simple dual semigroup S can be partitioned into two 0-disjoint subsets, namely, $\mathcal{A}$, the subset of all null cells of S and $\mathcal{B}$, the subset of all group cells of S such that $S = \mathcal{A} \dot{\cup} \mathcal{B}$. Moreover, $\overline{\overline{\mathcal{A}}} = \overline{\overline{E}} \times (\overline{\overline{E-1}}), \overline{\overline{\mathcal{B}}} = \overline{\overline{E}}$.*

Proof. According to Theorem 1.2, $S = \dot{\cup}_{e \in E} eS = \dot{\cup}_{f \in E} Sf$. Since all 0-simple dual semigroups are globally idempotent, so

$$\begin{aligned} S = S^2 = &\ \dot{\cup}_{e,f \in E}(eS)(Sf) \\ = &\ \dot{\cup}_{e,f \in E} eSf \\ = &\ (\dot{\cup}_{e \neq f} eSf) \dot{\cup} (\dot{\cup}_{e \in E} eSe). \end{aligned}$$

As $\{eSf : e \neq f\}$ is the subset of 0-disjoint null cells and $\{eSe\}$ is the subset of group cells of S, so $S = \mathcal{A} \dot{\cup} \mathcal{B}$. Because E excludes the zero of S and all group cells are mutually isomorphic, therefore the cordinality of $\mathcal{B}$ and E are equal, that is $\overline{\overline{\mathcal{B}}} = \overline{\overline{E}}$. Similarly, it can be shown that $\overline{\overline{\mathcal{A}}} = \overline{\overline{E}} \times \overline{\overline{(E-1)}}$.

§3. Application

The decomposition theorem can be used to clarify the structure of certain classes of semigroups.

Definition 3.1. A semigroup S is said to be reduced if S contains no non-zero nilpotent elements.

Theorem 3.2 *An 0-simple dual semigroup S is reduced if and only if S is a group adjointed with a zero.*

Proof. The sufficiency is trivial. For the necessity, we only need to consider the case $\overline{\overline{E}} \geq 2$. By Theorem 2.5, we have $\overline{\overline{\mathcal{A}}} = \overline{\overline{E}} \times \overline{\overline{(E-1)}}$ with $\mathcal{A} \geq 2$. This means that S contains at least one non-zero null cell, thus at least one non-zero nilpotent element. However, this contradicts that S is reduced. Hence $\overline{\overline{E}} = 1$. Consequently $\overline{\overline{A}} = 0$ and $\overline{\overline{B}} = 1$. So S must be a group adjointed with a zero.

Recall that a semigroup is a Brandt semigroup if it is completely 0-simple and inverse (see [1]). It was noticed by Schwarz[6] that a semigroup with 0 is a Brandt semigroup if and only if S is 0-simple and dual. By our decomposition theorem together with the result of Schwarz, we immediately obtain the following theorem.

Theorem 3.3. *If the cardinality of a finite Brandt semigroup S is $p+1$, where p a prime number, then S must be a cycle group adjointed with 0.*

References

[1] A.H. Clifford and G.B. Preston, *The algebraic theory of semigroups I and II*, Amer. Math. Soc., Providence, 1967.

[2] C.K. Lai and K.P. Shum, *Characterizations on dual semigroups*, Czechoslovak Math. J., 39(114)(1989), 578–584.

[3] C.K. Lai and K.P. Shum, *A Note on dual and inverse semigroups*, Czechoslovak Math. J. (to appear).

[4] M. Satyanarayana, *On dual semigroups*, Publications Math. Debrecen, 20(1973), 45–51.

[5] St. Schwarz, *On dual semigroups*, Czechoslovak Math. J., 10(1960), 201–230.

[6] St. Schwarz, *On the structure of dual semigroups*, Czechoslovak Math. J., 19(1969), 72–79.

[7] O. Steinfeld, *A characterization of dual semigroups without nilpotent ideals*, Semigroup Forum, 14(1977), 1–5.

[8] O. Steinfeld, *Quasi-ideals in Rings and semigroups*, Akademiai Kiado Budapest, 1978.

Proceedings of the Asian Mathematical
Conference, 1990

Annihilatingly Uniqueness of q-diwheels

K.S. LAM & C.K. LIM
Department of Mathematics
University of Malaya
59100 Kuala Lumpur, Malaysia

§1. Introduction

All graphs considered in this paper are directed, finite and without multiple arcs (however, directed loops are allowed). Undefined terms and notations can be found in [1].

The adjacency matrix $A(G) = (a_{ij})$ of a digraph G with vertex set $\{v_1, v_2, \cdots, v_n\}$ is a square matrix of order n, whose (i,j) entry, a_{ij}, is equal to the number of arcs starting at vertex v_i and terminating at the vertex v_j.

Let G be a digraph with n vertices and $A(G)$ its adjacency matrix. A monic polynomial $f(x)$ of degree $\leq n$ with $f(A(G)) = 0$ is called an annihilating polynomial of G. G is said to be annihilatingly unique if it possesses a unique annihilating polynomial.

§2. The q-diwheel $\mathbf{W}_{n,q}$ of Order n

The q-diwheel $\mathbf{W}_{n,q}$ of order n, consists of the cycle $\mathbf{C}_{n-q}$ with q additional vertices ($q < \frac{n}{2}$) so that each time an additional vertex is introduced, it is joined to all the $(n-q)$ vertices of $\mathbf{C}_{n-q}$ and the previously added vertices. We note that $\mathbf{W}_{n,1}$ is the diwheel $\mathbf{W}_n$. The digraph is illustrated below.

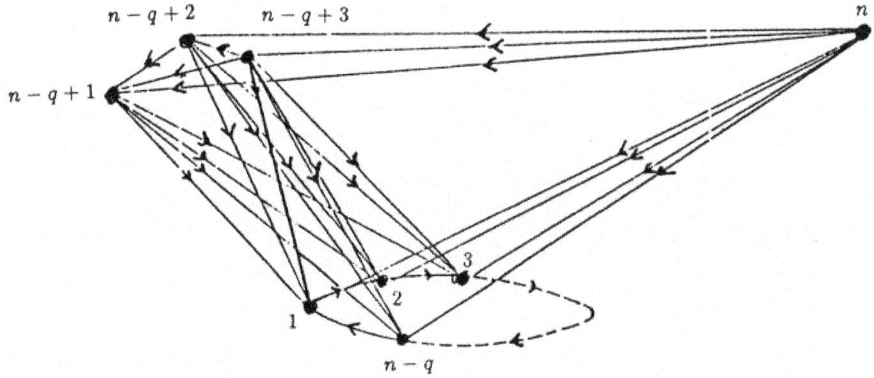

The adjacency matrix of $\mathbf{W}_{n,q}$ is given by

$$A(\mathbf{W}_{n,q}) = \begin{bmatrix} 0 & 1 & 0 & 0 & \cdots & \cdots & 0 & 0 & 0 & 0 & \cdots & \cdots & 0 \\ 0 & 0 & 1 & 0 & \cdots & \cdots & 0 & 0 & 0 & 0 & \cdots & \cdots & 0 \\ \vdots & & \ddots & \ddots & \ddots & & \vdots & \vdots & \vdots & \vdots & & & \vdots \\ \vdots & & & \ddots & \ddots & \ddots & 0 & \vdots & \vdots & \vdots & & & \vdots \\ 0 & \cdots & \cdots & \cdots & \ddots & \ddots & 1 & 0 & \vdots & \vdots & & & \vdots \\ 1 & 0 & \cdots & \cdots & \cdots & \cdots & 0 & 0 & 0 & 0 & & & \vdots \\ 1 & 1 & 1 & \cdots & \cdots & \cdots & 1 & 0 & 0 & 0 & & & \vdots \\ 1 & 1 & 1 & \cdots & \cdots & \cdots & 1 & 1 & 0 & 0 & & & \vdots \\ 1 & 1 & 1 & \cdots & \cdots & \cdots & 1 & 1 & 1 & 0 & \cdots & \cdots & 0 \\ \vdots & \vdots & \vdots & & & & \vdots & \vdots & \vdots & & & & 0 \\ \vdots & \vdots & \vdots & & & & \vdots & \vdots & \vdots & & & & 0 \\ 1 & 1 & 1 & \cdots & \cdots & \cdots & 1 & 1 & 1 & 1 & \cdots & 1 & 0 \end{bmatrix}$$

Expanding the determinant $|xI - A(\mathbf{W}_{n,q})|$, we have

$$\phi(x) = x^q|xI - A(\mathbf{C}_{n-q})| = x^q((x^{n-q} - 1) = x^n - x^q.$$

$A(\mathbf{W}_{n,q})$ is then partitioned as follow:

$$A(\mathbf{W}_{n,q}) = \begin{bmatrix} P & 0 \\ J^{(q)}_{n-q} & S \end{bmatrix},$$

where $P = A(\mathbf{C}_{n-q})$ and $J^{(q)}_{n-q}$ is the $q \times (n-q)$ matrix with all its entries equal to 1 and

$$S = \begin{bmatrix} 0 & 0 & 0 & \cdots & \cdots & \cdots & 0 \\ 1 & 0 & 0 & \cdots & \cdots & \cdots & 0 \\ 1 & 1 & 0 & \cdots & \cdots & \cdots & 0 \\ \vdots & \vdots & \ddots & & & & \vdots \\ \vdots & \vdots & & \ddots & & & \vdots \\ 1 & 1 & 1 & \cdots & \cdots & 1 & 0 \end{bmatrix} \quad \text{is an} \quad q \times q \quad \text{matrix.}$$

Let $S = (s_{ij})$ where $s_{ij} = \begin{cases} 0, & i \le j, \\ 1, & i > j. \end{cases}$ For $1 \le k \le q$, let $S^k = (s^{(k)}_{ij})$, we claim that

$$s^{(k)}_{ij} = \begin{cases} 0, & i < j + k, \\ 1, & i = j + k. \end{cases}$$

We prove this by induction on k. It is true for $k = 1$.

Assume that it is true for k.

For $i < j + k + 1$

$$s^{(k+1)}_{ij} = \sum_{l=1}^{q} s^{(k)}_{il} s_{lj} = \sum_{l=1}^{j} s^{(k)}_{il} s_{lj} + \sum_{l=j+1}^{q} s^{(k)}_{il} s_{lj}.$$

Note that

$$s_{lj} = 0 \quad \text{for} \quad l = 1, 2, \cdots, j$$

and

$$s_{il}^{(k)} = 0 \quad \text{for} \quad l = j+1, \cdots, q,$$

since

$$i < j + k + 1 \leq l + k \quad \text{so} \quad s_{ij}^{(k+1)} = 0 \quad \text{for} \quad i < j + k + 1.$$

For $i = j + k + 1$

$$\begin{aligned} s_{(j+k+1),j}^{(k+1)} = & \sum_{l=1}^{q} s_{(j+k+1),l}^{(k)} s_{lj} \\ = & \sum_{l=1}^{j} s_{(j+k+1),l}^{(k)} s_{lj} + s_{(j+k+1),j+1}^{(k)} s_{j+1,j} + \sum_{l=j+2}^{q} s_{(j+k+1),l}^{(k)} s_{lj}. \end{aligned}$$

Now, $s_{lj} = 0$ for $l = 1, 2, \cdots, j$

$$s_{(j+k+1),l}^{(k)} = 0 \quad \text{for} \quad l = j+2, \cdots, q$$

and $s_{j+k+1,j+1}^{(k)} = 1; s_{j+1,j} = 1$. Hence $s_{(j+k+1),j}^{(k+1)} = 1$.

The claim is thus proved. It follows that

(i) $S^q = 0_{q,q}$,

(ii) $S^k \neq 0$ for $k < q$.

We now claim that

$$A^k(\mathbf{W}_{n,q}) = \begin{pmatrix} P^k & 0 \\ \left(\sum_{i=0}^{k-1} S^i\right) J_{n-q}^{(q)} & S^k \end{pmatrix}, \quad k < q. \tag{2.1}$$

We prove this by induction on k. It is true for $k = 1$. Assume that it is true for k.

$$\begin{aligned} A^{k+1}(\mathbf{W}_{n,q}) = & A^k(\mathbf{W}_{n,q}) A(\mathbf{W}_{n,q}) \\ = & \begin{pmatrix} P^k & 0 \\ \left(\sum_{i=0}^{k-1} S^i\right) J_{n-q}^{(q)} & S^k \end{pmatrix} \begin{pmatrix} P & 0 \\ J_{n-q}^{(q)} & S \end{pmatrix} \\ = & \begin{pmatrix} P^{k+1} & 0 \\ \left(\sum_{i=0}^{k-1} S^i\right) J_{n-q}^{(q)} P + S^k J_{n-q}^{(q)} & S^{k+1} \end{pmatrix}. \end{aligned}$$

We note that $J_{n-q}^{(q)} P = J_{n-q}^{(q)}$

$$\begin{aligned} A^{k+1}(\mathbf{W}_{n,q}) = & \begin{pmatrix} P^{k+1} & 0 \\ \left(\sum_{i=0}^{k-1} S^i\right) J_{n-q}^{(q)} + S^k J_{n-q}^{(q)} & S^{k+1} \end{pmatrix} \\ = & \begin{pmatrix} P^{k+1} & 0 \\ \left(\sum_{i=0}^{k} S^i\right) J_{n-q}^{(q)} & S^{k+1} \end{pmatrix}. \end{aligned}$$

Thus it is true for $k+1$.

Various powers of $A(\mathbf{W}_{n,q})$ are computed accordingly and it is readily seen that

$$A^q(\mathbf{W}_{n,q}) = \begin{pmatrix} P^q & 0 \\ \left(\sum_{i=0}^{q-1} S^i\right) J_{n-q}^{(q)} & 0 \end{pmatrix} \tag{2.2}$$

$$A^{q+r}(\mathbf{W}_{n,q}) = \begin{pmatrix} P^{q+r} & 0 \\ \left(\sum_{i=0}^{q-1} S^i\right) J_{n-q}^{(q)} & 0 \end{pmatrix}, \quad r < n-2q \tag{2.3}$$

$$A^{n-q}(\mathbf{W}_{n,q}) = \begin{pmatrix} I_{n-q} & 0 \\ \left(\sum_{i=0}^{q-1} S^i\right) J_{n-q}^{(q)} & 0 \end{pmatrix} \tag{2.4}$$

$$A^{n-q+i}(\mathbf{W}_{n,q}) = \begin{pmatrix} P^i & 0 \\ \left(\sum_{i=0}^{q-1} S^i\right) J_{n-q}^{(q)} & 0 \end{pmatrix}, \quad i < q \tag{2.5}$$

$$A^n(\mathbf{W}_{n,q}) = A^q(\mathbf{W}_{n,q}). \tag{2.6}$$

The matrices $A(\mathbf{W}_{n,q}), A^2(\mathbf{W}_{n,q}), \cdots, A^{n-1}(\mathbf{W}_{n,q})$ are non-zero and distinct. We note that

$$\psi(x) = x^n - x^q = x^q(x^{n-q} - 1).$$

The polynomial $x^{n-q} - 1$ has $(n-q)$ distinct linear factors and Expression 2.4 indicates that $A^{n-q}(\mathbf{W}_{n,q}) - I \neq 0$. The minimum polynomial of $\mathbf{W}_{n,q}$ is thus of the form

$$x^r(x^{n-q} - 1) \quad \text{ie} \quad x^{n-q+r} - x^r, \quad r = 1, 2, \cdots, q.$$

Expressions 2.1 and 2.5 imply that

$$A^{n-q+r}(\mathbf{W}_{n,q}) - A^r(\mathbf{W}_{n,q}) \neq 0 \quad \text{for} \quad r = 1, 2, \cdots, q-1,$$

whereas Expression 2.6 implies that

$$A^n(\mathbf{W}_{n,q}) - A^q(\mathbf{W}_{n,q}) = 0.$$

The minimum polynomial of $\mathbf{W}_{n,q}$ is thus $x^n - x^q$ which is the same as the characteristic polynomial.

Theorem 2.1. *The annihilating polynomial of the q-diwheel $\mathbf{W}_{n,q}$ is unique and equals $x^n - x^q$.*

Reference

[1] D.M. Cvetkovic, M. Doob and H. Sachs., *Spectra of graphs*, Academic Press, New York, 1980.

Proceedings of the Asian Mathematical
Conference, 1990

Group Theoretical Properties of nth Order Ordinary Differential Equations*

P.G.L. LEACH†
Department of Mathematics and
Applied Mathematics
University of Natal
King George V Avenue
Durban, 4001, South Africa

§1. Mathematical Introduction

An nth order ordinary differential equation (node)

$$N(x, y, y', \cdots, y^{(n)}) = 0$$

has a Lie point symmetry

$$G = \xi(x, y)\frac{\partial}{\partial x} + \eta(x, y)\frac{\partial}{\partial y}$$

if

$$G^{[n]} N_{|_{N=0}} = 0,$$

where $G^{[n]}$ is the nth extension of G. For example

$$G^{[1]} = G + (\eta' - y'\xi')\frac{\partial}{\partial y'},$$

$$G^{[2]} = G^{[1]} + (\eta'' - 2y''\xi' - y'\xi'')\frac{\partial}{\partial y'},$$

where $\eta' = \frac{\partial \eta}{\partial x} + y'\frac{\partial \eta}{\partial y}$ etc.

The notion of symmetry can be extended to derivative dependent symmetries of the form

$$G = \xi(x, y, y', y'', \cdots, y^{(n-1)})\frac{\partial}{\partial x} + \eta(x, y, y', y'', \cdots, y^{(n-1)})\frac{\partial}{\partial y}$$

---

* Partially supported by the Foundation for research Development of South Africa.

† Associate member of the Centre for Nonlinear Studies, University of the Witwatersrand, Johannesburg.

although the only ones to have received significant attention are those for which the dependence is on y' or on derivatives of all orders. A subset of the former consists of contact symmetries for which

$$G = \xi(x,y,y')\frac{\partial}{\partial x} + \eta(x,y,y')\frac{\partial}{\partial y} + \varsigma(x,y,y')\frac{\partial}{\partial y'}.$$

All of these ideas can be have been extended to partial differential equations where the symmetries are used to obtain self-similar solutions.

§2. Historical Introduction

The study of symmetries of differential equations was initiated by Lie in the late nineteenth century. Lie used geometric arguments to prove that the maximum number of infinitesimal point transformations which leave an nth order ordinary differential equation invariant is 8 for $n = 2$ and $n+4$ for $n \geq 3$. In terms of practical developments (mathematically speaking!) not much happened until the seventies. In 1974 Anderson and Davison [1] and in 1976 Wulfman and Wybourne [2] showed that

$$y'' + y = 0$$

also had eight symmetries. In both cases the algebra of the symmetries was $sl(3,R)$.

A rush of papers followed extending this property to a number of linear sodes and it more or less became folklore that all linear sodes had $sl(3,R)$ symmetry although this was not proven until 1986 by F.M. Mahomed [3].

A more exciting idea was that nonlinear sodes could also have $sl(3,R)$ symmetry. A well known example [4] is the member of the Riccati hierarchy

$$y'' + 3yy' + y^3 = 0.$$

Since the algebra is invariant under transformation, this equation can be transformed to

$$Y'' = 0 \quad (\text{here} \quad ' = d/dX)$$

by the linearizing transformation

$$Y = -\frac{1}{2}x^2 + xy^{-1}, \quad X = x - y^{-1}.$$

§3. The Value of Symmetry

The existence of symmetry in a differential equation is not just a source of pleasure for the æsthetically minded mathematician.

1. Equivalence Classes of Differential Equations

If a given differential equation of order n has the symmetry algebra $\aleph$, it is a representative of an equivalence class of differential equations of order n related to it by a point transformation. If one can solve one member of the class, then one can solve all members of the class.

2. Linearization

Of particular practical importance is when a nonlinear node has the symmetry algebra of a linear node. Then there exists a point transformation which transforms the nonlinear equation to the linear equation and its solution is "trivialized". The savings in computer time are self evident.

§4. Developments at Wits 1985–1990

1. Classifiation of odes by their algebras [5, 6, 7]

Second order ordinary differential equations (sodes) can have symmetry algebras of dimension 0, 1, 2, 3, or 8, but not 4, 5, 6 or 7. The equivalence classes of sodes with algebras of dimension 1 and 2 have long been known. We determined all equivalence classes of sodes invariant under all the three-dimensional algebras (thereby completing the task begun by Lie).

Not all representations of three-dimensional algebras can be symmetries of a sode. For example the algebra $sl(2, R)$ is found in one representation for many equations (cf. $y''+y^n = 0$), but one representation

$$G_1 = \frac{\partial}{\partial y}, \qquad G_2 = y\frac{\partial}{\partial y}, \qquad G_3 = y^2\frac{\partial}{\partial y}$$

is not admitted by any sode.

In the case of nodes ($n \geq 3$) the situation is both simpler and more complex. The dimension of the algebra of a linear node depends upon the internal structure of the equation. Taking third order equations as an example

$$y''' + b(x)y' + \frac{1}{2}b'(x)y = 0$$

has 7 point symmetries.

$$y''' + b(x)y' + c(x)y = 0, \quad c \neq \frac{1}{2}b',$$

has 5 point symmetries.

$$y''' + a(x)y'' + a'(x)y' + c(x)y = 0, \quad c \neq 3a''/10 + 9a^2/100$$

has 4 point symmetries. In general a linear node can have $n+1, n+2$ or $n+4$ symmetries with the algebras $nA_1 \oplus_s A_1, nA_1 \oplus_s A_s$ and $nA_1 \oplus_s gl(2, R)$ respectively [8, 9]. The study of linear nodes throws light on the structure of the symmetry algebra. The nA_1 subalgebra corresponds to the n linearly independent solutions of the linear nodes.

In the case of sodes Lie had conjectured that any algebra associated with a sode must be a subalgebra of $sl(3,R)$. With the complete catalogue of all possible algebras that conjecture was verified by demonstration [3]. by contrast nonlinear nodes ($n>2$) can have algebras which are not subalgebras of the maximal algebra, $nA_1 \oplus_s gl(2,R)$, but there is no equation with $n+3$ symmetries with an algebra which is a subalgebra of the maximal algebra, e.g.

$$2y'''y' - y''^2 = 0$$

has the six-dimensional algebra $sl(2,R) \oplus sl(2,R)$, but this is not a subalgebra of $3A_1 \oplus_s gl(2,R)$. [10]

2. Linearization

Obviously, if a nonlinear node has the symmetry algebra associated with a linear node, there exists a point transformation which transforms the nonlinear equation into a linear one. If it does not, then it is not linearizable by means of a point transformation. However, a knowledge of all the symmetries is not necessary to determine whether a nonlinear node can be linearized. A necessary and sufficient condition for a node to be linearizable by a point transformation is that

1) for $n \neq 2$, it possesses the n-dimensional Abelian algebra nA_1 [9] and
2) for $n=2$, it possesses the 3-dimensional algebra [3]

$$(a) A_1 \oplus A_2, \quad (b) A^{\frac{1}{2}}_{3,5}, \quad (c) A_{3,3} \quad (d) A_{3,1} \quad \text{or} \quad (e) A_{3,9}.$$

The case $n=2$ is exceptional since there are two realizations of $2A_1$ one of which does not imply linearization.

3. Contact symmetries

Recall that a contact symmetry has the form

$$G = \xi(x,y,y')\frac{\partial}{\partial x} + \eta(x,y,y')\frac{\partial}{\partial y} + \varsigma(x,y,y')\frac{\partial}{\partial y'},$$

i.e., the y' dependence in ξ and η is such to produce no y'' dependence in ς since

$$\varsigma = \eta' - y'\xi.$$

G can be written in terms of a generating function $W(x,y,y')$ as

$$\begin{aligned} G = & -\frac{\partial W}{\partial y'}\frac{\partial}{\partial x} + (W - y'\frac{\partial W}{\partial y'})\frac{\partial}{\partial y} \\ & +(\frac{\partial W}{\partial x} + y'\frac{\partial W}{\partial y})\frac{\partial}{\partial y'}. \end{aligned}$$

In the case of sodes there is an infinite number of contact symmetries and the generator W is a function of the functionally independent first integrals of the equation [11].

Some examples are

$$y''=0 \quad W = xU(I,J)+V(I,J) \quad I=y' \quad J=y-xy'$$

$$y''=e^{-y'} \quad W = y'U(I,J)+V(I,J) \quad I=e^{y'}-x \quad J=(y'-1)e^{y'}-y.$$

The results for nodes are rather more sketchy. In the case of linear nodes, a tode can have 10 contact symmetries and for $n > 3$ there are no contact symmetries in addition to the point symmetries. The situation regarding nonlinearizable nodes remains a mystery.

References

[1] R.L. Anderson and S.M. Davison, *A generalization of Lie's "counting" theorem for second-order ordinary differential equations*, J. Math. Anal. Appl., 48(1974), 301–315.

[2] C.E. Wulfman and B.J. Wybourne, *The Lie group of Newton's and Lagrange's equations for the harmonic oscillator*, J. Phys. A 9(1976), 507–518.

[3] F.M. Mahomed, *Classification of nonlinear ordinary differential equations according to the their symmetries* (M. Sc. dissertation, University of the Witwatersrand, Johannesburg, 1986).

[4] F.M. Mahomed and P.G.L. Leach, *The linear symmetries of a nonlinear differential equation*, Quaest Math., 8(1985), 241–274.

[5] W. Sarlet, F.M. Mahomed and P.G.L. Leach, *Symmetries of nonlinear differential equations and linearization*, J. Phys. A 20(1987), 277-292.

[6] F.M. Mahomed and P.G.L. Leach, *The Lie algebra $sl(3,R)$ and linearization* Quaest Math., 12(1989), 121-139.

[7] F.M. Mahomed and P.G.L. Leach, *Lie algebras associated with scalar second order ordinary differential equations*, J. Math. Phys., 30(1989), 2770–2777.

[8] J. Krause and L. Michel, *Équations différentielles linéaires d'ordre $n > 2$ ayant une algèbre de Lie de symétrie de dimension $n+4$*, C. R. Acad. Sci. 307 Série 1(1988), 116–118.

[9] F.M. Mahomed and P.G.L. Leach, *Symmetry Lie algebras of nth order ordinary differential equations*, J. Math. Anal. Appl., 151(1990), 80–107.

[10] L. Michel, private communication, October 22, 1988.

[11] F.M. Mahomed and P.G.L. Leach, *Contact symmetries of second order differential equations* (Preprint CNLS-89-10, Centre for Nonlinear Studies, University of the Witwatersrand, Johannesburg, March, 1989).

Proceedings of the Asian Mathematical
Conference, 1990

Integration of Highly Oscillatory Functions in the Plane

LEE PENG YEE, & CHEW TUAN SENG
National University of Singapore
Republic of Singapore

§1. Introduction

It is known that the Henstock integral integrates functions which are not Lebesgue integrable. For example, let $F(x) = x^2 \sin x^{-2}$ when $x \neq 0$ and $F(0) = 0$. Note that F is differentiable everywhere and $F'(0) = 0$. The derivative F' of F is Henstock integrable on [0, 1] but not Lebesgue integrable there. The Henstock integral is also known as the Kurzweil integral or the generalized Riemann integral. Similarly, the Henstock integral in the plane integrates functions of two variables which are not Lebesgue integrable. For example, let $h(x, y) = f(x)g(y)$ in which $f(x) = F'(x)$ and F is the Function given above. Then h may be integrable on a rectangle containing the origin in the sense of Henstock but not of Lebesgue. In other words, the Henstock integral on the real line or in the plane is capable of integrating highly oscillatory functions where the Lebesgue integral fails.

The Henstock-Kurzweil integral was first defined by Kurzweil (1957) and independently by Henstock (see [2]). A general theory has also been developed by Henstock (see [7]). The integral in the Euclidean space has been studied by many. In particular, Ostaszewski [8] studied its various properties, Chew [1] gave a descriptive definition, Mahwin [6] and Pfeffer [9] investigated the divergence theorem, and Lee [4] proved the controlled convergence theorem (see also [3; p.130]). In this paper, we show that the controlled convergence theorem as proved for the Henstock integral in the Euclidean space, with slight modification, is in a sense the most natural.

§2. Generalized Absolute Continuity

The descriptive definition of the Henstock integral on the real line makes use of the ACG^* property. This property has been the stumbling block for extension to higher dimensions for decades. Recently, this property was used by Djvarsheishvili, and independently by Lee and Chew [3; p.172], to prove the controlled convergence theorem for the Henstock integral. Again, to extend the theorem to higher dimensions, the difficulty is to rephrase the ACG^* property so that it will lead to an easy extension to higher dimensions. Lee [3, 4] defined ACG^{**} and used it to prove the controlled convergence theorem in the Euclidean space. Lee

[5] showed that ACG^* and ACG^{**} are equivalent on the real line. However it is not clear that the primitive function of a Henstock integrable function in the plane is ACG^{**}. In this section, we discuss the property ACG^{**} and modify it in a way that we can still prove the controlled convergence theorem in the plane as well as that the primitive functions are ACG^{**}.

Let E be an interval in the n-dimensional Euclidean space and $X \subset E$. A division D of E, which is a finite collection of interval-point pairs (I,x) with the intervals non-overlapping and their union E, is said to be δ-fine if for each interval-point pair (I,x) we have $I \subset S(x,\delta(x))$ where x is a vertex of I and $S(x,r)$ denotes an open sphere with center x and radius r_n (see [3; p. 127]). A function F defined on E is said to be $AC^{**}(X)$ if for every $\varepsilon > 0$ there are $\delta(x) > 0$ and $\eta > 0$ such that for any two δ-fine partial divisions of E with the associated points in X, namely $D_1 = \{(I_1,x_1)\}$ and $D_2 = \{(I_2,x_2)\}$, with $x_1, x_2 \in X$, satisfying.

$$(D_1 \backslash D_2)\Sigma|I| < \eta \text{ we have } |(D_1 \backslash D_2)\Sigma F(I)| < \varepsilon.$$

Let E_1 be the union of the intervals in D_1 and E_2 the union of the intervals in D_2. Then $D_1 \backslash D_2$ denotes the collection of component elementary sets (consisting of adjacent intervals) in $E_1 - E_2$ and $(D_1 \backslash D_2)\Sigma$ means $(D_1)\Sigma - (D_2)\Sigma$. We assume always $D_2 \subset D_1$ in which D_2 may be void. Here $D_2 \subset D_1$ means that any interval from D_2 lies in some interval in D_1.

An interval function F defined on E is said to be ACG^{**} if $E = \cup_{i=1}^{\infty} X_i$ so that F is $AC^{**}(X_i)$ for each i. This definition differs from the one in [4] in three aspects. First, we no longer require X_i to be closed here. Secondly, we consider $|(D_1 \backslash D_2)\Sigma F(I)| < \varepsilon$ instead of $(D_1 \backslash D_2)\Sigma|F(I)| < \varepsilon$; and thirdly, we always assume $D_2 \subset D_1$.

We recall the definition of the integral as follows. A real number H is the value of the integral of f over an interval E if given $\varepsilon > 0$ there is $\delta(x) > 0$ such that

$$|(D)\Sigma f(x)|I| - H| < \varepsilon$$

for all δ-fine divisions D of E. The integral is uniquely determined. As shown in [3; p.128], the primitive F of f can be regarded as an interval function or a point function and they are convertable.

Theorem 1. *If f is integrable on E with primitive F, then F is ACG^{**}.*

We shall prove the theorem in a series of lemmas.

Lemma 2. *If f is integrable on E, then f is measurable.*

The Lemma follows from the corresponding result of Theorem 5.7 [3; p.21].

Lemma 3. *If f is a bounded measurable function on E, then f is integrable on E.*

The lemma follows from Theorem 21.6 [3; p.132] with $X = E, f_X = f$, and $f_{n,X}$ being a uniformly bounded sequence of step functions.

Lemma 4. *If f is integrable on E and bounded, then f is McShane integrable on E.*

We recall that a function is McShane integrable if it is integrable in which the associated point x does not necessarily lie in I for each (I,x) in the division. The lemma follows from Theorem 16.10 [3; p.109] which holds true in the Euclidean space and the fact that bounded integrable functions are absolutely integrable. We define a McShane division $D = \{(I,x)\}$

to be one that does not require the associated point x to belong to I, whereas an ordinary division does.

Proof of Theorem 1. Let $X = \{x; |f(x)| \leq N\}$ and $f_X(x) = f(x)$ when $x \in X$ and 0 otherwise. By Lemma 2, f is measurable and so is f_X. Then by Lemma 3 and 4, f_X is McShane integrable on E. In other words, for every $\varepsilon > 0$ there exists $\delta(x) > 0$ such that for every δ-fine division $D = \{(I, x)\}$ we have

$$(D)\Sigma|F(I) - f(x)|I|\,| < \varepsilon$$

and also for every δ-fine McShane division $D = \{(I, x)\}$ we have

$$(D)\Sigma|F_X(I) - f_X(x)|I|\,| < \varepsilon$$

in which F denotes the primitive of f and F_X that of f_X.

Now take any δ-fine partial division $D = \{(I, x)\}$ with $x \in X$ we have

$$(D)\Sigma|F(I) - F_X(I)| \leq (D)\Sigma|F(I) - f(x)|I|| + (D)\Sigma|F_X(I) - f_x(x)|I|| < 2\varepsilon.$$

Next, let D_1 and D_2 be δ-fine partial divisions with associated points in X and $D_2 \subset D_1$ such that $(D_1 \backslash D_2)\Sigma|I| < \eta$. Then we obtain

$$\begin{aligned} |(D_1\backslash D_2)\Sigma F(I)| \leq\ & (D_1)\Sigma|F(I) - F_X(I)| + (D_2)\Sigma|F(I) - F_X(I)| \\ & +|(D_1\backslash D_2)\Sigma\{F_X(I) - f_X(x)|I|\}| + |(D_1\backslash D_2)\Sigma f_X(x)|I|\,| \\ <\ & 5\varepsilon + N\eta. \end{aligned}$$

Note that $D_1 \backslash D_2$ is a δ-fine McShane partial division with x from D_1. Consequently, for every $\varepsilon > 0$ there is $\eta > 0$, in fact $N\eta < \varepsilon$, such that the condition in the definition of $AC^{**}(X)$ holds. Hence F is ACG^{**}.

§3. Controlled Convergence

We see that with the modified definition of ACG^{**} the following controlled convergence theorem still holds.

Theorem 5. *If the following conditions are satisfied:*
(i) $f_n(x) \to f(x)$ everywhere in E as $n \to \infty$ where each f_n is integrable on E;
(ii) the primitives F_n of f_n are $UACG^{**}$, i.e. ACG^{**} uniformly in n, then f is integrable on E and we have

$$\int_E f_n \to \int_E f \quad \text{as} \quad n \to \infty.$$

The proof follows as in Section 21 [3], and uses the remark of [3; p.136]. We sketch the additional step as follows. Suppose F_n are $UAC^{**}(X_i)$ for each i, i.e., $AC^{**}(X_i)$ uniformly in n for each i. Choose a sequence of closed sets $Y_i \subset X_i, i = 1, 2, \cdots$, such that $S = E - \cup_{i=1}^{\infty} Y_i$ is of measure zero. The following strong Lusin condition holds for F_n and F (see [5]). For every set $Z \subset E$ of measure zero and for every $\varepsilon > 0$ there is $\delta(x) > 0$ such that for any

δ-fine partial division $D = \{(I, x)\}$ of E with $x \in Z$ we have $|(D)\Sigma F_n(u, v)| < \varepsilon$ for all n, and $|(D)\Sigma F(I)| < \varepsilon$, in particular, when $Z = S$.

The proof follows as usual for $\cup_{i=1}^{\infty} Y_i$ in which Y_i is closed. In view of the strong Lusin condition, combining with $\delta(x) > 0$ defined on $\cup_{i=1}^{\infty} Y_i$ as in [3; p.135] and using the notation there we have

$$|(D)\Sigma\{F_{m(x)}(I) - F(I)\}| < 5\varepsilon$$

for all δ-fine division D and suitable $m(x)$ as required in Theorem 21.4 [3; p.130]. Hence Theorem 5 is now proved with the modified definition of $UACG^{**}$.

We remark that the statement of Theorem 6.22 [3; p.36] is faulty. The condition (N) there should be replaced by the strong Lusin condition. In other words, for the one-dimensional case a version of the descriptive definition states that f is Henstock integrable on $[a, b]$ if and only if its continuous primitive F satisfies the strong Lusin condition and that the derivative $F'(x) = f(x)$ almost everywhere. It is not known whether the corresponding n-dimensional version exists.

It is conjectured that if f is integrable on E then there is a sequence $\{\varphi_n\}$ of step functions satisfying the conditions of Theorem 5. The difficulty is in proving the primitives of φ_n having $UACG^{**}$. The one-dimensional proof [3; p.59] does not apply there because of its real-line dependence.

References

[1] T.S. Chew, *On the equivalence of Henstock-Kurzweil and restricted Denjoy integrals in* R^n, Real Analysis Exchange, 15(1989–90), 259–268.

[2] R. Henstock, *Lectures on the theory of integration*, World Scientific 1988.

[3] P.Y. Lee, *Lanzhou lectures on Henstock integration*, Wold Scientific, 1989.

[4] P.Y. Lee, *Generalized convergence theorems for Denjoy-Perron integrals*, Springer-Verlag LN 1419 New integrals pp. 97–109 (1990).

[5] P.Y. Lee, *On ACG* functions, Real Analysis Exchange*, 15(1989–90), 754–759.

[6] J. Mawhin, *Generalized multiple Perron integrals and the Green-Goursat theorem for differentiable vector fields*, Czechoslovak Math. J., 31(1981), 614–632.

[7] P. Muldowney, *A general theory of integraion in function spaces*, Pitman Research Notes in Maths 153, Longmans 1987.

[8] K.M. Ostaszewski, *Henstock integration in the plane*, AMS Memoirs, 353(1986).

[9] W.F. Pfeffer, *The divergence theorem*, Trans. Amer. Math. Soc., 295(1986), 665–685.

Proceedings of the Asian Mathematical
Conference, 1990

A Quasilinearization Method for Robust Parameter Estimation

RICHARD LEE
Department of Mathematics and Statistics University of New Brunswick Fredericton, N.B., Canada

In this paper we consider the problem of estimating the constant m-vector parameter γ in the differential equation

$$\frac{dx}{dt} = f(t, x, \gamma), \quad x \text{ in } R^n, \quad t \text{ in } [0, \infty), \tag{1}$$

by observing the system at a number of discrete times, say, $t_1, t_2, \cdots, t_L$, giving the observations

$$y_i = (\alpha_i, x(t_i)) + e_i, \quad i = 1, \cdots, L,$$

where e_i is the observation error at time t_i, α_i is a given n-dimensional vector, and $(\ ,\ )$ is the usual inner product. This is an important problem in using differential equations to model physical phenomena.

Traditionally, the vector γ is estimated by choosing γ so as to minimize

$$S_2 = \sum_{i=1}^{L} w_i(y_i - (\alpha_i, x(t_i)))^2,$$

where $w_i (i = 1, \cdots, L)$ is a weight factor. It does give reasonable results when the errors of observations are normally distributed. Alternatively, when the error distributions are not normal or are not known too well, Huber [3] suggested choosing γ to minimize the sum

$$S_H = \sum_{i=1}^{L} \rho((y_i - (\alpha_i, x(t_i)))/\sigma) \tag{2}$$

in which σ is a scale constant and $\rho(x)$ is a function defined by

$$\rho(x) = \begin{cases} (1/2)x^2, & \text{if } |x| \leq k, \\ kx - (1/2)k^2, & \text{if } |x| > k; \end{cases}$$

here, k is a positive constant, say, $k = 1.5$.

To solve this non-linear optimization problem, we propose a quasilinearization method. This is done by first adding the vector equation

$$\frac{d\gamma}{dt} = 0$$

to equation (1) to obtain the extended $N = n + m$ vector equation

$$\frac{dz}{dt} = g(t, z), \tag{3}$$

where

$$z = \begin{bmatrix} x \\ \gamma \end{bmatrix}, \quad g(t, z) = \begin{bmatrix} f(t, x, \gamma) \\ 0 \end{bmatrix}$$

so that

$$z(t_1) = \begin{bmatrix} x(t_1) \\ \gamma \end{bmatrix} \equiv \beta.$$

Thus, the constant vector parameter γ is estimated by estimating the initial vector β.

Assume an initial estimate β^0, and let $z^0(t)$ be the corresponding solution of (3). Linearize about $z^0(t)$ in the usual way to obtain an approximating linear equation for $z(t)$:

$$\frac{dz}{dt} = g(t, z^0(t)) + \frac{\partial g}{\partial z}(t, z^0(t))(z - z^0(t)).$$

If $\Phi(t) = [\phi_{ij}(t)]$ is the fundamental matrix of the homogeneous linear system of equations

$$\frac{dz}{dt} = \frac{\partial g}{\partial z}(t, z^0(t))z$$

such that $\Phi(t_1)$ is the identity matrix, and $p(t)$ is the solution of the initial value problem

$$\frac{dz}{dt} = g(t, z^0(t)) + \frac{\partial g}{\partial z}(t, z^0(t))(z - z^0(t)),$$
$$z(t_1) = 0,$$

then the general solution of (3) is given by

$$z(t) = \Phi(t)C + p(t).$$

Hence, $z(t_1) = C = \beta$.

The vector C is estimated by minimizing the sum (2) which may be rewritten as

$$S_H = \sum_{i=1}^{L} \rho((b_i - A_i C)/\sigma),$$

where

$$b_i = y_i - (\alpha_i, \tilde{p}(t_i)), \qquad A_i = \alpha_i^T \tilde{\Phi}(t_i),$$

and

$$\tilde{p}(t_i) = \begin{bmatrix} p_1(t_i) \\ \vdots \\ p_n(t_i) \end{bmatrix},$$

$$\tilde{\Phi}(t_i) = \begin{bmatrix} \phi_{11}(t_i) \cdots \phi_{1N}(t_i) \\ \vdots \\ \phi_{n1}(t_i) \cdots \phi_{nN}(t_i) \end{bmatrix}.$$

This robust linear regression problem can be solved by an iteratively reweighted least squares method [2], that is, by choosing C so as to minimize the sum

$$S = \sum_{i=1}^{L} w_i (b_i - A_i C)^2$$

in which

$$w_i = \begin{cases} \psi((b_i - A_i\overline{C})/\overline{\sigma})/((b_i - A_i\overline{C})/\overline{\sigma}), & \text{if } b_i \neq A_i\overline{C}, \\ 1, & \text{if } b_i = A_i\overline{C}, \end{cases}$$

$$\psi(u) = \partial\rho(u)/\partial u.$$

Here, $\overline{C}$ and $\overline{\sigma}$ are estimates of C and σ from a previous estimation, and

$$\sigma = \text{median}\{|b_i - A_i C|\}/0.6745.$$

Initially, the parameter C is estimated by minimizing the sum

$$S_1 = \sum_{i=1}^{L} |b_i - A_i C|.$$

It is well-known that this so called l_1-problem can be done by solving the linear programming problem

$$\text{Minimize} \quad \sum_{i=1}^{L} v_i,$$

subject to

$$v_i + A_i C \geq b_i, \quad i = 1, \cdots, L,$$

$$v_i - A_i C \geq -b_i, \quad i = 1, \cdots, L.$$

See [1].

Now let $C = \beta^1$ be the solution of the iteratively reweighted least squares problem. If β^1 is reasonably close to β^0, we accept β^1 as an estimate of β; otherwise, the procedure is repeated with β^0 replaced by β^1 to obtain a next approximation.

We have applied the method to an example of the logistic equation

$$\frac{dx}{dt} = \gamma_1(1 - (x/375)^{\gamma_2}), \quad 1 \leq t \leq 4,$$

$$x(0) = 5.0,$$

with two constant parameters γ_1 and γ_2. The system is observed through the equation

$$y_i = x(t_i) + e_i$$

with

$$t_i = 0.2 + (i - 1)(0.2), \quad i = 1, \cdots, 20.$$

It is assumed that the observation errors follow a contaminated normal distribution

$$0.90N(0, 1) + 0.10N(0, 100).$$

The following table summarizes the average results based on 40 sets of observations.

Results Based on	Average Values of Parameters $\hat{\gamma}_1$	$\tilde{\gamma}_2$	Average Value of RISE
S_2	2.3078	0.5016	1.3956
S_H	2.3087	0.5002	0.6689

True Values: $\gamma_1 = 2.309, \gamma_2 = 0.500$.

If we measure how good the estimates are with the root integral square error (RISE) criterion defined by

$$\text{RISE} = \int_0^4 [x(t,\gamma) - x(t,\hat{\gamma})]^2 dt$$

in which $\hat{\gamma}$ is the estimated value of γ, then it is seen from the table that the result based on Huber's ρ-function is much better than that obtained by the conventional least squares method. The root integral square error for the latter is about twice as large. We have also done many other numerical studies. Our conclusion is that if the error distributions are not known too well it is preferable to estimate the parameters using Huber's ρ-function.

References

[1] V. Chvatal, *Linear Programming*, W.H. Freeman, New York, 1980.

[2] R.V. Hogg, *Statistical Robustness: One View of Its Use in Application Today*, The American Statistician, 33(1979), 108–115.

[3] P.J. Huber, *Robust Statistics*, John Wiley and Sons, New York, 1981.

Proceedings of the Asian Mathematical
Conference, 1990

Positive Solutions to Certain Nonlinear Elliptic Biological Interacting Systems of Three Variables

LIGE LI & YAPING LIU
Department of Mathematics
Kansas State University
Manhattan, Kansas 66506, U.S.A.

§1. Introduction

In this short note we shall give some results on the existence of positive solutions to elliptic biological interacting systems involving 3 species in all possible ten interactions in terms of the combinations of competing, symbiosis and predator-prey types under the homogeneous Dirichlet boundary condition. This is to find the conditions for the existence of strict positive solutions $u, v, w \in C(\overline{\Omega}), u > 0, v > 0, w > 0$ of the following nonlinear elliptic system

$$\begin{cases} -\Delta u = uf(u,v,w), & \\ -\Delta v = vg(u,v) & \text{in } \Omega, \\ -\Delta w = wh(u,w), & \\ u = v = w = 0 & \text{on } \partial\Omega, \end{cases} \tag{1}$$

where Ω is a bounded domain in $\mathbf{R}^n$ with smooth boundary. The f, g, h are C^1 functions. The system (1) will model various biological interactions. Our results show that the existence of positive solutions of (1) can be characterized by the spectral property of certain Shrödinger operators and the equilibria of the equation (1), which reflects the dynamic aspect of the system.

§2. Preliminaries

In what follows, the $\lambda_1(L)$ will denote the first eigenvalue of a suitable linear operator L. First we define the class $\mathcal{F}$ by the following: $f \in \mathcal{F}$ if and only if $f \in C(\overline{\Omega} \times \mathbf{R}^+), f$ is C^1 is second component, $-L < f_u(x,u) < 0$ in $\Omega \times \mathbf{R}^+$ for some $L > 0$, and

$$\overline{\lim_{u \to \infty}} f(x,u) < L_1 \quad \text{uniformly for} \quad x \in \Omega.$$

Here $\lambda_1 = \lambda_1(-\Delta)$ is the first eigenvalue of $-\Delta$ under the homogeneous Dirichlet boundary condition.

Lemma 1. *Let $f \in \mathcal{F}$. Consider*

$$\begin{cases} -\Delta u = uf(x,u) & \text{in } \Omega, \\ u = 0 & \text{on } \partial\Omega. \end{cases} \tag{2}$$

(i) If $\lambda_1(\Delta + f(x,0)) > 0$ then equation (2) has a unique positive solution.

(ii) If $\lambda_1(\Delta + f(x,0)) \leq 0$ then (2) has no positive solution.

Proof. (i) Let $\lambda_1(\Delta + f(x,0)) > 0$. Define operator $A : C(\overline{\Omega}) \to C(\overline{\Omega})$ by $Au = (-\Delta + P)^{-1}(uf(x,u) + Pu)$ where $(-\Delta + P)^{-1}$ is under the boundary condition $u = 0. P > 0$ is to be chosen. $A'(0) = (-\Delta + P)^{-1}[f(x,0) + P]$ and $\lambda_1(\Delta + f(x,0)) > 0 \Longrightarrow r(A'(0)) > 1$. (See Lemma 2 [3]). $A'(0)$ is strongly positive in $C_e(\overline{\Omega})$ for $P > \max_{x\in\overline{\Omega}}\{-f(x,0)\}$ (see Amann [1]). So by Krein-Rutman Theorem and Lemma 7.5 of [1], $\exists \underline{u} > 0$ such that $A(\underline{u}) > \underline{u}$. Enlarge Ω to $\widetilde{\Omega}$ by $\widetilde{\Omega} = \{x \in \mathbf{R}^n : \text{dist}(x,\Omega) < \delta\}$ with $\delta > 0$. So $\widetilde{\lambda}_1 < \lambda_1$ where $\widetilde{\lambda}_1$ is the $\lambda_1(-\Delta)$ on the domain $\widetilde{\Omega}$. By the continuous dependence of λ_1 on domain Ω, we can choose δ so small such that

$$\overline{\lim_{u\to\infty}}\, f(x,u) < \widetilde{\lambda}_1 \quad \text{uniformly for} \quad x \in \Omega.$$

So $\exists\ C > 0$ such that $f(x,u) < \widetilde{\lambda}_1$, $\forall x \in \Omega$, $u > C$. Let u be a positive eigenfunction corresponding to $\widetilde{\lambda}_1$ in $\widetilde{\Omega}$. Then $u > 0$ on $\overline{\Omega}$. Let $\widetilde{u} = Ku$ where K can be chosen large enough such that $\widetilde{u} > C, \widetilde{u} > \underline{u}$ on Ω. Then $f(x,\widetilde{u}) < \widetilde{\lambda}_1$. So $-\Delta\widetilde{u} = \widetilde{\lambda}_1\widetilde{u} > \widetilde{u}f(x,\widetilde{u})(-\Delta + P)\widetilde{u} > \widetilde{u}f(x,\widetilde{u}) + P\widetilde{u}$. Thus $\widetilde{u} > (-\Delta + P)^{-1}[\widetilde{u}f(x,\widetilde{u}) + P\widetilde{u}] = A\widetilde{u}$. Let $C_0 = \max_{x\in\overline{\Omega}}\widetilde{u}(x)$ and $P > 2\max\{\sup_{(x,u)\in\overline{\Omega}\times[0,C_0]} \mid f(x,u) \mid, C_0L\}$. Then A is an increasing, compact operator on the order interval $[[\underline{u},\widetilde{u}]]$. By Corollary 6.2 [1], A has a maximal fixed point $\overline{u} \in [[\underline{u},\widetilde{u}]]$. So $\overline{u}$ is a positive solution.

Let u_1, u_2 be two positive solution of (2). From the above argument, we can choose $\widetilde{u}$ large enough such that $u_1, u_2 \leq \widetilde{u}$. Then $u_1, u_2 \leq \overline{u}$. Since $-\Delta u_1 = u_1 f(x,u_1), -\Delta\overline{u} = \overline{u}f(x,\overline{u})$, we have $\int_\Omega u_1\overline{u}[f(x,u_1) - f(x,\overline{u})]dx = \int_\Omega -\overline{u}\Delta u_1 dx - \int_\Omega -u_1\Delta\overline{u}dx = 0$ by Green's Identity. So $f(x,u_1) - f(x,\overline{u}) \equiv 0$ by monotonicity of f. Then $u_1 = \overline{u}$ since f is strictly decreasing in u. Similarly, $u_2 = \overline{u}$. Therefore the positive solution is unique. for another proof of the uniqueness, see Lemma 2.5 [2].

(ii) See Lemma 3 [3]. □

Note. This Lemma generalizes Lemma 3 [3] since we do not assume that $\exists C_0 > 0$ such that $f(x,C_0) \leq 0$, $\forall x \in \Omega$.

In view of Lemma 1, $\forall f \in \mathcal{F}$, let u_f be the unique positive solution of (2) if $\lambda_1(\Delta + f(x,0)) > 0$ and $u_f \equiv 0$ otherwise.

Lemma 2. *If the constant L in the definition of the class $\mathcal{F}$ is independant of functions f, then*

(i) The mapping $f \to u_f$ is continuous in sense of $\mathcal{F} \to C^{1,\alpha}(\Omega \times \mathbf{R}^+)$ where $\alpha \in (0,1)$.

(ii) If $f_1 \geq f_2 \not\equiv f_1$, then either $u_{f_1} > u_{f_2}$ or $u_{f_1} \equiv u_{f_2} \equiv 0$.

Proof. See Lemma 4 [3]. □

We impose the following hypotheses on the functions f, g, h:

(H1) $f \in C^1(\mathbf{R}^+ \times \mathbf{R}^+ \times \mathbf{R}^+), g, h \in C^1(\mathbf{R}^+ \times \mathbf{R}^+)$. $f(\cdot, v, w), g(u, \cdot), h(u, \cdot) \in \mathcal{F}$ for any fixed $u, v, w \in C(\overline{\Omega})^+$. $f_u, f_v, f_w, g_w, g_v, h_u, h_w \neq 0$ and are all uniformly bounded on $\mathbf{R}^+ \times \mathbf{R}^+ \times \mathbf{R}^+$ and on $\mathbf{R}^+ \times \mathbf{R}^+$ respectively.

(H2) We always assume that the species u, v, w can survive by themselves in the absence of the other species, i.e., the following equations

$-\Delta u = uf(u, 0, 0)$ in $\Omega, u = 0$ on $\partial\Omega$,

$-\Delta v = vg(0, v)$ in $\Omega, v = 0$ on $\partial\Omega$,

$-\Delta w = wh(0, w)$ in $\Omega, w = 0$ on $\partial\Omega$

have positive solutions u_0, v_0, w_0 respectively. In View of Lemma 1, this would mean that $\lambda_1(\Delta + f(0,0,0)) > 0, \lambda_1(\Delta + g(0,0)) > 0, \lambda_1(\Delta + h(0,0)) > 0$.

In the following, u_0, v_0, w_0 will always denote the unique positive solution of the above equations respectively, without exception.

Similar argument can be made when not all of the u_0, v_0, w_0 exist. Our methods also apply to some of those cases and many other cases still remain open.

By **(H1)**, f_v, f_w, g_u, h_u do not change sign. So the interactions between u and v and between u and w have certain fashion: either predator-prey, or symbiosis, or competition. We will use the following notation:

$u \longrightarrow v$ if u, v are a predator-prey pair.

$u \multimap v$ if u, v cooperate.

$u \longleftrightarrow v$ if u, v compete.

So for example, $v \longleftarrow u \longrightarrow w$ means that (1) is a one predator (u) two prey (v, w) model whereas $v \multimap u \longleftrightarrow w$ means that u, v cooperate and u, w compete.

Let $\mathbf{X} = C(\overline{\Omega})$ and $\mathbf{K} = C(\overline{\Omega})^+$ it's positive cone. Define operators $S, T : C(\overline{\Omega}) \to C(\overline{\Omega})$ by the following: for $u \in C(\overline{\Omega})$, Su is the unique positive solution of equation

$$\begin{cases} -\Delta v = vg(u, v) & \text{in } \Omega \\ v = 0 & \text{on } \partial\Omega \end{cases} \tag{3}$$

if $\lambda_1(\Delta + g(u, 0)) > 0$ and $Su \equiv 0$ otherwise. Tu is determine similarly from the equation

$$\begin{cases} -\Delta w = wh(u, w) & \text{in } \Omega \\ w = 0 & \text{on } \partial\Omega \end{cases} \tag{4}$$

By Lemma 2, it is easy to see that S and T are continuous operators and if u, v cooperate or v is a predator on u, then S is a strictly increasing operator in the sense that $u_1 \not\equiv u_2 \geq u_1 \Rightarrow Su_2 > Su_1$ and if u, v compete or v is a prey for u, then S is a strictly decreasing operator in the same sense. Similar conclusion is true for T.

Now define $A : C(\overline{\Omega}) \to C(\overline{\Omega})$ by $Au = (-\Delta + P)^{-1}[uf(u, Su, Tu) + Pu]$, where $P > 0$ is a constant. Let $A_\theta = \theta A$ for $0 < \theta \leq 1$. Clearly, $\overline{u}$ is a fixed point of A in $\mathbf{K}$ if and only if $(\overline{u}, S\overline{u}, T\overline{u})$ is a nonnegative solution of (1). We are looking for the positive fixed point $\overline{u}$ of A with $S\overline{u} > 0, T\overline{u} > 0$.

Lemma 3. *Given any $C > 0$. A is a positive, compact operator on $[[0, C]]$ for large P and if no predation is involved in system (1), then A is also increasing for large P.*

Proof. It is easy to verify by using Lemma 2. □

Lemma 4. *If $\lambda_1(\Delta + f(0, v_0, w_0)) > 0$ and $\exists R > 0$ such that A_θ has no fixed point on $\partial B_R(0)$ in $\mathbf{K}$ for $0 < \theta \leq 1$, then A has a positive fixed point in $[[0, R]]$.*

Proof. Since $A'(0) = (-\Delta+P)^{-1}[f(0, v_0, w_0)+P]$ and $\lambda_1(\Delta+f(0, v_0.w_0)) > 0$, we have $r(A'(0)) > 1$. By Krein-Rutman Theorem, $r(A'(0))$ is an eigenvalue of $A'(0)$ with a positive eigenvector and $A'(0)$ has no other eigenvalues with positive eigenvectors. $\theta Au = A_\theta u \neq u$ for $u \in \partial B_R(0) \cap \mathbf{K}, \theta \in (0, 1] \Longrightarrow Au \neq \lambda u, \forall \lambda \in [1, \infty)$. Then apply Theorem 13.2 [1]. □

Lemma 5. *The positive fixed points of A_θ have a-priori bound which is independant of $\theta \in (0, 1]$ if no symbiosis relation is involved in system (1).*

Proof. Let $A_\theta(u_\theta) = u_\theta$. Then $-\Delta u_\theta = \theta_{u_\theta} f(u_\theta, Su_\theta, Tu_\theta) + P(\theta - 1)u_\theta$. Consider the following cases:

(1) Both u, v and u, w are of predator-prey type. Then $f(u, Su, Tu) \leq f(u, C_2, C_3)$ where $C_2 = \max_{x\in\overline{\Omega}} v_0(x), C_3 = \max_{x\in\overline{\Omega}} w_0(x)$.

(2) u, v is predator-prey, u, w is not. Then $f(u, Su, Tu) \leq f(u, C_2, 0)$.

(3) u, v and u, w compete or u is prey. Then $f(u, Su, Tu) \leq f(u, 0, 0)$.

In any case, $\exists \overline{C}_2, \overline{C}_3 \geq 0$, independant of θ such that $f(u, Su, Tu) \leq f(u, \overline{C}_2, \overline{C}_3)$ for any $u \in \mathbf{K}$. So $-\Delta u_\theta \leq \theta u_\theta f(u_\theta, \overline{C}_2, \overline{C}_3)$. Let $u_\theta(x_0) = \max_{x\in\overline{\Omega}} u_\theta(x)$ for some $x_0 \in \Omega$. Then $0 \leq -\Delta u_\theta(x_0) \leq \theta u_\theta(x_0) f(u_\theta(x_0), \overline{C}_2, \overline{C}_3)$. So $f(u_\theta, \overline{C}_2, \overline{C}_3) \geq 0$. Since $\theta \in (0, 1]$, we have $-\Delta u_\theta \leq u_\theta f(u_\theta, \overline{C}_2, \overline{C}_3)$. So u_θ is a lower solution of

$$\begin{cases} -\Delta u = uf(u, \overline{C}_2, \overline{C}_3) & \text{in } \Omega, \\ u = 0 & \text{on } \partial\Omega. \end{cases} \tag{5}$$

By (**H1**) and proof of Lemma 1, (5) has arbitrarily large upper solution $\tilde{u} > u_\theta$. So (5) has unique positive solution $\overline{u} \in [[u_\theta, \tilde{\theta}]]$. Then $u_\theta \leq \overline{u}, \overline{u}$ does not depend on θ, therefore $K = \max_{x\in\Omega} \overline{u}(x)$ is the a-priori bound. □

Theorem 1. *Assume that no symbiotic interaction is involved in system (1). Then A has a positive fixed point if $\lambda_1(\Delta + f(0, v_0, w_0)) > 0$.*

Proof. Apply Lemma 4 and Lemma 5. □

§3. Main Result

In this section we give the conditions for the existence of positive solutions of the system (1) on all possible ten interactions of 3 species in terms of the combinations of competition, symbiosis and predation.

For $u_i, v_i, w_i \in C(\overline{\Omega})$ or $= \infty, i = 1, 2$, let $\ll (u_1, v_1, w_1), (u_2, v_2, w2) \gg := \{(u, v, w) \in C(\overline{\Omega}) \oplus C(\overline{\Omega}) \oplus C(\overline{\Omega}) : (u_1, v_1, w_1) < (u, v, w) < (u_2, v_2, w_2) \text{ in } \Omega\}$.

Theorem 2. *Let f, g, h satisfy* (**H1**) *and* (**H2**). *Then the following are true.*

Case 1 $v \to u \longleftarrow w$. The system (1) has a positive solution if and only if $\lambda_1(\Delta + f(0, v_0, w_0)) > 0$.

Case 2 $v \multimap u \multimap w$. If system (1) has a positive equilibrium (C_1, C_2, C_3),then it has a positive solution in $\ll (u_0, v_0, w_0), (C_1, C_2, C_3) \gg$.

Case 3 $v \longleftrightarrow u \longleftrightarrow w$. Assume $\lambda_1(\Delta + f(0, v_0, w_0)) > 0, \lambda_1(\Delta + g(u_0, 0)) > 0, \lambda_1(\Delta + h(u_0, 0)) > 0$. Then (1) has a positive solution in $\ll (0, 0, 0), (u_0, v_0, w_0) \gg$.

Note. If one of $\lambda_1(\Delta + g(u_0, 0))$ and $\lambda_1(\Delta + h(u_0, 0))$ is 0, the conclusion is still true.

Case 4 $v -\!\!\circ\!\!- u \longrightarrow w$. Assume $\lambda_1(\Delta + f(0, v_0, w_0)) > 0$ and the subsystem

$$\begin{cases} -\Delta u = uf(u, v, 0) \\ -\Delta v = vg(u, v) \end{cases} \tag{6}$$

has a positive equilibrium (C_1, C_2), i.e., $f(C_1, C_2, 0) = 0, g(C_1, C_2) = 0$. Then (1) has a positive solution in $\ll (0, v_0, w_0), (C_1, C_2, \infty) \gg$.

Case 5 $v \longleftrightarrow u \longleftarrow w$. Assume $\lambda_1(\Delta + f(0, v_0, w_0)) > 0, \lambda_1(\Delta + g(u_0, 0)) \geq 0$. Then (1) has a positive solution in $\ll (0, 0, w_0), (u_0, v_0, \infty) \gg$.

Case 6 $v -\!\!\circ\!\!- v \longleftrightarrow w$. Assume that

$$\begin{cases} -\Delta u = uf(u, v, 0) \\ -\Delta v = vg(u, v) \end{cases} \tag{7}$$

has positive equilibrium (C_1, C_2) and $\lambda_1(\Delta + f(0, v_0, w_0)) > 0, \lambda_1(\Delta + h(C_1, 0)) \geq 0$. Then (1) has a positive solution in $\ll (0, v_0, 0), (C_1, C_2, w_0) \gg$.

Note. The above result generalizes the existence result of Theorem 2.7 in §2 [4] in which the condition $\alpha_1(1 - K_2) > \lambda_0 d_1 \Longrightarrow \lambda_1(\Delta + f(0, 0, w_0)) > 0$ in our term and $\alpha_2 - \gamma K_1 > \lambda_0 d_2 \Longrightarrow \lambda_1(\Delta + h(C_1, 0)) > 0$.

Case 7 $v \longleftarrow v \longrightarrow w$. Let u_1 be the unique positive solution of

$$\begin{cases} -\Delta u = uf(u, v_0, w_0) & \text{in } \Omega \\ u = 0 & \text{on } \partial\Omega \end{cases} \tag{8}$$

(Existence is guranteed by Lemma 1 since $\lambda_1(\Delta + f(0, v_0, w_0)) \geq \lambda_1(\Delta + f(0, 0, 0)) > 0$.) If $\lambda_1(\Delta + g(u_1, 0)) \geq 0$ and $\lambda_1(\Delta + h(u_1, 0)) \geq 0$, then (1) has a positive solution in $\ll (u_0, 0, 0), (u_1, v_0, w_0) \gg$.

Case 8 $v -\!\!\circ\!\!- u \longrightarrow w$. Let $C = \max_{x \in \Omega} w_0(x)$ and

$$\begin{cases} -\Delta u = uf(u, v, C) \\ -\Delta v = vg(u, v) \end{cases} \tag{9}$$

has a positive equilibrium (C_1, C_2) and assume $\lambda_1(\Delta + h(C_1, 0)) \geq 0$. Then (1) has a positive solution in $\ll (u_0, v_0, 0), (C_1, C_2, w_0) \gg$.

Case 9 $v \longleftarrow u \longleftrightarrow w$. Assume $\lambda_1(\Delta + f(0, v_0, w_0)) > 0$ and $\lambda_1(\Delta + g(u_0, 0)) \geq 0$. Let u_1 be the unique positive solution of

$$\begin{cases} -\Delta u = uf(u, v_0, 0) & \text{in } \Omega \\ u = 0 & \text{on } \partial\Omega \end{cases} \tag{10}$$

(Existence is guranteed by Lemma 1 since $\lambda_1(\Delta + f(0, v_0, 0)) \geq \lambda_1(\Delta + f(0, 0, 0)) > 0$.) Suppose $\lambda_1(\Delta + h(u_1, 0)) \geq 0$. Then (1) has a positive solution in $\ll (0, 0, 0), (u_1, v_0, w_0) \gg$.

Case 10 $v \longleftarrow u \longleftarrow w$. Assume $\lambda_1(\Delta + f(0, v_0, w_0)) > 0$ and $\lambda_1(\Delta + g(u_0, 0)) \geq 0$. Then (1) has a positive solution in $\ll (0, 0, w_0), (\infty, v_0, \infty) \gg$.

Note. The condition $\lambda_1(\Delta + g(v_0, 0)) \geq 0$ can be weakened to $\lambda_1(\Delta + g(u_1, 0)) \geq 0$ where u_1 is the unique positive solution of

$$\begin{cases} -\Delta u = uf(u, 0, w_0) & \text{in } \Omega \\ u = 0 & \text{on } \partial\Omega \end{cases} \tag{11}$$

if $\lambda_1(\Delta+f(0,0,w_0))>0$ and $u_1\equiv 0$ otherwise. Thus for example, if $\lambda_1(\Delta+f(0,0,w_0))\le 0$ then $\lambda_1(\Delta+f(0,v_0,w_0))>0$ would imply the existence of positive solution of (1).

Proof.

Case 1 Let $\lambda_1(\Delta+f(0,v_0,w_0))>0$. by Theorem 1, A has a fixed point $\overline{u}>0.S\overline{u}>S0=v_0>0, T\overline{u}>T0=w_0>0$. So $(\overline{u},S\overline{u},T\overline{u})$ is a positive solution of (1).

Conversely, let $(\overline{u},\overline{v},\overline{w})$ be a positive solution. Then $\overline{v}=S\overline{u}>S0=v_0, \overline{w}=T\overline{u}>T0=w_0$. So $f(0,v_0,w_0))>f(0,\overline{v},\overline{w})$. Therefore $\lambda_1(\Delta+f(0,v_0,w_0))>\lambda_1(\Delta+f(0,\overline{v},\ \overline{w}))>0$ since

$$\begin{cases}-\Delta u=uf(u,\overline{v},\overline{w}) & \text{in } \Omega\\ u=0 & \text{on } \partial\Omega\end{cases}\tag{12}$$

has positive solution $\overline{u}$.

Case 2 $-\Delta u_0=u_0f(u_0,0,0)<u_0f(u_0,Su_0,Tu_0)\Longrightarrow u_0<Au_0$. Equation

$$\begin{cases}-\Delta v=vg(C_1,v) & \text{in } \Omega\\ v=0 & \text{on } \partial\Omega\end{cases}\tag{13}$$

has unique positive solution $v_1=SC_1.C_2$ is upper solution of (13) and θv_1 is lower solution of (13) for $\theta\in(0,1]$, when θ is small, we have $\theta v_1<C_2$, so $\exists$ positive solution $\widetilde{v}$ of (13) in $[[\theta v_1,C_2]]$. By uniqueness, $\widetilde{v}=v_1$. This shows that $SC_1\le C_2$. Similarly, $SC_1\le C_3$. So $-\Delta C_1=0=C_1f(C_1,C_2,C_3)\ge C_1f(C_1,SC_1,TC_1)$. Thus $C_1\ge AC_1$. By the similar argument as above, we see that $u_0\le C_1$. Corollary 6.2 [1] and Lemma 3 $\Longrightarrow$ A has a positive fixed point $\overline{u}\in[[u_0,C_1]]$. Therefore $(\overline{u},S\overline{u},T\overline{u})$ is a positive solution of (1). Also $C_1\not\equiv\overline{u}\le C_1\Longrightarrow v_0=S0<S\overline{u}<SC_1\le C_2, w_0=T0<T\overline{u}<TC_1\le C_3\Longrightarrow u_0<\overline{u}<C_1$ by Lemma 2.

Case 3. By Theorem 1, A has positive fixed point $\overline{u}$. By Lemma 2, $\overline{u}\le u_0$. So $v_0=S0>S\overline{u}\ge Su_0>0$ and $w_0=T_0>T\overline{u}\ge Tu_0>0$. Then $\overline{u}<u_0.(\overline{u},S\overline{u},T\overline{u})$ is a positive solution of (1).

Note Case 3 can also be proved by applying Theorem 7.6, [1] since $r(A'(0))>1$ and $Au_0<u_0$.

Case 4 Let $R=C_1$ and $A_\theta u=u$ with $u\in\partial B_R(0)\cap\mathbf{K},\theta\in(0,1]$. Then $u(x_0)=\max_{x\in\Omega}u(x)=C_1$ for some $x_0\in\Omega$ and $-\Delta u=\theta uf(u,Su,Tu)+P(\theta-1)u, u\le C_1\Longrightarrow Su\le C_2$. Since $Tu>T0=w_0>0$, we have $0\le-\Delta u(x_0)\le\theta u(x_0)f(u(x_0),Su(x_0),Tu(x_0))<\theta u(x_0)f(u(x_0),C_2,w_0(x_0))<\theta C_1f(C_1,C_2,0)=0$, a contradiction. So A_θ has no fixed point on $\partial B_R(0)\cap\mathbf{K}$ for $0<\theta\le 1$. By Lemma 4, A has positive fixed point $\overline{u}\in[[0,C_1]]$. Now $(\overline{u},S\overline{u},T\overline{u})$ is a positive solution of (1). $T\overline{u}>T0=w_0$. $C_1\not\equiv\overline{u}\le C_1\Longrightarrow v_0=S0<S\overline{u}<SC_1\le C_2$. So $\overline{u}<C_1$.

Case 5 By Theorem 1, A has a positive fixed point $\overline{u}$. since $T\overline{u}>T0=w_>0$, we have, by Lemma 2, $\overline{u}<u_0$. So $v_0=S0>S\overline{u}>Su_0\ge 0$. Therefore $(\overline{u},S\overline{u},T\overline{u})$ is in $\ll(0,0,w_0),(u_0,v_0,\infty)\gg$.

Case 6 $\lambda_1(\Delta+f(0,v_0,w_0))>0\Longrightarrow r(A'(0))>1$. Since $SC_1\le C_2$ and $TC_1\ge 0$, we have $-\Delta C_1=0=f(C_1,C_2,0)\ge f(C_1,SC_1,TC_1)$. So $C_1\ge AC_1$. By Lemma 3 and Theorem 7.6 [1], A has a positive fixed point $\overline{u}\in[[0,C_1]].\overline{u}\ne C_1\Longrightarrow 0\le TC_1<T\overline{u}<T0=w_0$ and $0<v_0=S0<S\overline{u}<SC_1<C_2\Longrightarrow\overline{u}<C_1$. So $(\overline{u},S\overline{u},T\overline{u})$ is in $\ll(0,v_0,0),(C_1,C_2,w_0)\gg$.

Case 7 By Theorem 1, A has a positive fixed point $\overline{u}$. Since $S\overline{u} < S0 = v_0, T\overline{u} < T0 = w_0$, we have $f(0, S\overline{u}, T\overline{u}) < f(0, v_0, w_0)$. So by Lemma 2, $\overline{u} < u_1$. Thus $S\overline{u} > Su_1 \geq 0, T\overline{u} > Tu_1 \geq 0$. So $u_0 < \overline{u}$. $(\overline{u}, S\overline{u}, T\overline{u})$ is a positive solution of (1).

Case 8 Similarly as in Case 4, A_θ has no fixed point on $\partial B_R(0) \cap \mathbf{K}$ for $0 < \theta \leq 1$. By Lemma 4, A has a positive fixed point $\overline{u} \in [[0, C_1]]$ and $\overline{u} \neq C_1$. So $v_0 = S0 < S\overline{u} < SC_1 \leq C_2$ and $w_0 = T0 > T\overline{u} > TC_1 \geq 0$. Thus $u_0 < \overline{u} < C_1$. $(\overline{u}, S\overline{u}, T\overline{u})$ is a positive solution of (1).

Case 9 By Theorem 1, A has a positive fixed point $\overline{u}.S\overline{u} < S0 = v_0, T\overline{u} \geq 0 \Longrightarrow f(0, S\overline{u}, T\overline{u}) < f(0, v_0, 0) \Longrightarrow \overline{u} < u_1$. So $0 \leq Tu_1 < T\overline{u} < T0 = w_0$. Assume $S\overline{u} = 0$. Then $\overline{u}$ is the positive solution of

$$\begin{cases} -\Delta u = uf(u, 0, T\overline{u}) & \text{in } \Omega \\ u = 0 & \text{on } \partial\Omega \end{cases} \tag{14}$$

$f(0, 0, T\overline{u}) < f0, 0, 0) \Longrightarrow \overline{u} < u_0$ by Lemma 2. So $S\overline{u} > Su_0 \geq 0$, a contradiction. Therefore $S\overline{u} > 0$. $(\overline{u}, S\overline{u}, T\overline{u})$ is a positive solution of (1).

Case 10 By Theorem 1, A has a positive fixed point $\overline{u}.S\overline{u} < S0 = v_0, T\overline{u} > T0 = w_0 > 0$. Similarly as in Case 9, we can show that $S\overline{u} > 0$. So $(\overline{u}, S\overline{u}, T\overline{u})$ is a positive solution of (1) in $\ll (0, 0, w_0), (\infty, v_0, \infty) \gg$. □

References

[1] H. Amann, *Fixed point equations and nonlinear eigenvalue problems in ordered Banach spaces.* SIAM Rev., 18(1976), 620–709.

[2] L. Li, *On the uniqueness and ordering of steady-states of predator-prey systems*, Proc. Roy. Soc. Edinburgh, 110A, (1988), 295–303.

[3] L. Li and A. Ghoreishi, *On positive solutions of general nonlinear elliptic symbiotic interacting systems*, to appear in Appl. Anal., 1991.

[4] Z. Li, Z. Yang and Q. Ye, *Some persistence and stability result for a competitor-competitor-mutualist model*, Universitatis Pekinensis, 24(1)(1988), 1–12.

Proceedings of the Asian Mathematical
Conference, 1990

On the Geometry of Infinite Dimensional Teichmüller Spaces*

ZHONG LI
Department of Mathematics
Peking University
Beijing 100871, China

§1. Introduction

It is known that a finite dimensional Teichmüller space is a straight geodesic space in the sense of Buseman (see S.Kravetz [1]),e.i., a locally compact metric space in which there is always an unique geodesic joining any two points and such a geodesic can be extended as a straight line (an isometric embedding of $\mathbb{R}$). However, in the infinite dimensional Teichmüller space, the situation is more complicate. In fact, we do not know whether or not the geodesic joining two points in an infinite dimensional Teichmüller space is unique. For instance, let P be a point in the universal Teichmüller which is different from the original. Then P can be represented an equivalence class of a Betrami differential μ. If there are two extremal Betrami differentials μ_1 and μ_2 in $[\mu]$ (the first example of such a point was given by K.Strebel [2]), then $t\mu_1$ and $t\mu_2$ $(0 \leq t \leq 1)$ represent the geodesics joining the original to P. But we do not know whether or not the path $t\mu_1$ is the same as $t\mu_2$. This is an open problem for a period of time (see, for example, F.P. Gardiner [3]).

The purpose of this paper is to give a negative answer to this problem. Two extremal Beltrami differentials μ_1 and μ_2 with the following properties will be constructed in §3: μ_1 and μ_2 represent a same point P in the universal Teichmüller space, while the paths $t\mu_1$ and $t\mu_2$ $(0 \leq t \leq 1)$ represent different geodesics joining the original to the point P. Moreover, it will be shown that one can get infinitely many geodesics joining two points in such a way (§4).

By these results, one can conclude that the universal Teichmüller space is not a straight geodesic space.

In the last paragraph (§5), a sufficient condition that the geodesic joining two given points is unique is given.

Remark. most of the proofs of the lemmas and theorems in this paper will be omitted. They will be published some where else.

*Work done with support from China NSF grant.

§2. Notations and Lemmas

Let S be a strip in the plane:

$$S = \{z = x + iy : -\infty < x < \infty, 0 < y < 1\}.$$

Denote by $\mathcal{F} = \{f\}$ the set of all quasiconformal mappings of S onto itself which keep the boundary points $0, i$ and ∞ fixed. The Beltrami differential (or complex dilatation) of f is denoted by μ_f. Let M be the set of all Beltrami differentials of $f \in \mathcal{F}$. Two Beltrami differentials μ_1 and $\mu_2 \in M$ are called to be equivalent,if and only if $f_1 |_{\partial S} = f_2 |_{\partial S}$ where $f_j \in \mathcal{F}$ such that $\mu_{f_j} = \mu_j, j = 1, 2$.

With these notations, the universal Teichmüller space of S is defined as the set of all equivalence classes of Beltrami differentials of M. The equivalence class of μ is denoted by $[\mu]$.

A Beltrami differential μ is called to be extremal, if and only if

$$\| \mu \|_\infty \leq \| \mu' \|_\infty \quad \text{for all} \quad \mu' \in [\mu].$$

It is known that if μ is extremal then $t\mu$ is extremal $(0 \leq t < 1/ \| \mu \|_\infty)$.

Througout this paper, K is real number larger than e^π (The constant e^π is not essential. We just need K is sufficiently large).

As usual, we use the constant k determined by K:

$$k = \frac{K-1}{K+1}.$$

The following boundary correspondence of S will play an important role in what follows:

$$h_K : \begin{cases} x \mapsto Kx, \quad x + i \mapsto Kx + i, & \text{as } x \leq 0; \\ x \mapsto x/K, \quad x + i \mapsto x/K + i, & \text{as } x > 0. \end{cases}$$

Obviously, h_K allows a quasiconformal extension of S onto itself, for instance, the following mapping:

$$f_K : \begin{cases} x + iy \mapsto Kx + iy, & \text{as } x \leq 0; \\ x + iy \mapsto x/K + iy, & \text{as } x > 0. \end{cases}$$

Thus h_K determines a point of the universal Teichmüller space of S.

Lemma 1. *Let f be any quasiconformal mapping of S onto itself with the boundary correspondence h_K and the Beltrami differential of f has the estimate*

$$| \mu_f(z) | \leq k, \quad \text{for almost } z \in S.$$

Then f is an extremal mapping, namely, μ_f is extremal.

The proof of this lemma is based on some estimates of the Grötzsch function ([4],[5]).

Let us look at the rectangles

$$R = \{z = x + iy : \ -1 < x < 1, 0 < y < 1\}$$

and

$$R_K = \{z = x + iy : -K < x < \frac{1}{K}, 0 < y < 1\}.$$

A boundary homeomorphism of ∂R onto ∂R_K is given as follows:

$$\omega_K : \begin{cases} \omega_K(x) = h_K(x),\ \omega_K(x+i) = h_K(x+i), & \text{as } -1 \le x \le 1; \\ \omega_K(-1+iy) = -K+iy, \omega_K(1+iy) = \frac{1}{K}+iy, & \text{as } 0 \le y \le 1. \end{cases}$$

Obviously, ω_K allows a quasiconformal extension into R.

Lemma 2. *There is a Teichmüller mapping of R onto R_K with the given boundary correspondence ω_K, The maximal dilatation of which is smaller than K.*

The proof of the lemma is related to K.Strebel [6], R.Fehlmann [7], Zhong Li [8] and E.Reich and K.Strebel [9].

Lemma 3. *Let μ_0 and μ_1 be two Beltrami differentials of M such that*

$$[t\mu_0] = [t\mu_1], \quad \text{for all} \quad t \in [0,1].$$

Then we have

$$[t\mu_0] = [t\mu_1], \quad \text{for all} \quad t \in [o, 1/\tilde{k}),$$

where $\tilde{k} = \max\{\| \mu_0 \|_\infty, \|\mu_1\|_\infty\}$.

§3. Nonuniqueness of Geodesics

From Lemma 1, the mapping f_K defined in the previous section is an extremal mapping for boundary correspondence h_K. Now we construct another mapping for h_K.

Let g_K^* be the Teichmüller mapping in Lemma 2. We define g_K as follows:

$$g_K(z) = \begin{cases} g_K^*(z), & \text{as } z \in R; \\ f_K(z), & \text{as } z \in S - R. \end{cases}$$

By Lemma 1, g_K is also extremal.

The Beltrami differentials of f_K and g_K will be denoted by μ_0 and μ_1 respectively. We claim that

$$[t\mu_0] \ne [t\mu_1], \quad \text{for some} \quad t \in (0,1).$$

We now suppose this is not true, namely,

$$[t\mu_0] = [t\mu_1], \quad \text{for all } t \in [0,1].$$

A contradiction will be deduced.

By Lemma 3, we know that

$$[t\mu_0] = [t\mu_1], \quad \text{for all } t \in [0, 1/k)$$

(in our case, $\tilde{k} = k$).

Let f_t and g_t be the quasiconformal mappings in $\mathcal{F}$ with the complex dilatation $t\mu_0$ and $t\mu_1$, respectively. With these natations, $f_1 = f_K$ and $g_1 = g_K$.

One can easily get the explicit expression of f_t. In fact, the following mapping has the complex dilatation $t\mu_0$ and keeps $0, i$ and ∞ fixed:

$$\begin{cases} x+iy \mapsto K_t x + iy, & \text{as } x \le 0; \\ x+iy \mapsto x/K_t + iy, & \text{as } x > 0, \end{cases}$$

where $K_t = (1+tk)/(1-tk)$. Therefore this mapping is f_t.

Let us consider the mapping

$$W_t(z) = g_t \circ f_t^{-1}(z).$$

It is easily seen that $W_t(z)$ is conformal in $S - R_t$, where R_t is the rectangle

$$\left\{z = x + iy :\ -K_t < x < \frac{1}{K_t}, 0 < y < 1\right\}.$$

Moreover, we see

$$f_t(\xi) = g_t(\xi), \quad \text{for } \xi \in \partial S, \quad t \in [0, 1/k),$$

and hence $W_t(\xi)$ is an identical correspondence on ∂S. Consequently $W_t(z)$ is an identical mapping in $S - R_t$ for every $t \in [0, 1/k)$. This means

$$f_t(z) = g_t(z), \quad \text{for } z \in S - R_t, \quad t \in [0, 1/k).$$

On the other hand, the expression of f_t is known. From the expression of f_t, one can conclude that $g_t|_R$ is a quasiconformal mappings of R onto R_t. But the maximal dilatation of $g_t|_R$ is $(1+tl)/(1-tl)$, where l is the constant smaller than K. Thus we get

$$\begin{aligned}\frac{1+tl}{1-tl} &\geq \mathrm{Mod}(R_t)/\mathrm{Mod}(R)\\ &= \frac{1}{2}(K_t + K_t^{-1})\\ &= \frac{1+t^2k^2}{1-t^2k^2}, \quad t \in [0, 1/k).\end{aligned}$$

As t tends to $1/k$, the right hand side of the above inequality tends to $+\infty$, but the left hand side tends to $(k+l)/(k-l)$ (recall $l < k$). This is a contradiction.

The following theorem is proved:

Theorem 1. *There are two extremal Beltrami differentials μ_0 and μ_1 such that they represent a same point in the universal Teichmüller space, while $t\mu_0$ and $t\mu_1 (0 \leq t \leq 1)$ represent different geodesics.*

§4. Infinitely Many Geodesics

A natural questions is to ask how many geodesics joining two given points one can get in such a way. To answer this question, we have the following theorem:

Theorem 2. *There is a set of Beltrami differentials,*

$$\{\mu_\alpha : \alpha \in (0, +\infty)\},$$

with the following properties:

(i) each μ_α is extremal;

(ii) all of μ_α represent a same point in the universal Teichmüller space;

(iii) each geodesic determined by μ_α is different from all of the others.

The proof of this theorem is similar to that of theorem 1, but more complicated.

§5. Uniqueness of Geodesics

We have seen that for some special pair of points in the universal Teichmüller space, the geodesic joining them is not unique. A natural question is to ask when the geodesic joining two points is unique. The following theorem gives a sufficient condition.

Theorem 3. *Suppose $h : \partial S \to \partial S$ is a boundary homeomorphism for which there is an unique extremal mapping f and $|\mu_f|$ is constant. Then the geodesic joining $[0]$ to $[\mu_f]$ is unique.*

The following question is still open:

Does the uniqueness of the extremal Beltrami differential of $[\mu]$ always imply the uniqueness of the geodesic joining $[0]$ to $[\mu]$?

References

[1] S.Kravetz, *On the geometry of Teichmüller spaces and the structure of their modular groups*, Ann. Acad. Sci. Fenn., 278(1959), 1–35.

[2] K.Strebel, *Zur Frage der Eindeutigkeit extremaler quasikonforme Abbildungen des Einheitskreises*, Comment. Math. Helv., 36(1962), 306–323.

[3] F.P.Gardiner, *On completing triangles in infinite dimensional Teichmüller spaces*, Complex Variables, Vol. 10(1988), 237–247.

[4] O.Lehto and K.Virtanen, *Quasiconformal Mappings in the Plane*, Springer-Verlag, Berlin and New York, 1973.

[5] Zhong Li, *Quasiconformal Mappings and its Applications to the Theory of Riemann Surfaces*, Science Press, Beijing, 1988.

[6] K.Strebel, *On the existence of extremal Teichmüller mappings*, Journal d'analyse Math., 30(1976), 464–480.

[7] R.Fehlmann, *Ueber extremal quasikonforme Abbildungen*, Comment. Math. Helv. 56, Fasc., 4(1981), 558–580.

[8] Zhong Li, *On the existence of extremal Teichmüller mappings*, Comment. Math. Helvetici, 57(1982), 511–517.

[9] E.Reich and K.Strebel, *Extremal quasiconformal mappings with given boundary values*, in *Contributions to Analysis*, Academic Press, 1974, 375–391.

Proceedings of the Asian Mathematical
Conference, 1990

Approximation-extension Properties of Functions

FON-CHE LIU
Institute of Mathematics Academia Sinica, Taipei
and Department of Mathematics, Taiwan University
Taipei, China

§1. Introduction

Let $n \geq 1$ be an integer. We denote by μ the Lebesgue outer measure in $\mathbf{R}^n$. By a function we shall always understand an extended real valued function defined on a Lebesgue measurable subset of $\mathbf{R}^n$ and is finite almost everywhere. A classical theorem of N. Luzin states that a function f defined on D is measurable if and only if, given $\varepsilon > 0$, there is a closed subset F of D such that $\mu(D\backslash F) < \varepsilon$ and the restriction of f to F is continuous. This is an approximation-extension type characterization of measurable functions. Another characterization of measurable functions is in terms of approximate continuity. a function f with domain D is said to be approximately continuous at $x \in D$ if there is a measurable subset S of D with density one at x such that the restriction of f to $S \cup \{x\}$ is continuous at x. It is a consequence of Luzin's Theorem and Density Theorem that a function is measurable if and only if it is approximately continuous almost everywhere. Approximate continuity is a kind of weak form of smoothness; our concern in this paper is with approximation extension properties of Luzin type for functions which satisfy weak forms of smoothness stronger than approximate continuity. Regarding approximate continuity we note that f is approximately continuous at x if and only if $\text{aplim}_{y \to x} f(y) = f(x)$, where aplim stands for approximate limit and is defined as follows : $\text{aplim}_{y \to x} f(y) = l$ if and only if

$$\lim_{r \to 0} \frac{\mu[B(x,r) \cap \{|f - l| > \varepsilon\}]}{\mu[B(x,r)]} = 0,$$

for every $\varepsilon > 0$, where $B(x,r)$ is the ball in $\mathbf{R}^n$ with center x and radius r and for a function h with domain D, $\{h \text{ satisfies property } P\}$ stands for $\{y \in D : h(y) \text{ satisfies property } P\}$.

§2. A Theorem of H. Whitney

A function f is said to be approximately totally differentiable at a point x of its domain D of definition if there is $u \in \mathbf{R}^n$ such that $\text{aplim}_{y \to x} \frac{|f(y) - f(x) - u \cdot (y - x)|}{|y - x|} = 0$. The following theorem of Whitney [12] relates functions which are approximately totally differentiable almost everywhere to continuously differentiable functions in the way Luzin's Theorem relates measurable functions to continuous functions.

(2.1) Theorem. *Let f be a measurable function with domain D. Then f is approximately totally differentiable almost everywhere if and only if for every given* $\varepsilon > 0$ *there exist a closed subset F of D and a continuously differentiable function g on* $\mathbf{R}^n$ *such that* $\mu(D\backslash F) < \varepsilon$ *and f(x)=g(x) for all x in F.*

Some remarks concerning this theorem are now in order. The "only if" part of the theorem was first observed by H. Federer [4, p.442] for f being totally differentiable almost everywhere. One may obtain quite easily from Theorem (2.1) a classical theorem of H. Rademacher [10] which asserts that a Lipschitz function is totally differentiable almost everywhere. For this the reader is refered to Corollary (2.15) bellow. Our final remark concerns approximately α-Lipschitz continuous functions, $0 < \alpha \leq 1$. A measurable function f is called approximately α-Lipschitz continuous if there is $M > 0$ such that for almost every x in its domain of definition the set $\{y : |f(y) - f(x)|/|y - x|^\alpha \leq M\}$ has density 1 at x. It is clear from the proof of Theorem (2.1) in [12] that the following theorem holds:

(2.2) Theorem. *A measurable function f with domain of definition D is approximately* α*-Lipschitz continuous if and only if for every given* $\varepsilon > 0$ *there is a closed subset F of D and an* α*-Lipschitz continuous function g on* $\mathbf{R}^n$ *such that* $\mu(D\backslash F) < \varepsilon$ and $f(x) = g(x)$ *for all x in F.*

We now consider generalization of Theorem (2.1) to the situation of higher degree of approximate differentiability. Let D be a measurable subset of $\mathbf{R}^n$ and let $k \geq 0$ be an integer. A function f on D is said to be in the class $\mathrm{ap}^k(x)$ with $x \in D$ if and only if there is a polynomial $P(y)$ of degree at most k such that $\mathrm{aplim}_{y \to x} \frac{|f(y) - P(y)|}{|y - x|^k} = 0$. If x is a point of density of $D, P(y)$ is uniquely determined; in this case, we write

$$P(y) = \sum_{|\alpha| \leq k} \frac{f_\alpha(x)}{\alpha!} \cdot (y - x)^\alpha,$$

where α denotes a lattice point $(\alpha_1, \cdots, \alpha_n)$ with nonnegative components, $|\alpha| = \sum_{i=1}^n \alpha_i, \alpha! = \prod_{i=1}^n \alpha_i!$, and for $z = (z_1, \cdots, z_n), z^\alpha = \prod_{i=1}^n z_i^{\alpha_i}$. Thus each f_α is defined almost everywhere on D if $f \in \mathrm{ap}^k(x)$ for almost everywhere $x \in D$; if furthermore each f_α is measurable and is of class $\mathrm{ap}^{k-|\alpha|}(x)$ for almost every x, then we write $f \in \mathrm{ap}^k(D)$. We now prove the following generalization of Theorem (2.1):

(2.3) Theorem. $f \in \mathrm{ap}^k(D)$ *if and only if for every given* $\varepsilon > 0$ *there exist a closed subset F of D and a* C^k*-function g on* $\mathbf{R}^n$ *such that* $\mu(D\backslash F) < \varepsilon$ *and f(x)=g(x) for all x in F.*

Proof. We may assume that D is bounded. Suppose that f is in $\mathrm{ap}^k(D)$. For each α with $|\alpha| \leq k$. Let

$$P_\alpha(y; x) = \sum_{\alpha+\beta| \leq k} \frac{f_{\alpha|+\beta}(x)}{\beta!} \cdot (y - x)^\beta, \quad x \in D, \quad y \in \mathbf{R}^n.$$

Then it is clear that for almost all x in D the following hold:

$$\text{ap}\lim_{y\to x}\frac{|f_\alpha(y)-P_\alpha(y;x)|}{|y-x|^{k-|\alpha|}}=0. \tag{2.4}$$

Let now $\varepsilon>0$ be given. By Egoroff theorem for continuous parameter [5, (10.2.64), p.124], there is a compact subset S of D with $\mu(D\backslash S)<\frac{\varepsilon}{2}$ such that

$$\lim_{r\to 0}\frac{\mu[B(x,r)\cap W^\alpha(x;j)]}{\mu[B(x,r)]}=1 \quad \text{uniformly for} \quad x\in S, \tag{2.5}$$

where $W^\alpha(x;j)$ is the set of all those points y such that $\frac{|f_\alpha(y)-P_\alpha(y;x)|}{|y-x|^{k-|\alpha|}}\le\frac{1}{j}, |\alpha|\le k, j=1,2,3,\cdots$. Now choose a closed subset F of S with $\mu(S\backslash F)<\varepsilon/2$ such that

$$\lim_{r\to 0}\frac{\mu[S\cap B(x,r)]}{\mu[B(x,r)]}=1 \quad \text{uniformly for} \quad x\in F. \tag{2.6}$$

Let $\sigma>0$ be given, choose and fix j with $1/j\le\sigma$. From (2.5), (2.6) we can find $\delta>0$ such that $W^\alpha(x;j)\cap B(x,r)\cap W^\alpha(y;j)\cap B(y,r)\cap S\neq\emptyset$ whenever x and y are in F and $r=|x-y|\le\delta$. Hence for any x,y in F with $|x-y|\le\delta$ there is z in S with $|z-x|,|z-y|\le\delta$ such that

$$\frac{|f_\alpha(z)-P_\alpha(z;t)|}{|z-t|^{k-|\alpha|}}\le\frac{1}{j}, \quad \text{for} \quad t=x,y \quad \text{and} \quad |\alpha|\le k.$$

From this, since $P_\alpha(z;x)-P_\alpha(z;y)=P_\alpha(z;x)-f_\alpha(z)+f_\alpha(z)-P_\alpha(z;y)$ and $|x-z|,|y-z|\le|x-y|$, we have

$$\frac{|P_\alpha(z;x)-P_\alpha(z;y)|}{|x-y|^{k-|\alpha|}}\le\frac{2}{j}, \quad |\alpha|\le k. \tag{2.7}$$

If we expand the polynomial $P_\alpha(t;x)-P_\alpha(t;y)$ in t at z:

$$P_\alpha(t;x)-P_\alpha(t;y)=\sum_{|\alpha+\beta|\le k}[P_{\alpha+\beta}(z;x)-P_{\alpha+\beta}(z;y)]\cdot(t-z)^\beta,$$

it follows from (2.7) by letting $t=y$ that

$$|f_\alpha(y)-P_\alpha(y;x)|\le\frac{2c}{j}\le 2c\sigma, \tag{2.8}$$

where c is the number of all those α with $|\alpha|\le k$. Since (2.8) holds for arbitrary x and y in F satisfying $|x-y|\le\delta, f$ is a C^k-function on F in the sense of Whitney [11] and hence by a theorem of Whitney in [11] f can be extended from F to be a C^k-function on $\mathbf{R}^n$ this proves one direction of the theorem. The other direction is obvious. □

We point out here that the space $\text{ap}^k(D)$ is introduced first in [6] in a slightly different way and that Theorem (2.3) is also proved there; but the proof given here is much more transparent.

In connection with the space $\text{ap}^k(D)$ we introduce now the classes of functions that have been considered in [3] and have their origin in [8] when $n=1$ in questions concerning convergence of Fourier series. For $x \in G$, where G is an open subset of $\mathbf{R}^n$, let $t^k(x)$ be the space of all those functions f integrable on compact neighborhoods of x in G such that there is a polynomial $P(y)$ of degree at most k with the properties:

$$\sup \frac{1}{r^k} \fint_{B(x,r)} |f(y) - P(y)|dy = M(f,x) < +\infty; \tag{2.9}$$

$$\lim_{r\to 0} \frac{1}{r^k} \fint_{B(x,r)} |f(y) - P(y)|dy = 0, \tag{2.10}$$

where the supremum is taken over all $r > 0$ such that $B(x,r) \subset\subset G$, and $\fint_A h(y)dy$ stands for $\frac{1}{\mu(A)} \int_A h(y)dy$. Since $P(y)$ is uniquely determined we write as before

$$P(y) = \sum_{|\alpha|\le k} \frac{f_\alpha(x)}{\alpha!} \cdot (y-x)^\alpha,$$

and denote henceforth $P(y)$ by $P_x(y)$ to indicate the dependence of $P(y)$ on x. If $f \in t^k(x)$ for almost every $x \in G$, then $f_\alpha, |\alpha| \le k$, are measurable on G. To see this we consider some preliminaries. Let φ be a nonnegative spherically symmetric C^∞-function on $\mathbf{R}^n$ with support contained in the unit ball and with $\int \varphi = 1$. for $\varepsilon > 0$, let

$$f_\varepsilon(x) = \varepsilon^{-n} \int \varphi(\frac{x-y}{\varepsilon}) f(y)dy, \tag{2.11}$$

if $B(x,\varepsilon) \subset\subset G$. Then it is well-known that f_ε is smooth and

$$D^\alpha f_\varepsilon(x) = (D^\alpha f)_\varepsilon(x), \tag{2.12}$$

if the distributional derivative $D^\alpha f$ of f is locally integrable, where as usual

$$D^\alpha = \frac{\partial^{|\alpha|}}{\partial x_1^{\alpha_1} \cdots \partial x_n^{\alpha_n}}.$$

Any such a function φ will be called a mollifying function and fixing a mollifying function φ, we will call the function f_ε defined by (2.11) the mollified function of f. Now let $f \in t^k(x)$ for almost every x, by a known result of Calderon and Zygmund [3, Lemma 2.6] there is a mollifying function φ such that the mollified function P_ε of any polynomial P of degree $\le k$ is P itself. Fixing such a mollifying function φ, writing $f(y) - P_x(y) = R_x(y)$, and letting

$$I = \varepsilon^{-n-|\alpha|} \int (D^\alpha \varphi)(\frac{x-y}{\varepsilon}) R_x(y)dy,$$

we have

$$\begin{aligned} I &= \varepsilon^{-n-|\alpha|} \int (D^\alpha \varphi)(\frac{x-y}{\varepsilon})[f(y) - P_x(y)]dy \\ &= (D^\alpha f_\varepsilon)(x) - D^\alpha (P_x)_\varepsilon(x) \\ &= (D^\alpha f_\varepsilon)(x) - D^\alpha P_x(x) = (D^\alpha f_\varepsilon)(x) - f_\alpha(x) \end{aligned}$$

from which we obtain

$$\lim_{\varepsilon\to 0} D^\alpha f_\varepsilon(x) = f_\alpha(x)$$

for almost every x, because the integral I is dominated by

$$C\varepsilon^{K-|\alpha|}\frac{1}{\varepsilon^k}\int_{B(x,\varepsilon)}|R_x(y)|dy$$

which tends to 0 for almost every x as $\varepsilon \to 0$ according to (2.10). Since the f_α's are limits of smooth functions, they are measurable. Hence $M(f,x)$ is also measurable in x.

We denote by $t^k(G)$ the space of all those functions which belong to $t^k(x)$ for almost every x in G.

(2.13) Proposition. $t^k(G) \subset \mathrm{ap}^k(G)$.

Proof. Let $f \in t^k(G)$. For $N > 0$, let S be the set of all those x for which $M(f,x) \le N$. Choose a mollifying φ such that $P_\varepsilon = P$ for every polynomial P of degree $\le k$. For x, z in S, we write

$$f(y) = P_x(y) + R_x(y)$$

and

$$f(y) = P_z(y) + R_z(y);$$

and let $\varepsilon = |x - z|$. Following [3], consider

$$I = \varepsilon^{-n-|\alpha|}\int (D^\alpha\varphi)(\frac{z-y}{\varepsilon})f(y)dy.$$

Using the two expressions above for $f(y)$, we get

$$\begin{aligned} I &= \sum_{|\alpha|\le k}\frac{1}{\alpha!}f_\alpha(x)D^\alpha(z-x)^\alpha + \varepsilon^{-n-|\alpha|}\int (D^\alpha\varepsilon)(\frac{z-y}{\varepsilon})R_x(y)dy \\ &= \sum_{|\gamma|\le k-|\alpha|}\frac{1}{\gamma!}f_{\alpha+\gamma}(x)(z-x)^\gamma + \varepsilon^{-n-|\alpha|}\int (D^\alpha\varphi)(\frac{z-y}{\varepsilon})R_x(y)dy \\ &= f_\alpha(z) + \varepsilon^{-n-|\alpha|}\textstyle\int (D^\alpha\varphi)(\frac{z-y}{\varepsilon})R_z(y)dy. \end{aligned}$$

Consequently,

$$\begin{aligned} f_\alpha(z) = &\sum_{|\gamma|\le k-|\alpha|}\frac{1}{\gamma!}f_{\alpha+\gamma}(x)(z-x)^\gamma \\ &+\varepsilon^{-n-|\alpha|}\int (D^\alpha\varphi)\left(\frac{z-y}{\varepsilon}\right)[R_z(y) - R_x(y)]dy. \end{aligned}$$

Since the last integral is dominated by

$$\begin{aligned} &C\varepsilon^{-n-|\alpha|}[\int_{B(x,2\varepsilon)}|R_x(y)|dy + \int_{B(z,\varepsilon)}|R_z(y)|dy] \\ \le\ & C[M(f,x) + M(f,z)]\cdot\varepsilon^{k-|\alpha|} = 2CN\varepsilon^{k-|\alpha|}, \end{aligned}$$

we see right away that if x is a point of density of S, then $f_\alpha \in ap^k(x), |\alpha| \le k$. This is sufficient to conclude the proof. □

For an application, we consider the Sobolev space $W^k_{loc}(G)$ of equivalent classes of functions which are locally integrable in G together with all its distributional partial derivatives of order $\leq k$ with G being an open subset of $\mathbf{R}^n$.

(2.14) Corollary. *If $f \in W^k_{loc}(G)$, then for every given $\varepsilon > 0$ there exist a closed subset F of G and a C^k-function g on $\mathbf{R}^n$ such that $\mu(G\backslash F) < \varepsilon$ and $f(x) = g(x)$ for $x \in F$.*

Proof. If we show that $W^k_{loc}(G) \subset t^k(G)$, then the corollary follows from Proposition (2.13) and Theorem (2.3). For this purpose consider first f in $C^k(G)$ and let $F(t) = f(x + (y - x)t)$, where $x \in G, y \in B(x,r) \subset G$ are fixed. Then

$$f(y) = F(1) = \sum_{j=0}^{k} \frac{1}{j!} F^{(j)}(0) + \frac{1}{(k-1)!} \int_0^1 [F^{(k)}(t) - F^{(k)}(0)](1-t)^{k-1} dt,$$

where $F^{(j)}$ denotes the j-th derivative of F. Since

$$F^{(j)}(t) = \sum_{|\alpha|=j} \frac{j!}{\alpha!} f_\alpha(x + (y-x)t)(y-x)^\alpha,$$

where $D^\alpha f$ is denoted by f_α for notational simplicity, it follows that

$$f(y) = P_x(y) = k \cdot \int_0^1 [\sum_{|\alpha|=k} \frac{1}{\alpha!} \{f_\alpha(x + (y-x)t) - f_\alpha(x)(y-x)^\alpha](1-t)^{k-1} dt,$$

and hence on integrating w.r.t. y over $B(x,r)$, we have

$$\begin{aligned} \int_{B(x,r)} |f(y) - P_x(y)| dy \leq \ & kr^k \int (1-t)^{k-1} \sum_{|\alpha|=k} \int_{B(x,r)} |f_\alpha(x + (y-x)t) - f_\alpha(x)| dy dt \\ = \ & kr^k \int_0^1 (1-t)^{k-1} dt \cdot \sum_{|\alpha|=k} \int_0^{rt} \fint_\Omega |f_\alpha(x + \rho\theta) - f_\alpha(x)| \frac{\rho^{k-1}}{t^k} d\theta d\rho \\ = \ & kr^k \mu[B(x,r)] \int_0^1 (1-t)^{k-1} [\sum_{|\alpha|=k} \int_{B(x,rt)} |f_\alpha(y) - f_\alpha(x)| dy] dt, \end{aligned}$$

or

$$\frac{1}{r^k} \fint |f(y) - P_x(y)| dy \leq k \sup_{0 \leq t \leq 1} \sum_{|\alpha|=k} \fint_{B(x,rt)} |f_\alpha(y) - f_\alpha(x)| dy,$$

which also holds for almost every x if $f \in W^k_{loc}(G)$ and $B(x,r) \subset\subset G$. This shows that $f \in t^k(x)$ if x is a Lebesgue point of all $f_\alpha, |\alpha| \leq k$, and if the maximal functions of all f_α are finite at x. □

(2.15) Corollary (Rademacher). *A Lipschitz function is totally differentiable almost everywhere.*

Proof. Let f be Lipschitz on D. We may assume that D is bounded. Given $\varepsilon > 0$, by Corollary (2.14) there exist a compact set $F \subset D$ and a continuously differentiable function g on $\mathbf{R}^n$ such that $\mu(D\backslash F) < \varepsilon$ and $f = g$ on F. Now let $x \in F$ be a point of density of F.

For $y \in D$, choose a point $z(y)$ in F with $|y - z(y)|$ being the distance between y and F. Then

$$\lim_{y \to x} \frac{|y - z(y)|}{|y - x|} = 0.$$

We have then

$$\begin{aligned} & |f(y) - f(x) - \nabla g(x) \cdot (y - x)| \\ \leq \; & |f(y) - f[z(y)]| + |g[z(y)] - g(x) - \nabla g(x) \cdot [z(y) - x]| + M \cdot |y - z(y)| \\ \leq \; & |g[z(y)] - g(x) - \nabla g(x) \cdot [z(y) - x]| + 2M \cdot |y - z(y)|, \end{aligned}$$

where ∇g is the gradient of g and M is the Lipschitz norm of f. It follows from the preceding inequalities that f is totally differentiable at x. Since almost all points of F are points of density of F and since $\varepsilon > 0$ is arbitrary, the proof is complete. □

§3. Approximation-extension Property of Sobolev Functions

The functions we have considered so far satisfy properties which are of local character. We consider now functions which also satisfy global conditions, e.g., integrability conditions. In this respect the Sobolev space $W_p^k(G)$ of all those functions on G which are p-integrable together with all their distributional partial derivatives of order $\leq k$, seems to be the natural sequel to the spaces $ap^k(K)$. In [7] we prove the following theorem:

(3.1) Theorem. *Let G be a strongly Lipschitz domain in $\mathbf{R}^n$, $1 \leq p < +\infty$. A measurable function f on G is in $W_p^k(G)$ if and only if there is $M > 0$ such that for every given $\varepsilon > 0$ there exist a compact subset K of G and a C^k-function on G such that $\mu(G \backslash K) < \varepsilon$, $|g|_{k,p} \leq M$, $|g|_{k,p;G\backslash F} < \varepsilon$ and $f(x) = g(x)$ for all $x \in K$, where $|\cdot|_{k,p}$ and $|\cdot|_{k,p;G\backslash K}$ are the Sobolev norms on $W_p^k(G)$ and $W_p^k(G\backslash K)$ respectively.*

For a set D in $\mathbf{R}^n$ we denote by $C_b^k(D)$ the space of all those functions f in $ap^k(D)$ such that $|f_\alpha(x)| \leq M, x \in D, |\alpha| \leq k$ for some $M > 0$ and such that for any given $\varepsilon > 0$ there is $\delta > 0$ so that whenever x, y in D with $|x - y| < \delta$ we have

$$|f_\alpha(y) - P_\alpha(y; x)| \leq \varepsilon \cdot |y - x|^{k-|\alpha|}, |\alpha| \leq k.$$

For f in $C_b^k(D)$ the smallest $M > 0$ for which $|f_\alpha(x)| \leq M$ for $x \in D$ and $|\alpha| \leq k$ is called C_b^k-norm of f and is denoted by $|f|_k$. The functions in $C_b^k(D)$ are the C^k-functions on D as defined in [11]. Behind Theorem (3.1) is the following characterization of functions in $W_p^k(G)$:

(3.2) Theorem. *Let G be a strongly Lipschitz domain. There is $C > 0$ depending only on n and G such that a measurable function f on G is in $W_p^k(G), 1 \leq p < +\infty$, if and only if there exists $N > 0$ such that whenever $\varepsilon > 0$ is given, there is $\lambda_0 > 0$ with the property that whenever $\lambda \geq \lambda_0$ there is a C_b^k-function g on G with the following properties:*

$$|g|_{k,p} \leq N; \tag{3.3}$$

(3.4) $$|g|_k \leq \lambda;$$

and

(3.5) $$\mu\{f \neq g\} \leq \frac{CN}{\lambda^p} \cdot \varepsilon.$$

Since the proof of Theorem (3.2) essentially follows that of Theorem 3 in [7], we omit its proof. It follows easily from Theorem (3.2) the following corollary:

(3.6) Corollary. *Let G be a strongly Lipschitz domain. There is $C > 0$ such that for $f \in W_p^k(G)$ and $\lambda > 0$, there is a compact set F contained in G so that $f \in C_b^k(f)$ with $|f|_{k;F} \leq \lambda$ and $\lambda^p \cdot \mu(G \backslash F) \leq C \cdot |f|_{k,p}$.*

We remark here that in the case $k = 1$, Corollary (3.6) proves to be useful in questions of semicontinuity and regularity in Calculus of Variations for multiple integrals [1], [2].

Finally we point out that Theorem (3.1) has been extended to the case of arbitrary open domain G with the Lebesgue measure μ replaced by appropriate capacities by Michael and Ziemer when the approximating g ins C^1 with $1 < k$ [9] (see also [13, Chapter 3]).

References

[1] E. Acerbi and N. Fusco, *Semicontinuity problems in the calculus of variations*, Arch. Rational Mech. Anal., 86(1984), 125–145.

[2] E. Acerbi and N. Fusco, *A regularity theorem for minimizers of quasiconvex integrals*, Arch. Rational Mech. Anal., 99(1987), 261–281.

[3] A.P. Calderon and A. Zygmund, *Local properties of solutions of elliptic partial differential equations*, Studia Math., 20(1961), 171–225.

[4] H. Federer, *Surface area II*, Trans. Amer. Math. Soc., 55(1944), 438–456.

[5] H. Hahn and A. Rosenthal, *Set functions*, The University of New Mexico Press, Albuquerque, New Mexico, 1947.

[6] F.C. Liu, *On a theorem of Whitney*, Bull. Inst. Math. Acad. Sinica, 1(1973), 63–70.

[7] F.C. Liu, *A Luzin type property of Sobolev functions*, Indiana Univ. J. Math., 26(1977), 645–651.

[8] J. Marcinkiewicz, *Sur la sommabilite forte des series de Fourier*, J. London Math. Soc., 14(1939), 162–168.

[9] J. Michael and W.P. Ziemer, *A Lusin type approximation of Sobolev functions by smooth functions*, Contemporary Mathematics, AMS, 42(1985), 135–167.

[10] H. Rademacher, *Uber partielle und totale Differenzierbarkeit I*, Math. Ann., 79(1919), 340–359.

[11] H. Whitney, *Analytic extensions of differentiable functions defined in closed sets*, Trans. Amer. Math. Soc., 36(1934), 63–89.

[12] H. Whitney, *On totally differentiable and smooth functions*, Pacific J. Math., 1(1951), 143–159.

[13] W.P. Ziemer, *Weakly differentiable functions*, Springer-Verlag, New York, Berlin, Heidelberg, 1989.

Proceedings of the Asian Mathematical
Conference, 1990

On the Dynamics of Resource-consumer-toxicant Systems: Focus on the Reproductive Effort of an Organism

JOSE T. DE LUNA
Department of Mathematics
Mindanao State University
Marawi City, Philippines

§1. Introduction

Recent studies on the determination of effects of toxicants on populations have focused on the individual organism. Hallam et al.([4]) proposed that a proper focal level for eco-toxicology must account for the individual organism. This is because chemical effects occur at the individual level and not at the population level ([5]). Hallam et al. ([4]) further elaborated that the underlying premise is that the "effects of a chemical at the population level result entirely from the effects of the chemical on the individuals that compose the population".

This article presents a dynamic theory of the reproductive effort of an individual organism and the resulting offspring. We are interested firstly, in the quantitative relationship between ingested energy, allocation of ingested energy, maintenance energy, and reproductive energy of an individual organism. Secondly, since "a population must adjust its reproductive effort to variations in its environment"([1]), we are also interested in determining the effects of a toxic substance on the reproductive effort of an exposed organism. Thirdly, we would like to predict reproductive success from knowledge of prevailing environmental conditions.

§2. Model Formulation

We start our modelling effort by considering an organism as an input-output system that converts food into energy which is then utilized in the process of body maintenance , activity, growth, and reproduction.

Let $B = B(t)$ be the reproductive effort (in joules J) of a reproducing organism of size $W(mg)$: $I = I(B,C,W)$, the rate of food intake, C food concentration; $M = M(B,W)$, the rate of energy utilization for maintenance, growth, and activity. The general equation governing the reproductive effort B of an organism is given by

$$dB/dt = -k_1 M(B,W) + bk_2 I(B,C,W). \tag{1}$$

The biological significance of equation (1) is evident from its formulation. Food is ingested at a rate $I(B,C,W)$ and is assimilated at the rate k_2. A portion of assimilated energy is used in reproduction according to the allocation policy $b(0 \le b \le 1$, a pure number). A portion of reproductive energy, however, is utilized for maintenance of reproductive organs and tissues (eggs, fetuses, etc.) and hence is also a part of the organism's maintenance energy. Thus, the first term in equation (1) represents that portion of reproductive energy used in maintenance and activity, which the second term represents the portion of assimilated energy used in reproduction.

The second equation in our dynamical system governs the number of offspring that results from the organism's reproductive effort B. Let $R = R(t)$ be the number of offspring that will survive to reproductive age. Then from life history theory, R will depend on the energy invested per offspring (effort per offspring) and on the juvenile survival rate. Thus, we can write:

$$dR/dt = \frac{B}{E}\{k_3[1 - \exp(-k_4(E - E_m))]\} - \beta R. \tag{2}$$

The biological significance of equation (2) can be easily explained. The organism's reproductive effort B results in a maximum number of offspring $N = B/E$, if each offspring receives the same effort $E(J, 0 \le E \le B)$. The expression inside the braces represents offspring survival rate that depends on the effort E that each offspring receives, k_3 (time^{-1}) is the maximum attainable survival rate, $k_4(J^{-1})$ is a coefficient setting the rate of increase in offspring survival with increasing E ([13]), and $E_m(J)$ is the minimum effort per offspring. The last term in equation (2) represents losses due to natural causes such as overcrowding, predation, etc., with β (time^{-1}) representing the loss rate.

The dynamical system governing both the reproductive effort of an individual organism and the resulting offspring is given by the pair of equations:

$$dB/dt = -k_1 M(B,W) + bk_2 I(B,C,W)$$

$$dR/dt = \frac{B}{E}\{k_3[1 - \exp(-k_4(E - E_m))]\} - \beta R$$

with initial conditions $B(0) \ge 0$ and $R(0) \ge 0$.

§3. The Energy Utilization Index

We define the function

$$g(B,C,W) = \frac{I(B,C,W)}{M(B,W)}. \tag{3}$$

In order to see the biological significance of the function $g(B,C,W)$, it is necessary to specify some mathematical properties of $M(B,W)$ and $I(B,C,W)$. It is firstly reasonable to assume that $I(0,C,W) > 0$ and $M(0,W) > 0$. These inequalities imply that even if the organism is not reproducing, it still needs food for maintenance and activity. Further, we may assume that I is a bounded nondecreasing function of B, while M is an unbounded nondecreasing function of B. The above assumptions are based on biological considerations. For an organism's food intake, even when food is abundant, is limited by its gut capacity, by the time of retention of ingested food in the digestive tract, and by the rate at which digestion takes place. Moreover, a reproducing organism needs additional energy and obtains the required energy by increasing its food consumption. Hence, I must be a bounded, nondecreasing function of B.

On the other hand, reproduction is an activity that demands a lot of energy from the organism and this energy demand increases with B. It has been observed in some instances that energy demand for reproduction can be so high that it can affect the organism's survival ([3], [10], [11]). Thus, if the organism's reproduction effort grows without bound, then its energy for maintenance and activity must likewise increase without bound.

From our hypotheses on I and M, we see that

$$\lim_{B\to\infty} g(B,C,W) = \lim_{B\to\infty} \frac{I(B,C,W)}{M(B,W)} = 0. \tag{4}$$

The limit in (4) implies that if the reproductive effort of an organism increases without bound, then energy for maintenance and activity becomes so large that the return per unit of food intake becomes negligible. Note that if $g(B,C,W) < 1$, then the organism is using much energy for maintenance and activity that it exceeds the rate of energy intake. Since $g(B,C,W) \to \infty$ as $B \to \infty$, then $g(B,C,W) < 1$ for B sufficiently large. Hence, for fixed W, high energy investment in reproduction results in small values of $g(B,C,W)$. But in as much as high energy investment in reproduction does increase mortality ([10], [11]) then such high level of energy investment must subject the organism to some stress. The smaller the value of $g(B,C,W)$, the more stress on the organism. Hence, we can use $g(B,C,W)$ as a measure of stress on an organism resulting from its investment on reproduction. In this context, we will call $g(B,C,W)$ the organism's energy utilization index.

§4. The Threshold Reproductive Effort of an Organism

We now consider the qualitative behavior of $g(B,C,W)$ as a function of the organism's reproductive effort B. We observe from our assumption on I and M that $g(0,C,W) > 0$. Also, since $g(B,C,W) \to \infty$ as $B \to \infty$, then it is reasonable to assume that there is a level of reproductive effort, which we denote by B_M, such that $g(B_M,C,W)$ is maximum, that is B_M is that reproductive effort that is least stressful to the organism. From biological considerations, B_M must be unique.

A typical graph of $g(B,C,W)$ for fixed W is shown in Figure 1. Note that it increases initially until it reaches its maximum at B_M and then decreases asymptotically to zero as $B \to \infty$.

To appreciate the biological significance of B_M, it is helpful to consider the equilibrium solution of system (1)-(2). Let (B_*, R_*) be an equilibrium solution. We see in Figure 1 that a feasible equilibrium exists if $g(B_M,C,W) \leq k_1/bk_2$. It is also apparent from Figure 1 that if $g(B_m,C,W) < k_1/bk_2$, then there are two equilibrium solutions (B_-,R_-) and (B_+,R_+) with $B_- < B_M$ and $B_+ > B_M$. It is easily shown by linearization techniques that (B_-,R_-) is unstable while (B_+,R_+) is a stable equilibrium solution.

Consider now the case where $g(B_M, C, W) = k_1/bk_2$. Here, (B_M, R_M) is the only equilibrium solution. From our previous result, we see that (B_M, R_M) is stable from the right. Since in general, $g(B_*, C, W) < k_1/bk_2$, the stable equilibrium reproductive effort B_+ satisfies the inequality $B_+ > B_M$. Keeping B_+ "close" to B_M would allow for near optimal energy utilization index. Hence the organism if better of if B_+ is equal to B_M. But since $B_+ > B_M$ and (B_M, R_M) is stable only from the right, then B_M is the lower limit of the stable reproductive effort. In this sense, we call B_M the threshold reproductive effort of an organism. Our result suggests that since greater reproductive effort results in higher mortality rates ([10], [11]), then individual organisms should aim at lower reproductive effort which could result in lower mortality rates. Since the organism can then enjoy longer lives, it can "still have a lifetime reproductive success comparable to that of individuals with higher than average reproductive effort but consequently, shorter life span." ([10]).

§5. Effects of a Toxic Chemical on Reproduction

It has been long recognized that chemicals released in the environment may have adverse effects on exposed organisms. In order to incorporate into the model the physiological changes in an exposed organism, it is necessary to modify equations (1) and (2). We will introduce modifications on the above equations separately.

Let $c_0 (mg/g, 0 \le c_0 \le 1)$ be the concentration of the toxic compound in the organism (body burden), which is assumed to be constant. Let $M_2(c_0)$ be the increase in maintenance energy and $MT(B, W, c_0)$ the rate of utilization of energy for maintenance and activity of the exposed organism. A form of $MT(B, W, c_0)$ that will allow us to see toxicant effects on an organism's reproductive effort is given by

$$MT(B, W, c_0) = M(B, W) \cdot M_2(c_0) \tag{5}$$

such that $M_2(0) = 1$ and $dM_2/dc_0 \ge 0$. We can rewrite equation (1) in the form

$$dB/dt = -k_1 MT(B, W, c_0) + k_2 bI(B, C, W). \tag{6}$$

We now incorporate modifications on equation (2) so as to reflect effects of a toxic compound on the offspring. It will be assumed that because of sensitive early life stage ([8]), exposure may result in higher mortality rate. We rewrite equation (2) in the form

$$dR/dt = \frac{B}{E}\{k_3 S(c_0)[1 - \exp(-k_4(E - E_m))]\} - \beta R, \tag{7}$$

where $S(c_0)$ is a function of c_0 such that $S(0) = 1, 0 \le S \le 1$ and $dS/dc_0 \le 0$. Equations (6) and (7) together represent the dynamics of the organism's reproductive effort and its resulting offspring when the organism is exposed to a toxic chemical.

§6. Stress Effects on Reproductive Effort

We define the function $gT(B,C,W,c_0) = I(B,C,W)/MT(B,W,c_0)$ as the organism's stressed energy utilization index. Since $M_2(c_0)$ is a nondecreasing function of c_0 then $M_2(c_0) \geq 1$, so that using equations (3) and (5), we see that $gT(B,C,W,c_0) \leq g(B,C,W)$. Thus, one effect of exposure to a toxic substance is a decrease in the organism's utilization index. Typical graphs of $g(B,C,W)$ and $gT(B,C,W,c_0)$ are shown in Figure 2.

Let B_1 be such that

$$gT(B_*,C,W,c_0) = g(B_1,C,W) \tag{8}$$

where (B_*, R_*) is a feasible equilibrium solution of system (6)–(7). It is apparent from Figure 2 that $B_1 > B_+$. Since B_+ is closer to B_M as compared to B_1, then at first glance it seems that the organism is better off when exposed to a toxic compound. But from equation (8), we see that the stress on the exposed organism with reproductive effort B_+ is the same as the stress on an unexposed organism with reproductive effort B_1. It is also apparent from Figure 2 that $gT(B,C,W,c_0) < g(B,C,W)$ for $B > B_M$. Thus, exposure to a toxic compound may result in a lower reproductive effort that is in fact closer to B_M, but the stress on the exposed organism is the same as that of an unexposed organism with higher reproductive effort.

We define the reproductive index (RI) of an organism by setting

$$RI = \ln(gT(B_M,C,W,c_0)).$$

Since $M_2(c_0)$ is a nondecreasing function of c_0, then $gT(B_M,C,W,c_0)$ is a nonincreasing function of c_0. Let c_0^m be that minimum body burden and reproduction fails. Define RI_m by setting

$$RI_m = \ln(gT(B_m,C,W,c_0^m)).$$

Then RI_m is the largest value of RI that corresponds to reproductive failure. Thus, as long as $RI > RI_m$, the organism can reproduce successfully, perhaps at a lower rate. If $RI < RI_m$, reproduction fails.

§7. Numerical Studies

The ideas in our theoretical model are illustrated using Daphnia as test organism. The organism has been studied quite extensively, hence its bioenergetics is more or less known. Moreover, Daphnia has also been used as test organism in numerous toxicity studies, so that most of the parameter values used in this study have been estimated or are computable. Parameter estimates used in this study are taken from the growth models of Taylor ([12]), Kooijman and Metz ([6]), Paloheimo et al. ([9]), and Gabriel ([2]).

The test compounds used are the metals arsenic and cadmium. The choice of these compounds is based on available information on the test organism's LC_{50}-values (that concentration of a toxic compound calculated to cause the mortality of 50% of a test population) and on the relative sensitivity of Daphnia to these metals. Generally, arsenic is relatively less toxic to Daphnia, requiring in excess of 1 mg/li to cause mortalities in short-duration exposure, whereas cadmuim requires only a 4-day LC_{50} of 0.005 mg/li ([7]). Thus, the test compounds are chosen such that the test organism is more sensitive to one of the compounds.

§8 Simulation Results

Only a few significant results will be mentioned here for lack of space. We first perform simulation runs for different values of $B_M(2.35J, 3.10J, 3.45J)$ when the system is free from toxic chemicals. The results suggest that an individual Daphnia tends to reproduce more rapidly when B_M is small. Moreover, the results also show that when the organism's reproductive effort is large, its energy utilization index is small.

When a toxic compound has invaded the system our simulation results show two possible ways in which an organism may fail to reproduce. For the relatively less toxic chemical arsenic, the organism fails to reproduce when the body burden c_0 is high (8.8mg/g). This is due to the fact that when c_0 is large then $M_2(c_0)$ is also large. Hence from equation (5), the energy utilization for maintenance and activity is high and hence energy for reproduction is low. Moreover, toxicant-induced mortality reduces the already small number of offspring to almost nil. When the organism is sensitive to a toxic chemical, it may fail to reproduce even if its reproductive effort is still relatively large. For cadmium, a toxicant concentration of 0.01 mg/g still allows for substantial amount of reproductive energy, but toxicant-induced motality on the offspring leads to reproductive failure. Further, the organism can successfully reproduce if $RI > 6.76$ for arsenic and $RI > 6.86$ for cadmium.

References

[1] Bronson, F.H., *The Reproductive Ecology of the House Mouse*, The Quarterly Review of Biology, 54(1979), 265–299.

[2] Gabriel, W., *Modelling Reproductive Strategies of Daphnia*, Arch. Hydrobiol., 95(1982), 69–80.

[3] Gadgil, M. and Bossert, W.H., *Life Historical Consequences of Natural Selection*, Amer. Natur., 104(1970), 1–24.

[4] Hallam, T.G., Lassiter, R.R., Li, Jia, and Mckinney, W., *Toxicant-induced Mortality in Models of Daphnia Populations*, Environ. Tox. and Chem., 9(1990), 597–621.

[5] Kooijman, S.A.L.M., Hanstveit, A.D. and van de Hoeven, N., *Research on the Physiological Basis of Population Dynamics in Relation to Ecotoxicology*, Wat. Sci. Tech., 19(1987), 21–37.

[6] Kooijman, S.A.L.M. and Metz, J.A.J., *On the Dynamics of Chemically Stressed Populations: The Deduction of Population Consequences from Effects on Individuals*, Ecotox. and Environ. Safety, 8(1984), 254–274.

[7] Mance, G., *Pollution Threat of Heavy Metals in Aguatic Environments*, Elsevier Applied Science, London and New York, 1987.

[8] Mount, D. R., Ingersoll, C.G., Gulley, D.D., Fernandez, J.D., LaPoint, T.W. and Bergman, H.L., *Effects of Long-term Exposure to Acid, Aluminum, and Low Calcium on Adule*

Brook Trout (Salvelinus fontinalis). 1. Survival, Growth, Fecundity, and Progeny Survival, Can. J. Fish. Aquat. Sci., 45(1988), 1623–1632.

[9] Paloheimo, J.E., Crabtree, S.J., and Taylor, W.D., *Growth Model of Daphnia*, Can. J. Fish. Aquat. Sci., 39 (1982), 598–606.

[10] Pianka, E.R., *Natural Selection of Optimal Reproductive Tactics*, Amer. Zool., 16 (1976), 775–784.

[11] Stearns, S.C., *Life History Tactics: A Review of the Ideas*, The Quarterly Review of Biology, 51 (1976), 3–47.

[12] Taylor, B.E., *Effects of Food Limitation on Growth and Reproduction of Daphnia*, To appear in Arch. Hydrobiol. Beih. Ergeben. Limnol. Copy of manuscript obtained from the author at ICTP in 1988.

[13] Winkler, D.W. and Wallin, K., *Offspring Size and Number: A Life History Model Linking Effort per Offspring and Total Effort*, Amer. Natur., 129 (1987), 708–720.

Proceedings of the Asian Mathematical
Conference, 1990

Some Methods of Estimation of Slope Change Points

MIAO BAIQI
Department of Mathematis
University of Science and Thchnology of China
Hefei 230026 Anhui, China

In this paper some methods are developed to detect slope change points when the number of change points is unknown. These methods give rise to estimates which are strongly consistent. The rate of convergence is of the order $O(h_N \log N/N)^{1/3}$. For simplicity, we state results for one dimensional processes only, while these results can be easily extended to multi-demensional processes. The complexity of computation of our algorithms is approximately of the order $O(N^{5/3}\log^{1/3} N)$.

Let $x(t)$ be a stochastic process defined by

$$x(t) = f(t) + w(t) \quad \text{for} \quad 0 \le t \le 1, \tag{1}$$

where $w(t)$ is a white noise with mean 0 and (unknown) variance σ^2. Assume that $f(t) = E(x(t))$ has a continuous derivative at all points except finitely many points $t_1, \cdots, t_q, 0 < t_1 < \cdots < t_q < 1$, and $f'(t_j - 0) \ne f'(t_j + 0), f(t_j - 0) = f(t_j + 0) = f(t_j)$ for $j = 1, \cdots, q(q$ unknown). The problem considered in this paper is to suggest a method to estimate the number q and the points $t_1, \cdots, t_q$, based on a sample $x(k/N), k = 0, \cdots, N$. We need to impose another condition on the function $f(t)$, namely, there exists a positive constant M such that

$$|f'(t) - f'(s)| < M(t-s) \tag{2}$$

if no t_i lies in the interval $[s,t]$.

In this paper we propose two methods to estimate q and $t_1, \cdots, t_q$. The first method is nonparametric one, and the second is based on maximum likelihood ratio test. Let $m = m_N$ and C_N satisfy the following conditions.

$$\lim_{N\to\infty} N^{2/3}\log^{1/3} N/NC_N = \lim_{N\to\infty} NC_N/m = \lim_{N\to\infty} m/N = 0. \tag{3}$$

For fixed k, let

$$\overline{X}_L = m^{-1}\sum_{i=1}^{m} x((k-i)/N), \tag{4}$$

$$\overline{X}_R = m^{-1}\sum_{i=1}^{m} x((k+i)/N), \tag{5}$$

$$D_{Nk} = m^{-2}\left\{\sum_{i=1}^{m} i[x((k-1)/N) - \bar{x}_L] - \sum_{i=1}^{m} i[x((k+i)/N) - \bar{x}_R]\right\}, \tag{6}$$

and

$$Q_N = \{k/N : D_{Nk} > C_N\}.$$

For large N, the intervals defined by Q_{Nj}, where $Q_{Nj} = [t_j - 4m/N, t_j + 4m/N]$, for $j = 1, \cdots, q$, are mutually disjoint. Now we state our first result.

Theorem 1. *For large N, the following statements are true with probability one.*

(a) $Q_{Nj} \cap Q_N$ *are not empty for* $j = 1, \cdots, q$.

(b) $Q_N = \bigcup_{j=1}^{q} Q_{Nj}$.

From this theorem for large N, it is easy to see that Q_N splits into q groups with the following properties:

(i) Two points $k/N, k^1/N$ belong to the same group if and only if $|k - k'| < 4m$, and

(ii) For $i < j$, the number of points in Q_{Ni} are less than that of Q_{Nj}.

Now we construct an estimator $\hat{q}_N$ of q as follows. Let $NQ_N = \{k_1, \cdots, k_p\}, 1 \le k_1 \le \cdots \le k_p \le N$. We group these number by using the following rule: Let k_1 belong to the first group M_1. If $k_r \in M_i$, then $k_{r+1} \in M_i$ if $k_{r+1} - k_r < 4m$; $k_{r+1} \in M_{i+1}$ otherwise. At last, we get $\hat{q}_N$ groups $M_1, \cdots, M_{\hat{q}_N}$. This $\hat{q}_N$ will be taken as an estimator for q. Take any element of M_j, say k_{aj}, then $k_{aj}/N = \hat{t}_j$ may be taken as an estimator of t_j for $j = 1, \cdots, \hat{q}_N$. Using Theorem 1 it is easy to see that $q_n \to q$ and $|\hat{t}_j - t_j| = 2m/N$ for any j. The convergence rate of the estimators $\hat{t}_j$ is given by $2m/N$. Here we can take $m/N = O(h_N(\log N/N)^{1/3})$ with $H_N \to \infty$ as $N \to \infty$. We call $\hat{q}_N$ and $\hat{t}_j$ the non-parametric estimators of q and t_j respectively.

Next we propose a parametric method of estimating q and $t_1, \cdots, t_q$ based on maximum likelihood ratio procedure. Here, we use the fact that $w(t)$ is a white noise with mean 0 and variance $\sigma^2 > 0$ for $0 \le t \le 1$. To simply out notation, let us denote

$$x_i = x(i/N), \quad f_i = f(i/N), \quad \text{and} \quad w_i = w(i/N) \quad \text{for} \quad i = 1, \cdots, N.$$

$$\bar{x} = (\bar{x}_L + \bar{x}_R)/2,$$

$$\bar{w} = (2m)^{-1}\left[\sum_{i=1}^{m}(w_{k-i} + w_{k+i})\right],$$

$$\sum_L = \sum_{i=1}^{m} i x_{k-i}, \quad \sum_R = \sum_{i=1}^{m} i x_{k+i},$$

where $\bar{x}_L$ and $\bar{x}_R$ are defined in (4) and (5). Using Taylor's expansion for $f(t)$ at t_j, it follows

$$f(t) = f(t_j - 0) + (t - t_j)f'(t - 0) + o(t - t_j), \quad t_j < t < t_{j+1}$$

or

$$f(t) = f(t_j + 0) + (t - t_j)f'(t + 0) + o(t - t_j), \quad t_j < t < t_{j+1}.$$

Take N equal-spaced samples sequentially from the process (1), say $x_1, \cdots, x_N$. Assume that $k = k_N(j)$ satisfies the condition $|k/N - t_j| < 1/N$, then

$$x_{k-i} = -C_L i/N + \mu + o(1/N) + w_{k-i},$$

$$x_{k+i} = C_R i/N + \mu + o(1/N) + w_{k+i}, \quad i = 1, \cdots, m,$$

where $C_L = f'(t_j - 0)$, $C_R = f'(t_j + 0)$, $C_L \neq C_R$ and $\mu = f(t_j)$. The logarithm of the likelihood function based on $x_{k-m}, \cdots, x_{k-1}, x_{k+1}, \cdots, x_{k+m}$ is

$$\begin{aligned} & L(x_{k-m}, \cdots, x_{k-1}, x_{k+1}, \cdots, x_{k+m}, \sigma^2, C_L, C_R, \mu) \\ = \; & -m \log(2\pi\sigma) - (2\sigma^2)^{-1} \left[\sum_{i=1}^{m} (x_{k-i} + C_L i/N - \mu + i\theta_{k-i}/N)^2 \right. \\ & \left. + \sum_{i=1}^{m} (x_{k+i} - C_R i/N - \mu + \theta_{k+i} i/N)^2 \right], \end{aligned}$$

where $|\theta_j| < 1, k = k-m, \cdots, k-1, k+1, \cdots, k+m$, and $\lim_{N \to \infty} \theta_j = 0$. Using this expression, estimators of C_L, C_R, μ and σ^2 are given by

$$\widehat{C}_L = \frac{-3(5m+1)N}{m(m+1)(m-1)(2m+1)} \sum_L - \frac{9(m+1)N}{m(m+1)(m-1)(2m+1)} \sum_R + \frac{6N}{m-1} \bar{x}, \tag{7}$$

$$\widehat{C}_R = \frac{9(m+1)N}{m(m+1)(m-1)(2m+1)} \sum_L + \frac{3(5m+1)N}{m(m+1)(m-1)(2m+1)} \sum_R + \frac{6N}{m-1} \bar{x}, \tag{8}$$

$$\widehat{\mu} = \frac{2(2m+1)}{m-1} \bar{x} - \frac{3}{m(m-1)} \left(\sum_l + \sum_R \right), \tag{9}$$

$$\begin{aligned} \widehat{\sigma}^2 = \; & \frac{1}{m} \left[\sum_{i=1}^{m} x_{k-i}^2 + \sum_{i=1}^{m} x_{k+1}^2 - \frac{3}{m(m+1)(2m+1)} \left(\sum_R - \sum_L \right)^2 \right. \\ & \left. -2m\bar{x}^2 - \frac{6}{m(m-1)(m+1)} m(m+1)x - \left(\sum_L + \sum_R \right)^2 \right]. \end{aligned} \tag{10}$$

Going along the same line as above it follows that under the null hypothesis $H_0 : C_L = C_R$ the maximum likelihood estimators of C, μ and σ^2 are given by

$$\widehat{C} = \frac{3N}{m(m+1)(2m+1)} \left(\sum_R - \sum_L \right), \quad \widehat{\mu} = \bar{x},$$

$$\widehat{\sigma}^2 = \frac{1}{m} \left\{ \left(\sum_{i=1}^{m} x_{k-1}^2 + \sum_{i=1}^{m} x_{k+i}^2 \right) - \frac{3}{m(m+1)(2m+1)} \left(\sum_R - \sum_L \right)^2 - 2m\bar{x}^2 \right\}.$$

Thus the logarithm of the likelihood ratio test statistic for testing the null hypothesis $H_0 : C_L - C_R$ against the alternative $H_1 : C_L \neq C_R$ is given by

$$G_N(k) = -m \log(1 - u/v),$$

where

$$u = 6\left[\sum_{i=1}^{m} i(x_{k-i} - \bar{x}_L) + \sum_{i=1}^{m} i(x_{k+i} - \bar{x}_R)\right],$$

$$v = m(m^2-1)\left[\sum_{i=1}^{m} i(x_{k-i}^2 + x_{k+i}^2) - \frac{3}{m(m+1)(2m+1)}\left(\sum_R - \sum_L\right)^2 - 2m\bar{x}^2\right].$$

Now, choose $m = m_N$ such that

$$\lim_{N\to\infty} N^{2/3}\log^{1/3} N/m = 0, \quad \lim_{N\to\infty} \log N/B_N = \lim_{N\to\infty} B_N N^2/m^3 = 0.$$

Let

$$Q_N = \{k/N : G_N(k) > B_N, k = m, \cdots, N-m\} \equiv \{k_1/N, \cdots, k_p/N\}.$$

Again, by using the procedure given in Theorem 1 the set $\{k_1, \cdots, k_p\}$ can be split into some groups $M_1, \cdots, M_{\widehat{q}_N}$. Take any element from M_j, say k_{aj}, and set $\widehat{t}_j = k_{aj}/N, j = 1, \cdots, \widehat{q}_N$. Our estimators of q and t_j are $\widehat{q}_N$ and $\widehat{t}_j$ respectively. Then we have

Theorem 2. $(\widehat{q}_N, \widehat{t}_1, \cdots, \widehat{t}_{\widehat{q}_N})$ *is a strongly consistent estimator of* $(q, t_1, \cdots, t_q)$.

In the sequel, we discuss the case when the white noise $w(t)$ has mean 0 and variances $\sigma_j^2 > 0$ in the interval $(t_{j-1}, t_j]$ and $\sigma_j^2 \neq \sigma_{j-1}^2$ for $j = 1, \cdots, q+1$. It is easily seen that the logarithm of the likelihood function based on $x_{k-m}, \cdots, x_{k-1}, x_{k+1}, \cdots, x_{k+m}$ is

$$\begin{aligned} & L(x_{k-m}, \cdots, x_{k-1}, x_{k+1}, \cdots, x_{k+m}, \sigma_L^2, \sigma_R^2, C_L, C_R, \mu) \\ = & -m\log(2\pi\sigma_L\sigma_R) - (2\sigma_L^2)^{-1}\textstyle\sum_{i=1}^{m}(x_{k-i} + C_L i/N - \mu + \theta_{k-1} i/N)^2 \\ & -(2\sigma_R^2)^{-1}\textstyle\sum_{i=1}^{m}(x_{k+i} - C_R i/N - \mu + \theta_{k+i} i/N)^2, \end{aligned}$$

where $\sigma_L^2 = \text{Var}(x_{k-i}), \sigma_R^2 = \text{Var}(x_{k+i})$ for $i = 1, \cdots, m, |\theta_{k-i}| < 1, |\theta_{k+i}| < 1$, and $\lim_{N\to\infty} \theta_{k-i} = \lim_{N\to\infty} \theta_{k+i} = 0$ for $i = 1, \cdots, m$.

Proceeding the same way as in (7)–(10), one can get estimators of $\sigma_L^2, \sigma_R^2, C_L, C_R$ and μ as

$$\begin{aligned} \widehat{C}_L = & -6N[3(m+1)(\alpha\sum_L + \beta\sum_R)]/m(m^2-1)(2m+1) + 6N\bar{x}/(m-1), \\ \widehat{C}_R = & \; 6N[3(m+1)(\alpha\sum_L + \beta\sum_R) + (m-1)\sum_L]/m(m^2-1)(2m+1) - 6N\bar{x}/(m-1), \\ \widehat{\mu} = & \; 2(2m+1)\bar{x}/(m-1) - 6(\alpha\sum_L + \beta\sum_R)/m(m-1), \\ \sigma_L^2 = & \; m^{-1}\sum_{i=1}^{m}(x_{k-i} + iC_L/N - \widehat{\mu})^2, \\ \sigma_R^2 = & \; m^{-1}\sum_{i=1}^{m}(x_{k+i} - iC_R/N - \widehat{\mu})^2, \end{aligned}$$

where $\alpha = \widehat{\sigma}_L^2/(\widehat{\sigma}_L^2 + \widehat{\sigma}_R^2)$, $\beta = \widehat{\sigma}_R^2/(\widehat{\sigma}_L^2 + \widehat{\sigma}_R^2)$.

Thus the logarithm of the likelihood ratio statistic for testing the null hypothesis H_o : $C_L = C_R, \sigma_L^2 = \sigma_R^2$ against the alternative $H_1 : C_L \neq C_R$ with arbitary σ_L^2 and σ_R^2 is given by

$$G_N(k) = m \log \widehat{\sigma}^2 - (m/2)(\log \widehat{\sigma}_L^2 + \log \widehat{\sigma}_R^2),$$

where σ^2 is defined by (10).

Take a sequence of integers $m = m_N$ and B_N such that $\lim_{N\to\infty} B_N/m = \lim_{N\to\infty} \log^2 N/B_N = 0$. Defining $Q_N, k_{aj}, \widehat{t}_j, j = 1, \cdots, \widehat{q}_N$ in the same way as in Theorem 2, we have

Theorem 3. $(\widehat{q}_N, \widehat{t}_1, \cdots, \widehat{t}_{\widehat{q}_N})$ *is a strongly consistent estimator of* $(q, t_1, \cdots, t_q)$.

For more results of this type, one can referred to Krishnaiah and Miao (1986), Krishnaiah, Miao and Zhao (1987) and Miao, Subramanyam and Zhao (1988).

References

[1] Hudson, D.V., *Fitting segmented curves whose joint points have to be estimated*, Jour er. Statist. Assoc., 61(1966), 1097–1129.

[2] Krishnaiah, P.R. and Miao, *Review about estimates of change points*, Handbook of Statistics, 7(1989), 375–402.

[3] Krishnaiah, P.R., Miao, B.Q. and Zhao, L.C., *Local likelihood method in the problems related to change points*, Chin. Ann. of Math., 11B:3(1990), 363–375.

[4] Miao, B.Q., Subramanyam, K. and Zhao, L.C., *On detection and estimation of change points*, Technical Report No. 88–08, Center for Multivariate Analysis, University of Pittsburgh (1988).

Proceedings of the Asian Mathematical
Conference, 1990

Invariant Probability Measures on a Compact Right Topological Group*

PAUL MILNES
University of Western Ontario
Canada

§0. Introduction

On an example of a compact right topological group, we construct a probability measure that is invariant under all right translations, is unique as such, and is also invariant under all continuous left translations. The construction is done in such a way as to indicate how to construct such a measure in the general case, starting from the structure theorem for compact right topological groups. The details for the general case are given in [4, 2, 3]. The measure on the example is also uniquely determined by invariance under all continuous left translations, a conclusion known not to be valid in general.

§1. A Compact Right Topological Group

The enveloping semigroups of distal flows are compact right topological groups. Here is an example [5, 2; 1.3.40 (b)].

Let $\mathbb{T}$ be the circle group and let E be the set of all endomorphisms of $\mathbb{T}$ (including the discontinuous ones). E may be viewed as the almost periodic compactification of $\mathbb{Z}, E = (\mathbb{T}_d)^\wedge = ((\mathbb{Z}^\wedge)_d)^\wedge$. (Here $\mathbb{T}_d$ means $\mathbb{T}$ with the discrete topology). The example is the (compact) product space $G = \mathbb{T} \times E$ with multiplication

$$(w', h')(w, h) = (w'wh' \circ h(e^{2i}), h'h).$$

Note that the formula involves both the product and composition of endomorphisms. G is right topological, i.e., all right translations $(w', h') \longmapsto (w', h')(w, h), G \to G$ are continuous. Also

$$\Lambda(G) := \{s \in G \mid t \longmapsto st, G \to G \text{ is continuous}\} = \{(w, (\ )^n) \mid n \in \mathbb{Z}\}$$

* This research was supported in part by NSERC grant A7857.

(where $(\ )^n : v \longmapsto v^n$) is dense in G. This last condition of having the set of elements that translate continuously on the left be dense in G characterizes the compact right topological groups that come from distal flows. (We remark that flow $(\mathbb{Z}, \mathbb{T}^2)$ with action $n : (u,v) \longmapsto (e^{2ni}u, e^{n^2 i}u^n v)$ is distal, and its enveloping semigroup may be identified with $G, n \in \mathbb{Z}$ corresponding to $(e^{n^2 i}, (\ )^n) \in G$. See [5]) G is an extension of $\mathbb{T}$ by E, since $\mathbb{T}(= \mathbb{T} \times \{1\})$ is a normal subgroup of G and $G/\mathbb{T} \cong E(= \{1\} \times E)$; G is not a split extension of $\mathbb{T}$ by E, since E is not a subgroup of G.

§2. Construction of the Invariant Measure μ

Haar measure on compact topological groups is the starting point of the construction. We remind the reader that
(*) every compact topological group H has a probability measure called Haar measure, which possesses all of the following properties and is uniquely determined by any one of them: left invariance; right invariance; left invariance by members of a dense subset of G; and right invariance by members of a dense subset of G.

The measure μ is defined as a linear functional (bounded, positive and of norm 1) on $C(G)$. (We often identify a measure with the corresponding linear functional). We define it first on

$$C_{\mathbb{T}}(G) := \{f \in C(G) \mid f \text{ is constant on the cosets } s\mathbb{T}, s \in G\},$$

which is isomorphic to $C(G/\mathbb{T})$ via $f \longmapsto \overline{f}, f(s) = \overline{f}(\overline{s})$. We use the shorter notation $\overline{s} := s\mathbb{T}$, when convenient, and also the notation $L_s f(t) = f(st) = R_t f(s)$ for left and right translates. Since $G/\mathbb{T} \cong E$ is a compact topological group with Haar measure μ_1, say, we define $\mu(f) = \mu_1(\overline{f})$ for $f \in C_{\mathbb{T}}(G)$. To extend the domain of definition of μ to all functions $f \in C(G)$, we first write $s = (w,h) \in G$ and $t = (v,1) \in \mathbb{T}$ and use Haar measure μ_2 on $\mathbb{T} \times \{1\} \cong \mathbb{T}$ to define F on G by

$$F(s) = \int_{\mathbb{T}} f(st)\,d\mu_2(t) = \int_{\mathbb{T}} L_s f\,d\mu_2.$$

Since

$$(s,t) = ((w,h),(v,1)) \longmapsto (w,h)(v,1) = (wv,h) = st, \quad G \times \mathbb{T} \to G$$

is (jointly) continuous, the integrand here, $L_s f|_{\mathbb{T}}$ is continuous (although $L_s f$ might not be continuous, perhaps not even measurable, on all of G) ; moreover, F is continuous. By the invariance of $\mu_2, F \in C_{\mathbb{T}}(G)$ on which μ is already defined. So $\mu(f)$ is defined to be $\mu(f) = \mu_1(\overline{F})$. For $f \in C_{\mathbb{T}}(G) \subset C(G)$, this definition of μ agrees with the one given first, since $f = F$ in this case.

§3. Properties of μ

(i) Right invariance. For $f \in C(G)$ and $r \in G$, we must show that $\mu(R_r f) = \mu(f)$. If $f \in C_{\mathbb{T}}(G)$, then $\overline{R_r f} = R_{\bar{r}} \bar{f}$, since $\mathbb{T}$ is normal in G; so

$$\mu(R_r f) = \mu_1(R_{\bar{r}} \bar{f}) = \mu_1(\bar{f}) = \mu(f)$$

by the invariance of μ_1. For a general $f \in C(G)$, we have $\mu(R_r f) = \mu(F_1)$, where

$$F_1(s) := \int_{\mathbb{T}} L_s(R_r f) d\mu_2 = \int_{\mathbb{T}} f(srr^{-1}tr) d\mu_2(t) \int_{\mathbb{T}} L_{sr} f(r^{-1}tr) d\mu_2(t).$$

The automorphism $\theta : t \longmapsto r^{-1}tr$ of $\mathbb{T}$ is the identity here. The continuity of $L_{sr} f$ and F_1 (and of θ in the general case) is a consequence of the joint continuity consideration in 2 and the fact that G is right topological. (For the general case we note also that $(*)$ of 2 implies that a continuous automorphism of a compact topological group leaves Haar measure invariant.) So now

$$F_1(s) = \int_{\mathbb{T}} L_{sr} f d\mu_2 = F(sr),$$

and

$$\mu(R_r f) = \mu_1(\overline{F}_1) = \mu_1(R_{\bar{r}} \overline{F}) = \mu_1(\overline{F}) = \mu(f).$$

(ii) Uniqueness. Suppose m is a right invariant probability measure on G. We will show that $m = \mu$. First, the restriction m_1 of m to $C_{\mathbb{T}}(G) \cong C(E)$ is right invariant, so $(*)$ of 2 forces $m_1 = \mu_1$, i.e., $m = \mu$ on $C_{\mathbb{T}}(G)$. If $f \in C(G)$,

$$\begin{aligned} m(f) = & \int_G f(s) dm(s) = \int_{\mathbb{T}} \left(\int_G f(s) dm(s) \right) d\mu_2(t) \\ = & \int_{\mathbb{T}} \left(\int_G f(st) dm(s) \right) d\mu_2(t) = \int_G \left(\int_{\mathbb{T}} f(st) d\mu_2(t) \right) dm(s). \end{aligned}$$

The last integral in brackets in $F(s)$ as in 2. F is in $C_{\mathbb{T}}(G)$ on which m agrees with μ, so $m(f) = m(F) = \mu(F)$, which is the definition of $\mu(f)$. Thus $m = \mu$.

(iii) Left invariance. For $f \in \Lambda(G)$, the linear functional $\mu \circ L_s$ is right invariant, so must be just μ.

(iv) Uniqueness again. Suppose m' is a probability measure on G that is invariant under left translation by members of $\Lambda(G)$. To show $m' = \mu$, we need to modify the proof in (ii) only slightly. Replacing "right" with "left", and also taking note of $(*)$ in 2, we see that m' agrees with μ on $C_{\mathbb{T}}(G)$. $\mathbb{T}$ is in the center of G as well as in $\Lambda(G)$. Thus $f(ts)$ and $f(st)$ will be interchangeable in the last two integrals in (ii), so $m' = \mu$.

Note. After (i) and (ii) we see that μ is just the product of the Haar measures $\mu_2 \times \mu_1$ on $G = \mathbb{T} \times E$.

§4. The General Case

The Furstenberg-Ellis-Namioka structure theorem for a compact right topological group G with $\Lambda(G)$ dense in G is as follows [4, 2; Appendix C, 3].

G has a strong normal system $\{L_\xi \mid \xi \leq \xi_0\}$ of subgroups. It is indexed by the set of ordinals less than or equal to an ordinal ξ_0 and satisfies

(i) each L_ξ is a closed normal subgroup of G, $L_0 = G$ and $L_{\xi_0} = \{e\}$;

(ii) for $\xi \leq \xi_0$, $L_\xi \subset L_{\xi+1}$ and the function

$$(sL_{\xi+1}, tL_{\xi+1}) \longmapsto stL_{\xi+1}, \quad G/L_{\xi+1} \times L_\xi \times L_{\xi+1} \to G/L_{\xi+1}$$

is continuous; and

(iii) for each limit ordinal $\xi \leq \xi_0$, $L_\xi = \cap_{\nu<\xi} L_\nu$.

Each quotient G/L_ξ is a compact right topological group, of which $L_{\xi-1}/L_\xi$ is a compact topological normal subgroup if ξ is a successor ordinal.

For the example, $\xi_0 = 2$ and $L_1 = \mathbb{T}$. The construction of the measure in 2 and the proofs of right invariance and uniqueness all proceed by an induction process of length 2. In the general case, each of these requires a (perhaps transfinite) induction process of length ξ_0. The example shows how to deal with successor ordinals. For example, after μ is defined on $C_{L_1}(G) \cong C(G/L_1)$ using Haar measure on the compact topological group G/L_1, it is defined on $C_{L_2}(G) \cong C(G/L_2)$ using the compact topological group L_1/L_2 in the same way that $\mathbb{T}$ is used in 2. For limit ordinals ξ, $L_\xi = \cap_{\nu<\xi} L_\nu$, which implies that $C_{L_\xi(G)} = (\cup_{\nu<\xi} C_{L_\nu}(G))^-$ by the Stone-Weierstrass theorem, so these ordinals are dealt with by limit arguments.

N.B. The uniqueness under left invariance for the measure of the example is proved (in 3(iv)) using the fact that $\mathbb{T}$ is in the intersection of $\Lambda(G)$ and the center of G. In general, left invariance is insufficient to specify a unique probability measure even if the compact right topological group contains a dense copy of $\mathbb{Z}$[1].

We mention as well that a compact right topological group G_1 is given in [3] that has a probability measure with the invariance properties (i)–(iii) of 3 here, but G_1 does not have a strong normal system of subgroups. (Of course, G_1 does not come from a distal flow and $\Lambda(G_1)$ is not dense in G_1.)

References

[1] L. Auslander and F. Hahn, *Real functions coming from flows on compact spaces and concepts of almost periodicity*, Trans. Amer. Math. Soc., 106(1963), 415–426.

[2] J.F. Berglund, H.D. Junghenn and P. Milnes, *Analysis on Semigroups; Function Spaces, Compactifactions, Representations*, Wiley, New York, 1989.

[3] P. Milnes and J. Pym, *Haar measure for compact right topological groups* (to appear).

[4] I. Namioka, *Right topological groups, distal flows and a fixed point theorem*, Math. Systems Theory, 6 (1972), 193–209.

[5] I. Namioka, *Ellis groups and compact right topological groups*, in Contemporary Math. 26, Amer. Math. Soc., 1984, 295–300.

Proceedings of the Asian Mathematical
Conference, 1990

What Is a Semifield?

SIDNEY S. MITCHELL
Chulalongkorn University
Bangkok 10330 Thailand

SUTEP SUANTAI
Chiengmai University
Chiengmai 50002 Thailand

§1. Introduction

The theory of semirings is a moderately well developed theory of algebra and recently it has been expanding greatly due to its applications to computer science, see [2], [3], [4], [5], [6] and [12]. Just as a field is a very important and interesting special case of a ring, a semifield is also a very important and interesting special case of a semiring. However, the theory of semifields is still in its infancy. It represents a good opportunity for mathematicians to do research. As of the moment, there is no universally agreed on definition of a semifield, see [1], [10], [11], [13] and [9] for some of the different definitions used so far. The purpose of this paper is to give the most general definition of a semifield which covers all the cases studied so far so that mathematicians can have a basis for developing the theory.

§2. Definitions and Basic Properties of Semifields

Definition 2.1. A semiring is a triple $(S, +, \cdot)$ where $(S, +)$ is a commutative semigroup, $(S, \cdot)$ is a semigroup and multiplication distributes over addition both on the left and the right. A semiring is called a commutative semiring if $(S, \cdot)$ is commutative semigroup. Since a field is a commutative ring $(R, +, \cdot)$ having the property that $(R, \cdot)$ is a 0-group it would seem natural to say that a semifield is a commutative semiring $(S, +, \cdot)$ such that $(S, \cdot)$ is a0–group. And some authors have done this, see for example [7] and [8]. However, there is an equivalent way of defining a field. A field is a commutative ring $(R, +, \cdot)$ having the property that there exists an element $a \in R$ such that $(R \backslash \{a\}, \cdot)$ is a group. Of course $a = 0$ and nothing new is obtained by using the second definition. But, if we use the second definition in the case of a semiring, we do not get an equivalent definition, instead we get a definition which contains the first definition as a special case. This second definition we feel should be the one taken as the definition of a semifield because it includes all other semifields studied to date as special cases. So to spell things out we shall take the following definition as our proposal for the most general definition of a semifield.

Definition 2.2. A semifield is a commutative semiring $(S, +, \cdot)$ having the property that there exists an element $a \in S$ such that $(S \backslash \{a\}, +, \cdot)$ is a group. The element a is called a special element of the semifield.

Theorem 2.3. *Let $(K,+,\cdot)$ be a semifield and $a \in K$ a special element of K. Then either 1) $ax = a$ for all $x \in K$ or 2) $ax = x$ for all $x \in K$ or 3) $a^2 \neq a$ and $a \cdot 1 \neq a$ where 1 is the identity of the group $(K\backslash\{a\},\cdot)$.*

Proof. **Case 1** Suppose that $a^2 = a$ and $a \cdot 1 = a$. We shall prove that $ax = a$ for all $x \in K$. Let $x \in K$. Suppose that $ax \neq a$. Since $(K\backslash\{a\},\cdot)$ is a group, there is a $y \in K\backslash\{a\}$ such that $xy = 1$. Hence $a = a \cdot 1 = a(xy) = (ax)y$ which is a contradiction since $(K\backslash\{a\},\cdot)$ is closed.

Case 2. $a \cdot 1 \neq a$ and $a^2 = a$. Since $(a\cdot 1)(a \cdot 1) = a(1\cdot(a\cdot 1)1) = a(a\cdot 1) = a^2 \cdot 1 = a \cdot 1$, we see that $a \cdot 1$ is an idempotent in the group $(K\backslash\{a\},\cdot)$. Hence $a \cdot 1 = 1$. Let $x \in K\backslash\{a\}$. $ax = a(1\cdot x) = (a\cdot 1)x = 1 \cdot x = x$.

Case 3. $a^2 \neq a$ and $a \cdot 1 \neq a$. In this case we have 3) in the statement of the theorem.

Case 4. $a^2 \neq a$ and $a \cdot 1 = a$. First we want to show that $ax = a$ for all $x \in K\backslash\{a\}$. Let $x \in K\backslash\{a\}$. If $ax \neq a$ that there is a $y \in K\backslash\{a\}$ such that $xy = 1$. Then $a = a \cdot 1 = a(xy) = (ax)y$ so $(K\backslash\{a\},\cdot)$ is not closed, a contradiction, so we must have that $ax = a$ for all $x \in K\backslash\{a\}$. Since $a^2 \neq a$ there is a $z \in K\backslash\{a\}$ such that $a^2z = 1$. Then $1 = a^2z = a(az) = a^2$. Hence a has an inverse so $(K,\cdot)$ is a group. Suppose that $|K| > 2$. Let $x \in K\backslash\{a,1\}$. Then $x = 1\cdot x = a^2x = a(ax) = a^2 = 1$ which is a contradiction. Hence $|K| = 2$. But this contradicts satz 2, §3 of [9] which states that if $(S,+,\cdot)$ is a finite semiring having the property that $(S,\cdot)$ is a group then $|S| = 1$. Hence this case cannot occur. ♮

Definition 2.4. Given a semifield K and a special element $a \in K$, if 1) in Theorem 2.3 occurs then we shall say that K is a Type I semifield with respect to a, if 2) in Theorem 2.3 occurs then we shall say that K is Type II semifield with respect to a and if 3) in Theorem 2.3 occurs then we shall say that K is a Type III semifield with respect to a.

Note that in the proof of Case 4 in Theorem 2.3 we encountered a semiring $(S,+,\cdot)$ where $(S,\cdot)$ is a group. Such semirings have been called "division semirings" or "semifields" in the literature. However, we do not like these names because a division ring is never a division semiring and a field is never a semifield. Hence we prefer to call them ratio semirings. To make things clear we shall give the following definition.

Definition 2.5. A semiring $(S,+,\cdot)$ is called a ratio semiring if $(S,\cdot)$ is group. If $(S,\cdot)$ is abelian then S is called a commutative ratio semiring.

The next natural question to ask is how unique is the special element of a semifield.

Theorem 2.6. *Let $(K,+,\cdot)$ be a semifeld and $a \in K$ a special element. 1) If $|K| > 2$ and K is a Type I semifield with respect to a then a is the only special element of K. 2) If $|K| > 2$ and K is a Type II semifield with respect to a then a is the only special element of K. 3) If K is a Type III semifield with respect to a then a is the only special element of K.*

Proof. We shall first prove 1) and 2). Suppose that $b \in K$ is a special element different from a. Since $a^2 = a \in K\backslash\{b\}$ and $(K\backslash\{b\},\cdot)$ is a group, a must be the identity of $(K\backslash\{b\},\cdot)$. Let $x \in K\backslash\{b,a\}$. Then there is a $y \in K\backslash\{b\}$ such that $xy = a$. If $y = a$ then $a = xy = xa = x$, a contradiction. Hence $y \neq a$. Now $x, y \in K\backslash\{a\}$ and $xy = a$ so $(K\backslash\{a\},\cdot)$ is not closed, a contradiction. Hence a is the only special element of K.

Now we shall prove 3). Suppose that $b \in K$ is a special element different from a. Let 1 denote the identity of $(K\backslash\{a\},\cdot)$ and e denote the identity of $(K\backslash\{b\},\cdot)$. Hence $ae = a$.

Since $a^2 \neq a, e$ is the only idempotent of K therefore $e = 1$ so $a = ae = a \cdot 1$, a contradiction since K is a Type III semifield with respect to a, so $a \cdot 1 \neq a$. Hence a is the only special element of K. ♯

Now we shall show that if $|K| = 2$ then the special element may not be unique.

Examples 2.7. 1) Let $K = \{0, 1\}$ with addition and multiplication given by the following tables:

$\cdot$	0	1
0	0	0
1	0	1

$+$	0	1
0	0	1
1	1	1

We call this the Boolean semifield. K is a semifield of Type I with respect to 0 and a semifield of Type II with respect to 1.

2) Let $K = \{\infty, 1\}$ with addition and multiplication given by the following tables:

$\cdot$	∞	1
∞	∞	∞
1	∞	1

$+$	∞	1
∞	∞	∞
1	∞	1

K is a semifield of Type I with repect to ∞ and a semifield of Type II with respect to 1.

These are the only possible cases, up to isomorphism, where the special element of a semifield is not unique. So disregarding these two trivial cases we can speak of Type I, Type II and Type III semifields realizing that the special element is unique.

§3. Type I Semifields

In a Type I semifield $(K, +, \cdot)$ the special element is a multiplicative zero so that $(K, \cdot)$ is a 0-group. Type I semifields are the ones used in [7] and [8]. We shall now show that Type I semifields can be divided into two different types.

Theorem 3.1. *Let K be a Type I semifield and $a \in K$ the multiplicative zero. Then either 1) $a + x = x$ for all $x \in K$ or 2) $a + x = a$ for all $x \in K$.*

Proof. First we shall show that $a + a = a$. Let $y = a + a$. Then $a = ay = a(a + a) = a^2 + a^2 = a + a$.

Case 1. There exists an $x \in K\backslash\{a\}$ such that $x + a = a$. Let $y \in K$. Then $yx^{-1}(x + a) = yx^{-1}$ so $y + a = a$. Since y is arbitrary, $y + a = a$ for all $y \in K$.

Case 2. For all $x \in K\backslash\{a\}$, $x + a \neq a$. Then $1 + a \neq a$. Let $u = 1 + a$. Then $u \neq a$ and $u + a = (1 + a) + a = 1 + (a + a) = 1 + a = u$. Let $v \in K$ be arbitrary. Then $vu^{-1}(u + a) = vu^{-1}u$ so $v + a = v$. Since $v \in K$ is arbitrary, $v + a = v$ for all $v \in K$. ♯

From Theorem 3.1 we see that there are two kinds of Type I semifields. If 1) of Theorem 3.1 holds then the multiplicative zero is also an additive identity so it behaves like the zero element of a ring and therefore we shall always denote it by 0 and call such a semifield a 0-semifield. Every field is a 0-semifield. If 2) of Theorem 3.1 holds then the multiplicative zero is also an additive zero so it behaves like infinity in the extended real line and therefore we shall always denote it by ∞ and call such a semifield an ∞-semifield.

Examples 3.3. 1) $(\mathcal{Q}_+ \cup \{0\}$ and $\mathcal{R}_+ \cup \{0\}$ (where $\mathcal{Q}_+$ and $\mathcal{R}_+$ are the positive rational and positive real numbers respectively) are examples of 0-semifields.

2) $\mathcal{Q}_+ \cup \{\infty\}$ and $\mathcal{R}_+ \cup \{\infty\}$ are examples of ∞-semifields.

3) The Boolean Semifield Example 2.7 is an example of a 0-semifield with two elements.

4) Let $(G, \cdot)$ be an abelian group. Let ∞ be a symbol not representing any element of G. On $G \cup \{\infty\}$ define $x \cdot \infty = \infty \cdot x = x$ for all $x \in G$. Define $x + y = \infty$ for all x and y in $G \cup \{\infty\}$. Then $G \cup \{\infty\}$ is an ∞-semifield called the ∞-semifield with the trivial addition.

5) Taken everything the same as in the preceding example except define

$$x + y = \begin{cases} \infty & \text{if } x \neq y, \\ x & \text{if } x = y. \end{cases}$$

Then $G\{\infty\}$ is an ∞-semifield called the ∞-semifield with the almost trivial addition. The second example in Example 2.7 is an example of an ∞-semifield with the almost trivial addition containing two elements.

Theorem 3.4. *Let K be a 0-semifield. Then either 1) every element in $K\backslash\{0\}$ has an additive inverse or 2) no element in $K\backslash\{0\}$ has an additive inverse.*

Proof. Suppose that $x \in K\backslash\{0\}$ has an additive inverse y. Let $z \in K$ be arbitrary then $x+y = 0$ so $zx^{-1}(x+y) = zx^{-1} \cdot 0$ and therefore $z + zx^{-1}y = 0$ hence z has an additive inverse. Since $z \in K$ is arbitrary we are done. ♮

If 1) of Theorem 3.4 holds then K is a field. So we see that in a 0-semifield which is not a field 0 is the only element having an additive inverse. In such semifields, $(K\backslash\{0\}, +, \cdot)$ is a commutative ratio semiring.

Definition 3.5. Let $(S, +, \cdot)$ be a semiring and $x \in S$. Then x is called an additively cancellative element if $x + y = x + z$ implies that $y = z$. S is called additively cancellative if every element of S is additively cancellative.

Theorem 3.6. *Let K be a 0-semifield. If one nonzero element is additively cancellative then K is additively cancellative.*

Proof. Suppose that $x \in K\backslash\{0\}$ is additively cancellative. Let $y \in K\backslash\{0\}$ be arbitrary and suppose that $y+u = y+v$. Then $xy^{-1}(y+u) = xy^{-1}(y+v)$ so $x+xy^{-1}u = x+xy^{-1}v$. Since x is additively cancellative, $xy^{-1}u = xy^{-1}v$. Multiplying both sides by yx^{-1} we get that $u = v$ so y is additively cancellative. Since $y \in K\backslash\{0\}$ is arbitrary, K is additively cancellative. ♮.

Since the only finite ratio semirings have one element, we see that the only finite 0-semifields are the finite fields and the Boolean semifield. From Theorem 4.6 of [7] we get that the every finite ∞-semifield has either the trivial or almost trivial addition.

However there is a relation between some ∞-semifield and commutative ratio semirings.

Theorem 3.7. *Let K be an ∞-semifield without the trivial addition. Let $L = \{x \in K \mid 1 + x \neq \infty\}$. Then L is a commutative ratio semiring.*

Proof. First we shall show that L is not empty. Since K does not have the trivial addition, there exist $x, y \in K\backslash\{\infty\}$ such that $x + y \neq \infty$. Multiplying by x^{-1} we get that $1 + x^{-1}y \neq \infty$ so $x^{-1}y \in L$. Let $x, y \in L$. Then $1 + x \neq \infty$ and $1 + y \neq \infty$ so $(1 + x)(1 + y) \neq \infty$ and therefore $1 + x + y + xy \neq \infty$. From this we conclude that $x + y$

and xy are in L. Lastly, let $x \in L$. Then $1 + x \neq \infty$ so $x \neq \infty$. Multiplying by x^{-1} we get that $1 + x^{-1} \neq \infty$ so L is a commutative ratio semiring. ♯

Corollary 3.8. *Let K be an ∞-semifield. Then K had the trivial addition if and only if $1 + 1 = \infty$.*

Proof. Clearly if K has the trivial addition then $1 + 1 = \infty$. Suppose that $1 + 1 = \infty$. Let L be as in the statement of the preceding theorem. If K does not have the trivial addition then L is a commutative ratio semiring so $1 \in L$. Hence $1 + 1 \neq \infty$, a contradiction. Hence L has the trivial addition.

Another consequence of Theorem 3.7 is that an ∞-semifield without the trivial addition is constructed from a commutative ratio semiring D by taking an abelian extension group E if $(D, \cdot)$ and defining $x + y = \infty$ if x and y belong to different cosets of D.

§4. Type II Semifields

First we shall show that every Type II semifield is obtained from a commutative ratio semiring by adjoining one element.

Theorem 4.1. *Let K be a Type II semifield with respect to a. Then $K\backslash\{a\}$ is a commutative ratio semiring.*

Proof. Let $x, y \in K\backslash\{a\}$. Since $(K\backslash\{a\}, \cdot)$ is a group, $xy \in K\backslash\{a\}$. So we must show that $x+y \in K\backslash\{a\}$. Suppose that $x+y = a$. Hence $a = x+y = x\cdot 1+y\cdot 1 = (x+y)\cdot 1 = a\cdot 1 = 1$ which is a contradiction.

Corollary 4.2. *Let K be a finite Type II semifield with respect to a. Then $|K| = 2$.*

Proof. $K\backslash\{a\}$ is a finite commutative ratio semiring so $|K\backslash\{a\}| = 1$. ♯

If K is a Type II semifield with respect to a then we know that $ax = x$ for all $x \in K$. How about $a + x$? This is answered in the next theorem.

Theorem 4.3. *Let K be a Type II semifield with respect to a. Then either $a+a = 1+1$ or $a + a = a$ and if $a + a = a$ then $(K, +)$ is a band. In addition, for all $x \in K\backslash\{a\}$ either $a + x = a$ or $a + x = 1 + x$.*

Proof. Suppose that $a + a \neq a$. Then $a + a = (a + a)\cdot 1 = a\cdot 1 + a\cdot 1 = 1 + 1$. Suppose that $a + a = a$. Let $x \in K$ be arbitrary. Then $x + x = ax + ax = (a + a)x = ax = x$ so $(K, +)$ is a band.

Let $x \in K\backslash\{a\}$. If $x + a \neq a$ then $a + x = (a + x)\cdot 1 = a\cdot 1 + x\cdot 1 = 1 + x$. ♯

Definition 4.4. Let S be a semiring and $x \in S$. $y \in S$ is called an identity of x if $x + y = x$. We shall denote the set of identities of x in S by $I_S(x)$.

Theorem 4.5. *Let D be a ratio semiring such that $|D| > 1$. Then $I_D(x) \neq D$ for all $x \in D$.*

Proof. It is easy to show that for all $x \in D, I_D(x) = xI_D(1)$. Hence it suffices to show that $I_D(1) \neq D$. If $I_D(1)$ is empty or $\{1\}$ then we are done. Suppose $|I_D(1)| > 1$. Let $y \in I_D(1)\backslash\{1\}$. Then $y + 1 = 1$ so $y^{-1} + 1 = y^{-1}$. Therefore $y^{-1} \in I_D(1)$ so $I_D(1) \neq D$. ♯

Theorem 4.6. *Let K be a semifield of Type II with respect to a. Let $D = K\backslash\{a\}$ and $S = \{x \in D \mid a + x = a\}$. Then 1) either S is empty or S is a subsemigroup of $I_D(1)$, 2) if $1 \in S$ then $S = I_D(1)$ and 3) $D\backslash S$ is an ideal in $(D, +)$.*

Proof. 1) Suppose that S is nonempty. Let $x \in S$. Then $1 = a \cdot 1 = (a + x) \cdot 1 = a \cdot 1 + x \cdot 1 = 1 + x$ so $x \in I_D(1)$. Clearly S is closed with respect to addition.

2) Suppose that $1 \in S$. Let $x \in I_D(1)$. Then $x + 1 = 1$ therefore $a = a + 1 = a + (1 + x) = (a + 1) + x = a + x$ so $x \in S$. Hence $I_D(1) \subseteq S$. From 1) we get that $I_D(1) = S$.

3) Let $x \in D \backslash S$ and $y \in D$. $a + x \in D$ so $(a + x) + y \in D$ therefore $a + (x + y) \in D$ so $a + (x + y) \neq a$ yielding $x + y \in D \backslash S$. ♮

We shall now show how to construct all possible Type II semifields of order greater than 2.

Theorem 4.7. *Let D be a commutative ratio semiring, a a symbol not representing any element of D and let $S \subseteq I_D(1)$ be either empty or a subsemigroup of $(I_D(1), +)$ which also has the property that $D \backslash S$ is an ideal in $(D, +)$. Then we can extend the binary operations of D to $K = D \cup \{a\}$ making K into a Type II semifield with a as its special element by defining $a + x = a$ for all $x \in S, a + x = 1 + x$ for all $x \in D \backslash S$ and*

$$a + a = \begin{cases} a \text{ or } 1 & \text{if } 1 + 1 = 1, \\ 1 + 1 & \text{if } 1 + 1 \neq 1. \end{cases}$$

Proof. The proof of this theorem is a straightforward case by case verification that the associative and distributive laws hold. We shall omit the steps of this proof. ♮

§5. Type III Semifields

Using the exact same proofs used in Theorems 4.1 and Corollary 4.2 we get the following results.

Theorem 5.1. *Let K be a Type III semifield with respect to a. Then $K \backslash \{a\}$ is a commutative ratio semiring. (So every Type III semifield is obtained from a commutative ratio semiring by adjoining one element.)*

Corollary 5.2. *Let K be a finite Type III semifield. Then $|K| = 2$.*

We shall now give a theorem describing the multiplication of a Type III semifield.

Theorem 5.3. *Let K be a Type III semifield with respect to a. Then there exists a unique $d \in K \backslash \{a\}$ such that $ax = dx$ for all $x \in K$.*

Proof. Let $d = a \cdot 1$. Then, by the definition of a Type III semifield, $d \in K \backslash \{a\}$. Let $x \in K \backslash \{a\}$. Then $ax = a(1 \cdot x) = (a \cdot 1)x = dx$. Since $a^2 \neq a, a^2 = a^2 \cdot 1 = a(a \cdot 1) = ad$ so we get that $ax = dx$ for all $x \in K$. To prove uniqueness, suppose that $ax = d'x$ for all $x \in K$. Then $d' = d' \cdot 1 = a \cdot 1 = d \cdot 1 = d$. ♮

Next we shall describe the addition of a Type III semifield which is very similar to the addition of a Type II semifield.

Theorem 5.4. *Let K be a Type III semifield with respect to a. Let $d \in K \backslash \{a\}$ be such that $ax = dx$ for all $x \in K$. Then either $a + a = d + d$ or $a + a = a$ and if $a + a = a$ then $(K, +)$ is a band. In addition, for all $x \in K \backslash \{a\}$ either $a + x = a$ or $a + x = d + x$.*

Proof. Suppose that $a + a = a$. Let $x \in K \backslash \{a\}$. Since $a \cdot 1 \neq a$ there is a $z \in K \backslash \{a\}$ such that $(a \cdot 1)z = 1$. Now, $1 = (a \cdot 1)z = a(1 \cdot z) = az$ therefore $az = 1$. Hence $x + x = 1 \cdot x + 1 \cdot x = (az)x + (az)x = a(zx) + a(zx) = (a + a)zx = a(zx) = (az)x = 1 \cdot x = x$ so $(K, +)$ is a band.

Suppose that $a+a \neq a$. Then $a+a=(a+a)\cdot 1 = a\cdot 1 + a\cdot 1 = d\cdot 1 + d\cdot 1 = d+d$. Lastly, if $a+x\neq a$ then $a+x=(a+x)\cdot 1 = a\cdot 1 + x\cdot 1 = d\cdot 1 + x = d+x$. ♯

Theorem 5.5. *Let K be a semifield of Type III with respect to a, $D = K\backslash\{a\}, S = \{x\in K \mid a+x=a\}$ and let $d\in K\backslash\{a\}$ be such that $ax = dx$ for all $x \in K$. Then 1) either S is empty or S is a subsemigroup of $I_D(d)$, 2) if $d\in S$ then $S = I_D(d)$ and 3) $D\backslash S$ is an ideal in $(D,+)$.*

Proof. 1) Suppose that S is nonempty. Let $x\in S$. Then $a+x=a$ so $d=d\cdot 1 = a\cdot 1 = (a+x)\cdot 1 = a\cdot 1 + x\cdot 1 = d+x$ so $x\in I_D(d)$ and therefore $S\subseteq I_D(d)$. Clearly S is closed with respect to addition.

2) Suppose that $d\in S$. Then $a+d=a$. Let $x\in I_D(d)$ so $d+x=d$. Now $a=a+d=a+(d+x)=(a+d)+x=a+x$ so $x\in S$ therefore $I_D(d)\subseteq S$. From 1) we get that $S=I_D(d)$.

3) Let $x\in D\backslash S$ and $y\in D$. Since $K\backslash\{a\}$ is a commutative ratio semiring and $a+x, y\in K\backslash\{a\}$, we get that $(a+x)+y\in K\backslash\{a\}$ and therefore $a+(x+y)\neq a$ so $x+y\in D\backslash S$.

We shall now show how to construct all possible Type III semifields of order greater than 2. ♯

Theorem 5.6. *Let D be a commutative ratio semiring, a a symbol not representing any element in $D, d\in D$ and let $S\subseteq I_D(d)$ have the property that either S is empty or S is a subsemigroup of $(I_D(d),+)$ such that $D\backslash S$ is an ideal of $(D,+)$. Then we can extend the binary operations of D to $K = D\cup\{a\}$ making K into a Type III semifield with a as its special element by defining $ax=dx$ for all $x\in K$ and $a^2=d^2, a+x=a$ for all $x\in S$ and $a+x=d+x$ for all $x\in D\backslash S$ and*

$$a+a=\begin{cases} a \text{ or } d & \text{if } 1+1=1,\\ d+d & \text{if } 1+1\neq 1.\end{cases}$$

Proof. As in Theorem 4.6 the verification that the associative and distributive laws hold is a straight forward case by case consideration which we shall omit. ♯

References

[1] Bourne, S., *On multiplicative idempotents of a potent semiring*, Proc. Nat. Acad. Sci. USA, 42(1956), 623–638.

[2] Eilenberg, S., *Automata Languages and Machines*, Academic Press, New York, 1974.

[3] Golan, J., *The theory of Semirings with applications in mathematics and theoretical computer science* (to appear).

[4] Kuich, W. and Salomaa, A., *Semirings automata languages*, Springer-Verlag, New York, 1986.

[5] Lallement, G., *Semigroups and combinatorial applications*, John Wiley, New York, 1979.

[6] Maslov, V.P. and Volosov, K.A., *Mathematical Aspects of Computer Engineering*, Mir, Moscow, 1988 (English translation).

[7] Mitchell, S. and Fenoglio, P., *Congruence-free commutative semirings*, Semigroup Forum, 37(1988), 79–91.

[8] Mitchell, S. and Sinutoke, P., *The theory of semifields*, Kyungpook Math. Jour., 22(2)(1982), 352–348.
[9] Weinert, H.J., *Über halbringe und halbkörper I*, Acta Math. Acad. Sci. Hungar, 13(1962), 365–378.
[10] Weinert, H.J., *On 0-Simple semirings, semigroup semirings and two kinds of division semirings*, Semigroup Forum, 28(1984), 313–333.
[11] Wiegandt, R., *Über die struktursätze der halbringe*, Ann. Univ. Sci. Budapest Eötvös Sect. Math., 5(1962), 51–68.
[12] Wongseelashote, A., *Semirings and path spaces*, Discrete Math., 26(1979), 55–78.
[13] Zeleznekow, J., *Regular Semirings*, Semigroup Forum, 23(1981), 119–136.

Proceedings of the Asian Mathematical
Conference, 1990

On Edge-disjoint Union of Paths

THELMA C. MONTERO-GALLIGUEZ
Mathematics Department
Mindanao State University-Marawi
c/ P.O. Box 5527, Iligan City 9200
Philippines

§1. Introduction

Let $G_1 = (V_1, E_1)$ and $G_2 = (V_2, E_2)$ be finite simple graphs. The union of the graphs, denoted by $G_1 \cup G_2$, is the graph $(V_1 \cup V_2, E_1 \cup E_2)$, where $\cup$ represents set union [1]. Here, edges are denoted as 2-subsets of the vertex-set. Figure 1 below shows the non-isomorphic unions of the square graph and the path P_2. We observe that the union of graphs may involve identification of vertices or edges, and may be affected by the way vertices are labelled.

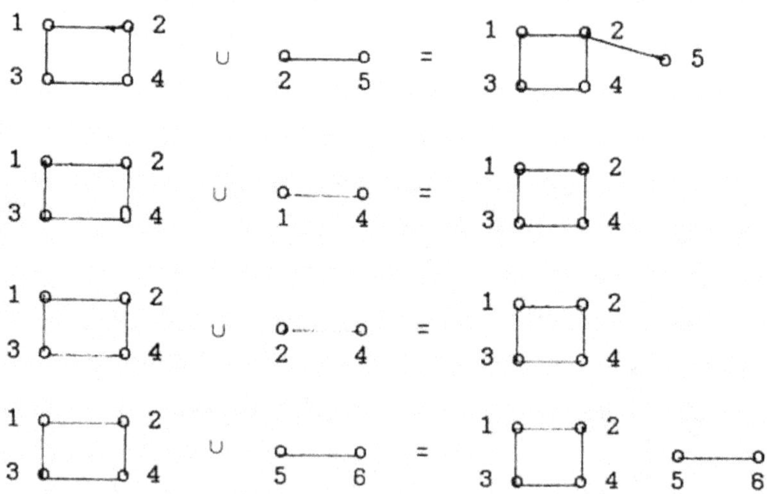

Figure 1. The non-isomorphic unions of C_4 and P_2

The disjoint union of sets X and Y, which we will represent by $X\dot{\cup}Y$, is the multiset obtained by putting together the elements of X and Y, where each element of X is considered distinct from each element of Y. For example, the disjoint union of $\{1,2,3\}$ and $\{1,2,4\}$ is the multiset $\{1,2,3,1,2,4\}$.

The disjoint union of graphs $G_1 = (V_1, E_1)$ and $G_2 = (V_2, E_2)$ is the graph $G_1\dot{\cup}G_2 = (V_1\dot{\cup}V_2, E_1\dot{\cup}E_2)$, where $\dot{\cup}$ denotes disjoint union of sets. Figure 2 illustrates the disjoint union of a K_3 and a P_2, where the label 3 on P_2 is distinguished from the label 3 on K_3. We note that the disjoint union of graphs is independent of the vertex-labelling: it may thus be considered as the union of unlabelled graphs.

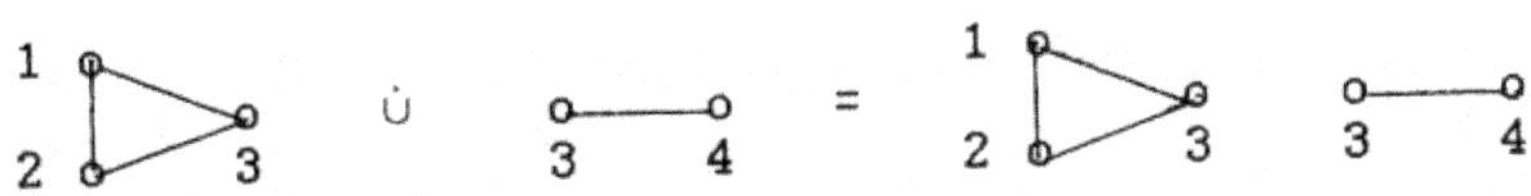

Figure 2. The disjoint union $K_3\dot{\cup}P_2$

We now extend the union and disjoint union operations on graphs as follows. Let $G_1 = (V_1, E_1)$ and $G_2 = (V_2, E_2)$ be simple finite graphs. The edge-disjoint union of graphs G_1 and G_2, denoted by $G_1\underline{\cup}G_2$, is the graph $G = (V, E)$ such that

i. $V = V_1 \cup V_2$;

ii. $E = E_1\dot{\cup}E_2$; and

iii. E_1 and E_2 are disjoint.

For example, the two non-isomorphic edge-disjoint unions of the path P_2 with itself, namely P_3 and $2P_2$, are shown in Figure 3.

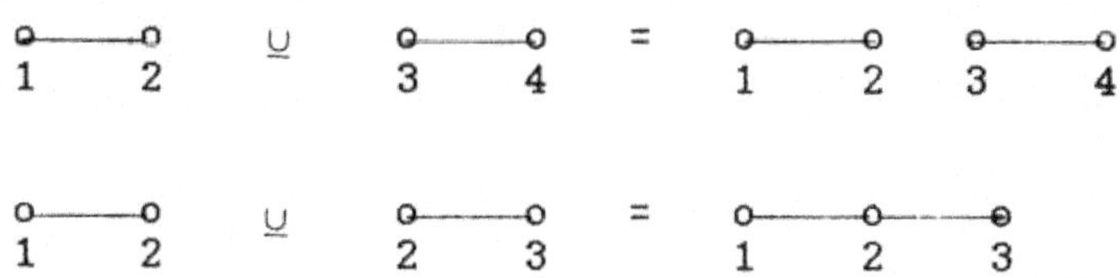

Figure 3. The non-isomorphic edge-disjoint unions of P_2 with itself

Condition (iii) of the definition is essential to yield simple graphs. If (iii) is relaxed, the operation $\underline{\cup}$ may produce a multigraph as given in Figure 4.

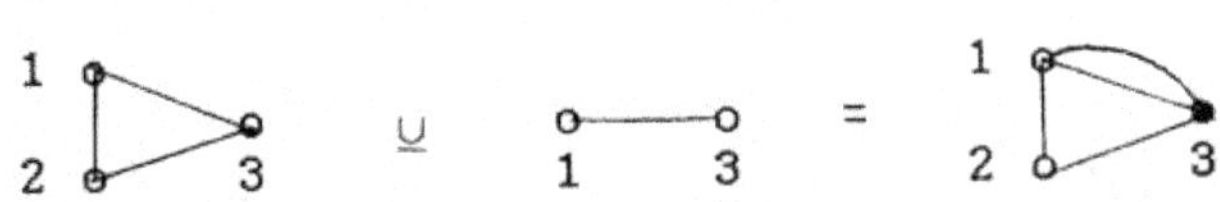

Figure 4. $\underline{\cup}$ may produce a multigraph when edge-sets are not disjoint

The edge-disjoint union of unlabelled graphs may be obtained through an appropriate vertex labelling such that the edge-sets of the graphs are disjoint. In particular, the edge-disjoint union will coincide with disjoint union when the vertex labels are such that the vertex-sets are disjoint.

If G_1 and G_2 are connected graphs, then the edge-disjoint union is connected provided we use an appropriate vertexlabelling so that G_1 and G_2 have at least one common vertex. In this sense, edge-disjoint union may be viewed as vertex-gluing of graphs [2].

This paper considers the following problem: Given a family $\mathcal{F}_n$ consisting of n copies of paths, determine the edge-disjoint union of the paths in $\mathcal{F}_n$ such that the resulting graph is a tree with minimum diameter. Here, diameter refers to the maximum distance between two vertices.

For the rest of the paper, the symbol $\underline{\cup}$ will denote a connected edge-disjoint union with minimum diameter. Diam (G) will represent the diameter of the graph G; and $d(c,v)$, the distance between the vertices c and v.

A vertex c is called a central vertex, or simply center, of a graph G if $\max\{d(c,v)\}$, for all vertices $v \in G$, is minimum,

Lemma 1 [1]. *A tree has either exactly one center or two, adjacent, centers.*

Corollary 2. *The path $P_k = \{a_1, a_2, \cdots, a_k\}$ has the unique center $a_{\lceil k/2 \rceil}$ if k is odd, and has two (adjacent) centers $a_{k/2}$ and $a_{(k+2)/2}$ if k is even.*

A path that joins any two vertices of a tree is unique; thus, if vertices of two trees are labelled so that the trees share exactly one common vertex, then the edge-disjoint union of the trees is also a tree. Together with Corollary 2, this indicates that the edge-disjoint union of a family of paths is a tree of minimum diameter if vertex-identification occurs at central vertices.

§2. The Homogeneous Family of Paths

Suppose $\mathcal{F}_n = \{P_k^i\}, 1 \le i \le n$, is a family of $n \ge 2$ copies of paths $P_k, k \ge 2$. We denote by $\underline{\cup}^n P_k$ the edge-disjoint union which is a tree with minimum diameter of the paths in $\mathcal{F}_n$. Let $K_{1,n}$ represent the complete bipartite graph with bipartition X, Y such that $|X| = 1$ and $|Y| = n$. The graph $K_{1,n}$ is also called a star graph.

The next two theorems present initial results for families of paths P_k, for $k = 2, 3$.

Theorem 3. *For $n \ge 2, \underline{\cup}^n P_2 = K_{1,n}$ and has diameter 2.*

Proof. Let P_2^i denote the ith copy of P_2 in the family, with vertices labelled 0 and a_i as follows:

P_2^i :

Identification of all vertices labelled 0 results in the star graph $K_{1,n}$, which has diameter 2. □

Theorem 4. *For $n \ge 1, \underline{\cup}^n P_3 = K_{1,2n}$ and has diameter 2.*

Proof. Let P_3^i be the ith copy of P_3 with vertices labelled 0, a_i and b_i as follows:

P_3^i :

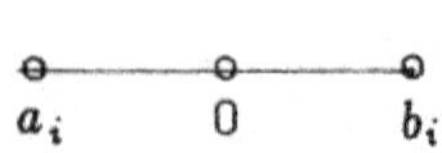

Identification of all vertices labelled 0 yields a $K_{1,2n}$, which has diameter 2. □

Let $m_1, m_2, \cdots, m_s$ and $r_1, r_2, \cdots, r_s$ be positive integers with $\sum_{i=1}^{n} r_i = n$. We define a generalized star graph as the graph obtained from the star graph $K_{1,n}$ by replacing each of the r_i endvertices of $K_{1,n}$ by a path P_{m_i} such that only an endvertex of P_{m_i} is joined to the center of $K_{1,n}$. The generalized star graph will be denoted by Star $(r_1 P_{m_1}, r_2 P_{m_2}, \cdots, r_s P_{m_s})$. Figure 5 shows how to obtain a generalized star graph from a $K_{1,5}$.

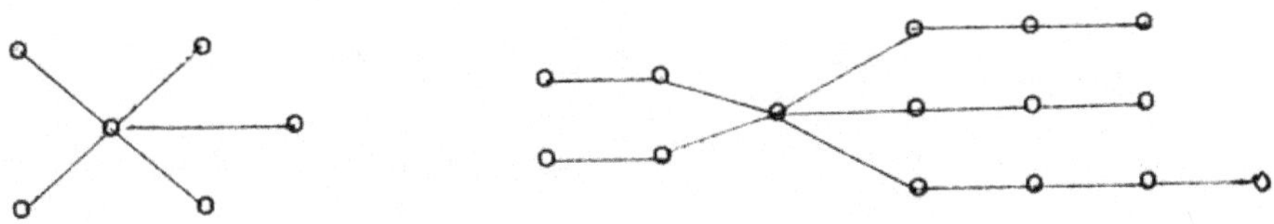

Figure 5. Star$(2P_2, 2P_3, P_4)$ constructed from $K_{1,5}$

Theorem 5. *i. For odd positive integers $k \geq 3$ and for $n \geq 1$.*

$$\underline{\cup}^n P_k = \text{Star}\ (2nP_{\lfloor k/2 \rfloor})$$

and has diameter $k-1$.

ii. For even positive integers k and for $n \geq 1$,

$$\underline{\cup}^n P_k = \text{Star}\ (nP_{(k/2-1)}, nP_{k/2})$$

and has diameter k if $n \geq 2$, or has diameter $k-1$ if $n = 1$.

Proof. For odd k, we represent P_k^i as

where 0 is the center of the path. By identifying all vertices with label 0, we obtain $\underline{\cup}^n P_k =$ Star$(2nP_{\lfloor k/2 \rfloor})$, which has diameter $k-1$. This is the minimum possible diameter since diam$(P_k) = k-1$.

The case when k is even is proved similarly. □

§3. The Nonhomogeneous Family of Paths

A family $\mathcal{F}$ of paths of type $(a_1, a_2, \cdots, a_n)$ is defined as a collection of paths comprising a_1 copies of P_2, a_2 copies of P_3, and so on, up to a_n copies of P_{n+1}, where $a_n \neq 0$. For example, the collection P_3, P_4, P_4, P_6 is a family of paths of type $(0, 1, 2, 0, 5)$.

Theorem 6. *Let $\mathcal{F}$ be a family of paths of type $(a_1, a_2, \cdots, a_n), n \geq 1$ and $a_n \neq 0$. Let $\underline{\cup}\mathcal{F}$ denote the edge-disjoint union of the paths in $\mathcal{F}$ such that the resulting graph is a tree with minimum diameter. Then*

$$\underline{\cup}\mathcal{F} = \mathrm{Star}(b_1 P_1, b_2 P_2, \cdots, b_{(n+1)/2} P_{\lfloor (n+1)/2 \rfloor}),$$

where $b_j = a_{2j-1} + 2a_{2j} + a_{2j+1}$ for $j = 1, 2, \cdots, \lfloor (n+1)/2 \rfloor$, and $a_{2j} = 0$ if $2j > n$ or $a_{2j+1} = 0$ if $2j+1 > n$.

Furthermore,

$$\mathrm{diam}(\underline{\cup}\mathcal{F}) = \begin{cases} n, & \text{if } n \text{ is even, or if } n \text{ is odd and } a_n = 1, \\ n+1, & \text{if } n \text{ is odd and } a_n \geq 2. \end{cases}$$

Proof. Let $\mathcal{F}_k$ be the subcollection (of $\mathcal{F}$) comprising all the a_{k-1} copies of P_k. For each k, let P_k^i denote the ith copy of $P_k, 1 \leq i \leq a_{k-1}$ and label a central vertex as 0. Choose an appropriate vertex-labelling such that all the vertices of all the paths P_k^i in $\mathcal{F}_k$, except the vertices labelled 0, have distinct labels. For each k, vertex-identification at 0 yields an edge-disjoint union which is a tree with minimum diameter; thus, by Theorem 5,

$$\underline{\cup}\mathcal{F}_k = \underline{\bigcup}^{a_{k-1}} P_k = \begin{cases} \mathrm{Star}(a_{k-1} P_{(k/2)-1}, a_{k-1} P_{k/2}), & \text{if } k \text{ is even,} \\ \mathrm{Star}(2a_{k-1} P_{\lfloor k/2 \rfloor}), & \text{if } k \text{ is odd.} \end{cases}$$

Now use an appropriate vertex labelling so that for $k = 2, 3, \cdots, n+1$, the vertices of the trees $\underline{\cup}^{a_{k-1}} P_k = \underline{\cup}\mathcal{F}_k$ will all have distinct labels except at their centers, which are all labelled 0. Identify the trees at the vertices labelled 0 to obtain the desired tree $\underline{\cup}\mathcal{F}$ of minimum diameter, which is a generalized star graph. The formula on the coefficients of the generalized star graph can be easily proved by induction on n.

Longest paths in $\underline{\cup}\mathcal{F}$ occur between endvertices on two paths $P_{\lfloor (n+1)/2 \rfloor}$ that are joined to 0. If n is even, then $\underline{\cup}\mathcal{F}$ contains a path P_n of odd order $n+1$. Thus, for even values of n.

$$\mathrm{diam}(\underline{\cup}\mathcal{F}) = \mathrm{length}(P_n) = n.$$

If n is odd, then $\underline{\cup}\mathcal{F}$ contains either two paths $P_{\lfloor (n+1)/2 \rfloor}$ that are joined to 0, or only one subpath P_{n+1}, depending on the value of a_n. Thus, for odd values of n

$$\mathrm{diam}(\underline{\cup}\mathcal{F}) = \begin{cases} \mathrm{length}(P_{n+1}) = n, & \text{if } a_n = 1, \\ n+1, & \text{if } a_n \geq 2. \end{cases}$$

This completes the proof of the theorem. □

To illustrate Theorem 6, consider a family $\mathcal{F}$ of paths of type $(0, 1, 2, 1, 2)$ as shown in Figure 6. Through identification

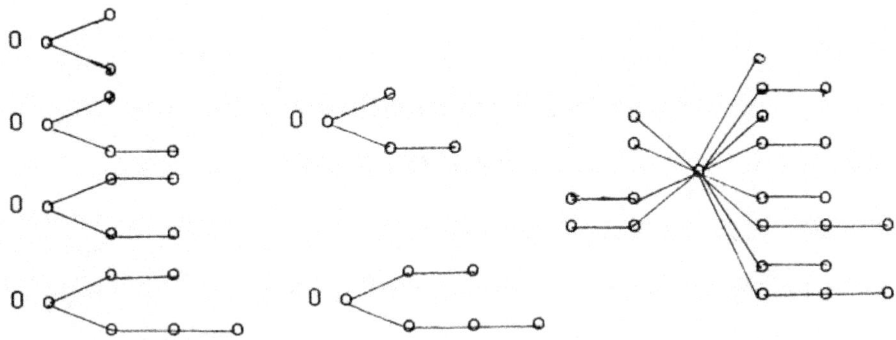

A type $(0, 1, 2, 1, 2)$ family $\mathcal{F}$ of paths

Figure 6. A family $\mathcal{F}$ of nonhomogeneous paths and the edge-disjoint union $\underline{\cup}\mathcal{F}$ of minimum diameter

of the vertices labelled 0, we obtain $\underline{\cup}\mathcal{F} = \mathrm{Star}(b_1P_1, b_2P_2, b_3P_3)$ where

$$b_1 = a_1 + 2a_2 + a_3 = 0 + 2 + 2 = 4,$$

$$b_2 = a_3 + 2a_4 + a_4 = 2 + 2 + 2 = 6,$$

$$b_3 = a_5 = 2.$$

References

[1] J.A. Bondy and U. S. R. Murty, *Graph theory with applications*, Macmillan Press, 1976.

[2] K.M. Koh and K.S. Poh, *Vertex-gluings of graphs with their closed-set lattices*, Paper presented at the Australia-Singapore Joint Conference, 1986.

Proceedings of the Asian Mathematical
Conference, 1990

Analytic Functionals and Their Transformations

MITSUO MORIMOTO
Department of Mathematics
Sophia University
Tokyo, Japan

§1. Analytic Functionals

An analytic functional is a continuous linear functional on a certain topological linear space of holomorphic functions. The notion of analytic functionals goes back to the Italian mathematician Luigi Fantappié (1943). But a systematic study was initiated by the doctoral thesis [6] of the French mathematician André Martineau (1963). The following are the spaces of holomorphic functions on which analytic functionals are usually defined.

We shall denote by $\mathcal{O}(U)$ the sapce of the holomorphic functions on an open subset U of C^{d+1}.$\mathcal{O}(U)$ is equipped with the topology of uniform convergence on every compact subset of U and it becomes a Fréchet space.

We shall denote by $\mathcal{O}(K)$ the space of the germs of holomorphic functions on K. For every open set $U \supset K$, there is the canonical linear mapping $\rho_{U,K} : \mathcal{O}(U) \to \mathcal{O}(K)$. We define the topology of $\mathcal{O}(K)$ to be the strongest locally convex topology such that $\rho_{U,K}$ is continuous for every open neighborhood U of K.

For $r \geq 0$ we define

$$\begin{aligned} Exp(\mathbf{C}^{d+1};r) = \quad & \{ f(\varsigma) \in \mathcal{O}(\mathbf{C}^{d+1});\ \forall \varepsilon > 0,\ \exists C \geq 0, \forall \varsigma \in \mathbf{C}^{d+1}, \\ & |f(\varsigma)| \leq C \exp((r+\varepsilon)(|\varsigma_1| + |\varsigma_2| + \cdots + |\varsigma_n|))\}. \end{aligned} \tag{1}$$

A function $f \in Exp(\mathbf{C}^{d+1};r)$ is called an entire function of exponential type $\leq r$. A theorem due to E. Borel reads as follows:

Theorem 1. *Let $f \in Exp(\mathbf{C}^{d+1};r)$. Then for every $\varepsilon > 0$ there is a mesure μ such that supp $\mu \subset \{z; |z_j| < r + \varepsilon\}$ and that*

$$f(\varsigma) = \int_{C^{d+1}} \exp(z\varsigma)\, d\mu(z). \tag{2}$$

The mesure μ is not unique even if ε is fixed. Theorem 1 can be formulated in the following manner using the notion of analytic functionals.

Let $\lambda \neq 0$ be a complex number. For $T \in \mathcal{O}'(K)$, we define

$$f_\lambda(\varsigma) = \langle T_z, \exp(i\lambda z \cdot \varsigma)\rangle, \tag{3}$$

where $z \cdot \zeta = z_1\zeta_1 + z_2\zeta_2 + \cdots + z_{d+1}\zeta_{d+1}$. We call the this transformation $T \longmapsto f_\lambda$ the Fourier-Borel transformation.

Theorem 2. *Let D_r be the polydisc with radius r:*

$$D_r = \{z \in \mathbf{C}^{d+1}; |z_j| \le r\}. \tag{4}$$

The Fourier-Borel transformation $T \longmapsto f_\lambda$ is a topological linear isomorphism of $\mathcal{O}'(D_r)$ onto $Exp(\mathbf{C}^{d+1}; |\lambda|r)$.

A. Martineau proved Theorem 2 can be generalized for every convex compact subset K of $\mathbf{C}^{d+1}$.

§2. Hyperfunctions

In 1951–1952, L. Schwartz published the book [1] and established the theory of distributions. Distributions are defined through the duality in the C^∞ category. The Japanese mathematician Mikio Sato prefered the analytic framework to the C^∞ framework and introduced the notion of hyperfunction in 1958 [4]. As there are no (non-trivial) analytic functions of compact support, the idea of duality does not work well for hyperfunctions..

Intuitively speaking, a hyperfunction of one real variable is a difference of two boundary values of holomorphic functions f_1 and f_2 defined on the upper (resp. lower) half plane:

$$f(x) = f_1(x + i0) - f_2(x - i0). \tag{5}$$

In more mathematical terminology, the space of hyperfunctions $\mathcal{B}(I)$ on an interval I is defined as follows:

$$\mathcal{B}(I) = \mathcal{O}(U \backslash I)/\mathcal{O}(U), \tag{6}$$

where U is a complex neighborhood of I. Gelfand and Shilov [3] studies many special distributions in term of the difference of two boundary values of holomorphic functions. But Sato's idea is to understand the hyperfunctions algebraically without the notion of limit.

But for several variables, the approach through boundary values becomes too complicated. Sato defines the space of hyperfunctions $\mathcal{B}(I)$ on an open set I of $\mathbf{R}^{d+1}$ as follows: Take a complex neighborhood U of I, i.e. U is an open set in $\mathbf{C}^{d+1}$ such that I is a closed subset of U and define $\mathcal{B}(I)$ to be the $(d+1)$-th space of relative cohomology of U with support in I and with coefficients in the sheaf $\mathcal{O}$ of the germs of holomorphic functions

$$\mathcal{B}(I) = H_I^{d+1}(U, \mathcal{O}). \tag{7}$$

Theorem 3. *(i) $\mathcal{B}(I)$ is defined independent of the choice of a complex neighborhood U of I.*

(ii) $I \longmapsto \mathcal{B}(I)$ is a flabby sheaf.

The definitition of hyperfunctions of several variables implies the theorem of Edge of the Wedge theorem. This idea was first formulated by A. Martineau [8] and was used to reformulate the notion of micro-functions [10]. The micro-local analysis is a great contribution to the theory of partial differential equations [11]. Let me mention here some names of mathematicians who worked in the area of PDE and hyperfunctions: T. Kawai, M. Kashiwara, A. Kaneko, K. Kataoka.

For a compact set K in $\mathbf{R}^{d+1}$, we denote by $\mathcal{B}[K]$ the space of hyperfunctions on $\mathbf{R}^{d+1}$ with support in K. Then we have the following theorem due to A. Martineau and R. Harvey.

Theorem 4.

$$\widetilde{\mathcal{B}}[K] \cong H_K^{d+1}(U; \mathcal{O}) \cong \mathcal{O}'(K). \tag{8}$$

It turns out that a hyperfunctions with compact support is nothing but an analytic functional and we can exploit the notion of duality in the study of hyperfunctions of compact support.

The duality (8) was one of the key lemmas when Hikosaburo Komatsu gave a solid foundation to Sato's theory by means of functional analytic method [7]. On the other hand, A. Martineau reconstructed hyperfunctions starting form his theory of analytic functionals [5]. Martineau's idea was followed by Pierre Schapira in his lecture note [9].

§3. Hyperfunctions on a Real Analytic Manifold

For every real analytic manifold X of dimension $d+1$, there exists a complexification $\widetilde{X}$ of X. The space $\mathcal{B}(X)$ of hyperfunctions on X is defined as follows:

$$\mathcal{B}(X) = H_X^{d+1}(\widetilde{X}, \mathcal{O}). \tag{9}$$

If X is a compact real analytic manifold, we have the isomorphism

$$\mathcal{B}(X) \cong \mathcal{A}'(X) \cong \mathcal{O}'(X). \tag{10}$$

We would like to study the case where X is the sphere. In this case, the natural complexification of X is the complex sphere.

§4. Circle and Complex Circle

In this 1958 paper [4], M. Sato considered hyperfunctions on the circle S^1 and studied their Fourier series expansion. An earlier work in this direction is Köthe's paper [2].

For $T \in \mathcal{O}'(S^1)$, we put

$$c_n = \langle T_n, z^{-n} \rangle \text{ for } n = 0, \pm 1, \pm 2, \cdots. \tag{11}$$

Then for $f(z) = \sum_{n=-\infty}^{\infty} a_n, z^n \in \mathcal{O}(S^1)$, we have

$$\langle T, f \rangle = \sum_{n=-\infty}^{\infty} c_{-n} a_n. \tag{12}$$

The formal Laurent series

$$\sum_{n=-\infty}^{\infty} c_n z^n \tag{13}$$

is called the fourier-Laurent series expansion of $T \in \mathcal{O}'(S^1)$. The space of hyperfunctions on ^{1} can be characterized by the growth condition on the Fourier-Laurent coefficients c_n:

$$\limsup_{n\to\pm\infty} \sqrt[|n|]{|c_n|} \leq 1. \tag{14}$$

The family of the annular sets

$$\check{S}^1(r) = \{z \in \mathbf{C}; 1/r < |z| < r\}, \quad r > 1 \tag{15}$$

forms a fundamental system of complex neighborhoods of the circle S^1. We may call $\check{S} = \mathbf{C}\backslash\{0\}$ the complex circle. The analytic functionals on the annular set $\check{S}^1(r)$ have the similar characterization by means of the growth condition of their Fourier-Laurent coefficients.

As the circle S^1 is not convex subset of $\mathbf{C}$, we cannot apply Theorem 2. We defined the Fourier-Borel transform of $T \in \mathcal{O}'(S^1)$ as follows:

$$f_\lambda(\varsigma) = \langle T_z, \exp(i\lambda z\varsigma)\rangle. \tag{16}$$

This simple Fourier-Borel transformation is not injective on $\mathcal{O}'(S^1)$.

Instead if we generalize the Fourier-Borel transform of T as follows:

$$f_\lambda(\varsigma_1, \varsigma_2) = \langle T_z, \exp(i\lambda(z\varsigma_1 + \frac{\varsigma_2}{z}))\rangle, \tag{17}$$

we will have the injectivity. See [12] for the details. This generalized Fourier-Borel transformation on $\mathcal{O}'(S^1)$ can be extended to the higher dimensional sphere.

§5. Sphere and Complex Sphere

We denote by $\widetilde{S}^d$ the complex sphere of complex dimension d:

$$\widetilde{S}^d = \{z \in \mathbf{C}^{d+1}; z^2 = 1\}, \tag{18}$$

where $z^2 = z \cdot z, z \cdot \varsigma = z_1\varsigma_1 + z_2\varsigma_2 + \cdots + z_{d+1}\varsigma_{d+1}$. The complex sphere $\widetilde{S}^d$ is a complexification of the sphere S^d:

$$S^d = \{x \in \mathbf{R}^{d+1}; \|x\| = 1\}, \tag{19}$$

where $\|x\| = (x^2)^{1/2}$ is the Euclidean norm on $\mathbf{R}^{d+1}$.

In order to define a fundamental system of complex neighborhoods of the sphere S^d, we introduce the Lie norm on $\mathbf{C}^{d+1}$.

$$\begin{aligned} L(z) = {} & L(x+iy) \\ = {} & [\|x\|^2 + \|y\|^2 + \{(\|x\|^2)\|y\|^2 - (x\cdot y)^2\}^{1/2}]^{1/2}. \end{aligned} \tag{20}$$

$$\widetilde{B}(r) = \{z \in \mathbf{C}^{d+1}; L(z) < r\} \tag{21}$$

is called the Lie ball of radius r.

$$\widetilde{S}(r) = \widetilde{S}^d(r) = \widetilde{S}^d \cap \widetilde{B}(r), \quad r > 1 \tag{22}$$

form a fundamental system of complex neighborhoods of the sphere S^d.

Remark that for $d = 1$, we have the complex analytic diffeomorphism of $\widetilde{S}^1$ onto $\check{S}^1$ via $(z_1, z_2) \longmapsto z = z_1 + iz_2$. $\widetilde{S}^1(r)$ and $\check{S}^1(r)$ can be identified by this correspondence.

We will need also the dual Lie norm defined as follows:

$$\begin{aligned} L^*(z) = & \sup\{|\varsigma \cdot z|; L(z) \leq 1\} \\ = & \frac{1}{\sqrt{2}}[\|x\|^2 + \|y\|^2 + \{(\|x\|^2 - \|y\|^2)^2 + 4(x \cdot y)^2\}^{1/2}]^{1/2}. \end{aligned} \tag{23}$$

Now we are going to consider holomorphic functions on $\widetilde{S}(r)$ and show some results obtained jointly by Ryoko Wada and M. Morimoto [14], [15], [16], [17], [18].

Theorem 5. *Suppose $\lambda \in \mathbf{C}$ satisfies the following condition:*

$$(\lambda/2)^{-n-(d-1)/2} J_{n+(d-1)/2}(\lambda) \neq 0 \text{ for all } n = 0, 1, 2, 3, \cdots. \tag{24}$$

Then the restiction mapping

$$\alpha_\lambda : \mathcal{O}_\lambda(\widetilde{B}(r)) \rightarrow \mathcal{O}(\widetilde{S}(r)) \tag{25}$$

is a topological linear isomorphism, where

$$\mathcal{O}_\lambda(U) = \{f \in \mathcal{O}(U); (\Delta + \lambda^2)f(z) = 0\}, \tag{26}$$

$$\Delta = \partial^2/\partial z_1^2 + \cdots + \partial^2/\partial z_{d+1}^2. \tag{27}$$

Note that $\lambda = 0$ satisfies the condition (24).

For an analytic functional $T \in \mathcal{O}'(\widetilde{S}(r))$, we define its Fourier-Borel transform $G_\lambda(z)$ as follows:

$$G_\lambda(z) = \langle T_\varsigma, \exp(i\lambda_z \cdot \varsigma)\rangle. \tag{28}$$

Theorem 6. *The Fourier-Borel transformation $\mathcal{F}_\lambda : T \longmapsto G_\lambda$ is a topological linear isomorphism:*

$$\mathcal{F}_\lambda : \mathcal{O}'(\widetilde{S}(r)) \rightarrow Exp_\lambda(\mathbf{C}^{d+1}; [|\lambda|r : L^*]), \tag{29}$$

where

$$Exp_\lambda(\mathbf{C}^{d+1}; [A : L^*]) = \mathcal{O}_\lambda(\mathbf{C}^{d+1}) \cap Exp(\mathbf{C}^{d+1}; [A : L^*]), \tag{30}$$

$$\begin{aligned} Exp(\mathbf{C}^{d+1}; \quad [A : L^*]) = \{f \in \mathcal{O}(\mathbf{C}^{d+1}); \forall\varepsilon > 0, \exists C \geq 0, \\ \forall z \in \mathbf{C}^{d+1}, |f(z)| \leq C\exp((A + \varepsilon)L^*(z))\}. \end{aligned} \tag{31}$$

We shall define the space of holomorphic functions of exponential type on the complex sphere $\widetilde{S}$.

$$\begin{aligned} Exp(\widetilde{S}^d; \quad [A : L^*]) = \{f \in \mathcal{O}(\widetilde{S}^d); \forall\varepsilon > 0, \exists C \geq 0, \\ \forall z \in \widetilde{S}^d, |f(z)| \leq C\exp((A + \varepsilon)L^*(z))\}. \end{aligned} \tag{32}$$

Remark that $f \in Exp(\widetilde{S}^d; [A : L^*])$ is defined only on the complex sphere $\widetilde{S}$ and satisfies the exponential growth condition only on the complex sphere. Then we have a theorem of the same type as Theorem 5.

Theorem 7. *Suppose $\lambda \in \mathbf{C}$ satisfies the condition (24). Then the restiction mapping*

$$\alpha_\lambda : Exp_\lambda(\mathbf{C}^{d+1}; [A : L^*]) \to Exp(\widetilde{S}; [A : L^*]) \tag{33}$$

is a topological linear isomorphism.

Combining the preceeding Theorems 6 and 7, we can find the Fourier-Borel isomorphism on the complex sphere.

Theorem 8. *Suppose $\lambda \in \mathbf{C}$ satisfies the condition (24). Then the Fourier-Borel transformation (25) composed with the restiction mapping (33) $T \longmapsto \langle \alpha_\lambda(G_\lambda)$ establishes a topological linear isomorphism:*

$$\alpha_\lambda \circ \mathcal{F}_\lambda : \mathcal{O}'(\widetilde{S}(r)) \to Exp(\widetilde{S}; [|\lambda| r : L^*]). \tag{34}$$

§6. Spherical Sphere and Complex Light Cone

Let me stress the similarity of Theorem 2 (Euclidean case) and Theorem 8 (spherical case). This kind of Fourier-Borel isomorphism can be constructed also for the complex light come.

We understand by the complex light come M the following analytic set:

$$M = \{z \in \mathbf{C}^{d+1}; z^2 = 0\} \tag{35}$$

and by the spherical sphere Σ the following set:

$$\Sigma = \{z = x + iy \in M; \|x\| = \|y\| = 1\}. \tag{36}$$

Σ is a compact real analytic manifold of dimension $2d - 1$ and complex analytically diffeomorphic to the unit cotangential bundle of the sphere S^d. See for the details [18].

References

[1] L. Schwartz, *Théorie des distributions*, Hermann, Paris, 1951–52.

[2] G. Köthe, *Dualität in der Funktiontheorie*, J. Reine Angew. Math., 191(1953), 30–49.

[3] I.M. Gelfand and G.E. Shilov, *Generalized Functions I*, Moscow, 1958.

[4] M. Sato, *On the theory of hyperfunctions*, Sugaku, 10(1958), 1–27.

[5] A. Martineau, *Les hyperfonctions de M. Sato*, Séminaire Bourbaki 1960–61, No. 214.

[6] A. Martineau, *Sur les fonctionnelles analytiques et la transformation de Fourier-Borel*, J. d'anal. Math. de Jérusalem, 11(1963), 1–163.

[7] H. Komatsu, *Hyperfunctions and partial differential operators with constant coefficients*, Todai Seminer Note 22, 1968.

[8] A. Martineau, *Le' edge of the wedge' en théorie des hyperfonctions de Sato*, Proc. Intern. Conf. on FA, 1969, Tokyo, Univ. of Tokyo Press, 1970.

[9] P. Schapira, *Théorie des hyperfonctions*, Lecture Notes in Math., 126, Springer, 1970.

[10] M. Morimoto, *Sur la décomposition du faisceau des germes de singularités d'hyperfonctions*, J. Fac. Sci. Univ. Tokyo, Sect. IA, 17(1970), 215–239.

[11] M. Sato, T. Kawai and M. Kashiwara, *Hyperfunctions and pseudo-differential equations Part II*, Lecture Notes in Math., 287, Springer, 1973, 263–529.

[12] M. Morimoto, *A generalization of the Fourier-Borel transformation for the analytic functionals with non-convex carrier*, Tokyo J. Math., 2(1979), 301–322.

[13] M. Morimoto, *Hyperfunctions on the Sphere, Sophia Kokyuroku in Mathematics 12*, Sophia Univ. Dept. Math., 1982 (in Japanese).

[14] M. Morimoto, *Analytic functionals on the sphere and their Fourier-Borel transformations*, Complex Analysis, Banach Center Publications 11. PWN-Polish Scientific Publishers, Warsaw, 1983, 223–250.

[15] R. Wada, *On the Fourier-Borel transformations of analytic functionals on the complex sphere*, Tohoku Math. J., 38(1986), 417–432.

[16] R. Wada and M. Morimoto, *A uniqueness set for the differential operator* $\Delta_z + \lambda^2$, Tokyo J. Math., 10(1987), 93–105.

[17] R. Wada, *Holomorphic functions on the complex sphere*, Tokyo J. Math., 11(1988), 205–218.

[18] M. Morimoto and R. Wada, *Analytic functionals on the complex light cone and their Fourier-Borel transformation*, Algebraic Analysis, Vol. 1, Academic Press, 1988, 439–455.

Proceedings of the Asian Mathematical
Conference, 1990

Presentations of Groups-computational Methods

DANIELA B. NIKOLOVA
Institute of Mathematics
Bulgarian Academy of Sciences
str. Acad. G. Bonchev, bl.8, 1113 Sofia, Bulgaria

§1. The Todd-Coxeter Method (TC)

In 1936 Todd and Coxeter proposed a practical method for enumerating cosets in a finite abstract group and used their method for hand calculations. In 1953 C.B. Haselgrove implemented for the first time the Todd-Coxeter method on a computer in Cambridge.

Let us be given a finite presentation for a group:

$$\begin{aligned} G = &\quad \langle g_1, g_2, \cdots, g_n | r_1(g_1, g_2, \cdots, g_n), r_2(g_1, g_2, \cdots, g_n), \cdots, r_m(g_1, g_2, \cdots, g_n) = 1\rangle \\ = &\quad \langle E|R\rangle, \end{aligned}$$

with

$$E = \{g_1, g_2, \cdots, g_n\}, E' = \{g_1, g_2, \cdots, g_n, g_1^{-1}, g_2^{-1}, \cdots, g_n^{-1}\},$$

$$R = \{r_i(g_1, g_2, \cdots, g_n) \mid i = 1, 2, \cdots, m\},$$

where g_i are called generators and r_j-relators for the group G.

The first natural question that arises is: Is the group G finite and if so, what is the order of G?

We may slightly generalize the situation. Let us be given a finite set of words in E':

$$S = \{s_1, s_2, \cdots, s_p\}$$

and let $U = gp(S)$ be the subgroup of G, generated by this set. Is the index $[G : U]$ finite and if so, can we determine it?

The Todd-Coxeter method attempts tro find that index (if finite), by enumerating cosets of U in G. The procedure is based on two simple facts:

(1) $\forall\ s(g_1, g_2, \cdots, g_n) \in S \Longrightarrow Us(g_1, g_2, \cdots, g_n) = U$.

(2) $\forall\ r(g_1, g_2, \cdots, g_n) \in R \Longrightarrow \forall\ h \in G,\ Uhr(g_1, g_2, \cdots, g_n) = Uh$.

On the base of the above stated facts we construct three types of tables:

(a) subgroup tables

(b) relations tables

(c) coset table.

We assign a number to each coset and determine its multiplication with every g_i or g_i^{-1}. Let us see how these tables look like.

For each $s = s(g_1, g_2, \cdots, g_n) = g_{i_1}, g_{i_2}, \cdots, g_{i_t} \in S$ the subgroup table consists of a single line:

	g_{i_1}		g_{i_2}	$\cdots$	g_{i_t}				g_{i_j}		
1	|		|	$\cdots$	|	1,	where	k	|	1	means $\mathbf{k}g_{i_j} = 1$.

(a) subgroup table

For each relator $r = r(g_1, g_2, \cdots, g_n) = g_{j_1}, g_{j_2}, \cdots, g_{j_q}$ the relation table is organized analogously:

	g_{j_1}		g_{j_2}	$\cdots$	g_{j_q}	
1	|		|	$\cdots$	|	1
2	|		|	$\cdots$	|	2
$\cdots$	|	$\cdots$	|	$\cdots$	|	$\cdots$
k	|		|	$\cdots$	|	k
$\cdots$	|	$\cdots$	|	$\cdots$	|	$\cdots$

(b) relation table

The coset table represents a multiplication table of the cosets with the generators and their inverses:

	g_1	$\cdots$	g_n	g_1^{-1}	$\cdots$	g_n^{-1}
1						
2						
$\cdots$	$\cdots$	$\cdots$	$\cdots$	$\cdots$	$\cdots$	$\cdots$
k						
$\cdots$	$\cdots$	$\cdots$	$\cdots$	$\cdots$	$\cdots$	$\cdots$

(c) coset table

We define new numbers in the coset table systematically by the rule:

$$1 := \mathbf{k}g, \quad \mathbf{k} < 1, \quad g \in E'.$$

It follows, of course, that $\mathbf{k} = 1g^{-1}$ too. Thus we initialize the l rows in the tables (b) and (c). Then we fill in all possible vacant places, going from left to right and from right to left. When it happens that a certain line closes up, then we get an equality of the type $\mathbf{k}g = 1, g \in E'$, called a deduction. Deductions could lead to a coincidence. It could happen that $\mathbf{A} = \mathbf{k}g$ and $\mathbf{B} = \mathbf{k}g$. Then, of course, $\mathbf{A} = \mathbf{B}$ and $\mathbf{A}g = \mathbf{B}g$ for each $g \in G$. If $\mathbf{A} < \mathbf{B}$, we replace $\mathbf{B}$ by $\mathbf{A}$ in all tables. When all m lines of all tables close simultaneously. Then $[G : U] = m$ and all cosets are represented by different numbers [7].

Example. Let us consider the presentation of the alternating group A_5 [2]:

$$A_5 = \langle R, S | R^2 = S^3 = (RS)^5 = 1 \rangle, \quad \text{where} \quad R = (12)(45), \quad S = (134)$$

and let $U = gp(RS)$.

We start by the definition $2 : 1R \Longrightarrow 2R^{-1} = 1$.

Then follow the deductions

$$2S = 1 \Longrightarrow 1S^{-1} = 2, \quad 2R = 1 \Longrightarrow 1R^{-1} = 2.$$

The number 3 is defined at the next vacant place in the coset table: $3 := 1S \Longrightarrow 3S^{-1} = 1$ and s.o.

The complete picture of the Todd-Coxeter procedure can be seen in the Appendix.

The order of filling in the tables is not important. It could only influence the efficiency. There exist several algorithms: Felsch algorithm, HLT-method, the "lookahead" algorithm ...

§2. Presentations of Subgroups

Let a group G has a final presentation: $G = \langle E|R\rangle$. Let U be a subgroup of G of finite index $[G : U]$. We want to write down a presentation for the subgroup U.

1. The Reidemeister-Shreier method (RS)

A generating set for U we get immediately from the Schreier's theorem [6, p.88]. As there exists a transformation τ, called Reidemeister's rewriting, we could transform any word $u \in U$ in a word, written in the Schreier generators. We get a presentation for the subgroup U from a theorem of Reidemeister [6, p.94]. This was used by G. Havas [3] to develop the program RS, which constructs a presentation of U in terms of Schreier generators from the coset table of G mod U.

2. The Tietze transformation program (TZ)

This program serves to simplify the presentation using the so called Tietze transformations [1, p.69–73].

3. The Modified Todd-Coxeter algorithm (MTC)

The problem is that usually Schreier generators are too many. Therefore, it is desirable to get a presentation in terms of a given set of generators $s_j(g_1, g_2, \cdots, g_n) \in U$. This can be obtained considering numbers to express coset representatives, and not cosets themselves [1, p.27–32]. Thus

$$\forall s = s(g_1, g_2, \cdots, g_n) = g_{i_1}, g_{i_2}, \cdots, g_{i_t} \Longrightarrow 1s = s1 (\text{and not } 1s = 1).$$

The subgroup table has the form:

$$\begin{array}{ccccccc} & g_{i_1} & & g_{i_2} & \cdots & g_{i_t} & \\ -- & -- & -- & -- & -- & -- & -- . \\ 1 & | & & | & \cdots & | & s1 \end{array}$$

In the coset table deductions look like:

$$1g = v_{1.g}(s_j)\mathbf{k}.$$

this table enables us to get relators in terms of s_j.

MTC, of course, states problems of space and time. It is possible to use tree structures for a recursive storing of these words. Now more efficient methods are discussed all the time. Their purpose is on one hand, to get as few as possible addition generators and on the other hand, to organize coincidences in the shortest possible way.

§3. Determination of All Subgroups Up to a Given Index

Let

$$G = \langle g_1, g_2, \cdots, g_n | r_1(g_i) = 1, \cdots, r_m(g_i) = 1 \rangle.$$

We want to find all subgroups $U \leq G : [G, U] \leq z$, with $z \in \mathbf{N}$. We have seen that each subgroup determines a complete coset table with concrete properties. On the other hand, we can assign to such a table a subgroup (i.e. one could get the Schreier generators from it). This correspondence is not bijective, but there exist bijective methods of constructing subgroups from coset tables. The task is finite. In order to get all subgroups of a given index z, it suffices to construct all coset table with z rows. The method is proposed by C.Sims [9]. there exist some further refinements to it [7].

§4. Method for Abelian Groups (IMD)

There exists a program [4], which enables us to get the structure of U/U' from the presentation of U. The algorithm represents a constructive proof of the Main theorem of finitely generated abelian groups. The matrix obtained by the powers of the elements in the relators of $\overline{U} = U/U'$ has to be diagonalized in the form $D = (d_1, d_2, \cdots, d_n, 0 \overset{t}{\cdot\cdot}, 0)$, so that $d_1 | d_2 | \cdots | d_n$. Then the abelian group $\overline{U} \cong C_{d_1} C_{d_2} \cdots C_{d_n} C_\infty^t$.

§5. The Nilpotent Quotient Algorithm (NQ)

Consider a finitely presented group $G = \langle E | R \rangle$ and a prime p. Using NQ we can construct the lower exponent p-central series of G [5].

§6. Heineken's Group

The $SPAS$ system, originating from Canberra, St. Andrews and Aachen, includes the main coset table methods for finitely presented groups. The " stand-alone" programs mentioned above, have been implemented on an IBM 4341 in the Institute of Mathematics with Computing Center of the Bulgarian Academy of Sciences, We want now to illustrate them on a concrete problem, put by H. Heineken. In the attempt to solve it, we used all those methods, and of course, the $CAYLEY$ system- the most complete system of group-theoretical programs, elaborated by J.J. Cannon [1, p. 145–183].

Let

$$H = \langle x_1, x_2, x_3 | [X_1, [x_1, x_2]] = x_3, [x_2, [x_2, x_3]] = x_1, [x_3, [x_3, x_1]] = x_2 \rangle,$$

where

$$[x, y] = x^{-1}y^{-1}xy.$$

What can we say about the order of this group; is it finite or infinite? As a matter of fact, the TC-procedure showed that

$$H_0 = \langle x_1, x_2, x_3 | [x_1, x_2] = x_3, [x_2, x_3] = x_1, [x_3, x_1] = x_2 \rangle = \langle 1 \rangle.$$

We began by enumerating cosets of given by us subgroups of H. This trial did not succeed because using TC, after 5000 definitions only two coincidences were found. With the subgroup $U = \langle 1 \rangle, TC$ produced 18712 coset numbers without any coincidence$\cdots$ Is the group H in fact infinite?

Then J. Neubueser found an epimorphism of H onto A_5. H. Heineken found even an epimorphism of H onto $SL(2,5)$. This was the starting point. We explored further the group H in the following way:
Step 1. Find $U \leq H : [H : U] \leq 6$. We found two such subgroups; U_2 and U_3.
Step 2. Use IMD for U_2 and U_3. It turned out that $U_3/U_3' \cong C_3$ therefore $U_3 \supseteq N$, where $U_3/N \cong A_4$. but $U_2/U_2' \cong C_4$, which means that $U - 2 \not\supseteq N$.
Step 3. Use NQ for $N \Longrightarrow \exists C : [N : C] = 2^4$. From the list of perfect groups [8] of order $\leq 10^4$, we got $H/C = 960_2$.
Step 4. We got that $H/L_1 \cong 1920_1$ is also factor of H (from the list [8]).
Step 5. -Use $CAYLEY$ to find 2 generators h_1, h_2 of H/L_1 such that from one side, the relations of Heineken R_1, R_2, R_3 hold for them and from the other, they generate H/L_1.
-Use MTC to find a new presentation for H/L_1:

$$H/L_1 = \langle h_1, h_2, \mid r_1, r_2, r_3, r_4, r_5 \rangle.$$

As $H = \langle x_1, x_2, x_3 | \; R_1, R_2, R_3 \rangle$, we have also

$$H/L_1 = \quad \langle h_1, h_2 \mid R_1, R_2, R_3, r_1, r_2, r_3, r_4, r_5 \rangle$$

$$H/L_1 \leq H/G_i = \quad \langle h_1, h_2 \mid R_1, R_2, R_3, r_1, \cdots, \bar{r}_i, \cdots, r_5 \rangle,$$

where $\bar{r}_i$ means that we omit the relation r_i.

Use TC to fine $[H/G_4 : 1] = 3840$. Thus $H/S = H/G_4 \cong 3840_5$. From this fact it follows also that $\exists L_2.L_3 : H/L_2 \cong 1920_6$ and $H/L_3 \cong 1920_7$, because they are contained into 3840_5.
Step 6. Use NQ to fine X-the maximal 2-nilpotent factor of $N.[N : X] = 2^{24}$.

Thus we get the picture shown in the Appendix. This group was worked out by J. Neubueser, V. Felsch, A. Luccini and the author in Lehrstuhl D fuer Mathematik. Aachen 1985–1986. Unfortunately, at this point we had to stop our investigation of the group H. As several unsuccessful tries didn't bring us new information. The question whether H is finite or infinite is still open. The exploration of it shows on one side how useful the methods of Computational Group Theory could be and on the other, that they certainly have their limits, leaving maybe the final decision to abstract theorists. As I mentioned before, the Sydney $CAYLEY$ system includes most of the known algorithms for computing with finite groups. Recently, a new programming system, called *GAP*, designed for Computational Group Theory was developed in Aachen (M. Schoenert, J. Meir, A. Niemeyer and W. Nickel). Its main difference from *CAYLEY* is that it enables the user to implement his own algorithms quite easily.

Appendix

The Appendix contains the result of the Todd-Coxeter procedure upon the group A_5 (Fig. 1) as well as the picture of our calculations for the Heineken's group (Fig. 2).

We used the following notations New numbers are underlined in the coset table at the places they were defined: deductions are underlined in the other tables with a full line, if they bring a new information and by a dotted line otherwise; the numbers below the underlining show the sequence in which they occur.

References

[1] *Computational Group Theory* (1984). Proc. L.M.S.Symposium, Durham 1982, Ed. M.D. Atkinson Academic Press, London.

[2] W.R. Hamilton, *Memorandum Respecting a New System of Roots of Unity*. Phil. Mag. 12(1856), 446.

[3] G. Havas, *A Reidemeister-Schreier Program* (Proc. Second Internat. Conf. Theory of Groups, Canberra 1973, Ed. M.F. Newman), Lecture Notes in Mathematics, Springer-Verlag, Berlin, 372(1974), 347–356.

[4] G. Havas, L. S Sterling, *Integer Matrices and Abelian Groups* in Symbolic and Algebraic Computation (Proc. EUROSAM, 79, Marseille, Ed. W.N.G. Edward), Lecture Notes in Computer Sciences, Springer-Verlag, Berlin, 72(1979), 431–451.

[5] I.D. Macdonald, *A Computer Application to Finite p-Groups*. J. Austral. Math. Soc. 17(1974), 102–112.

[6] W. Magnus, A. Karrass, D. Solitar, *Combinatorial Group Theory: Presentations of Groups in Terms of Generators and Relations*, Pure and Appl. Math., Interscience, New York 13(1966)

[7] J. Neubueser, *An Elementary Introduction to Coset Table Methods in Computational Group Theory* (Proc. Groups St. Andrews, 1981, Eds. C.M. Campbell E.F. Robertson). L.M.S. Lecture Notes. Cambridge University Press 71(1982) 1–45.

[8] G. Sandloebes, *Perfekte Gruppen bis Ordnung 10* in, Schriften zur Informatik und angewandten Mathematik, Eds. K. Indermark, R. Kaerkes, J. Merkwitz. W. Oberschelp, RWTH Aachen (1977)

[9] C.C. Sims, *Group-Theoretical Algorithms* in Topics in Algebra, Ed. M.F. Newman. Lecture Notes in Mathematics, Springer-Verlag, Berlin, 697(1978) 108–124.

$A_5 = \langle R, S, \mid R^2 = S^3 = (RS)^5 = 1 \rangle, \quad u = gp\langle RS \rangle$

$R = (12)(45), \qquad S = (134)$

Subgroup table

	R	S
1	2	1
	(1)	

Relation tables

	R	R
1	2	1 (2)
2	1	2 (3)
3	4	3 (8)
4	3	4 (9)
5	7	5 (13)
6	8	6 (15)
7	5	7 (14)
8	6	8 (18)
9	10	9 (25)
10	9	10 (26)
11	12	11 (31)
12	11	12 (32)

	S	S	S
1	3	2 (5)	1
2	1	3	2 (6)
3	2	1	3 (7)
4	5	6 (10)	4
5	6	4 (11)	5
6	4	5	6 (12)
7	8	9 (22)	7
8	9	7 (23)	8
9	7	8	9 (24)
10	11	12 (27)	10
11	12	10 (28)	11
12	10	11	12 (29)

	R	S	R	S	R	S	R	S	R	S
1	2	1	2	1	2	1	2	1	2	1 (4)
2	1	3	4	5	7	8 (17)	6	4	3	2
3	4	5	7	8	6 (18)	4	3	2	1	3
4	3	2	1	3	4	5	7	8	6 (19)	4
5	6	8	6 (20)	4	3	2	1	3	4	5
6	8	9	10	11	12 (30)	10	9	7	5	6
7	5	6	8	9	10	11	12	10 (33)	9	7
8	6	4	3	2	1	3	4	5	7	8 (21)
9	10	11	12	10 (34)	9	7	5	6	8	9
10	9	7	5	6	8	9	10	11	12	10 (35)
11	12	10 (36)	9	7	5	6	8	9	10	11
12	11	12	11	12	11	12	11	12	11	12 (37)

The coset table

x	R	S	R	S
1	2	3	2	2
2	1	1	1	3
3	4	2	4	1
4	3	5	3	6
5	7	6	7	4
6	8	4	8	5
7	5	8	5	9
8	6	9	6	7
9	10	7	10	8
10	9	11	9	12
11	12	12	12	10
12	11	10	11	11

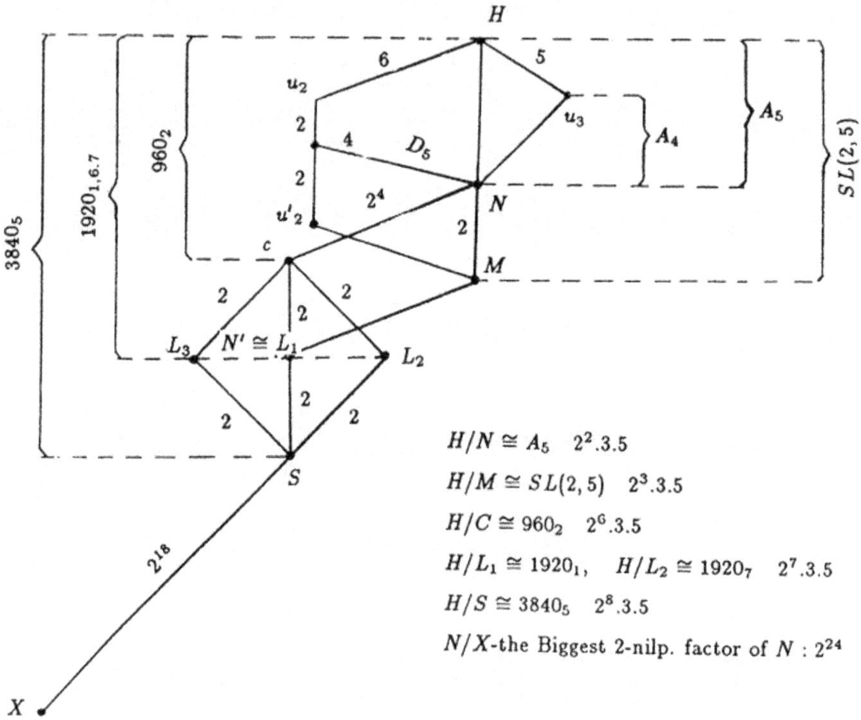
H
N
M
S
X
u2
u3
u'2
c
L3
L2
N' ≅ L1
D5
A4
A5
SL(2,5)
9602
19201,6.7
38405
218
H/N ≅ A5 2².3.5
H/M ≅ SL(2,5) 2³.3.5
H/C ≅ 9602 2⁶.3.5
H/L1 ≅ 19201, H/L2 ≅ 19207 2⁷.3.5
H/S ≅ 38405 2⁸.3.5
N/X-the Biggest 2-nilp. factor of N : 2²⁴

Proceedings of the Asian Mathematical
Conference, 1990

Adaptive Mesh Refinements with Iterated Defect Corrections in One Dimensional Partial Differential Equations

FERNANDA OLIVEIRA*
University of East Asia
P.O.Box 3001
Macau

PAULA OLIVEIRA*
Department of Math.
University of Coimbra
3000 Coimbra, Portugal

§1. Introduction

In this paper we describe a general method for the numerical solution of initial-boundary value problems of evolutionary (PDEs). The class of problems considered have the form:

$$\mu_t = L(\mu), \qquad x \in]0,1[, \quad t > 0 \tag{1}$$

$$\begin{cases} \mu(x,0) = f(x), & x \in]0,1[, \\ \mu(0,t) = \gamma_1(t), & t \in [0,t_f], \\ \mu(1,t) = \gamma_2(t), & t \in [1,t_f], \end{cases} \tag{2}$$

where L represents a linear or nonlinear spatial differential operator of the second order.

Along the last years several numerical adaptive mesh techniques have been proposed in engineering literature [1], [2], [3], etc., in order to solve (PDEs) that present sharp transitions in spatial derivatives, regions of larger solution variations, etc. We can say that the different known algorithms follow one of two grid adapting strategies: the first redistributes a fixed number of grid points, according to some criteria such that the overall error measured by some means is reduced; The second increases the number of grid points where a certain measure of the error indicates that refinement is needed.

Here, we follow the last strategy based on the known idea, if the exact value of the total local truncation error was included in the discretized equation, that corrected equation would give the exact solution of the given (PDE).

Since the exact value of total local truncation error T is not known, we define an iterated defect approximation for T together with one way of including that approximation in the discretized equation. The suggested numerical process is theoretically based on the

* This work has been supported by "Junta Nacional de Investigaçao Cientifica e Tecnológica", Portugal, under the research contract no. 87/191.

existence of an asymptotic expansion for the total local truncation error. Several authors have suggested the inclusion of the truncation error in the discretized equation. See Fung [3], Oliveira [4], etc.

§2. Discretized Equations

With the method of lines (1) (2) is space discretized on a uniform h-grid, where $h = x_i - x_{i-1} = \frac{1}{M}, i = 1(1)M$ and M is odd (only for convenience). That gives a system of ordinary differential equations (ODEs):

$$\begin{cases} \frac{d\mu(t)}{dt} = L_h\mu(t) + b_h + T_s(t;h), \\ \mu(0) = f, \quad t > 0, \end{cases} \tag{3}$$

where $\mu(t) = [\mu(x_1,t),\cdots,\mu(x_{M-1},t)]^T, f = [f(x_1),\cdots,f(x_{M-1})]^T, b_h$ and $T_s(t;h)$ are $M-1$ vectors; b_h carries the boundary conditions and $T_s(t;h)$ represents the local spatial truncation error along the lines (x_i,t), $i = 1(1)M-1$.

System (3) is then solved with a time integrator such as

$$C_k = \sum_{j=0}^{l} \alpha_j \mu_{n+j}^{hk} - k\sum_{j=0}^{l} \beta_j \mu_{n+j}^{hk} = 0, \tag{4}$$

where k is the time step size, α_j, β_j are real coefficients satisfying the classical conditions to guarantee stability and order $p (p \geq 1)$ and μ_{n+j}^{hk} is the approximation of the $M-1$ vector $\mu(t_{n+j})$ from (3).

In recent years several (MOL) packages have been developed for one-space dimensional (PDEs) system like (1) (2) which automatically adjust the time step. Throughout the computation we use such (ODEs) solvers coupling their time step control with our spatial step control.

The exact solution of (1) (2) at the hk-grid points satisfies the system

$$\sum_{j=0}^{l} \alpha_j \mu(t_{n+j}) - k\sum_{j=0}^{l} \beta_j [L_h\mu(t_{n+j}) + b_h + T_s(t_{n+j};h)] = T_t(t_{n+l};k), \quad n = 0,1,\cdots, \tag{5}$$

where $T_t(t_{n+l};k)$ is the time local truncation error of (4) at $t = t_{n+l}, n = 0,1,\cdots$. Then the total local truncation error $T(t_{n+l})$ is given by

$$T(t_{n+l}) = k\sum_{j=0}^{l} \beta_j T_s(t_{n+j};h) + T_t(t_{n+j};k), \quad n = 0,1,\cdots, \tag{6}$$

where we notice a sum of two terms, the first one carries the contribution of the spatial discretization error and the second one carries the time integrator error.

On a fixed spatial h-grid, starting with $n = 0$, we solve the system

$$\sum_{j=0}^{l} \alpha_j \mu_{n+j}^{hk} - k\sum_{j=0}^{l} \beta_j (L_h \mu_{n+j}^{hk} + b_h) = 0, \quad n = 0,1,\cdots \tag{7}$$

for the $M-1$ vector μ^{hk}_{n+j}. Then $\mu^{hk}_{ij} \approx \mu(x_i, t_l), i = 1, 2, \cdots, M-1$. An estimator TE of $T_s(t_l; h)$ is then defined.

Definition. Let the $\frac{M-1}{2}$ vector TE be defined by

$$TE = \sum_{j=0}^{l} \alpha_j \mu^{hk}_{n+j} - k \sum_{j=0}^{l} \beta_j (L_{2h} \mu^{hk}_{n+j} + b_{2h}). \tag{8}$$

Denoting by $\mu^{2h,k}$ the numerical solution corresponding to a spatial $2h$-grid we can write

$$\mu^{2h,k} = \mu^{hk} + \eta^{hk}$$

and see that TE is the numerical solution of the correction differential equation

$$\eta_t = L\eta, \quad \eta(0) = 0. \tag{9}$$

Localization of Spatial Subdomains where μ^{hk} is not Sufficiently Accurate.

At $t = t_l$, we localize spatial subdomains, where refinement is needed with the following test:

$$\text{Given a tolerance } \varepsilon > 0, \quad \text{if} \quad \exists\, i_0 : |i_0(TE)| > \varepsilon \tag{10}$$

then the accuracy of the i_0 component of μ^{hk}_l given by (8) is not acceptable, so we need to improve such accuracy. It should be notice that when TE is evaluated at $t = t_l$, the time truncation error has already been reduced by the time step control of the integrator.

§3. Spatial Refinements

Let us assume, for the sake of simplicity, that there is only one component of μ^{hk}, say $\mu^{hk}_{i_0}$, that implies (10).

We define now a spatial subdomain $]x_{i_0-1}, x_{i_0+1}[$ and a local associated problem

$$\begin{cases} \dfrac{\partial v(x,t)}{\partial t} = Lv(x,t), &]x_{i_0-1}, x_{i_0+1}[, t \in]0, t_l[, \\ v(x_{i_0-1}, t) = \mu^{hk}_{i_0-1}, & v(x_{i_0+1}, t) = \mu^{hk}_{i_0+1}, \\ v(x, 0) = P(x), \end{cases} \tag{11}$$

where $P(x)$ is an interpolation polynomial for the already known data

$$(x_{i_0-1}, \mu^{hk}_{i_0-1}), (x_{i_0}, \mu^{hk}_{i_0}) \quad \text{and} \quad (x_{i_0+1}, \mu^{hk}_{i_0+1}).$$

We will comment later that polynomial $P(x)$.

Then, a sequence of local refinements with $h = \frac{h}{N}, N = 2, 2^2, 2^3, \cdots$ is carried out together with:

a) the evaluation of the numerical approximations $v^{\frac{h}{N}k}$ for the solution of (11), and

b) the evaluation of the $M-1$ dimensional vectors RC^N given by

$$RC^N = \sum_{j=0}^{l} \alpha_j v^{\frac{h}{N}k}_{n+j} - k \sum_{j=0}^{l} \beta_j (L_h v^{\frac{h}{N}k}_{n+j} + b_h). \tag{12}$$

Obviously, the components of the vector $v^{\frac{h}{N}k}$ in (12) are given by

$$v_{n+j}^{\frac{h}{N}k} \quad \text{is the numerical solution of (11) for } x \in]x_{i_0-1}, x_{i_0+1}[, \quad t = t_l$$

and

$$v_{n+j}^{\frac{h}{N}k} = 0 \quad \text{elsewhere.}$$

Such local refinements, the correspondent solving process of (11) and (12) evaluation are carried out until (13) is verified

$$^{i}|RC^S - RC^R| < \varepsilon \quad \text{for } S > R, \quad i = 1, 2, \cdots, M-1. \tag{13}$$

Comparing to the total local truncation error (6) with (12) we can consider (12) an approximation of (6) and the inequalities (13) a "measure" of the improvement of the accuracy obtained with the spatial refinements.

In each subdomain localized by the test (10) we stop the refinements when (13) is verified. After that we add (12) in the discretized equation (7) and the next step is to solve the corrected equation (14):

$$\sum_{j=0}^{l} \alpha_j Z_{n+j}^{hk} - k \sum_{j=0}^{l} \beta_j (L_h Z_{n+j}^{hk} + b_h) - RC^N = 0, \quad n = 0, 1, \cdots. \tag{14}$$

Z_{n+l}^{hk} is now the improved numerical solution of (1) (2) at $t = t_l$ for the hk-grid. Then, we continue for $n = 1$, repeating the numerical process for $t = t_{l+1}$.

We have been assuming that, at $t = t_l$, only one component $\mu_{i_0}^{hk}$ of μ^{hk} was not sufficiently accurate. If it is not the case, we will perform one or more associated problems similar to (11), one for each subdomain.

§4. Diagram for the Algorithm

(1) set $n = 0$;
(2) Compute μ_{n+l}^{hk} with (7);
(3) Localize subdomains with (8) and (10);
(4) For each of those subdomains, solve a sequence of spatial refined problems (11);
(5) Stop (4) when (13) is verified;
(6) Solve (14) with RC^N given by (12);
(7) $n = n + 1$;
(8) Go to (2).

A PASCAL program for that algorithm is available from the authors. Our numerical results have been obtained with cubic spline interpolate polynomials $P(x)$. The program accepts other kind of interpolation polynomials $P(x)$ and also different kind of time integrators.

§5. Numerical Results

We consider the problem

$$\begin{cases} \mu_t = \gamma_1\mu_x + \gamma_2\mu_{xx} + \gamma_3(\mu - \mu_\omega) + \gamma_4 r(\mu) \\ \mu(x,0) = f(x), x \in [0,1]; \mu(0,0) = T_0; \mu(0,t) = r_1(t), t \geq 0 \\ \dfrac{\partial\mu}{\partial x}(1,t) = 0, t \geq 0 \end{cases} \tag{15}$$

with different real parameters $\gamma_1, \gamma_2, \gamma_3, \gamma_4$ and $r(\mu) = \mu^2$. That problem gives the energy balance of a fixed bed catalytic reactor, $f(x), T_0$ and $r(t)$ are assumed to be known functions.

We solve (15) for the values $\gamma_1 = -2, \gamma_2 = \gamma_3 = 0; \gamma_4 = -1; T_0 = 0.8; r_1 = 1.2$ and $f(x_i), i = 0(1)M-1$ given from an approximation of $\mu(x_i, 0)$ obtained by solving the stationary problem (15). Also we consider the boundary conditions $\mu_0 = 0.8$ and $\mu_{M-1} = \mu_M$.

For $h = k = 0.1 * 2^{-6}$ we solve the problem and considered such numerical solution as the exact one.

The algorithm presented here was applied then with $h = 5 * 10^{-2}$ with $N = 2$ and $N = 4$

	reference Solution	Solution with the Algorithm	
		$N = 2$	$N = 4$
Mean Error	1.6946E-2	9.2836E-3	8.0348E-3
Mean Number of Mesh Points	79	31	61

Acknowledgements: The Authors would like to thank Almiro M. Castro and A. Portugal, from the Chemical Dep. Faculty of Sc. and Tech., University of Coimbra and M.Ceu Lopes from Polytechnic. Institute. of Viseu for all the computations.

References

[1] M. Bieterman and I. Babuska, *An adaptive method of lines with error control for parabolic equations of the reaction-diffusion type*, J. Comp. Phys., 63(1986).

[2] J.G. Blom and P.A. Zegeling, *A moving-gird interface for systems of one-dimensional timedependent partial differential equations*, Report NM-R8904(1989). Center for Math. and Computer Science-1009 AB Amsterdam, The Netherlands.

[3] K.Y. Fung et al, *Adaptive refinement with truncation error injection*, Comp. Math Appl. Mech. Eng., 66(1989).

[4] P.Oliveira and F.A. Oliveira, *On a theoretical justification of adaptive gridding for finite difference approximations*. Int. Series of Numerical Maths., 86(1988).

[5] M.A. Revilla, *Simple time and space adaptation in one-dimensional evolutionary partial differential equations*, Int. Journal for Num. Methods in Eng., 23(1986).

[6] J.G. Verwer, J.G. Blom and J.M. Sanz-Serna, *An adaptive moving grid method for one-dimensional systems of partial differential equations*, Journal of Comp. Physics, 82(1989).

Proceedings of the Asian Mathematical
Conference, 1990

A Note for the Branching Character of Superprocesses

OU JINGQI
Department of Mathematics
Zhongshan University
Guangzhou 510275, China

Superprocesses are the analogue of branching Markov processes. The branching character $g(z)$ usually appears in the limit procedure of the construction of such processes. E.B.Dynkin expects that $g(z), z \geq 0$, has the form

$$g(z) = bz + cz^2 + \int_{0+}^{\infty} (e^{-zu} - 1 + \frac{zu}{1+u}) \frac{1+u^2}{u^2} m(du). \tag{$*$}$$

But it has never been proved. In this paper we shall prove the following.

Theorem. *Let $p_N = (p_{Nk}), N = 1, 2, \cdots$ be a sequence of probability distributions on $(k : k = 0, 1, \cdots)$. Let*

$$g_N(z) = N^2(\sum_{k=0}^{\infty}(1 - z/N)^k p_{Nk} - (1 - z/N)), \quad z \geq 0.$$

Assume that the following conditions hold:

$$\sum_{k=1}^{\infty} k^2 p_{Nk} \quad \text{converge uniforly in} \quad N. \tag{1}$$

$$\lim_{N\to\infty} \sum_{k=1}^{\infty} k p_{Nk} \quad \text{exists,} \tag{2}$$

$$\lim_{N\to\infty} g_N(z) = g(z), \quad z \geq 0. \tag{3}$$

Then the limit function $g(z)$ has the form $(*)$, wher b and c are constants; m is affinite measure on $(0, \infty)$.

Proof. Let $q_N(du)$ be a probability measure on the set $(k/N; k = 0, 1, \cdots)$ such that $q_N(\{k/N\}) = p_{Nk}$. It is easy to verify that

$$g_N(z) = Nz + N^2 \int_0^{\infty} ((1 - z/N)^{Nu} - 1) q_N(du). \tag{4}$$

Set

$$h_N(z) = Nz + N^2 \int_0^{\infty} (e^{-zu} - 1) q_N(du). \tag{5}$$

By Lagrange's theorem, we have

$$\begin{aligned} h_N(z) - g_N(z) &= N^2 \sum_{k=0}^{\infty} (e^{-zk/N} - (1 - z/N)^k) p_{Nk} \\ &= N^2 (e^{-z/N} - (1 - z/N)) \sum_{k=1}^{\infty} k p_{Nk} y_N^k, \quad 1 - z/N < y_N < e^{-z/N}. \end{aligned}$$

By (2) it is easy to see that

$$\lim_{N\to\infty} \sum_{k=1}^{\infty} k p_{Nk} y_N^k = \lim_{N\to\infty} \sum_{k=1}^{\infty} k p_{Nk} = c_1.$$

Therefore

$$\lim_{N\to\infty} h_N(z) = \lim_{N\to\infty} g_N(z) + \frac{1}{2} z^2 c_1 = g(z) + \frac{1}{2} z^2 c_1. \tag{6}$$

Set

$$m_N(du) = N^2 \frac{u^2}{1+u^2} q_N(du)$$

we are to prove that the family $(m_N; N = 1, 2, \cdots)$ is weakly precompact. Indeed, we have by (1):

$$\sup_N m_N(u > a) = \sup_N \sum_{k>a} \frac{N^2 k^2}{N^2 + k^2} p_{Nk} \leq \sup_N \sum_{k>a} k^2 p_{Nk} \to 0 \quad (a \to \infty),$$

and

$$\sup_N m_N(R_\pm) = \sup_N \sum_{k=1}^{\infty} \frac{N^2 k^2}{N^2 + k^2} p_{Nk} \leq \sup_N \sum_{k=1}^{\infty} k^2 p_{Nk} < \infty.$$

It has been proved that the family (m_N) is weakly precompact. For simplicity, we suppose that m_N converges weakly to a finite measure m. The function

$$f(u) = \left(e^{-2u} - 1 + \frac{2u}{1+u}\right) \frac{1+u^2}{u^2} \quad \left(\text{at u=0 it is defined as } \frac{1}{2} z^2\right)$$

is a bounded and continuous function, hence

$$\lim_{N\to\infty} \int_0^\infty f(u) m_N(du) = \int_0^\infty f(u) m(du). \tag{7}$$

Using (6) and (7) we have

$$\begin{aligned} g(z) = & -\tfrac{1}{2} z^2 + \int_0^\infty f(u) m(du) + \lim_{N\to\infty} N \left(z - N \int_0^\infty \frac{zu}{1+u} q_N(du)\right) \\ = & \; z \lim_{N\to\infty} N \left(1 - \sum_{k=1}^{\infty} \frac{Nk}{N+k} p_{Nk}\right) + \frac{1}{2} z^2 \left(m(\{0\}) - c_1\right) \\ & + \int_{0+}^\infty \left(e^{-zu} - 1 + \frac{zu}{1+u}\right) \frac{1+u^2}{u^2} m(du) = (*) \end{aligned}$$

where

$$b=\lim_{N\to\infty}N\left(1-\sum_{k=1}^{\infty}\frac{Nk}{N+k}p_{Nk}\right),\quad c=\frac{1}{2}(m(\{0\})-c_1).$$

The theorem is proved.

References

[1] Dynkin.E.B (1988), *Superprocesses and their linear additive functionals*, to appear in Trans. Amer. Math. Soc.

[2] Dynkin.E.B, (1988), Lecture notes in Nankai University.

[3] Dynkin. E.B, (1989), Lecture notes in Zhongshan University.

Proceedings of the Asian Mathematical
Conference, 1990

Three Families of Chromatically Unique Graphs

YEE-HOCK PENG
Department of Mathematics
Universiti Pertanian Malaysia
43400UPM SERDANG, West Malaysia

§1. Introduction

The graphs which we consider here are finite, undirected, simple, and loopless. For a graph G, let $V(G)$ denote its vertex set, $E(G)$ denote its edge set and $P(G;\lambda)$, or simply $P(G)$ if there is no likelihood of confusion, denote its chromatic polynomial. Two graphs X and Y are said to be chromatically equivalent if $P(X) = P(Y)$. A graph G is said to be chromatically unique if $P(G) = P(H)$ implies that H is isomorphic to G, denoted by $H \cong G$.

The search for infinite families of chromatically unique graphs is quite challenging and it is one of the topics on chromatic polynomials that has come to the fore in recent years. For an expository paper giving a survey on most of the works done on chromatically unique graphs, the reader is refer to Koh and Teo [4].

The complete bipartite graph in which the two partite sets consist of m and n vertices will be denoted by $K_{m,n}$. We denote by $K^1_{m,n}(s)$ (resp., $K^2_{m,n}(s)$ and $K^3_{m,n}(s)$), the graph obtained from $K_{m,n} \cup P_s$ by identifying the end vertices of the path P_s on s vertices with two vertices of $K_{m,n}$ which are adjacent (resp., nonadjacent with degree n and nonadjacent with degree m). Note that $K^2_{m,n}(s) \cong K^3_{m,n}(s)$ if and only if $m = n$.

In [2] the author proved the following result.

Theorem 1. *The graphs $K^i_{2,3}(s)(i = 1,2,3)$ where $s \geq 3$ are chromatically unique.*

As a natural extension of Theorem 1, we will show that the graph $K^i_{2,4}(s)$ is also chromatically unique for $1 \leq i \leq 3$ and $s \geq 3$. Furthermore, in this paper, we will prove that the graphs $K^1_{3,3}(2s)$ and $K^2_{3,3}(2s-1)$ are chromatically unique for $s \geq 4$. We also show that a bipartite graph obtained from $(K_{3,3} - e) \cup P_s (s \geq 5)$ where e is an edge of $K_{3,3}$, by identifying the end vertices of the path P_s with two nonadjacent vertices of $K_{3,3} - e$ is chromatically unique.

For terms used but not defined here the reader is refer to [1].

§2. Preliminaries

This section contains known results which will be used to prove our main theorems in the next section. We need the following notations to state the first known result.

Let H be a nonempty graph with two nonadjacent vertices u and v. The graph denoted by H^* is obtained from H by identifying the vertices u and v. Let G_1 (resp., G_2) be the graph obtained from $H \cup P_s$ by identifying the end vertices of the path P_s with the vertices u and v (resp., any two adjacent vertices) of H. Then by applying Theorems 1 and 3 in [6], and by using the well known fact that for the cycle C_n with n vertices, $P(C_s;\lambda) = (\lambda-1)^s + (-1)^s(\lambda-1)$, we have

Theorem A(Read [7] and Chia [2]). *Let G_1, G_2, and H^* be the graphs defined as above. Then*

$$P(G_1) = P(G_2) + (-1)^{s-1}P(H^*).$$

Let $K^-_{m,n}$ be the graph obtained by deleting one edge from the complete bipartite graph $K_{m,n}$. $N_G(L)$ will denote the number of subgraphs of G isomorphic to L, and $I_G(L)$ will denote the number of induced subgraphs of G isomorphic to L (so that $N_G(L) = I_G(L)$ if L is complete, or if G is bipartite and L is complete bipartite).

The following necessary conditions for two graphs to be chromatically equivalent can be deduced and are well known (see, for examples [3] and [9]).

Theorem B. *Let G and H be two chromatically equivalent graphs. Then*

(i) $|V(G)| = |V(H)|$;

(ii) $|E(G)| = |E(H)|$;

(iii) $N_G(C_3) = N_H(C_3)$;

(iv) $I_G(C_4) - 2N_G(K_4) = I_H(C_4) - 2N_H(K_4)$;

(v) $(q-3)I_G(C_4) + N_G(K_{2,3}) - I_G(C_5) = (q-3)I_H(C_4) + N_H(K_{2,3}) - I_H(C_5)$ *if G has at most one triangle*;

(vi) $\chi(G) = \chi(H)$, *where $\chi(G)$ denotes the chromatic number of G*;

(vii) *G is connected if and only if H is connected*;

(viii) *G is 2-connected if and only if H is 2-connected.*

For bipartite graphs, we have another necessary condition for them to be chromatically equivalent.

Theorem C (Peng [5]). *Let G and H be two chromatically equivalent bipartite graphs. Then*

$$\begin{aligned} & I_G(C_6) + N_G(K_{2,4}) - I_G(K^-_{3,3}) - 4N_G(K_{3,3}) \\ = \; & I_H(C_6) + N_H(K_{2,4}) - I_H(K^-_{3,3}) - 4N_H(K_{3,3}). \end{aligned}$$

§3. Main Results

We shall show the following key result (see also [2]). First the definition:

Definition. A subgraph F of a graph G is called a chromatic subgraph of G if for any graph H with $P(H) = P(G), H$ contains F as a subgraph.

Theorem 2. *Let G be a 2-connected graph obtained from $F \cup P_s$ by identifying the end vertices of the path P_s with two distinct vertices of a graph F. Suppose that F is chromatic subgraph of G. If H is chromatically equivalent to G, then H is also a graph obtained from $F \cup P_s$ by identifying the end vertices of P_s with two distinct vertices of F.*

Proof. It is trivial if $s = 2$. Assume that $s \geq 3$. Since F is chromatic subgraph of G, H contains two subgraphs F and $Y = H - F$ with some edges connecting them. Thus by Theorem B(ii),

$$|E(Y)| + e(F,Y) \leq s - 1, \tag{1}$$

where $e(F,Y)$ denotes the number of edges joining vertices of F to vertices of Y. Now let c be the number of components of Y. Then

$$|E(Y)| \geq s - 2 - c, \tag{2}$$

since $|V(Y)| = s - 2$. Also $e(F,Y) \geq 2c$ because H is 2-connected by Theorem B(viii). Hence from (1) we have

$$|E(Y)| \leq s - 1 - 2c. \tag{3}$$

Therefore (2) and (3) imply that $c = 1$, i.e., Y is connected. Consequently, $|E(Y)| = s - 3$ and $e(F,Y) = 2$. Since H is 2-connected, Y must be the path P_{s-2} whose end vertices are join to two distinct vertices of F. □

Corollary 1. *Let G be a 2-connected graph obtained from $F \cup P_s$ by identifying the end vertices of the path P_s with two adjacent vertices of F. Suppose that F is chromatic subgraph of G. If F is edge-transitive, then G is chromatically unique.*

Proof. Let H be a graph with $P(H) = P(G)$. Then by Theorem 2, H is a graph obtained from $F \cup P_s$ by identifying the end vertices of P_s with two distinct vertices x and y of F. By Theorem A, x is adjacent to y. Since F is edge-transitive, we have $H \cong G$. □

Corollary 2. *If the complete bipartite graph $K_{m,n} (m, n \geq 2)$ is chromatic subgraph of $K^i_{m,n}(s)$ where $s \geq 2$, then $K^i_{m,n}(s)$ is chromatically unique for $1 \leq i \leq 3$.*

Proof. For $i = 1$, the result follows from Corollary 1 because $K_{m,n}$ is edge-transitive. If H is chromatically equivalent to $K^2_{m,n}(s)$ or $K^3_{m,n}(s)$, then by Theorem 2, H is also a $K^i_{m,n}(s)(i = 2, 3)$. If $m = n$, we are done. Otherwise, by Theorem A, $P(K^2_{m,n}(s)) \neq P(K^3_{m,n}(S))$ since $K_{m-1,n}$ is not chromatically equivalent to $K_{m,n-1}$ (see Teo and Koh [8]). Hence $K^i_{m,n}(s)$ is chromatically unique for $i = 2, 3$. □

We shall now give three infinite families of chromatically unique graphs. The first one is stated in the following theorem.

Theorem 3. *The graph $K^i_{2,4}(s)$ is chromatically unique for $s \geq 3$ and $1 \leq i \leq 3$.*

Proof. Let $G_1 = K^2_{2,4}(3), G_2 = K^3_{2,4}(3), G_3 = K^1_{2,4}(4), G_4 = K^2_{2,4}(4), G_5 = K^3_{2,4}(4)$, and $G_6 = K^1_{2,4}(5)$. By Corollary 2 to Theorem 2, we need only to show that $K_{2,4}$ is chromatic subgraph of $K^i_{2,4}(s)$ for $1 \leq i \leq 3$ and $s \geq 3$. We consider two cases:

Case (a). $K^i_{2,4}(s)(1 \le i \le 3) \not\cong G_j(1 \le j \le 6)$.

Let H be a graph with $P(H) = P(K^i_{2,4}(s))$ for $1 \le i \le 3$ and $s \ge 3$. We will show that $K_{2,4}$ is chromatic subgraph of $K^i_{2,4}(s)$, i.e., $N_H(K_{2,4}) \ne 0$. By Theorem B, we have $I_H(C_4) = 6$ and $N_H(K_{2,3}) \ge 4$. Suppose that $I_H(K^-_{3,3}) = 1$. Note that $I_{K^-_{3,3}}(C_4) = 5$ and $N_{K^-_{3,3}}(K_{2,3}) = 2$. If $N_H(K_{2,3}) \ge 3$ and $I_H(K^-_{3,3}) = 1$, then we must have $I_H(C_4) \ge 7$. Also if $I_H(K^-_{3,3}) \ge 2$, then it is easy to see that $I_H(C_4) \ge 7$ (see Figure 1 for the minimal example). Since $N_H(K_{2,3}) \ge 4$, the suppossition that $I_H(K^-_{3,3}) \ge 1$ leads to contradiction and hence we have $I_H(K^-_{3,3}) = 0$.

G :

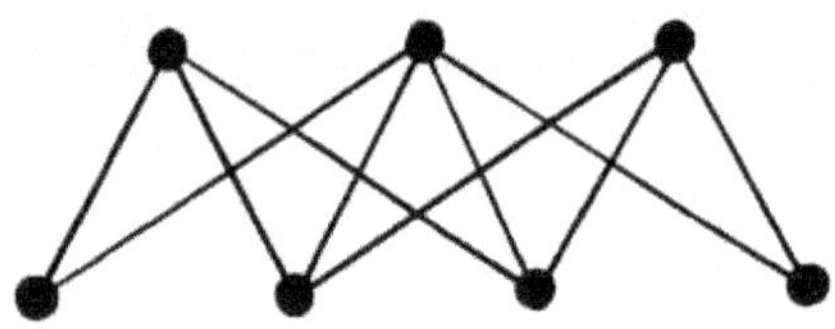

Figure 1. $I_G(K^-_{3,3}) = 2$ and $I_G(C_4) = 7$

We now assume that $N_H(K_{2,4}) = 0$ and we shall get a contradiction. Note that if $N_H(K_{2,3}) = 2$, then these two $K_{2,3}$'s have at most one edge in common, for otherwise, H contains $K_{2,4}$ or $K^-_{3,3}$ as an induced subgraph. So if $N_H(K_{2,3}) \ge 3$, we must have $I_H(C_4) \ge 7$, a contradiction. Therefore $N_H(K_{2,4}) \ne 0$ because $N_H(K_{2,3}) \ge 4$.

Case (b). $G_j(1 \le j \le 6)$.

The graph $G_1 \cong K_{2,5}$ is chromatically unique (see [8]). Let H be a graph with $P(H) = P(G_2)$ or $P(G_3)$. Then by Theorem B, H is a bipartite graph satisfying $10 \le |E(H)| \le 11$, $7 \le |V(H)| \le 8$, and $7 \le I_H(C_4) \le 8$. Thus $N_H(K_{3,3}) = 0$ and $I_H(C_6) = 0$, for otherwise, $I_H(C_4) \ge 9$ or $I_H(C_4) \le 5$ which is a contradiction. So by Theorem C, we have $N_H(K_{2,4}) - I_H(K^-_{3,3}) = 1$ which implies $N_H(K_{2,4}) \ne 0$.

We now consider $G_j(4 \le j \le 6)$. Let H be a graph chromatically equivalent to G_j for any $4 \le j \le 6$. Then by Theorem B, H contains 6 induced subgraphs C_4 but has no triangles, and H is not a bipartite graph.

We first consider G_6. By Theorem B (v), we have $N_H(K_{2,3}) - I_H(C_5) = 3$, i.e., $N_H(K_{2,3}) \ge 3$. By using the arguments similar to Case (a), we can show that $N_H(K_{2,4}) \ne 0$.

We next consider G_5. Theorem B (v) gives us $N_H(K_{2,3}) - I_H(C_5) = 2$. We claim that $I_H(C_5) \ge 1$, for otherwise, $I_H(C_7) = 1$ because $|V(H)| = 8$ and H is not a bipartite graph; but then there is no graph H of order 8 and size 11 such that $N_H(C_3) = I_H(C_5) = 0$ and $I_H(C_7) = 1$. Thus $N_H(K_{2,3}) \ge 3$. By applying the arguments used in Case (a), we can again show that $N_H(K_{2,4}) \ne 0$.

We now consider the last graph G_4. For this graph, we have $N_H(K_{2,3}) = I_H(C_5)$ by Theorem B. As in the case of G_5 above, $I_H(C_5) \ne 0$. If $I_H(C_5) = N_H(K_{2,3}) = 1$, then H must contain a spanning subgraph which is obtained from $K_{2,3} \cup C_5$ by identifying an edge of C_5 with an edge of $K_{2,3}$ because $|V(H)| = 8$. Since H has no triangles, it can easily be confirmed that $|E(H)| = 10$, which is a contradiction. So we must have $I_H(C_5) = N_H(K_{2,3}) \geq 2$. Thus H contains J_1 or J_2 of Figure 2 as a subgraph.

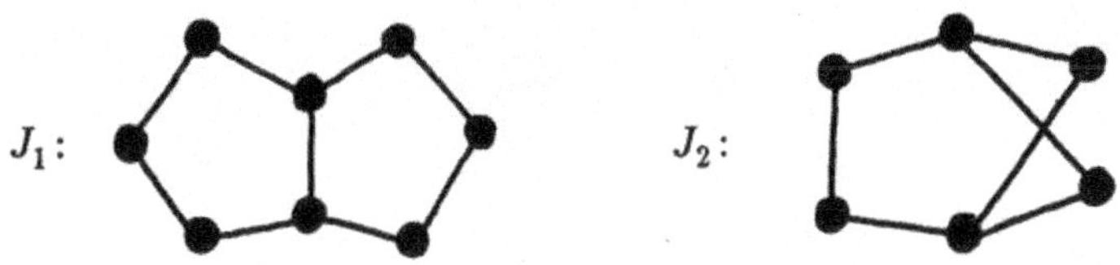

Figure 2.

Since $|V(H)| = 8, |E(H)| = 11$, and $I_H(C_4) = 6$, it is not difficult to check that J_1 can not be a subgraph of H. So H contains two vertex disjoint subgraphs J_2 and L, where L is a graph of order 2. Since $|E(H)| = 11$, and $N_H(K_{2,3}) \geq 2$, there are only three possibility for H. We show these three graphs in Figure 3. But $H \not\cong H'$ or G_5 because $I_{H'}(C_4) = 5$ and $P(G_5) \neq P(G_4)$. Thus G_4 is chromatically unique.

The proof is now complete. □

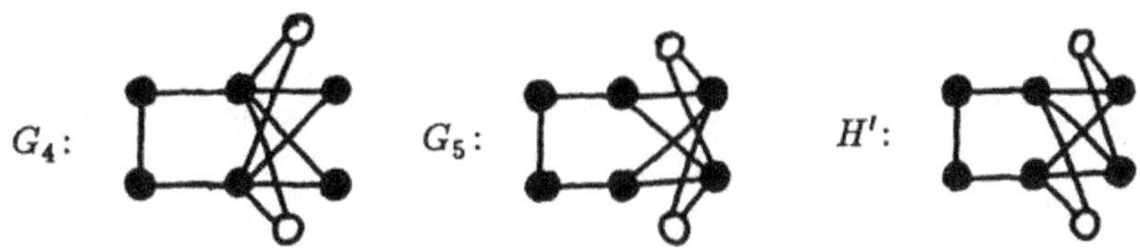

Figure 3.

Remark. The graph $K^3_{2,4}(2)$ is also chromatically unique.

Our second and third infinite families of chromatically unique graphs are stated in the following theorems.

Theorem 4. *The graphs $K^1_{3,3}(2s)$ and $K^2_{3,3}(2s-1)$ are chromatically unique for $s \geq 4$.*

Proof. Let H be a graph with $P(H) = P(K^1_{3,3}(2s))$ or $P(K^2_{3,3}(2s-1))$ where $s \geq 4$. Then by Theorem B, we have $I_H(C_4) = 9$. Since $K^1_{3,3}(2s)$ and $K^2_{3,3}(2s-1)$ are bipartite graphs, and $s \geq 4$, we have $I_H(C_6) + N_H(K_{2,4}) - I_H(K^-_{3,3}) - 4N_H(K_{3,3}) = -4$ by Theorem C. If $N_H(K_{3,3}) = 0$, then $I_H(K^-_{3,3}) \geq 4$. Note that one $K^-_{3,3}$ contains 5 induced subgraphs C_4, and if $I_H(K^-_{3,3}) \geq 4$. then it is not difficult to see that $I_H(C_4) \geq 11$ (see Figure 4 for the minimal case). Therefore $N_H(K_{3,3}) \neq 0$ and by Corollary 2 to Theorem 2, the graphs $K^1_{3,3}(2s)$ and $K^2_{3,3}(2s-1)$ are chromatically unique for $s \geq 4$. □

Remark. The graph $K^1_{3,3}(4)$ is also chromatically unique.

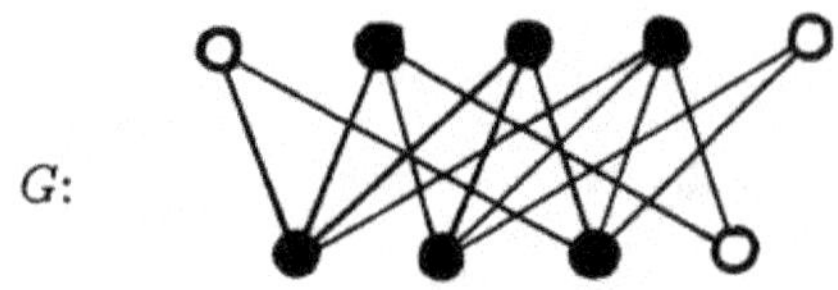

Figure 4. $N_G(K_{3,3}) = 0, I_G(K_{3,3}^-) = 4$ and $I_G(C_4) = 11$

Theorem 5. *The bipartite graph obtained from $K_{3,3}^- \cup P_s (s \geq 5)$ by identifying the end vertices of P_s with two nonadjacent vertices of $K_{3,3}^-$ is chromatically unique.*

Proof. Let G be a bipartite graph defined as in the theorem and let H be a graph with $P(H) = P(G)$. Then by Theorem B and C, we have $I_H(C_6) - I_H(K_{3,3}^-) = -1$ since $I_H(C_4) = I_G(C_4) = 5$. Thus $I_H(K_{3,3}^-) \neq 0$, i.e., the graph $K_{3,3}^-$ is chromatic subgraph of G. Therefore by Theorem 2, H must be a graph obtained from $K_{3,3}^- \cup P_s$ by identifying the end vertices of P_s with two distinct vertices of $K_{3,3}^-$. By Theorem A, it is not difficult to verify that H is isomorphic to G. □

In view of Theorems 3 and 4 it is tempting to state the following conjectures:

Conjectures.

(1) The graph $K_{2,n}^i(s)$ is chromatically unique for $1 \leq i \leq 3$ and $s, n \geq 3$.

(2) The graph $K_{m,m}^1(2s)$ and $K_{m,m}^2(2s-1)$ are chromatically unique for $m \geq 2$ and $s \geq 4$.

References

[1] M. Behzad, C. Chartrand, L. Linda-Lesniak, *Graphs and Digraphs*, Wadsworth, Belmont, Calif. (1979).

[2] G.L. Chia, *Personal communication.*

[3] E.J. Farrell, *On chromatic coefficients*, Discrete Math., 29(1980), 257–264.

[4] K.M. Koh and K.L. Teo, *The search of chromatically unique graphs*, Graphs and Combinatorics (to appear).

[5] Y.H. Peng, *On the chromatic coefficients of a bipartite graph*, (submitted).

[6] R.C. Read, *An introduction to chromatic polynomials*, J. Comb. Theory, 4(1968), 52–71.

[7] R.C. Read, *Broken wheels are SLC*, Ars Combinatoria, 21-A(1986), 123–128.

[8] C.P. Teo and K.M. Koh, *The chromaticity of complete bipartite graphs with at most one edge deleted*, J. Graph Theory, 14(1990), 89–99.

[9] E.G. Whitehead Jr and L.C. Zhao, *Cutpoints and the chromatic polynomials*, J. Graph Theory, 8(1984), 371–377.

[10] H. Whitney, *The colouring of graphs*, Ann. of Mathematics, 33(1932), 688–718.

Proceedings of the Asian Mathematical
Conference, 1990

Groups and Designs

CHERYL E. PRAEGER
Department of Mathematics
University of Western Australia
Nedlands WA 6009, Australia

§1. Introduction

To a combinatorial mathematician or to an algebraist, a design is an incidence structure, while to a statistician a design is an experimental layout. The theory of designs developed in these two mathematical cultures in response to different sets of mathematical problems. Perhaps the first systematic study of a design was undertaken by Euler [21] in 1782 in his analysis of the problem of the 36 officers: select, from each of 6 regiments, 6 officers - one of each of 6 ranks - and parade the 36 officers in a 6×6 formation so that each row and each column contains one officer from each regiment and one of each rank. Then in the 19th century Plücker [40, 41] studied a class of 2-designs with blocks of size 3, which were related to third degree curves in a plane. Woolhouse [48] and Kirkman [33, 34] studied the same designs, and Steiner [44] was led to the study of these and other t-designs, $t \geq 3$ (now called Steiner systems) by some problems concerning double tangents to fourth degree curves in the complex plane. The designs studied by these people all belong to the class of designs we now call t-designs.

Definition 1.1. A t-(v,k,λ) design $D = (\Omega, \mathcal{B})$ (or simply a t-design) consists of a set Ω of v points and a set $\mathcal{B}$ of k-element subsets of points, called blocks, with the property that every t-element subset of points is contained in exactly λ blocks.

One of the first significant interactions between group theory and design theory concerns the Mathieu groups. In 1861, as a consequence of some investigations in analysis, Emile Mathieu [37] defined several multiply transitive permutation groups. Recall that a permutation group $G \leq \mathrm{Sym}(\Omega)$ on a set Ω is said to be transitive on Ω (or 1-transitive on Ω) if, for all points $\alpha, \beta \in \Omega$ there is an element $g \in G$ such that $\alpha^g = \beta$; and, for $k \geq 2, G$ is said to be k-transitive on Ω, if G is transitive on the set of ordered k-tuples of distinct points of Ω. Mathieu defined 5-transitive permutation groups M_{12} and M_{24} on sets of 12 and 24 points respectively. He gave generating sets of permutations for M_{12} and M_{24}. However his computations were very complicated and hence his assertions about M_{12} and M_{24} were difficult to check. In 1938 Ernst Witt [46, 47] constructed a 5-$(12,6,1)$ design and a 5-$(24,8,1)$ design and showed that their automorphism groups were M_{12} and M_{24} respectively. Recall that an automorphism of a t-design $\mathcal{D}$ is a permutation of the set Ω of points which maps blocks to blocks. The set of all automorphisms forms the automorphism group Aut $\mathcal{D}$ of $\mathcal{D}$, a subgroup of $\mathrm{Sym}(\Omega)$. The structure of the 5-designs gave a new and better understanding of the Mathieu groups.

The fields of finite group theory, and design theory, have each contributed to the other, new constructions and new insights. We shall discuss some of these interactions, concentrating on designs with automorphism groups acting transitively on the set of blocks, that is on block-transitive designs.

§2. 2-Transitive Groups and Simple Groups

Up to 1965 the only known finite simple groups which did not fit into an infinite family of simple groups were the five Mathieu groups: M_{12} and M_{24} together with the subgroups M_{11} and M_{23} fixing a point, and the stabilizer M_{22} of two points in M_{24}. In the 1960's several new so-called sporadic simple groups were discovered, some of them involving the Witt designs. D.G. Higman and C.C. Sims [29] used the 2-$(22,6,1)$ design admitting M_{22} to construct a transitive group HS on a set of 100 points with one-point stabilizer M_{22}. J. Conway [15] constructed several new simple groups as groups of automorphisms of the Leech lattice; and the construction of the Leech lattice itself involved the Witt 5-$(24,8,1)$ design admitting M_{24}.

Also, among the infinite families of simple groups and their automorphism groups, several families were known to have representations as 2-transitive permutation groups, for example the projective groups $\mathrm{PSL}(d,q)$, the 3-dimensional unitary groups $\mathrm{PSU}(3,q)$, the Suzuki groups $Sz(q)$, and the Ree groups $\mathrm{Ree}(q)$. And several families of these 2-transitive simple groups were known to occur as groups of automorphisms of 2-$(v,k,1)$ designs, namely the projective, unitary, and Ree groups. Thus group theorists had an interest in 2-designs, hoping that the study of the designs arising would give a fresh understanding of the groups and, perhaps, new constructions of designs and groups.

The finite simple group classification was completed around 1980 and one of its first consequences was the complete classification of the finite 2-transitive groups (see [6, 25, 26, 35]). Kantor [31] used this latter classification to obtain a classification of all 2-$(v,k,1)$ designs admitting an automorphism group which was 2-transitive on points –the so-called 2-transitive designs.

Theorem 2.1 (Kantor). *Let $\mathcal{D}=(\Omega,\mathcal{B})$ be a 2-$(v,k,1)$ design with an automorphism group G which is 2-transitive on the point set Ω. Then $\mathcal{D}$ and G are one of the following:*

(a) $\mathcal{D}=PG(d,q)$, the design of points and lines of a d-dimensional projective geometry over $GE(q), d\geq 2$, and either $\mathrm{PSL}(d,q)\leq G\leq \mathrm{P\Gamma L}(d,q)$, or $(d,q)=(3,2)$ and $G=A_7$.

(b) $\mathcal{D}=AG(d,q)$, the design of points and lines of a d-dimensional affine space over $GE(q), d\geq 2$, and G is a 2-transitive subgroup of $\mathrm{A\Gamma L}(d,q)$.

(c) $\mathcal{D}$ is a Hermitian unital, that is the design of q^3+1 totally isotropic points (1-spaces) and nondegenerate lines (2-spaces) of a 3-dimensional unitary space over $GF(q^2)$, and $\mathrm{PSU}(3,q)\leq G\leq \mathrm{P\Gamma U}(3,q)$.

(d) $\mathcal{D}$ is Ree unital, that is $\mathrm{Ree}(q)={}^2G_2(q)\leq G\leq \mathrm{Aut\ Ree}\ (q)$, where $q=3^{2e+1}\geq 3$, Ω is a set of q^3+1 points on which G acts, and the blocks of $\mathcal{D}$ are the sets of fixed points of involutions of G.

(e) $\mathcal{D}$ *is one of two non-Desarguesian affine planes of order* k=27 *(the Hering plane [24], or see [20, p. 236]) or* k=9 *(a nearfield plane [22] or see [20, p.236]).*

(f) $\mathcal{D}$ *is one of two* 2-$(3^6, 3^2, 1)$ *designs [27], and* $G = Z_3^6.SL(2,13)$.

§3. Flag-transitive Designs

A flag of a design $\mathcal{D} = (\Omega, \mathcal{B})$ is a pair (α, B) where $\alpha \in \Omega, B \in \mathcal{B}$ and $\alpha \in B$. The design $\mathcal{D}$ is said to be flag-transitive if its automorphism group is transitive on the set of flags of $\mathcal{D}$. Most interest in flag-transitive designs has focussed on the case of 2-(v, k, λ) designs $\mathcal{D}$ with $\lambda = 1$. A result of higman and McLaughlin [28] in 1961 showed that in this case the automorphism group is primitive on the point set Ω, that is $\mathcal{D}$ is pointprimitive. (A permutation group G on Ω is said to preserve a partition P of Ω if, for every part $a \in P$ and every $g \in G$, the set a^g is also a part of P. Each transitive group G on Ω preserves the two so-called trivial partitions $\{\Omega\}$ and $\{\{\alpha\} \mid \alpha \in \Omega\}$; if these are the only partitions preserved by G then G is said to be primitive on Ω, while if G preserves a nontrivial partition of Ω then G is said to be imprimitive on Ω). Then several people showed that the socle of the automorphism group of a flag-transitive 2-$(v, k, 1)$ design is either elementary abelian or a nonabelian simple group (see [13] and [3, 12, 49]). Using this reduction Buekenhout and others have completed the mammoth task of classifying all flag-transitive 2-$(v, k, 1)$ designs apart from those with automorphism group a 1-dimensional affine group. This theorem is reported in [4] and is being prepared for publication [5]. Note that all the 2-transitive designs in Theorem 2.1 are examples of flag-transitive 2-$(v, k, 1)$ designs.

Theorem 3.1. *Let* $\mathcal{B} = (\Omega, \mathcal{D})$ *be a* 2-$(v, k, 1)$ *design with an automorphism group* G *which is transitive on the set of flags of* $\mathcal{D}$. *Then one of the following holds.*

(a) $\mathcal{D}$ *is one of the designs listed in Theorem 2.1.*

(b) $\mathcal{D}$ *is a Witt-Bose-Shrikhande space, that is* $\mathrm{PSL}(2, q) \leq G \leq \mathrm{P\Gamma L}(2, q)$, *where* $q = 2^n \geq 8, \Omega$ *is a set of* $q(q-1)/2$ *points on which* G *acts (with the stabilizer of a point in* $\mathrm{PSL}(2, q)$ *being a subgroup* $D_{2(q+1)}$*), and the blocks of* $\mathcal{D}$ *are the sets of fixed points of involutions in* $\mathrm{PSL}(2, q)$.

(c) $\mathcal{D}$ *is the design of points and lines of a non-Desarguesian Lüneburg affine plane [36] of order* $k = 2^{2e+1} \geq 8$.

(d) v *is a prime power and* $G \leq \mathrm{A\Gamma L}(1, v)$.

There are many designs arising under (d) but they have not been completely classified. Among them are the generalized Netto systems [4, 41] and the Kantor systems [4, 4.2].

The theorem has important applications to finite geometry and has already been used by Delandtsheer and Doyen [17, 19] to study semiovals and maximal arcs in finite projective planes.

§4. Block-transitive, Point-imprimitive Designs

In the area of block-transitive designs there has been some recent work done on 2-(v, k, λ) designs for general λ which is interesting in its own right, as well giving new results in the geometrically interesting special case of $\lambda = 1$. It follows from a result of Block [2] (or see [20, 2.3.1]), that if a group of automorphisms of a t-design $\mathcal{D}$ is transitive on the blocks of $\mathcal{D}$ then it is also transitive on the points of $\mathcal{D}$. Now by [28], a flag-transitive 2-$(v, k, 1)$ design must be point-primitive. But what can be said about a block-transitive 2-$(v, k, 1)$ design with automorphism group acting imprimitively on points (other than the fact that it cannot be flag-transitive)? Recently Delandtsheer and Doyen [18] showed that, for a bloc-transitive, point-imprimitive 2-(v, k, λ) design, the number v of points cannot be too large relative to the block size k.

Theorem 4.1[18]. *Let $\mathcal{D}$ be a block-transitive, point imprimitive 2-(v, k, λ) design. Then*

$$v \leq \left(\binom{k}{2} - 1\right)^2.$$

It follows from a result of Ray-Chaudhuri and Wilson [43] that the automorphism group of a block-transitive 4-design is 2-homogeneous, and hence primitive on points. Thus, for a block-transitive, point-imprimitive t-design, the parameter t is at most 3. In [8] it was shown that for a block-transitive, point-imprimitive 3-(v, k, λ) design $\mathcal{D}, v \leq \binom{k}{2} + 1$ if Aut$\mathcal{D}$ preserves a partition with either 2 parts or parts of size 2; this bound was also shown to hold (and was often attained) whenever $k \leq 70$ by a computer computation. It is an open question whether or not this better holds bound in general for 3-designs.

Question 4.2[8]. For a block-transitive, point-imprimitive 3-(v, k, λ) design, is it true that $v \leq \binom{k}{2} + 1$?

The question of when the Delandtsheer-Doyen bound for 2-designs is attained has been almost completely answered. This was done in [8] for $k \neq 3, 4, 5$ or 8.

Theorem 4.3[8] *Let $\mathcal{D}$ be a 2-(v, k, λ) design with $k \neq 3, 4, 5$ or 8 such that $v = m^2 = (\binom{k}{2} - 1)^2$, and with automorphism group G acting transitively on blocks and imprimitively on points. Then for some 2-transitive subgroups T_1, T_2 of S_m, either*

(a) $T_1 \times T_2 \leq G \leq \text{Aut } T_1 \times \text{Aut } T_2$

or

(b) $T_1^m \leq G \leq \text{Aut } T_1 \text{ wr } S_m$.

Moreover the possible designs $\mathcal{D}$ arising are completely determined in [8]. In particular, if the only 2-transitive groups of degree m are A_m and S_m, then there are up to isomorphism just three possible designs $\mathcal{D}$ (and by [7] this is the case for "almost all" m). Note that 3, 4, 5, 8 are precisely the values of k for which $v = (\binom{k}{2} - 1)^2$ is a prime power and in this case the proof in [8] does not work when G has an abelian socle. The classification is an open problem for these values of k.

Problem 4.4[8]. *Determine all block-transitive, point-imprimitive 2-(v, k, λ) designs with $v = (\binom{k}{2} - 1)^2$, and $k = 3, 4, 5, 8$.*

A little more is known. In the special case where $\lambda = 1$ the classification is complete and the result is quite surprising. All of the designs constructed from Theorem 4.3 are such that the parameter λ is very large, and so the only possible 2-$(v, k, 1)$ designs satisfying the hypotheses and bound of Theorem 4.1 must have $k = 3, 4, 5$ or 8. In fact we showed that k must equal 8 and there are lots of examples.

Theorem 4.5[38,39]. *Let $\mathcal{D}$ be a block-transitive, point-imprimitive 2-(v, k, λ) design with $v = (\binom{k}{2} - 1)^2$. Then $\mathcal{D}$ is a 2-$(729, 8, 1)$ design, the automorphism group is of the form $N.Z_{13}$, where $|N| = 729$, and one of the following holds:*

(a) *$N = Z_3^6$, and there are up to isomorphism 27 such designs,*

(b) *$N = Z_9^3$, and there are up to isomorphism 13 such designs,*

(c) *N is the free nilpotent class 2, 3 generator, exponent 3 group, and there are up to isomorphism 427 such designs.*

Apart from these examples, block-transitive, point-imprimitive 2-$(v, k, 1)$ designs seem quite rare. There are several such 2-$(91, 6, 1)$ designs the automorphism groups of which are analysed in [11], but these and the ones in Theorem 4.5 are the only ones known to the author. It was shown by Camina and di Martino [10, 12] that there are no further examples with $k \leq 6$. Further, it follows from Camina and Gagen [13] that k does not divide v, and information about the structure of the automorphism groups of such designs can be found in [14].

§5. Block-primitive Designs

There has been some work done on 2-$(v, k, 1)$ designs with an automorphism group primitive on blocks. First Kantor considered the special case of projective planes.

Theorem 5.1[32]. *Let $\mathcal{D}$ be a finite projective plane of order q, that is a 2-$(q^2 + q + 1, q + 1, 1)$ design, with a group G of automorphisms acting primitively on blocks. Then either*

(a) *$\mathcal{D}$ is desarguesian and $\mathrm{PSL}(3, q) \leq G \leq \mathrm{P\Gamma L}(3, q)$,*

or

(b) *$v = q^2 + q + 1$ is prime and G is cyclic of order v or a Frobenius group of order dividing $v(q+1)$ or vq.*

Then Delandtsheer considered such designs which are not projective planes and showed that their automorphism groups are almost simple.

Theorem 5.2[16]. *Let $\mathcal{D}$ be a 2-$(v, k, 1)$ design which is not a projective plane, with a group G of automorphisms acting primitively on blocks. Then $T \leq G \leq \mathrm{Aut}\ T$ for some nonabelian simple group T.*

Problem 5.3. Classify all block-primitive 2-$(v, k, 1)$ designs which are not projective planes.

The connection between block-primitive and point-primitive 2-$(v, k, 1)$ designs is not completely clear. Delandtsheer [16] showed that for small block sizes block-primitivity implies point-primitivity.

Theorem 5.4[16]. *Let $\mathcal{D}$ be a 2-$(v, k, 1)$ design with $k < 30$ and with a group G of automorphisms acting primitively on blocks. Then G acts primitively on the points of $\mathcal{D}$.*

This result supports the conjecture:

Conjecture 5.5(Delandtsheer). If $\mathcal{D}$ is a block-primitive 2-$(v,k,1)$ design then $\mathcal{D}$ is point-primitive.

The paper [16] contains a lot of information about the automorphism group of any design which is a counterexample to this conjecture, and in particular it is shown that $k^2 < v \le (k^2-k-4)(k^2-k-2)/8$.

§6. Block-transitive, Point-primitive Designs.

By Theorem 4.1 (of Delandtsheer and Doyen), most "large" block-transitive t-designs are point-primitive, $t \ge 2$. Of course for every $v \ge k \ge t \ge 2$, we have the trivial t-(v,k,λ) design for which every k-element subset of points is a block, $\lambda = \binom{v-t}{k-t}$, and the group S_v is transitive on blocks and primitive on points. All other t-designs will be called nontrivial, and for nontrivial t-designs, one question we might ask is:

how large can t be?

Nontrivial t-designs were constructed for every integer t by Tierlinck [45], but for nontrivial block-transitive t-designs, the value of t is every restricted. It follows from [43] that the automorphism group of a nontrivial block-transitive t-design is transitive on $[t/2]$-element subsets of points. That is, it is $[t/2]$-homogeneous on points. A little extra work shows:

Theorem 6.1[9]. *There are no nontrivial block-transitive t-designs for $t \ge 8$. There are no nontrivial flag-transitive t-designs for $t \ge 7$.*

We belive that a much stronger result holds, namely:

Conjecture 6.2[9]. There are no nontrivial block-transitive 6-designs.

Work is currently underway on this by Cameron, O'Keefe, Penttila and the author.

Returning now to the general problem of constructions of block-transitive t-designs for small $t \ge 2$, it was observed in [8, Proposition 1.1]:

Proposition 6.3[8]. *If G is a group of automorphisms of a t-(v,k,λ)* design $\mathcal{D} = (\Omega, \mathcal{B})$ which is transitive on the block set $\mathcal{B}$, and if $G < H \le \mathrm{Sym}(\Omega)$, then $\mathcal{D}^* = (\Omega, \mathcal{B}^*)$, where

$$\mathcal{B}^* = \{B^h \mid B \in \mathcal{B}, h \in H\}$$

is also a t-(v,k,λ^*) design admitting H as a block-transitive group of automorphisms for some $\lambda^* \ge \lambda$.

The design $\mathcal{D}^*$ will be nontrivial if and only if H is not k-homogenous on points (that is H is not transitive on k-element subsets of points). Thus:

Corollary 6.4. *For a given triple $(v,\ k,\ t)$ of positive integers, with $2 \le t \le k \le v$, there exists a nontrivial block-transitive t-(v,k,λ) design for some $\lambda \ge 1$, if and only if there exists such a design with automorphism group maximal in S_v subject to being not k-homogeneous.*

Note that for $k \ge 6$ the only k-homogeneous subgroups of S_v are A_v and S_v (see [6]). (This Corollary could be used even by those only interested in designs with $\lambda = 1$ as a first step in their search for such block-transitive designs.) Now, if $2 \le t < k$ and a subgroup

H of Sym(Ω), $|\Omega| = v$, is t-homogeneous but not k-homogeneous, then for any k-element subset B of Ω, $\mathcal{D} = (\Omega, B^H)$, where $B^H = \{B^h \mid h \in H\}$, is a nontrivial t-(v, k, λ) design for some λ. So the natural problem arising from Corollary 6.4 is the following.

Problem 6.5 Given $2 \leq t < v = |\Omega|$, and given a subgroup H of Sym(Ω) which is transitive on Ω and is maximal subject to being not t-homogeneous on Ω, decide whether, for some k and λ, a nontrivial t-(v, k, λ) design can exist. (and if so give a construction.)

This problem was considered in 1971 by W.O. Alltop [1] for the class of affine groups AGL$(d, 2), d \geq 3$, and also more recently by the author [42] for several other classes of subgroups $H \leq S_v$ which are maximal, non t-homogeneous. Divisibility conditions involving v, k and t, which are necessary conditions for the existence of such designs were given, and some interesting constructions were found. But many open questions on existence and design construction remain. One important tool in this work which has not been mentioned yet, is the following observation which provides a means of deciding, for a given k-element subset B of points and a given $G \leq$Sym (Ω), whether (Ω, B^G) is a t-design.

Proposition 6.6 (see [30, pp. 146–147]). *Let $2 \leq t < v = |\Omega|$, and let $H \leq$ Sym(Ω) be transitive but not t-homogeneous on Ω. Let $B \subseteq \Omega$ and let $Q_1, \cdots, Q_m$ be the H-orbits on t-element subsets of Ω. Then (Ω, B^H) is a (nontrivial block-transitive) $t - (v, |B|, \lambda)$ design if and only if*

$$\frac{q_1}{|Q_1|} = \cdots = \frac{q_m}{|Q_m|} = \frac{\binom{k}{t}}{\binom{v}{t}},$$

where, for $i = 1, \cdots, m$, q_i is the number of t-element subsets of B in Q_i.

Finally we give an illustration of the types of divisibility conditions and construction problems which arise by considering the affine groups $H =$AGL(d, p) where $d \geq 2$ and p is a prime. If p is odd then H is a homogeneous but not 3-homogeneous so in this case let us consider the problem of the existence of nontrivial block-transitive 3-(p^d, k, λ) designs admitting H. There are two H-orbits on 3-element subsets of $\Omega = V(d, p)$ (a d-dimensional vector space over $GF(p)$), namely collinear triples and non-collinear triples. If Q_1 is the set of all collinear triples then, by Proposition 6.6, we have a construction of a 3-(p^d, k, λ) design from any subset of k vectors which contains exactly

$$q_1 = \frac{k(k-1)(k-2)(p-2)}{6(p^d-2)}$$

collinear triples (see [42, Lemma 2.1]). In particular this number q_1 must be an integer. Gamble [23] has found, using the Chinese Remainder Theorem, the exact number of values of k (for which q_1 is an integer) in the range $[3, p^d/2]$ in terms of the number of distinct primes dividing $(p^d - 2)$. (Note that this is the range of interest for nontrivial designs; for we may assume $k \leq v/2$, since if (Ω, B^H) is a t-design then so also is $(\Omega, (\Omega \backslash B)^H)$.) For $p = 3$, the smallest value of d for which there are integer values for q_1, with $3 \leq k \leq 3^d/2$, is $d = 7$, and in this case there are 6 values of k in the range of interest and the smallest two of these are $k = 115, (q_1 = 113)$ and $k = 116, (q_1 = 116)$. For these two values of k Gamble [23] has constructed many non-isomorphic block-transitive 3-$(3^7, k, \lambda)$ designs with automorphism group AGL$(7, 3)$.

The group H =AGL$(d,2)$, where $d \geq 3$, is 3-homogeneous, but not 4-homogeneous on $\Omega = V(d,2)$ so we are interested in constructing 4-designs admitting H. There are two H-orbits on 4-element subsets of Ω, namely the set of affine planes in Ω and the set of noncoplanar 4-set. For $B \subseteq \Omega, |B| = k$, it follows from Proposition 6.6 (see [1, Theorem]) that (Ω, B^H) is a block-transitive 4-design admitting H if and only if the number of affine planes in B is

$$\frac{k(k-1)(k-2)(k-3)}{24(2^d-3)}.$$

Moreover, Alltop [1] proved the surprising result that in this situation (Ω, B^H) is a 4-design if and only if (Ω, B^H) is a 5-design. The smallest value of d for which the expression above can be a positive integer is $d = 8$, and for this value of d there are 10 possible values of k in the range $[3, 2^8-3]$. To obtain a flag-transitive 5-design the block size k must divide $|\mathrm{AGL}(8,2)|$ and there are only 3 values of k with this property, namely 24, 25 and 210. Alltop [1] found a construction of a 5-$(2^8, 24, \lambda)$ design with automorphism group AGL$(8,2)$, Cameron and the author rediscovered this construction making use of the Witt design associated with M_{24} and showed that the design is flag-transitive. (The value of λ is $2^{24}.3^2.5^2.7.31$).

References

[1] Alltop, W.O., *5-designs in affine spaces*, Pacific J. Math., 39(1971), 547–551.

[2] Block, R.E., *On the orbits of collineation groups*, Math. Zeit, 96(1967), 33–49.

[3] Buekenhout, F., Delandtsheer, A. and Doyen, J., *Finite linear spaces with flag transitive automorphism groups*, J. Combin. Theory, Series A, 49(1988), 268–293.

[4] Buekenhout, F., Delandtsheer, A., Doyen, J., Kleidman, P.B., Liebeck, M.W. and Saxl, J., *linear spaces with flag-transitive automorphism groups*, preprint, 1989.

[5] Buekenhout, F., Delandtsheer, A., Doyen, J., Kleidman, P.B., Liebeck, M.W. and Saxl, J., *The classification of flag-transitive finite linear spaces*, in preparation, 1990.

[6] Cameron, P.J., *Finite permutation groups and finite simple groups*, Bull. London Math. Soc., 13(1981), 1–22.

[7] Cameron, P.J., Neumann, P.M. and Teague, D.N., *On the degrees of primitive permutation groups*, Math. Z., 180(1982), 141–149.

[8] Cameron, P.J. and Praeger, C.E., *Block-transitive designs, I: point-imprimitive designs*, Discrete Math. (submitted) (1989).

[9] Cameron, P.J. and Praeger, C.E., unpublished (1990).

[10] Camina, A.R. and Di Martino, L., *Block designs on 196 points*, Arch. Math., 53(1989), 414–416.

[11] Camina, A.R. and Di Martino, L., *The group of automorphisms of a transitive 2-(91, 6, 1) design*, Geom. Ded., 31(1989), 151–164.

[12] Camina, A.R. and Di Martino, L., *Some imprimitive block designs*, (preprint).

[13] Camina, A.R. and Gagen, T.M., *Block transitive automorphism groups of block designs*, J. Algebra, 86(1984), 549–554.

[14] Camina, A.R. and Seimons, J., *Block transitive automorphism groups of 2-$(v,k,1)$ designs*, J. Combin. Theory, Series A, 51(1989), 268–276.

[15] Conway, J.H., *Three lecture on exceptional groups*, in Finite Simple Groups Eds. M. Powell and G. Higman, pp. 215–247(Academic Press, New York, 1971).

[16] Delandtsheer, A., *Line-primitive automorphism groups of finite linear spaces*, European J. Combinatorics, 10(1989), 161–169.

[17] Delandtsheer, A., *Finite flag-transitive semiovals*, to appear.

[18] Delandtsheer, A. and Doyen, J., *Most block-transitive t-designs are point primitive*, Geom. Ded., 29(1988), 397–410.

[19] Delandtsheer, A. and Doyen, J., *A classification of line-transitive maximal (v,k)-arcs in finite projective planes*, to appear.

[20] Dembowski, P., *Finite Geometries*, Springer-Verlag, New York, 1968.

[21] Euler, L., *Recherches sur une nouvelle espece de quarres magiques*, Verhandlingen Zeeuwach Genootschap Wetenschappen Vlissengen 9(1782), 85–239, or see Leonardi Euleri Opera Omnia, Serie 1, 291–392, (1923).

[22] Foulser, D.A., *Solvable flag-transitive affine groups*, Math. Zeit., 86(1964), 191–204.

[23] Gamble, G., *Block-transitive 3-designs with affine automorphism group*, unpublished mannscript, 1990.

[24] Hering, C., *Eine nicht-desarguessche zweifach transitive affine Ebene der Ordnung 27*, Abh. Math. Sem. Hamburg, 34(1969), 203–208.

[25] Hering, C., *Transitive linear groups which contain irreducible subgroups of prime order*, Geom. Ded, 2(1974), 425–460.

[26] Hering, C., *Transitive linear groups and linear groups which contain irreducible subgroups of prime order, II*, J. Algebra, 93(1985), 151–164.

[27] Hering, C., *Two new sporadic doubly transitive linear spaces*, in Finite Geometries, Lecture Notes in Pure and App. Math., Eds. C.A. Baker and L.M. Batten, pp. 127–129 (Marcel Decker, New York, 1985).

[28] Higman, D.G. and McLaughlin, J.E., *Geometric ABA-groups*, Illinois J. Math., 5(1961), 382–397.

[29] Higman, D.G. and Sims, C.C., *A simple group of order 44,352,000*, Math. Zeit. 105(1968), 110–113.

[30] Hughes, D.R. and Piper, F.C., *Design Theory*, (Cambridge University Press, Cambridge, 1985).

[31] Kantor, W.M., *Homogeneous designs and geometric lattices*, J. Combin. Theory Series A 38(1985), 66–74.

[32] Kantor, W.M., *Primitive permutation groups of odd degree and an application to finite projective planes*, J. Algebra, 106(1987), 15–45.

[33] Kirkman, T.P., *On a problem in combinatorics*, Cambridge and Dublin Mathematics Journal, 2(1847), 191–204.

[34] Kirkman, T.P., *Note on an unanswered prize question*, Cambridge and Dublin Mathematics Journal, 5(1850), 255–262.

[35] Liebeck, M.W., *The affine permutation groups of rank 3*, Proc. London Math. Soc. (3) 54(1987), 477–516.

[36] Luneburg, H., *Translation Planes*, Springer-Verlag, Berlin, 1980.

[37] Mathieu, E., *Mémoire sur l'étude des fonctions de plusieurs quantités*, J. Math. Pures et App. (2) 6(1861), 241–323.

[38] Nickel, W., Niemeyer-Nickel, A., O'Keefe, C.M., Penttila, T. and Praeger, C.E., *The block-transitive, point-imprimitive 2-(729, 8, 1) designs*, preprint (1990).

[39] O'Keefe, C.M., Penttila, T., and Praeger, C.E., *The block-transitive, point-imprimitive designs with* $\lambda = 1$, submitted (1990).

[40] Plücker, J., *System der analytischen Geometrie, auf neue Betrachtung-weisen gegrundet, und insbesondere eine ausfuhrliche Theorie der Curven dritter Ordung enthaltend*, (Duncker and Humbolt, Berlin, 1835).

[41] Plücker, J., *Theorie der algebraischen Curven, gegrundet auf eine neue Behandlungweise der analytischen Geometrie*, (Marcus, Bonn, 1839).

[42] Praeger, C.E., *Block-transitive designs and maximal subgroups of finite Symmetric groups*, Australas, J. Combin., 1(1990), 193–205.

[43] Ray-Chaudhuri, D.K. and Wilson, R.M., *On t-designs*, Osaka J. Math., 12(1975), 737–744.

[44] Steiner, J., *Combinatorisches Aufgabe*, Journal fur die reine und angewandt Mathematik, 45(1853), 181–182.

[45] Tierlinck, L., *Non-trivial t-designs without repeated blocks exist for all t*, Discrete Math., 65(1987), 301–311.

[46] Witt, E., *Die 5-fachen transitiven Gruppen von Mathieu*, Abhandlungen aus dem Math. Sem der Univ. Hamburg, 12(1938), 256–264.

[47] Witt, E., *Uber Steinersche Systeme*, Abhandlungen aus dem Math. Sem der Hamburg, 12(1938), 265–275.

[48] Woolhouse, W.S.B., *Prize Question 1733*, in Lady's and Gentlemen's Diary (1844).

[49] Zieschang, P.H., *Flag transitive automorphism groups of 2-designs with* $(r, \lambda) = 1.$, J. Algebra, 118(1988), 369–375.

Proceedings of the Asian Mathematical
Conference, 1990

Para–Fourier Integral Operators and Their Application to the Propagation of Nonlinear Singularities*

QIU QINGJIU
Department of Mathematics
Nanjing University
Nanjing 210008, China

In this paper we develop the concept of para–Fourier integral operators and use them to establish the Egorov theorem for paradifferential operators, which is applied then to reduce microlocally paradifferential opeators. Thus, we may study the propagation of singularities of solutions to paradifferential equations and the corresponding non-linear equations in the "lower frequency range". Here we only discuss two kinds of operators: principal type operator and one with constant multiple characteristics.

This paper is a summary and revison of my papers [4], [5], [6] and [7]. We refer to those papers for details.

Let $h(x,\eta) \in C^{\rho+1}(\mathbb{R}^n \times \mathbb{R}_n \backslash \{0\})$, with $\rho > 0$ an integer, be real positive–homogeneous of degree 1 in η, with nonsingular Hessian matrix. In this paper we stipulate that $h \in C^{\rho+1}$ implies $h_\eta(x,\eta) \in C_*^\rho$ (Zygmund class), which is homogeneous of degree zero in η. $a(x,\eta) \in S^m_{1,0}$ with compact support in x for simplicity. By Euler theorem and the technique of dyadic decomposition and homogeneous extension in η, we get

$$h(x,\eta) = \sum_{k=-1}^{\infty} H_k(x,\eta), \tag{1}$$

where $H_k(x,\eta)$ is C^∞ and positively homogeneous of degree 1 in η.

Consider the following Fourier integral operator:

$$F_k u(x) = \int e^{iS_k h(x,\eta)} a(x,\eta)\hat{u}(\eta) \quad d\eta, \tag{2}$$

where $S_k h(x,\eta) = \sum_{j=-1}^{k-1} H_j(x,\eta)$.

We may assume that k is large enough througout this paper without loss of generality. Hence, if $\rho > 2|s-m| + 2n + 4$, the norm of operator $\|F_k\|_{s\to s-m}$ is independent of k([4]).

* Project supported by the National Science Foundation of China.

Take the dyadic decomposition of $u(x), \sum_{k\geq -1} u_k(x)$, whose corresponding dyadic rings are $\{\mathbb{C}_k\}$. We call the operator $F_{(p)}$ defined by

$$F_{(p)}u = \sum_{k\geq 0} [F_k u_k]_k \tag{3}$$

a para-Fourier integral operator with order m, where the symbol $[\ \cdot\ ]_k$ denotes $[v]_k = \sum_{|k-j|\leq N} v_i$ if the dyadic decomposition of v is $\sum_{j\geq -1} v_i$ and the corresponding dyadic rings are $\{\mathbb{C}'_j\}, 2N$ is the number of $\mathbb{C}'_j$ intersected by the fixed $\mathbb{C}'_k$, which does not depend on k. It is not hard to know that the above definition is well-defined. $F_{(p)}$ is a bounded operator $: H^s \to H^{s-m}$ when $\rho > 2n+4$.

Take the adjoint Fourier integral operaor F_k^* of F_k, in which symbol $b(z,\eta) \in S^m_{1,0}$ with compact support in z. We can similarly define an adjoint para-Fourier intergal operator $F^*_{(p)}$. Henceforth, we only need the case $m = 0$ in $F_{(p)}$ and $F^*_{(p)}; b(z,\eta)$ in $F^*_{(p)}$ will be non-regular ([6]).

In this paper, we apply the symbol class $\sum^0{}^m_\sigma$ (rewrite it as $\sum(m,\sigma)$ later) introduced by Boulkhemair in [1]. We may also define $\sum(m,\sigma,\Omega), \sum(m,\sigma;\mathbb{R}^n\times\mathbb{R}_{n-1})$ and $\sum(m,\sigma;\Omega\times\mathbb{R}_{n-1}), \Omega$ being open set in $\mathbb{R}^n$. Denote by T_f the paradifferential operator T_f corresponding to $f\in\sum(m,\sigma)$. T_f is a bounded one : $H^s \to H^{s-m}$.

Operator class $\tilde{op}\sum(m,\sigma;\Omega)$ is defined in [1] for $s\in\mathbb{R}$. We can similarly define $\tilde{op}'\sum(m,\sigma;\Omega)$ by modifying for $s\geq 0$ in the definition of $\tilde{op}\sum(m,\sigma;\Omega)$. Obviously, it makes sense and $\tilde{op}\sum(m,\sigma;\Omega)\subset\tilde{op}'\sum(m,\sigma;\Omega)$.

The properties of $\tilde{op}\sum(m,\sigma,\Omega)$ are discussed in [1] in detail. Since s is restricted as $s\geq 0$ in this paper, $\tilde{op}'\sum(m,\sigma;\Omega)$ shares the same properties as $\tilde{op}\sum(m,\sigma,\Omega)$ essentially later.

For $G\in\tilde{op}\sum(m,\sigma;\Omega)(\tilde{op}'\sum(m,\sigma;\Omega))$, we can easily define its principal symbol; and, obviously, the principal symbol is not unique. However, the homogeneous principal symbol is unique if it exists.

Consider the transformation $K : (y,\eta)\to(x,\xi)$ defined by $x = X(y,\eta), \xi = \Theta(y,\eta)$ given by $y = h_\eta(x,\eta)$ and $\xi = h_x(x,\eta), K(y_0,\eta_0) = (x_0,\xi_0)$. When $\rho > 2$, this is a non-regular canonical transformation. Since our discussion is microlocal, we may confine the discussion within the conic neighborhoods of these two points, i.e., we suppose that the conic neighborhood $V\times\Delta$ of (y_0,η_0) is simplectically homeomorphic to the conic neighborhood $U\times\Gamma$ of (x_0,ξ_0), and assume that $h(x,\eta)$ is defined in $U\times\Delta$, the compact conic support of $a(x,\eta)$ is in $U\times\Delta, a(x_0,\eta_0)\neq 0$. Moreover, by [6], the compact conic supprort of $b(x,\eta)$ is also in $U\times\Delta, b(x_0,\eta_0)\neq 0$. Clearly , when $\rho > 3$ and k is large enough, $(x^{(k)},\xi^{(k)})$ given by $y = d_2S_kh(x,\eta)$ and $\xi = d_1S_kh(x,\eta)$ is in $U\times\Gamma$, for $(y,\eta)\in V\times\Delta$. $K_k : (y,\eta)\to(x^{(k)},\xi^{(k)})$ is also C^∞ canonical, $K_k(y_0,\eta_0) = (x_0^{(k)},\xi_0^{(k)})$. It is easy to know that $(x_0^{(k)},\xi_0^{(k)})\to(x_0,\xi_0)$ as $k\to\infty$.

Notice that , under the above assumption, the phase $h(x,\eta)$ defines only in a conic set $U\times\Delta$. In an obvious way , we also can define $F_{(p)}$ corresponding to such an $h(x,\eta)$.

For $u\in H^s_{x_0}\cap H^{s'}_{(x_0,\xi_0)}, s'\geq s\geq 0$, when $\rho > \max(4t+5n+3, 3t+2n-1+s'-s, 2t+3n+3+s'-s)$, the operator $F^*_{(p)}F_{(p)} - I$ is t-regular microlocally at $(x_0,\xi_0), F_{(p)}F^*_{(p)} - I$

has a similar result. Moreover, for any $u \in H^s_{y_0} \cap H^{s'}_{(y_0,\eta_0)}$, if $\rho > \max(2n+5, 2n+2+s'-s)$, then $F_{(p)}u \in H^s_{x_0} \cap H^{s'}_{(x_0,\xi_0)}$, $F^*_{(p)}$ also has a similar conclusion.

Theorem 1 (Egorov type). *For the above $K, h(x,\eta), a(x,\eta), b(x,\eta)$ and $f(x,\eta), a(x_0,\eta_0) \neq 0, f \in \sum(m,\sigma), u \in H^s, s \geq 0$. Then for provided $\rho > \max(4t+3n+3, 3t+7n+3, m+n+1), \sigma > \max(4t+2n+3, m)$, there exists an $\tilde{f} \in \sum(m,\alpha), \alpha = \min([\rho]-n-1, \sigma)$, such that $R = F^*_{(p)}T_f F_{(p)} - T_{\tilde{f}}$ is t-regular, and the principal symbol $\tilde{f}_0 = f \circ K$ in the neighborhood of (y_0,η_0), where t is taken as $m-1 < t < \frac{1}{2}[\alpha]-1$.*

Since this paper is devoted to the study of the nonlinear differential operator (cf. (7)) with which the associated paradifferential operator has homogeneous principal symbol (cf. (8)). So , we may confine the original principal symbol $g \in \sum(m,\sigma;\Omega)$ to the positively homogeneous function in ξ, later, in microlocal reduction. Thus by the Darboux type theorem (Lemma 4 in [6]) , we obtain the following microlocal reduction lemma for paradifferential operators of the principla type .

Theorem 2. *For the properly supported operator $Q \in \tilde{op}\sum(1,\sigma;\Omega)$ of principal type, whose real principal symbol is homogeneous in ξ, there are the above $K, F_{(p)}$ and $F^*_{(p)}$ such that $\tilde{Q} = F^*_{(p)}\mu' Q\mu F_{(p)} \in \tilde{op}'\sum(1,\tilde{\sigma};\Omega)$ with principal symbol $q_1 \circ K = \eta_n$ in the conic neighborhood of (y_0,η_0), if $\sigma > \max(4\tilde{\sigma}+3n+4, 3\tilde{\sigma}+7n+4), \tilde{\sigma} > 2$, where q_1 is the principal symbol of $Q, \mu(x) \in C_0^\infty(\Omega)$ which is equal to 1 on $\overline{U}, \mu'(x) \in C_0^\infty(\Omega)$ and $= 1$ on support μ.*

Let $P \in \tilde{op}\sum(m,\sigma;\Omega))(\tilde{op}'\sum(m,\sigma;\Omega))$ with symbol $p(x,\xi)$. We also assume that $p(x,\xi)$ is associated with nonlinear differential operator, and so P has a unique real homogeneous principal symbol $p_m(x,\xi)$. We say that $P, \sigma > 1$, is with constant multiple characteristics if $p_m(x,\xi)$ has the follwoing expression:

$$p_m(x,\xi) = q_1^{r_1}(x,\xi)\cdots q_s^{r_s}(x,\xi), \tag{4}$$

where $r_1,\cdots,r_s \geq 1$ are integer numbers, $q_1,\cdots,q_s$ are the real homogeneous symbols of principal type, $q_i^{-1}(0) \cap q_j^{-1}(0) = \emptyset, i \neq j$. Without loss of generality, we assume that $q_j \in \sum(1,\sigma;\Omega), j = 1,\cdots,s, r_1+\cdots+r_s = m$.

Lemma 1. *Let $f(x,\xi) \in \sum(m,\sigma;\Omega)), m,\sigma > 0, a(x,\eta) \in S^0_{1,0}(\Omega \times I\!R_{n-1})$ with compact support in $x, \varphi(x,\eta') \in C^\alpha(\Omega \times I\!R_{n-1}\backslash\{0\})$ be a real nondegenerate phase function, $\eta = (\eta',\eta_n), d_1\varphi \neq 0$ for $(x,\eta') \in$ conic supp $a, \eta' \neq 0$. If $\alpha > \max(2n+[\frac{n}{2}]+\frac{\tau}{2}+4, 2n+\tau+3, 2n+t+2)$ for some $0 < \tau < \sigma$ and t . then $g = e^{-i\varphi}T_f(ae^{i\varphi})$ can be decomposed into $g = g^{(1)} + g^{(2)}$, where $g^{(1)} \in \sum(m,\tau;\Omega \times I\!R_{n-1})$ and $g^{(1)} = \sum_{j=0}^N g_j^{(1)} + r^{(1)}, g_j^{(1)} \in \sum(m-j, \tau-2j;\Omega \times I\!R_{n-1})$, N is the maximal integer less than $\frac{\tau}{2}, r^{(1)} \in \sum(m-\frac{\tau}{2}+s, \varepsilon;\Omega \times I\!R_{n-1}), \forall\varepsilon > 0, g^{(2)}$ is φ-Fourier regular with order t, that is , the Fourier integral operator $F_{(\varphi,g^{(2)})}$ with phase φ and amplitude $g^{(2)}$ is a bounded one: $H^s_{comp}(\Omega) \to H^{s-m+t}_{loc}(\Omega)$.*

For any $(x_0,\xi_0) \in P_m^{-1}(0)$, there is a unique q_j such that $q_j(x_0,\xi_0) = 0$. We may assume that $(\partial q_j/\partial x_n)(x_0,\xi_0) \neq 0$. Let $\varphi_j(x,\xi')$ be such a local solution of the following initial value problem

$$\begin{cases} q_j(x, d_1\varphi_j) = 0, \\ \varphi_j(x,\xi')\mid_{x_n = x_n^0} = \langle x',\xi'\rangle \end{cases} \tag{5}$$

that $d_1\varphi_j \neq 0$ for $(x,\xi') \in$ conic supp $a(x,\xi')$ and for any $a(x,\xi') \in S^0_{1,0}(\Omega \times I\!R_{n-1})$

whose compact support in x is near x_0. Due to Lemma 1, we say that P satisfies the Levi condition at (x_0, ξ_0) if we have $\mu > 0, \sigma > \max(2n+[n/2]+\mu/2+5, 2n+\mu+4, 2n+r+3), r = \max(r_1, \cdots, r_s)$, such that

$$e^{-i\varphi_j}T_p(ae^{i\varphi_j} \in \sum(m - r_j, \mu; \Omega \times I\!R_{n-1}), \tag{6}$$

under Mod φ–Fourier regular with order r_j in the sense of Lemma 1.

We can also define the Levi condition on Ω.

For $\sigma > r$, (6) does not depend on the choice of the symbol p in the definition of $\tilde{op}\sum(m, \sigma; \Omega)(\tilde{op}'\sum(m, \sigma; \Omega))$ and of the para-truncated function in the definition of T_p. Hence the above definition makes sense.

Since K is homogeneous, if P is with constant multiple characteristics, then so does $\tilde{P}$ microlocally by Theorem 2. Let $\tilde{p}_m = \tilde{q}_1^{r_1}, \cdots, \tilde{q}_s^{r_s}$ be the homogeneous principal symbol of $\tilde{P}$ near $(y_0, \eta_0), \tilde{q}_j \in \sum(1, \tilde{\sigma}), j = 1, \cdots s$. By following the idea provided in [2], [3] and a careful discussion, we know that if P satisfies the Levi condition in the conic neighborhood of (x_0, ξ_0), then $\tilde{P} = F^*_{(p)}\mu' P \mu F_{(p)}$ also satisfies the Levi condition in the conic neighhborhood of (y_0, η_0), provided $\sigma > \max(4\sigma + 3n + 4, 3\sigma + 7n + 4, 2n + [\frac{n}{2}] + \frac{\mu}{2} + 5, 2n + \mu + 4), \sigma > 2n + 6 + r, \mu > n + 1$.

Now we can obtain the singularity propagation theorems of those two kinds of operators.

Theorem 3. *Suppose that $P \in \tilde{op}\sum(m, \sigma; \Omega)$ of principal type has compact support , p_m is its real principal symbol, $n \geq 2, u \in H^s_{loc}(\Omega), s \geq 0$. If $\sigma_0 - 3 + s > s', \sigma_0 = \min(\frac{1}{4}\sigma - \frac{5}{4}n - 1, \frac{1}{3}\sigma - \frac{7}{3}n - \frac{4}{3})$, then $\{WF_{s'}(u) \backslash WF_{s'-m+1}(Pu)\} \subset p_m^{-1}(0)$, and it is invariant under the Hamiltonian flow of p_m.*

Theorem 4. *For the properly supported $P \in \tilde{op}\sum(m, \sigma; \Omega)$ with constant multiple characteristics, which has real homogeneous principal symbol $p_m(x, \mu) = q_1^{r_1} \cdots q_s^{r_s}$, we assume that P satisfies the Levi condition on Ω with $\mu > \max(n + 1, 2r), r = \max r_j, u \in H^s_{loc}(\Omega), s \geq 0, n \geq 2$, If $\sigma > 4\tilde{\sigma} + 5n + 4, \tilde{\sigma} > \max(2n + r(r - 1) + \mu + 4, 2n + [\frac{n}{2}] + r(r - 1) + \frac{\mu}{2} + 5), s < s' < s + \tilde{\sigma} - r^2 + 1$, then*

$$S_j \equiv \{WF_{s'}(u) \backslash WF_{s'-m+r_j}(Pu)\} \subset p_m^{-1}(0),$$

for any $j = 1, \cdots, s$ and $S_j \cap q_j^{-1}(0)$ is invariant under the Hamiltonian flow of q_j.

Consider the nonlinear differential equation in an open set $\Omega \subset I\!R_n$:

$$F(x, u, \cdots, \partial^\alpha u, \cdots)_{|\alpha| \leq m} = 0, \tag{7}$$

where $F \in C^\infty$. Suppose that $u(x) \in H^s_{loc}(\Omega)$ is a real solution, $s > m + n, n \geq 2$. The principal symbol P_m of (7) corresponding to this solution is

$$P_m(x, \xi) = \sum_{|\alpha|=m} (\partial F/\partial y_\alpha)(x, u(x), \cdots, \partial^\alpha u(x), \cdots)(i\xi)^\alpha, \tag{8}$$

and $P_m(x, \xi) \in \sum(m, s - m - n; \Omega)$ is homogeneous in ξ.

By paralinearization, (7) is equivalent to the following paradifferential equation:

$$Pu \equiv \sum_{|\alpha| \leq m} T_{(\partial F/\partial y_\alpha)(x, u(x), \cdots, \partial^\alpha u(x), \cdots)} \partial^\alpha u = R, \tag{9}$$

where $R \in H_{loc}^{2s-2m-n/2}$, $P \in \tilde{op}\sum(m, s-m-n;\Omega)$. We say this opertoar P is associated with the nonlinear differential operator F. Since $p_m(x,\xi)$ in (8) is also the principal symbol of P in (9), we can say that the nonlinear differential equation (7) correspoinding to that solution $u(x)$ is of principal type, or with constant multiple characteristics and satisfies the Levi condition. Therefore, by Theorems 3 and 4, we can easily obtain theorems for singularity propagation of solution to (7) in the "lower frequency range":

Theorem 5. *For the equation (7) of principal type, $u \in H_{loc}^{s}(\Omega)$ is a real solution, $n \geq 2, s > \frac{3}{2}n + m + 1, s_0 = \min\{\frac{1}{8}(2s - 2m - 13n - 10), \frac{1}{6}(2s - 2m - 17n - 10)\}$. If $s + s_0 - 3 > s'$, then $WF_{s'}(u)$ is invariant under the Hamiltonian flow of p_m.*

Theorem 6. *For the equation (7) with constant multiple characteristics and satisfying the Levi condition , $u(x) \in H_{loc}^{s}$ is a real solution, $n \geq 2, s > 4\tilde{\sigma} + 6n + m + 4, \tilde{\sigma} > \max(3n + 8, 2n + r(r-1) + \mu + 4), \mu > \max(n+1, 2r), r = \max\ r_j, j = 1, \cdots, s$. If $s < s' < s + \tilde{\sigma} - r^2 + 1$, then for any $j = 1, \cdots, s, WF_{s'}(u) \subset P_m^{-1}(0)$, and $WF_{s'}(u) \cap q_j^{-1}(0)$ is invariant under the Hamiltonian flow of q_j.*

References

[1] Boulkhemair, A., *Opérateurs Paradifférentiels et Conjugaison par des Opérateurs Intégraux de Fourier, Thése de 3'éme Cycle*, Univ. Paris XI (Orsay.), 1984.

[2] Chazarain, J., *Opérateurs Hyperboliques a Charactéristiques de Multiplicité Constante*, Ann. Inst. Fourier, Grenoble, 24:1 (1974), 137–202.

[3] Chazarain, J., *Propagation des Singularités pour une Class D'opérateurs a Charactéristiques Multiples et Résobulité Locale*, Ann. Inst. Fourier, Grenoble, 24:1 (1974), 203–223.

[4] Qiu Qingjiu, *Para-Fourier Integral Operators*, Science in China, Ser. A, XXXII: 9 (1989), 1036–1047.

[5] ——, *On Egorov Theorem for Paradifferential Operators*, Science in China, Ser. A, XXXIII: 6(1990), 663–676.

[6] ——, *Application of Para-Fourier Integral Operators to the Propagation of Nonlinear Singularities*, Science in China, Ser. A, XXXIII: 9 (1990), 1060–1071.

[7] ——, *Propagation of Singularities for Nonlinear Differential Equations with Constant Multiple Characteristics*, preprint.

Proceedings of the Asian Mathematical
Conference, 1990

Global Small Solutions to Nonlinear Evolution Equations

REINHARD RACKE†
Institute of Applied Mathematics
University of Bonn Wegelerstr
10, 5300 Bonn, Germany

§1. Introduction

We consider nonlinear evolution equations like

- the system of nonlinear thermoelasticity,
- fully nonlinear damped wave equations or
- parabolic equations of higher order,

and we present global existence theorems for small smooth solutions for the Cauchyproblem (thermoelasticity) and for exterior domains.

Fully nonlinear problems have extensively been studied in the past years, e.g. wave equations, the equations of elasticity or heat equations, mainly for the case of the Cauchyproblem, i.e. the unknown functions are functions of $t \geq 0$ and $x \in \Omega = I\!R^n$. In this case several sufficient and/or necessary conditions for the existence of global, small, smooth solutions are known, see e.g. [10, 13, 14, 16, 27, 35]. For the case of an exterior domain, i.e. Ω is the complement of a bounded domain, three are only few results into this direction, e.g. for nonlinear wave equations in [30, 31].

One general scheme to obtain a global existence theorem is the following:

- To prove decay rates for L^q-norms of solutions to the associated linearized problem.
- To prove a local existence theorem.
- To prove a priori estimates for the "energy" of the solution V-usually in terms of $W^{s,2}(\Omega)$-norms of the local solution with sufficiently large $s-$ and for weighted norms of lower order, typically: $\sup(1+t)^d \cdot \|V(t)\|_{W^{m,q}(\Omega)}$. Here the sup is to be taken on the interval of local existence, m is small, $d > 0, q \geq 2$ are chosen depending on the special problem, exploiting the information obtained in the first step.

We shall show how this procedure works for the system of equations of nonlinear thermoelasticity for the Cauchyproblem. This system is of hyperbolic-parabolic type and has different types of nonlinearities (with respect to the order of vanishing near zero).

† Supported by the Deutsche Forschungsgemeinschaft.

The first step in the general scheme above – i.e. to obtain decay rates for solutions of the linearized problems – is difficult for the case of exterior domains even if the governing differential operator has constant coefficients. Here we shall also be interested in problems with variable coefficients. A method will be introduced which uses the so-called generalized eigenfunction expansion (or generalized Fourier transform) in order to obtain decay rates of solutions to some linear damped problems. It is a very simple straightforward approach for domains with star-shaped complement. (In contrast cf. the more delicate approach by Vainberg [33] and Eidus [5] for more general domains.) Finally this will lead to global existence theorems for the following typical equations.

A nonlinear parabolic equation of fourth-order:

$$\left.\begin{array}{ll} u_t + \Delta^2 u = f(u, \nabla u, \nabla^2 u, \nabla^3 u, \nabla^4 u) & \text{in } [0,\infty)\times\Omega, \\ u = \Delta u = 0 & \text{on } [0,\infty)\times\partial\Omega. \end{array}\right\} \text{(NPE)}$$

A non-homogeneous second-order damped wave equation:

$$\left.\begin{array}{ll} u_{tt} - \partial_i a_{ik}(x)\partial_k u + u_t = h(u_t, \nabla u, \nabla u_t, \nabla^2 u) & \text{in } [0,\infty)\times\Omega, \\ u = 0 & \text{on } [0,\infty)\times\partial\Omega. \end{array}\right\} \text{(NDWE)}$$

(See Section 4 for further assumptions.)

The paper is organized as follows: Section 2 discusses the Cauchyproblem for the equations of nonlinear thermoelasticity. Section 3 presents the general ansatz via generalized eigenfunction expansions to obtain decay rates in exterior domains for the linearized problems. Finally, Section 4 gives the global existence theorem for (NPE) and (NDWE) as typical applications and examples.

Notation

$W^{m,p}(\Omega), W_0^{m,p}(\Omega)$: usual Sobolev spaces with norm $\|\cdot\|_{m,p}$ (cf. [1]); $W^{0,p}(\Omega) \equiv L^p(\Omega)$ with norm $\|\cdot\|_p, m \in \mathbb{N}_0, 1 \leq p \leq \infty$.

$C_0^\infty(\Omega)$: C^∞–functions with compact support in Ω, Ω a domain in $\mathbb{R}^n$.

$\|\cdot\|_{m,p(G)}, \|\cdot\|_{p(G)}$: Norm in $W^{m,p}(G)$ and $L^p(G)$ resp., G a domain.

$B(x,r)$: ball with center x and radius r;

$\Omega_r : \Omega \cap B(0,r)$.

$\partial_i = \partial/\partial x_i, i = 1, \cdots, n, \partial_t = \partial/\partial t, \nabla^\alpha = (\partial_1^{\alpha_1}, \cdots, \partial_n^{\alpha_n}), \alpha \in \mathbb{N}_0^n$, similar for D^α, where $D = (\partial_t, \nabla)$.

$$\overline{\nabla}^j u = (\nabla^\alpha u)_{0\leq|\alpha|\leq j}, \quad j \in \mathbb{N}_0.$$

$$\|\nabla^l u\|_{\dots} = \left(\sum_{|\alpha|=L} \|\nabla^\alpha u\|_{\dots}^2\right)^{1/2}, \quad \text{similar for } \|D^L u\|_{\dots},$$

$$\|\overline{\nabla}^L u\|_{\dots} = \left(\sum_{0\leq|\alpha|\leq L} \|\nabla^\alpha u\|_{\dots}^2\right)^{1/2}, \quad \text{similar for } \|\overline{D}^L u\|_{\dots}, \quad L \in \mathbb{N}_0.$$

§2. Nonlinear Thermoelasticity

The system of equations of nonlinear thermoelasticity is a hyperbolic-parabolic coupling of the elasticity equations and of the heat equation. In three space dimensions the equations for the displacement $U = U(t,x)$ and the temperature difference $\theta = \theta(t,x)$ can be written in the following form:

$$\begin{aligned} & U_{tt} - ((2\mu+\lambda)\nabla\nabla' - \mu\nabla\times\nabla\times)\, U + \gamma\nabla\theta \\ = \;& q_1(\nabla U, \nabla^2 U, \theta, \nabla\theta) + K_1(\nabla U, \nabla^2 U, \theta, \nabla\theta), \end{aligned} \tag{2.1}$$

$$\begin{aligned} & \theta_t - \kappa\Delta\theta + \gamma\nabla' U_t \\ = \;& \theta\Delta\theta + q_2(\nabla U, \nabla^2 U, \nabla U_t, \theta, \nabla\theta, \nabla^2\theta) + K_2(\nabla U, \nabla^2 U, \nabla U_t, \theta, \nabla\theta, \nabla^2\theta), \end{aligned} \tag{2.2}$$

$$U(t=0) = U^0, \quad U_t(t=0) = U^1, \quad \theta(t=0) = \theta^0, \tag{2.3}$$

where q_j is quadratic and K_j is cubic near the origin, $j = 1,2$. μ and λ are the elastic moduli. They are constants as well as $\gamma \neq 0$ and $\kappa > 0$. U^0, U^1, θ^0 are prescribed initial data.

Remark. One-dimensional models have been discussed in [6, 8, 11, 12, 25, 32, 36], proving global existence for small data; the development of singularities is shown for large data in [4].

The equations (2.1)–(2.3) are discussed by appropriately modifying the general scheme from above, taking into account the different nonlinearites and the different type of equations leading to different kinds of asymptotic behaviour for the components of the solution (U,θ). Notation: $W^{m,p} \equiv W^{m,p}(\mathbb{R}^3)$.

Theorem 2.1 [21]. *If $q_j = 0, j = 1,2$, then there are an integer $s > 0$ and a $\delta > 0$ such that for $(\nabla U^0, U^1, \theta^0) \in W^{s,2} \cap W^{s,1}$ with norm less than δ, there exists a unique solution (U,θ) of (2.1)–(2.3),*

$$(\nabla U, U_t) \in C^1([0,\infty), W^{s-1,2}) \cap C^0([0,\infty), W^{s,2}),$$

$$\theta \in C^1([0,\infty), W^{s-2,2}) \cap C^0([0,\infty), W^{s,2}).$$

Moreover

$$\|(\nabla U, U_t, \theta)\|_\infty = O(t^{-5/9}), \quad \|(\nabla U, U_t, \theta)\|_{s,2} = O(1) \quad \text{as } t \to \infty.$$

Remark. The assumptions on q_j are assumptions on the underlying medium. The term $\theta\Delta\theta$ always appears. The assumptions on the initial data can be sharpened to: "There are integers s_0, s_2, s_4, s_6 such that for $(\nabla U^0, U^1, \theta^0) \in W^{s_0,2} \cap W^{s_2,9/7} \cap W^{s_4,13/11} \cap W^{s_6,26/15} \cap W^{s_6,18/11}$, with norm less than δ, there is a global solution".

Proof.

1. Decay estimates for the linearized system: Decompose U into $U = U^s + U^{po}$ corresponding to the orthogonal decomposition

$$L^2 = D_0 + \overline{\nabla W^{1,2}} \tag{2.4}$$

(D_0: fields with divergence zero).
This leads to a wave equation for U^s (not coupled to θ), hence behavior is known. U^{po} and θ are coupled in a way that one can describe the asymptotic behaviour using the Fourier transform in an easy way. One obtains decay rates like those for solutions to the heat equation - also with better decay rates for the derivatives.

2. Transformation to a first order system: Essentially $V = (\nabla U, U_t, \theta)$.
3. Local existence: This follows from the results in [9, 11].
4. The following energy estimate is proved for a local solution:

$$\|V(t)\|_{s,2} \leq C\|V(0)\|_{s,2} \cdot \exp\left\{C \int_0^t (\|(V^1, V^3, V_t^3, \nabla V)\|_\infty^2 + \|(\nabla V^3, \nabla^2 V^3)\|_\infty^{\frac{1}{\infty}})(r)dr\right\},$$

$t \in [0,T]$, where C depends only on s, not on T.

Remark. One should notice the different exponents.

5. The following weighted a priori estimate is proved:

$$\sup_{0\leq t\leq T} \{(1+t)^{5/9}\|(V^1,V^2)(t)\|_{s_1,9/2} + (1+t)^{27/26}(\|V^3(t)\|_{s_3,13/2} + \|\nabla^2 V^3(t)\|_{s_5,26/11} + \|\nabla^2 V^3(t)\|_{s_5,18/7})\} < \infty,$$

if $V(0)$ is small.

6. The a priori estimates in the last two steps allow to continue a local solution as usual. □

By an elaborate analysis, using the stronger decay of θ and its derivatives as well as the special form of the nonlinearities, one can show that there is still a global solution if the assumption $q_j = 0$ is changed to: $K_j = 0, j = 1,2$, and q_j does not contain any terms purely quadratic in (derivatives of) U, see [20]. The assumption that there are no "purely hyperbolic" quadratic terms in U in general cannot be omitted, one has to expect a blow-up, see [22]; the idea there is to transfer the decomposition (2.4) to the nonlinear problem.

§3. Generalized Eigenfunction Expansions

In order to obtain existence theorems in exterior domains, we observe that the decay of solutions to the heat equation $u_t - \Delta u = 0$ in $\mathbb{R}^n$ can easily be obtained by using the Fourier transform and exploiting the fact that the kernel of this transform, namely $(2\pi)^{-n/2}e^{ix\xi}$, is uniformly bounded with respect to x and ξ. This leads to the following ansatz.

Let $\Omega \subset \mathbb{R}^n$ be an exterior domain ($n \geq 3$ will always be assumed), that is, Ω is a domain and $\mathbb{R}^n \backslash \Omega$ is bounded, and let $\partial\Omega$ be smooth.

Let $A: D(A) \subset L^2(\Omega) \to L^2(\Omega)$,

$$D(A) := \{v \in W_0^{1,2}(\Omega) \mid \partial_i a_{ik}(\cdot)\partial_k v \in L^2(\Omega)\},$$

$Av = -\partial_i a_{ik}(\cdot)\partial_k v$, where $a_{ik} = a_{ki}$ is a smooth function of $x \in \overline{\Omega}$, $a_{ik}(x) = \delta_{ik}$ (Kronecker symbol) for $|x| > r_0 > 0$ for some fixed $r_0 > 0$ with $\partial\Omega \subset B(0,r_0)$. (We use the Einstein summation convention, $i,k = 1,\cdots,n$.) We assume for $x \in \overline{\Omega}$, $\xi \in \mathbb{R}^n$: $a_{ik}(x)\xi_i\xi_k \geq a_0|\xi|^2, a_0 > 0$ fixed.

It is well-known that there is a generalized eigenfunction expansion (also called generalized Fourier transform) $\mathcal{F}_+ : L^2(\Omega) \to L^2(\mathbb{R}^n)$, $\mathcal{F}_+$ being unitary, with the property

$$\mathcal{F}_+(\varphi(A)w)(\xi) = \varphi(|\xi|^2)(\mathcal{F}_+w)(\xi) \tag{3.1}$$

for functions $\varphi(A)$ of A defined by the spectral theorem for the selfadjoint operator A (cf. [15, 34]). Moreover:

$$(\mathcal{F}_+w)(\xi) = \int_\Omega \overline{\psi(x,\xi)}w(x)dx \equiv \hat{w}(\xi),$$

and

$$(\mathcal{F}_+^*\hat{w})(x) = \int_{\mathbb{R}^n} \psi(x,\xi)\hat{w}(\xi)d\xi.$$

The kernel ψ is uniquely determined by the following conditions (3.2)–(3.6): Let $j \in C^\infty(\mathbb{R}), j \geq 0, j(r) = 0$ for $r \leq r_1, j(r) = (2\pi)^{-n/2}$ for $r \geq r_1 + 1, r_1 > r_0$ fixed.

$$\psi(x,\xi) = j(|x|)e^{ix\xi} + \psi'(x,\xi), \tag{3.2}$$

$$\forall \xi \in \mathbb{R}^n : \quad (1 - j(|\cdot|)\psi(\cdot,\xi)) \in D(A), \tag{3.3}$$

$$\forall \xi \in \mathbb{R}^n : j(|\cdot|)\psi(\cdot,\xi) \in W^{2,2}(\Omega_r) \text{ for all } r > 0, \tag{3.4}$$

$$\forall \xi \in \mathbb{R}^n : (\partial_i a_{ik}(\cdot)\partial_k + |\xi|^2)\psi(\cdot,\xi) = -(\Delta + |\xi|^2)(j|\cdot|e^{i\cdot\xi}), \tag{3.5}$$

$\forall \xi \in \mathbb{R}^n \backslash \{0\} : \psi'(\cdot,\xi)$ satisfies the outooing radiation condition:

$$\begin{aligned} &\frac{\partial\psi'(x,\xi)}{\partial|x|} - i|\xi|\psi'(x,\xi) = O(|x|^{-(n+1)/2}), \\ &\psi'(x,\xi) = O(|x|^{-(n-1)/2}), \quad \text{as } |x| \to \infty. \end{aligned} \tag{3.6}$$

Remark. Such a generalized eigenfunction expansion was given first for the Schrödinger operator in $\mathbb{R}^3$ by Ikebe ([7]). Later on this was extended to higher dimensions and to perturbations of the Laplace operator, also for exterior domains, see e.g. [2, 17, 28, 29] and also [18].

In view of (3.1) and the remarks from the beginning of this Section, we are now interested in pointwise estimates on ψ' resp. ψ.

Theorem 3.1 [19, 26]. *Let $A = -\Delta$, i.e. $a_{ik}(x) = \delta_{ik}$ for all $x \in \overline{\Omega}$, let $\mathbb{R}^n\backslash\Omega$ be star-shaped, $n \geq 3$. Then we have*

$$\exists m \in \mathbb{N} \quad \exists c > 0 \quad \forall x \in \overline{\Omega} \quad \forall \xi \in \mathbb{R}^n : |\psi(x,\xi)| \leq c \cdot (1 + |\xi|)^m. \tag{3.7}$$

This essentially follows from the results in [19], see [26] for the necessary estimates near the boundary of Ω.

Using (3.1), the factor $(1 + |\xi|)^m$ appearing in (3.7) will finally turn into differentiation of the initial data only.

For A with variable coefficients we obtain the following theorem.

Theorem 3.2 [3, 23]. *Let $A = -\partial_i a_{ik}(\cdot)\partial_k$ be given as defined above, let $\mathbb{R}^3\backslash\Omega$ be star-shaped, moreover let*

$$\min_{x\in\overline{\Omega}} \left(2 \min_{|\xi|=1} a_{ik}(x)\xi_i\xi_k - \max_{i,k=1,2,3} |\nabla a_{ik}(x) \cdot x| \right) > 0. \tag{3.8}$$

Then we have

$$\forall r>0 \quad \exists c>0\ \forall x \in \Omega_r\ \forall \xi \in \mathbb{R}^3 : |\psi(x,\xi)| \le c \cdot (1+|\xi|)^2. \tag{3.9}$$

The proof uses the results in [3] for the high frequency estimate ($|\xi| \to \infty$) and the principle of limiting absorption for the low frequency estimates ($|\xi| \to 0$).

Remarks. The local character (with respect to x) in the estimate for ψ in the last Theorem makes it still possible to obtain a global decay estimate for solutions of the time-dependent problem (NDWE) below. It is sufficient to prove first a local energy decay rate which is large enough (which is necessary to combine it to the Cauchy problem, here: $A = -\Delta$ in $\mathbb{R}^3$, by using cut-off functions, following [31]). The more complicated technique for obtaining a local energy decay rate from [31], cf. also [33, 5], could lead to a removal of the star-shapedness assumption on the domain.

§4. Global Existence in Exterior Domains

As an application of the straightforward approach presented in the previous section, we obtain for example for solutions of

$$u_t + \Delta^2 u = 0, \quad u = \Delta u = 0 \text{ on the boundary}, \quad u(t=0) = u_0$$

an L^∞-decay which is of the order $t^{-n/4}$. This immediately follows from the representation

$$u(t,x) = \int_{\mathbb{R}^n} \psi(x,\xi) e^{-t|\xi|^4} \widehat{u}_0(\xi) d\xi$$

and Theorem 3.1. Together with appropriate energy estimates this finally leads to a global existence theorem for the following fourth-order nonlinear parabolic equation:

$$\left.\begin{array}{lr} u_t + \Delta^2 u = f(u, \nabla u, \nabla^2 u, \nabla^3 u, \nabla^4 u) & \text{in } [0,\infty) \times \Omega, \\ u = \Delta u = 0 & \text{on } [0,\infty) \times \partial\Omega, \\ u(t=0) = u_0 & \text{in } \Omega, \\ f(w) = O(|w|^2) \text{ near } w = 0, & \\ \mathbb{R}^n \backslash \Omega \text{ is star-shaped.} & \end{array}\right\} (NPE)$$

Theorem 4.1 [26]. *Let $n > 4, m \ge [n/2] + 1$. Then there is a $\delta > 0$ such that for $u_0 \in W^{4m+2,2}(\Omega) \cap W^{4m,1}(\Omega)$ with norm less than δ, satisfying the usual compatibility conditions, there exists a unique solution u of (NPE).*

$$u \in \bigcap_{k=0}^{m} C^{m-k}([0,\infty), W^{4k,2}(\Omega)).$$

Moreover:

$$\|u(t)\|_\infty + \|u(t)\|_2 = O(t^{-n/8}) \quad \text{as } t \to \infty.$$

Using Theorem 3.2 we first obtain a decay result for the linearized equations associated to the following problem and finally a global existence result for the non-homogeneous second-order damped wave equation (NDWE).

$$\left.\begin{array}{lr} u_{tt} - \partial_i a_{ik}(x)\partial_k u + u_t = h(u_t, \nabla u, \nabla u_t, \nabla^2 u) & \text{in } [0,\infty)\times\Omega, \\ u = 0 & \text{on } [0,\infty)\times\partial\Omega, \\ u(t=0) = u_0, u_t(t=0) = u_1 & \text{in } \Omega \\ h(w) = \mathcal{O}(|w|^3) \text{ near } w = 0, & \\ \mathbb{R}^3\backslash\Omega \text{ is star-shaped}, & \\ a_{ik} \text{ as given in Section 3, assuming (3.8).} & \end{array}\right\} \text{(NDWE)}$$

Theorem 4.2 [24]. *There are an integer $L_0 \geq 9$ and a $\delta > 0$ such that if $(u_0, u_1) \in W^{L,2}(\Omega) \cap W^{L,6/5}(\Omega)$, with $L > L_0$, L even, with norm less than δ, satisfy the usual compatibility condition (of order $L + 10$), three exists a unique solution u of (NDWE),*

$$u \in \left\{ \bigcap_{j=0}^{L-1} C^j([0,\infty), W^{L-j,2}(\Omega) \cap W_0^{1,2}(\Omega)) \right\} \cap C^1([0,\infty), L^2(\Omega)).$$

Moreover

$$\|(u(t))\|_\infty + \|u(t)\|_6 = \mathcal{O}(t^{-1}) \quad \text{as } t \to \infty.$$

References

[1] Adams, R., *Sobolev spaces.*, Academic Press. New York et al., 1975.

[2] Alsholm. P., Schmidt, G., *Spectral and scattering theory for Schrödinger operators*, Arch. Rat. Mech. Anal., 40(1971), 281–311.

[3] Bloom, C.O., *Estimates for solutions of reduced hyperbolic equations of the second order with a large parameter*, J. Math. Anal. Appl., 44(1973), 310–332.

[4] Dafermos, C.M., Hsiao. L., *Development of singularities in solutions of the equations of nonlinear thermoelasticity*, Quart. Appl. Math., 44(1986), 463–474.

[5] Eidus, D., *Asymptotical expansions of solutions of linear parabolic equations as $t \to \infty$*. J. Math. Anal. Appl., 130(1988), 155–170.

[6] Hrusa, W.J., Tarabek, M.A., *On smooth solutions of the Cauchy problem in one-dimensional nonlinear thermoelasticity*, Quart. Appl. Math., 47(1989), 631–644.

[7] Ikebe, T., *Eigenfunction expansions associated with the Schrödinger operators and their applications to scattering theory*, Arch. Rat. Mech. Anal., 5(1960), 1–34.

[8] Jiang, S., *Global existence of smooth solutions in one-dimensional nonlinear thermoelasticity*, To appear in: Proc. Roy. Soc, Edinburgh, 1990.

[9] John, F., *Finite amplitude waves in a homogeneous isotropic elastic solid*, Comm. Pure Appl. Math., 30(1977), 421–446.

[10] John, F., *Long time effects of nonlinearity in second order hyperbolic equations*, Comm. Pure Appl. Math., 39(S)(1986), 139–148.

[11] Kawashima, S., *Systems of a hyperbolic-parabolic composite type, with applications to the equations of magnetohydrodynamics*, Thesis, Kyoto University, 1983.

[12] Kawashima, S., Okada, M., *Smooth global solutions for the one-dimensional equations in magnetohydrodynamics*, Proc. Jap. Acad., Ser. A, 53(1982), 384–387.

[13] Klainerman, S., *Long-time behavior of solutions to nonlinear evolution equations*, Arch. Rat., Mech. Anal., 78(1982), 73–98.

[14] Klainerman, S., Ponce, G., *Global, small amplitude solutions to nonlinear evolution equations*, Comm. Pure Appl. Math., 36(1983), 133–141.

[15] Leis, R., *Initial boundary value problems in mathematical physics*, B.G. Teubner-Verlag, Stuttgart; John Wiley & Sons; Chichester et al., 1986.

[16] Li. T.-T., Chen, Y., *Initial value problems for nonlinear wave equations*, Comm. Part. Differential Equ., 13(1988), 383–422.

[17] Majda, A., *Outgoing solutions for perturbations of $-\Delta$ with applications to spectral and scattering theory*, J. Differential Equ., 16(1974), 515–547.

[18] Mochizuki. K., *Spectral and scattering theory for symmetric hyperbolic systems in an exterior domain*, Publ. RIMS, Kyoto University, 13(1977), 349–379.

[19] Morawetz, C.S., Ludwig, D., *An inequality for the reduced wave operator and the justification of geometrical optics*, Comm. Pure Appl. Math., 21(1968), 187–203.

[20] Ponce, G., Racke, R., *Global existence of solutions to the initial value problem for nonlinear thermoelasticity*, to appear in: J. Differential. Equ., (1990).

[21] Racke, R., *On the Cauchyproblem in nonlinear 3-d-thermoelasticity*, Math. Zeitschrift, 203(1990), 649–682.

[22] Racke, R., *Blow-up in nonlinear three-dimensional thermoelasticity*, Math. Meth. Appl. Sci., 12(1990), 267–273.

[23] Racke, R., *Decay rates for solutions of damped systems and generalized Fourier transforms*, to appear in: J.f.d. reine u. angewandte Math. (1990/1).

[24] Racke, R., *Non-homogeneous, nonlinear damped wave equations in exterior domains*, SFB Preprint 97, Universität Bonn (1990). (Submitted to: Math. Meth. Appl. Sci.)

[25] Racke, R., Shibata, Y., *Global smooth solutions and asymptotic stability in one-dimensional nonlinear thermoelasticity*, SFB Preprint 104, Universität Bonn, 1990.

[26] Racke, R., Zheng, S., *Global existence of solutions to a fully nonlinear fourth-order parabolic equation in exterior domains*, to appear in: Nonlin. Anal., T.M.A. (1990).

[27] Shatah, J., *Global existence of small solutions to nonlinear evolution equations*, J. Differential Equ., 46(1982), 409–425.

[28] Shenk, N.A., *Eigenfunction expansions and scattering theory for the wave equation in an exterior region*, Arch. Rat. Mech. Anal., 21(1966), 120–150.

[29] Shenk, N.A., Thoe, D., *Eigenfunction expansions and scattering theory for perturbations of $-\Delta$*, J. Math. Anal. Appl., 36(1971), 313–351.

[30] Shibata, Y., *On the global existence of classical solutions of second order fully nonlinear hyperbolic equations with first order dissipation in the exterior domain*, Tsukuba J. Math., 7(1) (1983), 1–68.

[31] Shibata, Y., Tsutsumi, Y., *On a global existence theorem of small amplitude solutions for nonlinear wave equations in an exterior domain*, Math. Zeitschrift, 191(1986), 165–199.

[32] Slemrod, M., *Global existence, uniqueness, and asymptotic stability of classical smooth solutions in one-dimensional non-linear thermoelasticity*, Arch. Rat. Mech. Anal.,

76(1981), 97–133.

[33] Vainberg, B.R., *On the short wave asymptotic behaviour of solutions of stationary problems and the asymptotic behaviour as $t \to \infty$ of solutions of non-stationary problems*, Russian Math. Surveys, 30(2) (1975), 1–58.

[34] Wilcox, C.H., *Scattering theory for the d'Alembert equation in exterior domains*, Lec. Notes Math., 442(1975).

[35] Zheng, S., *Remarks on global existence for nonlinear parabolic equations*, Nonlin. Anal., T.M.A., 10(1986), 107–114.

[36] Zheng, S., Shen, W., *Global solutions to the Cauchy problem of quasilinear hyperbolic parabolic coupled systems*, Sci. Sinica, Ser. A, 30(1987), 1133–1149.

Proceedings of the Asian Mathematical
Conference, 1990

Linear Hyperbolic Equations of the Fourth Order with Two Singular Lines

NUSRAT RAJABOV
Tajik State University
Dushanbe USSR

Let D denote a rectangle $D = \{0 < x < \alpha, 0 < y < \beta\}$. Further denote $\Gamma_i = \{0 < x \leq \alpha, y = 0\}, \Gamma_2 = \{x = 0, 0 < y \leq \beta\}$. In domain D we consider the equations

$$\begin{aligned} Lu \equiv \quad & x^2y^2\frac{\partial^4 u}{\partial x^2 \partial y^2} + xy^2 a_1(x,y)\frac{\partial^3 u}{\partial x \partial y^2} + x^2 y a_2(x,y)\cdot\frac{\partial^3 u}{\partial x^2 \partial y} \\ & + x^2 a_3(x,y)\frac{\partial^2 u}{\partial x^2} + xya_4(x,y)\frac{\partial^2 u}{\partial x \partial y} + y^2 a_5(x,y)\cdot\frac{\partial^2 u}{\partial y^2} \\ & + xa_6(x,y)\frac{\partial u}{\partial x} + ya_7(x,y)\frac{\partial u}{\partial y} + a_8(x,y)u = f(x,y), \end{aligned} \tag{1}$$

where $a_j(x,y), 1 \leq j \leq 8$, and $f(x,y)$ are given functions in D.

The corresponding non-model linear hyperbolic and elliptic equations of the second order of the systems with singularity coefficients have been already studied in the series of the author's articles [1]–[9].

In this paper on basis of the working out methods [1]–[9], for equation (1), we obtain integral representations of manifold of solutions. Independence of the roots of the characteristic equations corresponding to the model of the second order equation of types

$$\frac{\partial^2 V}{\partial x^2} + \frac{a_1(y,y)}{x}\frac{\partial V}{\partial x} + \frac{a_5(y,y)}{x^2}V = 0, \tag{2}$$

$$\frac{\partial^2 u}{\partial y^2} + \frac{a_2(x,x)}{y}\frac{\partial u}{\partial y} + \frac{a_3(x,x)}{y^2}u = 0, \tag{3}$$

that is dependence of the roots of algebrical equations:

$$\varphi^2(y) + (a_1(y,y) - 1)\varphi(y) + a_5(y,y) = 0, \tag{4}$$

$$\psi^2(x) + (a_2(x,x) - 1)\psi(x) + a_3(x,x) = 0, \tag{5}$$

or dependence of significance functions

$$\lambda^{\pm}(x,y) = 2^{-1}\left(a_1(x,y) \pm \sqrt{a_1^2(x,y) - 4a_5(x,y)}\right), \tag{6}$$

$$\mu^{\pm}(x,y) = 2^{-1}\left(a_2(x,y) \pm \sqrt{a_2^2(x,y) - 4a_3(x,y)}\right), \tag{7}$$

and the other coefficients, are obtained from the different integral representations of the manifold of solution in terms of four arbitrary functions of a single variable. The integral representations obtained are applied to the solutions of various boundary value problems.

We introduce the following designations:

$$\begin{aligned}
A_4(x,y) &= a_1a_2 + x\frac{\partial a_2}{\partial x},\\
A_6(x,y) &= a_1a_3 + 2x\frac{\partial a_3}{\partial x},\\
A_7(x,y) &= a_2a_7 + xa_1\frac{\partial a_2}{\partial x} + x^2\frac{\partial^2 a_2}{\partial x^2},\\
A_8(x,y) &= a_3a_5 + xa_1\frac{\partial a_3}{\partial x} + x^2\frac{\partial^2 a_3}{\partial x^2},\\
\omega_1^{\pm}(x,y) &= \int_{x_0}^{x}\frac{\lambda^{\pm}(t,y) - \lambda^{\pm}(0,y)}{t}dt,\\
\omega_2^{\pm}(x,y) &= \int_{y_0}^{y}\frac{\mu^{\pm}(x,s) - \mu^{\pm}(x,0)}{s}ds,\\
K_1(\psi_1,\psi_2,f) &\equiv \psi_2(y) + \psi_1(y)F_1(x,y) + \int_{x_0}^{x} M_1(x,y;s)f(s,y)ds,\\
F_1(x,y) &= \int_{x_0}^{x} t^{\lambda^-(0,y)-\lambda^+(0,y)}\exp[\omega_1^-(t,y) - \omega_1^+(t,y)]dt,\\
M_1(x,y;s) &= s^{\lambda^+(0,y)-2}y^{-2}\Big(\int_s^x \exp[\omega_1^-(t,y) - \omega_1^+(t,y)\\
&\quad +\omega_1^+(s,y)]t^{\lambda^-(0,y)-\lambda^+(0,y)^*}dt\Big),\\
K_2(\varphi_1,\varphi_2,V) &= \varphi_2(x) + \varphi_1(x)F_2(x,y) + \int_{y_0}^{y} M_2(x,y,\tau)V(x,\tau)d\tau\\
F_2(x,y) &= \int_{y_0}^{y} s^{\mu^-(x,0)-\mu^+(x,0)}\exp[\omega_2^-(x,s) - \omega_2^+(x,s)]ds,\\
M_2(x,y,\tau) &= \tau^{\mu+(x,0)-2}\int_{y_0}^{y}\exp[\omega_2^-(x,\xi) - \omega_2^+(x,\xi)\\
&\quad +\omega_2^+(x,y)]\xi^{\mu^-(x,0)-\mu^+(x,0)}d\xi,\\
P_x^- &\equiv \frac{\partial}{\partial x} + \frac{\lambda^-(x,y)}{x},\\
P_y^- &\equiv \frac{\partial}{\partial y} + \frac{\mu^-(x,y)}{y},
\end{aligned}$$

$$
\begin{aligned}
P_y \equiv\ & \frac{\partial^2}{\partial y^2} + \frac{a_2(x,y)}{y}\frac{\partial}{\partial y} + \frac{a_3(x,y)}{y^2}\\
P_y^0 \equiv\ & \frac{\partial^2}{\partial y^2} + \frac{a_2(x,x)}{y}\frac{\partial}{\partial y} + \frac{a_3(x,x)}{y^2},\\
Q_x \equiv\ & \frac{\partial^2}{\partial x^2} + \frac{a_1(x,y)}{x}\frac{\partial}{\partial x} + \frac{a_5(x,y)}{x^2},\\
B_4(x,y) =\ & a_1 \cdot a_2 + 2y\frac{\partial a_1}{\partial y},\\
B_6(x,y) =\ & a_1 a_3 + 2ya_2\frac{\partial a_1}{\partial y} + y^2\frac{\partial^2 a_1}{\partial y^2},\\
B_7(x,y) =\ & a_2 a_7 + 2y\frac{\partial a_5}{\partial y},\\
B_8(x,y) =\ & a_3 a_5 + ya_2\frac{\partial a_5}{\partial y} + y^2\frac{\partial^2 a_5}{\partial y^2},
\end{aligned}
$$

$$
\begin{aligned}
&\Omega_1(c_1(x), c_2(x), c_3(y), c_4(y), f)\\
\equiv\ & c_1(x)y^{\psi_1(x)} + c_2(x)y^{\psi_2(x)} + (\psi_2(x) - \psi_1(x))^{-1}\Big\{\int_{y_0}^{y}\Big[(\frac{y}{s})^{\psi_2(x)} - (\frac{y}{s})^{\psi_1(x)}\Big]\\
&\cdot(c_3(s)x^{\varphi_1(s)} + c_4(s)x^{\varphi_2(s)})s\,ds + \int_{y_0}^{y} ds\int_{x_0}^{x}\Big[(\frac{y}{s})^{\psi_2(x)} - (\frac{y}{s})^{\psi_1(x)}\Big]\\
&\cdot\Big[(\frac{x}{t})^{\varphi_2(s)} - (\frac{x}{t})^{\varphi_1(s)}\Big]\cdot(\varphi_2(s) - \varphi_1(s))^{-1}\cdot\frac{f(t,s)}{ts}dt\Big\}
\end{aligned}
$$

$$
Q_x^0 \equiv \frac{\partial^2}{\partial x^2} + \frac{a_1(y,y)}{x}\frac{\partial}{\partial x} + \frac{a_5(y,y)}{x^2}.
$$

The following assertion holds:

Theorem 1. *Let in equation (1) the coefficients and the right part are satisfying the following conditions: 1)* $a_2 \in C^2_{xx}(\overline{D}), a_3 \in C^2_{xx}(\overline{D}), a_1 \in C^1_x(\overline{D}), a_5 \in C^1_x(\overline{D}), a_j \in C^1_y(\overline{D}), j = 2,3, a_k \in C(\overline{D}), k = 4,6,8, f \in C(\overline{D})$; *2) The functions* $a_j(x,y), j = 4,6,7,8$ *with the other coefficients are connected by the aid of formulas:* $a_j(x,y) = A_j(x,y), j = 4,6,7,8$; *3)* $a_2 2 \geq 4a_3, a_1^2 \geq 4a_5$; *4) Functions* $\lambda^\pm(x,y), \mu^\pm(x,y)$ *satisfy the following conditions:*

$$
\lambda^+(0,y) > 1, \lambda^+(0,y) - \lambda^-(0,y) < 1, \mu^+(x,0) > 1, \mu^-(x,0) - \mu^+(x,0) < 1;
$$

5) In the neighborhood Γ_1, *function* $c_1(x,y) = x\frac{\partial\lambda^-(x,y)}{\partial x} - \lambda^-(x,y)$ *has zero of* α *order, that is*

$$
c_1(x,y) = O(|x|^\alpha), \alpha > 1 + \lambda^-(0,y) - \lambda^+(0,y).
$$

In the neighborhood of Γ_2, *function* $c_2(x,y) = y\frac{\partial\mu^-(x,y)}{\partial y} - \mu^-(x,y)$ *has zero of* β *order, that is* $c_2(x,y) = O(|y|^\beta), \beta > 1 + \mu^-(x,0) - \mu^+(x,0)$; *6) The functions* $\mu^\pm(x,y)$ *and* $\nu^\pm(x,y)$ *are satisfying the following conditions:*

$$
|\mu^\pm(x,y) - \mu^\pm(x,0)| \leq H_1^\pm|y|^{\delta_1}, \quad 0 < \delta_1 < 1,
$$

$$|\nu^{\pm}(x,y)-\nu^{\pm}(0,y)|\le H_2^{\pm}|x|^{\delta_2},\quad 0<\delta_2<1. \tag{8}$$

Then any solution of equation (1) of class $c^3(D)$ can be represented in the form:

$$\begin{aligned} u(x,y)= \;& \exp[w_2^-(x,0)-\mu_1^-(x,0)\ln y]\cdot[K_2(\varphi_1(x),\varphi_2(x),V)\\ &+\int_{y_0}^{y}\Gamma^2(x,y,\tau)K_2(\varphi_1,\varphi_2,V)d\tau], \end{aligned} \tag{9}$$

$$\begin{aligned} V(x,y)= \;& \exp[w_1^-(x,y)-\lambda^-(0,y)\ln x]\cdot[K_1(\psi_1(y),\psi_2(y),f)\\ &+\int_{x_0}^{x}\Gamma_1(x,y,t)K_1(\psi_1,\psi_2,f)dt], \end{aligned} \tag{10}$$

where $\varphi_1(x),\varphi_2(x),\psi_1(y),\psi_2(y)$ are arbitrary functions in Γ_1 and Γ_2; moreover $\varphi_j(x)\in c^2(\Gamma_1),\psi_j(y)\in c^2(\Gamma_2), j=1,2, \Gamma_1(x,y;t),\Gamma_2(x,y;\tau)$ are resolvents of Volterra integral equations of the second kind with the weak singularity.

Remark 1. The solution of type (9) has the following characteristics:

$$(y^{\mu^-(x,0)}u(x,y))_{y=y_0}=\varphi_2(x),(x^{\lambda^-(0,y)}P_y(u))_{x=x_0}=\psi_2(y),$$

$$[x^{\lambda^+(0,y)}P_x^-(P_y(u))]_{x=x_0}=\psi_1(y),(y^{\mu^+(x,0)}P_y^-(u))_{y=y_0}=\varphi_1(x).$$

Remark 2. The solution of type (9) in the neighborhood has the following characteristics:

$$u(x,y)=O(y^{-\mu^-(x,0)}). \tag{11}$$

If $\mu^-(x,0)<0$, then $u(x,y)$ in neighborhood Γ_1 has zero of $|\mu^-(x,0)|$ order; if $\mu^-(x,0)=0$, then $u(x,y)$ bounded and if $\mu^-(x,0)>0$, then $u(x,y)$ in unbounded and this order is determine from equality (2).

Remark 3. The assertion similar to Theorem 1 is given and in the case, when $a_1^2(x,y)-4a_5(x,y)<0, a_2^2(x,y)-4a_3(x,y)\ge 0; a_1^2(x,y)-4a_5(x,y)\ge 0, a_2^2(x,y)-4a_3(x,y)<0; a_1^2(x,y)-4a_5(x,y)<0, a_2^2(x,y)-4a_3(x,y)<0.$

Remark 4. The assertion similar to Theorem 1 and Remarks 1–3 is given in the case, when $a_4\in c_{yy}^2(\overline{D}), a_5\in c_{yy}^2(\overline{D}), a_1\in c_y^1(\overline{D})$, but functions $a_j(x,y), j=4,6,7$ with the other coefficients are connected by the aid of formuls:

$$a_j(x,y)=B_j(x,y),\quad 4\le j\le 8.$$

Remark 5. In the case, when $a_j(x,y)\ne A_j(x,y), a_j(x,y)\ne B_j(x,y), 4\le j\le 8$, we also obtain integral representations of the manifold of solutions in terms of four arbitrary functions of a single variable and the resolvents Volterra integral equations of the second kind with the weak singular kernel.

Theorem 2. *Let in equation (1) the coefficients and the right part are satisfying the conditions 1), 2) of Theorem 1 and $a_j(x,y)=a_j(y,y), j=1,5, a_k(x,y)=a_k(x,x), k=2;3;$ the roots of characteristic equation (4) $\varphi_1(y),\varphi_2(y)$ and the roots of characteristic* *equation (5) $\psi_1(y), \psi_2(y)$, be real and distinct, $\psi_j(x) < 2, j = 2$, the function $f(x,y)$ in the neighbourhood $\Gamma_1 \cup \Gamma_2$ has zero of $\delta + \gamma$ order, that is*

$$f(x,y) = O(x^\delta y^\gamma), \quad \delta > \varphi_j(y), \gamma > \psi_j(x), \quad j = 1,2.$$

Then any solution of equation (1) of class $C^3(D)$ can be represented in the form

$$U(x,y) = \Omega_1(C_1(x), C_2(x), C_3(y), C_4(y), f), \tag{12}$$

where $C_j(x), C_k(y), j,k = 1,2$–are arbitrary functions in Γ_1 and Γ_2, moreover $C_j(x) \in C^2(\Gamma_1), C_k(y) \in C^1(\Gamma_2), j = 1,2; k = 3,4$.

Remark 6. The integral representations, similar to (12) has been obtained in all others cases: when the roots of the equation (4) be real and distinct, the roots of the equation (5) be real and equal or complex-conjugates; when the roots of the equation (4) be real and equal, and the roots of the equation (5) be real and distincts, be real and equal or complex-conjugates; when the roots of the equation (4) complex-conjugates, and the roots equation of the (5) be real and distincts, be real and equal or complex-conjugates, in all 9 possible cases.

Remark 7. In the case, when the corresponding coefficients of equation (1) do not satisfy the conditions

$$a_j(x,y) = a_j(y,y), \quad j = 1,5, \quad a_k(x,y) = a_k(x,x), \quad k = 2,3.$$

We also obtain the integral representations of the manifold solutions which are depending on the roots of the characteristic equations (4), (5) and the resolvents of Volterra integral equations of the second kind with the weak singular kernel.

On the basis of the integral representations we are investigating the following boundary valued problems:

Problem H. Find the solution of equation (1) in the class $C^3(D)$ satisfying one of the following boundary conditions:

$$(y^{\mu^-(x,0)} u(x,y))_{y=y_0} = f_{11}(x), \quad 0 \le y_0 \le \beta$$

$$\left.\begin{array}{r}(y^{\mu^+(x,0)}P_y^-(u))_{y=y_0}=f_{12}(x),\quad 0\le y_0\le\beta\\(x^{\lambda^-(0,y)}P_y(u))_{x=x_0}=g_{11}(y),\quad 0\le x_0\le\alpha\\(x^{\lambda^+(0,y)}P_x^-(P_y(u)))_{x=x_0}=g_{12}(y),\quad 0\le x_0\le\alpha\end{array}\right\}\text{(H1)}$$

$$\left.\begin{array}{r}(x^{\lambda^-(0,y)}u(x,y))_{x=x_0}=g_{21}(y)\\(x^{\lambda^+(0,y)}P_x^-(u))_{x=x_0}=g_{22}(y)\\(y^{\mu^-(x,0)}Q_x(u))_{y=y_0}=f_{21}(x)\\(y^{\mu^+(x,0)}P_y^-(Q_x(u)))_{y=y_0}=f_{22}(x)\end{array}\right\}\text{(H2)}$$

$$\left.\begin{array}{c}u(x,y_0)=f_{31}(x),(P_y^0(u))_{x=x_0}=g_{31}(y)\\(\frac{\partial u}{\partial y})_{y=y_0}=f_{32}(x),(\frac{\partial}{\partial x}(P_y^0(u)))_{x=x_0}=g_{32}(y)\\0<x_0\le\alpha,\quad 0<y_0\le\beta\end{array}\right\}\text{(H3)}$$

$$\left.\begin{array}{c}u(x_0,y)=g_{41}(y),(Q_x^0(u))_{y=y_0}=g_{41}(y)\\(\frac{\partial u}{\partial x})_{x=x_0}=g_{42}(y),(\frac{\partial}{\partial x}(Q_x^0(u)))_{y=y_0}=g_{42}(y)\\0<x_0\le\alpha,\quad 0<y_0\le\beta\end{array}\right\}\text{(H4)}$$

$$\left.\begin{array}{c}[y^{\psi_1(x)}(u(x,y)\psi_2(x)-y\frac{\partial u}{\partial y})]_{y=y_0}=f_{51}(x)\\ [y^{\psi_2(x)}(u(x,y)\psi_1(x)-y\frac{\partial u}{\partial y})]_{y=y_0}=f_{52}(x)\\ [x^{\varphi_1(y)}(P_y^0(u)\varphi_2(y)-x\frac{\partial}{\partial x}(P_y^0(u)))]_{x=0}=g_{51}(y)\\ [x^{\varphi_2(y)}(P_y^0(u)\varphi_1(y)-x\frac{\partial}{\partial x}(P_y^0(u)))]_{x=x_0}=g_{52}(y)\\0\le x_0\le\alpha,\quad 0\le y_0\le\beta\end{array}\right\}\text{(H5)}$$

In the problems $(\mathrm{H}_j)1\le j\le 5, f_{ik}(x), g_{ik}(y), 1\le j\le 5, 1\le k\le 2$ are given functions in Γ_1 and Γ_2.

References

[1] Rajabov, N. *Integral representations and boundary value problems for some differential equations with singular line or Singular surfaces*, part 4. Dushanbe: 1985 (in Russian), -148.

[2] Rajabov, N., *Integral representations and boundary value problems for some differential equations with one and two singular lines*. Soviet. Math. Dokl., 218(3)(1985), 534–539.

[3] Rajabov. N., *Representation of the manifold of Solutions and boundary value problems for a class of linear second order hyperbolic Systems with an interior Singular line*, Soviet Math. Dokl., 37(3)(1988), 659–663.

[4] Rajabov, N.O., *Theory of linear elliptic systems second order for which influence the boundary is to appear singular line), Matematicki Vesnik. Third international symposium–Complex analysis and applications*, – Herceg-Novi, Jugoslavia, May 23–29, 1988, 40(1988), 301–307.

[5] Rajabov. N., *On the theory of linear hyperbolic equations with two super singular lines*. Dokl. Akad. Nauk Tajik SSR, 32(9)(1989), 573–577.

[6] Rajabov, N., *Integral representations for certain equations with two singular lines and their application*, Complex Analysis and Applications 85. Sofia, (1986), 548–555.

[7] Rajabov, N., *On one class of linear hyperbolic equations of the second order with a super singular point*, Dokl. Akad. Nauk Tajik SSR, 31(10)(1988), 630–634.

[8] Rajabov, N., *Integral representations and boundary value problems for some second order linear elliptic systems with regular and singular coefficients*, Complex Analysis and Applications, 87, Sofia 1989, p. 431–441.

[9] Rajabov N., *On the second order equation of hyperbolic type with two singular lines*, Differential Equations. Minsk, 24(12)(1988), 2129–2139.

Proceedings of the Asian Mathematical
Conference, 1990

Conformal Modules and Capacitance of a Transmission Line

MOHAMAD RASHIDI RAZALI
Department of Mathematics
Universiti Teknologi Malaysia
Locked Bag No. 791 80990 Johor Bahru, Malaysia

§1. Introduction

Conformal mapping has been a popular analytical tool used by researchers in various problems such as in the solution of Laplace's equation. Unfortunately, not all conformal mappings are analytically known. Only a few is listed ([17], [18]) and most of them need to be determined numerically ([7], [12], [36]).

A particular domain of significant interest is the quadrilateral which we need to map on simple domains such as the unit disc or the upper half-plane. This mapping is related to the conformal module M of the quadrilateral.

This conformal module has been shown useful in electromagnetic field theory ([16], [22], [35]), the measurement of diffusion coefficients of solid materials ([2], [21]), an area problem for quadrilateral ([1], [10]) and numerical grid generation ([24], [34]).

In this paper, we are only concerned with the application of conformal modules in the determination of the capacitance of a transmission line with quadrilateral cross-section. Here, we use a numerical method based on singular integral equations, which has appeared before ([30], [32]).

This paper is arranged as follows: In Section 2, we write the capacitance problem in the form of the Laplace's equation with mixed boundary conditions. The numerical method to solve the latter is discussed in Section 3. Some results are then given in Section 4 and are compared to those obtained by other researchers. The crowding phenomenon associated with slender or elongated quadrilaterals is next discussed with suggestions to avoid this serious problem.

§2. The Capacitance Problem

Let L be a closed contour bounding a simply-connected plane domain Ω, and let A, B, C, D be 4 points on it. The capacitance C_M between AB and CD is defined by

$$C_M = \frac{1}{4\pi}\int_{AB}\frac{\partial u}{\partial n}ds \tag{2.1}$$

where u satisfies the Laplace's equation:

$$\frac{\partial^2 u}{\partial x^2} + \frac{\partial^2 u}{\partial y^2} = 0 \quad \text{in} \quad \Omega \tag{2.2}$$

subject to the boundary conditions

$$u = 1 \quad \text{on} \quad CD, u = 0 \text{ and } DA, AB \text{ and } BC.$$

Here the derivative in (2.1) is taken to be directed into the given domain.

This capacitance problem has been studied by many researchers ([9], [20], [33]). Gaier [9] has given two expressions connecting the capacitance C_M and the conformal module M of the quadrilateral $ABCD$. This module is defined by a conformal mapping of Ω onto a rectangle

$$R = \{(\xi, \eta) : 0 < \xi < a, 0 < \eta < b\}$$

such that AB and CD map onto the horizontal sides and is given by the ratio b/a. The two expressions are

$$C_M = \frac{2}{\pi^2} \sum_{r=1}^{\infty} \{(2r-1)\sin h[(2r-1)\pi M]\}^{-1} \tag{2.3}$$

and

$$C_M = \frac{1}{2\pi^2} \ln \left[4\sqrt{q} \quad \Pi \left(\frac{1+q^{2r}}{1+q^{2r-1}} \right)^4 \right] \tag{2.4}$$

where $q = \exp(-\pi/M)$.

Symm ([33]) has pursued this problem further. Using the Dirichlet principle [5], he deduces that the conformal module M can be found from the following problem: to solve

$$M = -\int_{DA} \frac{\partial u}{\partial n} ds, \tag{2.5}$$

where u satisfies the Laplace's equation (2.2) subject to the boundary conditions:

$$u = 0 \quad \text{on} \quad BC, \tag{2.6}$$

$$u = 1 \quad \text{on} \quad DA, \tag{2.7}$$

$$\frac{\partial u}{\partial n} = 0 \quad \text{on} \quad AB \quad \text{and} \quad CD. \tag{2.8}$$

Using the Cauchy-Riemann equations, (2.5) can be likewise written as:

$$M = -(v(D) - v(A)), \tag{2.9}$$

where v is the conjugate harmonic function of u.

§3. A Numerical Method

In order to solve the mixed boundary value problem (2.2), (2.6), (2.7) and (2.8), we employ a numerical method based on singular integral equations which has been described in [30] and [32]. Here, we just give some highlights of the method. For details, the reader is advised to consult the above two references. Application of the method to solve a groundwater seepage problem has been given in [31].

Consider the following modified Volterra problem ([13], [15]): To obtain a complex function $\Phi(z) = u(x,y) + iv(x,y)$ holomorphic in the upper half plane $y \geq 0$ and bounded at infinity and constants $c_1, c_2, \cdots, c_p$ such that

$$v(x,0) = v_0(x) + c_j \quad \text{if} \quad x \in [a_j, b_j], \quad j = 1, \cdots, p, \tag{3.1}$$

$$u(x,0) = u_0(x) \quad \text{if} \quad x \in L_2, \tag{3.2}$$

where $u_0(x)$ and $v_0(x)$ are two given functions assumed to be smooth (satisfy a Holder condition), $L_1 = \cup_{j=1}^{p}[a_j, b_j]$, $L_2 = I\!R \backslash L_1$.

The above problem is obtained from the mixed boundary value problem using the Cauchy-Riemann equations and on integration. The c_j are thus constants due to integration.

Using the fact that $\phi(z)$ satisfies

$$\Phi(z) = \frac{1}{\pi i}\int_{-\infty}^{\infty}\frac{u(t,o)dt}{t-z} + iC, \tag{3.3}$$

where C is an arbitrary constant ([25]), and the Sokhotski-Plemelj formulas, the relevant singular integral equation is

$$-\frac{1}{\pi}\int_{L_1} u_0(t)\frac{dt}{t-s} - c_j = v_0(s) + \frac{1}{\pi}\int_{L_2} u_0(t)\frac{dt}{t-s} \tag{3.4}$$

which enables $u_0(s), s \in L_1$ to be calculated, where it is unknown.

The solution of (3.4) is done through the use of Chebyshev polynomials and collocation. The Hilbert transform of the Chebyshev polynomial of the second kind $U_n(t)$ is utilized:

$$-\frac{1}{\pi}\int_{-1}^{1}(1-t^2)^{1/2}U_n(t)\frac{dt}{t-s} = \begin{cases} T_{n+1}(s), & -1 \leq s \leq 1, \\ G_{n+1}(s), & s < -1, \\ H_{n+1}(s), & s > 1, \end{cases} \tag{3.5}$$

where $T_n(s)$ is the Chebyshev polynomial of the first kind,

$$G_n(s) = (s - (s^2-1)^{1/2})^{-n}, \quad H_n(s) = (s^2 + (s^2-1)^{1/2})^{-n}.$$

For details, see [30, p.120]. The collocation points are taken to be the zeros of $T_{n+2}(s)$.

For the mixed boundary value problem (MBVP) in Section 2, the rectangle is first transformed onto an upper half plane. This MBVP will lead to a singular integral equation on two arcs, and as such, we have to determine the constants c_1 and c_2 together with the unknown potentials. Equation (2.9) can simply be written as

$$M = |c_1 - c_2|. \tag{3.6}$$

The capacitance C_M is then calculated using equation (2.3).

§4. Some Numerical Results

We give here some numerical results obtained by the method given in Section 3.

Problem A ([33]). To compute the capacitance between opposite sides AB and CD of a 2×1 rectangle:

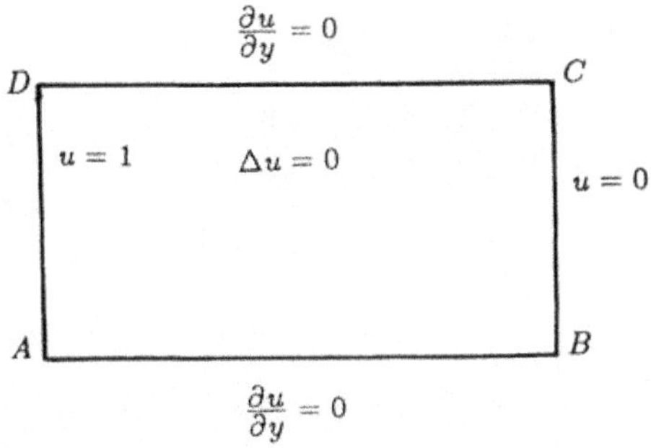

Here, it is immediately evident that $M = 0.5$ and hence $C_m = 0.08930$. Lampard ([20]) has shown that

$$C_M = \frac{1}{4\pi^2} \ln(17 + 12\sqrt{2}).$$

In table 4.1, we give the values of c_1, c_2 and M. Our results for M are better than those obtained in [33]. L denotes the number of equations used

Table 4.1

L	c_1	c_2	M
20	-0.211491	0.289829	0.501320
36	-0.210970	0.289106	0.500076
52	-0.210946	0.289074	0.500020
68	-0.210941	0.289067	0.500008
84	-0.210939	0.289065	0.500004

Problem B ([8], [26, p.77], [33]). To compute the capacitance C_M between the sides AB and CD of the L-shaped domain, formed by removing an unit square from a square of side 2.

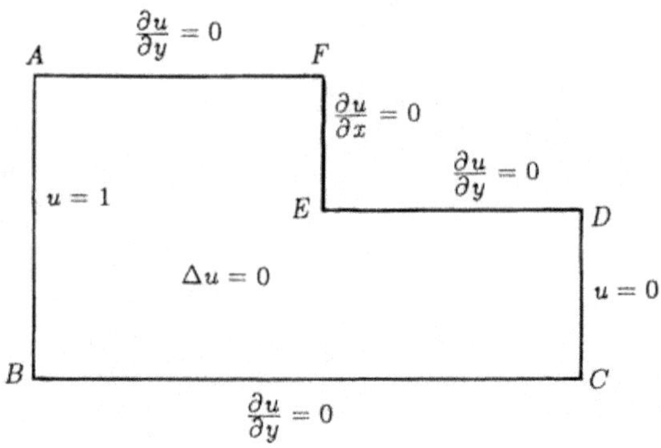

The true value of $m = 1/M$ is $1/\sqrt{3}$ ([8]). Hence $M = \sqrt{3}$ and $C_M = 0.001756$. Table 4.2 gives the values of c_1, c_2, m and M for various L. For comparison, Symm ([33]) has obtained $M = 1.7230$ by a modified boundary integral method using $n = 128$.

Table 4.2

L	c_1	c_2	$m = c_2 - c_1$	$M = 1/m$
20	-0.292020	0.285647	0.577667	1.731101
36	-0.291877	0.285507	0.577384	1.731950
52	-0.291862	0.285492	0.577354	1.732040
68	-0.291858	0.285488	0.577346	1.732064
84	-0.291856	0.285487	0.577343	1.732073

§5. The Crowding Phenomenon

An intrinsic property associated with conformal mapping is the crowding phenomenon which arises when we map elongated regions. Some authors give it a different name "the Geneva effect". The term "crowding" is due to Menikoff and Zemach ([23]). Other authors that have found this crowding problem in their works are [3], [4], [11], [26] [27] and [37].

The theory behind the occurence of this intrinsic problem has not been established yet. Let the points $\xi_j = f(z_j), j = 1,2,3,4$ divide the unit disc into four arcs $\gamma_j = (\xi_j, \xi_{j+1}), j = 1,2,3$ and $\gamma_4 = (\xi_4, \xi_1)$. Let $\phi_1 = \min\{\gamma_1, \gamma_2\}$ and $\phi_2 = \min\{\gamma_3, \gamma_4\}$. According to Papamichael [26], $\phi_1 \approx \exp(-\pi M/2)$ when the conformal module M is large and $\phi_2 \approx \exp(-\pi/(2M))$ when M is small. For these values of M, due to the architecture of the computer, the points $\xi_j, j = 1,2,3,4$ will no more be in the correct order. The quantity

$$Q = \begin{cases} 8\exp\{-\pi/2M\} & \text{if } M \text{ is small,} \\ 8\exp\{-\pi M/2\} & \text{if } M \text{ is large} \end{cases} \tag{5.1}$$

can be used as a measure for crowding ([27]).

To avoid the difficulties associated with crowding, a domain decomposition method has been suggested ([28], [29]) whereas Krikeles and Rubin ([19]) use a normalization for Schwarz-Christoffel mappings. Using a transformation originally devised in [6], Howell & Trefethen ([14]) use another elongated region, an infinite strip, as an intermediate domain during the Schwarz-Christoffel mappings.

References

[1] A. Acker, *Heat flow inequalities with applications to heat flow optimization problems*, SIAM J. Math. Anal., 8(1977), 604–618.

[2] G.R. Bell and J. Crank, *Influence of imbedded particles on steady-state diffusion*, J. C. S. Faraday Trans. II 70(1974), 1259–1273.

[3] J.P. Berrut, *Ueber integralgleichungen and Fourier methoden zur numeriche konforme abbildung*, Diss no 7754, Swiss Federal Institute of Technology, Zurich(1985).

[4] N.V. Challis and D.M. Burley, *A numerical method for conformal mapping*, IMA J. Numer Anal., 2(1982), 169–181.

[5] R. Courant, *Dirichlet's principle, conformal mapping and minimal surface*, Interscience, New York, 1950

[6] R.T. Davis, *Numerical methods for cordinate generation based on Schwarz-Christoffel transformations*, Proc. 4th AIAA Comput. Fluid Dynamics Conf., Williamsburg, VA, 1979.

[7] D. Gaier, *Konstruktive methoden der konformen Abbildung*, Springer, 1964.

[8] D. Gaier, *Ermittlung des konformen Modulus von Vierecken mit Differenzenmethoden*, Number. Math., 19(1972), 179–194.

[9] D. Gaier, *Capacitance and the conformal module of quadrilaterals*, J. Math. Anal. Appl., 70(1979), 236–239.

[10] D. Gaier, *On a area problem in conformal mapping*, Results in Math., 10(1986), 66–81.

[11] E. Grassmann, *Numerical experiments with a method of successive approximations for conformal mapping*, Z. Angew. Math. Phys., 30(1979), 873–884.

[12] P. Henrici, *Applied and Computational Complex Analysis*, Vol. III, Wiley, New York, 1986.

[13] D. Homentcovschi, *On the mixed boundary value problem for harmonic functions in plane domains*, J. Appl. Math., Phys. (ZAMP) 31, 352–366, 1980.

[14] L.H. Howell and L.N. Trefethen, *A modified Schwarz-Christoffel transformation for elongated regions*, to appear in SIAM J. Sci. Stat. Comp.

[15] C. Jacob, *Introduction mathematique a la mecanique des fluides*, Bucharest-Paris, 1959.

[16] H.B. Keller and D. Sachs, *Calculations of the conductivity of a medium containing cylindrical inclusions*, J. Appl. Phys., 35(1964), 537–538.

[17] H. Kober, *Dictionary of Conformal Representations*, Dover, New York, 1957.

[18] W. Koppenfels & F. Stallmann, *Praxis der Konformen Abbildung*, Springer-Verlag, Berlin, 1959.

[19] B.C. Krikeles and R.L. Rubin, *On the crowding of parameters associated with Schwarz-Christoffel transformations*, Appl. Math. Comp., 28(1988), 297–308.

[20] D.G. Lampard, *A new theorem in electrostatics with applications to calculable standards of capacitance*, Proc. I. E. E. Monograph No. 216M, (1957), 271–280.

[21] J. Larsen, *On the measuring of diffusion coefficients of solid materials*, Z. Angew Math. Phys., 32(1981), 229–231.

[22] P.A.A. Laura and L.E. Luisoni, *An application of conformal mapping to the determination of the characteristic impedance of a class of coaxial systems*, IEEE Trans. Microwave Theory Tech., 25(1977), 162–164.

[23] R. Menikoff and C. Zemach, *Methods for numerical conformal mapping*, J. Comput. Phys., 36(1979), 360–410.

[24] C.D. Mobley and R.J. Stewart, *On the numerical generation of boundary-fitted orthogonal curvilinear coordinate systems*, J. Comput. Phys., 34(1980), 124–135.

[25] N.I. Muskhelishvili, *Singular Integral Equations*, P. Noordhoff N.V., Groningen-Holland, 1953.

[26] N. Papamichael, *Numerical conformal mapping onto a rectangle with applications to the solution of Laplacian problems*, J. Comp. Appl. Math., 28(1989), 63–83.

[27] N. Papamichael, C.A. Kokkinos, and M.K. Warby, *Numerical techniques for conformal mapping onto a rectangle*, J. Comput. Appl. Math., 20(1987), 349–358.

[28] N. Papamichael and N.S. Stylianopoulos, *On a domain decomposition method for the computation of conformal modules*, Appl. Math. Letters, 1(1988), 277–280.

[29] N. Papamichael and N.S. Stylianopoulos, *A domain decomposition method for conformal mapping onto a rectangle*, Tech. Report TR/04/89, Dept. of Mathematics and Statistics Brunel University, Uxbridge, U.K., 1989.

[30] M.R. Razali, *Singular integral equations and mixed boundary value problem for harmonic functions*, Ph. D Thesis, Univ. of Southampton, UK, 1983.

[31] M.R. Razali, *The treatment of seepage problems by Cauchy singular integral equations. In: Numerical Solution of singular integral equations*, edited by A. Gerasoulis and R. Vichnevetsky, IMACS, 91–93, 1984.

[32] M.R. Razali and K.S. Thomas, *Singular integral equations and mixed boundary value problems for harmonic functions.* In: *Treatment of Integral Equations by Numerical Methods*, edited by C.T.H. Baker and G.F. Miller, Acad. Press, London, 1982.

[33] G.T. Symm, *Two methods for computing the capacitance of a quadrilateral*, Appl. Math. Modelling, 5(1981), 428–431.

[34] J.F. Thompson, Z.U.Z. Warsi and C.W. Mastin, *Numerical Grid Generation*, North-Holland, Amsterdam, 1985.

[35] L.N. Trefethen, *Analysis and design of polygonal resistors by conformal mapping*, Z. Angew. Math. Phys., 35(1984), 692–704.

[36] L.N. Trefethen, Ed. *Numerical Conformal Mapping*, North-Holland, Amsterdam, 1986 (reprinted froom J. Comput. Appl. Math., 14(1&2), 1986).

[37] C.Zemach, *A conformal map formula for difficult cases.* In: [36], 1986, 207–215.

Proceedings of the Asian Mathematical
Conference, 1990

Predictability and Behaviour of the Atmospheric Circulation

S.M. SHARFUDDIN
Advanced Centre for Mathematical Sciences and Mathematics Education 58 Lake Circus Kalabagan Dhaka-1205, Bangladesh

§1. Introduction

The aim of this paper is to obtain a better understanding of the dynamics of the atmospheric circulation in the midlatitudes (roughly between 30 degrees and 60 degrees latitudes), especially in relation to the problem of long range weather prediction. Modern weather forecasts are based on the results of detailed and complicated numerical model of European Centre for Medium Range Weather Forecasting (ECMWF) in Reading, England. It has long been assumed that the period over which the weather is predicted could be increased forever if the numerical models would be further improved and the initial state (determined by means of observations) better prescribed. However, now a days it is known that there is a fundamental limit to the period over which the weather can be predicted i.e. it cannot be enlarged by carrying out more and better observations. In most cases the predictability horizon of the atmosphere is encountered after a period of five to fourteen days. As a consequence, it is not possible to give forecasts for a period longer than about two weeks. But unfortunately there is a strong need for accurate long range weather predictions. This is because on time scales between a week and three months frequently climate anomalies occurs which have large socio-economic consequences such as drought, heat waves, tornado, floods etc.

In order to understand why forecasts fail on long term we first present some qualitative arguments. The weather as we experience (in midlatitudes) it is the result of day to day variations in the geographical distributions of high and low pressure belts. These so called synoptic scale eddies have typical horizontal dimensions of 1000km, a life span of about a week and are embeded in a belt of predominantly westerly winds. The latter results from an approximate balance between thermal forcing, due to equator pole temperature gradient, and the Coriolis force (induced by the rotation of the earth) acting on a moving fluid. Furthermore, due to the presence of large scale topography (to which, in particular the Himalays, Rocky mountain and the oceans contribute) and the thermal differences between land and ocean, ultra long quasi-stationary waves are generated which give the flow a meandering structure. The planetary waves have much larger dimensions (about 1000 km) and longer lives (of order of several months) than the synoptic scale eddies. Thus, the atmospheric circulation is characterised by two distinct scales of motion: a planetary scale and a synoptic scale. Little is known about the subtle interplay between these scale of motion. It appears that synoptic scale eddies develop spontaneously as initially small perturbations of the locally unstable planetary scale circulation. Moreover, the planetary-scale flow tends to steer and organise the eddies along preferent paths, which are the stormtracks. On the other hand the eddies themselves influence the evolution of the planetary waves. The consequences for the predictability of the atmospheric circulation were systematically studied by Lorentz [5]. He has shown that interaction between different scales of motion are the principal cause for limited predictability of the atmosphere.

The feed back between planetary waves and the synoptic scale eddies causes quasi-stable flow configuration which in its turn causes short range climate anomalies. The existence of such large scale preferent states of the atmospheric circulation (some times called weather regimes) has been known for a long time. They can be divided into three major types: (i) zonal (high index) states with strong western winds and small wave amplitudes; (ii) meridional (low index) states with large waves embeded in a weak zonal flow and (iii) transitional states which have characteristics of both high and low index states. Typical flow configurations in the European region are shown in Fig.1 (from Van Dijk et al [7]), and the same for Pacific Region are shown in Fig. 1 (from R.M. Dole: Thesis MIT [2]).

The atmosphere is a chaotic system which shows vacillatory behavior i.e. irregularly visits different preferent states. Quasi-stable flow patterns suddenly develop and disappear without a clear indication why (Dole [2], Reinhold [11]). Furthermore, the life span of the weather regimes is highly variable without having a preferent time scale. Within the frame work of long term weather predictions it is important to obtain a better understanding of the dynamics responsible for this vacillatory behaviour. A method for studying this problem consists of analysing highly simplified models which represents qualitative figures of the atmospheric circulation. Our purpose for doing this is that from the results, indications may be found how to consider more complicated models as well as real data. In this way we hope to enhance our understanding in the atmospheric dynamics.

§2. Quasi-geostrophic Atmospheric Dynamics

Here we use the simplified form of full equations of motions by the applications of scale analysis, which is a standard technique in Geophysical Fluid Dynamics (Pedlosky [6]). The method requires an a priori specification of the type of motion to be studied. Next it gives, by means of physical arguments, characteristic scales for the flow with which the equations of motion are written in a dimensionless form. The resulting system will contain several dimensionless parameters. The aim of this method is to find small parameters. Then, by means of standard perturbation techniques, simplified equations are derived which describe the type of motion under consideration.

Let us consider a flow near some central latitude $\phi = \phi_0$ on the Northern Hemisphere distant from equator and pole. Let it have a horizontal (parallel to the earth's surface) length scale k^{-1}, a vertical length scale H (which is the depth of the fluid) and a time scale σ^{-1}, such that

$$H << K^{-1} << r_0, \quad \sigma << f_0 = 2\Omega \sin \phi_0 \tag{2.1}$$

where f_0 is the Coriolis parameter at $\phi = \phi_0$, Ω the angular speed of rotation of the earth and r_0 the radius of the earth. The first condition implies that the flow is nearly horizontal and 2-dimensional. The second one means that to a first approximation the momentum equations reduce to a balance between the Coriolis force and pressure gardient force, which is the geostrophic balance. Obviously, (2.1) is satisfied for large-scale atmospheric motions near $\phi_0 = 45°N$ where $k^{-1} \sim 10^6 m, H \sim 10^4 m, \sigma^{-1} \sim 10^5 s, f_0 = 10^{-4} s^{-1}$ and $r_0 \sim 6.4.10^6 m$. Pedlosky [6] has shown that under these conditions the equation of motion reduces to one non-linear partial differential equation. With the additional assumption that the flow is barotropic (i.e. density is a function of pressure only) we get (in a dimensionless form)

$$\underset{(1)}{\frac{\partial}{\partial t}\nabla^2\psi} + \underset{(2)}{J(\psi,\nabla^2\psi)} + \underset{(3)}{\nu J(\psi,h)} + \underset{(4)}{\beta\frac{\partial\psi}{\partial x}} + \underset{(5)}{c\nabla^2(\psi-\psi^*)} = 0 \tag{2.2}$$

This is the barotropic vorticity equation. Here t is the time and $\psi(x,y,t)$ is stream function to which all the state variables (velocities, density, pressure and temperature) are related. At a fixed time the flow is along the stream line $\psi =$ constant. Furthermore,

$$\nabla = (\frac{\partial}{\partial x}, \frac{\partial}{\partial y}), \quad J(a,b) = (\frac{\partial a}{\partial x}\frac{\partial b}{\partial y} - \frac{\partial a}{\partial y}\frac{\partial b}{\partial x}), \tag{2.3}$$

$$dx = kr_0\cos\phi_0 d\lambda, \quad dy = kr_0 d\phi,$$

where λ is the longitude and ϕ is the latitude. In equation (2.2): term (1) represents the local change of relative vorticity $\nabla^2\psi$ (which is the vertical component of the curl of the velocity vector), (2) the advection of the vorticity by the flow itself, (3) the production of vorticity due to the presence of a large scale topography $h(x,y)$ (in particular the high mountains and the oceans) with characteristic amplitude h_0, (4) the planetary vorticity advection due to the variation of Coriolis parameter with latitude and (5) represents both a damping of vorticity and an external vorticity forcing (modelling the equator-pole temperature gradient indicated by the function $\psi^*(x,y)$). The dimensionless parameters are

$$\gamma = (f_0 h_0)/(\sigma H), \quad \beta = \beta_0/(\sigma k) = (2\Omega\cos\phi_0)/(\sigma k r_0), \quad C = (f_0\delta_E)/(2\sigma H), \tag{2.4}$$

where δ_E is the depth of the boundary layer near the earth's surface in which frictional effects are important. The flow described in (2.2) is called quasi-geostrophic because the small departures from the geostrophic balance determine the evolution of the flow (Pedlosky [6]).

§3. Spectral Models by Galerkin Projection Techniques

Because of nonlinear structure the barotropic vorticity/equation (2.2) is difficult to handle. In order to obtain approximate solutions we apply Galerkin projection techniques where explicit use is made of the boundary conditions to the equation (Voigt et al [8]). Consider equation (2.2) in a rectangular channel on β-plane of length $L(= 2\pi/k)$ and width $B(= bL/2)$. The dimensionless length is 2π in zonal x-direction and width πb in meridional y-direction. We investigate the existence of travelling wave solutions in the zonal x-direction. At the boundary $y = 0$ and $y = \pi b$ the meridional velocity component is assumed to be zero and it follows that mean zonal velocity component over these boundaries should be constant. Consequently the boundary conditions are

$$\left.\begin{aligned} &\psi(x+2\pi, y, t) = \psi(x, y, t), \\ &\frac{\partial\psi}{\partial x} = 0 \text{ and } \frac{\partial}{\partial t}\int_0^{2\pi}\frac{\partial\psi}{\partial y}dx = 0, \quad \text{at } y = 0,\ y = \pi b. \end{aligned}\right\} \tag{3.1}$$

Applying the spectral method, we expand the stream function $\psi(x, y, t)$ in a series of eigenfunctions $\{\Phi_j\}_j$ of the Laplace operator ∇^2 with corresponding eigenvalues λ_j, thus

$$\psi(x, y, t) = \sum_j \psi_j(t)\phi_j(x, y), \quad j = (j_1, j_2). \tag{3.2}$$

Each mode $\psi_j\phi_j$ satisfies the boundary conditions and the eigenfunctions are orthonormalised with respect to the domain average. In this case

$$\{\phi_j\} = \sqrt{2}\cos(j_2 y/b), \quad \lambda_j = j_2^2/b^2, \tag{3.3a}$$

(These functions describe $(0, j_2)$ zonal flow modes because they are independent of x.)

$$\{\phi_j\} = \sqrt{2}\exp(ij_1 x)\sin(j_2 y/b), \tag{3.3b}$$

(These functions describe $(|j_1|, j_2)$ wave modes.)

$$\lambda_j = j_1^2 + j_2^2/b^2, \quad |j_1|,\ j_2 = 1, 2, \cdots. \tag{3.3c}$$

The topography and forcing stress functions are represented by

$$\left.\begin{aligned} h(x, y) &= \cos(x)\sin(y/b), \\ \psi^*(x, y) &= \sqrt{2}\{\psi_{01}^*\cos(y/b) + \psi_{02}^*\cos(2y/b)\}, \end{aligned}\right\} \tag{3.4}$$

i.e. topography is given as a (1.1) mode, being the longest wave present in the model.

Projecting equation (2.2) on the eigenfunctions (3.3) which is called Galerkin projection, we obtain a spectral model

$$\begin{aligned} \lambda_j\dot{\psi}_j = &\ \tfrac{1}{2}\sum_l\sum_m c_{jlm}(\lambda_l - \lambda_m)\psi_l\psi_m + \nu\sum_l\sum_m c_{jlm}\psi_l h_m \\ &+ \sum_l b_{jl}\psi_l - c\lambda_j(\psi_j - \psi_j^*), \end{aligned} \tag{3.5}$$

consisting of an infinite number of coupled ordinary differential equation. This system describes the time evolution of the expansion $\{\psi_j\}_j$. The dot denotes differentiation with respect to time,

$$c_{jlm} = \langle\phi_j, J(\phi_l, \phi_m)\rangle, \quad b_{jl} = \beta\langle\phi_j, \frac{\partial\phi_l}{\partial x}\rangle, \tag{3.6}$$

are the interaction coefficients and $\langle A, B\rangle = \int_D AB^{cc}dr/\int_D d\bar{r}$ an inner product on the domain considered, cc denotes complex conjugate. It appears that non-linear contributions always occur as triads in which two modes interact and affect the evolution of a third mode. From (3.3) and (3.6) we find that there are two types of non-linear triads: one involving a zonal flow mode and two Rossby wave modes and the other involving three move modes. The latter are called barotropic wave triads. Furthermore, it appears that many non-linear interactions do not lead to resonant coupling. The underlying physical mechanism is discussed in Pedlosky [6]. The interaction coefficients obey the relations

$$c_{jlm} = c_{lmj}, \quad c_{jlm} = -c_{jml}, \quad b_{jl} = -b_{lj}. \tag{3.7}$$

§4. Truncated Spectral Model

In practice the expansion of (3.2) is truncated after a finite number of eigenfunctions. Only large scale modes are resolved since it is observed that most energy of quasi-geostrophic flow is contained in the long waves. We obtain a finite dimensional dynamical system of the type

$$x = f_\mu(x) + F(t) \quad \text{in } R^N. \tag{4.1}$$

Here N is the truncated number, R^N is the N-dimensional phase space, $x = (x_1, x_2, \cdots, x_N)$ real valued velocity amplitude of the modes (to be specified later on), $f_\mu(x)$ is an N-dimensional vector field depending on the resolved mode amplitudes and parameters $\mu = (\mu_1, \mu_2, \cdots, \mu_m)$. Furthermore, the N components of vector $F(t)$ are forcing terms which describe the effect of the neglected modes on the dynamics of retained modes. We also note that the original barotropic potential vorticity equation with boundary conditions is an approximation to the dynamics of the circulation. Thus, the forcing terms also account for the effect of physical process not incorporated in the model. In that case the $F(t)$ are defined by the conditions that the projection of the actual atmospheric circulation onto the N resolved modes is described by the solutions of (4.1). We remark that if (4.1) is used as a forecast model $F(t)$ should also account for the effect of the physical process and boundary conditions not (correctly) incorporated in the model. These forcing terms are unknown by definition.

Generally in (4.1) a priori the effect of forcing term is neglected ($F(t)$ is put equal to zero) and consider the properties of spectral models with increasing truncation number. The underlying motivation is that they will at least represent properties of the original vorticity equation. This is done because it follows from observations that generally most energy is contained in only a few modes (the long waves). We argue that this is due to an intrinsic property of 2-dimensional turbulent flow, as described by quasi-geostrophic theory: there is an energy cascade from the small scales to the larger scales (Pedlosky [6]). Some formal indication that this idea is correct have been found by Constantin et al [10]. They showed that for the spectral models of the Navier-Stokes equations minimum number $N_s(\geq (kL_D)^{-d})$ of eigenfunctions could be selected such that the solutions of the truncated spectral models with $N \geq N_s$ and $F(t) = 0$ have equal attractor properties as the solution of the priginal system. Here k^{-1} is the typical length scale of the flow, L_D is a dissipation length scale and d is the dimension of the flow. Although it is not clear whether these results are applicable to the barotropic vorticity equation, they at least suggest that it is useful to consider truncated spectral model.

In principle we would like to investigate the properties of (4.1) for arbitrary values of N. However, it is not possible to carry out such an analysis systematically since the system have a complicated dynamics due to the large number of nonlinear terms in the equations. Therefore as a first step, it becomes worthwhile to study low order spectral models, in which only a few modes are retained and investigate in what sense they reflect features like transitions between weather regimes and a flow with a limited predictability. An important advantage is that they can be analysed with the techniques originated from the theory of dynamical systems whereas from the results indications may be found how to study more complicated models as well as real data.

The structure of the vector fields of the spectral models discussed here is such that non-transient solutions are found in bounded subsets of the phase space. These can either be regular sets, including stationary points (equilibrium flow patterns), limited cycles (oscillating flow) and invariant tori (quasi-periodically oscillating flow), as well as irregular sets which are in fact strange attractors (chaotic flow). These sets of limit points are determined from a numerical bifurcation analysis of the spectral model.

A spectral model is assumed to give at least a qualitative description of the atmospheric circulation if trajectories irregularly visit different preferent regions in phase space. In this way the index cycle mentioned in the introduction is simulated. If this behavior does not occur the transition is apparently too severe and more modes should be included in the spectral expansions. Another possibility is to take account of the effect of the synoptic scale transient eddies on the dynamics of planetary-specific scale flow by adding specific forcing terms to the spectral equations. However, this requires a thorough understanding of the interactions between different scales of motion, this being one of the major problems in modern dynamic meteorology.

§5 Three and Six-component Model

Charney and Devore [1] in their three-component model retained only the (0, 1) zonal flow mode and (1, 1) wave mode. This implies that we assume ψ_{02} in (3.4) to be zero. The stationary point of this model can be calculated analytically. There are either one or three of them depending on the nodal parameters. As a characteristic situation let us consider a channel of length 5000km($= 2\pi/k$) and width 4000km centered at latitude $\phi_0 = 45°$, where $f_0 = 10^{-4}s^{-1}$ and $\beta_0 = 1.6.10^{-11}m^{-1}s^{-1}$. The vertical length scale is taken $H = 10^4 m$, the time scale $\sigma^{-1} = 10^5 s$ (about a day), the mountain amplitude $h_0 = 10^3 m$ and the dissipation time scale about 10 days. This yields the parameter values $b = 1.6, \beta = 1.25, \nu = 1$ and $c = 0.1$. In Fig. 2 the x_1-component of the stationary points (where $x_1 = \psi_{01}/b$) is presented as a function of a external forcing $x_1^* = \psi_{01}^*/b = U/U_0$, where U is the velocity for the forcing and $U_0 = \sigma/k = 7.8ms^{-1}$.

In Fig. 3 the stream function associated with the equilibria E_1, E_2 and E_3 occuring for $x_1^* = 4$ (zonal forcing velocity amplitude of 31.7ms^{-1}) are shown. Note their strong resemblance to the circulation pattern shown in Fig. 1. Based on this agreement Charney and Devore [1] suggest that equilibria of spectral models indicate large scale preferent states of the atmospheric circulation. The existence of multiple equilibria is the consequence of the presence of topography, forcing and dissipation. However, no flow of limited predictability and no vacillatory behavior (i.e. an index cycle) is found: the non-transient solutions are always stationary. The reason for this behavior is the lack of non-linear interaction in the model.

Therefore, we extend the model by including also the (0, 2) zonal flow mode and the (1, 2) wave mode in the spectral expansions, which gives a six-component model. In Fig. 4 the x_1- and $x_4(=\psi_{02}/b)$-component of its stationary points are shown as a function of x_1^* in case $x_4^*(=\psi_{02}^*/b)=0$ and all other parameter values similar as before. Clearly the equilibria of the 3-component model are also equilibria of the 6-component model but stability properties can be different because of the increased number of degrees of freedom. Furthermore, the model contains a new type non-linear interactions involving a zonal flow component and two different wave modes. As a result additional equilibria are found. However, the non-transient behavior can be more complicated. Grasman and De Swart [12] has shown that also stable periodic orbit exist, indicated by the presence of Hoff bifurcation points in Fig.4, as well as strange attractors. However, the latter have only a limited domain of attraction in phase space and the chaotic solutions remain in the low index regime forever. Thus no simulation of an index cycle is obtained. The conclusion remains unchanged if x_4^* is given non-zero values. It is due to the presence of one triad of nonlinear interactions in the model. Thus in order to obtain vaccilating solutions either forcing terms must be added to the equations or more modes should be included in the spectral expansions. Both possibilities is to be considered.

§6. Effect of Stochastic Perturbations on Low Order Spectral Models

Egger [3] was the first who tried to model the effect of the unresolved modes on the 3-component model. He added stochastic forcing terms of Gaussian white noise with a fixed small intensity to the equations. The noise forces the system to visit alternately the attraction domains of the two stable equilibria, thus in this way an index cycle is simulated. A justification for choosing this type of forcing was given by Egger and Schilling [3] who showed, using atmospheric data, that the forcing terms $F(t)$ in (4.1) can be modelled by colour noise process. These are stationary and Gaussian Markov processes and contain white noise as a limit for the corelation time tending to zero. For simplicity we only consider the effect of white noise forcing. The equation (4.1) became

$$dx_i = f_i(x)dt + \varepsilon dW_i(t) \quad \text{in} \quad R^3, \quad i = 1,2,3, \tag{6.1}$$

where the three components of $W(t)$ are independent Wiener processes and ε is the noise intensity which is assumed to be small ($\varepsilon << 1$). Let the stable equilibria E_1 and E_3 have attraction domains Ω_1 and Ω_3 with boundaries $d\Omega_1$ and $d\Omega_3$ respectively. We investigate the distribution of residence times $\tau(x)$ starting from a state $x \in \Omega_i (i = 1,3)$ of the system in these attraction domains. The expected value $\langle\tau(x)\rangle = T(x)$ gives a measure of the persistence of a large scale preferent state of the atmospheric circulation. It is shown that (Gordiner [13]) $T(x)$ obeys:

$$\left.\begin{array}{c}\frac{1}{2}\overset{\varepsilon^2}{\in}\nabla^2 T(x)+f_i(x)\cdot\nabla T(x)=-1 \text{ in } \Omega_i,\\ T=0 \text{ at } \delta\Omega_i \quad (i=1 \text{ or } 3).\end{array}\right\} \tag{6.2}$$

Matkowsky et al [14] by applying singular perturbation techniques derived an asymptotic solution of this differential equation. Outside a boundary layer near $\delta\Omega_i$ it reads

$$T_i \sim C_i e^{K_i/\epsilon^2}, \quad K_i = \lim_{x\to E_2} Q(x), \tag{6.3}$$

where $Q(x)$ is the solution of the eikonal equation

$$\frac{1}{2}(Q(x))^2 + f_i(x)\cdot\nabla Q(x) = 0, \quad Q(E_i) = 0, \tag{6.4}$$

which can be solved by means of ray method (Ludwig [15]). However, the computed residence times are order of months whereas from observations we expect life spans of weather regimes in the order of two weeks. We shall discuss the indiscrepancy later on. Furthermore, it is found that most probable region of exit from the attraction domain is an ε-neighborhood of the unstable equilibrium E_2. Here the system remains for characteristic time

$$T_2 \sim \frac{1}{\lambda}\log(1/\varepsilon) \tag{6.5}$$

where λ is the largest positive real part of the deterministic system linearised at E_2.

Once the stochastic dynamical system is in its statistical equilibrium it is characterized by the expected residence times in the different regimes. However, in this way no information is obtained about the time scale over which the effect of initial conditions is important. This can be investigated with a discrete-state Markov process model. For randomly forced spectral models discussed in this paper we can derive such a model with three states: (i) a zonal state, (ii) a transitional state and (iii) a meridianal state. Let Q_{ij} denote the transition probability per unit of time from state i to state j and let $p_i(t)$ denote the probability for the system to be in state i at time t. Then $p_i(t)$ satisfy

$$\left.\begin{array}{c}\dot{p}_1 = -(Q_{12}+Q_{21})p_1 - Q_{21}p_3 + Q_{21},\\ \dot{p}_3 = -Q_{23}p_1 - (Q_{32}+Q_{23})p_3 + Q_{23},\\ p_2 = 1 - p_1 - p_3,\end{array}\right\} \tag{6.6}$$

where

$$Q_{21} = Q_{23} = 1/(2T_2), \quad Q_{12} = 1/T_1, \quad Q_{32} = 1/T_3. \tag{6.7}$$

In Fig. 5 the probability function $p_i(t)$ are given for a process with $T_1 = 9, T_2 = 1$, and $T_3 = 31$ that starts in state 1, 2, 3 respectively. It is seen that once the initial state is given, the Markov model contains more information about the system than the stationary probability distribution for a period of about 50 days.

§7. Higher Dimensional Spectral Model (Deterministic Chaos and Vacillation)

In higher dimensional spectral models the system will exihibit irregular dynamics from itself. No stochastic forcing is needed to obtain vacillation between states with a zonal flow of different intensities. In this section we summerise some results. The purpose is to formulate a spectral model with the lowest dimension that still has chaotic behavior with two clearly different scales of motion (planetary/synoptic) and that has a zonal component (x_1) that varies over a sufficiently large realistic range. The model describes the evolution of two zonal flow profiles (a (0,1) and (0,2) mode) and four waves (the (1,1), (1,2), (2,1) and (2,2) modes) in a barotropic atmosphere. Compared to 6-component model the 10-component model contains a new type of non-linear interaction involving three waves: the (1,1), (1, 2) and (2,1) modes. In Fig. 6, the x_1-and x_4-component of the stationary points of this model are shown for the same parameter values as discussed in Section 5. Due to the presence of the wave triad, isolated branches of equilibria occurs. By letting x_4^* become non-zero all regular solution may be turned unstable. For $x_4 = -8$ the model represents a flow vacillating between three preferment regimes where the latter are actually unstable periodic solutions of the model. The wave triad provide for interaction between clearly distinct scales of motion: a planetary scale and synoptic scale. This behavior is similar to what is observed in the atmosphere. Fig. 7.

By computing the spectrum of Lyapunov exponents, using the method of Wolf et al [9] the existence of a global strange attractor is shown. Lyapunov exponents measure the average divergence between nearby orbits in phase space whereas the chaos is defined by at least one positive exponent. From the spectrum of Lyapunov exponents we can estimate the fractal dimension of the attractor which also yields an upper bound to the number of degrees of freedom of the chaotic flow (Farmer et al [4]).

In practice the initial conditions are never known with infinite precision. Thus small errors are introduced in the system which will grow during its evolution because of the chaotic dynamics. Consequently the predictability of the flow is limited: the time scale of the average prediction is given by the reciprocal of the sum of all positive Lyapunov exponents. However, of more interest to meteorologists is the dependence of predictability on the state of the system. It is argued that the local eigenvalues at each point of an orbit may determine the time evolution of small errors on this orbit. In that case the eigen vectors corresponding to eigenvalues with positive real part determine the geographical distribution of the error growth. However, this is on the the condition that the time scale of error growth is small compared to the time scale on which the flow itself evolves. This method can be applied to spectral models showing long periods of quasi-stationary behavior.

The impact of neglected short scale modes on a planetary scale model was studied by considering the chaotic ten-component model to represent the real atmosphere and the six-component model (for identical parameter values) to be a forecast model. For obtaining equivalance between solutions of the two systems forcing terms must be added to the equations of the forecast model. It appears that these forcing terms have an unpredictable nature and they cannot be modelled by the simple stochastic process used in Section 6.

§8. Some Remarks and Conclusions

In introduction we remarked that an accurate modelling of the feedback between quasi-stationary planetary-scale motion and transient synoptic scale eddies is important for the development of long range weather forecast models. Here we have argued that this problem may be studied by considering simplified models which still represent the chaotic properties and vacillatory behavior of the atmosphere. Next we investigated whether they provide clues on how to analyse more complicated models as well as real data.

Extremely low order spectral models show qualitative features of the atmospheric circulation. They possess multiple equilibria for a range of parameter values and the corresponding flow patterns resemble large scale preferent states of the atmospheric circulation. However, we remark that the existence of weather regime has never been convincingly demonstrated by a systematic data analysis; only recently some indications have been found.

It has been found that three- and six-component models cannot simulate a flow vacillating between different weather regimes. In order to meet this imperfection we added stochastic forcing terms to the spectral equations. They are assumed to account for the effect of neglected modes and physical process not incorporated in the model. A justification for choosing white noise and colored noise parameterizations is found from the data study of Egger [3]. The noise forces the system to visit alternately attraction domains of the stable equilibria. During a transition the system remains a characteristic time near an unstable equilibrium. This suggests that stable and unstable regular solutions of spectral model may have relevance for the dynamics of the atmospheric circulation. A method was discussed for computing expected residence times near the equilibria of the unperturbed system. Comparing the results with observational data it appears that the computed life spans of the weather regimes are a factor of ten times larger than those in the atmosphere.

A systematic way for investigating the effect of neglected modes on a truncated spectral model has been discussed. A 10-component model is discussed which is minimum order model representing a finitely predictable flow having two distinct scales of motion (a planetary and synoptic scale) and vacillating between two preferment regimes. We assume this model to represent the real atmosphere and considered a six-component subsystem as a forecast model. To the subsystem forcing terms were added such that its solutions are equivalent to those of the full model projected onto the modes which also belong to the subsystem. It was found that those forcing terms have an unpredictable nature and they cannot be modelled by color noise process. This result is in agreement with that of Lindenberg and West [6], who analysed explicit expression for the forcing terms representing the effect of the neglected modes on truncated spectral models of the barotropic vorticity equation. It does not contradict the results of Egger and Schilling [3] since these authors also include the effect of neglected physical process in their definition of the forcing terms.

We remark that the effect of the topography are overestimated in barotropic models since they act directly on the entire fluid column. Baroclinic multilevel models of the quasi-geostrophic potential vorticity give better results at this point. Again multiple equilibria are found and due to the presence of the baroclinic instability mechanisms vacillatory behaviour is even more easily produced. The two-level twenty-component model of Reinhold

[11] is probably the simplest model containing all basic physical mechanism: topographic, barotrophic and baroclinic instability as well as the occurance of wave triad interactions.

In general, the problems with low order spectral models is that unrealistically large external forcing values (corresponding to equator-pole temperature difference of more than 150°C) are required to produce vacillatory behavior. Moreover, the characteristic lives of the regimes in the models are much larger than those obtained from atmospheric data. These imperfections are probably due to the severe truncation in both the horizontal and vertical directions. A better description of the atmospheric circulation is expected from multilevel high resolution models. Since their structure is extremely complicated they are difficult to analyse. Alternately we can study lower dimensional spectral models which include an appropriate parametrization of the synoptic forcing terms. This problem remains to be investigated in more details.

Equations of the 10-componentmodel

$$
\begin{array}{llllllll}
\dot{x}_1 = & & \gamma^*_{11}x_3 & -C(x_1 - x^*_1), & & & \\
\dot{x}_2 = & -(\alpha_{11}x_1 - \beta_{11})x_3 & & -Cx_2 & -\delta_{11}x_4x_6 & -\rho_{11}(x_5x_8 - x_6x_7), & \\
\dot{x}_3 = & (\alpha_{11}x_1 - \beta_{11})x_2 & -\gamma_{11}x_1 & -Cx_3 & +\delta_{11}x_4x_5 & +\rho_{11}(x_5x_7 + x_6x_8), & \\
\dot{x}_4 = & & \gamma^*_{12}x_6 & -C(x_4 - x^*_4) & +\varepsilon_1(x_2x_6 - x_3x_5) & +\varepsilon_2(x_7x_{10} - x_8x_9), & \\
\dot{x}_5 = & -(\alpha_{12}x_1 - \beta_{12})x_6 & & -Cx_5 & -\delta_{12}x_3x_4 & +\rho_{12}(x_2x_8 - x_3x_7) & +\gamma'_{12}x_8, \\
\dot{x}_6 = & (\alpha_{12}x_1 - \beta_{12})x_5 & -\gamma_{12}x_4 & -Cx_6 & +\delta_{12}x_2x_4 & -\rho_{12}(x_2x_7 + x_3x_8) & -\gamma'_{12}x_7, \\
\dot{x}_7 = & -(\alpha_{21}x_1 - \beta_{21})x_8 & & -Cx_7 & -\delta_{21}x_4x_{10} & -\rho_{21}(x_2x_6 + x_3x_5) & +\gamma'_{21}x_6, \\
\dot{x}_8 = & (\alpha_{21}x_1 - \beta_{21})x_7 & & -Cx_8 & +\delta_{21}x_4x_9, & +\rho_{21}(x_2x_5 - x_3x_6) & -\gamma'_{21}x_5, \\
\dot{x}_9 = & -(\alpha_{22}x_1 - \beta_{22})x_{10} & & -Cx_9 & -\delta_{22}x_4x_8, & & \\
\dot{x}_{10} = & (\alpha_{22}x_1 - \beta_{22})x_9 & & -Cx_{10} & +\delta_{22}x_4x_7, & & \\
 & \text{advection} & \text{topo-graphy} & \text{forcing/dissipation} & \text{advection} & \text{wave triad} & \text{topo-graphy}
\end{array}
$$

where

$$
\alpha_{nm} = \frac{8\sqrt{2}n}{\pi}\frac{m^2}{4m^2-1}\frac{n^2b^2+m^2-1}{n^2b^2+m^2}, \qquad \beta_{nm} = \frac{\beta nb^2}{n^2b^2+m^2},
$$

$$
\delta_{nm} = \frac{64\sqrt{2}n}{15\pi}\frac{n^2b^2-(m^2-1)}{n^2b^2+m^2}, \qquad \gamma^*_{nm} = \frac{4m}{4m^2-1}\frac{\sqrt{2}nb\gamma}{\pi},
$$

$$
\varepsilon_n = \frac{16\sqrt{2}n}{5\pi}, \qquad \gamma_{nm} = \frac{4m^3}{4m^2-1}\frac{\sqrt{2}nb\gamma}{\pi(n^2b^2+m^2)},
$$

$$
\rho_{nm} = \frac{9}{2}\frac{(n-2)^2b^2-(m-2)^2}{n^2b^2+m^2}, \qquad \gamma'_{nm} = \frac{3b\gamma}{4(n^2b^2+m^2)}.
$$

It contains a six-component subsystem, denoted between the dashed lines, and a three-component subsystem denoted between the dotted lines. The relation between state variables and model amplitudes of the streamfunction is given.

The α_{nm}-terms describe interactions between the (0, 1) and (n, m) modes, the δ_{nm}-and ε_n-terms represent interactions between the (0, 2) $(n, 1)$ and $(n, 2)$ modes and the terms involving ρ_{nm} are interactions between the (1, 1), (1, 2) and (2, 1) modes. Furthermore the

β_{nm}-contributions represent planetary vorticity advection and the γ^{*}_{nm}, γ_{nm} and γ'_{nm}-terms the various couplings between flow and topography.

References

[1] Charney, J.G., Devore, J.G., *Multiple flow equilibria in the atmosphere and blocking*, J. Atmos. Sci., 36(1979), 1205–1216.

[2] Dole, R.M., *Persistent anomalies of the extratropical northern hemisphere wintertime circulation: Structure*, Mon. Wea. Rev., 114(1986), 178–207.

[3] Egger, J., Schilling, H.D., *On the theory of long term variability of the atmosphere*, J. Atmos. Sci., 40(1983), 1073–1085.

[4] Farmer, D.J., Ott, E. Yorke, J.A., *The dimensions of chaotic attractors*, Physica, 7D(1983), 153–180.

[5] Lorenz, E.N., *The predictability of a flow which possesses many scales of motion*, Tellus, 21(1969), 281–307.

[6] Pedlosky, J., *Geophysical Fluid Dynamics*, 2nd Edition, Springer Verlag, (1987).

[7] Van Dijk, W., Schmidt, F.H., Schuurmans, C.J.E., *Beschrijving en toepassings mogeligkheden van gemiddelde topografien van het 500 mb. valk in afhankelijkheid van circulatietypen*, KNMI, del Bilt, Wetenschappelijk Repport WR 74-3 (1974).

[8] Voigt, R.G., Gottalieb, D., Yousoff-Hussaini, M(eds), *Special Methods for partial differential equations*, SIAM, Philadelphia. (1984).

[9] Wolf, A., Swift, J.B., Swinney, H.L., Vastano, J.A., *Determining Lyapunov exponents from a time series*, Physica, 16D(1985), 285–317.

[10] Constantin, P., Foias, C., Manley, O.P., Teman, R., *Determining modes and fractal dimensions of turbulent flows*, J. Fluid Mech, 150(1985), 427–440.

[11] Reinhold, B.E. *Weather regimes: The challange in extended-range forecasting*, Science, 235(1987), 437–441.

[12] Grasman, J., and De Swart, H.E., *Low order spectral models of the atmospheric circulation*, Proceedings ICIAM 87(1987), Paris-La Villette June 29–July 3, 1987.

[13] Gordiner, C.W., *Hand book of stochastic methods for physics, chemistry and natural sciences*, Springer Verlag, Berlin, (1983).

[14] Matkowsky, B.J., Schuss, Z., Tier, C., *Diffusion across characteristic boundaries with critical points*, SIAM J. Appl. Math., 43(1983), 673–695.

[15] Ludwig D., *Persistence of the dynamical systems under random perturbations*, SIAM Rev., 17(1975), 605–640.

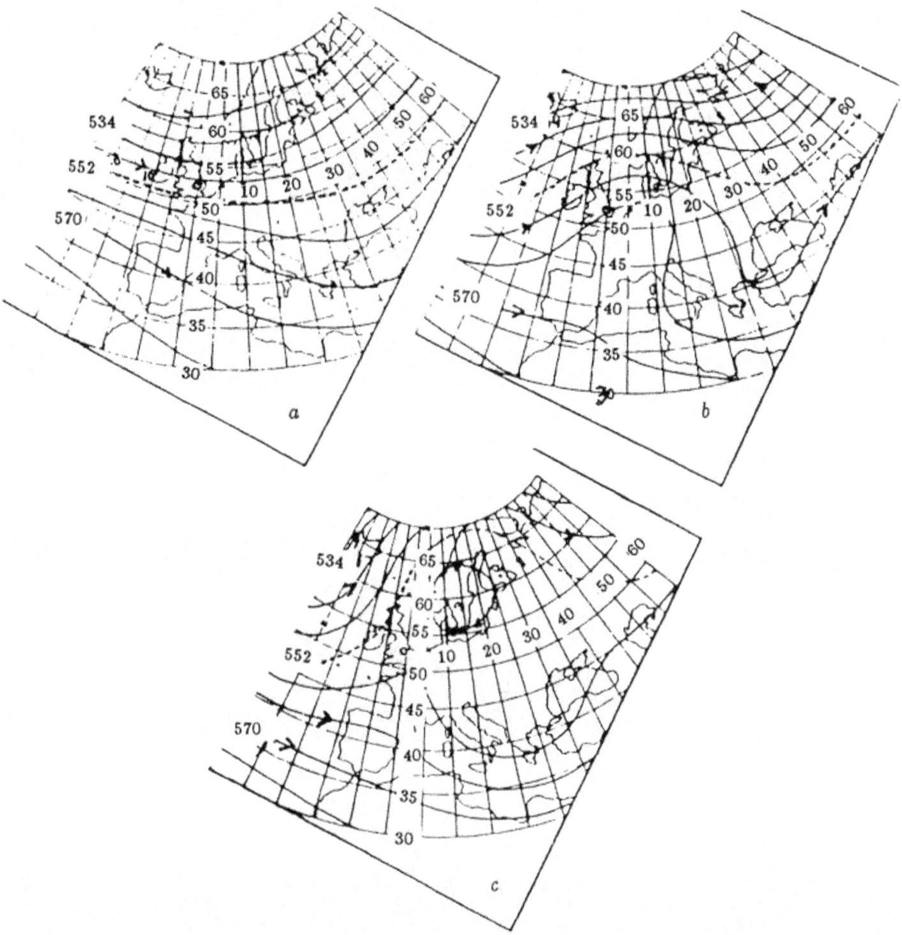

Figure 1. Geographical distribution of height (in geopotential decameter) of the 500 mb level for the (i) WZ, (ii) HM and (iii) HFa winter Grosswetterlagen, which are of zonal mixed and meridional type respectively. The isohyps are approximate streamlines of the flow; arrows indicate flow direction. From Van DIJK ET AL [7].

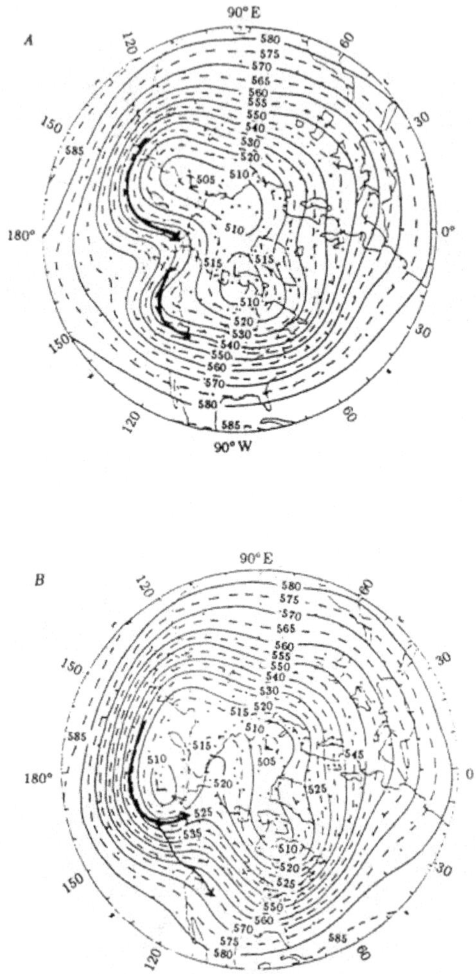

Fig1'. Polar stereographic projection of the composite 500-mbar flow panems for the (A) positive and (B) negative-persistent anomaly states for the Pacific region. The arrows on each figure represent the principal storm tracks associated with each pattern, ascertained by calculating the regions of maximum activity on the 2-to 6-day time scale. These figures show the different systematic persistures structures present in the atmosphere. [From R.M. Dole: Thesis, MIT, 1982].

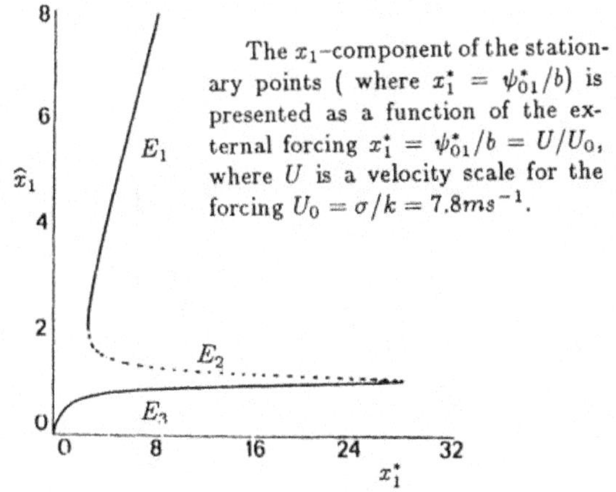

Figure 2. The x_1–component of the stationary points $\hat{x}$ of the three-component model for the parameter values discussed in the text. A solid line denotes that the solution is stable whereas a dashed line refers to an unstable solution.

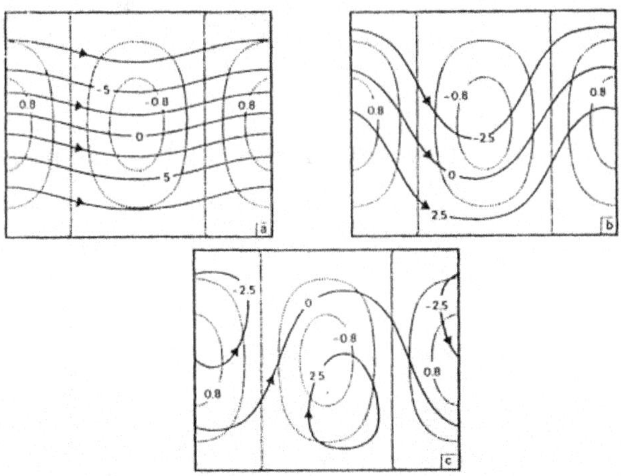

Figure 3. Nondimensional streamfunction patterns of the equilibria $E_1(a)$, $E_2(b)$ and $E_3(c)$ for the parameter values discussed in the text. The arrows indicate the flow direction which is along the streamilines ψ= constant. Here a difference $\Delta\psi = 1$ corresponds to a zonal transport of $2.6 \cdot 10^7 m^2 s^{-1}$. The dashed lines represent contours of the topolography $(10^3 m)$. Note the resemblace to circulation patterns with Fig.1.

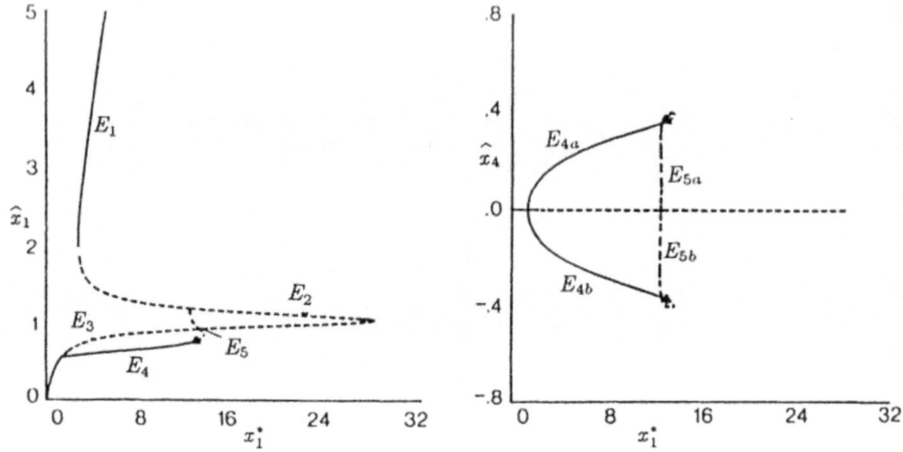

FIGURE 4. As Figure 2, but the x_1-component (left) and x_4-component (right) of the six-component model. A triangle denotes a Hopf bifurcation point.

The x_1- and $x_4(=\psi_{02}/b)$-component of its stationary points are shown as a function of x_1^* in case $x_4^*(=\psi_{02}^*/b)=0$ and all other parameter values similar as before. Clearly, equilibria of the three-

In Figure 5 the probability functions $p_i(t)$ are given for a process with $T_1=9$, $T_2=1$ and $T_3=31$ that starts in state 1, 2 and 3, respectively.

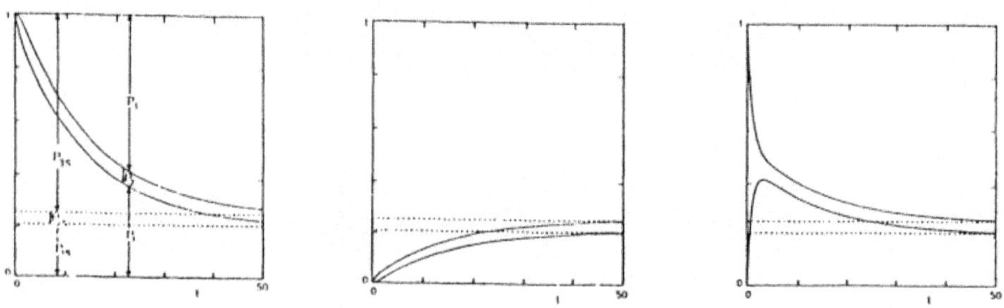

FIGURE 5. Evolution of the probability distribution of the Markov process starting in state 1 (left), 2 (middle) and 3 (right) respectively. The dotted lines represent the stationary distribution.

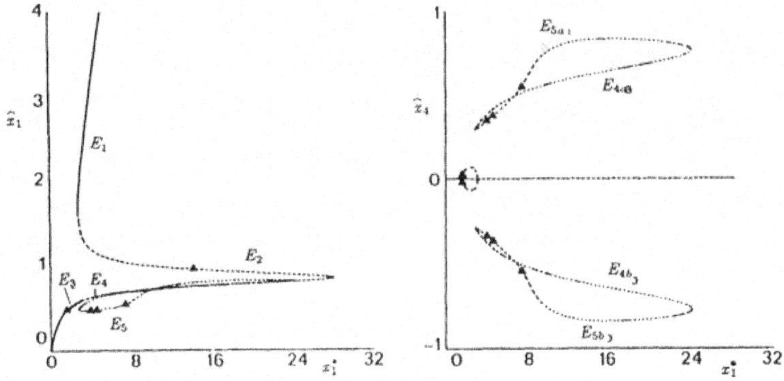

FIGURE 6. As Figure 4, but for the ten-component model.

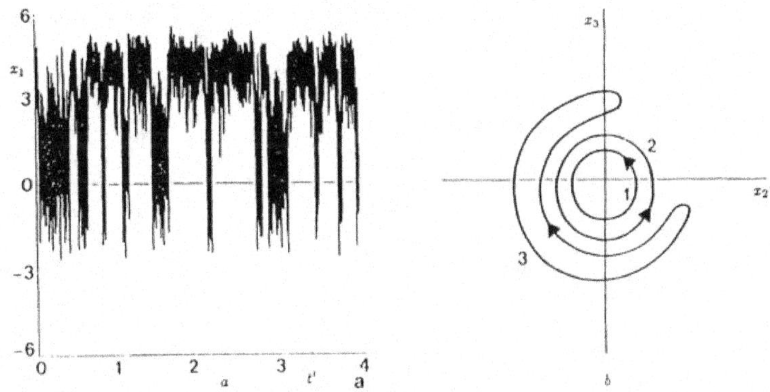

FIGURE 7a. The x_1-component of a chaotic solution of the ten-component model as a function of time. Here $t'=(t-1000)/500$ and the dimensional period is approximately six years.

b. Sketch of the unstable periodic orbits projected onto the x_2-x_3 plane. The preferent regions of the strange attractor are small tubes around these orbits.

Proceedings of the Asian Mathematical
Conference, 1990

Eigenvalue Problem for Quasilinear Elliptic Systems with Limiting Nonlinearity

SHEN YAOTIAN & YAN SHUSEN
Department of Applied Mathematics
South China University of Technology
Guangzhou 510641, China

In this paper, we consider the following eigenvalue problem:

$$\begin{aligned} & -D_\beta(a^{ij}_{\alpha\beta}(x,u)D_\alpha u^i) + \frac{1}{2}a^{im}_{\alpha\beta u^j}(x,u)D_\alpha u^i D_\beta u^m + h(x)u^j \\ = \;& \lambda |u|^{2^*-2}u^j, \quad j = 1,2,\cdots,N, \quad x \in \Omega, \end{aligned} \tag{1}$$

$$u \in H^1_0(\Omega, R^N) \cap L^\infty(\Omega, R^N), \tag{2}$$

where Ω is a bounded domain in R^N, $n \geq 3, 2^* = 2n/(n-2), a^{ij}_{\alpha\beta}(x,u) = a_{\alpha\beta}(x)g_{ij}(u)$, $a^{im}_{\alpha\beta u^j}(x,u) = \frac{\partial}{\partial u^j}a^{im}_{\alpha\beta}(x,u)$, $|u|^2 = \sum_{i=1}^N u_i^2$ and $h \in C(\overline{\Omega})$. Here we use the convention that repeated Latin letters are summed from 1 to N, while repeated Greek letters are summed from 1 to n. Throughout this paper, we set $\|v\|^2 = \int_\Omega |Dv|^2$ and $|v|_p^p = \int_\Omega |v|^p$.

Definition 1. If $\lambda \in R$ and $u \in H^1_0(\Omega, R^N) \cap L^\infty(\Omega, R^N)$ satisfy

$$\begin{aligned} & \int_\Omega (a^{ij}_{\alpha\beta}(x,u)D_\alpha u^i D_\beta \varphi^j + \frac{1}{2}\varphi^j a^{im}_{\alpha\beta u^j}(x,u)D_\alpha u^i D_\beta u^m + h(x)u^j\varphi^j) \\ = \;& \lambda \int_\Omega |u|^{2^*-2}u^j\varphi^j, \quad \forall \varphi \in H^1_0(\Omega, R^N) \cap L^\infty(\Omega, R^N), \end{aligned} \tag{3}$$

then (λ, u) is called the weak solution of (1)–(2).

It is well known that the usual way to solve (1)–(2) is to prove that the following minimization problem is achieved by some $u \in H^1_0(\Omega, R^N) \cap L^\infty(\Omega, R^N)$:

$$I_\lambda = \inf\{I(u) \mid u \in H^1_0(\Omega, R^N), |u|^{2^*}_{2^*} = \lambda\}, \tag{4}$$

where

$$I(u) = \int_\Omega (a^{ij}_{\alpha\beta}(x,u)D_\alpha u^i D_\beta u^j + h(x)|u|^2). \tag{5}$$

Similar problems were considered in the nonlimiting case in [2,3]. In the present situation, the difficulty in getting the existence result for (1)–(2) consists of two aspects: first, in order to show that I_λ is achieved by some $u \in H^1_0(\Omega, R^N)$, we need a concentration compactness principle similar to that of [1]; second, the method to prove that $u \in L^\infty(\Omega, R^N)$ seems to be not straightforward due to the nondifferentiability of the functional $I(u)$ in $H^1_0(\Omega, R^N)$ and the limiting growth of the nonlinearity.

The assumptions imposed on $a^{ij}_{\alpha\beta}(x,u)$ are the following:

(a_1) $a_{\alpha\beta}(x)=a_{\beta\alpha}(x)$, $a_{\alpha\beta}(x)\in C(\bar\Omega)$ $(\alpha,\beta=1,2,\cdots,n)$ and there are $a_2\geq a_1>0$, such that

$$a_1|\eta|^2\leq a_{\alpha\beta}(x)\eta_\alpha\eta_\beta\leq a_2|\eta|^2;$$

(a_2) $g_{ij}(u)\in C^1(R^N)$, $g_{ij}(u)=g_{ji}(u)$ $(i,j=1,2,\cdots,N)$ and there are $C_2\geq C_1>0$, such that

$$C_1|\xi|^2\leq g_{ij}(u)\xi_i\xi_j\leq C_2|\xi|^2;$$

(a_3) $|u^k g_{iju^k}(u)|\leq C$;

(a_4) There is an $a_3\in(0,1)$, such that

$$-a_{\alpha\beta}(x)u^k(g_{ij}(u)g^{jk}_{u^m}(u)+\frac{1}{2}g^{jk}(u)g_{imu^j}(u))\xi^i_\alpha\xi^m_\beta\leq a_3a_{\alpha\beta}(x)\xi^i_\alpha\xi^j_\beta,$$

where $(g_{ji}(u))=(g_{ij}(u))^{-1}$;

(a_5) $a_{\alpha\beta}(x)u^k(g_{ij}(u)g^{jk}_{u^m}(u)+\frac{1}{2}g^{jk}(u)g_{imu^j}(u))\xi^i_\alpha\xi^m_\beta\geq-\delta(L)|\xi|^2$, for $|u|\geq L$, where $\delta(L)\to0$ as $L\to+\infty$.

Set $I_0=\inf\{\int_\Omega|Dv|^2\mid v\in H^1_0(\Omega),|v|_{2^*}=1\}$, $\lambda_1=\varliminf_{|t|\to+\infty}\lambda_1(t)$, where $\lambda_1(t)$ is the least eigenvalue of $(g_{ij}(u))$. The main result of this paper is the following:

Theorem 1. *Suppose that* (a_1)–(a_5) *hold. If*

$$I_1<\lambda_1\inf_{x\in\bar\Omega}(\det(a_{\alpha\beta}(x)))^{1/n}I_0\triangleq I^\infty, \tag{6}$$

then I_{λ_0} is achieved by some $u\in H^1_0(\Omega,R^N)$, where

$$0<\lambda_0=\sup\{\tilde\lambda\mid I_1<I_\lambda+(1-\lambda)^{2/2^*}I^\infty,\ \forall\,0\leq\lambda<\tilde\lambda\}. \tag{7}$$

Furthermore, $u\in L^\infty(\Omega,R^N)$. In particular, (1), (2) have nontrivial weak solution (λ,u).

Now we sketch the proof of Theorem 1.1.

By the Ekeland variational principle, we can select a minimizing sequence $\{u_m\}$ of I_{λ_0} satisfying $\int_\Omega|u_m|^{2^*}=\lambda_0$ and

$$I(u_m)<I_{\lambda_0}+1/m, \tag{8}$$

$$I(w)\geq I(u_m)-\|w-u_m\|/m\quad\text{for}\quad w\in H^1_0(\Omega,R^N)\quad\text{with}\quad|w|^{2^*}_{2^*}=\lambda_0. \tag{9}$$

It is easy to see from (8) that there is a $C>0$, such that

$$\|u_m\|\leq C, \tag{10}$$

so we can assume that there is a $u\in H^1_0(\Omega,R^N)$ with

$$u_m\rightharpoonup u\quad\text{weakly in}\quad H^1_0(\Omega,R^N),$$
$$u_m\to u\quad\text{strongly in}\quad L^2(\Omega,R^N).$$

Lemma 1. *For fixed $x_0\in\bar\Omega$ and $\tau>0$, we have*

$$\int_{B_\rho(x_0)\cap\{|u_m|\leq\tau\}}|Du_m|^2\to0$$

as $\rho \to 0$, *uniformly for* $m = 1, 2, \cdots$.

Proof. For fixed $L \geq \tau$ and m, set $\phi^j = \eta g^{jk}(u_m)u_m^k |u_m|_L/|u_m| \in L^\infty(\Omega)$, where $\eta \in C_0^\infty(B_\rho(x_0)), \eta = 1$ for $x \in B_{\frac{1}{2}\rho}(x_0), 0 \leq \eta \leq 1$ and $|D\eta| \leq 2/\rho$.

Set

$$w = \lambda_0^{1/2^*}(u_m + t\phi)/|u_m + t\phi|_{2^*}.$$

Then $|w|_{2^*}^{2^*} = \lambda_0$. From (9), (10) and the Lebesgue dominated convergence theorem, we can prove

$$\begin{aligned}&\int_\Omega a_{\alpha\beta}(x)g_{ij}(u_m)D_\alpha u_m^i D_\beta \phi^j + \frac{1}{2}\int_\Omega a_{\alpha\beta}(x)g_{iju^k}(u_m)\phi^k D_\alpha u_m^i D_\beta u_m^j + \int_\Omega h(x)u_m^i\phi^i \\ = \;& \lambda_m \int_\Omega |u_m|^{2^*-2}u_m^i\phi^i + \mu_m,\end{aligned} \tag{11}$$

where $\mu_m \to 0$ as $m \to +\infty$ and

$$\begin{aligned}\lambda_m = \;& \lambda_0^{-1}\int_\Omega a_{\alpha\beta}(x)g_{ij}(u_m)D_\alpha u_m^i D_\beta u_m^j \\ & + \frac{1}{2\lambda_0}\int_\Omega a_{\alpha\beta}(x)g_{iju^k}(u_m)u_m^k D_\alpha u_m^i D_\beta u_m^j + \lambda_0^{-1}\int_\Omega h(x)|u_m|^2.\end{aligned}$$

From (11), we get that for a fixed $\tau > 0$,

$$\lim_{\rho\to 0}\int_{B_{\frac{1}{2}\rho}(x_0)\cap\{|u_m|\leq\tau\}} |Du_m|^2 = 0$$

uniformly for $m = 1, 2, \cdots$.

By Lemma 1.1 of [1], there are bounded nonnegative measures v, μ and $x_k \in \bar{\Omega}, v_k \geq 0$ $(k = 1, 2, \cdots)$, such that

$$|u_m|^{2^*} \rightharpoonup v = |u|^{2^*} + \sum_k v_k \delta_{x_k}, \tag{12}$$

$$|D|u_m||^2 \rightharpoonup \mu \geq |D|u||^2 + I_0 \sum_k v_k^{2/2^*}\delta_{x_k}. \tag{13}$$

Theorem 2. *Under the conditions of Theorem 1, I_{λ_0} is achieved, where λ_0 is defined by (7).*

Proof. By the definition of I_{λ_0}, we see that

$$I_1 = I_{\lambda_0} + (1-\lambda_0)^{2/2^*}I^\infty, \tag{14}$$

$$I_1 < I_\lambda + (1-\lambda)^{2/2^*}I^\infty \quad \text{for} \quad \lambda \in (0, \lambda_0). \tag{15}$$

First, we claim that

$$\begin{aligned}& a_{\alpha\beta}(x)g_{ij}(u_m)D_\alpha u_m^i D_\beta u_m^j \rightharpoonup \\ & \mu_1 \geq a_{\alpha\beta}(x)g_{ij}(u)D_\alpha u^i D_\beta u^j + I_0\textstyle\sum_k \lambda_1(\det(a_{\alpha\beta}(x_k)))^{1/n}v_k^{2/2^*}\delta_{x_k}.\end{aligned} \tag{16}$$

In fact, let $\varphi \in C_0^\infty(B_2(0))$ with $\varphi(x) = 1$ for $x \in B_1(0)$ and $0 \le \varphi \le 1$. Then, by Lemma 1, we get

$$\begin{aligned} &\int_\Omega \varphi\left(\frac{x-x_k}{\varepsilon}\right) a_{\alpha\beta}(x) g_{ij}(u_m) D_\alpha u_m^i D_\beta u_m^j \\ \ge\ & \gamma(\varepsilon,\tau) + (\lambda_1 - \delta_1(\tau)) \int_\Omega \varphi\left(\frac{x-x_k}{\varepsilon}\right) a_{\alpha\beta}(x) D_\alpha |u_m| D_\beta |u_m|, \end{aligned} \tag{17}$$

where $\gamma(\varepsilon,\tau) \to 0$ as $\varepsilon \to 0$ for a fixed $\tau > 0$.

By Lemma 1.4 of [1], letting $m \to +\infty$, then $\rho \to 0$, and finally $\tau \to +\infty$ in (17), we get

$$\mu_1(\{x_k\}) \ge \lambda_1 \det(a_{\alpha\beta}(x_k))^{1/n} I_0 v_k^{2/2^*} \tag{18}$$

which, together with the weak lower semicontinuity of the functional (5) in $H_0^1(\Omega, R^N)$, implies (16).

Now we are ready to prove that $|u|_{2^*}^{2^*} = \lambda_0$, and so u achieves I_{λ_0}.

Suppose that $\lambda = |u|_{2^*}^{2^*} < \lambda_0 \le 1$. Then by (14) and (16), we can prove

$$I_1 \ge I_\lambda + (1-\lambda)^{2/2^*} I^\infty,$$

which contradicts (15). Thus, $|u|_{2^*}^{2^*} = \lambda_0$. Q.E.D.

Lemma 2. *Suppose that* (a_1)*–*(a_4) *hold and* u *achieves* I_{λ_0}. *Then* $u \in L^\infty(\Omega, R^N)$.

Proof. Set $\psi^j = \eta^2 g^{jk}(u) u^k (|u| - q)^+/|u|$, where $q > 0, \eta \in C_0^\infty(B_\rho(x_0))$ with $\eta = 1$ for $x \in B_t(x_0), 0 < t < \rho, 0 \le \eta \le 1$ and $|D\eta| \le 2/(\rho - t)$. For this ψ, using the fact that u achieves I_0, we can prove that

$$\begin{aligned} &\int_\Omega a_{\alpha\beta}(x) g_{ij}(u) D_\alpha u^i D_\beta \psi^j + \frac{1}{2}\int_\Omega a_{\alpha\beta}(x) g_{iju^k}(u) \psi^k D_\alpha u^i D_\beta u^j + \int_\Omega h(x) u^i \psi^i \\ =\ & \lambda \int_\Omega |u|^{2^*-2} u^i \psi^i, \end{aligned} \tag{19}$$

where λ is a constant depending on u.

Using (19) and the reverse Hölder inequality, we can prove that $u \in L^\infty(\Omega, R^N)$.

Proof of Theorem 1. Theorem 1 is just a consequence of Theorem 2 and Lemma 2.

Finally, we give an example in which (a_1)–(a_5) hold.

Example. Let $(a_{\alpha\beta}(x))$ be a definitely positive matrix, $N = 2$, $g_{ij}(u) = \delta_{ij} + u_i u_j/(1 + |u|^4), i, j = 1, 2$.

By direct calculation, we have

$$\begin{aligned} f(u,\xi) &\triangleq -\left(\frac{1}{2} g_{jmu^j}(u) g^{jk}(u) + g_{ij}(u) g^{jk}_{u^m}(u)\right) u^k \xi_i \xi_m \\ &= (|u|^2 \delta_{im} - 2|u|^4 u^i u^m/(1+|u|^4)) \xi_i \xi_m / (1 + |u|^2 + |u|^4). \end{aligned}$$

Hence,

$$f(u,\xi) \le \frac{|u|^2}{1+|u|^2+|u|^4} |\xi|^2 \le \beta |\xi|^2,$$

where $\beta = \max_{t \ge 0} \frac{t^2}{1+t^2+t^4} < 1$, while

$$-f(u,\xi) \ge -\frac{|u|^2}{1+|u|^2+|u|^4} |\xi|^2 \ge -\delta(L)|\xi|^2, \quad \text{for} \quad |u| \ge L,$$

where $\delta(L) = \max_{t \ge L} \frac{t^2}{1+t^2+t^4} \to 0$ as $L \to +\infty$.

References

[1] Lions, P.L., *The concentration compactness principle in the calculus of variations, the limit case*, Part 1, Rev. Mat. Iberoamericana, Vol.1, No.1, 1985, 145–201.

[2] Shen Yaotian, Lu De, Cheng Baolong and Guo Xinkang, *Eigenvalue problems for quasi-linear elliptic systems* (in Chinese), Acta Math. Sinica, 31(1986), 845–849.

[3] Struwe, M., *Quasilinear elliptic eigenvalue problems*, Comment. Math. Helvetici, 58(1983), 509–527.

Proceedings of the Asian Mathematical
Conference, 1990

Identities of Euler and Finite Classical Groups

KEN-ICHI SHINODA
Department of Mathematics
Sophia University
7-1 Kioichō, Chiyodaku, Tokyo, Japan

§1.

Let $V = V_n$ be an n-dimensional vector space over a finite field $\boldsymbol{F}_q$ consisting of q elements and $G = G_n$ be a subgroup of $GL(V_n)$, the group of all linear automorphisms of V_n. Let $\boldsymbol{L} = \boldsymbol{L}_n$ be the set of all subspaces of V_n. Then $\boldsymbol{L}$ becomes a poset with respect to the inclusion, $\supset$, and moreover $\boldsymbol{L}$ is a G-poset, i.e. the canonical action of G on $\boldsymbol{L}$ preserves the inclusion relation. Let X be a subset of G. For $g \in X$, put $V^g = \{v \in V \mid g(v) = v\} = \mathrm{Ker}(g - 1_V)$. We consider the following problems.

Problem 1. For $W \in \boldsymbol{L}_n$, put $D_X(W) = \{g \in X \mid V^g = W\}$. Find $d_X(W) = |D_X(W)|$ and $p_X(W) = d_X(W)/|X|$.

We denote by [W] the G-orbit of W in $\boldsymbol{L}_n$ and we let $D_X([W])$ the union of $D_X(U)$ over all $U \in [W]$ and $d_X([W]) = |D_X([W])|$. Then $d_X([W]) = d_X(W)[G : N_G(W)]$, where $N_G(W) = \{g \in G \mid g(W) = W\}$. Therefore if we know explicitly the index $[G : N_G(W)]$, Problem 1 above is equivalent to find $d_X([W])$ and $p_X([W]) = d_X([W])/|X|$.

Problem 2. Suppose that for each natural number n we are given a subgroup G_n and its subset X_n in a natural manner. Let $p_{X_n}(m) = \sum_{\dim W = m} p_{X_n}(W)$. Find $p_{X_n}(m)$ and $\lim_{n\to\infty} p_{X_n}(m)$.

§2.

To consider the problems in §1, we apply the Möbius inversion formula for the functions over $\boldsymbol{L}_n$. Let $C_X(W) = \{g \in X \mid g(w) = w \text{ for all } w \text{ in } W\}$ and $c_X(W) = |C_X(W)|$. Then we have

$$c_X(W) = \sum_{U \supset W} d_X(U),$$

which is quivalent to

$$d_X(W) = \sum_{U \supset W} \mu(W, U) c_X(U),$$

where $\mu(W, U)$ is the Möbius function of the finite poset $\boldsymbol{L}$, given by

$$\mu(W,U) = \begin{cases} (-1)^k q^{k(k-1)/2}, & \text{if } W \subset U \text{ and } \dim\ U/W = k, \\ 0, & \text{if } W \not\subset U. \end{cases}$$

Then by the same method as in Section 2 of $[RS_2]$, we can show

2.1 Lemma. *Let $\{0\} = W_0, W_1, \cdots, W_N = V$ be a complete system of representatives of the G-orbits in L_n and define a subset $\Theta^{W_i}([W])$ of L_n by $\Theta^{W_i}([W]) = \{Y \in [W] \mid Y \subset W_i\}$. We may assume that if $\Theta^{W_i}([W]) \neq \phi$, then $W \subset W_i$. Then we have*

$$p_X([W]) = \frac{|G|}{|X|} \sum_{0 \leq i \leq N} \mu(W, W_i) \left|\Theta^{W_i}([W])\right| c_X(W_i) \left|N_G(W_i)\right|^{-1}.$$

§3.

In $[RS_2]$ complete answers for the problems above in the case of $X_n = G_n$ are given, where G_n is the general linear group $GL(V_n)$ or the group of all isometries $G(V_n, <,>)$ with respect to certain nondegenerate bilinear or sesquilinear form $<,>$ and hence G_n is the unitary, symplectic or orthogonal group. (To be precice, in the unitary case, V_n is an n-dimensional vector space over $\boldsymbol{F}_{q^2}$).

In particular , part of the answers for Problem 2 can be described as follows. Let $p_\infty^\Gamma(m) = \lim_{n\to\infty} p_{G_n}(m)$, where we denote by Γ the type of the series of groups $\{G_n\}$ and Γ is GL (resp. U,Sp or 0) when $\{G_n\}$ is the series of general linear (resp. unitary, symplectic or orthogonal) groups. Put $x = q^{-1}$ and for $|z| < 1$, define $(z)_k = \prod_{1\leq j\leq k}(1 - z^j)$ for $0 \leq k \leq \infty$.

3.1 Theorem($[RS_2]$). *For each nonnegative integer m, we have*

$$p_\infty^{GL}(m) = x^{m^2}(x)_\infty (x)_m^{-2},$$

$$p_\infty^{U}(m) = x^{m^2}\{(x^2)_m \prod_{n=0}^{\infty}(1 + x^{2n+1})\}^{-1},$$

$$p_\infty^{Sp}(m) = x^{m(m+1)/2}\{(x)_m \prod_{n=1}^{\infty}(1 + x^{n})\}^{-1},$$

$$p_\infty^{0}(m) = x^{m(m-1)/2}\{(x)_m \prod_{n=0}^{\infty}(1 + x^{n})\}^{-1}.$$

3.2 Remark. An obvious equation of probability , $\sum_{m=0}^{\infty} p_\infty^\Gamma(m) = 1$, where Γ denotes the type, is equivalent to the following famous identities of Euler (cf. [HW]) corresponding to each Γ.

$$(GL) \quad \sum_{m=0}^{\infty} x^{m^2}\{(x)_m\}^{-2} = (x)_\infty^{-1},$$

$$(U) \quad \sum_{m=0}^{\infty} x^{m^2}\{(x^2)_m\}^{-1} = \prod_{j=0}^{\infty}(1 + x^{2j+1}),$$

$$(Sp) \quad \sum_{m=0}^{\infty} x^{m(m+1)/2}\{(x)_m\}^{-1} = \prod_{j=1}^{\infty}(1 + x^{j}),$$

$$(O) \quad \sum_{m=0}^{\infty} x^{m(m-1)/2}\{(x)_m\}^{-1} = \prod_{j=0}^{\infty}(1 + x^{j}).$$

§4.

Now we consider the case when X_n is the set of all unipotent elements of G_n. For simplicity we let $G_n = GL(V_n)$.

Two subspaces U_1 and U_2 belong to a same G_n-orbit if and only if $\dim U_1 = \dim U_2$. Choose a chain of subspaces $\{0\} = W_0 \subset W_1 \subset \cdots \subset W_n = V_n$, where W_i is a subspace with $\dim W_i = i$ for each $i, 0 \le i \le n$. Then $\{W_i\}_{0 \le i \le n}$ is a complete system of representatives of G_n-orbits in $\boldsymbol{L}_n$. Therefore by (2.1) we have

$$(4.1) \qquad p_{X_n}(m) = \frac{|G_n|}{|X_n|} \sum_{i=0}^{n} \mu(W_m, W_i) \left|\Theta^{W_i}([W_m])\right| c_{X_n}(W_i) \left|N_{G_n}(W_m)\right|^{-1}.$$

For an indeterminant t and nonnegative integers i, k, let $\varphi_k(t) = \Pi_{j=1}^{k}(t^j - 1)$ and $\left[{k \atop i}\right]_t = \frac{\varphi_k(t)}{\varphi_i(t)\varphi_{k-i}(t)}$ if $0 \le i \le k$ and $\left[{k \atop i}\right]_t = 0$, otherwise, which is called a Gaussian polynomial. Then since in this case $\left|\Theta^{W_i}([W_m])\right|$ is the number of all m-dimensional subspaces of W_i, it is $\left[{i \atop m}\right]_q$. It is well known that $|G_n| = q^{n(n-1)/2}\varphi_n(q)$ and it can be shown that $|G_n| = q^{i(n-i)} |G_i| |G_{n-i}|$. Moreover by a theorem of Steinberg [St], we have $|X_n| = q^{n(n-1)}$ and $c_{X_n}(W_i) = q^{i(n-i)+(n-i)(n-i-1)}$. Substituting these values into (4.1) and using $x = q^{-1}$, we have

$$(4.2) \qquad p_{X_n}(m) = \frac{x^{m(m-1)}(x)_n}{(x)_m} \sum_{j=0}^{n-m} (-1)^j \frac{x^{mj+j(j-1)/2}}{(x)_j (x)_{n-m-j}}.$$

Now we recall the following Lemma, which is called q-binomical expansion (cf. [A]).

4.3 Lemma .

$$\prod_{j=0}^{k-1}(1 - zt^j) = \sum_{j=0}^{k}(-1)^j \begin{bmatrix} k \\ j \end{bmatrix}_t z^j t^{j(j-1)/2}.$$

Applying (4.3) to the right hand side of (4.2), we have

4.4 Theorem. $p_{X_n}(m) = \frac{x^{m(m-1)}(x)_n (x)_{n-1}}{(x)_m (x)_{n-m} (x)_{m-1}}$.

Since $\lim_{n\to\infty} p_{X_n}(m) = \dfrac{x^{m(m-1)}(x)_\infty}{(x)_m (x)_{m-1}}$, *the obvious equation* $\sum_{m=0}^{\infty} \lim_{n\to\infty} p_{X_n}(m) = 1$ *implies this time*

$$(GL, unip) \quad \sum_{m=0}^{\infty} \frac{x^{m(m-1)}}{(x)_m (x)_{m-1}} = (x)_\infty^{-1}.$$

A combinatorial proof of the above equation $(GL, unip)$ is easy to give. In fact we only need to consider the maximal $m \times (m-1)$ rectangle in the Young diagram instead of Durfee square which is used in a combinatorial proof of (GL). For detailes we refer p. 281 of $[HW]$.

For the other classical groups we can prove the following

4.5 Theorem ([S]). *Let X_n be the set of all unipotent elements of the group G_n. We write $p_{G_n,u}$ instead of $p_{X_n}(m)$. Then*

$$p_{U_n,u}(m) = \frac{x^{m(m-1)}(-x)_n(-x)_{n-1}}{(-x)_m(-x)_{n-m}(-x)_{m-1}}.$$

$$p_{Sp_{2\nu},u}(2\mu) = \frac{x^{2\mu^2}(x^2)_\nu(x^2)_{\nu-1}}{(x^2)_{\nu-\mu}(x^2)_\mu(x^2)_{\mu-1}}$$

and

$$p_{Sp_{2\nu},u}(2\mu+1) = \frac{x^{2\mu(\mu+1)}(x^2)_\nu(x^2)_{\nu-1}}{(x^2)_{\nu-\mu-1}\{(x^2)_\mu\}^2}.$$

If the characteristic of $\boldsymbol{F}_q$ is odd,

$$p_{O_{2\nu+1},u}(2\mu+1) = \frac{x^{2\mu^2}\{(x^2)_\nu\}^2}{(x^2)_{\nu-\mu}\{(x^2)_\mu\}^2} \quad \text{and} \quad p_{O_{2\nu+1},u}(2\mu) = 0.$$

$$p_{O^\epsilon_{2\nu},u}(2\mu) = \frac{x^{2\mu(\mu-1)}(x^2)_\nu(x^2)_{\nu-1}}{(x^2)_{\nu-\mu}(x^2)_\mu(x^2)_{\mu-1}} \quad \text{and} \quad p_{O^\epsilon_{2\nu},u}(2\mu+1) = 0.$$

If the characteristic of $\boldsymbol{F}_q$ is 2, then

$$p_{O^\epsilon_{2\nu},u}(2\mu) = \frac{x^{2\mu(\mu-1)}(x^2)_\nu(x^2)_{\nu-1}}{(x^2)_{\nu-\mu}(x^2)_\mu(x^2)_{\mu-1}}$$

and

$$p_{O^\epsilon_{2\nu},u}(2\mu+1) = \frac{x^{2\mu^2-1}\{(x^2)_{\nu-1}\}^2(1-\varepsilon x^\nu)}{\{(x^2)_\mu\}^2(x^2)_{\nu-\mu-1}},$$

where on the left hand side ε is + or - as in usual notation, while on the right hand side ε stands for $+1$ or -1 respectively.

4.6 Remark. As can be seen easily, the equation $\sum_{m=0}^{\infty} \lim_{n\to\infty} p_{G_n,u}(m) = 1$ in each of the above cases implies an identity of Euler of type (GL) or (GL, uniq), except for the orthogonal case of char $\boldsymbol{F}_q = 2$. He detailes of which will be discussed in [S].

§5.

In this section we consider a related problem. The problems given in section 1 concerns with enumerations in finite classical groups and so one can say it is a q-version of "problem de rencontres". As for symmetric group S_n, many elegant results in enumerative combinatorics are known. For example, let $f(n)$ be the number of involutions, i.e. elements σ such that $\sigma^2 = 1$, in S_n. Then

$$\sum_{n=0}^{\infty} f(n)t^n/n! = \exp(t + t^2/2). \tag{5.1}$$

Moreover it is known that $f(n)$ is the sum of degrees of all ordinary irreducible characters of S_n.

Returning to $G_n = GL(V_n)$, let $F(n)$ be the sum of degrees of all ordinary irreducible characters of G_n. Then we have

5.2 Proposition.

$$\sum_{n=0}^{\infty} F(n)t^n/\varphi_n(q) = (1+t)\Pi_{n=0}^{\infty}(1-t^2q^{-2n})^{-1}.$$

Proof. By a theorem of Klyachko [K], we have

$$F(n) = \sum_{0\le 2k\le n} [GL_n(q) : H_k],$$

where H_k is the subgroup $GL_n(q)$ consisting of all matrices of the form $\begin{pmatrix} U & * \\ 0 & A \end{pmatrix}$, where U is unipotent uppertriangular and $A \in Sp_{2k}(q)$. Taking into acount that

$$\sum_{k=0}^{\nu} q^{k(k-1)}(q^{2(k+1)}-1)(q^{2(k+2)}-1)\cdots(q^{2\nu}-1) = q^{\nu(\nu+1)},$$

which can be proved easily by induction on ν, we have

$$F(n) = \begin{cases} q^{\nu(\nu+1)}(q-1)(q^3-1)\cdots(q^{2\nu-1}-1), & \text{if } n = 2\nu, \\ q^{\nu(\nu+1)}(q-1)(q^3-1)\cdots(q^{2\nu+1}-1), & \text{if } n = 2\nu+1. \end{cases}$$

Therefore

$$\begin{aligned} \text{L.H.S. of (5.2)} &= (1+t)\textstyle\sum_{\nu=0}^{\infty} q^{\nu(\nu+1)}t^{2\nu}/\varphi_\nu(q^2) \\ &= (1+t)\textstyle\sum_{\nu=0}^{\infty} t^{2\nu}/(x^2)_\nu \qquad \text{(by putting } x = q^{-1}) \\ &= (1+t)\Pi_{n=0}^{\infty}(1-t^2x^{2n})^{-1} \qquad \text{(by (2.2.5) of [A]).} \end{aligned}$$

So the following question naturally arises: Are there any enumerative problems in G_n which have $F(n)$ as the solution? Unfortunately the author does not know an answer.

References

[A] G.E. Andrews, *The theory of partitions, Encyclopedia of mathematics and its applications 2*, Addison–Wesley, 1976.

[HW] G.H. Hardy and E.M.Wright, *An introduction to the theory of numbers*, Oxford, 1968.

[K] A.A.Klyachko, *Models for the complex representations of the groups $GL(n,q)$*, Math. USSR–Sb., 48 (1984), 365–379.

[RS_1] A.Rudvalis and K.Shinoda, *On counting some elements in finite classical groups* (preprint, 1988).

[RS_2] ——, *An enumeration in finite classical groups* (tentative title, in preparation).

[S] K. Shinoda, *Counting some unipotent elements in finite classical groups* (in preparation).

[St] R.Steinberg, *Endomorphisms of linear algebraic groups*, Memoirs of A.M.S., 80 (1968).

Proceedings of the Asian Mathematical
Conference, 1990

Differential-difference Inequalities with Unbounded Delays and Their Applications*

SI LIGENG & MA WANBIAO
Department of Mathematics
Inner Mongolia Normal University
Huhhot 010022, China

§1. Introduction

It is well known that in the stability analysis of large-scale dynamical systems, the methods of vector liapunov functions and scalar liapunov functions are the main and powerful methods. But, generally, great difficulties arise in applying these methods to large-scale dynamical systems with unbounded delays. Particularly, for the stability analysis of neutral-type large-scale dynamical systems with unbounded delays, very few results have been obtained.

In this paper, first, we establish two very important differential-difference inequalities with unbounded delays. Second, by making use of these inequalities, we study the global stability and instability of several classes of large-scale dynamical systems with unbounded delays (including neutral-type), and obtain some simple and practical sufficient criteria.

§2. Differential-difference Inequalities with Unbounded Delays

Definition 1. An $n \times n$ matrix $C = (c_{ij})_{n\times n}$ is an M-matrix if $c_{ij} \le 0 \quad i \ne j, i, j = 1, 2, \cdots, n$, and all principal minors of C are positive.

Theorem 1. *Assume that the function $p(t) = \mathrm{col}(p_1(t) \cdots p_n(t)) : R \to R^n_+$ is continuous and satisfies the inequality*

$$K\dot{p}(t) \le g(t)(Ap(t) + B\overline{p}(t)) \quad \text{for} \quad t \ge t_0,$$

where $K = \mathrm{diag}(k_1 \cdots k_n), k_i$ are nonnegative constants; $g(t) = \mathrm{diag}\ (g_1(t) \cdots g_n(t))$ is continuous such that $g_i(t) \ge \sigma > 0$ for $t \ge t_0$, is constant; $A = (a_{ij})_{n\times n}$ and $B = (b_{ij})_{n\times n}$ are constant matrices such that $a_{ii} < 0, a_{ij} \ge 0, i \ne j, b_{ij} \ge 0, i, j = 1, 2, \cdots, n$, and that -$(A+B)$ is an M-matrix; $\overline{p}(t) = \mathrm{col}(\overline{p}_1(t) \cdots \overline{p}_n(t)), \overline{p}_i(t) = \sup_{-\Delta(t)\le s\le 0} p_i(t+s), \Delta(t) : [t_0, +\infty] \to [0, +\infty)$ is continuous such that $\Delta(t) < t$ and $t - \Delta(t) \to +\infty$ as $t \to +\infty$. Then there exists

---

*This research was supported by the National Natural Science Foundation of China.

a constant $M_1 \geq 1$ *such that* $p_i(t) \leq M_1\|p\|$ *for* $t \geq t_0$ *and* $\lim_{t\to+\infty} p_i(t) = 0, i = 1, 2, \cdots, n$, *where* $\|p\| = \sum_{i=1}^n \sup_{(-\infty,\ t_0]} p_i(s) < +\infty$.

Remark I. 1. If $\Delta(t)$ in Theorem 1 is bounded, by making use of the method of [10], we can obtain the following exactly estimation:

$$\sum_{i=1}^{n} p_i(t) \leq \|p\|_\Delta M_2 \exp(-\alpha(t-t_0)) \quad \text{for} \quad t \geq t_0,$$

where $M_2 \geq 1$ and $\alpha > 0$ are constants, $\|p\|_\Delta = \sum_{i=1}^n \sup_{-\Delta\leq s\leq 0} p_i(t_0+s), \Delta$ is the bound of $\Delta(t)$.

2. We can obtain the following non-linear differential-difference inequality by making use of the similar methods:

Assume that the functions $p_i(t) \quad (i = 1, 2, \cdots, n) : R \to R_+$ are continuous and satisfy the non-linear inequality

$$k_i\dot{p}_i(t) \leq g_i(t)f_i\{p_1(t), \cdots, p_n(t); \overline{p}_1(t), \cdots, \overline{p}_n(t)\};$$

$$\int_0^\infty A_1(u)p_1(t-u)du, \cdots, \int_0^\infty A_n(u)p_n(t-u)du \quad \text{for} \quad t \geq t_0, \quad i = 1, 2, \cdots, n,$$

where k_i are nonnegative constants; $g_i(t) > 0$ are continuous and $\lim_{t\to+\infty} \int_{t_0}^t g_i(t)dt = +\infty$ are uniform for $t \geq t_0; \overline{p}_i(t)$ are defined in Theorem 1; $A_i(u) \geq 0$ are continuous for $u \geq 0$ such that $\int_0^\infty A_i(u)du < +\infty$ the functions f_i are continuous such that

$$f_i(x_1, \cdots, x_n; y_1, \cdots, y_n; z_1, \cdots, z_n) \leq f_i(\overline{x}_1, \cdots, \overline{x}_n; \overline{y}_1, \cdots, \overline{y}_n; \overline{z}_1, \cdots, \overline{z}_n),$$

for $x_i = \overline{x}_i, x_j \leq \overline{x}_j, i \neq j, y_j \leq \overline{y}_j, z_j \leq \overline{z}_j, j, i = 1, 2, \cdots, n$. In addition, there exist a continuous function $\overline{f}(\theta)$ and constants $\alpha_i > 0$ such that $\overline{f}(0) = 0$ and

$$f_i(\alpha_1\theta, \cdots, \alpha_n\theta; \alpha_1\theta, \cdots, \alpha_n\theta; \alpha_1 S_1\theta, \cdots, \alpha_n S_n\theta) \leq -\overline{f}(\theta) < 0$$

for $\theta \neq 0$ and $S_i = \int_0^\infty A_i(u)du, i = 1, 2, \cdots, n$. Then there exists constant $M_3 \geq 1$ such that $p_i(t) \leq M_3\|p\|$ for $t \geq t_0$ and $\lim_{t\to+\infty} p_i(t) = 0, i = 1, 2, \cdots, n$, where $\|p\| = \max\{\sup_{\theta\leq 0} p_s(t_0+\theta), s = 1, 2, \cdots, n\}$.

Theorem 2. *If the function* $p(t)=\mathrm{col}(p_1(t) \cdots p_n(t)) : R \to R_+^n$ *is continuous and satisfies the inequality*

$$\dot{p}(t) \geq f(t)(Ap(t) + B\overline{p}(t)) \quad \text{for} \quad t \geq t_0$$

or

$$\dot{p}(t) \geq f(t)p_*^{\frac{1}{2}}(t)(Ap^{\frac{1}{2}}(t) + B\overline{p}^{-\frac{1}{2}}(t)) \quad \text{for} \quad t \geq t_0,$$

where $p_i(t)$ *are positive and non-decreasing on* $(-\infty, t_0], i = 1, 2, \cdots, n; f(t) = \mathrm{diag}\ (f_1(t) \cdots f_n(t))$ *is positive and continuous for* $t \geq t_0$, *such that* $\int_{t_0}^\infty f_i(t)dt = +\infty, i = 1, 2, \cdots, n; p_*^{\frac{1}{2}}(t) = \mathrm{diag}\ (p_1^{\frac{1}{2}}(t) \cdots p_n^{\frac{1}{2}}(t)), p^{\frac{1}{2}}(t) = \mathrm{col}(p_1^{\frac{1}{2}}(t) \cdots p_n^{\frac{1}{2}}(t)), \overline{p}^{\frac{1}{2}}(t) = \mathrm{col}(\overline{p}_1^{\frac{1}{2}}(t) \cdots \overline{p}_n^{\frac{1}{2}}(t)), \overline{p}(t)$ *and* $\overline{p}_i(t)$ *are defined in Theorem 1;* $A = (a_{ij})_{n\times n}$ *and* $B = (b_{ij})_{n\times n}$ *are constant matrices such that* $a_{ii} > 0, a_{ij} \leq 0, i \neq j, b_{ij} \leq 0, i, j = 1, 2, \cdots, n$, *and that* $A+B$ *is an M-matrix. Then* $\underline{\lim}_{t\to+\infty}(p_1(t) + \cdots + p_n(t)) = +\infty$.

§3. Stability of Neutral-type Large-scale Dynamical Systems with Unbounded Delays

Let us consider the following composite system of neutral-type

$$\dot{x}(t) = A(t)x(t) + f(t, x(t), x(t-\Delta(t)), \dot{x}(t-\Delta(t))) \quad \text{for} \quad t \geq t_0, \tag{3.1}$$

with decomposition

$$\dot{x}_i(t) = A_i(t)x_i(t) + f_i(t, x(t), x(t-\Delta(t)), \dot{x}(t-\Delta(t))) \tag{3.2}$$

for $t \geq t_0, i = 1, 2, \cdots, r$, where $x = \mathrm{col}(x_1 \cdots x_r) \in R^n, x_i \in R^{n_i}, n_1 + \cdots + n_r = n$; $A(t) = \mathrm{diag}(A_1(t) \cdots A_r(t))$, $A_i(t)$ are $n_i \times n_i$ continuous matrices, $f = \mathrm{col}(f_1 \cdots f_r), f_i : R \times R^n \times R^n \times R^n \to R^{n_i}$ and $\Delta(t) : [t_0, +\infty) \to [0, +\infty)$ are continuous such that $f_i(t, 0, 0, 0) = 0, \Delta(t) < t$ and $t - \Delta(t) \to +\infty$ as $t \to +\infty, i = 1, 2, \cdots, r$.

The initial condition of the systems (3.1) and (3.2) are

$$x_i(t) = \varphi_i(t), \quad \dot{x}_i(t) = \dot{\varphi}_i(t) \quad \text{for} \quad t \in E_{t_0}, \quad i = 1, 2, \cdots, r, \tag{3.3}$$

where $\varphi_i(t)$ and $\dot{\varphi}_i(t)$ are n_i-dimensional continuous functions, $i = 1, 2, \cdots, r$. $E_{t_0} = \{t_0\} \cup \{s : s = t - \Delta(t) \leq t_0\}$ is the initial interval.

Set

$$\|\varphi\| = \sum_{i=1}^{r} \sup_{t \in E_{t_0}} (\|\varphi_i(t)\| + \|\dot{\varphi}_i(t)\|) < +\infty, a_i = \sup_{t \geq t_0} \|A_i(t)\| < +\infty, \quad i = 1, 2, \cdots, r.$$

Definition 2. The trivial solution of (3.2) is said to be stable in the metric space C_1 if for any $\varepsilon > 0$ there exists $\delta = \delta(\varepsilon, t_0) > 0$ such that whenever $\|\varphi| \leq \delta, \sum_{i=1}^{r}(\|x_i(t)\| + \|\dot{x}_i(t)\|) < \varepsilon$ for $t \geq t_0$.

Definition 3. The trivial solution of (3.2) is said to be uniformly stable in the metric space C_1 if it is stable in the space C_1 and the δ is independent of t_0.

Definition 4. The trivial solution of (3.2) is said to be globally asymptotically stable in the metric space C_1 if it is stable in the space C_1 and every solution of (3.2) satisfies

$$\sum_{i=1}^{r}(\|x_i(t)\| + \|\dot{x}_i(t)\|) \to 0 \quad \text{as} \quad t \to +\infty.$$

Condition I. Assume that the solution $Y_i(t, s)$ of the systems:

$$\begin{cases} \partial Y_i(t, s)/\partial t = A_i(t)Y_i(t, s), & \text{for } t_0 \leq s \leq t, \\ Y_i(s, s) = E_i(n_i \times n_i \text{ unit matrices}), & i = 1, 2, \cdots, r, \end{cases} \tag{3.4}$$

satisfies $\|Y_i(t, s)\| \leq M_i \exp(-r_i(t-s))$ for $t_0 \leq s \leq t$, where $M_i \geq 1$ and $r_i > 0$ are constants, $i = 1, 2, \cdots, r$.

Condition II. Assume that the functions f_i satisfies

$$\|f_i(t, x, y, z) - f_i(t, \bar{x}, 0, 0)\| \leq \sum_{j=1}^{r}(a_{ij}\|x_j - \bar{x}_j\| + b_{ij}\|y_j\| + c_{ij}\|z_j\|,$$

for $t \geq t_0$, where a_{ij}, b_{ij} and c_{ij} are nonnegative constants, $i, j = 1, 2, \cdots, r$.

Condition III. Assume that there exist differentiable functions $V_i(t, x_i)$ such that (1) $(\alpha_i \|x_i\|)^2 \leq V_i(t, x_i) \leq (\beta_i \|x_i\|)^2$; (2) $\|\partial V_i(t, x_i)/\partial x_i\| \leq 2\delta_i \sqrt{V_i(t, x_i)}$; (3) $(\partial V_i(t, x_i)/\partial x_i) \cdot (A_i(t)x_i + f_i(t, x, 0, 0)) + \partial V_i(t, x_i)/\partial t \leq -2r_i V_i(t, x_i)$, where $\alpha_i, \beta_i, \delta_i$ and r_i are positive constants, $i = 1, 2, \cdots, r$.

Theorem 3. *Assume that the Condition I and II hold, and that the matrix Q is an M-matrix, then the trivial solution of (3.2) is uniformly stable and globally asymptotically stable in the metric space C_1, where*

$$Q = -\left(\delta_{ij} - c_{ij} - \frac{a_{ij} + b_{ij} + a_{j,ij}}{r_j + a_j M_j} M_j\right)_{r \times r}$$

Theorem 4. *Assume that the Condition II and III hold, and that the matrix Q_1 is an M-matrix, then the conclusions of Theorem 3 hold, where*

$$Q_1 = -\begin{pmatrix} ((b_{ij}\delta_i/\alpha_i) - r_i\delta_{ij})_{r \times r} & (c_{ij}\delta_i/\alpha_i)_{r \times r} \\ (a_{ij} + b_{ij} + a_i\delta_{ij})_{r \times r} & (c_{ij} - \delta_{ij})_{r \times r} \end{pmatrix}_{2r \times 2r}$$

Remark II. If let $r = 1$ in the system (3.2), we can obtain some simple sufficient criteria from Theorems 3 and 4.

§4. Stability and Instability of Retarted-type Large-scale Dynamical Systems with Unbounded Delays

Consider the following retarted-type large-scale systems:

$$\dot{x}_i(t) = A_i(t)x_i(t) + f_i(t, x(t), x(t - \Delta(t))), \tag{4.1}$$

for $t \geq t_0, i = 1, 2, \cdots, r$, where $x, x_i, A_i(t)$ and $\Delta(t)$ are defined as in (3.2); $f_i : R \times R^n \times R^n \to R^{n_i}$ are continuous such that $f_i(t, 0, 0) = 0, i = 1, 2, \cdots, r$.

The initial condition of (4.1) is $x_i(t) = \varphi_i(t)$ for $t \in E_{t_0}, i = 1, \cdots, r$, where $\varphi_i(t)$ are n_i-dimensional continuous functions, E_{t_0} is the initial interval defined as in Section 3.

Set

$$\|\varphi\| = \sum_{i=1}^{r} \sup_{t \in E_{t_0}} \|\varphi_i(t)\| < +\infty.$$

Condition IV. Assume that the solutions $Y_i(t, s)$ of (3.4) satisfy $\|Y_i(t, s)\| \leq M_i \exp(-\int_s^t r_i(u)du)$ for $t_0 \leq s \leq t$, where $r_i(t)$ are continuous such that $r_i(t) \geq \sigma > 0$ for $t \geq t_0, M_i \geq 1$ and σ are constants, $i = 1, 2, \cdots, r$.

Condition V. Assume that the functions f_i satisfy

$$\|f_i(t, x, y) - f_i(t, \bar{x}, 0)\| \leq \sum_{i=1}^{r} (a_{ij}(t)\|x_j - \bar{x}_j\| + b_{ij}(t)\|y_j\|),$$

for $t \geq t_0$, where $a_{ij}(t)$ and $b_{ij}(t)$ are nonnegative and continuous for $t \geq t_0, i = 1, 2, \cdots, r$.

Condition VI. Assume that there exist differentiable functions $V_i(t, x_i)$ such that (1) and (2) of Condition III, and

$$\frac{\partial V_i(t,x_i)}{\partial x_i}(A_i(t)x_i + f_i(t,x,0)) + \frac{\partial V_i(t,x_i)}{\partial t} \le -2r_i(t)V_i(t,x_i)$$

hold, where $r_i(t)$ are continuous such that $r_i(t) \ge \sigma > 0$ for $t \ge t_0, \sigma$ is constant, $i = 1, 2, \cdots, r$.

Condition VII. Assume that there exist differentiable functions $V_i(t, x_i)$ such that (1) and (2) of Condition III, and

$$\frac{\partial V_i(t,x_i)}{\partial x_i}(A_i(t)x_i + f_i(t,x,0)) + \frac{\partial V_i(t,x_i)}{\partial t} \ge 2r_i(t)V_i(t,x_i)$$

hold, where $r_i(t) > 0$ are continuous such that $\int_{t_0}^{+\infty} r_i(t)dt = +\infty, i = 1, 2, \cdots, r$.

Condition VIII. Assume that the smallest real parts of the eigenvalues of the matrices $H_i(t) = \frac{1}{2}(A_i(t) + A_i^T(t))$, denoted by $\mathrm{Re}\lambda(H_i(t))$, satisfy $\mathrm{Re}\lambda(H_i(t))r_i(t) > 0$ for $t \ge t_0$, where $A_i^T(t)$ are the transpose of $A_i(t), r_i(t)$ are defined as in Condition VII, $i = 1, 2, \cdots, r$.

Set $\bar{a}_{ij} = \sup_{t\ge t_0}(a_{ij}(t)/r_i(t)) < +\infty, \bar{b}_{ij} = \sup_{t\ge t_0}(b_{ij}(t)/r_i(t)) < +\infty, i, j = 1, 2, \cdots, r$.

Theorem 5. *Assume that the Conditions Iv and V hold, and that the matrix Q_2 is an M-matrix, then the trivial solution of (4.1) is uniformly stable and globally asymptotically stable, where $Q_2 = (\delta_{ij} - M_i(\bar{a}_{ij} + \bar{b}_{ij}))_{r\times r}$.*

Theorem 6. *Assume that the conditions V and VI hold, and that the matrix Q_3 is an M-matrix, then the conclusions of Theorem 5 hold, where $Q_3 = (\delta_{ij} - \delta_i\bar{b}_{ij}/\alpha_j)_{r\times r}$.*

Theorem 7. *Assume that the Conditions V and VII hold, and that the matrix Q_4 is an M-matrix, then the trivial solution of (4.1) is instable, where $Q_4 = (\delta_{ij} - \delta_i\bar{b}_{ij}/\alpha_j)_{r\times r}$.*

Theorem 8. *Assume that the Conditions V and VIII hold, and that the matrix Q_5 is an M-matrix, then the trivial solution of (4.1) is instable, where $Q_5 = (\delta_{ij} - \bar{a}_{ij} - \bar{b}_{ij})_{r\times r}$.*

Example. Let us consider the linear system:

$$\dot{x}(t) = -(a + (t\sin t)^2)x(t) + bx(t - r(t)), \tag{4.3}$$

where $a > 0$ and b are constants, $r(t)$ is continuous such that $0 \le r(t) < t$ and $t - r(t) \to +\infty$ as $t \to +\infty$.

By Theorem 5 we can easily obtain that the trivial solution of (4.3) is uniformly stable and asymptotically stable if $a > |b|$. Obviously, the condition $a > |b|$ is better than the condition $a^2(1-\alpha) > b^2$ given in [6], where $r'(t) \le \alpha < 1$.

References

[1] V. Lakshmikantham and S. Leela, *Differential and Integral Inequalities*, 2(1969), Academic Press, New York.

[2] J.K. Hale, *Theory of Functional Equations*, Springer-Verlag, New York, 1977.

[3] Qin Yuanxun and Liou Iongqing and Wang Lian, *The Motion Stability of Dynamical Systems with Delays*, Academic Press, Beijing, 1963.

[4] A. Halanay, *Differential Equations: Stability, Oscillations, Time-lags*, Academic Press, New York, 1966.

[5] V.B. Kolmanovsky and V.R. Nosov, *Stability of neutral-type functional differential equations*, Nonlinear Analysis: Theory, Methods & Applications, 6(1982), 873–910.

[6] T.A. Burton, *Stability theory for delay equations*, Funkcial Ekvac, 22(1979), 67–76.

[7] Si Ligeng, *The boundedness and stability of nonlinear neutral differential equations with variable delay*, Acta Mathematics Sinica, 17(1974), 197–204.

[8] Si Ligeng and Zhang Yi, *The stability of neutral differential systems with variable delay*, Kexue Tongbao, 32(1987), 1009–1013.

[9] D.D. Siljak, *Large-Scale Dynamic Systems: Stability and Structure*, Noth-Holland, New York, 1978.

[10] Ma Wanbiao, *Differential-difference inequalities and their applications*, Journal of Inner Mongolia Normal University, 3(1986), 37–46.

[11] R.D. Driver, *Existence and stability of solutions of delay-differential systems*, Arch. Rationat Mech. Anal., 10(1962), 401–426.

[12] A. Berman and R.J. Plemmons, *Nonnegative Matrices in the Mathematical Sciences*, Academic Press, New York, 1977.

Proceedings of the Asian Mathematical
Conference, 1990

Comparison Theorems with Isometric Actions in Riemannian Manifolds

HUKUM SINGH
Department of Education in Science and Mathematics
National Council of Educational Research and Training
Sri Aurobindo Marg: New Delhi-110016, INDIA

The purpose of this paper is to prove the following theorems.

Theorem A. *Let M be complete connected Riemannian manifold of class C^∞ and G a compact connected Lie group whose elements are isometries of M and such that its canonical action α is of class C^∞. Let the orbit $G(x)$ of $x \in M$ be principal and the isotropy subgroup G_x is of maximal rank. Let $c : [0,a] \to M$ be the geodesic defined by the initial conditions $c(0) = x, \dot{c}(0) = w$ where $w \in N_x$ is a unit vector. If there is no focal point of $G(x)$ on c then to any vector field V defined along c there is a $G(x)$-Jacobifield X such that $X(a) = V(a)$ and in this case $I(X,X) \leq I(V,V)$, where the equality occures if and only if $X = V$.*

Theorem B. *Let $\alpha : G \times M \to M$ be an isometric action of a compact connected Lie group G on a complete connected Riemannian manifold M such that there is a point $x \in M$ for which the orbit $G(x)$ is principal and the isotropy subgroup G_x is of maximal rank. If $c : [a,\infty) \to M$ is the geodesic defined by the initial conditions $c(a) = x$ and $\dot{c}(a) = w \in N_x$ is a unit vector, and $c(b)$ a focal point of $G(x)$ for some $b > 0$. Then $c \restriction [a,d]$ is not a minimal geodesic for $d > b$.*

Theorem C. *Let M, M_0 be Riemannian manifolds such that $\dim M \leq \dim M_0, G, G_0$ compact connected Lie groups whose elements are isometries of M, M_0 respectively such that the isotropy subgroups of their actions $\alpha : G \times M \to M, \alpha_0 : G_0 \times M_0 \to M_0$ are of maximal rank. Let $G(x), G_0(x_0)$ be principal orbits and $c : [0,a] \to M, c_0 : [0,a] \to M_0$ geodesics such that $c(0) = x, c_0(0) = x_0; \dot{c}(0) = u \in N_xG(x), \dot{c}_0(0) = u_0 \in N_{x_0}(G_0(x_0))$, where $\|u\| = \|u_0\| = 1$. Assume that $K(P) \leq K_0(P_0)$ holds for sectional curvatures corresponding to 2-dimensional subspaces spanned by $X, \dot{c}(t) \in T_{c(t)}M, X_0, \dot{c}_0(t) \in T_{c_0(t)}M_0$ where $t \in [0,a]$ and that no $c_0(t)$ is a focal point of the orbit $G_0(x_0)$. Let V, V_0 be Jacobifields along c, c_0* such that $\|V(0)\| = \|V_0(0)\|$,

$$\langle V_0(0), \dot{c}_0(0) \rangle = \langle V(0), \dot{c}(0) \rangle; \quad V'(0) = d\dot{c}(0),$$

$V'_0 = d_0\dot{c}_0(0)$. *Then* $\|V(t)\| \geq \|V_0(t)\|$ for all $t \in [0,a]$.

Introduction

Let M be a complete connected n-dimensional Riemannian manifold of class C^∞ and G a compact connected Lie group whose elements are isometries of M. Let $\alpha : G \times M \to M$ be an action of class C^∞ such that every transformation $\alpha_g : M \to M, g \in G$ defined by $\alpha_g(x) = \alpha(g,x), x \in M$ is an isometry of M then α is called an isometric action. The space $G(x) = \{g(x) : g \in G\}$ is called the orbit of x. If x and y are points of M, then $G(x)$ and $G(y)$ are either disjoint or equal so that M is partitioned by the orbits. Denote by M/G the set of orbits, and let $p : M \to M/G$ be the natural projection assigning each point of M to its orbits. If we topologize M/G with the identification topology, i.e., $A \subset M/G$ is open if and only if $p^{-1}(A)$ is open in M, then p is continuous and open.

As is usual, we shall call $G_x = \{g \in G : g(x) = x\}$ the isotropy group at x. A group G is said to operate effectively if only the identity leaves every point of M fixed; that is, if the isotropy groups contain no common normal subgroups. We say G acts transitively if there is just one orbit. Besides, the action α is said to be closed if every orbit $G(x), x \in M$ is closed subset of M. Let now α be an isometric closed action. The orbit $G(x)$ of a point $x \in M$ being a compact connected submanifold of class C^∞ in M has its normal bundle $N(x)$ which is a subbundle of class C^∞ of the tangent bundle TM. The action α induces an action $T\alpha : G \times TM \to TM$ of class C^∞ by the following definition: $T\alpha(g,v) = T_x\alpha_g v$ for $g \in G$ and $v \in T_xM \subset TM$. The normal bundle $N(x)$ is obviously an invarient subset of TM for the action $T\alpha$. Consequently the restriction of $T\alpha$ to $N(x)$ yields an action $\beta : G \times N(x) \to N(x)$ of class C^∞. Since the action α is isometric the restricted exponential map $e : N(x) \to M$ is equivarient for the action β and α. This fact implies that the orbit of a focal point of $G(x)$ consists entirely of focal points of $G(x)$ both in $N(x)$ and in M. In fact, focal point of orbit may be defined as follows:

If the tangent linear map $T_{v^0} : T_vN(x) \to T_{e(v)}M$ of e at $v \in N(x)$ is not injective it is said that v is a focal point of $G(x)$ [1]. In this case the point $e(v)$ is called as well focal point of $G(x)$. The set of such focal points v is called the focal locus of $G(x)$ in $N(x)$ while the set of focal points $e(v)$ is called the focal locus of $G(x)$ in M.

The focal points of orbit in a Riemannian manifold are closely related to some Jacobi fields. Consider the second fundamental tensor $H_z : T_zG(x) \times T_zG(x) \to N_z$ of the orbit $G(x)$ at a point $z \in G(x)$ and fix a unit vector $w \in N_z$ then a symmetric linear map $A_w : T_zG(x) \to T_zG(x)$ is obtained by postulating the validity of $\langle A_w(Y), Z\rangle = \langle w, H(Y,Z)\rangle$ for $Y, Z \in T_zG(x)$ where $T_zG(x)$ and N_z are tangent and normal spaces of $G(x)$ respectively at $z \in G(x)$. Let $c : R \to M$ be the geodesic defined by the initial conditions $c(0) = z, \dot{c}(0) = w$. If $X : R \to TM$ is Jacobi field defined along the geodesic c then X is said to be an $G(x)$-Jacobifield provided that the following conditions are satified:

(1) $\langle X(t), \dot{c}(t)\rangle = 0$ for $t \in R$;

(2) $X(0) \in T_zG(x)$;

(3) $A_wX(0) - D_wX(0) \in N_z$,

where D is the Riemannian connection. The set of $G(x)$-Jacobifields forms a vector space of dimension $n-1$.

The role of $G(x)$-Jacobifields in studying the focal points of $G(x)$ is due to the following fact: A point $t_0w \in N_z$ is a focal point of $G(x)$ if and only if there is a non-trivial $G(x)$-Jacobifield X along geodesic c such that $X(t_0) = 0$ holds. In this case $y = e(t_0w)$ is said to be focal point of $G(x)$ in M. Moreover, the dimension of the vector space formed by these $G(x)$-Jacobifields X defined along c for which $X(t_0) = 0$ holds is called the order of the focal points t_0w and $e(t_0w)$.

Further, let $G(x), G(y); x, y \in M$ be orbits of x and y respectively in Riemannian manifold M, and $C : [a,b] \to M$ be geodesic perpendicular to $G(x)$ at $c(a)$ and to $G(y)$ at $c(b)$ and U the vector space of broken C^∞-vector fields $V : [a,b] \to TM$ along c such that $V(a)$ is tangent to $G(x)$ and $V(b)$ is tangent to $G(y)$. Then the bilinear form

$$I(V,W) = -[\langle A_{\dot{c}(t)}V(t), W(t)\rangle]_a^b + \int_a^b (\langle V'(t), W'(t)\rangle - \langle R(V(t),\dot{c}(t))\dot{c}(t), W(t)\rangle)dt$$

is called the index form of c with respect to $G(x)$ and $G(y)$.

Proof of Theorem A. Let $U(G(x),c)$ and $J(G(x),c)$ denote the set of vector fields and set of $G(x)$-Jacobifields defined along geodesic c respectively. Let $X_1, \cdots, X_{n-1}$ be a base of $J(G(x),c)$, then there is a unique expression $V + \sum_{i=1}^{n-1} f_i X_i$ where f_i are broken C^∞-functions. At points where V' exists let

$$W = \sum_{i=1}^{n-1} f_i' X_i; \quad Z = \sum_{i=1}^{n-1} f_i X_i'.$$

Then $V' = W + Z$ and following are valid

$$\langle V, \sum_{j=1}^{n-1} f_i' X_i' \rangle = \sum_{i,j=1}^{n-1} f_i f_j' \langle X_i, X_j \rangle = \sum_{i,j=1}^{n-1} f_i f_j' \langle X_i', X_j \rangle = \langle W, Z \rangle.$$

Consequently the following holds as well

$$\begin{aligned}
\langle V, Z\rangle' &= \langle V', Z\rangle + \langle V, Z'\rangle \\
&= \langle W, Z\rangle + \langle Z, Z\rangle + \langle V, \sum_{i=1}^{n-1} f_i' X_i'\rangle + \langle V, \sum_{i=1}^{n-1} f_i X''_i\rangle \\
&= \langle W, Z\rangle + \langle Z, Z\rangle + \langle W, Z\rangle - \langle V, \sum_{i=1}^{n-1} f_i R(X_i, \dot{c})\dot{c}\rangle \\
&= 2\langle W, Z\rangle + \langle Z, Z\rangle - \langle V, R(V, \dot{c})\dot{c}\rangle.
\end{aligned}$$

Therefore the integrand in $I(V,V)$ is the following:

$$\langle V', V'\rangle - \langle R(V,\dot{c})\dot{c}, V\rangle = \langle W + Z, W + Z\rangle - \langle R(V,\dot{c})\dot{c}, V\rangle = \langle W, W\rangle + \langle V, Z\rangle'.$$

Hence the value of $I(V,V)$ can be expressed as follows

$$\begin{aligned}
I(V,V) &= -[\langle A_{\dot{c}(t)}V(t), V(t)\rangle]_0^a + [\langle V(t), Z(t)\rangle]_0^a + \int_0^a \langle W, W\rangle dt \\
&= \langle Z(a) - A_{\dot{c}(a)}V(a), V(a)\rangle + \int_0^a \langle W, W\rangle dt
\end{aligned}$$

in consequence of the fact that the following is valid

$$A_{\dot{c}(0)}V(0) - Z(0) = \sum_{i=1}^{n-1} f_i(0)(A_{\dot{c}}X_i(0) - X_i'(0)) - V(0).$$

Let now $X(t), t \in [0,a]$ be the $G(x)$-Jacobifield with $X(a) = V(a)$. Then $X(t) = \sum_{i=1}^{n-1} c_i X_i(t)$ where $c_i = f_i(a), i = 1, \cdots, n-1$. Put $\bar{W} = \sum_{i=1}^{n-1} c_i' X_i = 0; \bar{Z} = \sum_{i=1}^{n-1} c_i X_i'$, then an analogue of the above calculation yields that $I(X,X) = \langle Z(a) - A_{\dot{c}(a)}V(a), V(a)\rangle$. Consequently $I(V,V) - I(X,X) = \int_0^a \langle W, W\rangle dt \geq 0$ i.e., $I(V,V) \geq I(X,X)$, and equality occures if and only if $W \equiv 0$ which is equivalent to $f_i' \equiv 0, i = 1, \cdots, n-1$. But then $V = X$.

Proof of Theorem B. Let X be a non-trivial $G(x)$-Jacobifield along c with $X(b) = 0$ and consider $V \in U(G(x), c \restriction [a,d])$ given by $V \restriction [a,b] = X \restriction [a,b], V \restriction [b,d] = 0$. Then

$$\begin{aligned} I(V,V) = & -[\langle A_{\dot{c}(t)}V(t), V(t)\rangle]_a^b + \int_a^b (\langle V', V'\rangle - \langle R(V,\dot{c})\dot{c}, V\rangle)dt \\ & + \int_b^d (\langle V', V'\rangle - \langle R(V,\dot{c})\dot{c}, V\rangle)dt = 0. \end{aligned}$$

Choose an infinitesimal deformation such that b is not in the subdivision and W is a vector field along deformed geodesic, then

$$I(W,W) < I(V,V) = 0$$

i.e., $I(W,W) < 0$. But then by the second variation formula the assertion follows.

Proof of Theorem C. Since c_0 does not contain focal points, V_0 does not vanish. Therefore it is sufficient to show that $\frac{d}{dt}(\frac{\|V\|^2}{\|V_0\|^2}) \geq 0$ holds considering the fact that $\frac{\|V(0)\|}{\|V_0(0)\|} = 1$. In fact, the inequality is equivalent to the following one

$$\frac{\langle V(t), V(t)\rangle}{\langle V(t), V(t)\rangle} \geq \frac{\langle V_0'(t), V_0(t)\rangle}{\langle V_0(t), V_0(t)\rangle}.$$

Fix $t_0 \in [0,a)$ and consider the following vector fields:

$$W(t) = \frac{V(t)}{\|V(t_0)\|}, \quad W_0(t) = \frac{V_0(t)}{\|V_0(t_0)\|}.$$

Then following are valid for the vector field introduced:

$$\begin{aligned} \frac{\langle V(t), V(t)\rangle}{\langle V(t), V(t)\rangle} &= \frac{\langle W'(t), W(t)\rangle}{\langle W(t), W(t)\rangle}, \\ \frac{\langle V_0'(t), V_0(t)\rangle}{\langle V_0(t), V_0(t)\rangle} &= \frac{\langle W_0'(t), W_0(t)\rangle}{\langle W_0(t), W_0(t)\rangle}. \end{aligned}$$

Consequently the following is obtained by using the Theorem A:

$$\begin{aligned}
\frac{\langle V_0'(t), V_0(t)\rangle}{\langle V_0(t), V_0(t)\rangle} &= \langle W'(t_0), W(t_0)\rangle \\
&= \int_0^{t_0} \langle W'(t), W(t)\rangle' dt + \langle W'(0), W(0)\rangle \\
&= \langle W'(0), W(0)\rangle + \int_0^{t_0} (\langle W'(t), W'(t)\rangle - \langle R(W(t), \dot{c}(t))\dot{c}(t), W(t)\rangle) dt \\
&= \langle W'(0), W(0)\rangle + \int_0^{t_0} (\langle W'(t), W'(t)\rangle - K(P)\|W(t)\|^2) dt,
\end{aligned}$$

where P is the plane section spanned by $\dot{c}, W(t_0)\|W(t_0)\|^{-1}$.

Let P'_{-c} denote parallel translation along c with the opposite parameterization and P'_{c_0} parallel translation along c_0. Let $I : T_{c(0)}M \to T_{c_0(0)}M_0$ be an injection preserving inner product and define $I_t : T_{c(t)}M \to T_{c_0(t)}M_0$ by $I_t(X) = P'_{c_0} \cdot I \cdot P'_{-c}(X)$. Assume that I has been chosen so that $I_t(\dot{c}) = \dot{c}_0$ and $I_{t_0}(W(t_0)) = W_0(t_0)$. Define a field $\widehat{W}_0(t_0)$ by $\widehat{W}_{0t_0}(t) = I_t(W_{t_0}(t))$. Since $\widehat{W}_{0t_0}$ has the same expression in term of parallel frames as W_{t_0}, clearly

$$\langle W_{t_0}(t), W_{t_0}(t)\rangle = \langle \widehat{W}_{0t_0}(t), \widehat{W}_{0t_0}(t)\rangle,$$

$$\langle W'_{t_0}(t), W_{t_0}(t)\rangle = \langle \widehat{W}'_{0t_0}(t), \widehat{W}_{0t_0}(t)\rangle.$$

Using this and our assumption on the curvatures gives

$$\begin{aligned}
&\langle W(0), W(0)\rangle + \int_0^{t_0} (\langle W'(t), W'(t)\rangle - \langle R(W, \dot{c}(t))\dot{c}(t), W\rangle) dt \\
\geq\ &\langle \widehat{W}_0'(0), \widehat{W}_0(0)\rangle + \int_0^{t_0} (\langle \widehat{W}_0'(t), \widehat{W}_0'(t)\rangle - \langle R_0(\widehat{W}_0, \dot{c}(t))\dot{c}(t), \widehat{W}_0\rangle) dt \\
\geq\ &\langle W_0'(0), W_0(0)\rangle + \int_0^{t_0} (\langle W_0'(t), W_0'(t)\rangle - \langle R_0(W_0, \dot{c}(t))\dot{c}(t), W_0\rangle) dt \\
=\ &\langle W_0'(t_0), W_0(t_0)\rangle \\
=\ &\frac{\langle V_0'(t_0), V_0(t_0)\rangle}{\langle V_0(t_0), V_0(t_0)\rangle}.
\end{aligned}$$

Since t_0 is arbitrary, the assertion of the theorem follows.

References

[1] Bishop, R.L. and Crittendon, D.J., *Geometry of manifolds*, Academic Press, New York, 1964.

[2] Bredon, G.E., *Introduction to compact transformation groups*, Academic Press, New York, 1972.

[3] Cartan, É., *Groupes simples clos et ouverts et géométric riemonnionne*, J. Math. Pures Appl., 8(1929), 1–33.

[4] Esiang, W.Y., *Cohomology theory of topological transformation groups*, Springer-Verlag, New York, 1975.

[5] Kobayashi, S. and Nomizu, K., *Foundations of differential geometry, I, II*, John Wiely and Sons, New York, 1963–1969.

[6] Steenrod, N.E., *The topology of fibre bundles*, Princeton University Press, Princeton, N.J., 1951.

[7] Warner, F.W., *Fxtensions of the Rauch comparison theorem to submanifolds*, Tran. Amer. Math. Soc., 122(1966), 341–356.

Proceedings of the Asian Mathematical
Conference, 1990

Algebraic Fuzzy Systems

UPPASETTY MADANA SWAMY
Andhra University Post Graduate Centre
Etcherla-532 402
Srikakulam (Dt), Andhra Pradesh, India

A class S of subsets of a set X is called an algebraic closure set system in X if S contains intersections of arbitrary subclasses of S and unions of upward directed subclasses of S. The classes of (normal) subgroups of a group, Ideals of a ring,, subspaces of a Vector space, Ideals of a lattice, congruences of an universal algebra and equivalence relations on a set are all algebraic closure set systems. In fact, it is known that any algebraic closure set system can be identified with the class of all subuniverses of an universal algebra and Vice-Versa ([3]). Ever since Zadeh ([19]) proposed fuzzy set theory, the concepts of a fuzzy ideal of a ring ([8] and [16]), Fuzzy subgroup of a group ([1]), [9] and [10]), Fuzzy normal subgroup of a group ([17]), Fuzzy subspace of a vector space ([6]), fuzzy subsemigroup, fuzzy bi-ideals of a semigroup ([4]), fuzzy ideal of a lattice ([13]), Fuzzy equivalence relation ([12]) and so on were introduced and studied by several authors. Here, we have developed a general theory of fuzzy systems by introducing the notion of a fuzzy S-subset of a set X corresponding to an algebraic closure set system S in X. The fuzzy sets here assume truth values (as suggested by Goguen ([5])) in a complete Browerian lattice. This talk is a brief survey of the work done by us in our papers ([12] to [16]).

By a complete Browerian lattice ([2]) we mean a complete lattice (L, V, Λ) satisfying the infinite meet distributive law:

$$a \wedge (\sup M) = \sup\{a \wedge m \mid m \in M\} \quad \text{for all } a \in L \quad \text{and} \quad M \subseteq L.$$

For example, any complete chain and, in particular, the closed unit interval [0, 1] and its powers are complete Browerian lattices.

Let X be any nonempty set and L a complete Browerian lattice. We write $\leq, \vee$ and $\wedge$ for the partial order, supremums and infimums respectively in L. Let 0 and 1 be the elements in L such that $0 \leq a \leq 1$ for all $a \in L$. By an L-fuzzy subset (or simply fuzzy subset when there is no ambiguity about L) of X we mean a mapping of X into L. If $\{F_i\}$ is a class of fuzzy subsets of X, then we write $\vee_i F_i$ and $\wedge_i F_i$ for the point-wise supremum and infimum of $\{F_i\}$ respectively. Fuzzy subsets form a partially ordered set under point-wise order; i.e., $F \leq G$ if and only if $F(x) \leq G(x)$ for all $x \in X$.

Definition 1. Let S be a class of subsets of X. A fuzzy subset F of X is said to be a fuzzy S-subset of X if and only if, for each $a \in L$, the set $F_a := \{x \in X | a \leq F(x)\}$ is a member of S.

Let us write $\mathcal{F}S$ for the class of all fuzzy S-subsets of X. We call $\mathcal{F}S$ the fuzzy S-system on X. It can be easily seen that $\mathcal{F}S$ is nonempty if and only if $X \in S$.

Definition 2. A class $\mathcal{F}$ of fuzzy subsets of X is said to be a closure fuzzy set system if and only if $\mathcal{F}$ contains the (point-wise) infimums of arbitrary subclasses of $\mathcal{F}$. A closure system $\mathcal{F}$ is said to be algebraic if and only if it contains (point-wise) supremums of upward directed subclasses of $\mathcal{F}$.

Theorem 1. *For any class $\mathcal{S}$ of subsets of X, $\mathcal{FS}$ is a closure fuzzy set system if and only if $\mathcal{S}$ is a closure set system.*

If $\mathcal{S}$ is a closure set system in X, for any (fuzzy) subset F of X, let us write $\overline{F}$ for the smallest (fuzzy) $\mathcal{S}$-subset of X containing F; note that $\bar{F}$ is the (point-wise infimum) intersection of all (fuzzy) $\mathcal{S}$-subsets of X containing F. $\bar{F}$ is said to be generated by F.

Theorem 2. *The following are equivalent for a closure set system $\mathcal{S}$.*

(1) $\mathcal{S}$ is algebraic;

(2) $\mathcal{FS}$ is algebraic;

(3) For any fuzzy subset F of $X, F \in \mathcal{FS}$ if and only if $\bigwedge_{y \in A} F(y) \leq F(x)$ for all finite subsets A of X and $x \in \overline{A}$.

Theorem 3. *Let $\mathcal{S}$ be an algebraic closure set system and F a fuzzy subset of X. Then the fuzzy $\mathcal{S}$-subset $\overline{F}$ generated by F is given by*

$$\overline{F}(x) = \vee \left\{ \bigwedge_{y \in A} F(y) \mid A \subseteq X, A \text{ is finite and } x \in \overline{A} \right\}$$

for any $x \in X$.

In what follows, let $\mathcal{S}$ always stand for an algebraic closure set system in a set X and $\mathcal{FS}$ for the fuzzy $\mathcal{S}$-system in X.

Theorem 4. *$\mathcal{FS}$ is a complete lattice under point-wise order in which, for any $\{F_i \mid i \in I\}$, $\bigwedge_{i \in I} F_i$ is the infimum and the supremum is given by*

$$\left(\overline{\bigvee_{i \in I} F_i}\right)(x) = \vee \left\{ \bigwedge_{y \in A} (\bigvee_{i \in I} F_i(y)) \mid A \subseteq X, A \text{ is finite and } x \in \overline{A} \right\}$$

Clearly $\mathcal{S}$ is also a complete lattice in which, for any $\{A_i \mid i \in I\}, \cap_{i \in I} A_i$ and $\overline{\cup_{i \in I} A_i}$ are infimum and supremum respectively. Also, the correspondence $A \longmapsto \psi_A$ establishes an embedding of the complete lattice $\mathcal{S}$ into the complete lattice $\mathcal{FS}$, where ψ_A is the characteristic map with respect to A.

Theorem 5. *$\mathcal{FS}$ is a modular (distributive) lattice if and only if $\mathcal{S}$ is so.*

Whenever $\mathcal{FS}$ is a distributive lattice, by its algebraic nature, it satisfies the infinite meet distributive law and hence $\mathcal{S}$ is Browerian if and only if $\mathcal{FS}$ is Browerian. The following theorem provides a method for constructing fuzzy $\mathcal{S}$-subsets of X from certain subclasses of $\mathcal{S}$.

Theorem 6. *Let $\{S_a \mid a \in L\} \subseteq \mathcal{S}$ such that $S_{supM} = \cap_{a \in M} S_a$ for all $M \subseteq L$. Define, for any $x \in X$,*

$$F(x) = \vee\{a \in L \mid x \in S_a\}.$$

Then $F \in \mathcal{FS}$. Conversely, any fuzzy $\mathcal{S}$-subset of X can be uniquely obtained as above .

If we adjoin the empty set ϕ to S (if $\phi \notin S$), then the resulting class is also an algebraic closure set system; let us denote this by S_ϕ. In the following we describe all the members of $\mathcal{F}S$ for certain known classes S.

Examples.

(1) If S is the class of subgroups of a group G, then

$$\mathcal{F}S_\phi = \{F : G \to L \mid F(x) \wedge F(y) \leq F(xy^{-1}) \text{ for all } x, y \in G\}$$

and $\mathcal{F}S = \{F \in \mathcal{F}S_\phi \mid F(e) = 1\}$, where e is the identity in G. Also, if S_1 is the class of normal subgroups of G, then

$$\mathcal{F}S_1 = \{F \in \mathcal{F}S \mid F(y) \leq F(xy, x^{-1}) \text{ for all } x, y \in G\}.$$

The members of $\mathcal{F}S$ are called fuzzy subgroups of G ([1], [9] and [10]) and members of $\mathcal{F}S_1$ are called fuzzy normal subgroups of G ([17] and [18]) .

(2) If S is the class of ideals of a ring R, then

$$\mathcal{F}S_\phi = \{F : R \to L \mid F(x) \wedge F(y) \leq F(x - y) \text{ and } F(x) \vee F(y) \leq F(xy) \text{ for all } x, y \in R\}.$$

Members of this $\mathcal{F}S_\phi$ are called fuzzy ideals of R ([8] and [16]).

(3) If S is the class of subspaces of a Vector space V over a field R, then

$$\mathcal{F}S_\phi = \{F : V \to L \mid F(x) \wedge F(y) \leq F(sx + ty) \text{ for all } x, y \in V, s, t \in R\}.$$

Members of this $\mathcal{F}S_\phi$ are called fuzzy subspaces of V ([6]).

(4) If S is the class of ideals of a lattice D with zero then

$$\mathcal{F}S_\phi = \{F : D \to L \mid F(x \vee y) = F(x) \wedge F(y) \text{ for all } x, y \in D\}$$

whose members are called fuzzy ideals of D ([13]).

(5) If $\mathcal{F}$ is the class of equivalence relations on a set X, then

$$\begin{aligned} \mathcal{F}S = \{ \quad & F : X \times X \to L \mid F(x, x) = 1, F(x, y) = F(y, x) \text{ and} \\ & F(x, y) \wedge F(y, z) \leq F(x, z) \text{ for all } x, y, z \in X\}. \end{aligned}$$

Whose members are called fuzzy equivalences on X ([12]).

Other than the usual lattice theoretic operations on S, there may be some other binary operations; for instance, we have the multiplication of ideals of a ring. Now, we demonstrate a procedure to extend certain binary operations on S to $\mathcal{F}S$ and discuss the irreducibilities with respect to these operations.

Definition 3. Let $\circ$ be a binary operation on S. For any $F, G \in \mathcal{F}S$ define $F \bar{\circ} G : x \to L$ by

$$(F \bar{\circ} G)(x) = \vee \left\{ \bigwedge_{y \in A} F(y) \wedge \bigwedge_{z \in B} F(z) \mid A, B \in S \quad \text{and } x \in A \circ B \right\}$$

A binary operation $\circ$ is said to be inclusion invariant if $A \subseteq B$ implies $A \circ C \subseteq B \circ C$ and $C \circ A \subseteq C \circ B$ for all $A, B, C \in \mathcal{S}$.

Theorem 7. *Let o be an inclusion invariant binary operation on $\mathcal{S}$. Then the following hold.*

(1) *$\bar{o}$ is inclusion invariant and is an extension of o to $\mathcal{FS}$;*

(2) *$\bar{o}$ is commutative if and only if o is commutative;*

(3) *$\bar{o}$ is idempotent if and only if o is idempotent;*

(4) *whenever o distributes over unions of upward directed subclasses of $\mathcal{S}$, $\bar{o}$ is associative if and only if o is associative.*

It is interesting to note that the lattice operations $\vee$ and $\wedge$ in $\mathcal{FS}$ are precisely the extensions of those in $\mathcal{S}$. In the following, we assume that o is an inclusion invariant binary operation on $\mathcal{S}$ and $\bar{o}$ is its extension to $\mathcal{FS}$. We also assume that $A \circ B \subseteq A \cap B$ for all $A, B \in \mathcal{S}$. Let us recall that an element $a \in L$ is said to be (meet) irreducible if $a \neq 1$ and, for any $b, c \in L, b \wedge c \leq a$ implies $b \leq a$ or $c \leq a$.

Definition 4. $P \in \mathcal{S}$ is said to be irreducible if $P \neq \emptyset, P \neq X$ and, for any $A, B \in \mathcal{S}$, $AoB \subseteq P$ implies $A \subseteq P$ or $B \subseteq P$. Analogously, a non-constant fuzzy $\mathcal{S}$-subset F is said to be irreducible if, for any $G, H \in \mathcal{FS}, G\bar{o}H \leq F$ implies $G \leq F$ or $H \leq F$.

The following theorem characterizes the irreducible members of $\mathcal{FS}$ in terms of those in $\mathcal{S}$ and L.

Theorem 8. *Let F be a fuzzy subset of X. Then F is an irreducible member in $\mathcal{FS}$ if and only if there exist an irreducible member P in $\mathcal{S}$ and an irreducible element $a \in L$ such that, for any $x \in X, F(x) = 1$ or a depending on whether $x \in P$ or $x \notin P$ respectively.*

It follows from the above that $\mathcal{I}(\mathcal{FS})$ and $\mathcal{I}(\mathcal{S}) \times \mathcal{I}(L)$ are bijective, where $\mathcal{I}(\mathcal{FS}), \mathcal{I}(\mathcal{S})$ and $\mathcal{I}(L)$ denote the set of irreducible members in $\mathcal{FS}$, $\mathcal{S}$ and L respectively. If we topologise these sets with the hult-Kernel topologies ([11]), then the correspondence in the above theorem becomes a homeomorphism of $\mathcal{I}(\mathcal{FS})$ onto the product space $\mathcal{I}(\mathcal{S}) \times \mathcal{I}(L)$.

Let us call $F \in \mathcal{FS}(A \in \mathcal{S})$ maximal if it is maximal among nonconstant members of $\mathcal{FS}$ (respectively, among members of $\mathcal{S}$ other than X). Also, let us recall that maximal elements in $L - \{1\}$ are called dual atoms.

Theorem 9. *Let F be a fuzzy subset of X. Then F is maximal in $\mathcal{FS}$ if and only if there exist a maximal A in $\mathcal{S}$ and a dual atom a in L such that, for and $x \in X, F(x) = 1$ or a depending on whether $x \in A$ or $x \notin A$ respectively.*

An algebraic lattice ([3]) is known to be distributive if and only if every one of its elements is an infimum of irreducible elements. Even though an algebraic closure set system $\mathcal{S}$ is always an algebraic lattice ([3]), the algebraic closure fuzzy set system $\mathcal{FS}$ need not be, in general, an algebraic lattice. However, in certain cases, the distributivity of $\mathcal{S}$ (and hence of $\mathcal{FS}$) is a necessary and sufficient condition to express any element of $\mathcal{FS}$ as an infimum of irreducible elements.

Theorem 10. *Suppose every element of L is an infimum of irreducible elements. Then $\mathcal{FS}$ is distributive if and only if every member of $\mathcal{FS}$ is an infimum of irreducible members.*

References

[1] J.M. Anthony and H. Sherwood, *A Characterization of fuzzy subgroups*, Fuzzy Set and Systems, 7(1982), 297–305.

[2] G. Birkhoff, *Lattice Theory*, Amer. Math. Soc. Colloq. Publication, Vol. XXV, (AMS, Providence, RI, 1967).

[3] S. Burris & H.P. Sankappanavar, *A course in Universal Algebra*, Springer-Verlag, New York, 1981.

[4] T.K. Dutta, *On Fuzzy pseudo ideals in semigroups*, Proc. of the Seminar on Fuzzy systems and Non-standard Logic, Calcutta, (1984).

[5] J.A. Goguen, *L. Fuzzy sets*, J. Math. Anal. Appl., 18(1967), 145–174.

[6] A.K. Katsaras and D.B. Liu, *Fuzzy Vector spaces and Fuzzy topological Vector spaces*, J. Math. Anal. Appl., 58(1977), 135–146.

[7] N. Kuroki, *Fuzzy ideals and Fuzzy bi-ideals in Semigroups*, Fuzzy Sets and Systems, 5(1981), 203–215.

[8] T.K. Mukherjee and M.K. Sen, *On Fuzzy ideals of a ring (1)*, Proc. Seminar on Fuzzy Systems and Non-standard Logic, Calcutta, (1984).

[9] A. Rosenfield, *Fuzzy groups*, J. Math. Anal. Appl., 35(1971) , 512–517.

[10] P. Siva Rama Krishna Das, *Fuzzy groups and level subgroups*, J. of Math. Anal. Appl., 84(1981), 264–269.

[11] M.H. Stone, *Topological representation of distributive lattices and Browerian logics*, Casopis, Pest. Math. Fyz., 67(1937), 1–25.

[12] K.L.N. Swamy and U.M. Swamy, *Fuzzy equivalences*, Partitions and congruences, communicated.

[13] U.M. Swamy and D.V. Raju, *Fuzzy ideals of Lattices*, communicated.

[14] U.M. Swamy and D.V. Raju, *Algebraic Fuzzy systems*, to appear in Fuzzy Sets and Systems.

[15] U.M. Swamy and D.V. Raju, *Irreducibility in Algebraic fuzzy systems*, to appear in Fuzzy Sets and Systems.

[16] U,M, Swamy and K.L.N. Swamy, *Fuzzy prime ideals of rings*, J. of Math. Anal. Appl., 134(1988), 94–103.

[17] Wang Jin-Liu, *Fuzzy invarient subgroups and fuzzy ideals*, Fuzzy Sets and Systems, 8(1982), 133–139.

[18] Wang Ming, *Normal fuzzy subgroups*, Fuzzy Math., 1(1) (1981), 21–30.

[19] L.A. Zadeh, *Fuzzy sets*, Information and Control, 8(1965), 338–353.

Proceedings of the Asian Mathematical
Conference, 1990

On Some Error Bounds in Information Theory

CHOON PENG TAN
Department of Mathematics
University of Malaya
59100 Kuala Lumpur, Malaysia

§ 1. Introduction

Let X be a discrete random variable taking on n values $x_1, x_2, \cdots, x_n$ with probabilities $p_1, p_2, \cdots, p_n$ respectively. We consider the experimental determination of the entropy of X, $H(X) = -\sum_{i=1}^n p_i \log p_i$ based on N independent observations on the occurrence of the events $\{X = x_i\}$. If q_i is the relative frequency of the event $\{X = x_i\}$ for $i = 1, 2, \cdots, n$, then $H(Y) = -\sum_{i=1}^{n} q_i \log q_i$ is an estimate of $H(X)$ based on the relative frequencies. If $|p_i - q_i|$ is sufficiently small for all $i = 1, 2, \cdots, n$, we expect $H(Y)$ to be close to $H(X)$, i.e., we expect $H(Y)$ to be a good estimate of $H(X)$. Pfaffelhuber [4] determined an upper bound on $|H(X) - H(Y)|$ which depends on $\max \{|p_i - q_i| : i = 1, 2, \cdots, n\}$ and applied the result to determine the smallest number N of independent observations required such that $P[|H(X) - H(Y)| \leq \delta] \geq 1 - \gamma$, where $\delta > 0$ and $0 < \gamma < 1$ are given numbers. Csiszár and Hörner [2] derived another bound on $|H(X) - H(Y)|$ which depends on $\sum_{i=1}^n |p_i - q_i|$ and this bound improves upon the Pfaffelhuber bound. In this paper, we shall derive upper bounds on $|H(X) - H(Y)|$ which depend on the ℓ_r norm of $\underset{\sim}{p} - \underset{\sim}{q}$ where $\underset{\sim}{p} = (p_i), \underset{\sim}{q} = (q_i)$ are probability distributions and $1 \leq r \leq \infty$. These bounds improve upon the Pfaffelhuber and Csiszár -Körner bounds whenever $|p_i - q_i|$ is sufficiently small for $i = 1, 2, \cdots, n$ and a positive lower bound on $\min \{p_1, \cdots, p_n, q_1, \cdots, q_n\}$ is available. Extension of these bounds to the divergence of two probability distributions will be discussed.

§ 2. Entropy Error Bounds

If X is a random variable taking on n values $x_1, x_2, \cdots, x_n$ with probabilities $p_1, p_2, \cdots, p_n$ respectively, we denote the entropy of X by

$$H(X) = H(p) \triangleq -\sum_{i=1}^{n} p_i \log p_i, \tag{1}$$

where $\underset{\sim}{p} = (p_i)$ is a probability distribution. The set of relative frequencies q_i of the events $\{X = x_i\}$ for $i = 1, 2, \cdots, n$ form a probability distribution and we denote the corresponding entropy by

$$H(Y) = H(\underset{\sim}{q}) \triangleq -\sum_{i=1}^{n} q_i \log q_i. \tag{2}$$

For any vector $\xi = (\xi_i) \in R^n$, we define the ℓ_r norm of ξ as

$$\|\xi\|_r = \left(\sum_{i=1}^{n} |\xi_i|^r\right)^{1/r} \quad \text{for} \quad 1 \le r < \infty \tag{3}$$

and

$$\|\xi\|_\infty = \max\{|\xi_i| : i = 1, 2, \cdots, n\}. \tag{4}$$

Suppose we estimate the entropy $H(\underset{\sim}{p})$ by $H(\underset{\sim}{q})$. Upper bounds on the error $\left|H(\underset{\sim}{p}) - H(\underset{\sim}{q})\right|$ are given by :

Proposition 1. *Let* $\underset{\sim}{p} = (p_i) > \underset{\sim}{0}$ *and* $\underset{\sim}{q} = (q_i) > \underset{\sim}{0}$ *be two probability distributions.* Then

(i) $$\left|H(\underset{\sim}{p}) - H(\underset{\sim}{q})\right| \le \max\left\{\left|\sum_{i=1}^{n}(q_i - p_i)\log q_i\right|, \left|\sum_{i=1}^{n}(p_i - q_i)\log p_i\right|\right\},$$

(ii) $$\left|H(\underset{\sim}{p}) - H(\underset{\sim}{q})\right| \le \left\|\underset{\sim}{p} - \underset{\sim}{q}\right\|_r \max \left\{\min\left\{\left\|\underset{\sim 1}{\beta} - c_1\underset{\sim}{1}\right\|_s : c_1 \in R\right\}, \min\left\{\left\|\underset{\sim 2}{\beta} - c_2\underset{\sim}{1}\right\|_s : c_2 \in R\right\}\right\}$$

where $\underset{\sim 1}{\beta} = (\log p_i), \underset{\sim 2}{\beta} = (\log q_i), r^{-1} + s^{-1} = 1, 1 \le r, s \le \infty$ and $\underset{\sim}{1}$ is the vector consisting of all 1's .

Proof. (i) By the divergence inequality for two probability distributions (see for example, [1], Lemma 1.4.1), $\sum_{i=1}^{n} p_i \log p_i \ge \sum_{i=1}^{n} p_i \log q_i$ and hence

$$H(\underset{\sim}{p}) - H(\underset{\sim}{q}) = \sum_{i=1}^{n} q_i \log q_i - \sum_{i=1}^{n} p_i \log p_i \le \sum_{i=1}^{n}(q_i - p_i)\log q_i. \tag{5}$$

Similarly,

$$H(\underset{\sim}{q}) - H(\underset{\sim}{p}) \le \sum_{i=1}^{n}(p_i - q_i)\log p_i. \tag{6}$$

(5) and (6) imply that

$$\left|H(\underset{\sim}{p}) - H(\underset{\sim}{q})\right| \le \max\left\{\left|\sum_{i=1}^{n}(q_i - p_i)\log q_i\right|, \left|\sum_{i=1}^{n}(p_i - q_i)\log p_i\right|\right\}. \tag{7}$$

(ii) Let $\underset{\sim}{\beta} \in R^n$ be any given vector. Then $\max\{\left|\underset{\sim}{\alpha}'\underset{\sim}{\beta}\right| / \left\|\underset{\sim}{\alpha}\right\|_r : \underset{\sim}{\alpha} \in R^n, \underset{\sim}{\alpha}'\underset{\sim}{1} = 0\} = \min\{\left\|\underset{\sim}{\beta} - c\underset{\sim}{1}\right\|_s : c \in R\}$ by the well-known result [3], p.119, where $r^{-1} + s^{-1} = 1$ and $1 \le r, s \le \infty$.

Now let $\underset{\sim}{\alpha} = \underset{\sim}{p} - \underset{\sim}{q}$. Then $\underset{\sim}{\alpha}' \cdot \underset{\sim}{1} = 0$ and hence

$$\left|\underset{\sim}{\alpha}'\underset{\sim}{\beta}_1\right| = \left|\sum_{i=1}^{n}(p_i - q_i)\log p_i\right| \leq \left\|\underset{\sim}{\alpha}\right\|_r \min\{\left\|\underset{\sim}{\beta}_1 - c_1\underset{\sim}{1}\right\|_s : c_1 \in R\}. \tag{8}$$

Similarly,

$$\left|\underset{\sim}{\alpha}'\underset{\sim}{\beta}_2\right| \leq \left\|\underset{\sim}{\alpha}\right\|_r \min\{\left\|\underset{\sim}{\beta}_2 - c_2\underset{\sim}{1}\right\|_s : c_2 \in R\}. \tag{9}$$

The desired result follows from (7) , (8) and (9) .

For the next result, we use the following notation $\beta_{\max}, \beta_{\min}$ and β_{med} to denote the maximum, minimum and median respectively of the n components of the vector $\underset{\sim}{\beta} = (\beta_i) \in R^n$, namely:

$$\beta_{\max} = \max\{\beta_i : i = 1, 2, \cdots, n\},$$

$$\beta_{\min} = \min\{\beta_i : i = 1, 2, \cdots, n\} \quad \text{and}$$

β_{med} is defined to be the middle value of the n components of $\underset{\sim}{\beta}$ if n is odd and either one of the 2 middle values if n is even. We have the following explicit entropy error bounds for Proposition 1 when $r = 1, 2, \infty$:

Proposition 2. *Let $\underset{\sim}{p} = (p_i) > \underset{\sim}{0}$ and $\underset{\sim}{q} = (q_i) > \underset{\sim}{0}$ be two probability distributions.* Then

(i) $$\left|H(\underset{\sim}{p}) - H(\underset{\sim}{q})\right| \leq \frac{1}{2}\left\|\underset{\sim}{p} - \underset{\sim}{q}\right\|_1 \max\left\{\log\frac{p_{\max}}{p_{\min}}, \log\frac{q_{\max}}{q_{\min}}\right\},$$

(ii) $$\left|H(\underset{\sim}{p}) - H(\underset{\sim}{q})\right| \leq \left\|\underset{\sim}{p} - \underset{\sim}{q}\right\|_2 \max\left\{\left[\sum_{i=1}^{n}\left(\log\frac{p_i}{p_{gm}}\right)^2\right]^{\frac{1}{2}}, \left[\sum_{i=1}^{n}\left(\log\frac{q_i}{q_{gm}}\right)^2\right]^{\frac{1}{2}}\right\},$$

where $p_{gm} = (p_1p_2\cdots p_n)^{1/n}$ and $q_{gm} = (q_1q_2\cdots q_n)^{1/n}$,

(iii) $$\left|H(\underset{\sim}{p}) - H(\underset{\sim}{q})\right| \leq \left\|\underset{\sim}{p} - \underset{\sim}{q}\right\|_\infty \max\left\{\sum_{i=1}^{n}\left|\log\frac{p_i}{p_{\text{med}}}\right|, \sum_{i=1}^{n}\left|\log\frac{q_i}{q_{\text{med}}}\right|\right\}.$$

Proof. The proof follows by invoking the following result: for any vector $\underset{\sim}{\alpha} \in R^n$ such that $\underset{\sim}{\alpha}'\underset{\sim}{1} = 0$,

(i) $$\left|\underset{\sim}{\alpha}'\underset{\sim}{\beta}\right| \leq \left\|\underset{\sim}{\alpha}\right\|_1 (\beta_{\max} - \beta_{\min})/2,$$

(ii) $$\left|\underset{\sim}{\alpha}'\underset{\sim}{\beta}\right| \leq \left\|\underset{\sim}{\alpha}\right\|_2 \left[\sum_{i=1}^{n}(\beta_i - \bar{\beta})^2\right]^{1/2}, \quad \text{where} \quad \bar{\beta} = \sum_{i=1}^{n}\beta_i/n,$$

(iii) $$\left|\underset{\sim}{\alpha}'\underset{\sim}{\beta}\right| \leq \left\|\underset{\sim}{\alpha}\right\|_\infty \sum_{i=1}^{n} |\beta_i - \beta_{\text{med}}|.$$

Since $p_{\min} \leq p_{gm} \leq p_{\max}$ and $q_{\min} \leq q_{gm} \leq q_{\max}$, we may replace (ii) in Proposition 2 by the following:

Corollary 1.

$$\left|H(\underset{\sim}{p}) - H(\underset{\sim}{q})\right| \leq \left\|\underset{\sim}{p} - \underset{\sim}{q}\right\|_{2n^{\frac{1}{2}}} \max\left\{\log \frac{p_{\max}}{p_{\min}}, \log \frac{q_{\max}}{q_{\min}}\right\}.$$

Proposition 3. *Let* $0 < a \leq \min\{p_1, \cdots, p_n, q_1, \cdots, q_n\}$ *and* $\left\|\underset{\sim}{p} - \underset{\sim}{q}\right\|_r \leq \varepsilon_r$ *for* $r = 1, 2, \infty$. *Then*

(i) $$\left|H(\underset{\sim}{p}) - H(\underset{\sim}{q})\right| < 2^{-1}\varepsilon_1 \log a^{-1},$$

(ii) $$\left|H(\underset{\sim}{p}) - H(\underset{\sim}{q})\right| < n^{1/2}\varepsilon_2 \log a^{-1},$$

(iii) $$\left|H(\underset{\sim}{p}) - H(\underset{\sim}{q})\right| \leq n\varepsilon_\infty \log(na)^{-1},$$

where $\underset{\sim}{p} = (p_i)$ *and* $\underset{\sim}{q} = (q_i)$ *are two probability distributions.*

Proof. (i), (ii) Exploiting the fact that $(p_{\max}/p_{\min}) < a^{-1}$ and $(q_{\max}/q_{\min}) < a^{-1}$, we obtain the desired result from Proposition 2 (i) and Corollary 1. (iii) Note that $\sum_{i=1}^{n} |\beta_i - \beta_{\text{med}}| = \min\{\sum_{i=1}^{n} |\beta_i - c| : c \in R\}$ (see for example [5] , Proposition 2) . Hence $\sum_{i=1}^{n} \left|\log \frac{p_i}{p_{\text{med}}}\right| = \min\{\sum_{i=1}^{n} |\log \frac{p_i}{c}| : 0 < c < \infty\} \leq \sum_{i=1}^{n} |\log \frac{p_i}{a}|$. By the inequality of the arithmetic and geometric means [1], p.25, $(p_1p_2\cdots p_n)^{1/n} \leq (1/n)$ and therefore $\sum_{i=1}^{n} \left|\log \frac{p_i}{p_{\text{med}}}\right| \leq n \log\left[(p_1p_2\cdots p_n)^{1/n}/a\right] \leq n\log(na)^{-1}$.

Similarly, $\sum_{i=1}^{n} \left|\log \frac{q_i}{q_{\text{med}}}\right| \leq n\log(na)^{-1}$ and the desired result follows from Proposition 2 (iii).

Suppose that $0 < \left\|\underset{\sim}{p} - \underset{\sim}{q}\right\|_r \leq \varepsilon_r \leq e^{-1}$ for $r = 1, \infty$. Then the Pfaffelhuber bound [4], p.51 states that

$$\left|H(\underset{\sim}{p}) - H(\underset{\sim}{q})\right| \leq n\left\|\underset{\sim}{p} - \underset{\sim}{q}\right\|_\infty \log \left\|\underset{\sim}{p} - \underset{\sim}{q}\right\|_\infty^{-1} \leq n\varepsilon_\infty \log \varepsilon_\infty^{-1}$$

and the Csiszár–Körner bound [2], Lemma 2.7 states that

$$\left|H(\underset{\sim}{p}) - H(\underset{\sim}{q})\right| \leq \left\|\underset{\sim}{p} - \underset{\sim}{q}\right\|_1 \log(n\left\|\underset{\sim}{p} - \underset{\sim}{q}\right\|_1^{-1}) \leq \varepsilon_1 \log(n\varepsilon_1^{-1}).$$

We observe that the bound in Proposition 3 (i) improves upon the Csiszár–Körner bound if and only if $\varepsilon_1 < na^{1/2}$. A similar result holds for the bound in Proposition 3 (iii) compared with the Pfaffelhuber bound if and only if $\varepsilon_\infty < na$. Thus , whenever $\left\|\underset{\sim}{p} - \underset{\sim}{q}\right\|_r$ is sufficiently small, sharper bounds on the error $\left|H(\underset{\sim}{p}) - H(\underset{\sim}{q})\right|$ are provided by Proposition 3.

§ 3.Divergence Error Bounds

The divergence of two probability distributions $\underset{\sim}{p} = (p_i)$ and $\underset{\sim}{q} = (q_i)$ is defined as $D(\underset{\sim}{p}, \underset{\sim}{q}) \triangleq \sum_{i=1}^n p_i \log(p_i/q_i)$. Divergence is also known as the Kullback–Leibler information number. If $\underset{\sim}{u}$ is an estimate of $\underset{\sim}{p}$ and $\underset{\sim}{v}$ is an estimate of $\underset{\sim}{q}$, we wish to derive some upper bounds on the divergence error $\left|D(\underset{\sim}{p}, \underset{\sim}{q}) - D(\underset{\sim}{u}, \underset{\sim}{v})\right|$.

Proposition 4. *Let $\underset{\sim}{p}, \underset{\sim}{q}, \underset{\sim}{u}$ and $\underset{\sim}{v}$ be four probability distributions where all the components of the vectors are strictly positive. Then*

$$\begin{aligned}\left|D(\underset{\sim}{p}, \underset{\sim}{q}) - D(\underset{\sim}{u}, \underset{\sim}{v})\right| \le \left\|\underset{\sim}{p} - \underset{\sim}{u}\right\|_r \max\ &\left\{\min\left\{\left\|\underset{\sim 1}{\beta} - c_1\underset{\sim}{1}\right\|_s : c_1 \in R\right\},\right.\\ &\min\left\{\left\|\underset{\sim 2}{\beta} - c_2\underset{\sim}{1}\right\|_s : c_2 \in R\right\},\\ &\left.\min\left\{\left\|\{\underset{\sim 3}{\beta} - c_3\underset{\sim}{1}\right\|_s : c_3 \in R\right\}\right\}\\ + \log\left[1 + \left\|\underset{\sim}{v} - \underset{\sim}{q}\right\|_\infty\right.\ &\left.\max\{v_1^{-1}, \cdots, v_n^{-1}, q_1^{-1}, \cdots, q_n^{-1}\}\right]\end{aligned}$$

where $\underset{\sim 1}{\beta} = (\log p_i), \underset{\sim 2}{\beta} = (\log u_i), \underset{\sim 3}{\beta} = (\log v_i), r^{-1} + s^{-1} = 1$ *and* $1 \le r, s \le \infty$.

Proof.

$$\begin{aligned}\left|D(\underset{\sim}{p}, \underset{\sim}{q}) - D(\underset{\sim}{u}, \underset{\sim}{v})\right| &= \left|\sum_{i=1}^n p_i \log(p_i/q_i) - \sum_{i=1}^n u_i \log(u_i/v_i)\right|\\ \le\ &\left|\sum_{i=1}^n p_i \log p_i - \sum_{i=1}^n u_i \log u_i\right| + \left|\sum_{i=1}^n (u_i - p_i)\log v_i\right| + \left|\sum_{i=1}^n p_i \log(v_i/q_i)\right|\\ \le\ &\left\|\underset{\sim}{p} - \underset{\sim}{u}\right\|_r \max\left\{\min\left\{\left\|\underset{\sim 1}{\beta} - c_1\underset{\sim}{1}\right\|_s : c_1 \in R\right\}, \min\left\{\left\|\underset{\sim 2}{\beta} - c_2\underset{\sim}{1}\right\|_s : c_2 \in R\right\}\right\}\\ &+ \left\|\underset{\sim}{u} - \underset{\sim}{p}\right\|_r \min\left\{\left\|\underset{\sim 3}{\beta} - c_3\underset{\sim}{1}\right\|_s : c_3 \in R\right\} + \max\{|\log(v_i/q_i)| : i = 1, \cdots, n\}\end{aligned}$$

by Proposition 1. Using the elementary inequality:

$$|\log(v_i/q_i)| \le \log\left[1 + \left\|\underset{\sim}{v} - \underset{\sim}{q}\right\|_\infty \max\{v_i^{-1}, q_i^{-1}\}\right] \quad \text{for} \quad i = 1, \cdots, n,$$

we have the desired result.

Remark. By interchanging $\underset{\sim}{p}$ and $\underset{\sim}{u}$ in Proposition 4 as well as $\underset{\sim}{q}$ and $\underset{\sim}{v}$, it may be possible to obtain a smaller upper bound.

Explicit upper bounds in terms of the entries of $\underset{\sim}{p}, \underset{\sim}{q}, \underset{\sim}{u}, \underset{\sim}{v}$ are given by the next proposition for $r = 1, 2, \infty$.

Proposition 5. *Let $\underset{\sim}{p}, \underset{\sim}{q}, \underset{\sim}{u}$ and $\underset{\sim}{v}$ be four probability distributions where all the components of the vectors are strictly positive. Then*

$$\text{(i)}\quad \begin{aligned} &\left|D(\underset{\sim}{p},\underset{\sim}{q}) - D(\underset{\sim}{u},\underset{\sim}{v})\right| \\ &\le \tfrac{1}{2}\left\|\underset{\sim}{p} - \underset{\sim}{u}\right\|_1 \max\{\log \tfrac{p_{\max}}{p_{\min}}, \log \tfrac{u_{\max}}{u_{\min}}, \log \tfrac{v_{\max}}{v_{\min}}\} \\ &\quad + \log\left[1 + \left\|\underset{\sim}{v} - \underset{\sim}{q}\right\|_\infty \max\{v_1^{-1},\cdots,v_n^{-1},q_1^{-1},\cdots,q_n^{-1}\}\right], \end{aligned}$$

$$\text{(ii)}\quad \begin{aligned} &\left|D(\underset{\sim}{p},\underset{\sim}{q}) - D(\underset{\sim}{u},\underset{\sim}{v})\right| \\ &\le \left\|\underset{\sim}{p} - \underset{\sim}{u}\right\|_2 \max\left\{\left[\sum_{i=1}^n \left(\log \frac{p_i}{p_{gm}}\right)^2\right]^{\frac{1}{2}}, \left[\textstyle\sum_{i=1}^n \left(\log \frac{u_i}{u_{gm}}\right)^2\right]^{\frac{1}{2}},\right. \\ &\quad \left.\left[\textstyle\sum_{i=1}^n \left(\log \frac{v_i}{v_{gm}}\right)^2\right]^{\frac{1}{2}}\right\} \\ &\quad + \log\left[1 + \left\|\underset{\sim}{v} - \underset{\sim}{q}\right\|_\infty \max\{v_1^{-1},\cdots,v_n^{-1},q_1^{-1},\cdots,q_n^{-1}\}\right], \end{aligned}$$

where $p_{gm} = (p_1p_2\cdots p_n)^{1/n}, u_{gm} = (u_1u_2\cdots u_n)^{1/n}$ *and* $v_{gm} = (v_1v_2\cdots v_n)^{1/n}$,

$$\text{(iii)}\quad \begin{aligned} &\left|D(\underset{\sim}{p},\underset{\sim}{q}) - D(\underset{\sim}{u},\underset{\sim}{v})\right| \\ &\le \left\|\underset{\sim}{p} - \underset{\sim}{u}\right\|_\infty \max\left\{\textstyle\sum_{i=1}^n \left|\log \frac{p_i}{p_{\text{med}}}\right|, \sum_{i=1}^n \left|\log \frac{u_i}{u_{\text{med}}}\right|, \sum_{i=1}^n \left|\log \frac{v_i}{v_{\text{med}}}\right|\right\} \\ &\quad + \log\left[1 + \left\|\underset{\sim}{v} - \underset{\sim}{q}\right\|_\infty \max\{v_1^{-1},\cdots,v_n^{-1},q_1^{-1},\cdots,q_n^{-1}\}\right]. \end{aligned}$$

A similar generalization of Proposition 3 to the case of divergence error bounds is straightforward and we shall omit the details. Let X and Y be two random variables having the probability distributions $\{p(x)\}$ and $\{p(y)\}$ respectively and joint distribution $\{p(x,y)\}$. The average mutual information between X and Y is defined as :

$$I(X;Y) \triangleq \sum_x \sum_y p(x,y) \log \frac{p(x,y)}{p(x)p(y)}.$$

Error bounds for $I(X;Y)$ are given by Propositions 4 and 5 by regarding $\{p(x,y)\}$ as the distribution vector $\underset{\sim}{p}$ and $\{p(x)p(y)\}$ as the vector $\underset{\sim}{q}$.

References

[1] R.Ash, *Information theory,* Wiley Interscience, New York, 1965.

[2] I.Csiszár and J. Körner, *Information theory: coding theorems for discrete memoryless systems,* Academic, New York, 1981.

[3] D.G. Luenberger, *Optimization by vector space methods,* Wiley New York, 1969.

[4] E.Pfaffelhuber, *Error estimation for the determination of entropy and information rate from relative frequencies,* Kybernetik, 8 (1971), 50–51.

[5] C.P.Tan, *A functional form for a particular coefficient of ergodicity,* J. Appl. Prob., 19 (1982), 858–863.

Proceedings of the Asian Mathematical
Conference, 1990

A Unified Numerical Approximation for Linear Differential Equations

S. TANGMANEE
Department of Mathematics
Faculty of Science Kasetsart University
Bangkok 10900, Thailand

§1. Continuous Problem

Let Ω be an open bounded domain of R^n, with a smooth boundary $\partial\Omega$ (at least continuous, and piecewise continuously differentiable). The closure of Ω will be denoted by $\overline{\Omega}$. Symmetric positive differential equations (K.O. Friedrichs [3]) are defined as follows.

We let the $A_i's$, $1 \leq i \leq n$, be symmetric matrics of $L(R^p)$, Lipschitz continuous in x for $x \in \Omega$, and we let $K(x)$ be a matrix of $L(R^p)$ whose coefficients are bounded in Ω.

We want to find a function $u : x \in \Omega, \quad u(x) \in R^p$ such that

$$\sum_{i=1}^{n} A_i(x)\frac{\partial u}{\partial x_i} + K(x)u = f \quad \text{for} \quad x \in \Omega, \tag{1.1}$$

$$(B - M)(x)u = 0 \quad \text{for} \quad x \in \Omega, \tag{1.2}$$

where $B = \sum_{i=1}^{n} v_i A_i$, the $v_i's$ being the components of the outer normal on $\partial\Omega$.

The matrix $M(x)$ belongs to $L(R^p)$ and is defined for $x \in \partial\Omega$. We let M^* denote the adjoint of M and assume that

$$\begin{aligned} &\text{(i)} \qquad M(x) \quad \text{is continuous in} \quad x \quad \text{for} \quad x \in \partial\Omega, \\ &\text{(ii)} \qquad M(x) + M^*(x) \geq 0 \quad \text{for all} \quad x \in \partial\Omega, \\ &\text{(iii)} \quad \text{Ker}(B - M)(x) + \text{Ker}(B + M)(x) = R^p \quad \text{for all} \quad x \in \partial\Omega. \end{aligned} \tag{1.3}$$

Let $A \in L(R^p) : A \geq 0$ simple means $(Au, u) \geq 0, \forall u \in R^p$.

Remark 1.1. Hypotheses (1.3) on the matrix $M(x)$ imply that boundary condition (1.2) is admissible in the sense of [3] and we have the following results.

Lemma 1.1. *Assume that B is positive (resp. negative) semi definite, the only matrix M satisfying hypotheses (1.3) is given by*

$$M = B \quad (\text{resp.} \quad M = -B), \tag{1.4}$$

We let the operator A be defined by

$$Au(x) = \sum_{i=1}^{n} A_i(x)\frac{\partial u}{\partial x_i}(x) + K(x)u(x) \quad \text{for} \quad x \in \Omega. \tag{1.5}$$

We say that the operator A is positive if there exists a constant $c_0 > 0$ such that

$$c - c_0 I = K + K^* - \sum_{i=1}^{n} \frac{\partial A_i}{\partial x_i} - c_0 I \geq 0 \quad \text{for all} \quad x \in \Omega, \tag{1.6}$$

where I denotes the matrix identity.

The adjoint operator A^* of A is defined by

$$A^*u(x) = -\sum_{i=1}^{n} \frac{\partial (A_i(x)u)}{\partial x_i} + K^*(x)u(x) \quad \text{for} \quad x \in \Omega. \tag{1.7}$$

We say that v satisfies the adjoint boundary conditions if

$$(B + M^*)v = 0 \quad \text{for} \quad x \in \partial\Omega. \tag{1.8}$$

We have the following Green formulas.

Lemma 1.2. *For all $u, v \in H^1(\Omega)$, we have*

$$(Au, v)_{L^2(\Omega)} = (u, A^*v)_{L^2(\Omega)} + \int_{\partial\Omega} (Bu, v)\, ds. \tag{1.9}$$

Let (,) denote the Euclidean inner product in R^p, we let

$$(f, g)_{L^2(\Omega)} = \int_{\Omega} (f(x), g(x))\, dx.$$

Lemma 1.3. *For all $u \in H^1(\Omega)$, we have*

$$2(Au, u)_{L^2(\Omega)} = (cu, u)_{L^2(\Omega)} + \int_{\partial\Omega} (Bu, u)\, ds. \tag{1.10}$$

Lemma 1.4. *For all u (resp. v) $\in H^2(\Omega)$ and satisfying boundary conditions (1.2) (resp. (1.8)) we have*

$$(Au, v)_{L^2(\Omega)} = (u, A^*v)_{L^2(\Omega)}.$$

Proof. This Lemma is a direct consequence of equality (1.9).

We just have to consider the boundary term:

$$\int_{\partial\Omega} (Bu, v)\, ds = \int_{\partial\Omega} \left(\frac{B - M}{2}u, v\right) ds + \int_{\partial\Omega} \left(u, \frac{B + M^*}{2}v\right) ds.$$

Definition 1.1. We say that $u \in L^2(\Omega)$ is a weak solution of problem (1.1), (1.2) if for all $v \in H^1(\Omega)$ and satisfying boundary conditions (1.8) we have

$$(u, A^*v)_{L^2(\Omega)} = (f, v)_{L^2(\Omega)}.$$

We have the following result. For more details see [4].

Theorem 1.1. *Assume that $f \in L^2(\Omega)$, and that hypotheses (1.3), (1.6) are satisfied. Then problem (1.1), (1.2) has a weak solution $u \in L^2(\Omega)$.*

Remark 1.2. Uniqueness is insured if we look for a solution $u \in H^1(\Omega)$. In that case, we have

$$Au = f \quad \text{in} \quad \Omega, \tag{1.11}$$

$$(B - M)u = 0 \quad \text{on} \quad \Omega. \tag{1.12}$$

Multiplying equation (1.11) by u and using Lemma 1.3, we get

$$(f, u)_{L^2(\Omega)} = \frac{1}{2}(cu, u)_{L^2(\Omega)} + \frac{1}{2}\int_{\partial\Omega} (Bu, u)\, ds. \tag{1.13}$$

Then, we have

$$\int_{\partial\Omega}(Bu, u)\, ds = \int_{\partial\Omega}((B - M)u, u)\, ds + \int_{\partial\Omega}(Mu, u) ds,$$

$$\int_{\partial\Omega}(Bu, u)\, ds = \tfrac{1}{2}\int_{\partial\Omega}((M + M^*)u, u)\, ds \geq 0.$$

Using the last inequality and the positivity of the operator A, we get

$$|u|^2_{L^2(\Omega)} \leq \frac{2}{c_0}(f, u)_{L^2(\Omega)},$$

and

$$|u|_{L^2(\Omega)} \leq \frac{2}{c_0}|f|_{L^2(\Omega)}. \tag{1.14}$$

This last inequality implies the uniqueness of u.

Definition 1.2. We say that $u \in L^2(\Omega)$ is a strong solution of problem (1.1), (1.2) if there exists a sequence $\{u_j\}$ such that

$$u_j \in H^1(\Omega), \quad (B - M)u_j = 0, \tag{1.15}$$

$$\lim_{j \to \infty} (\|u - u_j\|_{L^2(\Omega)} + \|Au_j - f\|_{L^2(\Omega)}) = 0. \tag{1.16}$$

The following result holds ([3]).

Theorem 1.2 *Assume that the boundary $\partial\Omega$ is twice continuously differentiable, that hypotheses (1.3) and (1.6) are satisfied and that the subspace $Ker(B(x) - M(x))$ of R^p is generated by vectors which are continuous functions of x for $x \in \partial\Omega$. Then for all $f \in L^2(\Omega)$, Problem (1.1), (1.2) has a unique strong solution.*

In the sequel we shall assume that the problem (1.1), (1.2) has a unique solution $u \in H^1(\Omega)$. We have shown in [6] that classical ordinary and partial differential equations can be transformed into a unified form (symmetric positive differential equation with an admissible boundary condition). Particular examples we want to give here are the Tricomi equation and the equation of seismic wave.

Tricomi equation is a model problem of transonic flow and is applicable to aeronautic engineering problems. A unified numerical method for these problems was constructed. The error estimates were also studied.

We consider the Tricomi equation

$$y\frac{\partial^2\Phi}{\partial x^2} - \frac{\partial^2\Phi}{\partial y^2} = f \quad \text{in} \quad \Omega \tag{1.17}$$

with the boundary conditions:

$$\begin{aligned} n_2\frac{\partial\Phi}{\partial x} - n_1\frac{\partial\Phi}{\partial y} &= 0 \qquad \text{on} \quad \Gamma_0, \\ \frac{\partial\Phi}{\partial x} &= a\frac{\partial\Phi}{\partial y} \quad \text{on} \quad \Gamma_1, \\ \frac{\partial\Phi}{\partial y} &= b\frac{\partial\Phi}{\partial x} \quad \text{on} \quad \Gamma_2. \end{aligned} \tag{1.18}$$

Where Ω is the bounded domain in the xy plane bounded by the boundary Γ (see Fig. 1), a and b are constants satisfying to Γ. $|a| \leq \sqrt{3}/2$ and $|b| \leq \sqrt{3}/2, n = (n_1, n_2)$ is the exterior normal.

Let

$$\Omega^- = \{(x, y) \in \Omega, y \leq 0\},$$

$$\Omega^+ = \{(x, y) \in \Omega, 0 < y < 1\}.$$

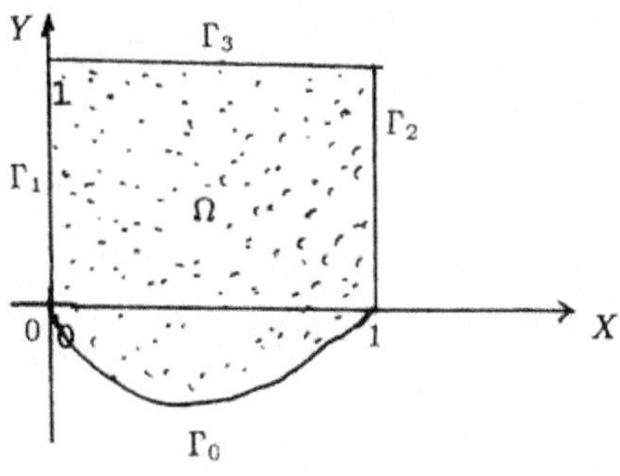

Fig. 1

In [7] we have shown that the equation (1.17) and (1.18) can be transformed into a unified form.

Consider the following equation of motion for seismic waves [8], in a quarter (x, y) plane:

$$\begin{cases} u_{tt} = \alpha^2 u_{xx} + (\alpha^2 - \beta^2)v_{xy} + \beta^2 u_{yy}, \\ v_{tt} = \beta^2 v_{xx} + (\alpha^2 - \beta^2)u_{xy} + \alpha^2 v_{yy}, \end{cases} \tag{1.19}$$

where u and v are the displacements in the x and y directions respectively.

α and β are dilatational and shear velocities respectively, and are defined by

$$\alpha^2 = \frac{\lambda + 2\mu}{\rho},$$

$$\beta^2 = \frac{\mu}{\rho},$$

where λ and μ are the Lamé constants and ρ is the density of the medium.

For a constant $\lambda > 0$, let

$$U = e^{-\lambda t}(u_t, v_t, u_x, v_x, u_y, v_y)^T,$$

and let the matrices A_1, A_2, A_3 be

$$A_1 = \begin{pmatrix} 1 & 0 & 0 & 0 & 0 & 0 \\ 0 & 1 & 0 & 0 & 0 & 0 \\ 0 & 0 & \alpha^2 & 0 & 0 & \left(\frac{\alpha^2-\beta^2}{2}\right) & 0 \\ 0 & 0 & 0 & \beta^2 & \left(\frac{\alpha^2-\beta^2}{2}\right) & \\ 0 & 0 & 0 & \left(\frac{\alpha^2-\beta^2}{2}\right) & \beta^2 & 0 \\ 0 & 0 & \left(\frac{\alpha^2-\beta^2}{2}\right) & 0 & 0 & \alpha^2 \end{pmatrix},$$

$$A_2 = \begin{pmatrix} 0 & 0 & -\alpha^2 & 0 & 0 & \frac{-(\alpha^2-\beta^2)}{2} \\ 0 & 0 & 0 & -\beta^2 & \frac{-(\alpha^2-\beta^2)}{2} & 0 \\ -\alpha^2 & 0 & 0 & 0 & 0 & 0 \\ 0 & -\beta^2 & 0 & 0 & 0 & 0 \\ 0 & \frac{-(\alpha^2-\beta^2)}{2} & 0 & 0 & 0 & 0 \\ \frac{-(\alpha^2-\beta^2)}{2} & 0 & 0 & 0 & 0 & 0 \end{pmatrix},$$

$$A_3 = \begin{pmatrix} 0 & 0 & 0 & \frac{-(\alpha^2-\beta^2)}{2} & -\beta^2 & 0 \\ 0 & 0 & \frac{-(\alpha^2-\beta^2)}{2} & 0 & 0 & -\alpha^2 \\ 0 & \frac{-(\alpha^2-\beta^2)}{2} & 0 & 0 & 0 & 0 \\ \frac{-(\alpha^2-\beta^2)}{2} & 0 & 0 & 0 & 0 & 0 \\ -\beta^2 & 0 & 0 & 0 & 0 & 0 \\ 0 & -\alpha^2 & 0 & 0 & 0 & 0 \end{pmatrix},$$

then, the equation (1.19) can be written in the symmetric positive form:

$$A_1 \frac{\partial}{\partial t} U + A_2 \frac{\partial}{\partial x} U + A_3 \frac{\partial}{\partial y} U + \lambda U = 0. \tag{1.20}$$

Note. If the boundary conditions, corresponding to the equation (1.20) are given in an appropriate form, then it is possible to express them as an admissible boundary condition.

§2. A Rayleigh Ritz Galbrkin Method for Solving Problem (1.1), (1.2)

We let v_h be a finite dimensional subspace of $(H^1(\Omega))^p$, where h is a parameter which will tend to zero, and let x_h be a subspace of v_h. Starting from the definition of a weak solution, we may define an approximate solution $u_h \in X_h$ as follows.

To find $u_h \in X_h$ such that, for all $v_h \in X_h$,

$$\frac{1}{2}((Au_h, v_h)_{L^2(\Omega)} + (u_h, A^* v_h)_{L^2(\Omega)}) + \frac{1}{2}\int_{\partial\Omega} (Mu_h, v_h)\, ds = (f, v_h)_{L^2(\Omega)}. \tag{2.1}$$

Theorem 2.1. *The approximate problem (2.1) has a unique solution $u_h \in X_h$ satisfying*

$$|u_h|_{L^2(\Omega)} \le \frac{2}{c_0}|f|_{L^2(\Omega)}. \tag{2.2}$$

Proof. In equality (2.1), one may choose $v_h = u_h$. Using Lemma 1.3 and Hypothesis (1.6), we get

$$\begin{aligned} &\tfrac{c_0}{2}|u_h|^2_{L^2(\Omega)} + \tfrac{1}{2}\int_{\partial\Omega}(Mu_h, u_h)\, ds \\ \le\; &\tfrac{1}{2}((A + A^*)u_h, u_h)_{L^2(\Omega)} + \tfrac{1}{2}\int_{\partial\Omega}(Mu_h, u_h)\, ds \\ =\; &(f, u_h)_{L^2(\Omega)}. \end{aligned}$$

According to hypothesis (1.3) (ii), we have

$$\int_{\partial\Omega}(Mu_h, u_h)\, ds = \frac{1}{2}\int_{\partial\Omega}(M + M^*)u_h, u_h)\, ds \ge 0.$$

The last two inequalities imply the existence of a unique solution $u_h \in x_h$ satisfying the inequality

$$\frac{c_0}{2}|u_h|^2_{L^2(\Omega)} \le (f, u_h)_{L^2(\Omega)}$$

from which we deduce inequality (2.2).

We show the following general estimate.

Theorem 2.2. *Let $u \in H^1(\Omega)$ be the solution of problem (1.1), (1.2)* and *$u_h \in X_h$ be the solution of problem (2.1). We have*

$$\begin{aligned} |u - u_h|_{L^2(\Omega)} \le\; & C \inf_{v_h \in X_h}\Bigg\{\|u - v_h\|_{H^1(\Omega)} \\ & + \sup_{w_h \in X_h} \frac{\int_{\partial\Omega} |u - v_h|\,|w_h|\, ds}{|w_h|_{L^2(\Omega)}}\Bigg\}, \end{aligned} \tag{2.3}$$

where C is a constant > 0 and independent of h.

Proof. For any $v_h \in X_h$, we let

$$E_h = \frac{1}{2}((A + A^*)(u_h - v_h), u_h - v_h)_{L^2(\Omega)} + \frac{1}{2}\int_{\partial\Omega} (M(u_h - v_h), u_h - v_h)\, ds.$$

According to Lemma 1.3 and hypothesis (1.3) (ii), we have

$$E_h \geq \frac{c_0}{2}|u_h - v_h|^2_{L^2(\Omega)}. \tag{2.4}$$

Relation (2.1), combined with Lemma 1.2, implies that

$$E_h = (f, u_h - v_h)_{L^2(\Omega)} - (Av_h, u_h - v_h)_{L^2(\Omega)} + \int_{\partial\Omega} \left(\frac{B\,X\,M}{2} v\ , u_h - v_h\right) ds.$$

Using equation (1.1) and boundary condition (1.2), we get

$$E_h = (A(u - v_h), u_h - v_h)_{L^2(\Omega)} + \int_{\partial\Omega} \left(\frac{M - B}{2}(u - v_h, u_h - v_h\right) ds. \tag{2.5}$$

Relation (2.4) and (2.5) imply that

$$|u_h - v_h|_{L^2(\Omega)} \leq C\|u - v_h\|_{H^1(\Omega)} + \sup_{w_h \in X_h} \frac{\int_{\partial\Omega} |u - v_h|\,|w_h|\, ds}{|w_h|_{L^2(\Omega)}}$$

for all $v_h \in X_h$.

Inequality (2.3) follows immediately.

§3. Continuous Finite Element Methods for Solving Problem (1.1), (1.2)

What we need now is to define the finite dimensional space X_h. In the sequel, we assume that the domain Ω is a polygonal. For the sake of simplicity, we consider only problem in two (or one)-dimension([5]). We let T_h be a triangulation of $\overline{\Omega}$ into triangular elements K (we could as well consider quadrilateral elements [4]). We let

$$h_K = \text{diameter of} K,$$

$$\rho_K = \sup\,(\text{diameter of the spheres included in } K),$$

$$h = \sup\{h_K; K \in T_h\}.$$

We assume that (T_h) is a regular family of triangulations, i.e.

$$\frac{h_K}{\rho_K} \leq C \quad \text{for all} \quad K \in T_h, \tag{3.1}$$

where the constant C is independent of K.

Let P_K be the space of polynomials of the form

$$P_k = \sum_{i+j\leq k} c_{ij} x^i y^j. \tag{3.2}$$

Any triangle K is the image of the reference triangle $\widehat{K}$ by an invertible mapping $F_K \in (P_1)^2$

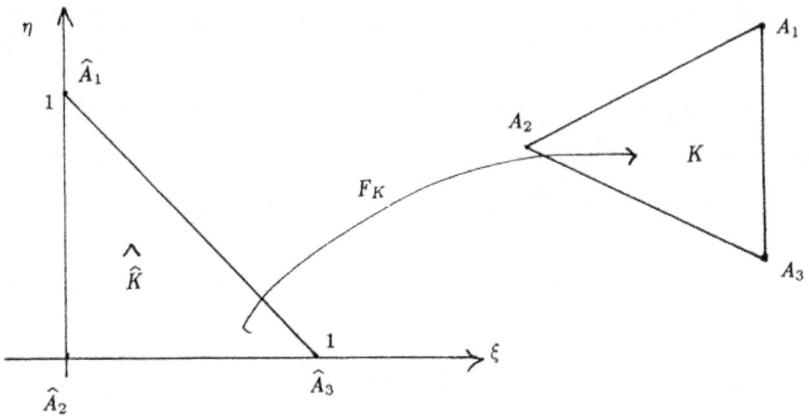

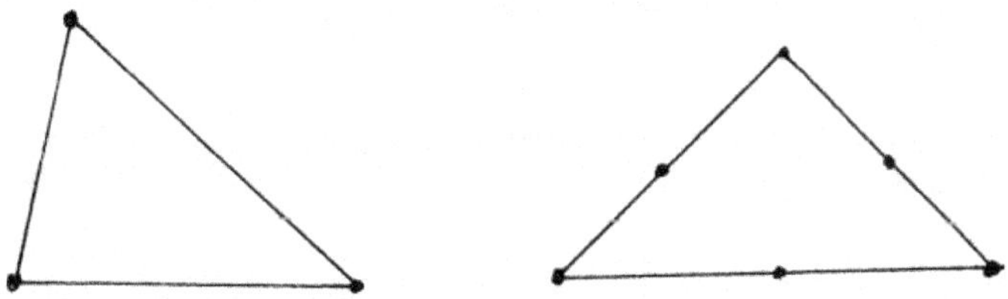

$k = 1$ $k = 2$

The space P_K of the shape functions is defined by

$$P_K = \{p = \widehat{p} \circ F_K^{-1}; \forall \widehat{p} \in P_K\} \quad \text{with} \quad k \geq 1. \tag{3.3}$$

We let

$$V_h = \{v_h \in C^0(\Omega); v_h| K \in P_K\}. \tag{3.4}$$

It is shown ([2] and bibliography of this paper) that if the finite elements are triangles of type k, the degrees of freedom of the space V_h can be chosen as the values of v_h at the nodes of the triangles.

We consider the possible choice for the space X_h, i.e.

$$X_h = \{v_h \in V_h; (B - M)v_h = 0 \quad \text{at the nodes} \quad \in \partial\Omega\}, \tag{3.6}$$

To be able to apply Theorem 2.2 and get some error bounds, we need to define a V_h-interpolate $r_h u$ of u and to introduce some notations.

Given a function $\widehat{u} \in C^0(K)$, its interpolate $\widehat{\pi}\widehat{u}$ is the unique polynomial of P_k equal to $\widehat{u}$ at the nodes of $\widehat{K}$. Now given a function $u \in C^0(K)$, its interpolate $\pi_K u \in P_K$ is defined by

$$\pi_K u = \widehat{\pi}\widehat{u} \circ F_K^{-1}, \qquad \text{where} \tag{3.7}$$

$$\widehat{u} = u \circ F_K.$$

(In fact $\pi_K u$ is the unique function of P_K equal to u at the nodes of K).

Now let $W^{m,p}(\Omega) = \{v \in L^p(\Omega), \partial^\alpha v \in L^p(\Omega) \text{ for } |\alpha| \le m\}$. We consider the following seminorm and norm on $W^{m,p}(\Omega)$.

$$\begin{aligned}
|v|_{m,p,\Omega} &= \left(\sum_{|\alpha|=m} |\partial^\alpha v|^p_{L^p(\Omega)}\right)^{\frac{1}{p}}, \\
\|v\|_{m,p,\Omega} &= \left(\sum_{l=0}^{m} |v|^p_{l,p,\Omega}\right)^{\frac{1}{p}}, \\
|v|_{m,\infty,\Omega} &= \max\{|\partial^\alpha v|_{L(\Omega)}; |\alpha| = m\}, \\
\|v\|_{m,\infty,\Omega} &= \max\{|\partial^\alpha v|_{L^\infty(\Omega)}; |\alpha| \le m\}.
\end{aligned}$$

We are now able to prove the following result.

Theorem 3.1. *Assume that the solution u of problem (1.1), (1.2) belongs to $H^{k+1}(\Omega)$, and let $u_h \in X_h$ be the solution of the approximate problem (2.5). Then we have*

$$|u - u_h|_{L^2(\Omega)} \le Ch^k |u|_{k+1,2,\Omega}. \tag{3.8}$$

Proof. We have the inclusion $H^{k+1}(\Omega) \subset C^0(\Omega)$, since $k \ge 1$. The interpolate $r_h u$ of u belongs to the space X_h since $(r_h u - u)$ is equal to zero at the nodes belonging to $\partial\Omega$, and we have $(B - M)(r_h u) = 0$ at those nodes. In the general estimation (2.7), we can replace v_h by $r_h u$ and we get

$$|u - u_h|_{L^2(\Omega)} \le C\Bigg\{\|u - r_h u\|_{1,2,\Omega} + \sup_{w_h \in X_h} \frac{\int_{\partial\Omega} |u - r_h u|\, |w_h|\, ds}{|w_h|_{L^2(\Omega)}}\Bigg\} \tag{3.9}$$

Using classical results of [1], we have

$$\|u - r_h u\|_{1,2,\Omega} \le Ch^k |u|_{k+1,2,\Omega}. \tag{3.10}$$

Then, as will be proved in the following two lemmas, for any side S of an element K, we have

$$|w_h|_{L^2(S)} \leq \; Ch_K^{-\frac{1}{2}}|w_h|_{L^2(K)},$$

$$|u - r_h u|_{L^2(S)} \leq \; Ch_K^{k+\frac{1}{2}}|u|_{k+1,2,K}.$$

Those two inequalities imply that we have

$$\sup_{w_h \in X_h} \frac{\int_{\partial\Omega} |u - r_h u|\,|w_h| ds}{|w_h|_{L^2(\Omega)}} \leq Ch^k |u|_{k+1,2,\Omega}. \tag{3.11}$$

Combining inequalities (3.9), (3.10) and (3.11), we get inequality (3.8).

Lemma 3.1. *For any $K \in T_h$, any $S \subset \partial K$ and any $w_h \in V_h$, we have*

$$|w_h|_{L^2(S)} \leq Ch_K^{-\frac{1}{2}}|w_h|_{L^2(K)}. \tag{3.12}$$

Proof. Let J_K denote the Jacobian of F_K and J'_K denote the Jacobian of the restriction of F_K to the side $\widehat{S}$, image of S by F_K^{-1}. We have

$$|w_h|^2_{L^2(S)} = \int_S w_h^2 \, ds = J'_K \int_{\widehat{S}} (\widehat{w}_h)^2 \, d\widehat{s}. \tag{3.14}$$

The expressions $(\int_{\widehat{s}} (\widehat{w}_h)^2 d\widehat{s})^{\frac{1}{2}}$ and $(\int_{\widehat{K}} (\widehat{w}_h)^2 d\widehat{\Omega})^{\frac{1}{2}}$ are respectively a semi norm and a norm on the space of the restrictions of the polynomials of P_K to the triangle $\widehat{K}$. Since P_K is a finite dimensional space, we get

$$\int_{\widehat{S}} (\widehat{w}_h)^2 d\widehat{s} \leq C \int_{\widehat{K}} (\widehat{w}_h)^2 d\widehat{\Omega}. \tag{3.15}$$

One may write:

$$\int_{\widehat{K}} (\widehat{w}_h)^2 d\widehat{\Omega} = J_K^{-1} \int_K w_h^2 d\Omega. \tag{3.16}$$

Combining relations (3.14), (3.15) and (3.16) we get

$$\int_S (w_h)^2 ds = J'_K J_K^{-1} \int_K (w_h)^2 d\Omega.$$

It is an easy matter to show that we have

$$J'_K \leq Ch_K, \tag{3.17}$$

$$J_K^{-1} \leq C\rho_K^{-2}. \tag{3.18}$$

The last three relations, combined with hypothesis 3.1 yield inequality (3.12).

Lemma 3.2. *For any $K \in T_h$, any $S \subset \partial K$ and any $v \in H^{k+1}(K)$, we have*

$$|v - r_h v|_{L^2(S)} \leq Ch_K^{k+\frac{1}{2}}|v|_{k+1,2,K}. \tag{3.19}$$

Proof. Using the same notations as in Lemma 3.1, we write

$$|v - r_h v|_{L^2(S)} = (J'_K)^{\frac{1}{2}} |\hat{v} - \hat{\pi}\hat{v}|_{L^2(\hat{S})} \tag{3.20}$$

For any $\hat{w} \in L^2(\hat{S})$, we have

$$|(\hat{v} - \hat{\pi}\hat{v}, \hat{w})_{L^2(\hat{S})}| \le C \|\hat{v}\|_{k+1,2,\hat{K}} |\hat{w}|_{L^2(\hat{S})}$$

and

$$(\hat{v} - \hat{\pi}\hat{v}, \hat{w})_{L^2(\hat{S})} = 0, \quad \text{for all} \quad \hat{v} \in P_k.$$

Applying the Bramble Hilbert Lemma ([1]) we get

$$|(\hat{v} - \hat{\pi}\hat{v}, \hat{w})_{L^2(\hat{S})}| \le C |\hat{v}|_{K+1,2,\hat{K}} |\hat{w}|_{L^2(\hat{S})}. \tag{3.21}$$

Combining relations (3.20) and (3.21), we have

$$|v - r_h v|_{L^2(S)} \le C (J'_K)^{\frac{1}{2}} |\hat{v}|_{k+1,2,K}.$$

It is shown in [2] that we have

$$|\hat{v}|_{k+1,2,\hat{k}} \le c h_k^{k+1} (J_k)^{-\frac{1}{2}} |v|_{k+1,2,k}.$$

The last two inequalities and inequalities (3.17), (3.18) lead to estimate (3.19).

References

[1] Ciarlet, P.G. et Raviart, P.A., *General Lagrange and Hermite interpolation in R^n with applications to finite element methods.* Arch. Rational Mech. Anal., 46(1972), 177–199.

[2] Ciarlet, P.G. et Raviart, P.A., *Interpolation theory over curved elements with applications to finite element methods.* Computer Methods in Applied Mechanics and Engineering, 1,(1972), 217–249.

[3] Friedrichs, K.O., *Symmetric positive differential equations,* Comm. on pure and appl. math., 11(1958), 333–418.

[4] Lesaint, P., *Sur la résolution des systems hyperboliques du premier order par des méthodes d'éléments finis,* Thesis, 1975.

[5] Lesaint P., *Finite Element methods for symmetric hyperbolic equations,* Numer. Math., 21(1973), 244–255.

[6] Tangmanee, S., *A unified numerical method for classical differential equations,* Conference in Mathematics, Chulalongkorn University, Bangkok, 1987.

[7] Tangmanee, S., *Existence and uniqueness of solution for a model problem of transonic flow,* Regional Conference in Mathematics, Yogyakarta, 1989.

[8] Alterman, Z.S. and Lowenthal, D., *Seismic wave in a quater and three-quarter plane* , Geophy. J.R. astr. Soc., 20(1970), 101–126.

Proceedings of the Asian Mathematical
Conference, 1990

The CES Method and Its Applications in Computing and Statistical Contexts

VENUGOPAL, N. & SUBRAHMANYAM, V.
Dept. Of Statistics
S.V. University
Tirupati-517 502, India

§1. General

An elegant and simple method is developed to obtain the sum of a given number of entries, based on the technique of cumulated summation. We call this technique as CES (Cumulated Entry Sum) method. Further, extended results for repeated summation upto a given order are derived.

Although the sums of binomial coefficients (of the type $\sum_{x=k}^{n}\binom{x}{k}, 0 < k < n, n \in N^{+}$) are discussed in earlier literature in some detail we not only provide alternative and elegant proofs for the formulae using the CES method but also various useful ramifications involving certain repeated sums are discussed.

Further, the use of the CES method is highlighted: (i) in the context of "Numerical Summation", based on the use of some known interpolation formulae and (ii) in determining the degree of a polynomial based on the limited information of the values of its rth order differences.

§2. Introduction

In Algebra and Numerical Analysis varied methods have been developed to obtain the sum of a given finite number of entries (see forexampl e, [1ppx]x[4]]x Thus, the summation techniques for a finite series of values have always been a dominantly interesting field of research.

The motivation for the present work stemmed from the interest to deal with certain types of sums and their important ramifications, with some freshness and comprehensivenss in approach. Specifically, towards this end, we develop the CES (Cumulated Entry Sum) method (see [6]).

Using the method, we derive some results not only general in character and content but are also direct, Simple and elegant in terms of computing labour. In some detail, we elaborate on these aspects in the following.

The formula (see, for example, Chung [2]) for $\sum_{x=k}^{n}\binom{x}{k}$ is not only derived afresh through our methods but also the other ramifications (useful, for example, in Blaire-Pascal's Negative Binomial Law) are also successfully dealt with. Further, the formulae for sums of factorials (see, for example, Hildebrand [4]) are extended for other important contexts.

Further results, more general and elegant in nature are derived to obtain:

(i) numerical sums of entries of polynomials of different degrees;

(ii) numerical sums of repeated orders upto a given, $r \in N^+ = 1, 2, 3, \cdots$;

(iii) numerical sums commencing from an intermediatory stage of argument;

(iv) numerical sums involving entries with arguments not equally spaced and so on, using the known results like Newton-Gregory Formulae, Newton-Gauss Formulae, Sterling's, Bessel's, Laplace-Everett's, Lagrange's Formulae and so on.

We developed a methodology called BCS (Binomial Coefficients Sum) method leading to a direct formula to determine a polynomial in x, say y_x of nth degree (upto 'r' constants) given its rth order differences $\Delta^r y_x, r \in N^+$ (polynomials of $(n-r)$th degree).

§3. The CES Method and Results

Let $y_i = f(x_i)$ be the entry at x_i, $i = 1, 2, \cdots, n$ and y_i follow a polynomial of nth degree.

Let $Y_t^{(1)} = \sum_{i=0}^{t} y_i$, so that $Y_0^{(1)} = y_0, Y_1^{(1)} = y_0 + y_1, Y_2^{(2)} = y_0 + y_1 + y_2$ and so on.

Now,

$$\Delta Y_i^{(1)} = y_{i+1}, \Delta^{j+1} Y_i^{(1)} = \Delta^j y_{i+1} = 0 \quad \text{for} \quad j \geq n+1. \tag{1}$$

Theorem 1.

$$Y_t^{(1)} = \binom{t+1}{1} y_0 + \binom{t+1}{2}\Delta y_0 + \binom{t+1}{3}\Delta^2 y_0 + \cdots + \binom{t+1}{n+1}\Delta^n y_0, \tag{2}$$

where $t = \frac{x_n - x_0}{h}$ *and* $h =$ *incremental interval in* x_i*'s,* $i = 0, 1, \cdots, n$.

Proof.

$$\begin{aligned} Y_t^{(1)} = {} & E^t Y_0^{(1)} = (1+\Delta)^t Y_0^{(1)} \\ = {} & Y_0^{(1)} + \binom{t}{1}\Delta Y_0^{(1)} + \binom{t}{2}\Delta^2 Y_0^{(1)} + \cdots + \binom{t}{n+1}\Delta^{n+1} Y_0^{(1)}, \quad \text{by (1)} \qquad (3) \\ = {} & y_0 + \binom{t}{1} y_1 + \binom{t}{2}\Delta y_1 + \binom{t}{3}\Delta^2 y_1 + \cdots + \binom{t}{n+1}\Delta^n y_1, \quad \text{by (1)}. \end{aligned}$$

Replacing y_1 by Ey_0 in (3) we have:

$$\begin{aligned} Y_t^{(1)} = {} & y_0 + \binom{t}{1} Ey_0 + \binom{t}{2}\Delta(Ey_0) + \binom{t}{3}\Delta^2(Ey_0) + \cdots + \binom{t}{n+1}\Delta^n(Ey_0) \\ = {} & \left[\binom{t}{0} + \binom{t}{1}\right] y_0 + \left[\binom{t}{1} + \binom{t}{2}\right]\Delta y_0 + \left[\binom{t}{2} + \binom{t}{3}\right]\Delta^2 y_0 + \cdots \\ & + \left[\binom{t}{n} + \binom{t}{n+1}\right]\Delta^n y_0, \end{aligned}$$

by replacing E by $(1+\Delta)$ and then rearranging like terms.

$$Y_t^{(1)} = \binom{t+1}{1} y_0 + \binom{t+1}{2} \Delta y_0 + \binom{t+1}{} \Delta^2 y_0 + \cdots + \binom{t+1}{n+1} \Delta^n y_0. \quad \square$$

This theorem is extended (using similar manipulations as above) to obtain "Repeated Summation": $Y_t^{(r)} = \sum_0^t \cdots \sum_0^j y_i$ in the result

$$Y_t^{(r)} = E^t Y_0^{(r)} = \binom{t+r}{r} y_0 + \binom{t+r}{r+1} \Delta y_0 + \cdots + \binom{t+r}{r+n} \Delta^n y_0. \qquad (4)$$

ILLUSTRATIONS

Example 1. Find the sum $x^3 + (x+1)^3 + (x+2)^3 + \cdots + (x+n-1)^3$.

Solution. Treating $y_i = (x+i)^3, i = 0, 1, 2, \cdots, (n-1)$ we obtain:

$$y_0 = x^3, \quad \Delta y_0 = \ x^2 + \ x + 1, \quad \Delta^2 y_0 = 6x + 6 \quad \text{and} \quad \Delta^3 y_0 = 6.$$

Then,

$$\sum_{i=0}^{n-1} y_i = Y_{n-1}^{(1)} = \binom{n}{1} x^3 + \binom{n}{2} (x\ ^2 + x\ + 1) + \binom{n}{} (6x + 6) + \binom{n}{4} 6, \quad \text{using (2)}.$$

Further,

$$Y_{n-1}^{(r)} = \binom{n+r-1}{r} x^3 + \binom{n+r-1}{r+1} (x\ ^2 + x\ + 1) \ + \binom{n+r-1}{r+2} (6x + 6) + \binom{n+r-1}{r+3} 6,$$

using (4).

Example 2. Sum to 'n' terms the series whose xth term is $x(x-1)(x-2)\cdots(x-\overline{m-1}\,) = x^{(m)}$, (see ( )) where the case of $m = 4$ only is dealt with.

Solution. $x(x-1)(x-2)\cdots(x-\overline{m-1}\,) = x^{(m)}$ in factorial notation. Let $y_x = x^{(m)}$, then $y_0 = \Delta y_0 = \Delta^2 y_0 = \cdots \Delta^{m-1} y_0 = 0, \Delta^m y_0 = m!$. and $t = n-1$.

Now

$$Y_{n-1}^{(1)} = \binom{n}{m+1} m!, \qquad Y_{n-1}^{(r)} = \binom{n+r-1}{m+r} m!, \quad \text{using (2) and (4) respectively.}$$

Example 3. Find $\sum_{i=1}^n i^4$ (see also [1]).

Solution. Let $y_i = i^4, i = 1, 2, \cdots, n$. Then, $y_1 = 1, \Delta y_1 = 15, \Delta^2 y_1 = 50, \Delta^3 y_1 = 60, \Delta^4 y_1 = 24$. Now,

$$Y_{n-1}^{(1)} = \frac{n}{0}(6n^4 + 15n^3 + 10n^2 - 1), \quad \text{using (2)}$$

and

$$Y_{n-1}^{(r)} = \binom{n+r-1}{r} + \binom{n+r-1}{r+1} 15 + \binom{n+r-1}{r+2} 50$$

$$+ \binom{n+r-1}{r+} 60 + \binom{n+r-1}{r+4} 24, \quad \text{using (4)}.$$

§4. The CES Method for General Sums of Binomial Coefficients

In Chung [2] and elsewhere a method is given to obtain $\sum_{x=k}^{n}\binom{x}{k}$. In the present work employing the 'CES Method', $\sum_{x=k}^{n}\binom{x}{k}$ is directly obtained, by noting $y_x = \binom{x}{k}$ as a polynomial of kth degree. After some calculations,

$$\begin{aligned} Y_t^{(1)} = \sum_{x=0}^{t} y_x = \sum_{x=k}^{n}\binom{x}{k} &= 0+0+\cdots+0+\binom{n+1}{k+1} \quad \text{using (2)} \\ &= \binom{n+1}{k+1}. \end{aligned} \tag{5}$$

Further

$$\sum_{x=1}^{n}\binom{-x}{k} = (-1)^k\sum_{x=1}^{n}\binom{x+k-1}{k} = (-1)^k\binom{n+k}{k+1} \quad \text{using (5).} \tag{6}$$

Result (6) is useful in Pascal's probability distribution,

$$P(x) = \binom{-r}{x}p^r(-q)^x, \quad x = 0,1,2,\cdots,$$

where negative binomial coefficients $\binom{-r}{x}$ occur.

We can further determine the sum of the binomial coefficients $\binom{c+ih}{k}$, $i = 0,1,2,\cdots,(n-1)$ with the interval 'h' with the following theorem.

Theorem 2.

$$\sum_{i=0}^{n-1}\binom{c+ih}{k} = n\binom{c}{k}+\binom{n}{2}\Delta\binom{c}{k}+\binom{n}{3}\Delta^2\binom{c}{k}+\cdots+\binom{n}{k+1}h^k, \tag{7}$$
$$l \le c \le n; \quad n > k; \quad c,\ k,\ n,\ h \in N^+.$$

Proof. Let $x_i = c+ih$, $i = 0,1,2,\cdots,(n-1)$ and $y_{x_i} = \binom{c+ih}{k}$ be an entry of a polynomial of kth degree.

After some calculations,

$$y_{c+ih} = E^i y_c = (1+\Delta)^i y_c = \binom{i}{0}y_c + \binom{i}{1}\Delta y_c + \binom{i}{2}\Delta^2 y_c + \cdots + \binom{i}{k}\Delta^k y_c. \tag{8}$$

Since $\Delta^{k+j}y_c = 0$ for $j \ge 1$.

Summing both sides of (8) from $i = 0$ to $i = t = n-1$, we obtain

$$\sum_{i=0}^{n-1} y_{c+ih} = y_c\sum_{i=0}^{n-1}\binom{i}{0} + \Delta y_c\sum_{i=0}^{-1}\binom{i}{1} + \Delta^2 y_c\sum_{i=0}^{n-1}\binom{i}{2} + \cdots + \Delta^k y_c\sum_{i=0}^{n-1}\binom{i}{k}$$

i.e.

$$\begin{aligned} \sum_{i=0}^{n-1}\binom{c+ih}{k} &= \binom{n}{1}y_c + \binom{n}{2}\Delta y_c + \binom{n}{3}\Delta^2 y_c + \cdots + \binom{n}{k+1}\Delta^k y_c \quad \text{using (2)} \\ &= n\binom{c}{k}+\binom{n}{2}\Delta\binom{c}{k}+\binom{n}{3}\Delta^2\binom{c}{k}+\cdots+\binom{n}{k+1}h^k. \quad \square \end{aligned}$$

Remark: On similar lines we can obtain the sum of $\binom{c-ih}{k}, i = 0, 1, 2, \cdots, (n-1)$ by taking $(c-ih), i = 0, 1, 2, \cdots, (n-1)$ as arguments, $y_x = \binom{c-ih}{k}, i = 0, 1, 2, \cdots, (n-1)$ as entries in a kth degree polynomial as follows:

$$\sum_{i=0}^{n-1} \binom{c-ih}{k} = n\binom{c}{k} + \binom{n}{2}\Delta\binom{c}{k} + \binom{n}{}\Delta^2\binom{c}{k} + \cdots + (-1)^k h^k \binom{n}{k+1}. \quad (9)$$

The results are easily extended to repeated sums also (after some calculations, straightforward enough).

§5. Sums of Factorials

Sums of factorials are important in combinatorial problems and these are briefly discussed in [4]. We add to these results by exploiting the sums of factorials in different contexts.

(A)

$$\sum_{x=k}^{n} x^{(k)} = k! \sum_{x=k}^{n} \binom{x}{k} = k!\binom{n+1}{k+1} = \frac{(n+1)^{(k+1)}}{(k+1)!} \quad \text{using (5).} \quad (10)$$

(B)

$$\begin{aligned} \sum_{x=c}^{n} x^{(k)} &= \sum_{x=0}^{n} x^{(k)} - \sum_{x=0}^{c-1} x^{(k)} = k!\left[\binom{n+1}{k+1} - \binom{c}{k+1}\right] \quad \text{using (10)} \\ &= \frac{(n+1)^{(k+1)} - c^{(k+1)}}{(k+1)}, \quad k \le c \le n. \end{aligned} \quad (11)$$

(C)

$$\sum_{x=1}^{n} (-x)^{(k)} = (-1)^k \sum_{x=1}^{n} (x+k-1)^{(k)} = (-1)^k \frac{(n+k)^{(k+1)}}{k+1} \quad \text{using (10).} \quad (12)$$

(D)

$$\begin{aligned} &\sum_{i=0}^{n-1} (c+ih)^{(k)} = k! \sum_{i=0}^{n-1} \binom{c+ih}{k} \\ &= k!\left[n\binom{c}{k} + \binom{n}{2}\Delta\binom{c}{k} + \binom{n}{}\Delta^2\binom{c}{k} + \cdots + \binom{n}{k+1}h^k\right], \quad \text{using (7)} \end{aligned} \quad (13)$$

using (7).

(E)

$$\left.\begin{array}{r} r\text{th order cumulated} \\ \text{sum of } x^{(k)} \text{ from } x = k \text{ to } x = n \end{array}\right\} = k!\binom{n+r}{k+r}, \quad \text{follows by remark (see (9)).} \quad (14)$$

On same lines we can also obtain rth order cumulated sums of $(-x)^{(k)}, x^{(k)}$ from $x = c$ to $x = n$, $(c+ih)^{(k)}$ and $(c-ih)^{(k)}$.

ILLUSTRATIONS

Example 4. Find $\sum_{x=10}^{17}\binom{x}{5}$ and $\sum_{x=10}^{17}x^{(5)}$.

Solution.

$$\sum_{x=10}^{17}\binom{x}{5}=\sum_{x=5}^{17}\binom{x}{5}-\sum_{x=5}^{9}\binom{x}{5}=\binom{18}{6}-\binom{10}{6}=18354,\quad \text{using (5)}$$

and

$$\sum_{x=10}^{17}x^{(5)}=\frac{18^{(6)}-10^{(6)}}{6}\quad \text{using (B)}.$$

Example 5. Compute $\sum_{x=1}^{100}\binom{-x}{50}$ and $\sum_{x=1}^{100}(-x)^{50}$.

Solution.

$$\sum_{x=1}^{100}\binom{-x}{50}=\binom{150}{51},\quad \text{using (6)}.$$

We can also get $1Y_{100}^{(r)}=\binom{149+r}{50+r}$.

$$\sum_{x=1}^{100}(-x)^{(50)}=\frac{(150)^{51}}{51}\quad \text{using (C)}.$$

§6. Numerical Summation

Numerical summation is proposed and developed, on the lines of numerical differentiation and numerical integration, to obtain the sum of polynomial entries of different degrees from the first argument, intermediate argument, ending argument, unequal arguments and so on.

The basic technique is as follows: We first recast polynomial $f(x)$ in factorial notation $(x^{(\cdot)})$ by constructing difference table and applying suitable interpolation formula and then summing the resulting binomial coefficients.

Theorem 3.

$$\begin{aligned}\sum_{i=0}^{t}f(x_0+ih)=&\binom{t+1}{1}f(x_0)+\binom{t+1}{2}\Delta f(x_0)+\binom{t+1}{3}\Delta^2 f(x_0)\\&+\cdots+\binom{t+1}{n+1}\Delta^n f(x_0),\end{aligned}\tag{15}$$

where $\frac{x_t-x_0}{h}=t\in N^+$, *using Newton-Gregory Forward Formula.*

Proof: We construct the difference table for $f(x_0+ih)$'s and use it in the Newton-Gregory Forward Formula given by

$$f(x_0+ih)=f(x_0)+\binom{i}{1}\Delta f(x_0)+\binom{i}{2}\Delta^2 f(x_0)+\cdots+\binom{i}{n}\Delta^n f(x_0),\tag{16}$$

where $i=\frac{x-x_0}{h}$ (since $\Delta^j f(x_0)=0$ for $j>n$ if $f(\cdot)$ is a polynomial of nth degree).

Now, applying the operator $\sum_{i=0}^{t}$ on both sides of (16) we obtain

$$\begin{aligned}\sum_{i=0}^{t} f(x_0+ih) = & \; f(x_0)\sum_{i=0}^{t}\binom{i}{0} + \Delta f(x_0)\sum_{i=0}^{t}\binom{i}{1} + \Delta^2 f(x_0)\sum_{i=0}^{t}\binom{i}{2} \\ & + \cdots + \Delta^n f(x_0)\sum_{i=0}^{t}\binom{i}{n}, \\ = & \binom{t+1}{1} f(x_0) + \binom{t+1}{2}\Delta f(x_0) + \binom{t+1}{}\Delta f(x_0) \\ & + \cdots + \binom{t+1}{n+1}\Delta^n f(x_0) \quad \text{using (5).} \qquad \square\end{aligned}$$

Corrollary. *The rth order repeated sum* $\sum^{(r)}$ *is given by*

$$\begin{aligned}\sum_{i=0}^{t}{}^{(r)} f(x_0+ih) = & \binom{t+r}{r} f(x_0) + \binom{t+r}{1+r}\Delta f(x_0) + \binom{t+r}{2+r}\Delta^2 f(x_0) \\ & + \cdots + \binom{t+r}{n+r}\Delta^n f(x_0),\end{aligned} \tag{17}$$

by applying the operator $\sum_{i=0}^{t}{}^{(r)}$ *to both sides of (16) and the remark under (9).*

We further observe that the results (15) and (17) can also be derived directly through the CES method (using (2) and (4)).

Theorem 4.

$$\begin{aligned}\sum_{i=0}^{t} f(x_0+ih) = & \sum_{i=0}^{n-1} f(x_0+ih) + \Big[\binom{t}{1} f(x_n) + \binom{t+1}{2}\nabla f(x_n) \\ & + \binom{t+2}{}\nabla^2 f(x_n) + \cdots + \binom{t+n}{n+1}\nabla^n f(x_n)\Big],\end{aligned} \tag{18}$$

where $\frac{x_t - x_n}{h} = t \in N^+$ *using Newton-Gregory Backward Formula. Proof follows on similar lines as in Theorem 3.*

Corrollary. *We can further derive:*

$$\begin{aligned}\sum_{i=0}^{t}{}^{(r)} f(x_n+ih) = & \binom{t+r}{r} f(x_n) + \binom{t+r}{1+r}\nabla f(x_n) + \binom{t+r+1}{2+r}\nabla^2 f(x_n) \\ & + \cdots + \binom{t+r+n-1}{n+r}\nabla^n f(x_n).\end{aligned} \tag{19}$$

In the following we state further theorems, omitting the proof for reasons stated under Theorem 4.

Theorem 5.

$$\begin{aligned}\sum_{i=0}^{t} f(x_0+ih) = & \sum_{i=0}^{c-1} f(x_0+ih + \Big[\binom{t+1}{1} f(0) + \binom{t+1}{2}\Delta f(0) \\ & + \binom{t+1}{}\Delta^2 f(-1) + \binom{t+2}{4}\Delta^3 f(-1) + \binom{t+2}{5}\Delta^4 f(-2) + \cdots \\ & + \binom{t+m}{2m}\Delta^{2m-1} f(-m+1) + \binom{t+m}{2m+1}\Delta^{2m} f(-m) + \cdots\Big],\end{aligned} \tag{20}$$

by Newton-Gauss Forward Formula, where $t = \frac{x_t - x_c}{h}$,

x_t = *the argument upto which sum is required,*

x_c = *the middle (central) argument.*

Theorem 6.

$$\sum_{x=0}^{t} f(x) = \sum_{j=0}^{n} j!\, H_j \binom{t+1}{j+1}, \quad \text{by Newtons' Divided-Difference Formula (NDDF).}$$

Theorem 7.

$$\sum_{j=0}^{n} (1+x)^j = \binom{n+1}{1} + \binom{n+1}{2} x + \binom{n+1}{3} x^2 + \cdots + x^n.$$

SOME EXAMPLES:

Example 6. Compute the sum of $y_x = 2x^3 - x^2 + 5$ at $x = 5, 9, 13, 17, 21, \cdots$ to 15 terms. Further, find $\sum_{x=0}^{61} y_x$ by Newton-Gregory Forward Formula.

Solution. The given entries are $y_{(5+4i)}; i = 0, 1, 2, \cdots, 14$ whose values are $y_5 = 230$, $y_9 = 1382$, $y_{13} = 4230$, $y_{17} = 9542$, $y_{21} = 18086, \cdots$. From the difference table we obtain $y_5 = 230$, $\Delta y_5 = 1152$, $\Delta^2 y_5 = 1696$ and $\Delta^3 y_5 = 768$ = constant.

Setting $t = \frac{(5+4\times 14)-5}{4} = 14$ and using (15),

$$\sum_{i=0}^{14} y_{5+4i} = \binom{15}{1} 230 + \binom{15}{2} 1152 + \binom{15}{3} 1696 + \binom{15}{4} 768 = 1944410.$$

Again,

$$y_x = 2x^3 - x^2 + 5 = 2x^{(3)} + 5x^{(2)} + x + 5 = 12\binom{x}{3} + 10\binom{x}{2} + \binom{x}{1} + 5\binom{x}{0},$$

using synthetic division.

Now $\sum_{x=0}^{61} y_x = 7074541$, using (2).

Remark. In the same way we can apply and obtain Numerical Summation by Central difference formula, NDDF and so on.

Example 7. Given that $f(1) = 6$, $f(3) = 50$, $f(6) = 401$ and $f(10) = 1905$, find (i) $\sum_{x=0}^{32} f(x)$ and (ii) $\sum_{x=0}^{32} {}^{(r)} f(x)$, using Lagrange's formula for numerical summation.

Solution. The intervals in the given arguments 1, 3, 6 and 10 are not uniform. We assume a polynomial relationship and obtain $f(x) = 2x^3 - x^2 + 5$, by Lagrange's formula

$$= 2x^{(3)} + 5x^{(2)} + x^{(1)} + 5 \quad \text{using "Synthetic Division"}$$

$$= 12\binom{x}{3} + 10\binom{x}{2} + \binom{x}{1} + 5\binom{x}{0}.$$

Now, $\sum_{x=0}^{32} f(x) = 514613$, using (2) and

$$\sum_{x=0}^{32} {}^{(r)} f(x) = 12\binom{32+r}{3+r} + 10\binom{32+r}{2+r} + \binom{32+r}{1+r} + 5\binom{32+r}{r} \quad \text{using (4).}$$

§7. BCS Method for FFD

Employing the results in Section 5 concerning Binomial Coefficients Sums (BCS), we derive the formula for Function Form Determination (FFD) of nth degree polynomial y_x (upto 'r' constants) given its rth order difference $\Delta^r y_x, r \in N^+$, to be a polynomial of $(n-r)$th degree, in the following theorem.

Theorem 8. *y_x, an unknown polynomial is determined upto a constant, y_0, in terms of a given $(n-1)$th degree polynomial, namely Δy_x, by the relation*]:

$$y_x = y_0 + \binom{x}{1}(\Delta y_0) + \binom{x}{2}\Delta(\Delta y_0) + \binom{x}{}\Delta^2(\Delta y_0) + \cdots + \binom{x}{n}\Delta^{n-1}(\Delta y_0).$$

Proof: Since Δy_x is given to be a known polynomial of $(n-1)$th degree, we have

$$\Delta^j(\Delta y_x) = 0, \quad \forall x, \quad j \geq n. \tag{21}$$

Further, y_x is a polynomial of nth degree, so that

$$\begin{aligned} \Delta^n y_x = &\ a \text{ constant}, \quad \forall x \\ = &\ \Delta^n y_0, \quad \text{say}. \end{aligned} \tag{22}$$

Let

$$Z_j = \sum_{x=0}^{j}(\Delta y_x), \quad j = 0, 1, 2, \cdots, n. \tag{23}$$

so that $Z_0 = y_0$,

$$\Delta Z_0 = Z_1 - Z_0 = (\Delta y_0 + \Delta y_1) - \Delta y_0 = \Delta y_1,$$

and in general

$$\Delta^n Z_0 = \Delta^n y_1, \quad n \in N^+. \tag{24}$$

Let $n < t \in N^+$. Then,

$$\begin{aligned} \Delta y_t = &\ \Delta E^t y_0 = (1+\Delta)^r \Delta y_0 \\ = &\ \Delta y_0 + \binom{t}{1}\Delta(\Delta y_0) + \binom{t}{2}\Delta^2(\Delta y_0) + \cdots + \binom{t}{n-1}\Delta^{n-1}(\Delta y_0), \quad \text{by (21)}. \end{aligned}$$

Also,

$$Z_t = E^t z_0 = (1+\Delta)^t Z_0 = Z_j + \binom{t}{1}\Delta Z_0 + \binom{t}{2}\Delta^2 Z_0 + \cdots + \binom{t}{n}\Delta^n y_0, \text{ using (22)}. \tag{25}$$

Alternatively,

$$\sum(\Delta y_t) = \Delta y_0 + \binom{t}{1}\Delta y_1 + \binom{t}{2}\Delta^2 y_1 + \cdots + \binom{t}{n}\Delta^n y_1, \tag{26}$$

using (2) and (24).

Subtracting (25) from (26) we obtain:

$$\begin{aligned}\sum(\Delta y_t) - \Delta y_t = & \binom{t}{1}[\Delta y_1 - \Delta^2 y_0] + \binom{t}{2}[\Delta^2 y_1 - \Delta^3 y_0] \\ & + \binom{t}{3}[\Delta^3 y_1 - \Delta^4 y_0] + \cdots + \binom{t}{n-1}[\Delta^{n-1} y_1 - \Delta^n y_0] + \binom{t}{n}\Delta^n y_1 \\ = & \binom{t}{1}\Delta[y_1 - \Delta y_0] + \binom{t}{2}\Delta^2[y_1 - \Delta y_0] + \cdots \\ & + \binom{t}{n-1}\Delta^{n-1}(y_1 - \Delta y_0) + \binom{t}{n}\Delta^n y_1 \\ = & \binom{t}{1}\Delta y_0 + \binom{t}{2}\Delta^2 y_0 + \cdots + \binom{t}{n-1}\Delta^{n-1} y_0 \\ & + \binom{t}{n}\Delta^n y_0, \quad \text{using (24)}.\end{aligned} \tag{27}$$

Adding y_0 to both sides of (27), we have:

$$\begin{aligned}\sum(\Delta y_t) - \Delta y_t + y_0 = & \binom{t}{0} y_0 + \binom{t}{1}\Delta y_0 + \binom{t}{2}\Delta^2 y_0 + \cdots \\ & + \binom{t}{n}\Delta^n y_0, \qquad \text{since } \binom{t}{0} = 1 \\ = & (1+\Delta)^t y_0 = E^t y_0 = y_t.\end{aligned} \tag{28}$$

Thus,

$$\begin{aligned}y_t = & y_0 + \binom{t}{1}(\Delta y_0) + \binom{t}{2}\Delta(\Delta y_0) + \binom{t}{3}\Delta^2(\Delta y_0) + \cdots \\ & + \binom{t}{n}\Delta^{n-1}(\Delta y_0) \quad \text{from (28)}.\end{aligned} \tag{29}$$

Since a polynomial $P_n(x)$ can be written as $P_n(x)\sum_{r=0}^{n} a_r x^r = \sum_{r=0}^{n} b_r \binom{x}{r}$, for any x, y_x is given (using (29)) as:

$$y_x = y_0 + \binom{x}{1}(\Delta y_0) + \binom{x}{2}\Delta(\Delta y_0) + \binom{x}{3}\Delta^2(\Delta y_0) + \cdots + \binom{x}{n}\Delta^{n-1}(\Delta y_0).$$

We now extend the result to the case of finding y_x from known $\Delta^r y_{x'}, r \in N^+$. This is stated in the following theorem, omitting the proof involving lengthy calculations.

Theorem 9. *$y_{x'}$ an unknown polynomial of nth degree is determined upto 'r' constants, $\Delta^j y_0, j = 0, 1, 2, \cdots, (r-1)$, in terms of a given $(n-r)$th degree polynomial $\Delta^r y_x, r \in N^+$, through the relation:*

$$\begin{aligned}y_x = & \binom{x}{n}\Delta^{n-r}(\Delta^r y_0) + \binom{x}{n-1}\Delta^{n-r-1}(\Delta^r y_0) + \binom{x}{n-2}\Delta^{n-r-2}(\Delta^r y_0) + \cdots \\ & + \binom{x}{r}(\Delta^r y_0) + \binom{x}{r-1}\Delta^{r-1} y_0 + \binom{x}{r-2}\Delta^{r-2} y_0 + \cdots + \binom{x}{1}\Delta y_0 + y_0.\end{aligned}$$

For operational and working convenience, the following algorithm proves useful.

The algorithm: Suppose we are given a function $\Delta^r y_x$, $r \in N^+$, as $(n-r)$th degree polynomial in x, say. Compute the values of $\Delta^r y_x$ at $x = 0, 1, 2, \cdots, (n-r)$. Construct the difference table for these values. The leading terms $\Delta^j(\Delta^r y_0), j = 0, 1, 2, \cdots, (n-r)$ are used in Theorem 9 to determine directly the functional form y_x, upto the 'r' constants $\Delta^j y_{0'}, j = 0, 1, 2, \cdots, (n-r)$.

Example 8. Determine the polynomial whose third difference is $12x^3 + 4x - 10$.

Solution. Let y_x be the function such that $\Delta^3 y_x = 12x^3 + 4x - 10$. Now, $\Delta^3 y_0 = -10, \Delta^3 y_1 = 6, \Delta^3 y_2 = 94, \Delta^3 y_3 = 326$.

Forming the difference table for these values as indicated in the algorithm we obtain:

$$\Delta^3 y_0 = -10, \Delta(\Delta^3 y_0) = 16, \Delta^2(\Delta^3 y_0) = 72 \text{ and } \Delta^3(\Delta^3 y_0) = 72.$$

Using theorem 9 for the case of $r = 3$ and $n = 6$, we directly obtain:

$$y_x = 72\binom{x}{6} + 72\binom{x}{5} + 16\binom{x}{4} - 10\binom{x}{3} + \binom{x}{2}(\Delta^2 y_0) + \binom{x}{1}(\Delta y_0) + y_0.$$

We notice that y_x is determined upto the constants $\Delta^j y_0, j = 0, 1, 2,$.

§8. Concluding Discussion

In conclusion, we remark that the CES method proves extremely useful in computing with ease, directness and simplicity such statistics as sample mean, variance, other higher order central moments, covariance, and correlation coefficient, from classified data related to equally spaced classes (as compared to the known formulae for these statistics, see for example [5]). Further defining or identifying the well known ARIMA models as ARMA models (after differencing), we have useful applications of our results reported in the above. These and other cognate results will be reported elsewhere.

References

[1] Barnard, S. and Child, J.M., *Higher Algebra, Macmillan & Co.* (1960)

[2] Chung, K.L. *Elementary Probability Theory with Stochastic Process*, Springer-Verlag. 1983.

[3] Freeman, Harry, *Finite Differences for Acturial Students*, Cambridge University Press., 1965.

[4] Hildebrand, *Introduction to Numerical Analysis*, McGraw-Hill., 1956.

[5] Mood, A.M. et al., *Introduction to the Theory of Statistics*, McGraw-Hill. 1974.

[6] Venugopal, N. and Subrahmanyam, V. *The CES Method and its Applications*–Prsented at the X ISPS Conference held at Dharwad, June, 1989.(1989)

[7] Venugopal, N. and Subrahmanyam, V., *The Product Rank Correlation Coefficient with SVSR for TIES*–Presented at the 77th ISCA held at Cochin, February 1990.

Proceedings of the Asian Mathematical
Conference, 1990

Smooth Curve and Surface Interpolation*

WANG RENHONG
Institute of Applied Mathematics
Dalian University of Technology
Dalian 116024, China

§1. Smooth Curve Interpolation by Rational Splines

Let H_n be the collection of all univariate polynomials of degree n, and $P_{r,l}$ the collection of all rational functions with the form $p(x)/q(x)$, where $p(x) \in H_r$, and $q(x) \in H_1$.

Denote by T the following partition of the interval $[a, b]$:

$$T: \qquad a = x_0 < x_1 < \cdots < x_{n-1} < x_n = b.$$

If a real function $r(x)$ defined on $[a, b]$ satisfies

1) $r(x) \in R_{r,1,}$ in each $[x_j, x_{j+1}], j = 0, \cdots, n-1$;

2) $r(x) \in C^s[a, b]$,

then $r(x)$ is said to be a rational spline or type $(r, 1)^s$ with respect to the partition T.

In this part, we shall discuss the rational splines of type $(2.1)^1$ and type $(2.1)^2$ with the forms

$$r(x) = p_{1j}(x) + \frac{(x - x_j)(x - x_{j+1})}{q_{1j}(x)}, \quad x_j \le x \le x_{j+1}, \quad j = 0, \cdots, n-1, \tag{1}$$

where $p_{1j}(x)$ and $q_{1j}(x) \in H_1$.

The interpolation conditions are

$$\begin{cases} r(x_j) = y_j, & r'(x_j) = y'_j, \\ r(x_{j+1}) = y_{j+1}, & r'(x_{j+1}) = y'_{j+1}; \end{cases} \tag{2}$$

$$\begin{cases} r(x_j) = y_j, \qquad r(x_{j+1}) = y_{j+1}, \\ r'(x_j) = (f(x_{j-1}, x_j) + f(x_j, x_{j+1}))/2, \\ r'(x_{j+1}) = (f(x_j, x_{j+1}) + f(x_{j+1}, x_{j+2}))/2; \end{cases} \tag{3}$$

and

$$\begin{cases} r(x_j) = y_j, & r(x_{j+1}) = y_{j+1}, \\ r'(x_j) = y'_j, & r''(x_j) = y''_j; \end{cases} \tag{4}$$

---

* Supported by The Rolf Nevanlinna Institute, Helsinki, Finland

respectively, where $f(x_j, x_{j+1}) = f_{j,j+1}$ denotes the divided difference of the first degree.

In [3] and [4], we have obtained the rational splines satisfying the interpolation conditions (1)–(2), (1)–(3), and (1)–(4):

$$\begin{aligned} r_{1j}(x) = & \ y_j + f(x_j, x_{j+1})(x - x_j) \\ & + \frac{(x-x_j)(x-x_{j+1})(y'_j - f_{j,j+1})(y'_{j+1} - f_{j,j+1})}{(x-x_j)(y'_j - f_{j,j+1}) - (x-x_{j+1})(y'_{j+1} - f_{j,j+1})}, \\ r_{2j}(x) = & \ y_j + f(x_j, x_{j+1})(x - x_j) \\ & + \frac{(x-x_j)(x-x_{j+1})(f_{j-1,j} - f_{j,j+1})(f_{j+1,j+2} - f_{j,j+1})}{2f_{j-1,j}(x-x_j) - 2f_{j+1,j+2}(x-x_{j+1}) - 2f_{j,j+1}(x_j - x_{j+1} - 2x)}, \end{aligned}$$

and

$$\begin{aligned} r_{3j}(x) = & \ y_j + f(x_j, x_{j+1})(x - x_j) \\ & + \frac{2(y'_j - f_{j,j+1})^2(x-x_j)(x_{j+1}-x)}{2(y'_j - f_{j,j+1})(x_{j+1}-x) - y''_j(x_j - x_{j+1})(x - x_j)}, \end{aligned}$$

respectively.

Denote by $r_i(x) (i = 1, 2, 3)$ the rational splines:

$$r_i(x) = \{r(x) \in (2,1)^1 | r(x)|[x_j, x_{j+1}] = r_{ij}(x),\ j = 0, \cdots, n-1\}, \quad i = 1, 2;$$

$$r_3(x) = \{r(x) \in (2,1)^2 | r(x)|[x_j, x_{j+1}] = r_{3j}(x),\ j = 0, \cdots, n-1\}.$$

By using the rational splines $r_1(x), r_2(x)$ and $r_3(x)$ we can get smooth curve interpolations with C^1 or C^2 continuities.

In the practical problems, however, an important problem is to analyse the singularity of rational spline interpolation. We have established the following theorems.

Theorem 1. *Let the interpolation function $y = f(x) \in C^2[a, b]$. If there exists a singular point of $r_i(x) (i = 1, 2)$ in the interval $[x_j, x_{j+1}]$, then the original interpolation function $y = f(x)$ has a inflection point in the open interval (x_j, x_{j+1}).*

Theorem 2. *Let the interpolation function $y = f(x) \in C^2[a, b]$. If $r_3(x)$ has a singular point in the interval $[x_j, x_{j+1}]$, then the original interpolation function $y = f(x)$ has inflection point in the open interval (x_j, x_{j+1}).*

Theorem 3. *For any given interpolation function $f(x) \in C^2[a, b]$ we can construct a partition T of the interval [a, b], such that rational splines $r_1(x), r_2(x)$, and $r_3(x)$ based on the partition T have no singularity.*

Otherwise we shall show a method for obtaining a C^1-rational interpolation which does not need to solve any equation.

Suppose that $y = l_j(x)$ and $y = r_j(x)$ are two curves passing through (x_{j-1}, y_{j-1}), $(x_j, y_j), (x_{j+1}, y_{j+1})$, and $(x_j, y_j), (x_{j+1}, y_{j+1}), (x_{j+2}, y_{j+2})$ respectively. By using so-called weight average

$$f_j(x) = W(x)\ l_j(x) + (1 - W(x))\ r_j(x), \quad x_j \le x \le x_{j+1}, \tag{5}$$

where $W(x)$ satisfies conditions

$$\begin{aligned} W(x_j) = 1, \quad W(x_{j-1}) = 0, \\ W'(x_j) = 0, \quad W'(x_{j-1}) = 0, \end{aligned} \tag{6}$$

we get a function

$$y = F(x) = \begin{cases} l_0(x), & a = x_0 \leq x \leq x_1, \\ f_j(x), & x_j \leq x \leq x_{j+1}, \quad j = 1, \cdots, n-2, \\ r_{n-2}(x), & x_{n-1} \leq x \leq x_n = b, \end{cases}$$

which has C^1-continuity and satisfied the following interpolation conditions

$$F(x_j) = y_j, \quad j = 0, \cdots, n.$$

Especially, $W(x)$ may be one of the following functions

$$1 - 3\underline{x}^2 + 2\_^3; \quad -3 + 2\_((1\!\!+\frac{1+\underline{x}}{1+\underline{x}}, \quad 9 - x\_ + 3\underline{x}^2 \quad \frac{8}{1+\underline{x}} \quad -8 \quad 3\underline{x}p + \frac{9+9\underline{x}}{1+\underline{x}+\underline{x}}, \quad \text{etc.},$$

where $\overline{x} = (x - x_j)/(x_{j+1} - x_j)$.

§2. Smooth Surface Interpolation

In this part, we shall present the localized methods for constructing smooth surface interpolation to a given scattered data.

Let $f(x,y)$ be know at only a limited number of points (scattered data). It is usually important in practice that the approximating function $p(x,y)$ has good localization properties. In other words, we require a form for $f(x,y)$ in any small region of the (x,y)-space with little or no influence on the value of $p(x,y)$ when (x,y) is far from the small region.

The first, by using the points of the scattered data and a certain optimality criteria, we can generate a triangulation T, such that the set of vertices of T is just the set of points of the scattered data.

Denote by T^1 the refinement of T which is obtained by subdividing each triangle of T into six triangles ([1]). By using the B-spline method given in paper [2]. We can generate a C^1-quadratic surface $S(x,y)$, provided the function $f(x,y)$ and its gradient at each vertex have been fixed. In fact, $S(x,y)$ satisfies interpolation conditions

$$\begin{cases} S(v_i) = f(v_i), \\ \text{Grad } S(v_i) = \text{Grad } f(v_i), \end{cases} \quad \text{for each vertex } v_i \text{ or } T,$$

and can be represented by the values of $f(x,y)$ and Grad f at the vertices of T.

A practical problem is how to determine the Grad f at all vrti ces of T. A natural way is to solve the minimization problem

$$\int\!\!\int_T \|\text{Grad } S(x,y)\|_2^2 dxdy = \text{Min}.$$

In general, however, we shall use the simpler local methods to determine those gradients.

For any given vertex v of T, for example, by using a suitable convex combination of Grad f on all triangles including v, we can obtain an approximation of Grad f at the vertex v.

Secondly, it is well known, Arner, Barnhill, Farin and Little have proposed an unsetted question if a C^2 scheme exists for the split depicted in Fig.1 which likes Morgan-Scott triangulation. Denote by T the refinement of T.

If $(\alpha,\beta,\beta,),(\beta,\alpha,\beta)$, and $(\beta,\beta,\alpha)(\alpha+2\beta=1)$ are the barycentric coordinates of the interior vertices B_1, B_2 and B_3 in the triangle $A_1A_2A_3$ respectively, we have proved

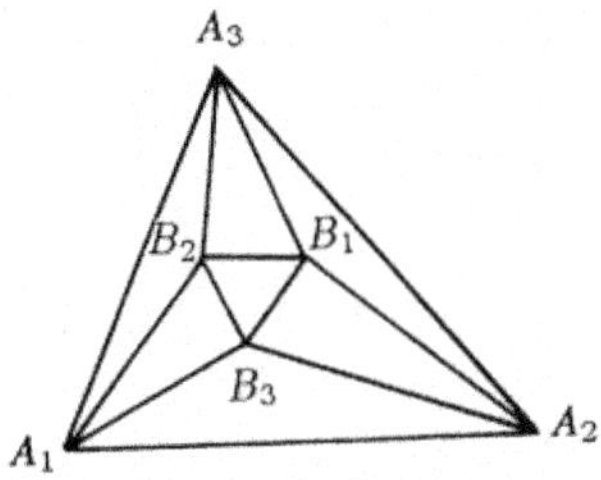

Theorem 4. *For any given parameters* $a_i^m (i=1,2,3,\ m=1,\cdots,6), b_i, d_i$ *and* $d_{i,j}(i=1,2,3, j=1,2,)$ *there exists a unique* C^2 *surfaces* $z=S(x,y)$ *of piecewise polynomial of degree 5 over* T^2 *such that*

$$\begin{cases} (1, D_x, D_y, D_{xx}, D_{xy}, D_{yy})S(A_i) = (a_i^1, a_i^2, \cdots, a_i^6), \\ S(B_i) = b_i, \\ D_{n_i}S((A_i + A_{i+1})/2) = d_i, \\ D_{n_i^2}S((j\cdot A_i + k\cdot A_{i+1})/3) = d_{i,j},\ k = 3-j,\ j=1,2, \end{cases}$$

where $A_4 = A_1$, *and* n_i *is the unit exterior normal vector on the edge of the triangle* $A_1A_2A_3$ *opposite to* A_i.

In [5] we have shown the method how to express this C^2 surface interpolation.

References

[1] M.J. Powell and M.A. Sabin, *Piecewise quadratic approximation on triangles*, ACM Trans. on Math. Software, 3(1977), 316–325.

[2] R.H. Wang, W.B. Wang, S.M. Wang and X.Q. Shi, *The C^1-quadratic spline space on triangulations*, Mathematica Applicata, 1(1988), 123–131.

[3] S.T. Wu and R.H. Wang, *On a problem of rational spline interpolation*, Numer. Math. J. Chinese Univ., 3(1981), 165–175.

[4] R.H. Wang and S.T. Wu, *On the rational spline functions*, Jour. Math. Res. and Exp., 4, (2) (1984), 31–36.

[5] R.H. Wang and Y.W. Zhan, *A bivariate C^2 Morgan-Scott scheme*, to appear.

Proceedings of the Asian Mathematical
Conference, 1990

Finite Element Approximations for Some Nonlinear Quasiconformal Mappings in Multiply Connected Domains

WEN GUO-CHUN
Department of Mathematics
Peking University
Beijing 100871, China

LI PING-QIAN
Department of Mathematics
Yantai University
Shandong 264005, China

§1. Reduction of Quasiconformal Mapping for Elliptic Complex Equation in a Multiply Connected Domain to a Boundary Value Problem.

The so-called quasiconformal mapping is showed to be a homeomorphic solution of the elliptic complex equation

$$w_{\bar z} f(z,w,w_z),\quad f = Q_1 w_z + Q_2 \bar w_{\bar z} = Q_0 w_z,\quad Q_j = Q_j(z,w,w_z),\quad j=0,1,2 \tag{1.1}$$

in a domain D. In this paper, we consider that D is an $(N+1)$-connected bounded domain with the boundary $\Gamma = \cup_{j=0}^{N}\Gamma_j \in C_{\mu}^{1}(0<\mu<1)$, where $\Gamma_j (j=0,1,\cdots,N)$ are simple closed curves, $\Gamma_j (j=1,\cdots,N)$ are located inside Γ_0, and $z=0\in D$. Suppose that the complex equation (1.1) satisfies Condition C, namely

(1) $Q_j(z,w,U)(j=1,2)$ are continuous in $w\in \mathbb{C}$ for almost every point $z\in D$ and $U\in \mathbb{C}$.

(2) $f(z,w,U)$ satisfies the following uniform ellipticity condition, i.e.

$$|f(z,w,U_1) - f(z,w,U_2)| \le q_0 |U_1 - U_2| \tag{1.2}$$

for almost every point $z\in D$ and $w, U_1, U_2 \in \mathbb{C}$, where $q_0 (0\le q_0<1)$ is a constant.

(3) For any function $v(z)\in W_{p_0}^1(D)$, there is

$$\left\| Q_1\left(z, \frac{1}{z}+v, -\frac{1}{z^2}+v^2\right)/z^2 + Q_2\left(z,\frac{1}{z}+v,-\frac{1}{z^2}+v_z\right)/\bar z^2\right\|_{L_p(D)} \le k_0, \tag{1.3}$$

where $p, p_0 (2<p_0<p<\infty), k_0(\ge 0)$ are constants.

Problem A. Find a homeomorphic solution of (1.1) which maps D onto a rectilinear slit domain G with the condition $w(0)=\infty$. The so-called rectilinear slit domain G means a domain whose boundary consists of $N+1$ rectilinear slits $L_j(j=0,1,\cdots,N)$ with oblique angles $\theta_j - \frac{\pi}{2} (j=0,\cdots,N)$ respectively.

In order to find a solution of Problem A, we introduce

Problem B. Find a function $W(z) \in C(\bar{D}) \cap H^1(D)$, such that

$$W_{\bar{z}} = F(z, W, W_z), \quad F = Q_1 W_z + Q_2 \bar{W}_{\bar{z}} + A,$$

$$Q_j = Q_j\left(z, \frac{1}{z} + W, -\frac{1}{z^2} + W_z\right), \quad j = 1, 2, \tag{1.4}$$

$$A = -Q_1/z^2 - Q_2/\bar{z}^2,$$

$$\mathrm{Re}[e^{-i\theta_j} W(t)] = -\mathrm{Re}[e^{-i\theta} j/t] + h(t), \quad t \in \Gamma, \tag{1.5}$$

$$LW = \mathrm{Im} \int_{\Gamma_0} e^{-i\theta_0} W(t) ds_0 = 0, \tag{1.6}$$

$$h(t) = h(1/t, W) = \begin{cases} 0, & t \in \Gamma_0, \\ \frac{1}{S_j} \int_{\Gamma_j} \mathrm{Re}[e^{-\theta_j}(\frac{1}{t} + W(t))], \ ds_j, & j = 1, \cdots, N, \end{cases} \tag{1.7}$$

where S_j is the arc length of $\Gamma_j, j = 0, 1, \cdots, N$.

Using the method as stated in [1], we can prove that there exists a solution of Problem B.

Theorem 1.1. *If $W(z)$ is a solution of Problem B, then $w(z) = \frac{1}{z} + W(z)$ is a solution of Problem A.*

Proof. Let $W(z)$ be a solution of Problem B. Obviously $w(z)$ satisfies the complex equation (1.1) and the boundary condition

$$\mathrm{Re}[e^{-i\theta_j} w(t)] = h(t), \quad t \in \Gamma. \tag{1.8}$$

According to the representation theorem of solutions of (1.1), we have $w(z) = \Phi[\varsigma(z)]$, where $\varsigma(z)$ is a homeomorphism, which maps D onto an $(N+1)$-connected circular domain H in the unit disk $|\varsigma| < 1$ with the condition $\varsigma(0) = 0$, $L_0 = \varsigma(\Gamma_0) = \{|\varsigma| = 1\}$, and $\Phi(\varsigma)$ is an analytic function in $H - \{0\}$. Similarly to Theorem 3.4 of Chapter 3 in [2], we can prove that $\Phi(\varsigma)$ is univalent. So $w(z)$ is a quasiconformal mapping which maps D onto G.

§2. Finite Element Solutions of Problem A

On the basis of Theorem 1.1, it is sufficient to find an approximate solution of Problem B by using finite element method. We use the regular triangular net to cut the domain D and denote by $\Delta_m (m = 1, \cdots, n)$ all triangular units, so that

$$\Delta = \bigcup_{m=0}^{n} \Delta_m \supset D, \quad \Delta_m \cap D \neq \phi, \quad h_m/d_m \leq k_0, \quad h/h_m \leq k_0, \quad 1 \leq m \leq n,$$

here $h = \max_{1 \leq m \leq n} h_m$, h_m is the miximum of lengths of three sides of Δ_m, d_m is the diameter of inner tangent circle in Δ_m. Denote by subspace E of $C(\bar{D}) \cap H^1(D)$ the set of all functions which are one-order polynomials in $z, \bar{z}$ on each triangular unit. Similarly to [3], we consider the following finite element approximation:

Problem C. Find $W_h \in E$, such that

$$J(W_h) = \min_{v\in E} J(v) = \min_{v\in E}\Big[\int\int_D |v_{\bar z} - F(z,v,v_z)|^2\, d\sigma_z \\ +(1/h)\int_\Gamma \{\mathrm{Re}[e^{-i\theta_j}(1/t+v(t))] - h(1/t,v)\}^2\, ds + (Lv)^2\Big]. \tag{2.1}$$

Lemma 2.1. *There exists a constant $M = M(q_0, h) > 0$, such that for any $v(z) \in E$,*

$$\begin{aligned} a(Q,v) = &\int\int_D |v_{\bar z} - Q(z)v_z|^2 d\sigma_z \\ &+\frac{1}{h}\int_\Gamma [\mathrm{Re}(e^{-i\theta_j}v(t)) - h(0,v)]^2 ds + (Lv)^2 \\ \geq\ & M\|v\|^2_{C(\bar D)\cap H^1(D)} \end{aligned} \tag{2.2}$$

holds, where $\|Q\|_{L_\infty(D)} \leq q_0 < 1, \|v\|_{C(\bar D)\cap H^1(D)} = \|v\|_{C(\bar D)} + \|v\|_{H^1(D)}$.

Proof. Let $A = \{(Q,v)|\ \|Q\|_{L_\infty(D)} \leq q_0, v \in E, \|v\|_{C(\bar D)\cap H^1(D)} = 1\}$, we are only to prove $\inf_A a(Q,v) > 0$. Otherwise, there exists $\{(Q_n, v_n)\} \subset A$ such that $a(Q_n, v_n) \to 0$ as $n \to \infty$. There is no harm in assuming that $\|v_n - v_0\|_{C(\bar D)\cap H^1(D)} \to 0, Q_n \xrightarrow[\text{weak}]{L_q(\bar D)} Q_0, q > 2$ as $n \to \infty$. Obviously $\|Q_0\|_{L_\infty(D)} \leq q_0 < 1$. Due to $v_0(z) \in E$, it can be seen that $v_{0\bar z}$ is a piecewise constant. Hence from $\|v_{n\bar z} - Q_n v_{nz}\|_{L_2(D)} = 0$, it follows that $v_{0\bar z} - Q_0 v_{0z} = 0, z \in D$, and $\mathrm{Re}(e^{-i\theta_j}v_0) = h(0,v), t \in \Gamma, L(v_0) = 0$. According to the method in the proof of Theorem 6.1, Chapter 5 in [1], we obtain $v_0(z) = 0$. However from $\|v_n\|_{C(\bar D)\cap H^1(D)} = 1$, it can be derived $\|v_0\|_{C(\bar D)\cap H^1(D)} = 1$. This contridiction shows that Lemma 2.1 is true.

Theorem 2.2. *Problem C has a solution.*

Proof. Noting $\min_{v\in E} J(v) = \min_{v\in E, J(v)\leq J(0)} J(v)$, we may only prove that $\min_{v\in E, J(v)\leq J(0)} J(v)$ exists. Setting $E_1 = \{v|v \in E, J(v) \leq J(0)\}$, noting the condition (1.3) and applying Lemma 2.1, for any $v(z) \in E_1$, we have

$$\begin{aligned} \|v\|^2_{C(\bar D)\cap H^1(D)} \leq\ & (1/M)\{\int\int_D |v_{\bar z} - F(z,v,v_z - Q_1/z^2 - Q_2/\bar z^2|^2\, d\sigma_z \\ &+(1/h)[\int_{\Gamma_j}[\mathrm{Re}[e^{-i\theta_j}(1/t+v(t))] - \mathrm{Re}(e^{-i\theta_j}/t) - h(1/t,v) \\ &+h(1/t,v) - h(0,v)]^2\, ds] + (Lv)^2\} \\ \leq\ & (1/M)\{2J(0) + k_0^2[\mathrm{mes}(D)]^{1-2/p} + 1/h \\ &\cdot\int_\Gamma [\mathrm{Re}(e^{-i\theta_j}/t) + h(0,v) - h(1/t,v)]^2\, ds\}. \end{aligned}$$

Since $h(0,v) - h(1/t,v)$ is independent of v, this shows that E_1 is a bounded set.

Next, suppose that for $v_n(z) \in E_1, n = 1,2,\cdots, v(z) \in E$, $\|v_n - v\|_{C(\bar D)\cap H^1(D)} \to 0$ as $n \to \infty$. Similarly to the proof of Theorem 3.1, Chapter 4 in [1], it can be obtained

$$S_n = \|f(z, 1/z + v_n, -1/z^2 + v_z) - f(z, 1/z + v, -1/z^2 + v_z)\|_{L_2(D)} \to 0$$

as $n \to \infty$. Therefore

$$\begin{aligned} S'_n = \quad & \|F(z, v_n, v_{nz}) - F(z, v, v_z)\|_{L_2(D)} \\ & \le q_0 \|v_{nz} - v_z\|_{L_2(D)} + S_n \to 0, \\ J(v_n) \to \quad & J(v) \quad \text{as} \quad n \to \infty. \end{aligned}$$

Thus $v(z) \in E_1$, and $J(v)$ is a continuous function on the bounded closed set E_1. Hence there exists $W_h(z)$ such that $J(W_h) = \min_{v \in E} J(v)$.

The solution $W_h(z)$ of Problem C will be called the numerical solution of Problem B. In the following, we shall give error estimates on $W_h(z)$.

§3. L_2 and H^1-Error Estimates of the Numerical Solution

Lemma 3.1. *Let* $T_0 f = \int_\Gamma \frac{f(\varsigma)}{\varsigma - z} d\varsigma, Tf = -\frac{1}{\pi} \iint_D \frac{f(\varsigma)}{\varsigma - z} d\sigma_\varsigma$. *Then for any* $f(z) \in H^{1/2}(\Gamma)$, *there are*

$$\|T_0 f\|_{H^{\frac{1}{2}}(\Gamma)} \le M(D) \|T_0 f\|_{H^1(\Gamma)} \le M(D) \|f\|_{H^{\frac{1}{2}}(\Gamma)}, \tag{3.1}$$

$$\|T_0 f\|_{L_2(D)} \le M(D) \|f\|_{H^{\frac{1}{2}}(\Gamma)}, \tag{3.2}$$

$$\|T_0 f\|_{H^{\frac{1}{2}}(\Gamma)} \le M(D) \|f\|_{H^{\frac{1}{2}}(\Gamma)}, \tag{3.3}$$

and for any $f(z) \in L_2(\bar{D})$, *we have*

$$\|Tf\|_{H^{-\frac{1}{2}}(\Gamma)} \le M(D) \|f\|_{H^{-1}(D)}, \tag{3.4}$$

$$\|Tf\|_{L_2(D)} \le M(D) \|f\|_{H^{-1}(D)}. \tag{3.5}$$

Proof. (3.1) and (3.2) are immediate consequences of Lemma 2.1 in [4]. Applying (3.1), (3.2) and according to the method used in Lemma 2.1 in [4], we can conclude (3.3)-(3.5).

Lemma 3.2. *Suppose that* $\Gamma \in C^2_\mu, W(z) \in C(\bar{D})$ *is analytic in* D *and satisfies the boundary condition*

$$\operatorname{Re}[e^{-i\theta_j} w(t)] = r(t) + h(t), \quad t \in \Gamma, \tag{3.6}$$

$$Lw = b = \text{constant}, \tag{3.7}$$

where $r(t) \in C_\alpha(\Gamma), 0 < \alpha, \mu < 1$. Then

$$\|w\|_{L_2(\bar{D})} \le M(D)[\|r(t)\|_{H^{-\frac{1}{2}}(\Gamma)} + |b|]. \tag{3.8}$$

Proof. There is no harm in assuming that D is a circular domain similar to H as stated in the proof of Theorem 1.1. In accordance with Theorem 5.2 in [2], the analytic function $w(z)$ can be expressed as

$$w(z) = \frac{1}{2\pi} \int_\Gamma P(z,t) r(t) d\theta + \left[b - \frac{1}{2\pi} L \left(\int_\Gamma P(z,t) r(t) d\theta \right) \right] \Phi_0, \tag{3.9}$$

where $\Phi_0(z)$ is an analytic function with the condition $\text{Re}[e^{-i\theta_j} \cdot \Phi_0(t)] = h(t), t \in \Gamma$ and $L\Phi_0 = 1$. For any $R(t) \in L_2(\Gamma)$, it is evident that

$$|LR| \le M(D)\|R\|_{H^{-\frac{1}{2}}(\Gamma)}. \tag{3.10}$$

From (3.9), (3.10), (3.2) and (3.3), we know that Lemma 3.2 is true.

Lemma 3.3. *If $\Gamma \in C^2_\mu$, and*

$$\begin{aligned} &\|Q_j\|W^1_{p_2}(D) \le k_1 < \infty (p_0 > 2), \quad \|A_j(z)\|_{H^1(D)} \le K_1 \quad (j = 1, 2), \\ &A_3(z) \in H^1(D), \quad r(t) \in H^{\frac{3}{2}(\Gamma)}, \quad p_2 = 2p_0/(p_0 - 2), \end{aligned} \tag{3.11}$$

then there exists a unique solution $w(z) \in H^2(D)$ of the following boundary value problem

$$w_{\bar z} = Q_1(z)w_z + Q_2(z)\overline{w}_{\bar z} + A_1(z) + A_2\overline{w} + A_3(z), \quad z \in D, \tag{3.12}$$

$$\text{Re}[e^{-i\theta_j} w(t)] = r(t) + h(t), \quad t \in \Gamma, \tag{3.13}$$

$$Lw = b, \tag{3.14}$$

and $w(z)$ satisfies the estimates

$$\|w\|_{H^2(D)} \le M(q_0, p_0, k_0, k_1, D)[\|r\|_{H^{\frac{3}{2}}(\Gamma)} + \|A_3\|_{H^1(D)} + |b|], \tag{3.15}$$

$$\|w\|_{L^2(\overline{D})} \le M(Q_1, Q_2, p_0, k_0, k_1, D)[\|r\|_{H^{-\frac{1}{2}}(\Gamma)} + \|A_3\|_{H^{-1}(D)} + |b|]. \tag{3.16}$$

Proof. (3.12)–(3.15) can be get from Lemma 2.3 and Theorem 3.2 in [3]. It remains to prove (3.16). Noting the condition (3.11) and using Lemma 1.2 of Chapter 1 in [1], we can transform (3.12) into a standard form, where $\varsigma(z) \in W^2_{p_2}(D)$. Hence there is no harm in assuming that $Q_1 = Q_2 = 0$ in (3.12) and D is a circular domain. Applying Theorem 1.1, Chapter 4 in [1], $w(z)$ can be expressed as

$$w(z) = [\Phi(z) + \psi(z)]e^{\varphi(z)}, \tag{3.17}$$

where $\Phi(z)$ is an analytic function in D and $\varphi(z), \psi(z)$ satisfy

$$\|\varphi\|_{W^1_{p_0}(D)} \le M(p_0, k_0, D), \quad \psi(z) = T[A_3 e^{-\varphi}]. \tag{3.18}$$

On account of $\|Tg\|_{H^1(D)} \le M(D)\|g\|_{L_2(\overline{D})}$, we obtain

$$\|\psi\|_{L_2(\overline{D})} \le M(D)\|A_3 e^{-\varphi}\|_{H^{-1}(D)} \le M(p_0, k_0, D)\|A_3\|_{H^{-1}(D)}, \tag{3.19}$$

$$\|\psi\|_{H^{-\frac{1}{2}}(\Gamma)} \le M(D)\|A_3 e^{-\varphi}\|_{H^{-1}(D)} \le M(p_0, k_0, D)\|A_3\|_{H^{-1}(D)}. \tag{3.20}$$

It is clear that $\Phi(z)$ satisfies the boundary condition

$$\begin{aligned} \text{Re}[e^{-i\theta_j} e^{\varphi}\Phi(t)] &= r(t) - \text{Re}[e^{-i\theta_j} e^{\varphi}\psi(t)] + h(t), \quad t \in \Gamma, \\ L(e^{\varphi}\Phi) &= b - L(e^{\varphi}\psi). \end{aligned}$$

Similarly to the proof of Lemma 3.2, there is

$$\|\Phi\|_{L_2(\overline{D})} \leq M(p_0, k_0, k_1, D)[\|r - \mathrm{Re}[e^{-i\theta_j+\varphi}\psi]\|_{H^{-\frac{1}{2}}(\Gamma)} + |b| + |L(e^{\varphi}\psi)|]. \quad (3.21)$$

From (3.19), (3.20), (3.4) and (3.5), it follows that (3.16) holds.

Lemma 3.4. *If $w(z) \in C_\alpha(\overline{D}) \cap H^1(D)(0 < \alpha < 1)$ is a solution of the boundary value problem (3.12)–(3.14), but the coefficients of (3.12)–(3.14) satisfy the following condition*

$$\begin{aligned} &\| |A_1| + |A_2| \|_{L_p(\bar{D})} \leq k_0 < \infty, \quad A_3(z) \in L_p(\bar{D}), \quad p > 2, \\ &\| |Q_1| + |Q_2| \|_{L_\infty(D)} \leq q_0 < 1, \\ &\|Q_j\| C_\alpha(\overline{D/D_\delta}) \leq k_1 < \infty, \quad j = 1, 2, \quad r(t) \in C_\alpha(\Gamma), \quad 0 < \alpha < 1, \end{aligned} \quad (3.22)$$

in which $D_\delta = \{z \mid z \in D,\ dist(z, \partial D) \geq \delta > 0\}$, k_0, k_1, q_0, α are constants, then the following estimates hold:

$$\|w\|_{L_2(D)} \leq M_1 k_2, \quad \|w\|_{H^1(D_\delta)} \leq M_2 k_2, \quad (3.23)$$

where $k_2 = \|A_3\|_{L_2(D)} + \|r\|_{L_2(\Gamma)} + |b|$, $M_j = M_j(q_0, p_0, k_0, k_1, \alpha, \delta, D), j = 1, 2,$ $2 < p_0 < p$.

Proof. We may assume that D is a circular domain as before. According to Theorem 1.1, Chapter 4 in [1], $w(z)$ can be written as

$$w(z) = [\Phi(\varsigma(z)) + \psi(z)]e^{\phi(z)}, \quad (3.24)$$

where $\varsigma(z)$ is a homeomorphism as stated in the proof of Theorem 1.1, $\|\psi(z), \|_{H^1(D)} \leq M_3\|A_3\|_{L_2(D)}, \|e^{\phi(z)}\|_{W^1_{p_0}(D)} \leq M_4, \|\varsigma(z)\|_{W^1_{p_0}}(D) \leq M_5, C^1_\alpha[\varsigma(z), \overline{D \backslash D\delta/2}] \leq M_6, C^1_\alpha[z(\varsigma), \varsigma(\overline{D \backslash D\delta/2})] \leq M_7, M_j = M_j(q_0, p_0, k_0, \alpha, \delta, k_1, D), j = 3, \cdots, 7$. Hence we have

$$\|\Phi(\varsigma)\|_{H^1[\varsigma(\overline{D\delta/2})]} \leq M_8 k_2, \quad \|\Phi(\varsigma)\|_{L_{p_2/2}[\varsigma(\overline{D\delta/2})]} \leq M_9 k_2, \quad (3.25)$$

where $M_j = M_j(q_0, p_0, k_0, k_1, \alpha, \delta, D), j = 8, 9$. On the basis of Theorem 4.1, Chapter 2 in [6], it can be obtained

$$\|\Phi[\varsigma(z)]\|_{H^1(D\delta/2)} \leq M_{10}[\|\Phi(\varsigma)\|_{L_{p_2}(\varsigma(\overline{D\delta/2}))} + \|\Phi_\varsigma\|_{L_2(\varsigma(\overline{D\delta/2}))}, \quad (3.26)$$

where $M_{10} = M_{10}(q_0, p_0, \delta, D)$. Combining (3.24)–(3.26), the second estimate in (3.23) is derived.

Noting that $\varsigma(z)$ and $z(\varsigma)$ are continuously differentiable in a neighborhood of the boundaries Γ and $L = \varsigma(\Gamma)$ respectively, and using the trace theorem, we choose $N + 1$ simple smooth closed curves Γ_j^*, such that

$$\|w\|_{H^{\frac{1}{2}}(\Gamma_j^*)} \leq M_{11} k_2, \quad 1 \leq j \leq N + 1, \quad (3.27)$$

where $M_{11} = M_{11}(q_0, p_0, k_0, k_1, \alpha, D)$. In a similar way used as before, it can be derived that

$$\|w\|_{L_2(D \backslash D\delta/2)} \leq M_{12} k_2,$$

where $M_{12} = M_{12}(q_0, p_0, k_0, k_1, \alpha, D)$. Due to for $H^1(D\delta/2)$ the second estimate in (3.23) still holds, and $H^1(D\delta/2) \hookrightarrow L_2(\overline{D\delta/2})$, thus the first estimate in (3.23) is obtained.

In what follows we shall give the error estimates for the linear case and nonlinear case of the complex equation (1.1) respectively.

Theorem 3.5. *Suppose that $\Gamma \in C^2_\mu, W(z) \in H^2(D)$, and for any $W_1(z)$ and $W_2(z) \in W^1_p(D)$, $F(z, W, W_z)$ satisfies*

$$F(z, W_1, W_{1z}) - F(z, W_2, W_{2z}) = Q_1(W_1 - W_2)_z + Q_2(\bar{W}_1 - \bar{W}_2)_{\bar{z}} + A(W_1 - W_2), \quad (3.28)$$

$$Q_j = Q_j(z, W_1, W_2, W_{1z}, W_{2z}), \quad z \in \bar{D},$$

$$Q_j = Q_j(z), \quad z \in \overline{D \backslash D_\delta}, \quad A + A(z, W_1, W_2, W_{1z}, W_{2z}),$$

where $\| \, |Q_1| + |Q_2| \, \|_{L_\infty(D)} \le q_0 < 1, C_\alpha(Q_j, \overline{D\backslash D_\delta}) \le k_0 < \infty, \quad j = 1, 2, \quad L_p(A, \bar{D}) \le k_0, p > 2$. *Then*

$$\|W - W_h\|_{L_2(D)} = O(h), \quad \|W - W_h\|_{H^1(D_\delta)} = O(h), \quad (3.29)$$

$$\|W - W_h\|_{L_2(D)} = O(h), \quad \|W - W_h\|_{H^1(D_\delta)} = O(h). \quad (3.30)$$

Proof. Noting the condition: $\Gamma \in C^2$ and $W(z) \in H^2(D)$, and using the extension theorem, there exists $W^*(z) \in H^2(R^2)$ such that $W^*(z) = W(z), z \in D, \|W^*\|_{H^2(R^2)} \le M(D)\|W\|_{H^2(D)}$. Denote by $\widehat{W}(z)$ the interpolation of $W^*(z)$, obviously $\widehat{W}(z) \in E$. By means of the interpolation approximation theorem, we have

$$\|W(z) - \widehat{W}(z)\|_{H^1(D)} \le \|W^*(z) - \widehat{W}(z)\|_{H^1(\Delta)} \le Mh\|w^*\|_{H^2(\Delta)} \le Mh\|W\|_{H^2(D)} = O(h).$$

Taking into account that $W(z)$ is a solution of Problem B, and using Condition C and (3.28), there is $J(\widehat{W}) = O(h)$, hence $J(W_h) \le J(\widehat{W}) = O(h^2)$. Noting that $W - W_h$ satisfies

$$\begin{aligned} (W - W_h)_{\bar{z}} = {} & F(z, W, W_z) - F(z, W_h, W_{hz}) + B_h(z) \\ = {} & Q_1(W - W_h)_z + Q_2(\bar{W} - \bar{W}_h)_{\bar{z}} + A((W - W_h) + B_h(z), \\ Q_j \; & = Q_j(z, W, W_h, W_z, W_{hz}), \quad j = 1, 2, \\ A \; & = A(z, W, W_h, W_z, W_{hz}), \quad \|B_h\|_{L_2(D)} = O(h). \end{aligned} \quad (3.31)$$

$$\mathrm{Re}[e^{-i\theta_j}(W - W_h)] = h(0, W - W_h) + r_h(t), \quad t \in \Gamma, \quad \|r_h\|_{L_2(\Gamma)} = O(h^{\frac{3}{2}}), \quad (3.32)$$

$$L(W - W_h) = b_h, \quad |b_h| = O(h). \quad (3.33)$$

By making use of Lemma 3.4, it is clear that (3.29) holds. Moreover, $W_h(z) = 1/z + W_h(z)$ possesses the error estimate (3.30).

Theorem 3.6. *If the complex equation (1.1) satisfies Condition C and is linear, where $Q_j = Q_j(z) \in W^1_{p_2}(D), j = 1, 2, \quad Q_1/z^2 + Q_2/\bar{z}^2 \in H^1(D), \Gamma \in C^2$, then the following error estimates hold:*

$$\|W - W_h\|_{L_2(D)} = O(h^2), \quad \|W - W_h\|_{H^1(D)} = O(h), \quad (3.34)$$

$$\|w - w_h\|_{L_2(\overline{D})} = O(h^2), \quad \|w - w_h\|_{H^1(D)} = O(h). \tag{3.35}$$

Proof. Using Lemma 3.3 and the similar method used in Chapter 8 of [3], we can prove the theorem.

Finally, we can similarly discuss the finite element approximation of quasiconformal mapping which maps D onto a sprial slit domain G with oblique angles $\theta_j - \frac{2}{\pi}(j = 0, 1, \cdots, N)$ such that $w(0) = \infty$, $w(a) = 0 (a \in D)$, where $\theta_j (j = 0, 1, \cdots, N)$ are real constants.

References

[1] Wen Guo-chun, *Linear and nonlinear elliptic complex equations*, Shanghai Scientific and Technical Publishers, 1986.

[2] Wen Guo-chun, *Conformal mappings and boundary value problems*, Higher Education Press, 1985.

[3] Wendland W., *Elliptic systems in the plane*, Pitman, 1979.

[4] Wen Guo-chun and Li Ping-qian, *The weak solution of the Riemann-Hilbert problem for elliptic complex equations of first order*, Complex Variables, Theory and Applications, 16(1991) (to appear).

[5] Wen Guo-chun and Begehr H., *Boundary value problems for elliptic equations and systems*, Longman Scientific and Technical Publishing Co., 1990.

[6] Wen Guo-chun, Yang Guang-wu and Huang Sha, *Generalized analytic functions and their generalizations*, Hebei Education Press, 1989.

Proceedings of the Asian Mathematical
Conference, 1990

Cone-absolutely Summing Mapping, Majorizing Mappings and Related Classes of Operators

YAU-CHUEN WONG
Department of Mathematics
The Chinese University of Hong Kong
Shatin, N.T. Hong Kong

Let E and F be locally convex spaces (abbreviated L C S). Recall that an operator $T : E \to F$ is absolutely summing if for any continuous seminorm q on F there is a disked o-neighbourhood V in E such that

$$\sum_{i=1}^{n} q(Tx_i) \leq \sup\left\{\sum_{i=1}^{n} \mid f(x_i) \mid \ : f \in V^0\right\} \quad \text{for any finite } \{x_1, \cdots, x_n\} \subset E.$$

Denote by $\mathcal{L}^s(E,F)$ the set of all absolutely summing mapping from E into F.

The notion of absolutely summing mapping considered first by the Great French mathematician *A*. Grothendieck in 1953 for studying the nuclearity of locally convex spaces; he also proved the following important well-known result:

Theorem 1 (Grothendick). *A LCS E is nuclear if and only if*

$$\mathcal{L}^s(E,F) = \mathcal{L}(E,F) \quad \text{for any LCS (or normed space) } F.$$

Denote by $E^{(IN)}$ the algebraic direct sum whose elements are denoted by $[x_n]_{(IN)}$ or $[x_n, n \in (IN)]$, and by $T^{(IN)} : E^{(IN)} \to F^{(IN)}$ the direct sum map of T (where $T : E \to F$ is linear), defined by

$$T^{(IN)}([x_n])_{(IN)} = [Tx_n]_{(IN)} \quad (\text{for any } [x_n]_{(IN)} \in E^{(IN)}).$$

Let p be a continuous seminorm on E and $V_p = \{x \in E : p(x) \leq 1\}$. We write, for nay $[x_n]_{(IN)} \in E^{(IN)}$, that

$$p_\pi[[x_n]_{(IN)}] = \sum_{(IN)} p(x_n);$$

$$p_\epsilon[[x_n]_{(IN)}] = \sup\left\{\sum_{(IN)} |f(x_n)| : f \in V_p^0\right\};$$

$$p_{(n)}[[x_n]_{(IN)}] = \sup\{p(x_n) : n \in (IN)\}.$$

Denoted by $E_\pi^{(IN)}$ the vector space $E^{(IN)}$ equipped with the topology $\{p_\pi : p \in I\!P_E\}$ (where $I\!P_E$ is the family of all continuous seminars on the LCS E). Similarly one can define $E_\epsilon^{(IN)}$ and $E_{(n)}^{(IN)}$. It can be shown that an operator $T : E \to F$ is absolutely summing (see Wong [1976]) if and only if

$$T^{(IN)} : E_\epsilon^{(IN)} \to F_\pi^{(IN)} \quad \text{is continuous.}$$

From this, it is natural to consider the following two classes of operators:

$$\text{(I)} \quad \mathcal{L}^{ss}(E.F) = \{T \in \mathcal{L}(E,F) : T^{(IN)} \in \mathcal{L}(E_{(n)}^{(IN)}, F_\pi^{(IN)})\} \text{ and}$$

$$\text{(II)} \quad \mathcal{L}^{qs}(E.F) = \{T \in \mathcal{L}(E,F) : T^{(IN)} \in \mathcal{L}(E_{(n)}^{(IN)}, F_\epsilon^{(IN)})\}.$$

It is easily seen that

$$\mathcal{L}^{ss}(E,F) \subset \mathcal{L}^{s}(E,F) \text{ and } \mathcal{L}^{ss}(E,F) \subset \mathcal{L}^{qs}(E,F),$$

so that elements in $\mathcal{L}^{ss}(E,F)$ are called temporarily strongly absolutely summing operators.

Because of Theorem 1, it is natural to ask the following question: (Q1) What kind of nice property of E is such that $I^{(IN)} \in \mathcal{L}(E_{(n)}^{(IN)},\ E_\pi^{(IN)})$?

Suppose now that $(X, X_+, I\!P_X)$ is a locally solid space, and that $p \in I\!P_X$ is a regular seminorm. Denote by $p_{\pi,D}$ (resp. $p_{\epsilon,D}$ and $p_{(n),D}$) the decomposable seminorm associated with p_π (resp. p_ϵ and $p_{(n)}$), that is, for any $[x_n]_{(IN)} \in X^{(IN)}$,

$$p_{\pi,D}([x_i]_{(IN)}) = \inf\{p_\pi[[u_i]_{(IN)}] + p_\pi[[v_i]_{(IN)}] : \quad [u_i], [v_i] \in X_+^{(IN)}, \} \\ [x_i] = [u_i] - [v_i].$$

Denoted by $X_{\pi,D}^{(IN)}$ the ordered vector space $(X^{(IN)}, X_+^{(IN)})$ equipped with the locally solid topology $\{p_{\pi,D} : p \in I\!P_X\}$; similarly one can define the locally solid spaces $X_{\epsilon,D}^{(IN)}$ and $X_{(N),D}^{(IN)}$.

Definition 2. Let $(X, X_+, I\!P_X)$ be a locally solid space, let $(F, I\!\!Q_F)$ be a LCS and suppose

$$\mathcal{L}^{l}(X,F) = \{T \in \mathcal{L}(X,F) : T^{(IN)} \in \mathcal{L}(X_{\epsilon,D}^{(IN)}, F_\pi^{(IN)})\}.$$

Elements in $\mathcal{L}^l(X,F)$ are called cone-absolutely summing mappings.

As the identity map $X_{\epsilon,D}^{(IN)} \to X_\epsilon^{(IN)}$ is continuous, it follows that $\mathcal{L}^s(X,F) \subset \mathcal{L}^l(X,F)$. Moreover, a $T \in \mathcal{L}(X,F)$ is cone-absolutely summing (see Wong [1974, 1976]) if and only if for any $q \in I\!\!Q_F$ there is a $p \in I\!P_X$ such that

$$\sum_{i=1}^{n} q(Tu_i) \le p(\sum_{i=1}^{n}(u_i)) \quad \text{for any finite } \{u_1, \cdots, u_n\} \subset X_+.$$

The concept of cone-absolutely summing mappings was first studied by Schaefer and his school in the Banach lattices case (see Schaefer [1972, 1974]) and has been extended by Walsh [1973], wong [1974, 1976] and others to the case of locally solid spaces.

The following result should be compared with Theorem 1.

Theorem 3 (Wong [1976]). *A locally solid space $(X, X_+, \mathbb{P}_X)$ is a generalized AL-space (i.e. $\mathbb{P}_X = \sigma_s(X, X')$) the topology uniform convergence on order-bounded subsets (X', X'_+) if and only if*

$$\mathcal{L}^l(X, F) = \mathcal{L}(X, F) \quad \text{for any LCS (or normed space) } F$$

Similar to operators of class (I) and (II), it is natural to consider the following two classes of operators:

$$\text{(III)} \quad \mathcal{L}^{sl}(X, F) = \{T \in \mathcal{L}(X, F) : T^{(IN)} \in \mathcal{L}(X^{(IN)}_{(n),D}, F^{(IN)}_\pi)\}$$

$$\text{(IV)} \quad \mathcal{L}^{ql}(X, F) = \{T \in \mathcal{L}(X, F) : T^{(IN)} \in \mathcal{L}(X^{(IN)}_{(n),D}, F^{(IN)}_\epsilon)\}$$

Also, operators in $\mathcal{L}^{oa}(E, X) = \{T \in \mathcal{L}(E, X) : T^{(IN)} \in \mathcal{L}(E^{(IN)}_\epsilon, F^{(IN)}_{\epsilon,D})\}$ are said to produce order-absolute convergence by Walsh [1973].

Let $(X, X_+, \mathbb{P}_X)$ be a locally solid space and $p \in \mathbb{P}_X$. For any $[x_n]_{(IN)} \in X^{(IN)}$, let

$$P_{\infty,s}([x_n]_{(IN)}) = \inf\{p(u) : -u \leq x_n \leq u(n \in (IN))\}.$$

Denote by $X^{(IN)}_\infty$ the ordered vector space $(X^{(IN)}, X^{(IN)}_+)$ equipped with the locally solid topology $\{P_{\infty,S} : p \in \mathbb{P}_X\}$. Operators in the following class

$$\mathcal{L}^m(E, X) = \{T\epsilon\mathcal{L}(E, X) : T^{(IN)}\epsilon\mathcal{L}(E^{(IN)}_{(n)}, X^{(IN)}_\infty)\}$$

are called majorizing operators.

It is clear that a $T \in \mathcal{L}(E, X)$ is a majorizing operator (see Wong [1977] and Schwarz [1984]) if and only if for any $p \in \mathbb{P}_X$ there is a continuous seminorm r on E such that

$$\inf\{p(w) : \quad -w \leq Tx \leq w(i = 1, \cdots, n)\} \leq \sup_{1 \leq i \leq n} r(x_i)$$
$$\text{for any finite set } \{x_1, \cdots, x_n\} \subset E,$$

and this is the case if and only if for any $p \in \mathbb{P}_X$ there is a continuous seminorm r on E such that

$$p(\sup_{1 \leq i \leq n} |Tx_i|) \leq \sup_{1 \leq i \leq n} r(x_i) \quad \text{for any finite } \{x_1, \cdots, x_n\} \subset E,$$

provided that (X, X_+) is an order-complete Riesz space.

The notion of majorizing operators was first considered by Schloterbeck [1973] in the Banach lattices cases (see Schaefer [1974]), and has been extended by Wahsh [1973], Wong [1977] and others. Moreover, the following result, due to Wong [1977], is a generalization of Schloterbeck's result (see Schaefer [1974, (iv.2.8)]) on the case of Banach lattice $(X, X_+, \|\cdot\|)$.

Theorem 4 (Wong). *A locally convex Riesz space $(X, X_+, \mathbb{P}_X)$ is an M-space if and only if*

$$\mathcal{L}(E, X) = \mathcal{L}^m(E, X) \quad \text{for any LCS } E.$$

References

[1970] Jameson, G.J.O., *Ordered linear spaces*, (Lecture Notes in Math. 104, Springer-Verlag.

[1972] Schaefer, H.H., *Normal tensor products of Banach lattices*, Israel J. Math. 12 400-15

[1974] Schaefer, H.H., *Banach lattices and positive operators*, Springer-verlag, Berlin.

[1984] Schwarz. H.U., *Banach lattices and operators*, Teubner-Texte zur Math. Band 71.

[1973] Walsh W., *Ordered vector sequence spaces and related classes of linear operators*, Math. Ann., 206, 89–139.

[1974] Wong Yau-Chuen, *Nuclear and L-nuclear spaces*, Yale University Press-Mimeographed.

[1976] Wong Yau-Chuen, *The topology of uniform convergence on order-bounded sets*, Lecture Notes in Math. 531, Springer-Verlag.

[1977] Wong Yau-Chuen. *On majorizing and cone-absolutely summing mappings*, J. Austral Math. Soc., 24(Series A), 245–51.

Proceedings of the Asian Mathematical
Conference, 1990

Admissibility of Simultaneous Estimation of Several Losses*

WU QIGUANG
Institute of Systems Science
Chinese Academy of Sciences
Beijing 100080, China

§1. Introduction

As many statisticians pointed out, to estimate an actual loss incurred by estimating unknown parameters is an important decision-theoretic problem. Let random vector X be distributed according to $P_\theta, \theta \in \Theta$. Suppose that $L_i(\theta, d_i(X))$ is the loss incurred in estimating $g_i(\theta)$ by a " good decision " $d_i(X), i = 1, \cdots, k$. One wishes to estimate $L(\theta, d(X)) = (L_1(\theta, d_1(X)), \cdots, L_k(\theta, d_k(X)))'$. To develop a decision-theoretic approach to this problem we shall be concerned with two kinds of loss functions. One is the usual loss function

$$L^*[L(\theta, d(X)), a] \tag{1.1}$$

assuming nonnegative real values, and then the risk of an estimator $\widehat{L}_d(X)$ of $L(\theta, d(X))$ is defined as

$$R_L^*(\theta, \widehat{L}_d) = E\{(L^*[L(\theta, d(X)), \widehat{L}_d(X)]\}.$$

Another is the matrix loss function

$$[a - L(\theta, d(X))][a - L(\theta, d(X))]', \tag{1.2}$$

and then the risk of an estimator $\widehat{L}_d(X)$ of $L(\theta, d(X))$ is defined as

$$M_L^*(\theta, \widehat{L}_d) = E\{[\widehat{L}_d(X) - L(\theta, d(X))][\widehat{L}_d(X) - L(\theta, d(X)]'\}.$$

The usual notations of admissibility, etc., can be applied to estimators $\widehat{L}_d(X)$ of $L(\theta, d(X))$ and their corresponding risks. For example, an estimator $\widehat{L}_d(X)$ of $L(\theta, d(X))$ is said to be inadmissible under the loss function (1.2) if there exists another estimator $h(X)$ which dominates $L_d(X)$ (that is,

$$E\{b'[h(X) - L(\theta, d(X))]\}^2 \leq E\{b'[\widehat{L}_d(X) - L(\theta, d(X))]\}^2$$

for all θ and $b \in R^k$ with strict inequality for some θ and b) and admissible if no such estimator $h(X)$ exists.

---

* Supported by National Natural Science Foundation of China .

Johnstone (1987) proved that the estimator , $\widehat{L}_X = p$, of the loss $(X-\theta)'(X-\theta)$, there $X \sim N_p(\theta, I)$, is admissible for $p \leq 4$, and inadmissible for $p \geq 5$ under the loss function $[a-(X-\theta)'(X-\theta)]^2$. Lu and Berger (1987), and Johnstone (1987) obtained the improved loss estimators based on several Stein-type estimators of θ.

In this paper, we shall give admissible estimators of losses of usual estimators for multivariate normal distribution mean, gamma parametes, binomial parameters, and Poisson parameters in Sections 1, 2, 3, 4 and 5 respectively. The admissibility of a loss estimator can be settled by using the usual methods for proving the admissibility of an estimator of parameters.

§ 2. Ammissibility of Unbiased Loss Estimator for Multivariate Normal Mean Estimators

In this section, we consistently assume that $X_1, \cdots, X_k$ are independent and $X_i \sim N_{n_i}(\theta_i 1, I)$, where $1 = (1, \cdots, 1)', i = 1, \cdots, k$. Put $Y_i = n_i^{-1} 1' X_i$ and $Y = (Y_1, \cdots, Y_k)'$. It is well known that for every $i \in \{1, \cdots, k\}$ Y_i is an admissible estimator and the uniformly minimum risk unbiased estimator (UMRUE), and the best equivariant estimator of θ_i with squared error loss $(d-\theta_i)^2$. Therefore one is interested in estimating the loss $L_y = ((Y_1-\theta_1)^2, \cdots, (Y_k-\theta_k)^2)'$. Clearly, $\widehat{L}_Y = (n_1^{-1}, \cdots, n_k^{-1})'$ is the UMRUE of L_Y with the loss function $(a-L_Y)'(a-L_Y)$, which leads us to discuss the admissibility of $\widehat{L}_Y$.

Theorem 2.1. *(i) The estimator $\widehat{L}_Y$ of the loss L_Y is admissible for $k \leq 4$, and is dominated by $(n_1^{-1} - 2cS^{-2}, \cdots, n_k^{-1} - 2cS^{-2})$, for $k \geq 5$, where $0 < c < 2k^{-1}(k-4)$ and $S^2 = \sum_{i=1}^k n_i^2 Y_i^2$, under the loss function $(a-L_Y)'(a-L_Y)$.*

(ii) Under the matrix loss function $(a-L_Y)(a-L_Y)$, $\widehat{L}_Y$ is an admissible estimator of L_Y for all k.

(iii) Under the loss function $\sum_{i=1}^k [1+|\theta_i|]^r [a_i - (Y_i-\theta_i)^2]^2$ with either $r \geq 1$ and $k \geq 1$, or $0 < r < 1$ and $k < (1-r)^{-1}(2-r)$, $\widehat{L}_Y$ is admissible for estimating L_Y.

§ 3. Estimating Losses of Estimators of Gamma Parameters

Let $X_1, \cdots, X_n$ be i.i.d. according to a gamma distribution with the density $[\Gamma(\alpha)\theta^\alpha]^{-1} \cdot y^{\alpha-1} \exp\{-\theta^{-1}y\} I_{(0,\infty)}(y)$, where $\alpha > 0$ is known, and $\theta \in (0, \infty)$ is a parameter. Denote $T(X) = \sum_{j=1}^n X_j$. As we know, $(n\alpha+1)^{-1}T(X)$ is admissible for θ under the loss $(d-\theta)^2$, and both minimax and best scale-equivariant under the loss $\theta^{-2}(d-\theta)^2$. We want to estimate losses $\theta^{-2}[(n\alpha+1)^{-1}T(X)-\theta]^2$ and $[(n\alpha+1)^{-1}T(X)-\theta]^2$.

Theorem 3.1. *Under the above assumptions and notations, we have*

(i) under the loss function $\{a - [(n\alpha+1)^{-1}T(X)-\theta]^2\}^2$, $cT^2(X)$, where $c \geq 0$, is an admissible estimator of $[(n\alpha+1)^{-1}T(X)-\theta]^2$ when and only when $c = (n\alpha+1)^{-2}(n\alpha+2)^{-1}$. As a special case, the UMRUE $(n\alpha+1)^{-2}(n\alpha)^{-1}T^2(X)$ is inadmissible.

(ii) under the loss function $\{a-\theta^{-2}[(n\alpha+1)^{-1}T(X)-\theta]^2\}^2, (n\alpha+1)^{-1}$ is not only the UMRUE but also an admissible minimax estimator of $\theta^{-2}[(n\alpha+1)^{-1}T(X)-\theta]^2$.

In the remainder of this section, we assume that

$$X_1 = (X_{11}, \cdots, X_{1n_1})', \cdots, X_k = (X_{k1}, \cdots, X_{kn_k})$$

are independent, and that for every $i \in \{1, \cdots, k\}$, $X_{i1}, \cdots, X_{in_i}$ are i.i.d. according to a gamma distribution with the density

$$[\Gamma(\alpha_i)\theta_i^{\alpha_i}]^{-1} y^{\alpha_i - 1} \exp\{-\theta_i^{-1} y\} I_{(0,\infty)}(y),$$

where $\alpha_i > 0$ is known, $\theta_i > 0$ is a parameter. Put $T_i(X_i) = 1'X_i$. We wish to estimate losses

$$L_{1T} = \left([(n_1\alpha_1 + 1)^{-1} T_1(X_1) - \theta_1]^2, \cdots, [(n_k\alpha_k + 1)^{-1} T_k(X_k) - \theta_k]^2\right)'$$

and

$$L_{2T} = \left(\theta_1^{-2}[(n_1\alpha_1 + 1)^{-1} T_1(X_1) - \theta_1]^2, \cdots, \theta_k^{-2}[(n_k\alpha_k + 1)^{-1} T_k(X_k) - \theta_k]^2\right)'.$$

Theorem 3.2. *(i) Under the loss function*

$$\sum_{i=1}^{k} \theta_i^{-4}(1 + |ln\theta_i|)^r \{a_i - [(n_i\alpha_i + 1)^{-1} T_i(X_i) - \theta_i]^2\}^2,$$

where either $r \geq 1$ and $k \geq 1$, or $0 < r < 1$ and $k < (1-r)^{-1}(2-r)$,

$$\widehat{L}_{1T} = ((n_1\alpha_1 + 1)^{-2}(n_1\alpha_1 + 2)^{-1} T_1^2(X_1), \cdots, (n_k\alpha_k + 1)^{-1}(n_k\alpha_k + 2)^{-1} T_k^2(X_k))'$$

is an admissible estimator of L_{1T}.

(ii) Under the matrix loss function $(a - L_{1T})(a - L_{1T})'$, $\widehat{L}_{1T}$ is an admissible estimator of L_{1T} for any k.

(iii) Under the loss function

$$\sum_{i=1}^{k} (1 + |ln\theta_i|)^r \{a_i - \theta_i^{-2}[(n_i\alpha_i + 1)^{-1} T_i(X_i) - \theta_i]^2\}^2,$$

with either $r \geq 1$ and $k \geq 1$, or $0 < r < 1$ and $k < (1-r)^{-1}(2-r)$, $\widehat{L}_{2T} = ((n_1\alpha_1 + 1)^{-1}, \cdots, (n_k\alpha_k + 1)^{-1})'$ is admissible for L_{2T}.

(iv) Under the matrix loss function $(a - L_{2T})(a - L_{2T})'$, $\widehat{L}_{2T}$ is an admissible estimator of L_{2T} for any k.

§ 4. Estimating Losses of Estimators of Binomial Parameters

Suppose that $X_1 = (X_{11}, \cdots, X_{1r_1})', \cdots, X_k = (X_{k1}, \cdots, X_{kr_k})'$ are independent, and that for every $i \in \{1, \cdots, k\}$ $X_{i1}, \cdots, X_{ir_i}$ are i.i.d. as the binomial distribution $B(N_i, \theta_i)$, where $\theta_i \in (0,1)$ is a parameter. Clearly, $T_i(X_i) = 1'X_i$ is distributed according to $B(n_i, \theta_i)$, where $n_i = r_i N_i$, and for every $i \in \{1, \cdots, k\}$ $n_i^{-1} T_i(X_i)$ is both the UMRUE and an admissible estimator of θ_i with squared error loss. So we shall discuss the problem of estimating the loss

$$L_T = ([n_1^{-1} T_1(X_1) - \theta_1]^2, \cdots, [n_k^{-1} T_k(X_k) - \theta_k]^2)'.$$

Note that $E\{[n_i^{-1}T_i(X_i)-\theta_i]^2\}=n_i^{-1}\theta_i(1-\theta_i)$. Hence there does not exist any unbiased estimator of $[n_i^{-1}T_i(X_i)-\theta_i]^2$ when $n_i=1$. If $n_i\geq 2$ for $i=1,\cdots,k$, then

$$\widehat{L}_T(X)=(n_1^{-2}(n_1-1)^{-1}T_1(X_1)[n_1-T_1(X_1)],\cdots,n_k^{-2}(n_k-1)^{-1}T_k(X_k)[n_k-T_k(X_k)])'$$

is the UMRUE of L_T, and hence its admissibility will be studied.

Theorem 4.1. *Under the above assumptions and notations if $n_i\geq 2$ for $i=1,\cdots,k$, then*

(i) under the loss function

$$\sum_{i=1}^{k}\theta_i^{-t_i}(1-\theta_i)^{-s_i}\{a_i-[n_i^{-1}T_i(X_i)-\theta_i]^2\}^2$$

with $t_i>1$ and $s_i>0, i=1,\cdots,k$, $\widehat{L}_T(X)$ is an sadmissible estimator of L_T for any k;

(ii) under the matrix loss function $(a-L_T)(a-L_T)'$ $\widehat{L}_T(X)$ is also an admissible estimator of L_T for any k.

Remark. Let $X_1,\cdots,X_N$ be independent and $X_i\sim B(n_i,\theta)$, where $\theta\in(0,1)$ is a parameter. If $n=\sum_{i=1}^{N}n_i\geq 2$, then

$$\widehat{L}_T(X)=n^{-2}(n-1)^{-1}T(X)[n-T(X)],$$

where $T(X)=\sum_{i=1}^{N}X_i$, is both an admissible minimax estimator and the UMRUE of the loss $L_T=[n^{-1}T(X)-\theta]^2$ under the loss function $\theta^{-2}(1-\theta)^{-2}(a-L_T)^2$.

§ 5. Estimating Losses of Estimators of Poisson Parameters

Let $X_1=(X_{11},\cdots,X_{1n_1})',\cdots,X_k=(X_{k1},\cdots,X_{kn_k})'$ be independent. Suppose that for every $i\in(1,\cdots,k)$, $X_{i1},\cdots,X_{in_i}$ i.i.d. as a Poisson distribution with the parameter λ_i. Denote $T_i(X_i)=1'X_i$. Observe that $n_i^{-1}T_i(X_i)$ is not only the UMRUE but also an admissible minimax estimator of λ_i under the loss $\lambda_i^{-1}(d-\lambda_i)^2$. Concerning the problems of estimating losses

$$L_{1T}=([n_1^{-1}T_1(X_1)-\lambda_1]^2,\cdots,[n_k^{-1}T_k(X_k)-\lambda_k]^2)'$$

and

$$L_{2T}=(\lambda_1^{-1}[n_1^{-1}T_1(X_1)-\lambda_1]^2,\cdots,\lambda_k^{-1}[n_k^{-1}T_k(X_k)-\lambda_k]^2)',$$

we have the following results.

Theorem 5.1. *(i) Under the loss function*

$$\sum_{i=1}^{k}\lambda_i^{-1}(1+|ln\lambda_i|)^r\{[n_i^{-1}T_i(X_i)-\lambda_i]^2-a_i\}^2,$$

where either $r\geq 1$ and $k\geq 1$, or $0<r<1$ and $k<(1-r)^{-1}(2-r)$, the UMRUE, $\widehat{L}_{1T}=(n_1^{-2}T_1(X_1),\cdots,n_k^{-2}T_k(X_k))'$, of L_{1T} is admissible.

(ii) Under the matrix loss function $(a-L_{1T})(a-L_{1T})'$, $\widehat{L}_{1T}$ is an admissible estimator of L_{1T} for any k.

(iii) Under the loss function

$$\sum_{i=1}^{k} \lambda_i(1+|ln\lambda_i|)^r\{a_i - \lambda_i^{-1}[n_i^{-1}T_i(X_i) - \lambda_i]^2\}^2$$

with either $r \geq 1$ *and* $k \geq 1$ *and* $k < (1-r)^{-1}(2-r)$, *the UMRUE* , $\widehat{L}_{2T} = (n_1^{-1}, \cdots, n_k^{-1})'$, *of* L_{2T} *is admissible.*

(iv) Under the matrix loss function $(a - L_{2T})(a - L_{2T})'$, $\widehat{L}_{2T}$ *is an admissible estimator of* L_{2T} *for any* k.

References

[1] Cheng, P. , *Admissibility of simultaneous estimation of several parameters,* J. Sys. Sci. & Math. Sci., 2 (1982), 176–195.

[2] Johnstone, I. , *On the inadmissibility of some unbiased estimates of loss, In Statistical Decision Theory and Related Topics iv* , 361–379, Gupta, S. S. and Berger, J.O. (Eds), Springer-Verlag New York (1987).

[3] Lu, K. L. & Berger, J. O. , *Estimation of means : the estimated loss approach,* In Sino-American Statistical Meeting Contributed Papers, (1987) 20–22.

[4] Rukhin, A. L., *Estimated loss and admissible loss estimators, In Statistical Decision Theory and Related Topics iv,* 409–417, Springer-Verlag New York, 1987.

[5] Wu, Q.G. & Chen, J.B., *All admissible linear estimates of regression coefficients under matrix loss,* Systems Science and Mathematical Sciences, 2(1989), 80–91.

[6] Zidek, J.V. , *Sufficient condition for the admissibility under squared error loss for formal Bayes estimators,* Ann. Math. Statist., 41 (1970), 446–465.

Proceedings of the Asian Mathematical
Conference, 1990

BV Solutions of Degenerate Quasilinear Parabolic Equations

WU ZHUOQUN, YIN JINGXUE
Department of Mathematics
Jilin University
Changchun 130023, China

§1. Introduction

When we study some kind of partial differential equations, it is important to choose a suitable class of functions in which we define solutions and discuss the solvability. Because of the degeneracy and nonlinearity, the solutions of general degenerate quasilinear parabolic equations of second order considered in this paper might be discontinuous. So, in treating such a kind of equations, one should choose a certain class of discontinuous functions. It was Vol'pert and Hudjaev [1] who first chose BV, the class of all functions of locally bounded variation, in which the Cauchy problem for such a kind of equations was studied and the existence was established. Afterwards, Wu and Zhao studied the first boundary value problem in the same framework and obtained corresponding results [2]-[4]. In [1], Vol'pert and Hudjaev also discussed the uniqueness of BV solutions. However, the uniqueness was established only for a very narrow subclass of BV functions. In the one-dimensional case

$$\frac{\partial u}{\partial t} - \frac{\partial}{\partial x}\left(a(u,t,x)\frac{\partial u}{\partial x}\right) + \frac{\partial}{\partial x}b(u,t,x) = c(u,t,x), \tag{1.1}$$

with $a(u,t,x) \geq 0$ they tried to prove a general result on uniqueness, but the proof is incorrect due to the adoption of a wrong form of discontinuity condition for the BV solutions. The root of mistakes is the neglect of the effect of the term $\frac{\partial}{\partial x}(a(u,t,x)\frac{\partial u}{\partial x})$, in which $a(u,t,x)\frac{\partial u}{\partial x}$ (for BV solution u) might not be a BV function. What can be known on $a(u,t,x)\frac{\partial u}{\partial x}$ is only that the partial derivative with respect to x is a regular measure. But nothing is known about the partial derivative with respect to t. This motivates us to make a deep investigation into a more general class of functions, called BV_x, and give an essential complement to the theory of BV functions. Some basic results on BV_x functions and BV functions will be presented in §2 and applied to the uniqueness in §3 and in §4 for the equation (1.1) and equations of other forms, for example,

$$\frac{\partial u}{\partial t} = \mathrm{div}\vec{\mathbf{F}}(\nabla A(u)) + \nabla\vec{\mathbf{B}}(u), \tag{1.2}$$

which arises in many applied problems such as the non-stationary flow of a compressible Newtonian fluid in a porous medium. We also discuss the existence of BV solutions both in §3 and in §4.

§ 2. Some Properties of BV Functions and BV_x Functions [5]

Let $Q_T = (0,T)\times \mathbb{R}$. By $BV(Q_T)$, we mean the class of all functions on Q_T of locally bounded variation. In other words, $u\in BV(Q_T)$ if and only if $u\in L^1_{\text{loc}}(Q_T)$ and $\frac{\partial u}{\partial t}$ and $\frac{\partial u}{\partial x}$ are regular measures on Q_T of locally bounded variation. We will frequently deal with a more general class of functions, denoted by $BV_x(Q_T)$, which consists of all functions $u\in L^1_{\text{loc}}(Q_T)$ with $\frac{\partial u}{\partial x}$ being regular measures on Q_T of locally bounded variation. It follows from [6] that if $u\in BV_x(Q_T)$, then for almost every $t\in(0,T), u(t,\cdot)\in BV(\mathbb{R})$ and for any fixed open interval I, $\int_I\left|\frac{\partial u(t,\cdot)}{\partial x}\right|$ is locally integrable on $(0,T)$. Using a delicate argument, we can prove the following

Theorem 2.1. *Let $u\in BV_x(Q_T)$ and $g(t,x)$ be an essentially bounded and measurable (with respect to $\frac{\partial u}{\partial x}$) function with compact support in Q_T. Then for almost every $t\in(0,T), g(t,x)$ is measurable with respect to $\frac{\partial u(t,\cdot)}{\partial x}$. Moreover, $\int_R g(t,x)\frac{\partial u(t,\cdot)}{\partial x}$ is integrable on $(0,T)$ and we have*

$$\iint_{Q_T} g(t,x)\frac{\partial u}{\partial x} = \int_0^T dt\int_R g(t,x)\frac{\partial u(t,\cdot)}{\partial x}. \tag{2.1}$$

As a formula reducing a double integral to a repeated integral, (2.1) is natural and reasonable. However, the proof is quite difficult. The difficulty lies in the fact that the two-dimensional measure $\frac{\partial u}{\partial x}$ might not be decomposible into a product of the Lebesgue measure dt and the measure $\frac{\partial u(t,\cdot)}{\partial x}$, since $\frac{\partial u(t,\cdot)}{\partial x}$ depends on t in general.

In applying the formula (2.1) for $u\in BV_x(Q_T)$, first of all, one needs to check the measurability of $g(t,x)$ with respect to $\frac{\partial u}{\partial x}$. Here we introduce a class of essentially bounded functions which are measurable and hence integrable with respect to $\frac{\partial u}{\partial x}$ for any $u\in BV_x(Q_T)$. The class, denoted by $\overline{Ca}(Q_T)$, is an extension of the class of all essentially bounded Caratheodory funcitons. Precisely, $g\in\overline{Ca}(Q_T)$ if and only if there exists a uniformly essentially bounded sequence $\{g_n(t,x)\}$ of Caratheodory functions, such that

$$\lim_{n\to\infty} g_n(t,x) = g(t,x)$$

for almost every $t\in(0,T)$ and any $x\in\mathbb{R}$. We recall that a function $g(t,x)$ is called a Caratheodory function, if for almost every $t\in(0,T), g(t,x)$ is continuous in x, and for any $x\in\mathbb{R}, g(t,x)$ is Lebesgue measurable in t.

The typical examples of functions in $\overline{Ca}(Q_T)$ are $\widetilde{u}(t,x), u^r(t,x)$ and $u^l(t,x)$ if $u\in L^\infty(Q_T)\cap BV(Q_T)$. Here $\widetilde{u}(t,x)$ denotes the symmetric mean value, $u^r(t,x)$ and $u^l(t,x)$ the right and left approximate limits of $u(t,x)$, as a function of x.

For $u\in BV(Q_T)$, we denote by Γ_u the set of all jump points of $u(t,x), \nu=(\nu_t,\nu_x)$ the normal to $\Gamma_u, u^+(t_0,x_0)$ and $u^-(t_0,x_0)$ the approximate limits of u at the point (t_0,x_0) from the side of the half planes $(t-t_0)\nu_t+(x-x_0)\nu_x>0$ and $(t-t_0)\nu_t+(x-x_0)\nu_x<0$ respectively. From the basic theory of BV functions (see [7]), the symmetric mean value $\overline{u}(t,x)$ exists and equals $\frac{1}{2}(u^+(t,x)+u^-(t,x))$ at any regular point $(t,x)\in\Gamma_u$.

We can prove that if $u \in BV(Q_T)$, then

$$u^+(t,x) = u^r(t,x)\mathrm{sgn}^+ v_x - u^l(t,x)\mathrm{sgn}^- v_x,$$

$$u^-(t,x) = u^l(t,x)\mathrm{sgn}^+ v_x - u^r(t,x)\mathrm{sgn}^- v_x$$

hold H-almost everywhere on $\Gamma_u^* = \{(t,x) \in \Gamma_u; v_x \neq 0\}$, where

$$\mathrm{sgn}^+ s = \begin{cases} 1, & s > 0 \\ 0, & s \le 0 \end{cases} \qquad \mathrm{sgn}^- s = \begin{cases} 0, & s \ge 0 \\ -1, & s < 0, \end{cases}$$

and H denotes the Hausdorff measure on Γ_u.

Now we state a useful theorem concerning the reduciton of double integrals or repeated integrals to curvilinear integrals, which supplements essentially the known results in the theory of BV functions.

Theorem 2.2. *Let $u \in L^\infty(Q_T) \cap BV(Q_T)$ and S be a bounded and measurable subset of Γ_u^*. Then for any bounded and measurable (with respect to $\frac{\partial u}{\partial x}$) function $g(t,x)$ with compact support in Q_T, we have*

$$\iint_S g(t,x)\frac{\partial u}{\partial x} = \int_S g(t,x)[u^r(t,x) - u^l(t,x)]\,|v_x|\,dH,$$

$$= \int_0^T dt \sum_{x \in S^t} g(t,x)[u^r(t,x) - u^l(t,x)]$$

$$\int_0^T dt \sum_{x\in S^t} g(t,x) = \int_S g(t,x)dH,$$

where $S^t = \{x; (t,x) \in S\}$.

Remark The results stated in this section can be easily extended to the multi-dimensional case.

§ 3. The One-dimensional Case

In this section we first consider the degenerate parabolic equations of the form

$$\frac{\partial u}{\partial t} - \frac{\partial}{\partial x}\left(a(u,t,x)\frac{\partial u}{\partial x}\right) + \frac{\partial}{\partial x}b(u,t,x) = c(u,t,x), \tag{3.1}$$

where functions a, b, c satisfy certain required smoothness conditions. It is known that the existence for both the Cauchy problem and the first boundary value problem has been successfully solved even in the most general case in which $a(u,t,x) \ge 0$ (see [1],[2]). However, this is not the case with the uniqueness. What has been obtained before are the results by Zhao in [8], which establish the uniqueness of solutions in the class of bounded and measurable functions under the condition that the function

$$A(u,t,x) = \int_0^u a(s,t,x)ds$$

is strictly increasing with respect to u. A result Vol'pert and Hudjaev tried to prove in [1] is the following

Theorem 3.1. *Let u_1 and u_2 be BV solutions of the Cauchy problem for the equation (3.1) with bounded and measurable initial functions u_1^0 and u_2^0 respectively (for definition of solutions see [1] or [5]). Then for any $t \in (0,T)$, we have*

$$\int_R |\overline{u}_1(t,x) - \overline{u}_2(t,x)|\omega_\lambda(x)dx \le e^{K_\lambda t}\int_R |u_1^0(x) - u_2^0(x)|\omega_\lambda(x)dx,$$

where $\omega_\lambda(x) = \exp(-\lambda\sqrt{1+x^2})$, $(\lambda > 0)$ and K_λ is a constant depending only on λ and the bounds of u_1 and u_2.

Vol'pert and Hudjaev's proof is based on the following discontinuity conditions: For the solution $u \in L^\infty(Q_T) \cap BV(Q_T)$,

$$[\mathrm{sgn}(u^+ - k) - \mathrm{sgn}(u^- - k)][(\overline{u} - k)v_t + (\overline{b(u,t,x)} - b(k,t,x))v_x] \le 0, \tag{3.2}$$

$$a(u,t,x) = 0, \quad u \in [u_*, u^*] \tag{3.3}$$

hold H-almost everywhere on Γ_u, for any $k \in \mathbb{R}$, where $u_* = \min(u^-, u^+), u^* = \max(u^-, u^+)$.

However, as is shown in [9], (3.2) does not hold in general. One can infer this from the fact that the equation considered is of scond order; however, the second order term dose not appear in (3.2). The correct version of the discontinuity conditions should be

$$(u^+ - u^-)v_t + (b(u^+,t,x) - b(u^-,t,x))v_x - \{w^r - w^l\}\,|v_x| = 0, \tag{3.4}$$

$$[\mathrm{sgn}(u^+ - k) - \mathrm{sgn}(u^- - k)][(\overline{u} - k)v_t + (\overline{b(u,t,x)} - b(k,t,x))v_x - \widetilde{w}v_x] \le 0, \tag{3.5}$$

where $w = \widehat{a}(u,t,x)\frac{\partial u}{\partial x}$, which can be derived by using the properties of functions in BV and BV_x presented in §2 (see [5]). Here the symbol $\widehat{g}(u,t,x)$ denotes

$$\widehat{g}(u,t,x) = \int_0^1 g(\lambda u^+ + (1-\lambda)u^-, t, x)d\lambda.$$

On the basis of (3.3), (3.4) and (3.5), we can prove the key integral inequality,

$$\begin{aligned} J(u_1,u_2,\varphi) \equiv \iint_{Q_T} &\mathrm{sgn}(u_1 - u_2)\Big[(c(u_1,t,x) - c(u_2,t,x))\varphi - (u_1 - u_2)\frac{\partial\varphi}{\partial t} \\ &+ (b(u_1,t,x) - b(u_2,t,x))\frac{\partial\varphi}{\partial x} - (w_1 - w_2)\frac{\partial\varphi}{\partial x}\Big]dtdx \ge 0, \end{aligned} \tag{3.6}$$

where $w_1 = \widehat{a}(u_1,t,x)\frac{\partial u_1}{\partial x}, w_2 = \widehat{a}(u_2,t,x)\frac{\partial u_2}{\partial x}, 0 \le \varphi \in C_0^\infty(Q_T)$ and u_1, u_2 are BV solutions of the equation (3.1). Once this is done, Theorem 3.1 follows just as in [1].

The fact that w_1 and w_2 are not in BV but only in BV_x generally makes the proof of (3.6) much more complicated. The integrals in (3.6) involving these two functions can not be directly transformed into curvilinear integrals on Γ_z with $z = u_1 - u_2$. We need first to reduce these integrals into repeated integrals by using Theorem 2.1 and then analyse carefully the interior integrals with respect to the space variable x by using the properties of one-dimensional BV functions. Finally, we transform the resulting repeated integrals into curvilinear integrals by using Theorem 2.2, and the desired inequality (3.6) follows from (3.3), (3.4) and (3.5).

Similarly, we can prove the uniqueness of BV solutions of the first boundary value problem for the equation (3.1) with $a(u,t,x) \geq 0$. Since the proof can be easily extended to more general nonlinear equations, namely, the equations with double-degeneracy, we will deal with the latter case. For simplicity of notations we consider equations of the form

$$\frac{\partial u}{\partial t} - \frac{\partial}{\partial x} F\left(\frac{\partial}{\partial x} A(u)\right) + \frac{\partial}{\partial x} b(u) = 0, \tag{3.7}$$

where

$$A(s) = \int_0^s a(\sigma)d\sigma, \quad F(s) = \int_0^s f(\sigma)d\sigma, \quad a(s) \geq 0, \quad f(s) \geq 0.$$

Theorem 3.2 ([10]). *Let u_1, u_2 be BV solutions of the first boundary value problem for (3.7) in the region $(0,T)\times(0,1)$ with initial data $u_1^0, u_2^0 \in L^1(0,1)$ respectively . Then*

$$\int_0^1 |\overline{u}_1(t,x) - \overline{u}_2(t,x)|\, dx \leq \int_0^1 |u_1^0(x) - u_2^0(x)|\, dx.$$

The proof is also based on the discontinuity conditions satisfied by the solutions considered. For the first boundary value problem there are two kinds of discontinuity conditions, the interior discontinuity conditions and the boundary discontinuity conditions.

As in the case of single degeneracy, the difficulty of the proof centres on the analysis of the function $F(\frac{\partial}{\partial x}A(u))$ which is only known to be in $BV_x((\delta,T)\times(0,1))$. The difference is that here we have to analyse the behaviour of the function $F(\frac{\partial}{\partial x}A(u))$ not only at the jump points of u bue also at the jump points of itself both in $(0,T)\times(0,1)$ and on the boundary.

For the existence of BV solutions of the first boundary value problem for (3.7) , we have the following

Theorem 3.3 ([11]). *Assume that*

$$\lim_{s\to\infty} F(s) = +\infty, \qquad \lim_{s\to-\infty} F(s) = -\infty.$$

If $u_0(x)$ is appropriately smooth on $[0,T]$ and satisfies certain compatibility conditions at $x = 0, 1$, then the first boundary value problem for (3.7) admits of a BV solution. Moreover, $A(u)$ is Hölder continuous provided that the initial value is appropriately smooth.

If, in addition, the function $A(s)$ is strictly increasing, then the solution is continuous.

Notice that the strict monotonicity of $A(s)$ is almost necessary for the existence of continuous solutions even in the special case where $F(s) = s$.

Our result is much more general than the one proved by Kalashnikov [12] for the equation

$$\frac{\partial u}{\partial t} = \frac{\partial}{\partial x} F(\frac{\partial}{\partial x} A(u)).$$

§ 4. The Multi-dimensional Case

Now we turn to the multi-dimensional equations of the form

$$\frac{\partial u}{\partial t} = \triangle A(u) + \nabla \vec{\mathbf{B}}(u), \tag{4.1}$$

where $A(s)$ is a nondecreasing function. The existence of BV solutions has been established for both the Cauchy problem and the first boundary value problem even for degenerate parabolic equations more general than (4.1). As for the uniqueness, an elegant work was done by Brezis and Crandall [13] who proved the uniqueness of solutions in the class of bounded and integrable funcitons for (4.1) with $\vec{\mathbf{B}}(u) \equiv \vec{\mathbf{O}}$. Some authors have been devoted to the case where $\vec{\mathbf{B}}(u) \not\equiv \vec{\mathbf{O}}$. However, the equations they considered have only one point of degeneracy, and in most of their papers a certain condition that $\vec{\mathbf{B}}(s)$ be dominated by $A(s)$ is required. In this paper, the only assumption, in addition to some smoothness conditions, is that the function $A(s)$ is strictly increasing. Much degeneracy is allowed. This and the existent term $\nabla\vec{\mathbf{B}}(u)$ make the proof of uniqueness more difficult. In our case we can not treat the uniqueness either by using the argument in [13], or on the basis of the discontinuity conditions as in §3. In fact, in multi-dimensional case, we can not conclude from the definition of generalized solutions that the second order derivatives of $A(u)$ with respect to the space variables are regular measures, although $\triangle A(u)$ is. This prevents us from deriving an analogue of discontinuity conditions in multi-dimensional case. However, in case $A(s)$ is strictly increasing, we can use some approximate process and apply the results in §2 to derive a basic inequality similar to (3.6), and hence prove the uniqueness of solutions.

Theorem 4.1 ([10]). *If $A(s)$ is strictly increasing and u_1, u_2 are solutions of the equation (4.1) with initial data $u_1^0, u_2^0 \in L^\infty(\mathbb{R}^n)$, then*

$$\int_{R^n} |\bar{u}_1(t,x) - \bar{u}_2(t,x)|\omega_\lambda(x)dx \le e^{K_\lambda t}\int_{\mathbb{R}^n} |u_1^0(x) - u_2^0(x)|\omega_\lambda(x)dx,$$

where $K_\lambda t$ is a constant depending only on λ and the bounds of u_1, u_2, $\omega_\lambda(x) = \exp(-\lambda\sqrt{1+|x|^2})$.

An analogous result can also be obtained for the following equation with double-degeneracy

$$\frac{\partial u}{\partial t} = \mathrm{div}\vec{\mathbf{F}}(\nabla A(u)).$$

The uniqueness is established in BV_t for the first boundary value problem.

The existence of BV_t solutions can be established by using the method of parabolic regularization under certain assumptions on $\vec{\mathbf{F}}(s)$ and $A(s)$. The Cauchy problem can be treated similarly. We are especially interested in the case where the initial data u_0 have compact support: supp $u_0 \subset B_M$, where B_M is a ball in $\mathbb{R}^n$. In this case we can obtain a unique solution $u \in BV_t$ with the property that for any $T > 0$, there exists an open ball $B_R \subset B_M$ such that for almost every $t \in (0,T)$, supp $u(t,\cdot) \subset B_R$ (see [14]). In the proof of this result, the Hardy inequality [15, Theorem 330] and the weighted Nirenberg interpolation inequality [16, Lemma 8] are used to establish a key differential inequality.

References

[1] Vol'pert, A.I and Hudjaev S.I., *Cauchy's problem for second order quasilinear degenerate parabolic equations*, Mat. Sb., 78 (120) (1969), 389–411.

[2] Wu Zhuoqun and Wang Junyu, *Some results on quasilinear degenerate parabolic equations of second order*, Proceedings of International Beijing Symposium on Differential Geometry and Partial Differential Equations, 1981.

[3] Wu Zhuoqun and Zhao Junning, *The first boundary value problem for quasilinear degenerate parabolic equations of second order in several space variables*, Chin. Ann. of Math., 4B (1) (1983), 57–76.

[4] Wu Zhuoqun and Zhao Junning, *Some general results on the first boundary value problem for quasilinear degenerate parabolic equations*, Chin. Ann. of Math., 4B (3) (1983), 319–328.

[5] Wu Zhuoqun and Yin Jingxue, *Some properties of functions in and their applications to the uniqueness of solutions for degenerate quasilinear parabolic equations*, Northeastern Math. J., 5(4) (1989), 395–422.

[6] Krickeberg, K., *Distributionen, Funktionen beschrankter Variation und Lebesguesher Inhalt nichtparametrischer Flachen*. Ann. mat. pura ed appl., 1957, 44.

[7] Vol'pert, A.I. and Hudjaev S.I., *Analysis in the Class of Discontinuous Functions and Equations of Mathematical Physics*, Izda., Nauka Moscwa, 1975.

[8] Zhao Junning, *Uniqueness of solutions of quasilinear degenerate parabolic equations*, Northeastern Math. J. 1(2) (1985), 153–165.

[9] Wu Zhuoqun, *A free boundary problem for degenerate quasilinear parabolic equations*, Nonlinear Analysis TMA, 9(9) (1985), 937–952.

[10] Yin Jingxue, *On the uniqueness and stability of BV solutions for nonlinear diffusion equations*, Comm. P.D.E., 1671–1683.

[11] Yin Jingxue, *On a class of quasilinear parabolic equations of second order with double–degeneracy*, J.PDEs, 3(4) (1990) , 49–64.

[12] Kalashnikov, A.S., *Cauchy problem for second order degenerate parabolic equations with nonpower nonlinearities*, J. Soviet Math., 33(1986), 1014–1025.

[13] Brezis, H. and Crandall, M.G., *Uniqueness of solutions of the initial–value problem for* $u_t - \Delta\varphi(u) = 0$, J. Math. Pures et Appl., 58 (1979) , 153–163.

[14] Yin Jingxue, *Solutions with compact support for nonlinear diffusion equation*, to appear in Nonlinear Analysis.

[15] Hardy, C.H., Littlewood, J.E. and Polya, G., *Inequalities*, Cambridge University Press, Cambridge, 1952.

[16] Bernis, F., *Compactness of the support for some nonlinear elliptic problems of arbitrary order in dimension* n , Comm. P.D. E. 9(1984), 271–312.

Proceedings of the Asian Mathematical
Conference, 1990

A Necessary and Sufficient Condition for Dual Algebras of Coalgebras to be Prime

XU YONGHUA, LU DIMING & ZHU HONGXIN
Institute of Mathematics
Fudan University
Shanghai 200433, China

§1. Introduction

A coalgebra over a field Φ is a triple (C, Δ, ε) where C is a vector space over $\Phi, \Delta : C \to C \otimes_\Phi C$ and $\varepsilon : C \to \Phi$ are linear maps such that $(I \otimes \Delta)\Delta = (\Delta \otimes I)\Delta$ and $(I \otimes \varepsilon)\Delta = I = (\varepsilon \otimes I)\Delta$, where I means identity. Let $C^* = \mathrm{Hom}_\Phi(C, \Phi)$. Then there is a natural algebra structure on C^* induced by the diagonal map Δ and the augmentation map ε of C. This means that for any $c^*, d^* \in C^*$ and $c \in C$ we can define $(c^*d^*)(c) = (c^* \otimes d^*)\Delta(c)$. It is easy to see that the counit ε of C is the identity of C^*. Let $(-, c^*) : C \to \Phi$ be a linear map. Then C can be defined as a left C^*-module by $c^*, c = (I \otimes (-, c^*))\Delta(c)$ for $c^* \in C^*$ and $c \in C$.

Since the multiplication of C^* is induced by the Δ of C, C^* has a special kind of algebra structure which is closely related to the structure of C. Therefore using the algebra structure of C^* one can always study coalgebra structure of C, for example, the decomposition of a coalgebra into a direct sum of its indecomposable components[1,5] and so on. Thus the dual algebra plays an important role in studying the structure of coalgebra. But in this paper we use the structure of coalgebra to describe some properties of its dual algebra. Our two main results are as follows:

Theorem A. *Let C be a coalgebra over a field Φ with infinite dimension and C^* its dual algebra. Suppose that there exists a non-zero element c^* of C^* such that $\dim c^* \cdot C^* < \dim C$. Then C^* is not prime.*

Note that the converse is false as the following Example 3 indicates, but we can obtain the following criterion for dual algebras to be prime.

Theorem B. *C^* is prime if and only if $C^*c^* \cdot C = C$ for any non-zero element c^* of C^*.*

§2. Examples

In this section we give three examples which illuminate the above results.

Example 1. Let C be a vector space with basis $\{c_i\}_{i=0,1,\dots}$. Define $\Delta(c_n) = \sum_{i=0}^{n} c_i \otimes c_{n-i}$ and $\varepsilon(c_n) = \delta_{0,n}$. Then (C, Δ, ε) is a coalgebra. Let X^i be define in C^* by $X^i(c_j) = \delta_{ij}$.

Note that $\varepsilon = 1_{C^*} = X^0$ is in C^* and that $(c_n, X^i X^j) = \delta_{i+j,n}$, so that $X^i X^j = X^{i+j}$. Then $C^* = \{\sum_{i=0}^{\infty} \lambda_i X^i | \lambda_i \in \Phi\}$ and this gives an isomorphism between C^* and the power series ring in one variable[2]. Thus C^* is an algebra without divisors of zero. Obviously C^* is prime. Choosing any non-zero element c^* of $C^*, c^* = \sum_{i=0}^{\infty} \lambda_i X^i$, we can easily calculate $c^* \cdot c_n = \sum_{i=0}^{n} \lambda_i c_{n-i}, n = 0, 1, \cdots$. Therefore $c^* \cdot C = C$ for any non-zero element $c^* \in C^*$, and clearly $\dim c^* \cdot C = \dim C$.

Example 2. Let C be a coalgebra with finite dimension. let C^* be its dual algebra. Let J denote the Jacobson radical of C^*. Then it is easy to see that $J^n = 0$ for some $n \in \mathbb{Z}^+$. Thus C^* is not prime. Furthermore we assume that C has a finite basis $\{s_1, \cdots, s_n\}$, and define $\Delta(s_i) = s_i \otimes s_i, \varepsilon(s_i) = 1 \in \Phi, i = 1, \cdots, n$. Then (C, Δ, ε) is a coalgebra. In C^* let s_i^* be defined by $s_i^*(s_j) = \delta_{ij}$, for $i, j = 1, \cdots, n$. Then C^* has a basis $s_1^*, \cdots, s_n^*$. For any element c of C we have $c = \sum_{i=1}^{n} \lambda_i s_i, \lambda_i \in \Phi$. It is easy to see that $s_i^* \cdot c = \sum_{j=1}^{n} \lambda_j s_j^*(s_j) s_j = \lambda_i s_i$. Thus $\Phi s_i = s_i^* \cdot C \neq C$. But C^* is not prime as mentioned above.

Example 3. Let C be a vector space with basis $T = \{(x,y) \in Z \times Z \mid x \leq y\}$. Define Δ by $\Delta(x,y) = \sum_{x \leq z \leq y} (x,z) \otimes (z,y)$, and ε by $\varepsilon(x,y) = \delta_{x,y}$ for any $(x,y) \in T$. Then (C, Δ, ε) is a coalgebra (cf. [2]). We take two non-zero elements $T_{(a,b)}, T_{(c,d)}$ in C^* as follows:

$$T_{(a,b)}(x,y) = \delta_{ax}\delta_{by},$$

$$T_{(c,d)}(x,y) = \delta_{cx}\delta_{dy}$$

for all $(x,y) \in T$. We may assume $d < a$. For any $c^* \in C^*, (x,y) \in T$, we have

$$\begin{aligned}
&(T_{(a,b)} c^* T_{(c,d)})(x,y) \\
= &\sum_{x \leq \alpha \leq \beta \leq y} (T_{(a,b)}(x,\alpha))(c^*(\alpha,\beta))(T_{(c,d)}(\beta,y)) \\
= &\sum_{x \leq \alpha \leq \beta \leq y} \delta_{ax}\delta_{b\alpha}\delta_{c\beta}\delta_{dy}(c^*(\alpha,\beta)) = 0.
\end{aligned}$$

It implies $T_{(a,b)} C^* T_{(c,d)} = 0$, i.e. C^* is not prime.

On the other hand, for any $D \neq c^* \in C^*$, there exists $(x',y') \in T$ such that $c^*(x',y') \neq 0$. For any $m < x'$, we have $c^*(m,y') = \sum_{m \leq z \leq y'} (m,z) c^*(z,y') \neq 0$, since $c^*(x',y') \neq 0$ and (m,z) is a basis element. Suppose $c^* \cdot C$ is a finite dimensional space with basis $\{c_1, \cdots, c_r\}$. We can find an element x_0 of $\mathbb{Z}$ such that $\{c_1, \cdots, c_r\} \subset W = \sum_{(x,y) \in T, x_0 \leq x} \Phi(x,y)$. Thus we obtain $c^* \cdot C$. If we take $m < \min\{x', x_0\}$, it follows that $0 \neq c^*(m,y') = (m,x')c^*(x',y') + \sum_{m \leq z \leq y', z \neq x'} (m,z) c^*(z,y') \overline{\in} W$. This is contrary to $c^* \cdot C \subset W$. Therefore $c^* \cdot C$ is infinite dimensional. Note that C has a countable basis. We obtain $\dim c^* \cdot C = \dim C$.

§3. Some Lemmas

For the sake of convenience, we first introduce some terminologies.

Let C and D be vector spaces over fields Φ and K respectively. A map S of C into D is called a semi-linear transformation if and only if

(i) S is a homomorphism of $(c, +)$ to $(D, +)$,

(ii) There exists an isomorphism ψ of Φ onto K such that $(\lambda c)S = \lambda^{\psi}(cS)$ for all c of C and λ of Φ.

We call ψ the isomorphism associated with S, and when we wish to indicate ψ explicitly we shall speak of the "semi-linear transformation (S, ψ)". Besides, if S is a 1-1 semi-linear transformation of C onto D, we call S a semi-linear isomorphism of C onto D, whose associated isomorphism is ψ. So the inverse mapping S^{-1} is also a semi-linear isomorphism of D onto C with associated isomorphism ψ^{-1}.

Definition 1. For any ring R, Let ${}_{\Phi}C_R$ and ${}_K D_R$ be (Φ, R)-bimodule and (K, R)-bimodule respectively. We say that C and D are (ψ, I)-isomorphisms if and only if there exists a mapping S of C onto D such that the following conditions are satisfied:

(i) S is a semi-linear isomorphism of ${}_{\Phi}C$ onto ${}_K D$ with the associated isomorphism ψ.

(ii) $(cr)S = (cS)r$ for all c of C and r of R.

Lemma 1. [3,4] *Let C be a vector space over a field Φ with a basis $\{u_i\}_{i\in\Gamma}$ and let $\{E_i\}_{i\in\Gamma}$ be the subset of Ω such that $u_j E_i = \delta_{ij} u_i$ for all $i, j \in \Gamma$. Let E_1 be an arbitrary element of $\{E_i\}_{i\in\Gamma}$ and denote $A = E_1\Omega$, $K = E_1\Omega E_1$. Then we have*

(i) $A = \sum_{i\in\Gamma} K\alpha_i$ can be expressed as a vector space over field K such that $\{\alpha_i\}_{i\in\Gamma}$ is a basis of A and $\Omega = \mathrm{End}_K A$. Denote the scale of Ω by $\mathcal{S}$. Then $\mathcal{S} = \Omega E_i \Omega$ for all $i \in \Gamma$. Moreover, if $\omega\mathcal{S} = 0$ for $\omega \in \Omega$ then $\omega = 0$.

(ii) There exists a (ψ, I)-isomorphism S of bimodules ${}_{\Phi}C_{\Omega}$ onto ${}_K A_{\Omega}$ such that $(\lambda c\omega)S = \lambda^{\psi}(cS)\omega$ for all $\lambda \in \Phi, c \in C$ and $\omega \in \Omega$, where ψ is the associated isomorphism with S.

(iii) Let $C^* = \mathrm{Hom}_{\Phi}(C, \Phi)$ and $A^* = \mathrm{Hom}_K(A, K)$. Then $A^* = \Omega E_1$. Moreover, there exists an (I, ψ^{-1})-isomorphism S' of ${}_{\Omega}A^*_K$ onto ${}_{\Omega}C^*_{\Phi}$ such that $(cS, a^*) = (c, S'a^*)^{\psi}$ for all $c \in C$ and $a^* \in A^*$. Note that the ring product of elements a and a^* of Ω is aa^*, $(a, a^*) = aa^*$. The mapping S' of A^* onto C^* is called an adjoint of S.

Remark. Lemma 1 indicates that the vector space C and the dual space C^* over Φ can be replaced by the spaces $A = E_1\Omega$ and $A^* = \Omega E_1$ over $K = E_1\Omega E_1$ respectively. Note that A and A^* are minimal right and left ideals of Ω respectively, which have ring structures. Moreover, the tensor product $C \otimes_{\Phi} C$ can be replaced by the tensor product $A \otimes_K A$ too. In fact. S is a semi-linear isomorphism by Lemma 1. As usual, we define $(S, S) : C \times C \to A \otimes_K A$ via $(S, S)(c, d) = cS \otimes dS$ for any $c, d \in C$. It is easy to see that (S, S) is Φ-balanced and there exists a unique $\mathbb{Z}$-homomorphism, dented by $S \otimes S$, such that $(S \otimes S)(c \otimes d) = cS \otimes dS$. Since (S, ψ) is a semi-linear isomorphism, $S \otimes S$ is also a semi-linear isomorphism of vector space $C \otimes_{\Phi} C$ over Φ onto the space $A \otimes_K A$ over K and ψ is also the associated isomorphism with $S \otimes S$.

According to the above description of semi-linear isomorphism S of C onto A, it is useful to introduce the concept of semi-linear coalgebra mapping which extends the concept of the usual coalgebra mapping in the following.

Definition 2. Let C and D be vector spaces over fields Φ and K respectively. A mapping S of C into D is called a semi-linear coalgebra mapping if and only if

(i) S is a semi-linear transformation of C into D and ψ is the associated isomorphism with S, where S may be denoted by $S = (S, \psi)$ as before,

(ii) The following two diagrams are commutative:

$$\begin{array}{ccccccc} C & \xrightarrow{S} & D & \quad & C & \xrightarrow{S} & D \\ \Delta_c \downarrow & & \downarrow \Delta_D & & \epsilon_c \downarrow & & \downarrow \epsilon_D \\ C \otimes_\Phi C & \xrightarrow{S \otimes S} & D \otimes_K D & & \Phi & \xrightarrow{\psi} & K \end{array}$$

Especially we say that S is a semi-linear coalgebra isomorphism of C onto D if S is an isomorphism.

Note that this concept coincides with the usual one of coalgebra mapping if $K = \Phi$ and $\psi = I$, where I is the identity.

After these preliminaries we can introduce a coalgebra structure on A in the following way.

Lemma 2. *Let C, A and S be as in Lemma 1. Then there exists a natural coalgebra structure on A and S is a semi-linear coalgebra isomorphism of C onto A.*

Proof. Let A be an arbitrary element of A. Since S is an isomorphism of C onto A, there is a unique $c \in C$ such that $cS = a$. Define

$$\Delta_A : A \to A \otimes_K A$$

via

$$\Delta_A(a) = (S \otimes S)\Delta_C(c), \quad a = cS. \tag{1}$$

First we prove that Δ_A is K-linear. In fact, we have $(K^{\psi^{-1}}c)S = K(cS) = Ka$ for $cS = a$. Since S is semi-linear, using (1) we have

$$\begin{aligned} \Delta_A(Ka) = & (S \otimes S)\Delta_c(K^{\psi^{-1}}c) = (S \otimes S)K^{\psi^{-1}}\Delta_c(c) \\ = & (S \otimes S)K^{\psi^{-1}}\sum_{(c)} c_{(1)} \otimes c_{(2)} = (S \otimes S)\sum_{(c)} K^{\psi^{-1}} c_{(1)} \otimes c_{(2)} \\ = & \sum_{(c)}(K^{\psi^{-1}} c_{(1)})S \otimes c_{(2)}S = \sum_{(a)} K(c_{(1)}S) \otimes c_{(2)}S \\ = & K(S \otimes S)\sum_{(c)} c_{(1)} \otimes c_{(2)} = K(S \otimes S)\Delta_C(c) \\ = & K\Delta_A(a). \end{aligned}$$

Since $\Delta_c(c) = \sum_{(c)} c_{(1)} \otimes c_{(2)}$, we obtain from (1) that

$$\Delta_A(a) = \sum_{(a)} a_{(1)} \otimes a_{(2)} \quad \in A \otimes_K A,$$

where $a_{(i)} = c_{(i)}S, i = 1, 2$.

For any $c \in C$ we have

$$(c, \varepsilon_c) = (aS^{-1}, \varepsilon_c) = (a, S'^{-1}\varepsilon_c)^{\epsilon^{-1}}. \tag{2}$$

Putting

$$\varepsilon_A = S'^{-1}\varepsilon_c, \quad \text{i.e.} \quad \varepsilon_c = S'\varepsilon_A, \tag{3}$$

we rewrite (2) as below:

$$(c, \varepsilon_c)^{\psi} = (cS, \varepsilon_A). \tag{4}$$

Similarly, we can prove that ε_A is K-linear.

Now we show that $(A, \Delta_A, \varepsilon_A)$ is a coalgebra. In fact, we observe that by (1),

$$\Delta_A S = (S \otimes S)\Delta_c, \tag{5}$$

and it follows from (5) that

$$\begin{aligned}(I \otimes \Delta_A)\Delta_A S = &\quad (I \otimes \Delta_A)(S \otimes S)\Delta_c = (IS \otimes \Delta_A S)\Delta_c \\ = &\quad (SI \otimes (S \otimes S)\Delta_c)\Delta_c = (S \otimes S \otimes S)(I \otimes \Delta_c)\Delta_c \\ = &\quad (S \otimes S \otimes S)(\Delta_c \otimes I)\Delta_c = ((S \otimes S)\Delta_c \otimes SI)\Delta_c \\ = &\quad (\Delta_A S \otimes IS)\Delta_c = (\Delta_A \otimes I)(S \otimes S)\Delta_c \\ = &\quad (\Delta_A \otimes I)\Delta_A S.\end{aligned}$$

Therefore $(I \otimes \Delta_A)\Delta_A = (\Delta_A \otimes I)\Delta_A$. Similarly, we have $(\varepsilon_A \otimes I)\Delta_A = I = (I \otimes \varepsilon_A)\Delta_A$. This proves that $(A, \Delta_A, \varepsilon_A)$ is a coalgebra.

Finally it follows immediately from (5) and (4) that S is a semi-linear coalgebra isomorphism. This completes the proof.

Lemma 3. *Let C and A be as before. Let C^* and A^* be the dual algebras of C and A respectively. Then the adjoint S' of S is a semi-linear algebra isomorphism of A^* onto C^*.*

Proof. By Lemma 1, we need only to show that

$$S'^{-1}(c^*d^*) = (S'^{-1}(c^*))(S'^{-1}(d^*))$$

for all $c^*, d^* \in C^*$. In fact, for any $c \in C$, we have

$$\begin{aligned}(c, c^*d^*)^{\psi} = &\quad \sum_{(c)}(c_{(1)}, c^*)^{\psi}(c_{(2)}, d^*)^{\psi} \\ = &\quad \sum_{(a)}(a_{(1)}S^{-1}, c^*)^{\psi}(a_{(2)}S^{-1}, d^*)^{\psi} \\ = &\quad \sum_{(a)}(a_{(1)}, S'^{-1}c^*)(a_{(2)}, S'^{-1}d^*) \\ = &\quad (\Delta_A a, S'^{-1}c^* \otimes S'^{-1}d^*) \\ = &\quad (a, (S'^{-1}(c^*))(S'^{-1}(d^*))).\end{aligned}$$

On the other hand, we have

$$(c, c^*d^*)^{\psi} = (aS^{-1}, c^*d^*)^{\psi} = (a, S'^{-1}(c^*, d^*)).$$

Hence for all $a \in A$,

$$(a, (S'^{-1}(c^*))(S'^{-1}(d^*))) = (a, S'^{-1}(c^*d^*)).$$

Thus we have

$$S'^{-1}(c^*d^*) = (S'^{-1}(c^*))(S'^{-1}(d^*))$$

for all $c^*, d^* \in C^*$. From (3) it follows immediately that S'^{-1} maps the identity ε_c of C^* to the identity ε_A of A^*. This completes the proof.

Remark. Lemma 3 indicates that the algebra C^* can be replaced by the algebra A^* when we study the algebra structure of C^*. This replacement is advantageous to proving our main results in the next section, because $A^* = \Omega E_1$ is a subring of Ω.

Analogous to the left C^*-module of C we can define a left A^*-module of coalgebra A by $a^* \cdot a = (I \otimes (-, a^*))\Delta_A(a)$ for all $a^* \in A^*$ and $a \in A$.

Lemma 4. *Let C, C^*, A, A^*, S and S' be as above. Then for any element $c^* \in C^*$ we have $(c^*, C)S = a^* \cdot A$ with $a^* = S'^{-1}c^*$.*

Proof. By definition,

$$\begin{aligned} c^* \cdot c = &\ (I \otimes (-, c^*))\Delta_c(c) = \sum_{(c)} c_{(1)}(c_{(2)}, c^*) \\ = &\ \sum_{(c)} (c_{(2)}, S'a^*)c_{(1)} = \sum_{(c)} (c_{(2)}S, a^*)^{\psi^{-1}} c_{(1)}. \end{aligned}$$

Thus we have

$$\begin{aligned} (c^* \cdot c)S = &\ \sum_{(c)} ((c_{(2)}S, a^*)^{\psi^{-1}} c_{(1)})S \\ = &\ \sum_{(c)} (c_{(2)}S, a^*)(c_{(1)}S) \\ = &\ (I \otimes (-, a^*))(S \otimes S)\Delta_c(c) \\ = &\ (I \otimes (-, a^*)(\Delta_A S))(c) = a^* \cdot (cS). \end{aligned}$$

Since S is an isomorphism, we obtain

$$(c^* \cdot C)S = a^* \cdot (CS) = a^* \cdot A.$$

§4. Main Results

From now on we will replace coalgebra $(C, \Delta_c, \varepsilon_c)$ by the coalgebra $(A, \Delta_A, \varepsilon_A)$. It will be seen in the sequel that this replacement is more convenient to our study. For the sake of simplicity we write Δ_A for Δ and ε_A for ε. We always regard A as a left A^*-module. As before we write $A = \sum_{i \in \Gamma} K\alpha_i$, $A^* = \Omega E_1$ and $\Omega = \mathrm{End}_K A$. Then for any element a^* of A^* and $\alpha_t \in \{\alpha_i\}_{i \in \Gamma}$ we have

$$\begin{aligned} &\Delta\alpha_i = \sum_i \alpha_i \otimes a_{it}, \\ &a^* \cdot \alpha_t = \sum_i \alpha_i \otimes (a_{it}, a^*) = \sum_i (a_{it}, a^*)\alpha_i, \\ &(a_{it}, a^*) = a^*(a_{it}) \in K \end{aligned} \tag{6}$$

for all $t \in \Gamma$. Hence there exists uniquely an element ω_{a^*} of Ω such that

$$\alpha_t \omega_{a^*} = a^* \cdot \alpha_t \tag{7}$$

for all $t \in \Gamma$. From (6) and (7) it follows that

$$\begin{aligned} b^* a^* \cdot \alpha_t = & \; b^* \cdot \sum_i (a_{it}, a^*) \alpha_i = \sum_i (a_{it}, a^*) b^* \cdot \alpha_i \\ = & \; \sum_i (a_{it}, a^*) \alpha_i \omega_{b^*} = \alpha_t \omega_{a^*} \omega_{b^*}, \end{aligned}$$

where $\omega_{b^*} \in \Omega, (a_{it}, a^*) \in K$. By (7) we have

$$\omega_{b^* a^*} = \omega_{a^*} \omega_{b^*}. \tag{8}$$

Theorem 1. *Let A, A^* be as above. Denote $W = \{\omega_{a^*} \mid a^* \in A^*\}$. Then there exists a ring anti-isomorphism $\omega : a^* \to \omega_{a^*}$ of A^* onto W such that $\omega_{a^*} \varepsilon = a^*$, where $\omega_{a^*} \varepsilon$ means the ring-product of elements ω_{a^*} and ε of Ω. Furthermore, ε is the inverse isomorphism of W onto A^* and $A^* = W\varepsilon$.*

Proof. by (8) W is closed by multiplication of its elements. Let $\omega_{a^*}, \omega_{b^*} \in W$. Then $(a^* + b^*) \cdot \alpha_t = a^* \cdot \alpha_t + b^* \cdot \alpha_t = \alpha_t(\omega_{a^*} + \omega_{b^*})$, which means

$$\omega_{a^* + b^*} = \omega_{a^*} + \omega_{b^*}. \tag{9}$$

It is easy to see that W is a ring. On the other hand, we have $\varepsilon \cdot \alpha_t = \alpha_t$; hence

$$\omega_\varepsilon = 1. \tag{10}$$

Thus W is a subring of Ω with the same identity 1 of Ω. Now we define $\omega : A^* \to W$ via $\omega : a^* \to \omega_{a^*}$. From (8), (9), (10) and the definition of W, it is enough to show that ω is injective. To do this, we prove that

$$\omega_{a^*} \varepsilon = a^* \tag{11}$$

for all $a^* \in A^*$. In fact, for any $\alpha_t \in \{\alpha_i\}_{i \in \Gamma}$, we have

$$\begin{aligned} \alpha_t(\omega_{a^*} \varepsilon) = & \; (\alpha_t \omega_{a^*}) \varepsilon = (a^*, \alpha_t) \varepsilon \\ = & \; \left(\textstyle\sum_i (a_{it}, a^*) \alpha_i \right) \varepsilon = \textstyle\sum_i (a_{it}, a^*) \varepsilon(\alpha_i) \\ = & \; \left(\textstyle\sum_i \varepsilon(\alpha_i) a_{it}, a^* \right) = (\alpha_t, a^*) = \alpha_t a^*. \end{aligned}$$

It implies (11).

Suppose $\omega(a^*) = \omega(b^*)$, that is, $\omega_{a^*} = \omega_{b^*}$. Then we have $a^* = \omega_{a^*} \varepsilon = \omega_{b^*} \varepsilon = b^*$, which proves that ω is injective. Also, from (11), it follows that $A^* = W\varepsilon$.

Now we are in a position to prove our first main result.

Theorem 2. *Let C be a coalgebra over a field Φ with infinite dimension and let C^* be its dual algebra. Suppose that there exists a non-zero element c^* of C^* such that* $\dim c^* \cdot C < \dim C$. *Then C^* is not prime.*

Proof. From Section 3, it is easy to see that $\dim c^* \cdot C < \dim C$ if and only if $\dim a^* A < \dim A$, where $a^* = S'^{-1}c^*$. In this case we only need to prove that A^* is not prime under the assumption of $\dim a^* \cdot A < \dim A$. But W and A^* are ring anti-isomorphisms; it is enough to show that W is not prime. In fact, since $a^* \cdot A = A\omega_{a^*}$, there exists a non-zero element $x = \omega_{a^*} \in W$ such that $\dim Ax < \dim A$. Denote $A = \sum_{i\in\Gamma} K\alpha_i$ and $Ax = \sum_{j\in I} K_{\beta_j}$. Then $|I| < |\Gamma|$, where $|I|$ means the cardinal number of I. On the other hand, $a^* \cdot \beta_j = \beta_j\omega_{a^*}$ by (7); hence $\beta_j W = \{\beta_j\omega_{a^*} \mid \omega_{a^*} \in W\} = A^* \cdot \beta_j$. It is easy to see that the dimension of the subspace $A^* \cdot \beta_j$ is finite. Since $|\Gamma|$ is infinite and the dimension $|I|$ of Ax is smaller that $|\Gamma|$, we have $\dim AxW < |\Gamma|$. Thus $AxW \neq A$. Putting $AxW = \sum_{j\in I'} K_{z_j}$, we have $|I'| < |\Gamma|$. Therefore A can be expressed by $A = AxW \oplus \sum_{j\in\Gamma-I'} K\lambda_j$ and has a basis $\{z_j\}_{j\in I'} \cup \{\lambda_j\}_{j\in\Gamma-I'}$. Now we choose an element e of Ω such that

$$\begin{aligned} z_j e &= 0 \qquad \text{for all} \quad j \in I', \\ \lambda_{i_0} e &= \lambda_{i_0}, \qquad i_0 \in \Gamma - I', \\ \lambda_i e &= 0, \qquad i \in \Gamma - I', \quad i \neq i_0. \end{aligned}$$

Since $e\Omega E_1 \neq 0$, there exists an element $w \in \Omega$ such that $b^* = ewE_1 \in A^*$ and $b^* \neq 0$. Thus we have

$$AxWb^* = 0.$$

By Theorem 1, $b^* = \omega_{b^*}\varepsilon$. Hence $AxW\omega_{b^*}\varepsilon = 0$. But $xW_{\omega_{b^*}}\varepsilon \subset \Omega$, therefore $xW\omega_{b^*}\varepsilon = 0$. Because ε is a ring anti-isomorphism of W onto A^* and $xW\omega_{b^*} \subset W$, it is clear that $xW\omega_{b^*} = 0, x \neq 0$ and $\omega_{b^*} \neq 0$. Thus W is not prime. This completes the proof.

From Example 3, we know the converse of Theorem 2 is not true; but we have the following

Theorem 3. *Let C be a coalgebra over a field Φ (Not necessarily of infinite dimension) and let C^* be its dual algebra. Then C^* is prime if and only if $C^*c^* \cdot C = C$ for any non-zero element c^* of C^*.*

Proof. As in the proof of Theorem 2, we only need to prove that A^* is prime if and only if $A^*a^* \cdot A = A$ for any non-zero element a^* of A^*, or equivalently that W is prime if and only if $AxW = A$ for any non-zero element x of W. We will prove the latter.

Suppose there exist two elements x, y of W with $x \neq 0$ such that $xWy = 0$. Then we have $0 = AxWy = Ay$ by assumption. Since $y \in \mathrm{End}_k A$, we obtain $y = 0$. Hence W is prime.

Conversely, if there exists a non-zero element x of W such that $AxW \neq A$, then A can be expressed by $A = AxW \oplus B$ as a vector space. We take the element $l \in \Omega$ satisfying $AxWl = 0$ and $bl = b$ for any $b \in B$. Then $l \neq 0$. Since $l\Omega E_1 \neq 0$, there exists an element $l' \in \Omega$ such that $a^* = ll'E_1 \in A^*$ and $a^* \neq 0$. Thus we have $xWa^* = 0$ and $xW\omega_{a^*}\varepsilon = 0$. Since $xW\omega_{a^*} \subset W$ and ε is a ring anti-isomorphism, we obtain $xW\omega_{a^*} = 0, x \neq 0, \omega_{a^*} \neq 0$. This proves that W is not prime.

As a consequence of Theorem 3, we have the following.

Corollary. *Let C be a coalgebra over a field Φ and let C^* be its dual algebra. Then we have*

(i) *C^* is semi-prime if and only if $c^*C^*c^* \cdot C = C$ for any non-zero element c^* of C^*,*

(ii) C^* *has no divisors of zero if and only if* $c^* \cdot C = C$ *for any non-zero element* c^* *of* C^*.

References

[1] T. Shudo and H. Miyamoto, *On the decomposition of coalgebras*, Hiroshima Math. J., 8(1978), 499–504.

[2] M.E. Sweedler, *Hopf Algebras*, W.A. Benjamin, INC. New York, 1969.

[3] Xu Yonghua, *On the structure of primitive rings having minimal one-sided ideals*, Scientia Sinica, 24(1981), 1056–1065.

[4] Xu Yonghua, *A theory of rings that are isomorphic to the complete rings of linear transformations I*, Acta Math. Sinica, 22(1979), 204–218.

[5] Xu Yonghua, *On the decomposition of coalgebras* (to appear).

Proceedings of the Asian Mathematical
Conference, 1990

Factorizations of Meromorphic Solutions of Differential and Functional Equations

CHUN-CHUN YANG
The Hong Kong University of Science and Technology

§1. Introduction

Let $f(z), g(z)$ and $h(z)$ be denoted as meromorphic functions in the complex plane. If every factorization of the form $f(z) = g(h(z))$ implies that either g or h is a linear fractional function (rational function), then f is called prime (pseudo-prime); g, h are called right- and left- factor of f, respectively. The definition of primemess of a function was first introduced in 1952 by P.C. Rosenbloom [11] in a quantitative study of six-points of composition of two entire functions. In [11] he stated without proof that $e^z + z$ is a prime function. Later on F. Gross presented a complete proof and extended the definition of factorization to meromorphic functions with meromorphic functions as the factors [4]. Since then the topic of factorization theory has been vigorously pursued by C. Yang, M. Ozawa, R. Goldstein, Prokopovich, etc. Based on the properties of the function $e^z + z$, many classes of prime and pseudo-prime functions have been constructed. Among them is that any meromorphic solution of a linear differential equation with rational functions as the coefficients must be pseudo-prime.

It is easy to see that a similar conclusion cannot be reached for non-linear algebraic differential equations. For instance, $w(z)$ =exp $e^z (= \exp z_0 e^z)$ is an entire solution of the equation $ww'' - ww' - (w')^2 = 0$. Note that $w(z)$ is of infinite order. However meromorphic solutions of various types of nonlinear algebraic differential equations that are pseudo-prime have been exhibited.

It has been shown [7] that under suitable conditions an entire solution of infinite order (with restricted growth) of a second order differential equation with periodic functions as the coefficients must be pseudo-prime.

The tools used in the study of factorization theory are mainly some classical theorems in complex function theory such as Picard's theorem, the theorem of Wiman-Valiron and the Nevanlinna's value-distribution theory of meromorphic functions (see e.g. [5]). Prokopovich also utilized Nevanlinna's theory for algebroidal functions to attack the factorization problem and obtained some significant results, see e.g. [10].

Recently Zheng and Yang [20] studied and defined the factorization of algebroidal functions, a generalized concept of meromorphic functions. Their results have shown some common and distinct properties between meromorphic functions and algebroidal functions.

In this paper we shall mainly review and announce some of the interesting results mentioned above (as theorems) and useful techniques which are employed in obtaining them (as lemmas).

§2. Preliminaries

Earlier Polya [9] showed that if a finite order entire function $F(z)$ has a factorization of the form $F(z) = f(g(z))$, then either f is of zero order or g is a polynomial.

Edrei and Fuchs [3] proved the extension of the above result for meromorphic functions as follows.

Lemma A. *Let $f(z)$ be a meromorphic function which is not of order zero, and let $g(z)$ be a transcendental entire function. Then $f(g(z))$ is of infinity order.*

Lemma B (Clunie [2]). *Let f be a transcendental meromorphic function and g be a transcendental entire function. Then*

$$\lim_{r\to\infty} \frac{T(r, f(g))}{T(r, g)} = \infty,$$

and

$$\lim_{r\to\infty} \frac{T(r, f(g))}{T(r, f)} = \infty, \quad \text{if} \quad f \text{ or } g \text{ is of finite order.}$$

Lemma C (Valiron-Wittich [18]). *Suppose that $w(z)$ is a meromorphic solution of a linear differential equation with rational as the coefficients. Then the order of $w(z)$ is a positive rational number.*

Lemma D (Strelitz-Simogljad [15]). *Every entire transcendental solution of a second order algebraic differential equation with rational coefficients is of a positive order.*

Lemma E (N. Steinmetz [12]). *Let F_j and $h_j (j = 0, 1, 2, \cdots, m)$ be two sets of $m + 1$ meromorphic functions not identically vanishing, and g be a nonconstant entire function satisfying*

$$\sum_{j=0}^{m} T(r, h_j) \leq KT(r, g) + S(r, g),$$

where K is a positive constant and $S(r, g) = oT(r, g)$outside a set of r values of finite measure.

If F_j and h_j $(j = 0, 1, 2, \cdots, m)$ satisfy the following functional equation

$$F_0(g(z))h_0(z) + F_1(g(z))h_1(z) + \cdots + F_m(g(z))h_m(z) \equiv 0,$$

then exist polynomials P_j and $Q_j (j = 0, 1, 2, \cdots, m)$ not all identically zero such that

$$P_0(g(z))h_0(z) + P_1(g(z))h_1(z) + \cdots + P_m(g(z))h_m(z) \equiv 0$$

and

$$F_0(z)Q_0(z) + F_1(z)Q_1(z) + \cdots + F_m(z)Q_m(z) \equiv 0.$$

In [13], the following result, useful to the factorization of meromorphic solutions algebraic differential equations, is derived.

Lemma F. *Let $w(z)$ be a meromorphic function satisfying an algebraic differential equation of the form $\Omega(z, w, w', \cdots, w^{(n)}) = 0$, where Ω is a polynomial in $w, w', \cdots, w^{(n)}$ with rational functions as the coefficients. Let $M[w]$ or $M_j[w]$ denote a differential monomial of the form*

$$a(z)(w)^{k_0}(w')^{k_1}\cdots(w^{(m)})^{k_m}$$

with $a(z)$ being a rational function $\not\equiv 0$, then $\Omega(z, w, w', \cdots, w^{(n)})$ can be expressed as $\sum_{j=0}^{t} M_j[w]$. Define

$$\gamma_{M_j} = k_0 + k_1 + \cdots + k_m \quad \text{and} \quad \Gamma_{M_j} = k_0 + 2k_1 + \cdots + (m+1)k_m.$$

Let

$$\gamma_\Omega = \max_{j=1}^{t} \gamma_{M_j} \quad \text{and} \quad \Gamma_\Omega = \max_{j=1}^{t} \Gamma_{M_j}.$$

If $w(z) = f(g)$, where f is transcendental meromorphic and g is entire, then f also satisfies an algebraic differential equation

$$\hat{\Omega}(z, w, w', \cdots, w^{(m)}) = 0 \ \text{with} \ \gamma_{\hat{\Omega}} \leq \gamma_\Omega \ \text{and} \ \Gamma_{\hat{\Omega}} \leq \Gamma_\Omega.$$

The following result is important in the study of factorization of algebroidal functions.

Lemma G(Zheng [19]). *Let $\phi(z, w) = A_0(z)w^n + \cdots + A_1(z)w + A_0(z)$ be irreducible, then $\phi(g(z), w)$ remains to be irreducible for any entire function $g(z)$.*

§3. Main Results

3.1. Solutions of linear equations

Theorem 1 (N. Steinmetz [13]). *Let $F(z)$ be a meromorphic solution of the linear differential equation of the form*

$$a_n(z)w^{(n)}(z) + a_{n-1}(z)w^{(n-1)}(z) + \cdots + a_1(z)w(z) + a_0(z) = 0, \tag{E$_1$}$$

where all the a_j's are polynomials. Then $F(z)$ must be pseudo-prime.

We shall use D to denote the class of meromorphic functions that satisfy the above type of differential equations (E_1). Then it was shown in [12] that $F(z) = \sum_{i=1}^{m} q_i(z)\psi_i(z) \in D$, where the q_1's are rational functions and $\psi_i \in D$. Recently by argument similar to that used in [11], Urabe and Yang [16] showed if $f \in D$ and $g \in D$, then the product fg also belongs to D. It is asked in [16]: Let $f \in D$ and $g \in D$ $(g \not\equiv 0)$, then is it true that f/g is pseudo-prime?

We note that any meromorphic solution of (E_1) is of finite order. For infinite order solutions we have the following.

Theorem 2 [7]. *Consider the differential equation*

$$w''(z) + P(e^z)w'(z) + Q(e^z)w(z) = 0, \tag{E$_2$}$$

where

$$P(z) = \sum_{j=0}^{p} p_j z^j, \quad Q(z) = \sum_{k=0}^{q} q_k z^k.$$

If there exist rational number r_1 and r_2 such that $p_0 = -(r_1 + r_2)$ and $q_0 = r_1 r_2$, then each subnormal solution of $(E_2)$is pseudo-prime. A function $w(z)$ is said to be subnormal if it satisfies

$$\overline{\lim_{r\to\infty}} \frac{\log T(r,w)}{r} = 0. \tag{C}$$

As generalization of the above, the following result has been obtained.

Theorem 3 (Urabe-Yang [16]). *Consider the following second order homogeneous differential equation*

$$w''(z) + P_1(e^z, e^{-z})w'(z) + P_2(e^z, e^{-z})w(z) = 0, \tag{D}$$

where $P_j(X,Y)(j = 1,2)$ is a polynomial of the two variables X and Y. If $w = h(z)$ is an entire and subnormal solution of equation (D), then h can be expressed as $h(z) = e^{\alpha z}\psi(e^z)$, where $\psi(z) = \psi_1(z) + \psi_2(1/z)$ and that $\psi_i(i = 1,2)$ is a polynomial. Moreover, $h(z)$ is pseudo-prime.

3.2. Meromorphic solutions of nonlinear differential equations

Theorem 4 (Mues [8]). *Let $F(z)$ be a meromorphic solution of the differential equation of Riccati type*

$$w'(z) = a_0(z) + a_1(z)w(z) + a_2(z)w^2(z),$$

where a_i's $(i = 0,1,2)$ are rational functions. Then $F(z)$ is pseudo-prime.

By using a different argument, Steinmetz derived the same conclusion in [14].

For first order nonlinear differential equations of the form

$$\left(\frac{dy}{dz}\right)^m = R(z,y), \tag{G}$$

where $R(z,y)$ is a rational function in z and y, N. Steinmetz [13], S. Bank and P. Kaufman [1] proved independently that there are essentially six canonical forms of (G) for the solution (or its linear fractional transform) to be a transcendental meromorphic function:

$$w' = a(z) + b(z)w(z) + c(z)w^2(z), \tag{G$_1$}$$

$$(w')^2 = a(z) + (w - b(z))^2 w(z), \quad b(z) \not\equiv 0, \tag{G$_2$}$$

$$(w')^3 = a(z) + (w-\beta)^2(w-\beta)^2, \quad \beta \text{ a complex number}, \tag{G$_3$}$$

$$(w')^2 = a_1(z)(w_1 - e_1)(w - e_2)(w - e_3), \tag{G$_4$}$$

where e_i's $(i = 1,2,3)$ are complex numbers satisfying $e_1 + e_2 + e_3 = 0$,

$$(w')^4 = a_2(z) + (w-\beta)^2 w^3, \quad \beta \text{ a complex number}, \tag{G$_5$}$$

$$(w')^6 = a_3(z) + (w-\beta)^3 w^4, \quad \beta \text{ a complex number}, \tag{G$_6$}$$

in the above, $a(z), b(z), c(z)$, and $a_j(z)(j = 1,2,3)$ are rational functions $\not\equiv 0$. Then the following three results have been obtained.

Theorem 5 (He-Lin [6]). *Each transcendental meromorphic solution of (G_2) or (G_3) is pseudo-prime.*

Theorem 6 (He-Lin [6], Theorem 3). *Let the Laurent's expansion of $a_1(z)$ in (G_4) around ∞ be*

$$a_1(z) = a_0^{(1)} z^{d_1} + a_1^{(1)} z^{d_1+1} + \cdots.$$

If $a_0^{(1)} \neq 0, d_1 = -2$, then each meromorphic solution of (G_4) is pseudo-prime.

Theorem 7 (He-lin [6], Theorem 4). *Let the Laurent's expansion of $a_2(z)$ in (G_5) around ∞ be*

$$a_2(z) = a_0^{(2)} z^{d_2} + a_1^{(2)} z^{d_2+1} + \cdots.$$

If $a_0^{(2)} \neq 0, d_1 = -4$, then each meromorphic solution of (G_5) is pseudo-prime.

Theorem 8. *Let the Laurent's expansion of $a_3(z)$ in (G_6) around ∞ be*

$$a_3(z) = a_0^{(3)} z^{d_3} + a_1^{(3)} z^{d_3+1} + \cdots.$$

If $a_0^{(3)} \neq 0, d_3 = -6$, then each meromorphic solution of (G_6) is pseudo-prime.

It is noted in [6] that if the coefficients of (G_4), (G_5), or (G_6) don't satisfy the conditions as stated in Theorems 6, 7 and 8 above, then the corresponding meromorphic solution of (G_4), (G_5) and (G_6) may not be pseudo-prime, for instance, if $a_1(z), a_2(z)$ and $a_3(z)$ are suitable constants.

Let $w_j(z)$ denote a meromorphic solution of equation (G_j), then the gradients to the proofs of the above three results are the facts that the order of $w_1(z)$ (or $w_2(z)$) is a positive integral multiple of 1/2, the order of $w_3(z)$ is an integral multiple of 1/3, and the order of w_4, w_5 and w_6 is equal to $d_1 + 2, d_2/2 + 2$ and $d_3/3 + 2$, respectively (see (Bank-Kaufman [1]) and Lemma E).

For second order nonlinear differential equations we have

Theorem 9 (Urabe-Yang [16]). *Consider the following differential equation with rational functions as the coefficients:*

$$P(z, w, w', w'') = 0. \tag{H}$$

Let $w = h(z)$ be entire and periodic solutions of (H). Suppose that

$$\overline{\lim_{r\to\infty}} \frac{\log\log T(r,h)}{\log r} = 0$$

(i.e., h is of hyper-order zero), or $h(z)$ is subnormal and of finite lower order, then h is pseudo-prime.

Along these lines, one can show similarly that any finite order entire solution of (H) is pseudo-prime. It remains to be answered whether or not a finite order entire solution of an arbitrary order algebraic differential equation is always pseudo-prime [16].

3.3. Factorization of algebroidal functions

Let $w(z)$ be a complex function defined by an irreducible equation of the form

$$\phi(z,w) = A_n(z)w^n + A_{n-1}(z)w^{n-1} + \cdots + A_1(z)w(z) + A_0(z) = 0, \qquad \text{(J)}$$

where $A_i(z)(i = 0, 1, 2, \cdots, n)$ are entire functions with $A_n \not\equiv 0$ and at least one of them being transcendental, then $w(z)$ is called an n-valued algebroidal function. If all the coefficients $A_i(z)s$ in equation (J) are polynomials with at least one of them being nonconstant, then $w(z)$ is called an n-valued algebraic function.

Based on Lemma G, Zheng [19] obtained the following result.

Theorem 10. *Let $w(z)$ be an n-valued algebroidal function defined by an irreducible equation of form (J). Then a necessary and sufficient condition for $w(z)$ to be expressed as the composition of an algebraic (algebroidal) function $w_0(z)$ with a nonlinear meromorphic (entire) function $g(z)$, i.e., $w(z) = w_0(g(z))$, is that $g(z)$ is a common right factor of the meromorphic functions $A_i(z)/A_n(z)(i = 0, 1, 2, \cdots, n)$; some of the $A_i(z)/A_n(z)$ can be constant.*

Definition 1 [20]. If an n-valued algebroidal function $w(z)$ can be expressed as

$$w(z) = w_0(g(z)), \qquad \text{(F)}$$

where $w_0(z)$ is an n_0-valued algebroidal function , $g(z)$ an entire function (g can be a meromorphic function if $w_0(z)$ is an n_0-valued algebraic function), then the form (F) will be referred to as a factorization of w, $w_0(z)$ and $g(z)$ are called left and right factors of $w(z)$, respectively. We also note that $n = n_0$ according to Lemma G.

Definition 2 [20]. Let $w(z)$ be an n-valued algebroidal function. If every factorization of w in the form (F) implies that $g(z)$ must be linear (w_0 an n-valued algebraic function and g is a polynomial), then $w(z)$ is called prime (pseudo-prime). Moreover, if, in (F), g can only be a polynomial, then $w(z)$ is called left-prime.

Based on Theorem 10, some equivalent definitions of prime, pseudo-prime, and left-prime for an algebroidal function can be established as follows.

Definition 3 [20]. Let $w(z)$ be an n-valued algebroidal function defined by an irreducible equation (J). Then

(i) $w(z)$ is called (left-) prime if all the functions $A_i(z)A_n(z)$ $(i = 0, 1, 2, \cdots, n-1)$ have no (transcendental) nonlinear entire functions as their common right factor.

(ii) $w(z)$ is called pseudo-prime if all the functions $A_i(z)/A_n(z)(i = 0, 1, 2, \cdots, n-1)$ have no nonlinear meromorphic functions as their common right factor, or otherwise, but each $A_i(z)/A_n(z)$ has only rational functions as its left factors.

It follows immediately from Definition 3 that $w(z)$ is pseudo-prime if all the $A_i(z)/A_n(z)(i = 0, 1, 2, \cdots, n-1)$ are.

There have been many studies on the factorizations of the product or sum of an entire function and a polynomial. For instance, it is known [17] that if $H(z)$ is a periodic entire function of finite lower order and if $Q(z)$ is a nonconstant rational function with at least one of its poles (if it exists) being simple, then $H(z) + Q(z)$ must be left-prime. The following is an analogous result for algebroidal functions.

Theorem 11 [20]. *Let $w(z)$ be an n-valued algebroidal function as defined by an irreducible equation of form (J). Assume further that $A_{n-1}(z)$ has at least one zero. Suppose that the lower order of w is finite. Then $w(z) - Q(z)$ is always left-prime, where $Q(z)$ is a rational function and has at least one of its poles (if it exists) being simple.*

The proof will use an extended Edrei-Fuchs theorem [19] on the order extimation of an algebroidal function composed with an entire function.

References

[1] S. Bank and P. Kaufman, *On the growth of meromorphic solutions of the differential equation $(y')^m = R(z,y)$*, Acta Math., 144(1980), 233–248.

[2] J. Clunie, *The composition of entire meromorphic functions*, McIntyre Memorial Volume, Ohio Univ. Press, 1970.

[3] A, Edrei and W.H.J. Fuchs, *On the zeros of $f(g(z))$, where f and g are entire functions*, J. Analyse Math., 12(1964), 243–255.

[4] F. Gross, *On factorization of meromorphic functions*, Trans. AMA Math. Soc., 131(1968), 215–222.

[5] W.K. Hayman, *Meromorphic functions*, Oxford Mathematical Monographs, Clarendon Press, 1964.

[6] Y. He and Y. Lin, *On factorization of meromorphic functions satisfying nonlinear ordinary differential equations*, preprint.

[7] Y. He, X. Zou and Y. Lin, *Meromorphic solutions of ordinary differential equations Research Memorandum*, Vol.3 No.10(1987), Institute of Mathematics, Academia Sinica, Beijing, China.

[8] E. Mues, *Über faktorisierbaren Lösungen von Riccatischen Differential gleichungen*, Math. Zeit., 121(1971), 145–156.

[9] G. Polya, *On an integral function of an integral function*, J. London Math. Soc., 1, 12(1926).

[10] G.S. Prokopovich, *On pseudo-simplicity of some meromorphic functions*, Ukrainskii Matematicheskii Zhurnal, Vol. 27, No. 2, March-April 1975, 261–273 (English transl. 219–222).

[11] P.C. Rosenbloom, *The fix-prints of entire functions*, Medd, Lunds Univ. Mat. Sem., Suppl. -Bd., M. Riesz, 186(1952).

[12] G.D. Song and C.C. Yang, *On pseudo-primality of the combination of meromorphic functions satisfying linear differential equations*, in value distribution theory and its applications (edited by C.C. Yang), Contemporary Math. Series, vol.25, American Math. Soc., Providence, RI, 1986.

[13] N. Steinmetz, *Eigenschaften eindeutigen Lösungen gewöhnlicher Differential gleichungen in Komplexen*, Karlsrube Dissertation, 1978.

[14] N. Steinmetz, *Über die faktoriserbaren Lösungen gewöhnlicher Differential gleichungen*, Math. Z., 170(1980), 168–180.

[15] Sh. Strelitz, *A remark on meromorphic solutions of algebraic differential equations*, Proc. Amer. Math. Soc., 65(1977), 255–261.

[16] H. Urabe and C.C. Yang, *On factorization of entire functions satisfying differential equations*, to appear in Kodai Math. J.

[17] H. Urabe, *On a factorization problem of certain entire functions and a theorem of Ostrovskii*, Bull. Kyoto Univ. of Education, Ser. 13, No.523 (1978), 1–9.

[18] H. Wittich, *Neuere Untersuchungen über eindentige analytische funktionen*, Berlin-Gottington-Heidelberg, 1955.

[19] C.H. Zheng, *The growth of a compositive algebroidal function and its factorization* (in Chinese), Nature Magazine, 12(1)(1988), 876–877.

[20] C.H. Zheng and C.C. Yang, *Factorization of algebroidal functions* (in Chinese), to appear in Chin. Ann. Math.

Proceedings of the Asian Mathematical
Conference, 1990

Total Deficiency of Meromorphic Functions and Their Derivatives*

LO YANG
Institute of Mathematics
Chinese Academy of Sciences
Beijing 100080, China

Suppose that $f(z)$ is a transcendental meromorphic function in the finite plane and a is a complex value which may be infinity. According to R. Nevanlinna [4, 6], if

$$\delta(a,f) = \liminf_{r\to\infty} \frac{m\left(r,\dfrac{1}{f-a}\right)}{T(r,f)} = 1-\limsup_{r\to\infty} \frac{N\left(r,\dfrac{1}{f-a}\right)}{T(r,f)}$$

is positive, a is called a deficient value of $f(z)$ and $\delta(a,f)$ is its deficiency. (When $a = \infty, m(r,\frac{1}{f-a})$ and $N(r,\frac{1}{f-a})$ in the definition of $\delta(a,f)$ should be replaced by $m(r,f)$ and $N(r,f)$ respectively.) The most important and classical result is that the set of all deficient values of $f(z)$ is at most countable and the total deficiency does not exceed two. (deficiency relation [4, 6]) The upper bound of two is sharp in general.

§1. Total Deficiency of Meromorphic Derivatives

Let k be a positive integer. Hayman [7] pointed out that the inequality

$$\sum_{a\in\mathbb{C}} \delta(a,f^{(k)}) \le \frac{k+2}{k+1}$$

holds for any transcendental meromorphic function $f(z)$. In 1971, Mues [5] improved this result to

$$\sum_{a\in\mathbb{C}} \delta(a,f^{(k)}) \le \frac{k^2+5k+4}{k^2+4k+2}. \tag{1.1}$$

*The research was partially supported by a grant from the National Science Foundation of China.

The following result was obtained in [7].

Theorem 1. *Let $f(z)$ be a transcendental meromorphic function in the finite plane and k be a positive integer. Then we have*

$$\sum_{a\in\mathbb{C}} \delta(a, f^{(k)}) \leq \frac{2k+2}{2k+1}, \tag{1.2}$$

where the summation is taken over all the finite complex numbers.

It is clear that for any positive integer k, the upper bound in Theorem 1 is always better than that in (1.1). Moreover, we improved more than last time. i.e.

$$\frac{k^2+5k+4}{k^2+4k+2} - \frac{2k+2}{2k+1} > \frac{k+2}{k+1} - \frac{k^2+5k+4}{k^2+4k+2}.$$

We also proved another estimate which includes $\delta(\infty, f^{(k)})$.

Theorem 2. *Let $f(z)$ be a transcendental meromorphic function of finite order in the finite plane and k be a positive integer. Then we have*

$$\sum_{a\in\overline{\mathbb{C}}} \delta(a, f^{(k)}) \leq 2 - \frac{2k(1-\Theta(\infty, f))}{1+k(1-\Theta(\infty, f))},$$

where $\Theta(\infty, f)$ is the ramification index of the infinity with respect to f, defined by

$$\Theta(\infty, f) = 1 - \limsup_{r\to\infty} \frac{\overline{N}(r, f)}{T(r, f)}.$$

Corollary 1. *If $\Theta(\infty, f) < 1$, then we have*

$$\lim_{k\to\infty} \left\{ \sum_{a\in\overline{\mathbb{C}}} \delta(a, f^{(k)}) \right\} = 0.$$

Corollary 2. *If $\Theta(\infty, f) = 0$, then we have*

$$\sum_{a\in\overline{\mathbb{C}}} \delta(a, f^{(k)}) \leq \frac{2}{k+1}.$$

Corollary 3. *If there exists a positive integer k_0 such that $\sum_{a\in\overline{\mathbb{C}}} \delta(a, f^{(k_0)}) = 2$, then we have $\Theta(\infty, f) = 1$.*

§2. Drasin's Problems

Let $f(z)$ be a transcendental meromorphic function of finite order in the finite plane and k be a positive integer. From Nevanlinna second fundamental theorem, it is easy to see that

$$\begin{aligned} \sum_{a\in C} \delta(a, f^{(k)}) \leq & \ (k+1-\Theta(\infty, f))\delta(0, f^{(k)}) \\ \leq & \ (k+1-\delta(\infty, f))\delta(0, f^{(k)}). \end{aligned} \tag{2.1}$$

Drasin [1] posed the following problems.

(A) If $\sum_{a\in\overline{\mathbb{C}}} \delta(a, f) = 2$ and $\delta(\infty, f) = 0$, do we must have

$$\sum_{b\in\overline{\mathbb{C}}} \delta(b, f') = \delta(0, f') = 1?$$

(B) Let $f(z)$ be transcendental meromorphic in the finite plane with $\delta(\infty, f) = 0$. Can we have

$$\sum_{a\in\overline{\mathbb{C}}} \delta(a, f) + \sum_{b\in\overline{C}} \delta(b, f') = 4? \tag{2.2}$$

(C) If the equality (2.2) can not be attended, what is the best upper bound of the left-hand side of (2.2)?

In order to answer these questions we have [8]

Theorem 3. *Let $f(z)$ be a transcendental meromorphic function in the finite plane and k be a positive integer. Then we have*

$$\sum_{a\in\mathbb{C}} \delta(a, f) + \sum_{b\in\overline{\mathbb{C}}} \delta(b, f^{(k)}) \leq 3. \tag{2.3}$$

The equality of (2.3) holds if and only if either

(i) $\Theta(\infty, f) = 1, \sum_{a\in\mathbb{C}} \delta(a, f) = 1, \sum_{b\in\overline{\mathbb{C}}} \delta(b, f^{(k)}) = 2;$

or

(ii) $k = 1, \Theta(\infty, f) = 0, \sum_{a\in\mathbb{C}} \delta(a, f) = 2, \sum_{b\in\overline{\mathbb{C}}} \delta(b, f') = 1.$

Theorem 3 together with the inequality (2.1) give a positive answer to the problem (A). It also gives a negative answer to the problem (B) and obtains the best upper bound for the problem (C).

If the condition $\delta(\infty, f) = 0$ is taken off from the problem (B), then the equality (2.2) is possible. In fact, we have

Theorem 4. *Let $f(z)$ be transcendental meromorphic and of finite order in the finite plane and k_0 be a positive integer. In order that*

$$\sum_{a\in\overline{\mathbb{C}}} \delta(a, f) + \sum_{b\in\overline{\mathbb{C}}} \delta(b, f^{(k_0)}) = 4,$$

the necessary and sufficient condition is

$$\delta(\infty, f) = 1, \quad \sum_{a\in\mathbb{C}} \delta(a, f) = 1$$

and

$$\delta(0, f^{(k)}) = \delta(\infty, f^{(k)}) = 1 \qquad (k = 1, 2, \cdots).$$

The problem (A) can be posed in a general way which has a positive answer, too.

Theorem 5. *Let $f(z)$ be transcendental meromorphic and of finite order in the finite plane. If $\sum_{a\in\mathbb{C}} \delta(a,f) = 2$, then we have*

$$\sum_{b\in\overline{\mathbb{C}}} \delta(b, f^{(k)}) = \delta(0, f^{(k)}) = \frac{2}{k+1} \quad (k = 1, 2, \cdots).$$

Similar to Theorem 3, we also have

Theorem 6. *With the assumptions of Theorem 3, we have*

$$\left(\sum_{a\in\mathbb{C}} \delta(a,f)\right)\left(\sum_{b\in\overline{\mathbb{C}}} \delta(b, f^{(k)})\right) \leq 2.$$

§3. Mues Conjecture

Let $f(z)$ be a transcendental meromorphic function in the finite plane and k be a positive integer. Mues [5] conjectured that the following relation seems true.

$$\sum_{a\in\mathbb{C}} \delta(a, f^{(k)}) \leq 1. \tag{3.1}$$

The inequality (3.1) is obviously true if $f(z)$ satisfies the condition $N(r,f) = o\{T(r,f)\}$. In the general case, if $f(z)$ satisfies

$$N(r,f) - k\overline{N}(r,f) = o\{T(r,f)\},$$

then the Mues conjecture is true by using the rewritten form of a lemma of Frank and Weissenborn [2]. Now we consider some quite different cases.

Theorem 7. *If the order of $f(z)$ is finite and*

$$\Theta(\infty, f) \leq \frac{(1-\frac{1}{k}) + (1+\frac{1}{k})\delta(\infty, f^{(k)})}{1+\delta(\infty, f^{(k)})}, \tag{3.2}$$

then the Mues conjecture holds. Especially, the condition (3.2) can be replaced by $\Theta(\infty,f) \leq 1 - \frac{1}{k}$.

Theorem 8. *If the order of $f(z)$ is finite and*

$$(1 + \delta(\infty, f^{(k)}))\left(\sum_{a\in\mathbb{C}} \delta(a,f)\right) \geq 2, \tag{3.3}$$

then the Mues conjecture holds. In particular, the condition (3.3) can be replaced by

$$\sum_{a\in\mathbb{C}} \delta(a,f) + \delta(\infty, f^{(k)}) \geq 2.$$

Quite recently, Wang Yuefei and I[9] proved the following general result.

Theorem 9. *Let $f(z)$ be a transcendental meromorphic function in the finite plane. Then the Mues conjecture is true for almost every positive integer k. i.e. There exists a positive number $K = K(f)$ such that (3.1) holds for every positive integer $k \geq K$.*

The positive number K in Theorem 9 can be chosen as $1/\{\limsup_{r \to \infty, r \bar{\in} e} \frac{\overline{N}(r,f)}{T(r,f)}\}$, where e is the union of unordinary intervals, when f has infinite order and empty, when f has finite order.

References

[1] Drasin D., *An introduction to potential theory and meromorphic functions*, complex Analysis and its Applications, Vol. 1, IAEA, Vienna, 1976, 1–93.

[2] Frank G. and Weissenborn G., *Rational deficient functions of meromorphic functions*, Bull. London Math. Soc., 18(1986), 29–33.

[3] Hayman W.K., *Picard values of meromorphic functions and their derivatives*, Ann. of Math., 70(1959), 9–42.

[4] Hayman W.K., *Meromorphic Functions*, Oxford, 1975.

[5] Mues E., *Über eine Defekt-und Verzweigungsrelation für die Ableitung meromorpher Funktionen*, Manuscripta Math., 5(1971), 275–297.

[6] Nevanlinna R., *Analytic Functions*, Springer-Verlag, 1970.

[7] Yang Lo, *Precise estimate of total deficiency of meromorphic derivatives*, J. d'Analyse Math., to appear. (A Chinese abstract appeared in Bull. Sci. (Beijing), 1990, No. 16, 1208–1210.)

[8] Yang Lo, *Some recent results and problems in the theory of value-distribution*, Lecture at Symposium on Theory of Value-distribution (held in April of 1990 at the Univ. of Notre Dame), to appear in the Proceedings.

[9] Yang Lo & Wang Yuefei, *Drasin's Problems and Mues Conjecture*, preprint.

Proceedings of the Asian Mathematical
Conference, 1990

Traveling Wave Solutions for Reaction-diffusion Systems*

YE QIXIAO
Department of Applied Mathematics
Beijing Institute of Technology
Beijing 100081, *China*

WANG MINGXIN
Institute of Systems Science
Chinese Academy of Sciences
Beijing 100080, *China*

LI ZHENGYUAN
Institute of Mathematics
Peking University
Beijing 100871, *China*

§1. Introduction

In this paper we will survey our work on the existence of traveling wave front solutions and periodical traveling ware solutions and explicit traveling wave front solutions for the following reaction-diffusion systems

$$\begin{cases} u_t - d_1 u_{xx} = f(u,v), \\ v_t - d_2 v_{xx} = g(u,v). \end{cases} \tag{1}$$

Example: A simplified mathematical model for Belousov-Zhabotinskii chemical reaction is

$$\begin{cases} u_t - u_{xx} = u(1-u-rv) + av, \\ v_t - v_{xx} = -buv + ev. \end{cases} \tag{2}$$

See [10, 14].

Solutions of (1) of the form

$$u(x,t) = u(x-ct), \quad v(x,t) = v(x-ct)$$

are called traveling wave solutions (TWS).

If $u(\xi), v(\xi) (\xi = x - ct)$ are both monotone, bounded, and nonconstant, they are called traveling wave front solutions (TWFS), for some c, the wave speed.

Set $\xi = x - ct$, $' = \frac{d}{d\xi}$.

By direct computation we have

$$\begin{cases} d_1 u'' + cu' + f(u,v) = 0, \\ d_2 v'' + cv' + g(u,v) = 0. \end{cases}$$

* Research Supported by the National Natural Science Foundation of China.

We seek for solutions of (1),

$$(u(\xi), v(\xi), c).$$

The existence of TWS is essentially a kind of nonlinear phenomenon. It is easy to see that for heat equation

$$u_t - a^2 u_{xx} = 0,$$

no bounded TWS exists.

But for a reaction diffusion equation $u_t - a^2 u_{xx} = f(u)$, for instance, we have Fisher equation (K-P-P equation)

$$u_t = u_{xx} + u(1-u). \tag{*}$$

It can be proved, by using phase plane method and shooting technique, that (*) has TWFS. More precisely we have

Lemma BVP.

$$\begin{cases} -u'' - cu' = (1-r)u(1-u), & 0 < r < 1 \\ u(-\infty) = 0, \quad u(+\infty) = 1, \end{cases}$$

has monotone solutions if and only if

$$c \le -2\sqrt{1-r}.$$

See [3], [12].

A quite a lot of researchers have studied the existence of TWS of (1) theoretically and numerically, for instance, see [1–2], [4], [6–9].

§2. A General Theorem for the Existence of TWFS

Assumptions:

(H_1) For $(u,v) \in (0,l) \times (0,l)$, $\quad l \ge 1$,

$$f(u,v) = uf_1(u,v), \quad f_1(0,0) > 0,$$

$$\frac{\partial f_1}{\partial u} < 0, \quad \frac{\partial f_1}{\partial v} > 0,$$

there exists an $l_0 \in (0,l)$ such that

$$g(u,v) > 0, \quad (u,v) \in (0,l) \times (0,l_0),$$

$$g(u,v) < 0, \quad (u,v) \in (0,l) \times (l_0,l),$$

equations $f(u,v) = 0, g(u,v) = 0$ only have solutions $(0,\alpha)(\alpha \in [0,l])$ and (1.1).

(H_2) There exist a pair of upper and lower solutions $(\tilde{u}, \tilde{v}), (\underset{\sim}{u}, \underset{\sim}{v})$ of the following BVP on R^1 for some c,

$$\begin{cases} d_1 u'' + cu' + f(u,v) = 0, \\ d_2 v'' + cv' + g(u,v) = 0, \\ u(-\infty) = v(-\infty) = 0, \quad u(+\infty) = v(+\infty) = 1 \end{cases} \tag{3}$$

such that

$$(0,0) < (\underset{\sim}{u}, \underset{\sim}{v}) < (\tilde{u}, \tilde{v}) < (l, l),$$

$$\tilde{u}'(\xi) > 0, \quad \underset{\sim}{u}'(\xi) > 0, \quad \tilde{v}'(\xi) > 0, \underset{\sim}{v}'(\xi) > 0,$$

$$\lim_{\xi \to -\infty} \tilde{v}(\xi) = 0, \quad \lim_{\xi \to -\infty} \underset{\sim}{v}(\xi) = 0.$$

We can prove

Theorem. *Under the assumptions* (H_1) *and* (H_2), (3) *has monotone, bounded, and nonconstant solutions.*

The proof consists of the following four steps.

1°. Consider a BVP on a finite interval $[-a, a]$ for $a > 0$ sufficiently large,

$$\begin{cases} d_1 u'' + cu' + f(u, v) = 0, & \\ d_2 v'' + cv' + g(u, v) = 0, & -a < \xi < a \\ u(-a) = \underset{\sim}{u}(-a), & u(a) = \tilde{u}(a), \\ v(-a) = \underset{\sim}{v}(-a), & v(a) = \tilde{v}(a), \end{cases} \tag{4}$$

2°. Using the upper and lower solution method we can prove that (4) has a strictly monotone solution

$$(\ u_a(\xi), v_a(\xi)\ )$$

for any $a > 0$ sufficiently large.

3°. We can get the a priori estimates as follows. Let m be any positive integer. For $a \geq m$

$$\|u_a(\xi)\|_{C^{2+\alpha}[-m,m]}, \ \|v_a(\xi)\|_{C^{2+\alpha}[-m,m]} \leq M_m,$$

$\alpha \in (0, 1)$ and M_m is independent of $a(m)$.

4°. There exists a subsequence $a_n \to \infty$, such that

$$\lim_{n \to \infty} u_{a_n}(\xi) = u(\xi), \quad \lim_{n \to \infty} v_{a_n}(\xi) = v(\xi).$$

Furthermore, we can prove that $u(\xi), v(\xi)$ are strictly monotone solution of (3)(see [5], [11], [13], [14]).

§3. Application

Consider

$$\begin{cases} u_t - u_{xx} = u(1 - u - rv), \\ v_t - v_{xx} = -buv, \\ u(-\infty) = v(+\infty) = 0, \quad u(+\infty) = v(-\infty) = 1. \end{cases} \tag{5}$$

By using the Theorem in Section 2 and the special forms in (5) we can prove the following results (see [5], [11], [13], [14]).

Theorem. *Suppose that* $0 < r < 1, 0 < b < \frac{1-r}{r}, c < -2\sqrt{1-r}$, *then (5) has strictly monotone solution* $(u(\xi), v(\xi))$.

Theorem. *Let $0 < r < 1$, if $(u(\xi), v(\xi))$ is a strictly monotone solution of (5), then*

$$c \leq -2\sqrt{1-r}, \quad \xi = x - ct.$$

§4. Nonmonotone (periodical) Traveling Wave Solutions

By using the following Hopf bifurcation theorem (see [5]) we can prove that (2) has periodic traveling wave solution with period close to $2\pi(\omega_0 = 1)$ when $|c| < \varepsilon$ (see [14–2]):

$$\frac{dx}{dt} = A(\lambda)x + a(\lambda, x), \quad a(\lambda, 0) = 0. \tag{**}$$

Theorem (Kozjakin-Krasnosel'skii). *Let the matrix $A(\lambda_0)$ has no eigenvalues of the form $0, \pm 2i\omega_0, \pm 3i\omega_0, \cdots$, and let $Re\mu(\lambda)$ of the eigenvalue $\mu(\lambda)$ takes values of opposite signs in every neighborhood of λ_0. Then for $\lambda = \lambda_0$ generation of small periodic solutions of (**) with periods close to $2\pi/\omega_0$ takes place.*

§5. Explicit Traveling Wave Front Solutions

We will obtain some explicit traveling wave front solutions for (5) in this section; see [15].

In this section, we will show that it is always possible to get some explicit traveling wave front solutions of the form $u(x,t) = u(x+ct)$, c being the wave speed, for the generalized Fisher equation

$$\begin{cases} u_t - Du_{xx} = pu(1-u^\alpha)(u^\alpha + q), \\ u(-\infty) = 0, \quad u(+\infty) = 1, \end{cases} \tag{6}$$

where D, p, α are positive constants, q is a real number.

Denoting $\xi = x + ct$, $\frac{d}{d\xi} = '$, we have

$$\begin{cases} -Du'' + cu' = pu(1-u^\alpha)(u^\alpha + q), \\ u(-\infty) = 0, \quad u(+\infty) = 1. \end{cases} \tag{7}$$

Setting

$$u' = \frac{du}{d\xi} = Ku(1-u^\alpha), \tag{8}$$

where K is an undetermined positive constant, one obtains

$$u'' = \frac{d^2u}{d\xi^2} = K^2(1-(1+\alpha)u^\alpha)u(1-u^\alpha).$$

Substituting them into (7) we get

$$-DK^2(1-(1+\alpha)u^\alpha)u(1-u^\alpha) + cKu(1-u^\alpha) = pu(1-u^\alpha)(u^\alpha + q).$$

Comparing coefficients, we have

$$\begin{cases} -DK^2 + cK = pq, \\ DK^2(1+\alpha) = p; \end{cases}$$

thus

$$K = \sqrt{\frac{p}{D(1+\alpha)}}, \quad c = \sqrt{\frac{pD}{1+\alpha}}[1 + q(1+\alpha)].$$

Integrating the Bernoulli equation (8), we get the explicit traveling wave front solutions

$$u(\xi) = (1 + e^{-\alpha K\xi})^{-1/\alpha} = \frac{e^{K\xi}}{(1 + e^{\alpha K\xi})^{1/\alpha}}$$

and $u'(\xi) > 0$, or

$$\begin{aligned} u(x,t) = \ & u(x+ct) = (1 + e^{-\alpha K(x+ct)})^{-1/\alpha} \\ = \ & \frac{e^{K(x+ct)}}{(1 + e^{\alpha K(x+ct)})^{1/\alpha}}. \end{aligned}$$

Remark. The method used here is quite useful. We can easily get the explicit solutions for the equations mentioned in [1, 2, 11].

In order to get the explicit wave front solutions for (5), it is reasonable to assume that there exists a explicit relation between the solutions u, v of (5), namely $F(u,v) = 0$.

If there exists an explicit function $v = f(u)$, then $f(1) = 0$ should hold.

We assume that

$$v = (\sqrt{u} - 1)(A\sqrt{u} - B) = Au - (A+B)\sqrt{u} + B,$$

where A and B are undetermined real constants.

Substituting it into the first equation of (5) we have

$$\begin{aligned} -Du'' + cu' = \ & u(1 - u - r(\sqrt{u} - 1)(A\sqrt{u} - B)) \\ = \ & (1 + rA)u(1 - \sqrt{u})\left(\sqrt{u} + \frac{1 - rB}{1 + rA}\right). \end{aligned}$$

So, there is an explicit solution

$$u(\xi) = (1 + e^{-N\xi})^{-2}$$

with $N = K/2$ and K, c being

$$K = \sqrt{\frac{2(1+rA)}{3D}}, \quad c = \sqrt{\frac{2(1+rA)^D}{3}}\left[1 + \frac{3}{2} \cdot \frac{1 - rB}{1 + rA}\right].$$

We will prove that it is possible to get explicit traveling wave front solutions of the form

$$v = B - (A+B)(1 + e^{-N\xi})^{-1} + A(1 + e^{-N\xi})^{-2}$$

by determining A and B.

$$\begin{cases} u(x,t) = (1+e^{-\frac{K}{2}(x+ct)})^{-2}, \\ v(x,t) = \left[1 - \dfrac{1}{1+e^{-\frac{K}{2}(x+ct)}}\right]^2 = \dfrac{e^{-K(x+ct)}}{(1+e^{-\frac{K}{2}(x+ct)})^2}. \end{cases} \tag{9}$$

Theorem. *If $D > 1/5$, then (5) has explicit traveling wave front solutions (9), provided the following relations hold:*

$$r = \frac{5D-1}{D+1}, \quad K = 2\sqrt{\frac{1}{D+1}}, \quad c = \frac{1}{\sqrt{D+1}}, \quad b = \frac{6}{D+1}.$$

Remark.

$$\lim_{D\to+\infty} r(D) = 5, \quad \frac{dr}{dD} > 0,$$

so

$$0 < r(D) < 5, \quad \text{for} \quad D > \frac{1}{5}.$$

In particular, for $D = 1$, then $r = 2 > 1, b = 3, K = \sqrt{2}, c = 1/\sqrt{2}$. This is an explicit traveling wave front solution for $r = 2 > 1$.

References

[1] Abdelkader, M.A., *Traveling wave solutions for a generalized Fisher equation*, J. Math. Anal. Appl., 85(1982), 287–290.

[2] Ablowitz, M.J. & A. Zeppet Ella, *Explicit solutions of Fisher's equation for a special wave speed*, Bull. Math. Biol., 41(1979), 835–840.

[3] Aronson, D.G. & Weinberger, H.F., *Nonlinear diffusion in population genetics, combustion and nerve pulse propagation*, Lecture Notes in Mathematics, 446 (1975), 5–49.

[4] Gibbs, R.G., *Traveling waves in the Belousov-Zhabotinskii reaction*, SIAM J. Appl. Math., 38(1980), 422–444.

[5] Kozjakin, V.S. & Krasnosel'skii, M.A., *The method of functionalization in the Hopf bifurcation problem*, Nonlinear Analysis TMA, 11(2)(1987), 149–161.

[6] Li, Z.Y., *The existence of traveling front solutions for reaction-diffusion system*, to appear.

[7] Manoranjan, V.S. & A.R. Mitchell, *A numerical study of the Belousov-Zhabotinskii chemical reaction using Galerkin finite element methods*, J. Math. Biol., 16(1983), 251–260.

[8] Murray, J.D., 1. *Mathematical Biology*, 1989, Springer, 56(1976), 329–353. 2. *On traveling wave solutions in a model for B-Z reaction*, J. Theor. Biol.

[9] Troy, W.C., *The existence of travelling wave front solutions of a model of the Belousov-Zhabotinskii reaction*, J. Diff. Eqns., 36(1980), 89–98.

[10] Tyson, J.J., *The Belousov-Zhabotinskii reaction*, Lecture Notes in Biomathematics, 10 1976, Springer.

[11] Wang, X.Y., *Exact soliton solutions of generalized Fisher equations* (in Chinese), Kexue Tongbao, 9(1988), 657–659.

[12] Li, Z.Y. & Ye, Q.X., *Travelling wave front solutions for reaction diffusion systems.* (accepted by the Journal of Partial Differential Equations).

[13] Ye, Q.X. & Li, Z.Y., *Introduction to reaction-diffusion equations* (in Chinese), Science Press, Beijing, 1990.

[14] Ye, Q.X. & Wang M.X., 1.*Travelling wave front solutions of Noyes-Field system for Belousov-Zhabotinskii reaction*, Nonlinear Analysis TMA, 11(1987), 1289–1302. 2.*A note on travelling wave solutions for Belousov-Zhabotinskii chemical reaction*, Journal of Beijing Institute of Techenology, 9 (4) (1989), 5–9.

[15] Wang, M.X., Xiong, S.L. and Ye, Q.X., *Explicit wave front solutions of Noyes-Field systems for Belousov-Zhabotinskii reaction*, to appear.

Proceedings of the Asian Mathematical
Conference, 1990

A Comparison Theorem on a Class of Reinhardt Domains*

YIN WEIPING
Department of Mathematics
University of Science and Technology of China
Hefei, 230026 Anhui, China

§1. Introduction

In this note we shall prove a comparison theorem for the Kobayashi and Bergman metrics on a class of Reinhardt domains in $\mathbb{C}^n$ of the following type

$$D(p_1,\cdots,p_n)=\left\{Z=(z_1,\cdots,z_n)\in\mathbb{C}^n:|z_1|^{2p_1}+|z_2|^{2p_2}+\cdots+|z_n|^{2p_n}<1\right\},$$

where $p_1\geq p_2\geq\cdots\geq p_n>0$.

If $p_1=k\neq 1, p_2=\cdots=p_n=1$, then $D(k,1,\cdots,1)$ is denoted by $D(k)$, that is

$$D(k)=\{Z=(z_1,z_2,\cdots,z_n)=(z_1,z)\in\mathbb{C}^n:|z_1|^{2k}+|z|^2<1\}.$$

More precisely, we have the following

Theorem. *Let $K_{D(p_1,\cdots,p_n)}$ and $B_{D(p_1,\cdots,p_n)}$ denote Kobayashi and Bergman metric of $D(p_1,\cdots,p_n)$ respectively. Then for each $Z\in D(p_1,\cdots,p_n)$ and $\chi\in\mathbb{C}^n$, there exists positive constant C, depending only on $D(p_1,\cdots,p_n)$ such that*

$$B_{D(p_1,\cdots,p_n)}(Z,\chi)\leq CK_{D(p_1,\cdots,p_n)}(Z<\chi).\tag{1}$$

In the case that $n=2$, similar results have already been obtained by K.T. Hahn and P.Pflug in [1]. For this case, the holomorphic sectional curvature of the Bergman metric of $D(p_1,1)$ is bounded from above by a negative constant [2], [3]. Then, by a well known result (see, for example [4]) there exists a constant $C_1>0$ such that

$$B_{D(p_1,1)}(Z,\chi)\leq C_1K_{D(p_1,1)}(Z,\chi)\tag{2}$$

for all $Z\in D(p_1,1)$ and $\chi\in\mathbb{C}^2$. Hahn and Pflug developed a method in [1] such that the form (1) with $n=2$ can be obtained from the form (2). We would like to mention that their method is valid not only for $n=2$ but also for $n>2$.

---

*Research supported by NSF of China

So, the key step is to estimate the holomorphic sectional curvature of the Bergman metric of $D(k)$. But such estimate is very complex even in the case that $n = 2$. My main idea is to construct a complete $\text{Aut}(D(k))$-invariant Kähler metric $Y(Z,\chi)$ such that

$$Y(Z,\chi) \geq B_{D(k)}(Z,\chi)$$

for all $Z \in D(k)$ and $\chi \in \mathbb{C}^n$. And prove that the holomorphic sectional curvature of $Y(Z,\chi)$ of $D(k)$ is bounded from above by a negative constant. Then by [4] there exists a constant $C_3 > 0$ such that

$$Y(Z,\chi) \leq C_3 K_{D(k)}(Z,\chi),$$

hence

$$B_{D(k)}(Z,\chi) \leq C_3 K_{D(k)}(Z,\chi) \tag{3}$$

for all $Z \in D(k)$ and $\chi \in \mathbb{C}^n$. By using the Hahn-Pflug method, we can get (1) from (3).

§2. Preparation

We will use the following results, which were proved in [5].

i. $X = X(Z,\bar{Z}) = |z_1|^2(1-|z|^2)^{-1/k}$ (where $Z = (z_1, z) \in D(k)$ is an $\text{Aut}(D(k))$ invariant function, i.e. for any $\gamma \in \text{Aut}(D(k))$, then $X(\gamma Z, \overline{\gamma Z}) = X(Z,\bar{Z})$.

ii. $K(Z,\bar{Z})$, the Bergman kernel of $D(k)$ can be expressed in terms of X, i.e.

$$K(Z,\bar{Z}) = F(X)(1-|z|^2)^{-(nk+1)/k}, \tag{4}$$

where

$$F(X) = \frac{k}{\pi^n}\sum_{j=0}^{n-1} b_j(1-X)^{-(n+1)+j} \tag{5}$$

and b_j is determined by the following equation of y

$$\prod_{j=0}^{n-1}(y+1+jk) = k^n \sum_{j=0}^{n-1} b_j \Gamma(y+n+1-j)[\Gamma(n+1-j)\Gamma(y+1)]^{-1}.$$

iii. Let

$$IM(D(k)) = \left\{G(X)(1-|z|^2)^{-(nk+1)/k} : G(X) \text{ satisfies condition (M)}\right\},$$

where the condition (M) is

$$(M)\begin{cases} G(X) \in C^\infty, & W = \log G(X), \\ X\dfrac{dW}{dX} + nk + 1 > 0, & X\dfrac{d^2W}{dX^2} + \dfrac{dW}{dX} > 0, \end{cases}$$

then any element of $IM(D(k))$ generates an $\text{Aut}(D(k))$-invariant Kähler metric, and the counterclaim is also true.

Remark 1. The condition (A)1 and (A)2 imply (A)3 in [5], so the counterclaim is also true.

iv. $\omega(Z, dZ)$, the holomorphic sectional curvature of the $\mathrm{Aut}(D(k))$-invariant Kähler metric of $D(k)$, has the following expression on points $(z_1, 0)$.

$$\begin{aligned}\omega(Z,dZ)|_{Z=(z_1,0)} = \; & [P|dz_1|^4 + Q|dz_1|^2(dz\overline{dz'}) + R(dz\overline{dz'})^2] \\ & \cdot[(XW''+W')|dz_1|^2 + k^{-1}(XW'+nk+1)dz\overline{dz'}]^{-2},\end{aligned}$$

where

$$\begin{aligned}P = \; & (XW''+W')^{-1}(XW'''+2W'')^2X - (XW'''+2W'') - X(XW'''+2W'')', \\ Q = \; & 4k^{-1}[X(XW''+W')^2(XW'+nk+1)^{-1} - (XW''+W') - X(XW'''+2W'')], \\ R = \; & -k^{-1}[2(XW'+nk+1) + 2k^{-1}X(XW''+W')], \\ W = \; & \log G(X), \quad W' = \frac{dW}{dX}, \quad W'' = \frac{d^2W}{dX^2} \quad \text{and so on.}\end{aligned}$$

Remark 2. There are some print errors in the English paper of [5]. For example, the R is missed. But the Chinese paper of [5] is correct.

§3. The Complete Aut($D(k)$)-invariant Kähler Metrics of $D(k)$

Any element of $IM(D(k))$ can be written as

$$G(X)(1-|z|^2)^{-(nk+1)/k} = \frac{G(X)}{F(X)}F(X)(1-|z|^2)^{-(nk+1)/k} = g(X)K(Z,\bar{Z}),$$

where

$$g(X) = G(X)/F(X)$$

and $K(Z, \bar{Z})$ is the Bergman kernel of $D(k)$. Then the metric $Y_G(Z, \chi)$ generated by $g(X)K(Z, \bar{Z})$ has the following form

$$Y_G(Z,\chi) = \chi\left(\frac{\partial^2 \log g(X)}{\partial z_\alpha \partial \bar{z}_\beta}\right)\bar{\chi}' + \chi\left(\frac{\partial^2 \log K(Z,\bar{Z})}{\partial z_\alpha \partial \bar{z}_\beta}\right)\bar{\chi}'.$$

The second term of the right hand side is the Bergman metric $B_{D(k)}(Z, \chi)$ of $D(k)$. If

$$\left(\frac{\partial^2 \log g(X)}{\partial z_\alpha \partial \bar{z}_\beta}\right) \geq 0, \tag{7}$$

then

$$Y_G(Z,\chi) \geq B_{D(k)}(Z,\chi)$$

for all $Z \in D(k)$ and $\chi \in \mathbb{C}^n$. But $B_{D(k)}(Z, \chi)$ is a complete metric, so is $Y_G(Z, \chi)$.

By calculations, we have

$$\left(\frac{\partial^2 \log g(X)}{\partial z_\alpha \partial \bar{z}_\beta}\right) = \begin{pmatrix} J_{11} & 0 \\ J_{21} & J_{22} \end{pmatrix} \cdot \begin{pmatrix} X\frac{d^2 \log g(X)}{dX^2} + \frac{d\log g(X)}{dX} & 0 \\ 0 & k^{-1}X\frac{d\log g(X)}{dX} I^{(n-1)} \end{pmatrix} \cdot \overline{\begin{pmatrix} J_{11} & 0 \\ J_{21} & J_{22} \end{pmatrix}}',$$

where

$$J_{11} = e^{i\varphi}(1-|z|^2)^{-1/2k}, \quad J_{21} = e^{i\varphi}k^{-1}z_1(1-|z|^2)^{-(2k+1)/2k}\bar{z}',$$

$$J_{22} = e^{i\varphi}(1-|z|^2)^{-1/2}Q, \quad (Q\bar{Q}')^{-1} = I^{(n-1)} - \bar{z}'z,$$

and

$$\begin{pmatrix} J_{11} & 0 \\ J_{21} & J_{22} \end{pmatrix} > 0.$$

So, if

$$\frac{d\log g(X)}{dX} \geq 0 \quad \text{and} \quad \frac{d^2 \log g(X)}{dX^2} \geq 0$$

then the (7) holds.

By calculations, we have

$$\frac{d\log g(X)}{dX} = \frac{G'(X)}{G(X)} - \frac{F'(X)}{F(X)}, \qquad \frac{d^2\log g(X)}{dX^2} = \left[\frac{G'(X)}{G(X)} - \frac{F'(X)}{F(X)}\right]'.$$

Therefore if $G(X)$ satisfies the condition

$$(CM) = \begin{cases} \frac{G'(X)}{G'(X)} - \frac{F'(X)}{F(X)} \geq 0, \\ \left[\frac{G(X)}{G(X)} - \frac{F'(X)}{F(X)}\right]' \geq 0, \end{cases}$$

then the $\mathrm{Aut}(D(k))$-invariant Kähler metric $Y_G(Z,\chi)$, which is generated by $G(X)(1-|z|^2)^{-(nk+1)/k}$ is complete.

Example. Let

$$G(X) = (1-X)^{-\lambda}, \quad \lambda \geq \max\{m_1, m_2\}, \quad m_1, m_2 < +\infty \tag{8}$$

then $G(X)(1-|z|^2)^{-(nk+1)/k}$ generates a complete $\mathrm{Aut}(D(k))$-invariant Kähler metric $Y_G(Z,\chi)$ and for all $Z \in D(k), \chi \in \mathbb{C}^n$, we have

$$Y_G(Z,\chi) \geq B_{D(k)}(Z,\chi),$$

where

$$m_1 = \max_{0\le X<1}\left\{\frac{F'(X)(1-X)}{F(X)}\right\},$$

$$m_2 = \max_{1\le X<1}\left\{\frac{[F(X)F''(X)-F'(X)^2](1-X)^2}{F(X)^2}\right\}.$$

Proof. By calculations, we have

$$\frac{G'(X)}{G(X)} - \frac{F'(X)}{F(X)} = \frac{\lambda(1-X)^{-1}F(X)-F'(X)}{F(X)},$$

$$\left[\frac{G'(X)}{G(X)} - \frac{F'(X)}{F(X)}\right]' = \frac{\lambda(1-X)^{-2}F^2(X)-F''(X)F(X)+F'(X)^2}{F^2(X)}.$$

Because $F(X)>0$ for $0\le X<1$. So, condition (CM) holds iff

$$\lambda(1-X)^{-1}F(X)-F'(X)\ge 0 \tag{9}$$

and

$$\lambda(1-X)^{-2}F^2(X)-F''(X)F(X)+F'(X)^2\ge 0. \tag{10}$$

Eq. (9) holds iff

$$\lambda\ge\max_{0\le X<1}\left\{\frac{F'(X)(1-X)}{F(X)}\right\}=m_1.$$

Because $F'(X)(1-X)$ and $F(X)$ are the real polynomial of $(1-X)^{-1}$ with same degree $(n+1)$. We have

$$\lim_{X\to 1}\frac{F'(X)(1-X)}{F(X)}=n+1.$$

So, m_1 exists and is finite. Eq. (10) holds iff

$$\lambda\ge\max_{1\le X<1}\left\{\frac{[F''(X)F(X)-F'(X)^2](1-X)^2}{F(X)^2}\right\}=m_2.$$

With the same reason, m_2 exists and is finite. So if

$$\lambda\ge\max\{m_1,m_2\},$$

then $G(X)=(1-X)^{-\lambda}$ satisfies the condition (CM). Thus we have completed our proof.

§4. The Estimate for the Holomorphic Sectional Curvature of $Y_G(Z,\chi)$ of $D(k)$

From §2. iv, we have

$$\omega(Z,dZ)|_{Z=(z_1,0)} = -C-\omega_1(z_1,dZ)\cdot\{(XW''+W')|dz_1|^2+k^{-1}(XW'+nk+1)dz\overline{dz'}\}^{-2},$$

where

$$\omega_1(z_1,dZ)=P_1|dz_1|^4+Q_1|dz_1|^2dz\overline{dz'}+R_1(dz\overline{dz'})^2$$

and

$$\begin{aligned} P_1 = &\ (XW''' + 2W'') + X(XW''' + 2W'')' \\ &-X(XW''' + 2W'')^2(XW'' + W')^{-1} - C(XW'' + W')^2, \\ Q_1 = &\ 4k^{-1}[XW'' + W' + X(XW''' + 2W'') \\ &-X(XW'' + W')^2(XW' + nk + 1)^{-1} - \frac{C}{2}(XW' + nk + 1)(XW'' + W')], \\ R_1 = &\ k^{-1}[2(XW' + nk + 1) + 2k^{-1}X(XW'' + W') \\ &-k^{-1}C(XW' + nk + 1)^2], \end{aligned}$$

where $C > 0$ is a constant.

Under the assumption of $G(X) = (1 - X)^{-\lambda} (\lambda > 0)$, we have

$$W' = \lambda(1 - X)^{-1}, \quad W'' = \lambda(1 - X)^{-2}, \qquad W''' = 2\lambda(1 - X)^{-3}, \quad W^{(4)} = 6\lambda(1 - X)^{-4}.$$

We place these in the expression of $\omega_1(z_1, dZ)$ and let $\lambda = m(nk + 1)$, then we have

$$\begin{aligned} P_1 = &\ m(nk + 1)[2 - Cm(nk + 1)](1 - X)^{-4}, \\ Q_1 = &\ 2k^{-1}m(nk + 1)(1 - X)^{-2}(XW' + nk + 1)^{-1}[m(nk + 1)(2 - Cm(nk + 1))(1 - X)^{-2} \\ &+2(1 - m)(nk + 1)(2 - Cm(nk + 1))(1 - X)^{-1} \\ &-(1 - m)(nk + 1)(2 + C(1 - m)(nk + 1))], \\ R_1 = &\ k^{-2}m(nk + 1)(2 - Cm(nk + 1))(1 - X)^{-2} \\ &+2k^{-1}m(nk + 1)(1 - k^{-1} - k^{-1}C(1 - m)(nk + 1))(1 - X)^{-1} \\ &+k^{-1}(1 - m)(nk + 1)[2 - k^{-1}C(1 - m)(nk + 1)]. \end{aligned}$$

It is easy see that $P_1 \geq 0$ if

$$C \leq \frac{2}{m(nk + 1)}. \tag{11}$$

Let

$$q(Y) = m[2 - Cm(nk + 1)]Y^2 + 2(1 - m)[2 - Cm(nk + 1)]Y + (m - 1)[2 - C(1 - m)(nk + 1)],$$

where $1 \leq Y < +\infty$.

$$q'(Y) = 2m[2 - Cm(nk + 1)]Y + 2(1 - m)[2 - Cm(nk + 1)].$$

Let

$$Y_0 = \frac{(m - 1)}{m},$$

then $q'(Y_0) = 0$. But $q''(Y_0) = 2m[2 - Cm(nk + 1)]$. If

$$C \leq \frac{2}{m(nk + 1)},$$

then $q''(Y_0) \geq 0$, and

$$q(Y_0) = \min q(Y) \quad \text{for} \quad Y \in (-\infty, +\infty).$$

But

$$\begin{aligned} q(Y_0) = & \frac{(m-1)^2}{m}[2 - Cm(nk+1)] \\ & + \frac{2(1-m)(m-1)}{m}[2 - Cm(nk+1)] + (m-1)[2 - C(1-m)(nk+1)] \\ = & \frac{2(m-1)}{m} \geq 0 \quad \text{if} \quad m \geq 1. \end{aligned}$$

Therefore, if

$$0 < C \leq \frac{2}{m(nk+1)}, \qquad m \geq 1 \tag{12}$$

then $Q_1 \geq 0$. Let

$$\begin{aligned} \gamma(Y) = & \; mk^{-1}[2 - Cm(nk+1)]Y^2 + 2m[1 - k^{-1} + k^{-1}C(m-1)(nk+1)]Y \\ & + (m-1)[2 - k^{-1}C(m-1)(nk+1)], \end{aligned}$$

then

$$\gamma'(Y) = 2mk^{-1}[2 - Cm(nk+1)]Y + 2m[1 - k^{-1} + k^{-1}C(m-1)(nk+1)]$$

where $1 \leq Y < +\infty$.

$$\gamma'(1) = 2m[k^{-1} + 1 - k^{-1}C(nk+1)].$$

If

$$C < \frac{k+1}{nk+1}, \tag{13}$$

then $\gamma'(1) > 0$. If

$$C < \frac{2}{m(nk+1)}, \tag{14}$$

then the coefficient of Y in $\gamma'(Y)$ is positive. So, under the condition (13) and (14) and $\gamma'(1) > 0$, we have

$$\gamma'(Y) > 0 \quad \text{for} \quad Y \geq 1.$$

That means $\gamma(Y)$ is increase for $Y \geq 1$. But $\gamma(1) = 2 - k^{-1}C(nk+1)$. If

$$C < \frac{2k}{nk+1},$$

we have $\gamma(1) > 0$, hence $\gamma(Y) > 0$ for $Y \geq 1$. Therefore, if

$$0 < C < \min\left\{\frac{2k}{nk+1}, \frac{2}{m(nk+1)}, \frac{k+1}{nk+1}\right\},$$

then $R_1 > 0$. And if

$$0 < C < \min\left\{\frac{2k}{nk+1}, \frac{k+1}{nk+1}, \frac{2}{m(nk+1)}\right\}, \qquad m > 1,$$

then $P_1 > 0, Q_1 > 0$ and $R_1 > 0$, hence

$$\omega(Z, dZ) < -C.$$

Therefore, if

$$m > \max\left\{\frac{m_1}{nk+1}, \frac{m_2}{nk+1}, 1\right\}$$

and

$$0 < C < \min\left\{\frac{2k}{nk+1}, \frac{k+1}{nk+1}, \frac{2}{m(nk+1)}\right\}$$

then

$$(1-X)^{-m(nk+1)}(1-|z|^2)^{-(nk+1)/k}$$

generates a complete $\mathrm{Aut}(D(k))$-invariant Kähler metric $Y(Z,\chi)$, such that

$$Y(Z,\chi) \geq B_{D(k)}(Z,\chi)$$

for all $Z \in D(k)$ and $\chi \in \mathbb{C}^n$, and the holomorphic sectional curvature of $Y(Z,\chi)$ of $D(k)$ is bounded from above by a negative constant $-C$. Therefore, by [4] there exist a constant $C_3 > 0$ such that

$$Y(Z,\chi) \leq C_3 K_{D(k)}(Z,\chi)$$

for all $Z \in D(k)$ and $\chi \in \mathbb{C}^n$. Thus, as mentioned in the end of Sec. 1, the Theorem is proved.

Acknowledgments. The author would like to thank Professor Abdus Salam, the International Atomic Energy Agency and UNESCO for hospitality at the International Centre for Theoretical Physics, Trieste. He would also like to thank Professor P. Pflug for suggesting this problem and also for sharing his knowledge on the problem, and Professor Pit-Mann Wong for the valuable discussions during his visit at the University of Notre Dame.

References

[1] K.T. Hann and P.Pflug, *The Kobayashi and Bergman metrics on generalized Thullen domains*, Procs. Amer. Math. Soc., 104(1)(1988), 207–214.

[2] K. Azakawa and M. Suzuki, *The Bergman metric on Thullen domain*, Nagoya Math. J., **89**(1983), 1–11.

[3] K. Azukawa, *Bergman metric on a domain of Thullen type*, Math. Rep. Toyama Univ., 7(1984), 41–65.

[4] M. Hens, *On a class of conformal metrics*, Nagoya Math. J., **21** (1962), 1–60.

[5] Weiping Yin, *Curvatures and invariant functions* (in Chinese), Sci. Sinica Ser., A 36(12)(1987), 1245–1257 or *Curvatures and invariant functions* (in English), Sci. Sinica Ser. A, 31(6)(1988), 675–686.

Proceedings of the Asian Mathematical Conference, 1990

The Monodromy of the Period Map of a 4-parameter Family of K3 Surfaces and the Hypergeometric Function of Type (3, 6)

MASAAKI YOSHIDA
Department of Mathematics
Faculty of Science
Kyushu University, Japan

In this note we present a system of linear differential equations in four variables which has several interesting properties that are familiar in the case of the hypergeometric equation of Gauss; and we explore new nice aspects of our system. This work was done in collaboration with professors K. Matsumoto and T. Sasaki.

The double cover of the projective plane branching along six lines in a K3 surface. We consider the 4-dimensional family of such surfaces; the configuration space X of six lines in the plane can be considered to be the algebraic moduli space of the family. On the other hand, the space of periods of the K3 surfaces is the transcendental modulo space of the family, which is isomorphic to a bounded symmetric domain modulo a discrete subgroup of the group of automorphisms of the domain. The equivalence between the algebraic moduli and the transcendental moduli is given by the period map; it turns out that the period map is described by the solutions of the hypergeometric system of type (3, 6), and that its monodromy group is the discrete subgroup above. This is the model we treat in this note.

The classical model is the family of elliptic curves; as is well known, the equivalence between its algebraic moduli and its transcendental moduli is given by elliptic integrals, which are solutions of the hypergeometric equation of Gauss. Thus our model can be regarded as a higher dimensional version of the classical model. We wish to show the reader that out theory is as rich as the classical theory.

We first explain the classical model in order to help understand our model. Among the hypergeometric differential equations $E(a,b,c): z(1-z)u'' + \{c-(a+b+1)z\}u' - abu = 0$ satisfied by the hypergeometric series

$$F(a,b,c;z) = \sum_{n=0}^{\infty} \frac{(a,n)(b,n)}{(c,n)(1,n)} z^n$$

where $(a,n) = a(a+1)\cdots(a+n-1)$, the equation $E(1/2,1/2,1)$ is especially interesting because of the following geometric relation with elliptic curves. Let $p = \{p_1,p_2,p_3,p_4\}$ be a system of four distinct points of the complex projective line $\mathbb{P}^1$, $C(p)$ the double cover of $\mathbb{P}^1$ branching at p, $\eta(p)$ a nonzero holomorphic 1-form, and $\gamma_i\,(i=1,2)$ be a basis of the first homology group $H_1(C(p), \mathbb{Z})$ such that the intersection matrix $(\gamma_i \cdot \gamma_j)$ takes the form $\begin{pmatrix} 0 & 1 \\ -1 & 0 \end{pmatrix}$. The integrals

$$\omega^i(p) = \int_{\gamma_i(p)} \eta(p), \qquad i = 1, 2,$$

give periods of the curve $C(p)$; they satisfy $\operatorname{Im}\omega^1(p)/\omega^2(p) > 0$, the monodromy group of the (multivalued) period map $p \longmapsto (\omega^1(p), \omega^2(p))$ is conjugate to the principal congruence subgroup

$$\Gamma(2) = \{g \in SL(2, \mathbb{Z}) \mid g \equiv I_2 \bmod 2\}$$

of the elliptic modular group. If we normalize the four points so that $p_1 = 0, p_2 = 1, p_3 = \infty, p_4 = z$, and write $\omega^i(z)$ instead of $\omega^i(p)$, then the periods $\omega^1(z)$ and $\omega^2(z)$ form a set of linearly independent solutions of the hypergeometric equation $E(1/2, 1/2, 1)$. Let $Z = \{z \in \mathbb{C} \mid z \neq 0, 1\}$ and $\mathbb{H} = \{\tau \in \mathbb{C} \mid \operatorname{Im}\tau > 0\}$. Then the projectivization $z \longmapsto \omega^1(z)/\omega^2(z)$ of the period map gives a multivalued map $X \to \mathbb{H}$ for which there is a single-valued inverse map that gives rise to the equivalence

$$Z \simeq \mathbb{H} \quad \text{modulo} \quad \Gamma(2).$$

Next we explain our model. We begin by considering a system l of six lines $l_1, \cdots, l_6$ in the complex projective plane $\mathbb{P}^2$ in general position. Let $S'(l)$ be the double cover of $\mathbb{P}^2$ branching along l which has fifteen singular points. Resolving the singularities we have a K3 surface $S(l)$, on which there is a nonvanishing holomorphic 2-form $\eta(l)$, unique up to multiplicative constants. Let $AC(l)$ be the $\mathbb{Z}$-submodule of the second homology group $H_2(S(l), \mathbb{Z})$ generated by the fifteen rational curves coming from the resolution of singularities and by the pull bake of a general line in $\mathbb{P}^2$. We can take a system of 2-cycles $c'_1(l), \cdots, c'_6(l)$ orthogonal to $AC(l)$ such that there exists another system $c_1(l), \cdots, c_6(l) \in H_2(S(l), \mathbb{Z})$ satisfying $c'_i(l) \cdot c_j(l) = \delta_{ij}$ and such that the intersection matrix $(c'_i(l) \cdot c'_j(l))$ takes a fixed form $A = (A_{ij})$ which is symmetric, integral and of signature $(2+, 4-)$. The vector $\omega(l) = (\omega^1(l), \cdots, \omega^6(l))$, where

$$\omega^i(l) = \int_{c_i(l)} \eta(l).$$

is called the period of $S(l)$; it satisfies the Riemann relation and the Riemann inequality

$$\sum A_{ij}\omega^i(l)\, \omega^j(l) = 0, \qquad \sum A_{ij}\omega^i(l)\, \overline{\omega}^j(l) > 0.$$

Now we let l vary in the space X formed by the systems of six lines in $\mathbb{P}^2$ modulo projective transformations, and let the cycles $c_j(l)$ depend continuously on l. Then the period map $l \longmapsto \omega(l)$ gives a multivalued map whose monodromy group is conjugate to the principal congruence subgroup of level 2

$$\Gamma_A(2) = \{Y \in GL(6, \mathbb{Z}) \mid {}^tYAY = A, f(Y) > 0, Y \equiv I_6 \bmod 2\},$$

where f is a certain quadratic form in Y. This group turns out to be a reflection group generated by twenty reflections. Here a reflection means the transformation

$$z \longmapsto z - 2\frac{(\alpha, z)}{(\alpha,\alpha)}\alpha$$

for some real vector α such that $(\alpha,\alpha) < 0$, where $(x,y) = (x,y)_A = xA^ty$. The projectivization of the period map has its image in the 4-dimensional classical bounded symmetric domain of type IV:

$$D := \{z = (z^1, \cdots, z^6) \in \mathbb{P}^5 \mid (z,z) = 0, (z,\bar{z}) > 0, \mathrm{Im}\frac{z^3}{z^1} > 0\},$$

on which the group $\Gamma_A(2)$ acts properly discontinuously. The inverse map gives the equivalence between X and $\Gamma_A(2)\backslash D_{\mathrm{reg}}$, where D_{reg} is the subset of D on which $\Gamma_A(2)$ acts freely.

The periods $\omega^j(l)(1 \le j \le 6)$ form a set of linearly independent solutions of a system of partial differential equations called the hypergeometric differential equation $E(3,6;1/2) = E(3,6;1/2,\cdots,1/2)$ of type $(3,6)$. In terms of local coordinates $x = (x^1,\cdots,x^4)$ on X, the system turns out to have the form

$$\frac{\partial^2 w}{\partial x^i \partial x^j} = g_{ij}\frac{\partial^2 w}{\partial x^1 \partial x^n} + \sum_{k=1}^{n} A_{ij}^k \frac{\partial w}{\partial x^k} + A_{ij}^0 w \quad (1 \le i,j \le 4),$$

in which the conformal structure $\Sigma g_{ij}dx^i dx^j$ plays a principal role.

By investigating the arrangement of the mirrors of reflections in $\Gamma_A(2)$, we know that a compactification $\overline{X}$ of X is isomorphic to the Satake compactification of the quotient space $\Gamma_A(2)\backslash D$. In the forthcoming paper, the inverse map $D \to X$ of the period map will be written in terms of theta functions on D.

We conclude this note by remarking that the family of curves of genus 2 can be considered as a 3-dimensional subfamily of our family of K3-surfaces and that periods of such curves are restrictions of solutions of the system $E(3,6;1/2)$.

To give a better understanding of this note, we give a list showing the correspondence between our theory and the theory of elliptic curves:

System l of 6 lines in $\mathbb{P}^2 \leftrightarrow$ system p of 4 points in $\mathbb{P}^1$,
Space X of $l \leftrightarrow$ Space $Y = \mathbb{C} - \{0,1\}$ of y,
Compactification $\overline{X}$ of $X \leftrightarrow$ compactification $\overline{Y} \simeq \mathbb{P}^1$ of Y,
K3-surface $S(l) \leftrightarrow$ Elliptic curve $C(p)$,
Holomorphic 2-form $\eta(l) \leftrightarrow$ holomorphic 1-form $\eta(p)$,
Transcendental cycles $c_j'(l) \leftrightarrow$ Standard bases $\gamma_j(p)$,
Intersection form $A \leftrightarrow$ Intersection form $\begin{pmatrix} 0 & 1 \\ -1 & 0 \end{pmatrix}$,
Periods $\omega_j(l) \leftrightarrow$ Periods $\omega_j(p)$,
Riemann inequality $\leftrightarrow$ $\mathrm{Im}\omega_1(p)/\omega_2(p) > 0$,
Domain D of type IV$\leftrightarrow \mathbb{H} = \{\tau \in \mathbb{C} \mid \mathrm{Im}\,\tau > 0\}$,
Period map $l \longmapsto \omega(l) \in D \leftrightarrow$ Period map $p \mapsto \omega_1(p)/\omega_2(p) \in \mathbb{H}$,
Congruence subgroup $\Gamma_A(2) \leftrightarrow$ Elliptic modular group $\Gamma(2)$,
System $E(3,6;1/2) \leftrightarrow$ Equation $E(1/2,1/2,1)$.

Proceedings of the Asian Mathematical
Conference, 1990

Statistical Models in Cross-Matching

H. J. ZAINODIN
Centre of Quantitative Studies
National University of Malaysia
43600 UKM Bangi, Selangor Darul Ehsan
MALAYSIA

§1. Introduction

Most growth trend is usually removed by some form of filtering process. The filtered sequence of indices is used rather than the raw tree-ring widths sequences by decomposing a sequence of yearly ring-width indices into signal and noise components. The noise represents the general climatic factors (including the general growth curve) that govern the growth of trees within a region. The signal for a year is supposed to be the same for all trees growing in the same climatic region but the signals received by the trees are different from year to year. A filtered sequence of indices may be decomposed into signal and noise components.

Dendrochronology is a well establised technique that uses the sequences of annual ring widths of samples of oak timbers, taken for example from historical buildings, to date the samples and then the buildings. One of the most tricky features arising from the dating of historical buildings is the contrast in the suitability for tree-ring dating purposes of timber from prestigious buildings. Thus any sequence formed is usually long and well replicated, and so is easily dated against the appropriate master chronology. On the other hand, the dating of a cruck-framed building is usually more difficult and sometimes impossible because there are few timbers to be sampled and often the number of rings per sample is smaller. However, the dating of these humble buildings is important as it is suspected that these were the dwellings of medieval peasants.

§2. The Model

A sequence of n indices is considered to be composed of a sequence of n signal components common to all oak trees growing in the same climatic region and a sequence of n noise components, the noise sequence being specific to a particular tree.

Consider five sequences of tree-ring width indices from modern trees from the Thoresby Estate, Nottinghamshire (grid reference SK 640685) and six from Bradgate Park, Leicestershire (grid reference SK 530108). The sample correlation coefficient matrix for the eleven sequences are shown in Table 1.

Table 1. The correlation coefficients (below the diagonal) at the corresponding correct relative positions (above the diagonal) for the five samples from Thoresby Estate and six samples from Bradgate Park

Sample	THO-B01A	THO-B02B	THO-B03B	THO-B04A	THO-B05A	BRA-P01C	BRA-P02B	BRA-P03A	BRA-P04A	BRA-P05A	BRA-P06B
THO-B01A		0	2	14	15	-211	-64	-78	-107	-112	-102
THO-B02B	0.42		2	14	15	-211	-64	-78	-107	-112	-102
THO-B03B	0.39	0.32		12	13	-213	-66	-80	-109	-114	-104
THO-B04A	0.42	0.49	0.38		1	-225	-78	-92	-121	-126	-116
THO-B05A	0.36	0.30	0.35	0.44		-226	-79	-93	-122	-127	-117
BRA-P01C	0.19	0.20	0.21	0.24	0.13		147	133	104	99	109
BRA-P02B	0.22	0.32	0.35	0.31	0.17	0.25		-14	-43	-48	-38
BRA-P03A	0.30	0.27	0.23	0.32	-.02	0.31	0.48		-29	-34	-24
BRA-P04A	0.14	0.21	0.10	0.16	0.23	0.35	0.37	0.39		-5	5
BRA-P05A	0.19	0.24	0.11	0.32	0.17	0.30	0.31	0.28	0.42		10
BRA-P06B	0.15	0.18	0.19	0.27	0.10	0.34	0.42	0.41	0.64	0.48	

It seems that the sample correlates between pairs of trees from Thoresby and between pairs of trees from Bradgate. In fact the median values of the correlation coefficients are very similar for both sets of samples from within the Thoresby Estate and within Bradgate Park (0.38 and 0.37 respectively) with the range of values of the correlation coefficients for the Bradgate Park samples being slightly wider than that for the Thoresby Estate trees. However, the value of the median correlation coefficients between trees from Thoresby and Bradgate is much lower at 0.27. This suggests that model without interaction for tree-ring indices for the region containing both the Thoresby Estate and Bradgate Park may be inadequate. The Thoresby Estate and Bradgate Park are about 55 kilometers apart roughly in the north-south direction and are in the same general climatic zone. But the above results suggest that within such a zone there are local differences as to how trees respond to the climate. Examination of other modern data sets reveals a similar pattern (for details see [7]). This suggests that a region with an overall similar climate should be subdivided into small regions, which we call subregions, within which the climate and its effect on tree growth is even more homogeneous.

The existence of an underlying signal of ring width indices is assumed to be $\theta_1, \theta_2, \cdots, \theta_n$ for years $1, 2, \cdots, n$, respectively. This signal is assumed to be common to all oak trees growing in a region which experiences similar climatic conditions, at least as regards to the growth of oaks. If δ_k is taken to be the effect of subregion k and γ_{ik} be the interaction effect for subregion k in year i, then the index for the year i for sample j and subregion k is given by

$$y_{ijk} = \theta_i + \delta_k + \gamma_{ik} + Z_{ijk}$$

for $i = 1, 2, \cdots, n$, $j = 1, 2, \cdots, J_k$ and $k = 1, 2, \cdots, K$ where θ_i and Z_{ijk} are the signal for the whole region and the noise respectively. The assumptions made for this model are

$$E(\theta_i)=0, \quad \mathrm{Var}(\theta_i)=P^2, \quad i=1,2,\cdots,n;$$

$$E(\delta_i)=0, \quad \mathrm{Var}(\delta_i)=B^2, \quad k=1,2,\cdots,K;$$

$$E(\gamma_{ik})=0, \quad \mathrm{Var}(\gamma_{ik})=C^2, \quad i=1,2,\cdots,n, \quad k=1,2,\cdots,K;$$

$$E(Z_{ijk})=0, \quad \mathrm{Var}(Z_{ijk})=\sigma^2, \quad i=1,2,\cdots,n, \quad j=1,2,\cdots,J,$$

$k=1,2,\cdots,K$; and all the terms are uncorrelated.

If we consider trees j and $j'(j'\neq j)$ from the same subregion k, then the correlation coefficient between the indices for the same year i is given by

$$\rho(y_{ijk},y_{ij'k})=\frac{P^2+B^2+C^2}{P^2+B^2+C^2+\sigma^2}=\frac{R_1+R_2+R_3}{R_1+R_2+R_3+1}=\rho_1 \quad \text{say,}$$

where $R_1=P^2/\sigma^2, R_2=B^2/\sigma^2$ and $R_3=C^2/\sigma^2$.

Now consider two trees j and j' from different subregions k and k', respectively. If the indices for year i are denoted by y_{ijk} and $y_{ij'k'}$, respectively, then the correlation coefficient between the two indices for year i is given by

$$\rho(y_{ijk},y_{ij'k'})=\frac{P^2}{P^2+B^2+C^2+\sigma^2}=\frac{R_1}{R_1+R_2+R_3+1}=\rho_2 \quad \text{say.}$$

Then the correlation coefficient matrix for trees from both of these two subregions is of the form

$$\begin{bmatrix} 1 & \rho_1 & \cdot & \rho_1 & \rho_2 & \rho_2 & \cdot & \rho_2 \\ \rho_1 & 1 & \cdot & \rho_1 & \rho_2 & \rho_2 & \cdot & \rho_2 \\ \cdot & \cdot & \cdot & \cdot & \cdot & \cdot & \cdot & \cdot \\ \rho_1 & \rho_1 & \cdot & 1 & \rho_2 & \rho_2 & \cdot & \rho_2 \\ \rho_2 & \rho_2 & \cdot & \rho_2 & 1 & 1 & \cdot & 1 \\ \rho_2 & \rho_2 & \cdot & \rho_2 & \rho_1 & 1 & \cdot & \rho_1 \\ \cdot & \cdot & \cdot & \cdot & \cdot & \cdot & \cdot & \cdot \\ \rho_2 & \rho_2 & \cdot & \rho_2 & \rho_1 & \rho_1 & \cdot & 1 \end{bmatrix}$$

which resembles the form of the sample correlation matrix for the Bradgate/Thoresby example given in Table 1 and observed in other data sets.

§3. Parameters Estimation of the Model

A model which appears to have some of the characteristics of tree-ring data sets originating from two or more subregions was proposed. To illustrate the differences in the parameter values and hence the implications for dating for different regions we have used eleven modern tree-ring data sets taken from sites in the United Kingdom (for further details see [1], [4] and [7]). The sites are defined into three regions.

Region I: Sites A, B, C, O and R. This corresponds to a fairly small regional system.

Region II: Sites G, M, P, R and S. This corresponds to a national system which is poorly replicated.

Region III: All sites. This corresponds to a national system but with better replication than Region II.

Table 2. Estimated values of R_1, R_2 and R_3 for regions I, II, and III together with the corresponding values of ρ_1 and ρ_2.

Region	R_1	R_2	R_3	ρ_1	ρ_2
I	0.35	0.00	0.40	0.43	0.20
II	0.20	0.00	0.40	0.38	0.13
III	0.25	0.00	0.38	0.39	0.15

For each region values of the ratios R_1, R_2 and R_3 were estimated. (Details of the filtering and estimation procedures used are given in [7].) The results suggest that typical values of R_1, R_2 and R_3 for the three regions are as given in Table 2. Interestingly the variation due to the effect of subregions is negligible. The important effects are the variation due to the year and the interaction between year and subregion. That is, there is a significant year-by-year variation in the signal for a region but in addition this variation varies from subregion to subregion due to local climatic and other factors. As expected the values of, ρ_1, are very similar. However, ρ_2, does vary with the extent of the overall region and is largest for Region I.

§4. Statistical Dating Against a Master Chronology

Circumstances as to when is it better to construct and use regional tree-ring chronologies rather than national chronologies is investigated. The method adopted based upon the work of [2] who showed that when a sequence tree-ring indices of length n_1 is compared with a master sequence of tree-ring indices of length $n_2 (n_1 < n_2)$ at all possible incorrect position then

$$Pr(\max r(p) \leq x) = \left\{\Phi\left(\frac{1}{2}\sqrt{n_1 - 3}\ln\left(\frac{1+x}{1+x}\right)\right)\right\}^{n_2 - n_1},$$

where $r(p)$ is the sample correlation between the two sequences at relative position $p, p = 1, 2, \cdots, n_2 - n_1$; $p = 0$ is taken as the correct relative position and $\Phi(x) = \int_{-\infty}^{x} 1/\sqrt{2\pi}$ $\exp(-t^2/2)dt$. In tree-ring dating both [5] and [6] stress the advantages of the z-value and thus this paper follows their recommendations where $z(p) = \frac{1}{2}\sqrt{n_1 - 3}\ln(\frac{1+r(p)}{1-r(p)})$ be the corresponding z-value at position p. If the z-values at the incorrect relative positions and let $F_M(x)$ be the distribution function of the maximum of the $z(p), p = 1, 2, \cdots, n_2 - n_1$, then following Kronberg's argument, we obtain

$$F_M(x) = [\Phi(x)]^{n_2 - n_1}.$$

At the correct relative position ($p = 0$) we assume that the underlying correlation is ρ and if we let $F_C(y)$ be the distribution function of the z-value at this position then

$$F_C(y) = \Phi\left(y - \frac{1}{2}\sqrt{n_1 - 3} \quad \ln\left(\frac{1+\rho}{1-\rho}\right)\right).$$

The probability that the z-value at the correct relative position ($p = 0$) is greater than the maximum of the z-values at the incorrect relative positions ($p \neq 0$) is given by

$$Pr\left[z(0) > \max_{p=1,2,\cdots,n_2-n_1}\{z(p)\}\right] = \int_{y=-\infty}^{y=\infty}\int_{x=-\infty}^{x=y} f_C(y) f_M(x)\,dx\,dy,$$

where $f_C(y)$ and $f_M(x)$ are the probability density functions corresponding to the distribution functions $F_C(y)$ and $F_M(x)$ respectively. This probability may be expressed as

$$(n_2 - n_1)\int_{-\infty}^{\infty} \phi(x)[\Phi(x)]^{n_2-n_1-1}[1 - \Phi(x-\mu)]dx,$$

where

$$\phi(x) = \frac{1}{\sqrt{2\pi}}\exp -\frac{1}{2}x^2 \quad \text{and} \quad \mu = \sqrt{n_1 - 3}\ln\left(\frac{1+\rho}{1-\rho}\right).$$

Thus for given ρ, n_1 and n_2 it is possible to evaluate the probability that the z-value at the correct relative position is greater than the z-values at all other incorrect positions and so the relative merits of using national or regional chronologies is to be examined.

§5. Conclusions and Comments

In tree-ring dating of a building samples are taken from timbers though to be contemporary. These samples are then prepared, measured and cross-matched against each other (for further details see [3]). The relationship among R_1, R_2 and R_3 are examined and the length and level of replication of the shortest average tree-ring sequence that can be dated successfully against a master chronology with at most a 1% chance of an error. For a fixed number (J^*) of samples in the average sequence from a new subregion within the overall region the shortest length (n^*) of an average sequence is computed such that the probability of the z-value at the correct relative position is greater than the maximum of the z-values at the incorrect positions is 0.99. (The choice of 0.99 is somewhat arbitrary and is used to illustrate this approach.) Assuming that the model holds, it can be shown that the correlation coefficient, ρ, at the correct relative position satisfies

$$\rho^2 = R_1^2 \Big/ \left\{[R_1 + R_2 + R_3 + J^{*-1}]J_{\text{TOT}}^{-2}\left\{(R_2 + R_3)\sum_{k=1}^{k} J_k^2 + J_{\text{TOT}} + R_1 J_{\text{TOT}}^2\right\}\right\},$$

where $J_{\text{TOT}} = \sum_{k=1}^{K} J_k$, K is the number of subregions forming the master chronology, J_k is the number of samples included in the master from subregion $k (k = 1, 2, \cdots, K)$ and J^* is the number of samples from a new subregion, the average of which is to be dated against the master chronology.

Plot of n^* against J^* for the typical values of R_1, R_2 and R_3 from the three different regions can be obtained. It is practical to take the number of samples for each subregion in the master chronology as ten. The number of subregions, K is taken to be five, ten or a hundred, the latter corresponds to an extremely well replicated master chronology for the region concerned. Between five to ten subregions is probably typical of most published master chronologies. We have assumed that the master chronology has 1100 indices, the length of the East Midlands master tree-ring chronology (see [3]).

As the number (J^*) of samples used to form the average sequence to be dated increases the least length (n^*) needed is reduced. This is true for all three regions when they contain either five, ten or a hundred subregions. The gain in terms of a lower value of n^* is about ten, or slightly more, indices when using a master chronology for Region I with ten subregions over a similar region with five subregions. This is true if the average sequence to be dated contains at least four samples. Hence a reasonably well replicated ($J^* \geq 5$) average sequence with at least eighty indices should be dated with some confidence. For Regions II and III, the gain is even more dramatic, about forty and thirty indices respectively. The average sequences should be both longer, say at least one hundred and twenty indices, and better replicated, say $J^* \geq 10$.

Hence to attempt to date buildings containing only a few oak timbers and/or timbers with relatively few rings (≤ 120), the use of local tree-ring chronologies rather than national chronologies is almost essential. However, it is unlikely that reliable dates will be feasible for sequences containing less than sixty indices, however well replicated.

References

[1] Briffa, K.R., *Tree-climate relationships and dendroclimatological reconstruction in British Isles.* PhD thesis (unpublished). University of East Anglia, Norwich, U.K. (1984).

[2] Kronberg, D., *Distribution of cross-correlations in two-dimensional time series with application to dendrochronology*, Research report no.72, Department of Theoretical Statistics, University of Aarhus, Denmark. (1981).

[3] Laxton, R.R. and Litton, C.D., *An East Midlands Master Tree-Ring Chronology and Its Use for Dating Vernacular Buildings*, Monograph Series III, Department of Classical and Archaeological Studies, University of Nottingham, Nottingham. (1988).

[4] Litton, C.D. and Zainodin, H.J., *Grouping Methods of Dendrochronology*, Science and Archaeology, 21(1987), 14–24.

[5] Okasha, M K M., *Statistical methods in dendrochronology*, PhD thesis (unpublished), University of Sheffield, Sheffield, U. K. (1987).

[6] Steward, S., *Statistical dendrochronology*, MSc thesis, University of Sheffield, Sheffield, U.K. (1983).

[7] Zainodin, H.J., *Statistical models and techniques for dendrochronology*, PhD thesis (unpublished), University of Nottingham, Nottingham, U.K. (1988).

Proceedings of the Asian Mathematical
Conference, 1990

A Criterion for Dependency of Algebraic Equations and Its Application to Automated Theorem Proving*

ZHANG JINGZHONG, YANG LU & HOU XIAORONG
Institute of Mathematical Sciences
Chinese Academy of Sciences
Chengdu 610015 Sichuan, China

§1. Introduction

Let $\{f_i(x_1, x_2, \cdots, x_n), i = 1, 2, \cdots, m\}$ be a set of polynomials over field K, and $g(x_1, x_2, \cdots, x_n)$ be a polynomial over field K. A very important problem is whether

$$C : g(x_1, x_2, \cdots, x_n) = 0 \tag{1.1}$$

holds or not under the condition

$$H : f_i(x_1, x_2, \cdots, x_n) = 0 \quad (i = 1, 2, \cdots, m). \tag{1.2}$$

The importance of the problem lies in the fact that it not only has many applications but also is one of the essential problems in the theory of the system of multivariate higher-degree polynomial equations. If all f_i in (1.2) and the g in (1.1) are of the first degree in $x_1, x_2, \cdots, x_n$, this problem is exactly the criterion problem for dependency of linear algebraic equations, one of the essential problems in linear algebra. Thus, the problem whether $H \Rightarrow C$ holds or not is likewisely essential in nonlinear algebra.

Let $\bar{K}$ be the algebraically closed field to K, and $x_j \in \bar{K}$. All $(x_1, x_2, \cdots, x_n)$ satisfying (1.2) and (1.1) form hypersurfaces Γ_H and Γ_C in $\bar{K}^n$, respectively. The above problem, in algebraic geometry, is whether $\Gamma_H \subset \Gamma_C$ or not ?

In recent years, owing to professor Wu Wentsun's outstanding contributions [1] and [3], this problem has made fundamental progress. Applying Wu-Ritt's well-ordering principle, we can convert (1.1) into the system of equations with triangular form, called ascending chain, as follows:

$$H : f_i(u, x_1, x_2, \cdots, x_i) = 0 \quad (i = 1, 2, \cdots, s), \tag{1.3}$$

* Supported by the NNSF of China

where $u = (u_1, u_2, \cdots, u_d)$ denotes some x_j in simplified form, $x_{i+1}, \cdots, x_s$ don't appear in f_i, and $x_1, x_2, \cdots, x_s$ are from changing some subscripts. Correspondingly, (1.2) is rewritten as follows:

$$C : g(u, x_1, x_2, \cdots, x_s) = 0. \tag{1.4}$$

On the basis, using remainder formula given in the [1], professor Wu gave a sufficient condition to decide whether $H \Rightarrow C$ or not. Applying the sufficient condition to mechanical geometry theorem proving, he has achieved universally acknowledged success. If the ascending chain $\{f_i\}$ in (1.3) is irreducible, the condition is necessary and sufficient. However, when $\{f_i\}$ in (1.3) is reducible, in order to treat this problem, also, one has to do decomposition[5] or seek the help of geometry[4]. Not long before, two methods, WE method and the resultant-matrix method, were shown to overcome reducibility difficult without doing decomposition beforehand[6]–[8]. However, in general, these methods eventually still cause doing decomposition. By the method suggested in the present paper, one can determine directly whether $H \Longrightarrow C$ or not without doing decomposition of ascending chain at all.

Now we need more notations. $ld(f_i)$ and I_i denote the degree and the leading coefficient of f_i in x_i, respectively. Generally, I_i belongs to $K[u, x_1, x_2, \cdots, x_{i-1}]$. Given the special case $\tilde{u} = (\tilde{u}_1, \tilde{u}_2, \cdots, \tilde{u}_d)$ of u, and if $I_1(\tilde{u}) \neq 0$, we can obtain the $m_1 = ld(f_1)$ roots of equation $f_1(\tilde{u}, x_1) = 0$ in $\bar{K}$, write $x_1^{(1)}, x_1^{(2)}, \cdots, x_1^{(m_1)}$. Taking $x_1 = x_1^{(1)}, x_1^{(2)}, \cdots, x_1^{(m_1)}$ in $I_2(\tilde{u}, x_1)$, respectively, if $I_2(\tilde{u}, x_1^{(i)}) \neq 0$ for all $i = 1, 2, \cdots, m_1$, we can obtain the $m_2 = ld(f_2)$ roots of equation $f_2(\tilde{u}, x_1^{(i)}, x_2) = 0$ for each i, write $x_2^{(i_1, i_2)} (i_1 = 1, 2, \cdots, m_1; i_2 = 1, 2, \cdots, m_2)$. Going on doing so, if

$$\{x_1^{(i_1)}, x_2^{(i_1,i_2)}, x_3^{(i_1,i_2,i_3)}, \cdots, x_j^{(i_1,i_2,\cdots,i_j)}\} = x^{(i_1,i_2,\cdots,i_j)} \tag{1.5}$$

have been determined for all $j < l$, and

$$I_{j+1}(\tilde{u}, x^{(i_1,i_2,\cdots,i_j)}) \neq 0 \quad (j = 1, 2, \cdots, l-1), \tag{1.6}$$

taking $u = \tilde{u}, x = x^{(i_1,i_2,\cdots,i_j)}$ in $f_l(u, x)$, we can get the $m_l = ld(f_l)$ roots of equation

$$f_l(\tilde{u}, x_1^{(i_1)}, \cdots, x_{l-1}^{(i_1,i_2,\cdots,i_{l-1})}, x_l) = 0 \tag{1.7}$$

for each set of superscripts $i_1, i_2, \cdots, i_{l-1}$, write $x_l^{(i_1,i_2,\cdots,i_l)} (i_l = 1, 2, \cdots, m_l)$.

If the above procedure can reach $l = s$, we get the $m = m_1 m_2 \cdots m_s$ sets of the solutions of the system of equations (1.3) for given $u = \tilde{u}$:

$$\begin{aligned} &\{x_1^{(i_1)}, x_2^{(i_1,i_2)}, \cdots, x_s^{(i_1,i_2,\cdots,i_s)}\} = x^{(i_1,i_2,\cdots,i_s)}, \\ &i_k = 1, 2, \cdots, m_k = ld(f_k); \quad k = 1, 2, \cdots, s \end{aligned} \tag{1.8}$$

and for all $x^{(i_1,i_2,\cdots,i_s)}$,

$$I_j(\tilde{u}, x^{(i_1,i_2,\cdots,i_s)}) \neq 0. \tag{1.9}$$

In this case, $\tilde{u}$ is called a normal point or a set of normal parameters of the ascending chain $\{f_i(u, x_i)\}$.

Now let K be any number field and $\bar{K}$ the complex number field. If $\widetilde{u}$ is a normal point, there exists a $\widetilde{u}$'s neighborhood consisting of normal points in d-dimension complex space. An ascending chain with the normal points is called a normal ascending chain. A set of solutions of the system of the equations (1.3), $(\widetilde{u}, \widetilde{x})$ is called a set of normal solutions if $I_j(\widetilde{u}, \widetilde{x}) \neq 0$ for all j. If $g(\widetilde{u}, \widetilde{x}) = 0$ for all normal solutions $(\widetilde{u}, \widetilde{x})$ of (1.2), we say that (1.4) generally holds under the condition (1.3). If $\widetilde{u}$ is a normal point, the $(\widetilde{u}, \widetilde{x})$ is a set of normal solutions. The reverse may not necessarily be true. For the normal ascending chain, we will give a necessary and sufficient condition which makes (1.4) generally hold under (1.3). In [7], the concept of proper ascending chain was induced, and a practical method with which an ascending chain is decomposed into some proper ascending chains was given. It is easy to prove that the proper ascending chain is just the normal ascending chain if K is a number field.

§2. Lemmas on Resultants

Let A be a commutative ring, $f, g \in A[v]$:

$$\left.\begin{array}{ll} f = a_n v^n + \cdots + a_0 & (a_n \neq 0) \\ g = b_k v^k + \cdots + b_0 & (a_i, b_i \in A) \end{array}\right\}. \tag{2.1}$$

Determinant

$$\mathrm{Res}(g, f, v) = \begin{vmatrix} a_n & a_{n-1} & \cdots & a_1 & a_0 & & 0 & \\ & a_n & a_{n-1} & \cdots & a_1 & a_0 & & \\ & & & \cdots & \cdots & & & \\ & 0 & & a_n & a_{n-1} & \cdots & a_1 & a_0 \\ b_k & b_{k-1} & \cdots & b_1 & b_0 & & 0 & \\ & b_k & b_{k-1} & \cdots & b_1 & b_0 & & \\ & & & \cdots & \cdots & & & \\ & 0 & & b_k & b_{k-1} & \cdots & b_1 & b_0 \end{vmatrix} \tag{2.2}$$

is called the resultant of polynomial g to f with respect to v. The real degree of polynomial g may not necessarily be k. The k is called the set degree of polynomial g in the resultant. The following lemmas show that the difference of set degrees doesn't impair the resultant on a certain meaning.

Lemma 2.1. *If $\lambda_1, \lambda_2, \cdots, \lambda_n$ are all roots of polynomial f, then*

$$\mathrm{Res}(g, f, v) = a_n^k \prod_{i=1}^{n} g(\lambda_i). \tag{2.3}$$

This is a well-known fact. By the Lemma 2.1, we have

Lemma 2.2. *If $f = f_1 f_2, f_1, f_2 \in A[v]$, then*

$$\mathrm{Res}(g, f, v) = \mathrm{Res}(g, f_1, v)\mathrm{Res}(g, f_2, v). \tag{2.4}$$

Given an ascending chain $\{f_1, f_2, \cdots, f_s\}$ and another polynomial g, putting

$$\left.\begin{aligned} R_{s-1} &:= \mathrm{Res}(g, f_s, x_s), \\ R_{s-2} &:= \mathrm{Res}(R_{s-1}, f_{s-1}, x_{s-1}), \\ &\cdots \\ R_0 &:= \mathrm{Res}(R_1, f_1, x_1), \end{aligned}\right\} \tag{2.5}$$

we call R_0 the resultant of ascending chain $\{f_1, f_2, \cdots, f_s)$ with respect to polynomial g and denote it simply by $Res(g, f_1, f_2, \cdots, f_s)$. Given the set degrees of g and $R_i (i = 1, 2, \cdots, s-1)$, we can get a unique R_0, which is a polynomial in u.

Applying the Lemma 2.1 and induction on s, we have

Lemma 2.3. *Given a normal point $\tilde{u}$ of ascending chain $\{f_1, f_2, \cdots, f_s\}$, let m_i be the degree of f_i in leading variable x_i, and the solutions of the system of equations (1.3) be (1.8), then*

$$\mathrm{Res}(g, f_1, f_2, \cdots, f_s)(\tilde{u}) = P(\tilde{u}) \prod_{j=1,\cdots,s}^{1\le i_j \le m_j} g(\tilde{u}, x^{(i_1, i_2, \cdots, i_s)}), \tag{2.6}$$

where $P(\tilde{u})$ is the value of a certain polynomial $P(u)$ in $u = \tilde{u}$, and $P(u) \neq 0$.

Proof. Induction on s. For $s := 1$, the proposition is the Lemma 2.1. Suppose that the proposition is true for $s := s-1$. For $s := s$, regard u and x_1 as independent variables. Consider the resultant of ascending chain $\{f_1, f_2, \cdots, f_s\}$ with respect to g. Given $\tilde{u}$ and corresponding $x_1^{(i_1)}$, by the induction hypothesis, we have

$$R_1 = \mathrm{Res}(g, f_2, \cdots, f_s)(\tilde{u}, x_1^{(i_1)}) = P^*(\tilde{u}, x_1^{(i_1)}) \prod_{j=2,\cdots,s}^{1\le i_j \le m_j} g(\tilde{u}, x^{(i_1, \cdots, i_s)}), \tag{2.7}$$

where $P^*(\tilde{u}, x_1^{(i_1)})$ is the value of polynomial $P(u, x_1)$ in $u = \tilde{u}, x_1 = x_1^{(i_1)}$, and $P(\tilde{u}, x_1^{(i_1)}) \neq 0$. Let $i_1 = 1, 2, \cdots, m_1$ in (2.7). Applying the Lemma 2.1, we have

$$\begin{aligned} \mathrm{Res}(g, f_1, \cdots, f_s)(u) =& \ \mathrm{Res}(R_1, f_1, x_1)(u) \\ =& \ (I_1(\tilde{u}))^N \prod_{i_1=1}^{m_1} \left(P^*(\tilde{u}, x_1^{(i_1)}) \prod_{j=2,\cdots,s}^{1\le i_j \le m_j} g(\tilde{u}, x^{(i_1, \cdots, i_s)}) \right) \\ =& \ (I_1(\tilde{u}))^N \left(\prod_{i_1=1}^{m_1} (P^*(\tilde{u}, x_1^{(i_1)}) \right) \prod_{j=2,\cdots,s}^{1\le i_j \le m_j} g(\tilde{u}, x^{(i_1, \cdots, i_s)}). \end{aligned} \tag{2.8}$$

Since $x_1^{(i_1)} (i_1 = 1, 2, \cdots, m_1)$ are all roots of polynomial $f_1(\tilde{u}, x_1), \Pi_{j=1}^{m_1} P^*(\tilde{u}, x_1)$ is a polynomial in u.

From the equality (2.6), we see that the value of resultant is zero if there is a set of solutions $(\tilde{u}, \tilde{x})$ satisfying $g(\tilde{u}, \tilde{x}) = 0$ for the normal point $\tilde{u}$, and vice versa. If we regard $g(\tilde{u}, x^{(i_1, \cdots, i_s)})$ as an algebraic function, then the resultant $Res(g, f_1, f_2, \cdots, f_s)$ identically equals zero if and only if some $g(\widehat{u}, x^{(i_1, \cdots, i_s)})$ in the left-hand side of (2.6) identically equals zero. We will discuss the problem in the following section.

§3. On λ-resultants

Let λ be independent from u, x. Consider the resultant of ascending chain $\{f_1, f_2, \cdots, f_s\}$ with respect to polynomial $g(u, x) + \lambda$

$$R(g+\lambda, f) = R(g+\lambda, f_1, f_2, \cdots, f_s). \tag{3.1}$$

By the Lemma 2.3, we have

$$R(g+\lambda, f) = P(\tilde{u}) \prod_{j=1,\cdots,s}^{1\le i_j \le m_j} (g(\tilde{u}, x^{(i_1,\cdots,i_s)}) + \lambda) \tag{3.2}$$

for any normal point u. Both sides in (3.2) are polynomials in $\tilde{u}$, and normal point set is a nonempty open set , so we have

Theorem 3.1. *Let $\{f_1, f_2, \cdots, f_s\}$ be a normal ascending chain. Condition (1.4) generally holds under the condition (1.3) if and only if*

$$R(g+\lambda, f)(u, \lambda) = P(u)\lambda^{m_1 m_2 \cdots m_s}. \tag{3.3}$$

Moreover, we can get more information from $R(g+\lambda, f)$. Given the system of equations (1.3), we can regard each set of solutions $x^{(i_1, i_2, \cdots, i_s)}$ as a set of the branches of algebric function of u. Each one is called a branch determined by (1.3). How many branches are there in the $m = m_1 m_2 \cdots m_s$ branches to make the equality (1.4) hold? By the (3.2), we have

Theorem 3.2. *Let $\{f_1, f_2, \cdots, f_s\}$ be a normal ascending chain. If $\mathrm{Res}(g+\lambda, f) = \lambda^k Q(u,\lambda), \lambda \nmid Q(u,\lambda), Q(u,\lambda) \in K[u,\lambda]$, then there are exactly k branches determined by (1.3) to satisfy (1.4).*

§4. Examples

Example 1.

$$\left.\begin{aligned} f_1 &= x_1 - u, \\ f_2 &= x_2^2 - 2x_1x_2 + u^2, \\ g &= x_2 - x_1. \end{aligned}\right\}$$

Solution. All u are normal points in this problem.

$$\mathrm{Res}(g+\lambda, f_2, x_2) = \begin{vmatrix} 1 & 1 & 0 \\ -2x_1 & \lambda - x_1 & 1 \\ u^2 & 0 & \lambda - x_1 \end{vmatrix} = -x_1^2 + \lambda^2 + u^2,$$

$$\mathrm{Res}(-x_1^2 + \lambda^2 + u^2, f_1, x_1) = \lambda^2.$$

Therefore $g = 0$ holds under $f_1 = 0, f_2 = 0$.

Example 2.

$$f_1 = x_1^2 + ux_1,$$

$$f_2 = x_2^2 + (x_1^2 - u)x_2 - u^2 x_1,$$

$$g = x_2^2 + x_1(1-u)x_2 - ux_1^2.$$

Solution. By Wu-method, we can get the final remainder (of g with respect to $\{f_1, f_2\}$) $\text{prem}(g, f_1, f_2) = (x_1 + u)x_2 = 0$. Taking $g := (x_1 + u)x_2$ and doing λ-resultant, we have

$$\begin{aligned}
& \text{Res}((x_1 + u)x_2 + \lambda, f_2, x_2) \\
= \;& \lambda^2 + (x_1 + u)^2 u^2 x_1 - \lambda(x_1 + u)(x_1^2 - u) \\
= \;& \lambda u x_1 + \lambda(\lambda + u^2) = R_1
\end{aligned}$$

(in the above calculation, we did reduction by $f_1 = 0$).

$$\text{Res}(R_1, f_1, x_1) = \lambda^3(\lambda + u^2)$$

so the ascending chain $\{f_1, f_2\}$ is reducible, and there is an ascending chain with the first degree to make the $g = 0$ false in general.

References

[1] Wu Wen-tsun, *On the Decision Problem and the Mechanization of Theorem Proving in Elementary Geometry*, Scientia Sinica, 21(1978), 157–179.

[2] L. Liu, Z.Y. Lu & D.M. Wang, *The Structure of LaSall's invariant set of Lotka – Volterra Systems*, (in Chinese) Scientia Sinica, 8, 33(1990), 799–804.

[3] Wu Wen-tsun, *Basic Principles of Mechanical Theorem Proving in Geometries* (in Chinese) Peking, 1984.

[4] Wu Wen-tsun, *On Reducibility Problem in Mechanical Theorem Proving of Elementary Geometries*, Chinese Quarterly J. Math. Vol. 2, No. 2(1987), 1–19.

[5] S. C. Chou, *Mechanical Geometry Theorem Proving*, D. Reidel Publishing Company, 1988.

[6] J. Z. Zhang & L. Yang, *WE – method of Mechanical Theorem Proving* (to appear).

[7] J. Z. Zhang & L. Yang, *The Resultant – Matrix Method of Mechanical Theorem Proving* (to appear).

[8] J. Z. Zhang & L. Yang, *A Method to Overcome the Reducibility Difficulty in Mechanical Theorem Proving*, ICTP preprint IC/89/263, (1989).

Proceedings of the Asian Mathematical
Conference, 1990

The Centralizer of Semigroup Rings

ZHANG MOUCHENG & CHEN YUQUN
Mathematics of Department
South China Normal University
Guangzhou 510631, China

§ 1. Introduction

Let R be a ring, G a semigroup. The semigroup ring of a semigroup G over a ring R is the ring $R[G]$ which consists of all finite formal sums of the form $\sum f(s)s, s \in G, f(s) \in R$, with the following rules of equality, addition and multiplication: Letting $f = \sum f(s)s, \quad g = \sum g(s)s$,

$$f = g \quad \text{if and only if} \quad f(s) = g(s) \quad \text{for all } s \in G;$$

$$f + g = \sum (f(s) + g(s))s \quad \text{for all } s \in G, f(s), \quad g(s) \in R;$$

$$f \cdot g = \sum \left(\sum_{hk=s} f(h)g(k) \right) s, \quad \text{for all} \quad s \in G, \quad f(s), \quad g(s) \in R.$$

Clearly $R[G]$ is a ring with the foregoing addition and multiplication. It is called the semigroup ring of a semigroup G over a ring R.

Let G be a semigroup, and let

$G' = \{s \in G \mid sa = sb \text{ or } as = bs \text{ implies } a = b \text{ for all } a, b \in G\}$. It is easy to see that, when $G' \neq \emptyset, G'$ is the subsemigroup of G. The element of G' is called the cancellation element of G. Moreover, if G satisfies the cancellation law then $G' = G$. Let R be a ring, A a subset of R with $A \neq \emptyset$, and let

$$C_r(A) = \{r \in R \mid ar = ra \quad \text{for any } a \in A\},$$

$C_R(A)$ is called the centralizer of A in R. In particular,

$C_R(R)$ is called the center of$R$and is denoted by $C(R)$.

Dauns, Krempa, Passman, Ponizovskii, Rudin and Schneider have done much work in this area. The purpose of this paper is to characterize the centralizer $C_{R[G]}(R[G'])$ of a semigroup ring $R[G']$ in $R[G]$. In particular, when $G = G'$, we obtain the center of $R[G]$. This generalizes Krempa's result in [2].

§ 2. Characterization of $C_{R[G]}(R[G'])$

Let G be a semigroup, A a nonempty subset of G satisfying $|A| < \infty$ and $x'A = Ax'$ for any $x' \in G'$, where G' is the cancellation subsemigroup of G. Let $\{A_\alpha\}_{\alpha \in I}$ be the family of all subsets of G satisfying the above property and denote the union of $\{A_\alpha\}_{\alpha\in I}$ by $\delta(G)$, i.e., $\delta(G) = \cup_{\alpha\in I} A_\alpha$.

Theorem 1. *Let G be a semigroup. If $\delta(G) \neq \emptyset$, then $\delta(G)$ is a subsemigroup of G.*

Proof. Suppose $s_1, s_2 \in \delta(G)$. By the definition of $\delta(G)$, there exist $\alpha_1, \alpha_2 \in I$ such that $s_1 \in A_{\alpha_1}$and $S_2 \in A_{\alpha 2}$, and then $s_1 s_2 \in A_{\alpha_1} A_{\alpha_2}$. Since $|A_{\alpha_1} A_{\alpha_2}| \leq |A_{\alpha_1}||A_{\alpha_2}| < \infty$, and $x' A_{\alpha_1} A_{\alpha_2} = A_{\alpha_1} x' A_{\alpha_2} = A_{\alpha_1} A_{\alpha_2} x'$ for any $x' \in G'$, we have $A_{\alpha_1} A_{\alpha_2} \in \delta(G)$. Therefore $s_1 s_2 \in \delta(G)$.

For any $x \in \delta(G)$, let $A^{(x)}$ be a subset of $\delta(G)$ satisfying the following conditions:

(a) $x \in A^{(x)}$;

(b) $gA^{(x)} = A^{(x)} g$ for any $g \in G'$;

(c) $|A^{(x)}|$ is the minimal for all subsets of $\delta(G)$ satisfying the foregoing conditions (a) and (b).

Theorem 2. *Let G be a semigroup. Then for any $x \in \delta(G)$, there exists a unique subset $A^{(x)}$ of $\delta(G)$.*

Proof. The existence of $A^{(x)}$ is clear by the definition of $\delta(G)$. Now we show the uniqueness. Suppose A_1 satisfies the conditions (a)–(c) . Then $x \in A^{(x)} \cap A_1$ and $A^{(x)} \cap A_1 \neq \emptyset$. For any $g \in G$ and any $a \in A^{(x)} \cap A_1$, we have $ga \in gA^{(x)} \cap gA_1$. Thus $g(A^{(x)} \cap A_1) \subseteq gA^{(x)} \cap gA_1$. Conversely, for any $y \in gA^{(x)} \cap gA_1$, we have $y = ga_1 = ga_2$, where $a_1 \in A^{(x)}, a_2 \in A_1$. Since g is a cancellation element of G, we have $a_1 = a_2$. Therefore, $y = ga_1 = ga_2 \in g(A^{(x)} \cap A_1)$, i.e., $g(A^{(x)} \cap A_1) = gA^{(x)} \cap gA_1 = A^{(x)} g \cap A_1 g = (A^{(x)} \cap A_1) g$. Since $|A^{(x)}|$ and $|A_1|$ satisfy the condition (c), we obtain $A^{(x)} = A^{(x)} \cap A_1 = A_1$.

Theorem 3. *Let G be a semigroup, x an arbitrary element of $\delta(G), y$ an element of G. Then $y \in A^{(x)}$ if and only if $A^{(x)} = A^{(y)}$.*

Proof. The sufficiency is clear . Now we show the necessity. Suppose $y \in A^{(x)}$. Then $y \in A^{(y)} \cap A^{(x)}$. Therefore for any $g \in G'$, we have $g(A^{(y)} \cap A^{(x)}) = gA^{(y)} \cap gA^{(x)} = A^{(y)} g \cap A^{(x)} g = (A^{(y)} \cap A^{(x)}) g$. Since $A^{(y)} \cap A^{(x)} \subseteq A^{(y)}$ and $A^{(y)} \cap A^{(x)}$ satisfies the conditions (a) and (b), and $A^{(y)}$ satisfies the conditions (a)–(c) of $A^{(x)}$, we have $A^{(y)} \cap A^{(x)} = A^{(x)}$. Thus $A^{(y)} \subseteq A^{(x)}$.

Suppose $x \notin A^{(y)}$. Let $B = A^{(x)} \backslash A^{(y)}$. Then $x \in B$. For any $g \in G'$ and any $b \in B$, since g is a cancellation element of G, we have $gb \in gA^{(x)}$ and $gb \notin gA^{(y)}$. Therefore $gb \in gA^{(x)} \backslash gA^{(y)} = A^{(x)} g \backslash A^{(y)} g = (A^{(x)} \backslash A^{(y)}) g = Bg$, i.e., $gB \subseteq Bg$. Symmetrically, $Bg \subseteq gB$. Thus $gB = Bg$. But $B \subseteq A^{(x)}$ and $B \neq A^{(x)}$. This is a contradiction. Thus $x \in A^{(y)}$ and $A^{(x)} \subseteq A^{(y)}$. Consequently $A^{(x)} = A^{(y)}$.

Now we show the main results in this paper.

Theorem 4. *Let G be a semigroup, G' the cancellation subsemigroup of $G, G' \cap C(G) \neq \emptyset, f \in R[G]$. Then $f \in C_{R[G]}(R[G'])$ if and only if $f = \sum_{x\in\delta(g)} \sum_{y\in A^{(x)}} f(x)y + f_0$, where $C(G)$ is the center of $G, f(X) \in C(R), f_0 \in N_R[G]$, and $N_R = \{r \in R \mid rR = Rr = 0\}$ is the annihilator of R.*

We stipulate $\sum_{x\in\delta(G)} \sum_{y\in A^{(x)}} f(x)y = 0$ if $\delta(G) = \emptyset$.

Proof. (1) The sufficiency is clear.

(2) Now we show the necessity. Let $f = \sum f(s)s = f_1 + f_2 \in C_{R[G]}(R[G'])$, where $f_1(s) \notin N_R, f_2 \in N_R[G]$ for any $s \in G$. Then for any $g \in G', r \in R$, we have $rgf_1 = rgf =$

$frg = f_1rg$, i.e., $f_1 \in C_{R[G]}(R[G'])$. Therefore we can suppose $f_2 = 0$, i.e., $f(s) \notin N_R$ for any $s \in G$.

(3) Since $G' \cap C(G) \neq \emptyset$, we can take $g \in G' \cap C(G)$. For any $r \in R$, we have $rgf = rg\sum f(s)s = \sum rf(s)gs$, and $frg = \sum f(s)srg = \sum f(s)rsg$. Therefore $\sum rf(s)gs = \sum f(s)rsg$. Since $gs = sg$ and g is a cancellation element, we have $rf(s) = f(s)r$ for any $r \in R$, i.e., $f(s) \in C(R)$ for any $s \in G$.

(4) For any element t of supp$f$, there exists an $r_t \in R$ such that $f(t)r_t \neq 0$ since $f(t) \notin N_R$. Since $f(t) \in C(R), r_t f(t) \neq 0$. Since $r_t gf = fr_t g$ for any $g \in G'$, we have $\sum r_t f(s)gs = \sum f(s)r_t sg$. Thus we obtain $gt \in$ supp $(r_t gf)$ by letting g to be a cancellation elemment and $r_t f(t) \neq 0$. Therefore $gt \in$supp$(fr_t g) \subseteq$ (supp$f)g$, i.e., $g($ supp $f) \subseteq$ (supp$f)g$. Comparing the cardinals on both sides, we obtain g(supp$f$) = (supp$f)g$. Thus supp $f \subseteq \delta(G)$. By Theorem 3 we have supp$f = \cup_{x \in \text{supp} f} A^{(x)}$. Therefore $\{A^{(x)} \mid x \in$supp$f\}$ constitutes a classification of supp$f$.

(5) Let $f_1 = f - \sum_{y \in A^{(x)}} f(x)y$. Since $x \in$ supp$f\backslash$suppf_1, supp$f_1 \subseteq$ supp$f$ and supp$f_1 \neq$ supp$f$ and $f_1 \in C_{R[G]}(R[G'])$. Let $f_3 = f - \sum_{y \in A^{(x)}} f(y)y$. Then $f_1 = f_3 + \sum_{y \in A^{(x)}}(f(y) - f(x))y$. Clearly $f_3 \in C_{R[G]}(R[G'])$, and suppf_3 = supp$f\backslash A^{(x)}$. Now we show $(f(y) - f(x)) \in N_R$, Otherwise, suppose there exists a $y \in A^{(x)}$ such that $f(y) - f(x) \notin N_R$. Let $f_1 = f_{11} + f_{12}$, where $f_{12} \in N_R[G], f_{11}(s) \notin N_R$ for any $s \in G$. Repeating the same process we obtain $A^{(y)} \subseteq$ supp$f_{11} \subseteq$ suppf_1. By Theorem 3, we have $x \in A^{(x)} = A^{(y)} \subseteq$ suppf_1. It contradicts $x \notin$ suppf_1. Therefore $f(y) - f(x) \in N_R$ for any $y \in A^{(x)}$. Then $f = f_1 + \sum_{y \in A^{(x)}} f(x)y = f_3 + \sum_{y \in A^{(x)}} f(x)y + \sum_{y \in A^{(x)}}(f(y) - f(x))y$. Repeating the same process for f_3, we obtain $f = \sum_{x \in \delta(G)} \sum_{y \in A^{(x)}} f(x)y + f_0$, where $f(x) \in C(R), f_0 \in N_R[G]$.

Corollary 5. *Let R be a ring, G a semigroup, G' the cancellation subsemigroup of $G, f \in R[G']$. Then $f \in C_{R[G]}(R[G'])$ if and only if $f = \sum_{x \in \delta(G)} \sum_{y \in A^{(x)}} f(x)y + f_0$, where $f(x) \in C_R(R), f_0 \in N_R[G]$.*

Proof. Assume $f \in C_{R[G]}(R[G'])$. Since $f \in R[G']$, supp$f \subseteq G$. Then for any $s \in$ supp$f$ and $r \in R, rsf = frs$. Comparing the coefficients of the terms of s^2 on both sides , we have $rf(s) = f(s)r$ for any $r \in R$. Thus $f(s) \in C(R)$, i.e., we needn't the condition $G' \cap C(G) \neq \emptyset$ for proving Corollary 5. Then Corollary 5 follows from Theorem 4.

§ 3. Characterization of $\triangle(G)$ and $S_G(g)$

Let G be a cancellation semigroup. Krempa in [2] defined the subset $\triangle(G)$ of G which consists of all those elements g for which there exists a finite subset $S_G(g) \subseteq G$ satisfying the following conditions:

(a) For any $x \in G$ there exists an $h \in S_G(g)$ such that $xg = hx$;

(b) For any $x \in S_G(g)$ there exists an $x \in G$ such that $xg = hx$.

Let G be a semigroup, $g \in G$. Dauns in [1] defined the left and right conjugate classes of g in G as follows:

$$\begin{aligned} lC(g) &= \{h \in G \mid \text{there exists an } x \in G \quad \text{such that } gx = xh\}; \\ rc(g) &= \{h \in G \mid \text{there exists an } x \in G \quad \text{such that } xg = hx\}. \end{aligned}$$

By definition, we have $h \in lc(g)$ if and only if $g \in rc(h)$.

We consider in this section the relationship between $\triangle(G)$ and $\delta(G), S_G(G)$ and $lc(g)(rc(g))$ and $A^{(x)}$.

Lemma 6. *Let G be a cancellation semigroup. Then the following statements are equivalent:*

(1) $g \in \triangle(G)$;

(2) $|rc(g)| < \infty$ and for any $x \in G, xg \in rc(g)x$;

(3) $|rc(g)| < \infty$ and for any $x \in G, xrc(g) = rc(g)x$;

(4) $g \in \delta(G)$;

(5) There exists a unique finite subset $A^{(g)}$ of $\delta(G)$ which satisfies

(a) $g \in A^{(g)}$;

(b) For any $x \in G, xA^{(g)} = A^{(g)}x$;

(c) $|A^{(g)}|$ is the minimal for all subsets of $\delta(G)$ satisfying the foregoing conditions (a) and (b).

Proof. (1)⇒(2). Let $g \in \triangle(G)$. By the definition of $\triangle(G)$, for any $h \in S_G(g)$, there exists an $x \in G$ such that $xg = hx$. By the definition of $rc(g)$, we have $h \in rc(g)$. Thus $S_G(g) \subseteq rc(g)$.

Conversely, for any $h \in rc(g)$ there exists an $x \in G$ such that $xg = hx$. Also by the definition of $\triangle(G)$, we have $xg \in S_G(g)x$. Then $h \in S_G(g)$. Therefore $S_G(g) \supseteq rc(g)$, i.e., $S_G(g) = rc(g)$. Consequently we have $|rc(g)| = |S_G(g)| < \infty$ and for any $x \in G, xg \in S_G(g)x = rc(g)x$.

(2)⇒(3). Suppose that $|rc(g)| < \infty$ and for any $x \in G$, $xg \in rc(g)x$. Then for any $h \in rc(g)$. there exists a $y \in G$ such that $hy = yg$. Thus $xhy = xyg = (xy)g \in rc(g)xy$. Therefore $xh \in rc(g)$ x, i.e., $xrc(g) \subseteq rc(g)x$. Since the cardinal of $xrc(g)$ is equal to that of $rc(g)x$, we obtain $xrc(g) = rc(g)x$.

(3)⇒(4). By the statement (3), we have $rc(g) \subseteq \delta(G)$. Since $g \in rc(g), g \in \delta(G)$.

(4)⇒(5). This follows from Theorem 2.

(5)⇒(1). Let $S_G(g) = A^{(g)}$. Then $|S_G(g)| < \infty$ and for any $x \in G, xg \in xA^{(g)} = A^{(g)}x = S_G(g)x$. Now we show the condition (b) of $S_G(g)$ holds. It suffices to show $A^{(g)} = rc(g)$. For any $h \in rc(g)$, there exists an $x \in G$ such that $hx = xg \in xA^{(g)} = A^{(g)}x$, i.e., $h \in A^{(g)}$. Therefore $rc(g) \subseteq A^{(g)}$. Conversely, for any $h \in rc(g)$, there exists a $g \in G$, such that $hy = yg$. Thus $xhy = xyg = (xy)g \in A^{(g)}xy$. Let $xyg = h'(xy)$, where $h' \in A^{(g)}$. By the definition of $rc(g)$, we have $h' \in rc(g)$. Since $xhy = h'xy$, we have $xh = h'x \in rc(g)x$. Therefore $xrc(g) \subseteq rc(g)x$. Since the cardinals of $xrc(g)$ and $rc(G)x$ are equal , we obtain $xrc(g) = rc(g)x$ for any $x \in G$. Since $g \in rc(g)$ and $A^{(g)}$ is unique, $rc(g) = A^{(g)}$, i.e., $S_G(g) = rc(g)$. The condition (b) of $S_G(g)$ holds and $g \in \triangle(G)$.

Corollary 7. *If $\triangle(G) \neq \emptyset$, then $\triangle(G)$ is the subsemigroup of G.*

Theorem 8. *Let G be a cancellation semigroup. Then*

(1) $\triangle(G) = \delta(G) = \cup\{A_\alpha \mid A_\alpha \subseteq G, |A_\alpha| < \infty$ and $xA_\alpha = A_\alpha x$ for any $x \in G\} = \{g \in G \mid |rc(g)| < \infty$ and $xg \in rc(g)x$ for any $x \in G\}$;

(2) If $g \in \triangle(G)$, then $S_G(g) = rc(g) = A^{(g)}$;

(3) For any $g \in \triangle(G), lc(g) = rc(g)$.

Proof. (1) and (2) follow from Lemma 6. Now we show (3). Let $\triangle'(G)$ be a subset of all elements g with the following properties: There exists a finite subset $S'_G(g)$ of G such that (a)' For any $x \in G$ there exists an $h \in S'(g)$ such that $xg = hx$;

(b)' For any $h \in S'_G(g)$ there exists an $x \in G$ such that $xg = hx$. By a process similar to that for Lemma 6, we have $\triangle(G) = \triangle'(G) = \delta(G)$ and $S'_G(g) = lc(g) = A^{(g)}$ for any $g \in \triangle(G)$. Thus $lc(g) = A^{(g)} = rc(g)$.

§ 4. Applications

Theorem 9. *Let R be a ring, G a cancellation semigroup. Then f is in the center of $R[G]$ if and only if*

$$f = \sum_{x \in \triangle(g)} \sum_{y \in A^{(x)}} f(x)y + \beta, \text{ where } f(x) \in C(R), \beta \in N_R[G].$$

We stipulate $\sum_{x \in \triangle(G)} \sum_{y \in A^{(x)}} f(x)y = 0$ if $\triangle(G) = \emptyset$.

Proof. This follows from Corollary 5 and Theorem 8.

Cororally 10 [1, Lemma 2.1]. *Let R be a ring with identity, G a cancellation monoid. Then $f \in C(R[G])$ if and only if $f = \sum_i a_i (\sum_{x \in S_G(g_i)} x)$, where $a_i \in C(R), g_i \in \triangle(G)$.*

Proof. Since G is a monoid, the identity e of G is in $\triangle(G)$. Then $\triangle(G) \neq \emptyset$. Also since R has an identity, we find $N_R = 0$. It follows from Theorem 9 that Corollary 10 holds.

Let R be a ring, G a group. In [4] , Rudin and Schneider defined the "class function": A function $f = \sum f(x)x$, where $f(x) \in R$ and $x \in G$, is called a class function if $f(xy) = f(yx)$ for any $x, y \in G$.

Corollary 11 [4, Theorem 1.3]. *Let R be a ring, G a group. Then the center of $R[G]$ consists of all forms $f = f_1 + f_2$, where $f_1 \in C(R)[G]$ is the class function and $f_2 \in N_R[G]$.*

Proof. By Theorem 9, $f \in C(R[G])$ if and only if $f = f_1 + f_2$, where $f_1 = \sum_{x \in \triangle(G)} \sum_{y \in A^{(x)}} f(x)y$, $f_2 \in N_R[G]$. Clearly, f_1 is a class function of $C(R)[G]$.

Let R be a ring with identity, G a cancellation semigroup. Consider a subset Ω of $R[G] : \Omega = \{\sum_{y \in A^{(x)}} y \mid x \in \triangle(G)\}$.

Corollary 12. *Let R be a ring with isentity, G a cancellation semigroup. Then Ω is a basis of $C(R)$-module $C(R[G])$.*

We stipultate $C(R[G]) = 0$ if $\Omega = \emptyset$.

Proof. Since $N_R = 0$ in the given conditions, Corollary 12 follows from Theorem 9.

In [5] Ponizovskii posed the Problem 9: For which semigroup rings is the supporting subsemigroup of any central idempotent finite?

Theorem 13. *Let R be a ring with $N_R = 0$, G a semigroup, G' the cancellation subsemigroup of G, f the central idempotent of $R[G]$. If $\text{supp} f \subseteq G'$ then the supporting subsemigroup <suppf> of G generated by supp f is the finite subgroup of G'.*

Proof. Since f is a central idempotent of semigroup ring $R[G']$ by using $\text{supp} f \subseteq G'$ and G' is the cancellation semigroup of G, $\langle \text{supp} f \rangle$ is the finite subgroup of G' by Theorem 2.5. in [2].

Let R be a ring , G a smigroup. The semigroup ring $R[G]$ is said to contain only trivial centers if , for any $f \in C(R[G])$, $|\text{supp} f| \leq 1$.

Theorem 14. *Let R be a ring with identity, G a cancellation semigroup.*

(1) If $\triangle(G) = \emptyset$, then $C(R[G]) = \{0\}$;

(2) If $\triangle(G) \neq \emptyset$, then $R[G]$ contains only trivial centers if and only if $\triangle(G) = C(G)$ and $|\triangle(G)| = 1$.

Proof. (1) Obvious.

(2) Suppose $R[G]$ contains only trivial centers. Then $|\Omega| = 1$ by Corollary 12; and , for any $g \in C(G), rc(g) = A^{(g)} = \{g\}$. Thus $g \in \Omega$, i.e., $|C(G)| \leq 1$. Also since $|\Omega| = 1$, there exists an $A^{(g)} \subseteq \triangle(G)$ such that $|A^{(g)}| = 1$. It implies $A^{(g)} = \{g\} \subseteq C(G)$ and $\triangle(G) = C(G) = \{g\}$.

Conversely, suppose $\triangle(G) = C(G) = \{g\}$. Then $A^{(g)} = \{g\}, \Omega = \{g\}$. By Corollary 12, $R[G]$ contains only trivial centers.

References

[1] Dauns, J., *Center of semigroup rings and conjugacy class*, Semigroup Forum, 38 (1989), 355–364.

[2] Krempa, J., *Special elements of semigroup ring*, Bull. Acad. Pol. Sci . Ser. Math., 28 (1980), 17–23.

[3] Passman, D.S., *The Algebraic structure of group ring,*, Wiley, 1977.

[4] Rudin, W. and H.Schneider, *Idempotents in group rings*, Duke Math. J., 31 (1964), 385-602.

[5] Ponizovskii, J.S., *Semigroup rings*, Semigroup Forum, 36 (1987), 1–46.

Proceedings of the Asian Mathematical
Conference, 1990

Groups Whose Proper Quotients are Finite-by-nilpotent*

ZHANG ZHIRANG
Chengdu Institute of Meteorology
Chengdu 610041, Sichuan, China

§ 1. Introduction

A group G is said to be an FN_c-group if the $(c+1)$th term $\gamma_{c+1}G$ of its lower central series is finite. It is easy to see that a group G is an FN_c-group if and only if it is finite-by-nilpotent of class $\leq c$ (or finite-by-N_c). A group G is called a JNFN_c-group if all its proper quotients are FN_c-groups, but G itself is not. The structure of JNFN_c-groups with $c = 1$ (JNFA-groups) has been described in [4]. In this paper we study JNFN_c-groups for an arbitrary c.

As often happens in the theory of just non-P-groups, with P being some group theoretical property, only JNFN_c-groups with non-trivial Fitting subgroup are studied, and they are naturally divided into those with non-trivial centre and those with trivial centre. We give a complete description of groups of both types in Sections 2 and 3.

It turns out that JNFN_c-groups with non-trivial centre are just the torsion-free nilpotent groups G of class $c+1$ such that $\gamma_{c+1}G$ is infinite cyclic and $Z(G)$ is a subgroup of rationals (see (2.1)). In Section 3 we deal with JNFN_c-groups with trivial centre and non-trivial Fitting subgroup A. It is shown that A is abelian and then A is a faithful just infinite module for the non-trivial FN_c-group $Q =: G/A$. The investigation also shows that it is significantly harder to describe JNFN_c-groups with $c \geq 2$ than with $c = 1$ in [4]; some interesting but complicated problems about groups, modules and cohomology and many new techniques are involved (see especially (3.2), (3.3), (3.6), (4.1), (4.7), (4.8) etc.).

In the final section we consider faithful just infinite modules over nilpotent FN_c-groups. For this purpose a well-known theorem of P. Hall (see [2]) is extended in the following form : If G is an FN_c-group, then $G/C_G(\gamma_c G)$ is residually finite and of finite exponent. Moreover, $G/Z_{2c-2}(G)$ is of finite exponent, provided $c \geq 2$ (see (4.1)). By using this proposition, necessary and sufficient conditions are obtained for a non-trivial nilpotent FN_c-group Q to have a faithful just infinite module or a simple faithful just infinite module. These results are markedly different from those in [4, Section 4].

---

* This project was suggested by Professor Derek J.S. Robinson while the author was a Vistiting Scholar at the university of Illinois in Urbana-Champaign, 1986-1987. To him I express my gratitude for many suggestions and comments.

§ 2. JNFN$_c$-groups with Non-trivial Centre

Theorem 2.1. *(i) If G is a $JNFN_c$-group with non-trivial centre, then G is torsion-free nilpotent of class $c+1$. Furthermore, $\gamma_{c+1}G$ is infinite cyclic and $Z(G)$ is isomorphic with a subgroup of the additive group of rational numbers.*

(ii) Conversely, a group G with the structure described in (i) is a $JNFN_c$-group with non-trivial centre.

Via Theorem (2.1) we are able to describe locally nilpotent JNFN$_c$-groups.

Theorem 2.2. *If G is a locally nilpotent $JNFN_c$-group, then G is nilpotent and so its structure is described in (2.1).*

A natural question arising from (2.1) is : What type of groups has the structure described in (i) of (2.1)? This appears to be a hard question in general, but we can give examples of JNFN$_c$-groups with non-trivial centre.

Example. Let G be $U(c+2, \mathbb{Z})$, the group of all $(c+2)\times(c+2)$ (upper) unitriangular matrices over $\mathbb{Z}$, that is, matrices with 1 on the diagonal and 0 below it. Then G is a torsion free nilpotent group of class $c+1$, and $\gamma_{c+1}G = Z(G) \cong \mathbb{Z}$. By (2.1), G is a JNFN$_c$-group with non-trivial centre.

§ 3. JNFN$_c$-groups with Non-trivial Fitting Subgroup and Trivial Centre

Proposition 3.1. *The Fitting subgroup of a $JNFN_c$-group with trivial centre is abelian. In the proof, the following two technical lemmas are made use of.*

Lemma 3.2. *Let F be a normal subgroup of group G. Write $B_i = \gamma_i G \cap F$. Then for $m \geq 1$,*

$$[F',{}_mG] \leq \prod_{\substack{t+s=m+2\\ t,s\geq 1}} [B_t, B_s].$$

Lemma 3.3. *Suppose that A and B are G-modules which have series of finite length whose factors have finite G-orbits. Then $A\otimes B$ has a series of this type.*

Our next object is to examine the structure of the Fitting subgroup of a JNFN$_c$-group, regarded as a module via conjugation. Here just infinite modules are the infinite modules whose proper quotients are all finite.

Theorem 3.4. *(i) Let G be a $JNFN_c$-group with trivial centre and non-trivial Fitting subgroups A. Then every non-trivial G-admissible subgroup A_0 of A is a faithful just infinite module for the non-trivial FN_c-group $Q : G/A$.*

(ii) Conversely, let A be a faithful just infinite module for a non-trivial FN_c-group Q. Then every extension of A by Q inducing the given module structure is a $JNFN_c$-group with trivial centre and Fitting subgroup equal to A.

Corollary 3.5. *Every non-trivial submodule of a faithful just infinite module over a finite-by-N_c group is itself faithful and just infinite.*

Since the underlying additive group of a just infinite module is either torsion-free or an elementary p-group for some prime p, we shall say that a JNFN$_c$-group with trivial centre and non-trivial Fitting subgroup has characteristic $p(\geq 0)$ if its Fitting subgroup has characteristic p.

Next, we consider the structure of a JNFN$_c$-group regarded as an extension of its Fitting subgroup. By using cohomology (see [3, Lemma 2.12]), we can prove

Theorem 3.6. *Let G be a $JNFN_c$-group with Fitting subgroup A. Then there is a subgroup X such that $X \cap A = 1$ and $|G : XA|$ is finite.*

Definition. Let G be a $JNFN_c$-group with trivial centre and non-trivial Fitting subgroup A; put $Q = G/A$.We shall call G a primitive JNFN$_c$-group if the following hold:

(i) G splits over A;

(ii) Q is nilpotent of class $\leq 2c$;

(iii) If Q is abelian-by-finite, then it is abelian.

An infinite group is called a just infinite group if all its proper quotients are finite. Just as in [4, Theorem (3.7)] we have

Theorem 3.7. *Let G be a $JNFN_c$-group with trivial centre and non-trivial Fitting subgroup. Assume that G is not just infinite. Then G embeds in a wreath product $W = \overline{G}$ wr U where (i) $\overline{G}$ is a primitive $JNFN_c$-group and U is finite;*

(ii) $\overline{G} \cong H/K$ where H is subnormal in G with defect at most 2;

(iii) The Fitting subgroup of $\overline{G}$ is a $\overline{G}$ -quotient of that of G;

(iv) The characteristics of G and $\overline{G}$ are equal.

In view of (3.7) we shall concentrate in Section 4 on faithful just infinite modules for nilpotent FN$_c$-groups.

Remark. Let G be a JNFN$_c$-group with trivial centre. Then G is also a JNFN$_{c+1}$-group with trivial centre. Hence JNFA-groups with trivial centre are also JNFN$_c$-groups of this type for any positive integer c.

§ 4. Faithful Just Infinite Modules for Finite-by-N$_c$ Groups

The main questions here are : Which (non-trivial) finite-by-N$_c$ groups possess faithful just infinite modules and how can such modules be constructed? If we are content to classify the primitive JNFN$_c$-groups, it is enough to solve these problems for nilpotent groups; this turns out to be possible on the basis of the following results on finite-by-N$_c$ groups.

Proposition 4.1. *If G is a finite-by-N_c group, then $G/C_G(\gamma_c G)$ is residually finite and of finite exponent. Moreover, $G/Z_{2c-2}(G)$ is of finite exponent, provided $c \geq 2$.*

(4.1) can be viewed as a generalization of the theorem of P. Hall on finite-by-nilpotent groups. Its proof is a variety of that of [5, 14.5.3]. By this and a similar argument to that of [4, Proposition (4.5)] we have

Proposition 4.2. *Let Q be a finite-by-N_c group. Assume that the torsion-free rank of $Z_{2c-2}(Q)$ is finite and positive, and $Z(Q)$ is not torsion when $p > 0$. If Q admits of a faithful just infinite module, then Q is finitely generated.*

This will play an essential role in the investigation of necessary and sufficient conditions for a nilpotent finite-by-N$_c$ group to have a faithful just infinite module. Our main results on this subject are

Theorem 4.3. *Let Q be a non-trivial nilpotent finite-by-N_c group with centre K.*

Then Q has a faithful just infinite module of characterisitic $p \geq 0$ if and only if the torsion-subgroup of K is a locally cyclic p'-group, and one of the following is true:

(i) The torsion-free rank of $Z_{2c-2}(Q)$ is infinite,

(ii) The torsion-free rank of $Z_{2c-2}(Q)$ is finite and positive, K is torsion and $p > 0$,

(iii) Q is finitely generated infinite and finite-by-abelian,

(iv) Q is an infinite torsion group and $p > 0$,

(v) Q is finite and $p = 0$.

Theorem 4.4. *Let Q be a non-trivial nilpotent finite-by-N_c group with centre K. Then Q has a faithful just infinite module of characteristic $p \geq 0$ which is simple if and only if the torsion-subgroup of K is a locally cyclic p'-group, and one of the following holds:*

(i) The torsion-free rank of $Z_{2c-2}(Q)$ is infinite;

(ii) The torsion-free rank of $Z_{2c-2}(Q)$ is finite and positive, K is torsion and $p > 0$,

(iii) Q is an infinite torsion group and $p > 0$.

The proofs depend on two propositions, the second in turn resting on [6], 2.8 and 5.1.

Proposition 4.5. *Let Q be a nilpotent group with centre K. If Q has infinite torsion-free rank and the torsion-subgroup of K is a locally cyclic p'-group ($p \geq 0$), then Q has a faithful just infinite module of characteristic p that is simple.*

Proposition 4.6. *Let Q be a finitely generated nilpotent group. If Q has a faithful just infinite module, then it is finite-by-abelian.*

In the proofs of these propositions two lemmas are used, the first one being similar to [1, Lemma 3.1].

Lemma 4.7. *Let G be a nilpotent group of infinite torsion-free rank. Then G has an abelian subgroup of infinite torsion-free rank.*

Lemma 4.8, *Let G be a finitely generated nilpotent group. If G is abelian-by-finite, then it is finite-by-abelian.*

Now Theorems (4.3) and (4.4) follow from Propositions (4.2), (4.5) and (4.6) , and the results in [4, Section 4]. Here the necessity of the conditions in (4.3) and (4.4) is essentially a consequence of (4.2) and (4.6); and the sufficiency of the conditions (i) and (ii) in (4.3) and (4.4) follows from (4.5) ,which is the crux of the matter. The rest of the results can be obtained via [4], (4.1), (4.2), (4.7) and (4.12).

References

[1] K. A. Brown, *Primitive group rings of soluble groups,* Arch. Math., 36 (1981), 404-413.

[2] P. Hall, *Finite-by-nilpotent groups,* Proc. Cambridge Philos. Soc., 52 (1956), 611–616.

[3] D. J. S. Robinson, *The vanishing of certain homology and cohomology groups,* J. Pure Appl. Algebra, 7 (1976), 145–167.

[4] D. J. S. Robinson and Z.R. Zhang, *Groups whose proper quotients have finite derived subgroups* , J. Algebra, (2) 118 (1988) , 346–368.

[5] D. J. S. Robinson, *A Course in the Theory of Groups,* Springer-Verlag, New York, 1982.

[6] D. J. S. Robinson and J. S. Wilson, *Soluble groups with many polycyclic quotients,* Proc. London Math. Soc., (3) 48 (1984), 193–229.

Proceedings of the Asian Mathematical
Conference, 1990

Expansion of Self-similar Functions*

ZHENG WEIXING
Mathematics Department
Nanjing University
Nanjing 210008, China

§1. Introduction

Let K be a p-series field[3], O its ring of integers, p its prime element, $|\cdot|$ the non-archimedean norm in $K, \mathcal{P}^k$ the sphere $\{x \in K : |x| \leq p^k\}$ for $k \in Z$, and $\{\chi_j\}_{j\in P}$ the complete orthonormal system on O. Denote $O/\mathcal{P} = \{\varepsilon_0, \varepsilon_1, \cdots, \varepsilon_{p-1}\} \cong Z_p$; and we will not distinguish the two sets Z_p and $\{\varepsilon_0, \varepsilon_1, \cdots \varepsilon_{p-1}\}$. For $x \in K$, we have $x = \sum_{\nu=-n}^{\infty} x_\nu p^\nu$, the formal series of x, where $x_\nu \in Z_p$. If we set $k = k_{-r}p^r + \cdots + k_{-1}p + k_0$ as a p-adic expression of $k \in P$, where $k_{-r} \neq 0$ and r depends on k, then

$$\chi_k(x) = \exp(2\pi i p^{-1}(k_0 x_0 + k_{-1}x_1 + \cdots + k_{-r}x_r)),$$

so that $\chi_k(x)$ is trivial on $\mathcal{P}^{r+1}$, and $\chi_0(x) = 1$ for any $x \in K$. Besides, we have

$$\chi_k(x) = \chi_{k-r}(x_r) \cdots \chi_{k-1}(x_1)\chi_{k_0}(x_0). \tag{1}$$

Assume that p is an odd prime number, and write $p = 2s + 1$ for some $s \in N$. A function f defined on O is said to be self-similar on O with parameters $A_0, A_1, \cdots, A_{2s}$, if for any coset $x_0 + \mathcal{P}, x_0 \neq \varepsilon_s$, there are constants α_{x_0} depending on x_0 and $t \in (0, 1)$ such that the following two conditions hold:

(i) $f(x_0 + z) = \alpha_{x_0} + tf(zp^{-1}),\ z \in \mathcal{P}$,

(ii) f takes constants $A_0, A_1, \cdots, A_{2s}$ on the cosets

$$\varepsilon_0 + \varepsilon_s p + \mathcal{P}^2, \cdots, \varepsilon_{s-1} + \varepsilon_s p + \mathcal{P}^2, \quad \varepsilon_s + \mathcal{P}, \varepsilon_{s+1} + \varepsilon_s p + \mathcal{P}^2, \cdots, \varepsilon_{2s} + \varepsilon_s p + \mathcal{P}^2,$$

respectively. If, moreover, the relation

$$\sum_{\gamma=0}^{2s} A_\gamma = (2s + 1)A_s \tag{2}$$

holds, we say that f is harmonic on O. A non-constant function f on O is called a p-adic Cantor function, if it is harmonic and continuous on O in the sense of K^+-topology.

---

* Supported by the National Natural Science Foundation of China.

As for the case of real field R, the definition of self-similarity [2] should be modified as follows. A function f defined on $[0,1)$ is called self-similar with parameters $A_0, A_1, \cdots, A_{2s}$, if there are constants α_γ and $t \in (0,1)$ such that the two conditions hold:

(i′) $f(x) = \alpha_\gamma + tf(p(x - x_\gamma))$, for $x \in [\gamma/p, (\gamma+1)/p], \gamma = 0, 1, \cdots, 2s$, where $x = \gamma/p, \gamma = 0, 1, \cdots, 2s$;

(ii′) f takes constants $A_0, A_1, \cdots, A_{2s}$ on the intervals

$$\left[\frac{1}{p^2}, \frac{2}{p^2}\right), \quad \frac{1}{p} + \left[\frac{1}{p^2}, \frac{2}{p^2}\right), \cdots, \frac{s-1}{p} + \left[\frac{1}{p^2}, \frac{2}{p^2}\right), \left[\frac{s}{p}, \frac{s+1}{p}\right),$$
$$\frac{s+1}{p} + \left[\frac{1}{p^2}, \frac{2}{p^2}\right), \cdots, \frac{p-1}{p} + \left[\frac{1}{p^2}, \frac{2}{p^2}\right),$$

respectively. The concepts of harmonic and Cantor function may be defined similarly.

For $\gamma \in \{0, 1, \cdots, 2s\}$, let

$$\Lambda(\gamma) = \begin{cases} -\omega^{-s\gamma}, & \gamma \neq 0, \\ 2s, & \gamma = 0, \end{cases} \tag{3}$$

where $\omega = \exp(2\pi i/p)$. The following equalities are easily verified:

$$\sum_{\gamma=0}^{2s} \Lambda(\gamma)\chi_\gamma(\varepsilon_\mu) = p(1 - \delta_{s,\mu}), \quad \delta_{s,\mu} : \text{Kronecker symbol}, \tag{4}$$

$$\sum_{\gamma=1}^{2s} \left(\sum_{l=0}^{2s} A_l \omega^{-l\gamma} \right) \chi_\gamma(\varepsilon_\mu) = p(A_{\varepsilon_\mu} - A_s). \tag{5}$$

§2. Expansion of Self-similar Functions on O

Let $k \in N, k = k_{-r}p^r + \cdots + k_0$ its p-adic expression, where $k_{-r} \neq 0$ and r depends on k.

Theorem 1. *Let f be a function on O which is periodic with period p^{-1}. The function f on O is harmonic iff its Fourier coefficients c_k with respect to $\{\chi_k\}$ satisfy the following recurrence formula:*

$$c_h = p^{-1}\left\{ \sum_{\gamma=0}^{2s} A_\gamma \omega^{-\gamma k_0} - tA_s\Lambda(k_0) \right\} \delta_{0,[k/p]} + tp^{-1}\Lambda(k_0)c_{[k/p]}, \qquad k \in P, \tag{6}$$

where $\delta_{\mu,\gamma}$ is the Kronecker symbol, and $[a]$ the integral part of a.

Proof. Assume that f is harmonic on O. Then for $k \in P$,

$$\begin{aligned} c_k &= \int_O f(x)\overline{\chi}_k(x)dx \\ &= A_s \int_{\varepsilon_s + P} \overline{\chi}_k(x)dx + \sum_{\substack{\nu=0 \\ (\nu\neq s)}}^{2s} \int_{\varepsilon_\gamma + P} (\alpha_{\varepsilon_\gamma} + tf((x - \varepsilon_\gamma)p^{-1}))\overline{\chi}_k(x)dx. \end{aligned} \tag{7}$$

Since

$$\int_{\varepsilon_\gamma+P} f((x-\varepsilon_\gamma)p^{-1})\overline{\chi}_k(x)dx = p^{-1}\omega^{-\gamma k_0}c_{[k/p]},$$

$$\alpha_\gamma = A_\gamma - tA_s, \gamma = 0,1,\cdots,s-1,s+1,\cdots,2s,$$

we have

$$c_k = p^{-1}\left\{A_s\omega^{-sk_0} + \sum_{\substack{\gamma=0\\(\gamma\neq s)}}^{2s}\alpha_p\omega^{-\gamma k_0}\right\}\delta_{0,[k/p]} + \sum_{\substack{\gamma=0\\(\gamma\neq s)}}^{2s}p^{-1}t\omega^{-\gamma k_0}c_{[k/p]}, \tag{8}$$

and consequently (5).

Conversely, if (6) holds, then so does (7). Let ξ_E denote the characteristic function of E. By (7) we can get the expression of c_k:

$$c_k = \int_O\left\{A_s\xi_{\varepsilon_s+P}(x) + \sum_{\substack{\nu=0\\(\nu\neq s)}}^{2s}(\alpha_\nu + tf(x-\varepsilon_\nu)p^{-1})\xi_{\varepsilon_\nu+P}(x)\right\}\overline{\chi}_k(x)dx. \tag{9}$$

By induction we can show from (6) that

$$c_k = t^r p^{-r-1}\Lambda(k_0)\cdots\Lambda(k_{-r+1})\left\{\sum_{\gamma=0}^{2s}A_\gamma\Omega^{-\gamma k_{-r}}\right\},\quad k\in N\quad (k = k_{-r}p^r+\cdots+k_0).$$

$$c_0 = \left\{\sum_{\gamma=0}^{2s}A_\gamma - 2stA_s\right\}/(p-2st) = A_s. \tag{10}$$

Thus $\sum_k |c_k|^2 < \infty$; and according to the Riesz-Fischer theorem we conclude that

$$f(x) = A_s\xi_{\varepsilon_s+P}(x) + \sum_{\substack{\gamma=0\\(\gamma\neq s)}}^{2s}(\alpha_\gamma + tf((x-\varepsilon_\gamma)p^{-1})\xi_{\varepsilon_\gamma+P}(x),\quad \text{a.e.},$$

that is, the relation (i) holds.

The formula (10) leads to the following theorem.

Theorem 2. *The p-adic Cantor function θ on O has the following expansion:*

$$\theta(x) = A_s + \sum_{k=1}^{\infty} t^r p^{-r-1}\Lambda(k_0)\cdots\Lambda(k_{-r+1})\left\{\sum_{\nu=0}^{2s}A_\nu\Omega^{-\nu k_{-r}}\right\}\chi_k(x), \tag{11}$$

where $k = k_{-r}p^r+\cdots+k_0$ is the p-adic expression of k, and the constants A_γ are (cf. [4]):

$$\begin{aligned} &A_s = 1/2,\\ &A_j = (2j+1)/(4s),\quad j=0,1,\cdots,s-1,\\ &A_j = (2j-1)/(4s),\quad j=s+1,\cdots,2s,\\ &t = 1/(2s). \end{aligned} \tag{12}$$

We note that Theorem 2 can be proved directly.

§3. Expansion of Self-similar Functions on [0, 1)

Let $a(k)$ be the Fourier coefficients of f with respect to the system $\{e^{2\pi ik}\}_{k\in Z}$:

$$a(k) = \int_0^1 f(x)e^{-2\pi ikx}dx, \quad k \in Z. \tag{13}$$

In the sequel we always extend $a(k)$ to the entire function $a(z) = \int_0^1 f(x)e^{-2\pi izx}dx$ of exponential type 2π, from Z to C. Then analogous to Theorem 1, we have

Theorem 3. *A function f is harmonic on [0, 1) iff*

$$\begin{aligned} a(k) = \sum_{\nu=0}^{2s} \quad & \alpha_\nu \gamma^\nu (1-\gamma)/(2\pi ik) \\ & + tp^{-1}(1+\gamma^{s+1})(1-\gamma^s)(1-\gamma)^{-1}a(k/p) \quad (\gamma = e^{-2\pi ik/p}). \end{aligned} \tag{14}$$

If we set $(\gamma(z) = e^{-2\pi iz/p})$

$$\begin{aligned} a(z) = \quad & \frac{1}{2\pi iz}\left\{\frac{\gamma(z)(1-\gamma(z)^s)(1+\gamma(z)^{s+1})}{2s(1-\gamma(z))} - \gamma(z)^{2s+1}\right\} \\ & + \frac{t(1-\gamma(z)^s)(1+\gamma(z)^{s+1})}{p(1-\gamma(z))} a\left(\frac{z}{p}\right), \end{aligned} \tag{15}$$

then for Cantor function θ on $[0,1)$ the Fourier coefficient $a(k)$ is $a(z)$ in (15) when $z = k$. Let

$$B_1(z) = (1-\gamma(z)^s)(1+\gamma(z)^{s+1})(1-\gamma(z))^{-1}, \tag{16}$$

$$A_1(z) = (2s)^{-1}\gamma(z)B_1(z) - \gamma(z)^{2s+1} \tag{17}$$

and

$$A_{j+1}(z) = A_j(z/p), \quad B_{j+1}(z) = B_j(z/p), \quad j \in N. \tag{18}$$

Then by induction, we have for any $j \in N$,

$$\begin{aligned} a(z) = \quad & (2\pi iz)^{-1}\{A_1(z) + tB_1(z)A_2(z) + \cdots + t^j B_1(z)\cdots B_j(z)A_{j+1}(z)\} \\ & + (tp)^{j+1}B_1(z)\cdots B_{j+1}(z)a(zp^{-j-1}). \end{aligned} \tag{19}$$

It is easy to show the estimates

$$|B_j(z)| \le 2s, \quad z \in R,$$

$$|A_j(z)| \le 2, \quad z \in R,$$

$$|tB_1(z)\cdots B_j(z)A_{j+1}(z)| \le 2\left|\cos\frac{\pi(s+1)z}{p}\cdots\cos\frac{\pi(s+1)z}{p^j}\right|,$$

so that

$$a(z) = (2\pi i z)^{-1} \lim_{j\to\infty} \{A_1(z) + tB_1(z)A_2(z) + \cdots + t^j B_1(z)\cdots B_j(z)A_{j+1}(z)\}.$$

This eventually reduces to

$$a(z) = \frac{1}{2\pi i z}\left\{-e^{-2\pi i z} + e^{-\pi i z}\prod_{j=1}^{\infty}\cos\frac{\pi(s+1)z}{p^j}\cdot\frac{\sin \pi s z/p^j}{s\cdot\sin \pi z/p^j}\right\} \tag{20}$$

The infinite product in (20) is absolutely convergent for any $z \in C$. In particular, we get for $z \in Z - \{0\}$,

$$a(n) = \frac{1}{2\pi i n}\left\{-1 + (-1)^n\prod_{j=1}^{\infty}\cos\frac{\pi(s+1)n}{p^j}\cdot\frac{\sin \pi s n/p^j}{s\cdot\sin \pi n/p^j}\right\} \tag{21}$$

and

$$a(0) = 1/2.$$

Theorem 4. *The Cantor function θ on $[0, 1)$ has the following expansion:*

$$\theta(x) = \frac{1}{2} + \sum_{n=1}^{\infty}\frac{-1}{\pi n}\left\{1 - (-1)^n\prod_{j=1}^{\infty}\cos\frac{\pi(s+1)n}{p^j}\cdot\frac{\sin \pi s n/p^j}{s\cdot\sin \pi n/p^j}\right\}\sin 2\pi n x. \tag{22}$$

For example, if $p = 3$, then the triadic Cantor function takes the following form:

$$\theta_3(x) = \frac{1}{2} + \sum_{n=1}^{\infty}\frac{-1}{\pi n}\left\{1 - (-1)^n\prod_{j=1}^{\infty}\cos\frac{2\pi n}{p^j}\right\}\sin 2\pi n x. \tag{23}$$

Comparing the two expansions (11) and (22), one can find the latter is more complicated to apply than the former. However, as an application, we can obtain, by integration by parts,

$$\int_0^1 e^{-2\pi i n x}d\theta(x) = (-1)^n\prod_{j=1}^{\infty}\frac{\cos(\pi(s+1)n/p^j)\sin(\pi s n/p^j)}{s\cdot\sin \pi n/p^j} \equiv \lambda_n. \tag{24}$$

Since when $n = p^\nu, p > 2$ and $\nu \in N, \lambda_n = (-1)^{(s+1)\nu}\lambda_1 \neq 0$, we see that there exist functions which are monotonic and continuous (Cantor functions) such that their Fourier-Stieltjes coefficients do not converge to zero as $n \to \infty$. For $p = 3$, cf. [1].

In the last part of this paper we give a remark to Theorem 4 on the calculation of the coefficients $a(k)$. For the sake of simplicity we only consider the case $p = 3$. We have from(15),

$$a(z) = \frac{i}{4\pi z}\{e^{-2\pi i z} - e^{-2\pi i z/3}\} + \frac{1}{6}(1 + e^{-4\pi i 2/3})a(z/3). \tag{25}$$

If we set $b(z) = za(z)$, then

$$b(z) = \frac{i}{4\pi}\{e^{-2\pi i z} - e^{-2\pi i z/3}\} + \frac{1}{2}(1 + e^{-4\pi i z/3})b(z/3). \tag{26}$$

Hence for $n \in P$,

$$\begin{aligned} b^{(n)}(z) = \; & \frac{i}{4\pi}\left\{(-2\pi i)^n e^{-2\pi i z} - \left(\frac{-2\pi i}{3}\right)^n e^{-2\pi i z/3}\right\} + \frac{1}{2}\left\{\left(-\frac{4\pi i}{3}\right)^n e^{-4\pi i z/3} b\left(\frac{z}{3}\right)\right. \\ & + \binom{n}{1}\left(\frac{-4\pi i}{3}\right)^{n-1} e^{-4\pi i z/3}\cdot\frac{1}{3}b'\left(\frac{z}{3}\right) + \cdots + \binom{n}{n-1}\left(\frac{-4\pi i}{3}\right)^1\left(\frac{1}{3}\right)^{n-1} b^{(n-1)}\left(\frac{z}{3}\right) \\ & \left. + (1 + e^{-4\pi i z/3})\left(\frac{1}{3}\right)^n b^{(n)}\left(\frac{z}{3}\right)\right\}, \end{aligned} \tag{27}$$

and consequently,

$$\begin{aligned} b^{(n)}(0) = \; & \frac{i}{4\pi}\{(-2\pi i)^n(1 - \frac{1}{3^n})\} + \frac{1}{2}\frac{1}{3^n}\{(-4\pi i)^n b(0) + \binom{n}{1}(-4\pi i)^{n-1} b'(0) \\ & + \binom{n}{2}(-4\pi i)^{n-2} + \cdots + \binom{n}{n-1}(-4\pi i) b^{(n-1)}(0) + 2b^{(n)}(0)\}. \end{aligned} \tag{28}$$

From (28) we can calculate the value of $b^{(n)}(0)$ for every $n \in P$. For example, the first four of $b^{(n)}(0)$ are:

$$b(0) = 0, \quad b'(0) = 1/2, \quad b''(0) = -5i/4, \quad b'''(0) = -11/4.$$

Therefore, we have

$$\begin{aligned} b(z) &= \frac{1}{2}z + \frac{1}{2!}(-\frac{5}{4}\pi i)z^2 + \frac{1}{3!}(-\frac{11\pi^2}{4})z^3 + \cdots, \\ a(z) &= \frac{1}{2} - \frac{5}{8}\pi i z - \frac{11\pi^2}{24}z^2 + \cdots, \end{aligned} \tag{29}$$

and $a(k)$ can be found by (29) with k replaced by z.

References

[1] E. Hille & J.D. Tamarkin, *Remarks on a known example of a monotone continuous function*, Amer. Math. Monthly, 36(1929), 255–264.

[2] B.B. Mandelbrot, *The fractal geometry of nature*, W.H. Freeman, San francisco, 1982.

[3] M.H. Taibleson, *Fourier analysis on local fields*, Math. Notes, No. 15, Princeton University Press, 1975.

[4] W.X. Zheng, *On p-adic Cantor functions*, to appear.

Proceedings of the Asian Mathematical
Conference, 1990

Locally Triangulable Spaces and Quasi-horseshoe Maps*

ZHOU ZUOLING
Department of Computer Science
Zhongshan University
Guangzhou 510275, China

§ I. Introduction and Main Result

It is well known that the compact topological manifold and finitely triangulable space are two important objects of study in topology. Unfortunately, there is no relations of inclusion between them. That is, a compact topological manifold is not always triangulable and conversely and a triangulable space is not necessarily a manifold, either. This is inconvenient for discussing self-map, because the results obtained on a compact topological manifold can not automatically be applied on a triangulable space, and vice versa. We hope to unify them in the sense of dynamics. For this, we introduce the notion of locally triangulable space. We shall form a class of self-maps, so called the quasi-horseshoe maps, and then use them and the approach of algebraic topology to discuss the discrete dynamics.

Let X be a compact metric space and denote by $C^0(X)$ the set of all continuous maps on X with the C^0-topology. Some notations and terminologies in [1–4] are used. The main result of this paper is the following:

Theorem. *Let X be a compact metrisable locally triangulable space (for the definition, see the next section). Then the following conclusions hold.*

(1) the entropy function ent: $C^0(X) \to R^+ \cup \{+\infty\}$ is surjective,

(2) the set $\{f \in C^0(X) \mid ent(f) > C\}$ contains an open everywhere dense subset of $C^0(X)$ for all $C > 0$,

(3) the set $\{f \in C^0(X) \mid f$ is chaotic in the Li-Yorke's sense modified$\}$ is dense everywhere in $C^0(X)$,

(4) ent is discontinuous at $f \in C^0(X)$, if ent $(f) < +\infty$,

(5) the set $\{f \in C^0(X) \mid$ ent is continuous at $f\}$ is a Baire residual set in $C^0(X)$,

(6) the set $\{f \in C^0(X) \mid$ ent $(f) = +\infty\}$ is a Barie residual set in $C^0(X)$,

(7) the entropy conjecture holds on an open everywhere dense subset of $C^0(X)$, if X is a compact manifold or a triangulable space.

For the entropy function and the entropy conjecture, see [5–6].

---

* Partially supported by National Natural Science Foundation of China and The Foundation of Zhongshan University Advanced Research Centre.

§ II. The Definitions and the Auxiliary Propositions

Definition 1. A topological space X is called locally triangulable, if for each $x \in X$ there is a neighborhood $V(x)$ of x such that $\overline{V(x)}$ is finitely triangulable or homeomorphic to a compact polyhedron.

Obviously, the manifolds and the triangulable space are locally triangulable. Throughgout this paper, let (X, d) be a locally triangulable compact metric space without any isolated points.

The C^0-topology of $C^0(X)$ is given by the following metric:

$$\rho(f,g) = \sup\{d(f(x), g(x)), \forall x \in X\}, \quad \forall f, g \in C^0(X).$$

Let $f \in C^0(X)$ and $\varepsilon > 0$. Set

$$\begin{aligned} M(f,\varepsilon) &= \{g \in C^0(X) \mid \rho(f,g) < \varepsilon\}, \\ \text{gra}(f) &= \{(x,y) \in X \times X \mid y = f(x)\} \end{aligned}$$

and

$$U(f,\varepsilon) = \{(x,y) \in X \times X \mid d(f(x), y) < \varepsilon\}.$$

Proposition 1. *Let $f, g \in C^0(X)$ and $\varepsilon > 0$. Then*

$$g \in N(f,\varepsilon) \quad \text{iff} \quad \text{gra}(f) \subset U(f,\varepsilon).$$

Proposition 2. *The set $\{f \in C^0(X) \mid P(f) \neq \emptyset\}$ is everywhere dense in $C^0(X)$.*

The proofs are simple and omitted.

Denote by $B^n(r)$ the n-dimensional solid sphere ($n \geq 1$) with the radius $r > 0$ and by $S^{n-1}(r)$ and $B^n(r)$ the boundary and the interior of $B^n(r)$, respectively. For $r = 1, r$ will be omitted.

Proposition 3. *Let $n > 0$ and $h : B^n \to B^n$ be continuous such that $h \mid S^{n-1} = id$. Then $h(B^n) = B^n$.*

This is a simple fact in algebraic topology (see [7]).

Proposition 4. *Let $n \geq 1$ and $0 < r_4 < r_3 < r_2 < r_1 < 1$. Also suppose that $h : B_n(r_2) - B^n(r_3) \to B^n$ satisfies $h(S^{n-1}(r_2)) \subset B^n - B^n(r_1)$ and $h(S^{n-1}(r_3)) \subset B^n(r_4)$, and $g : S^{n-1} \cup S^{n-1}(r_2) \to B^n - B^n(r_1)$ has a continuous extension*

$$G : B^n - B^n(r_2) \to B^n - B^n(r_1),$$

where $g \mid S^{n-1} = id$ and $g \mid S^{n-1}(r_2) = h \mid S^{n-1}(r_2)$. Then $h(B^n(r_2) - B^n(r_3)) \supset B^n(r_1) - B^n(r_4)$.

Proof. Define $f : B^n \to B^n$ such that

$$f(x) = \begin{cases} h(x), & \text{if } x \in B^n(r_2) - B^n(r_3), \\ th(g), & \text{if } x = tq, \quad q \in S^{m-1}(r_3), \quad t \in I, \\ G(x), & \text{if } x \in B^n - B^n(r_2). \end{cases}$$

Then f is well defined and continuous. Clearly, $f(x) = x$ for all $x \in S^{n-1}$. By Proposition 3, we have $f(B^n) = B^n$. From the above definition, it is easy to see that $f(B^n(r_3)) \subset B^n(r_4)$

and $f(B^n - B^n(r_2)) \subset B^n - B^n(r_1)$. Hence, we have $f(B^n(r_2) - B^n(r_3)) = h(B^n(r_2) - B^n(r_3)) \supset B^n(r_1) - B^n(r_4)$.

Next, let $N > 0, n > 0$ and $0 < r < 1$. Suppose $h : B^n(r) \to B^n$ be continuous.

Definition 2. Say that h is an N-type quasi-horseshoe map or h has N-type quasi-horseshoe structure, if

(1) there exist $r > r_0 > r_1 > \cdots > r_{N+1} > 0$ such that

$$\begin{cases} h(S^{n-1}(r_i)) \subset B^n - B^n(r), & \text{if } 0 \le i \le N \text{ and } i \equiv 0 \pmod 2, \\ h(S^{n-1}(r_i)) \subset B^n(r_{N+1}), & \text{if } 0 \le i \le N \text{ and } i \equiv 1 \pmod 2. \end{cases}$$

(2) for $0 \le i \le N$ and $i \equiv 0 \pmod 2$, the map

$$h_i : S^{n-1} \cup S^{n-1}(r_i) \to B^n - B^n(r)$$

has a continuous extension

$$H_i : B^n - B^n(r_i) \to B^n - B^n(r),$$

where $h_i \mid S^{n-1} = id$ and $h_i\, |S^{n-1}(r_i) = h|\, S^{n-1}(r_i)$.

Proposition 5. *The N-type quasi-horseshoe structure is preserved under the C^0-small perturbation. Furthermore, let $h : B^n(r) \to B^n$ be an N-type quasi-horseshoe map. Then there exist pairewise disjoint subsets $A_0, A_1, \cdots\cdots A_{N-1}$ of $B^n(r)$ such that $h(A_j) \supset \cup_{i=0}^{N-1} A_i, j = 0, 1, \cdots, N-1$.*

Proof. The proof of first conlusion is simple by Definition 2. By Proposition 4, we have $h(B^n(r_i) - B^n(r_{i+1})) \supset B^n(r) - B^n(r_{N+1}), i = 0, 1, \cdots, N-1$. Clearly, there exist a subset A_i of $B^n(r_i) - B^n(r_{i+1})$ such that $h(A_i) = B^n(r) - B^n(r_{N+1}), i = 0, 1, \cdots, N-1$. It is easy to see that $A_0, \cdots, A_{N-1}$ satisfy our requirement.

Defintion 3. $x \in X$ is called l-dimensional homogeneous, if there is a neighborhood $V(x)$ of x, such that $V(x)$ is homeomorphic to $B^l, l > 0$.

It is easy to prove that the set of all homogeneous points of X is dense everywhere in X.

Definition 4. Let $f \in C^0(X)$ and $N > 0$. Say that f is an N-type quasi-horseshoe map, if

(1) There is an l-dimensional homogeneous point x of X. Let $V(x)$ be a neighborhood of x and $\psi : V(x) \to B^l$ be homomorhpic. Denote by $V_r(x) = \psi^{-1}(B^l(r)), r \in (0, 1]$.

(2) there exist $k > 0$ and $r \in (0, 1]$ such taht $f^k(V_r(x)) \subset V(x)$,

(3) $\psi f^k \psi^{-1} : B^l(r) \to B^l$ is an N^k-type quasi-horseshoe map.

Denote by $C^0_N(X)$ the set of all N-type quasi-horseshoe map of $C^0(X), N > 0$.

Proposition 6. *$C^0_N(X)$ is open and dense everywhere in $C^0(X)$ and* ent $(f) \ge \log N$ *for all $N > 0$, if $f \in C^0_N(X)$.*

Proof. It follows from Proposition 5 that $C^0_N(X)$ is open in $C^0(X)$. Let $f \in C^0(X)$. Without loss of generality, we may assume that $\{p, f(p), \cdots, f^{n-1}(p)\}$ is an n-periodic orbit and $f^i(p)$ is l_i-dimensional homogeneous, $i = 0, 1, \cdots, n-1$, and $l_0 = \min\{l_i, 0 \le 1 \le n-1\}$. Using the absolute homotopy extension property (see [7]), we may form a map $g : X \to X$ such that $g \in N(f, \varepsilon)$ for given $\varepsilon > 0$ and $g \in C^0_N(X)$. This proves that $C^0_N(X)$ is dense everywhere in $C^0(X)$.

Let $\psi f^k \psi^{-1} : B^1(r) \to B^1, r \in (0,1]$, be an N^k-type quasi-horseshoe map. By Proposition 5, ther exist N^k pairewise disjoint subsets $A_1, \cdots, A_{N^k}$ of $B^l(r)$ such that $\psi f^k \psi^{-1}(A_j) \supset \cup_{i=1}^{N^k} A_i, j = 1, \cdots, N^k$. It is easy to see that $\psi^{-1}(A_1), \cdots, \psi^{-1}(A_{N^k})$ are pairewise disjoint and compact and $f^k\psi^{-1}(A_j) \supset \bigcup_{i=1}^{N_k} \psi^{-1}(A_i), j = 1, \cdots, N^k$. By [8] , there is a compact subset $\wedge$ of $\overline{V}_r\overline{(x)}$ such that $\sigma f^{\ k} \mid \wedge = h\sigma$, where $\sigma : \sum_{N^k} \to \sum_{N^k}$ is the one sided shift on the one sided symbolic space with N^k symbols and $h : \wedge \to \sum_{N^k}$ is surjective and continous. Hence, ent $(f^k) \geq \log N^k$ and ent $(f) \geq \log N$.

Proposition 7. *Let $f \in C^0(X)$ and $\varepsilon > 0$. Then there exist $g \in N(f, \varepsilon)$ and a single arc $\overline{ab} \subset U(f, \varepsilon)$ such that $\overline{ab}$ is invariant under $g^n, n > 0$, and ent $(g^n \mid \overline{ab}) > 0$.*

Proof is similar to [2, Th.2].

§ III. Proof of Theorem

Proof of (1) is similar to [2, Th. 1].

(2) and (3) are given by Proposition 6.

Let $f \in C^0(X)$ and $\varepsilon > 0$. Suppose that $g \in N(f, \varepsilon)$ is the same as in Proposition 7. g^n is a self-map of the interval. By [9], g^n is chaotic and so is g. (4) is completed.

It is easy to see that $\{f \in C^0(X) \mid \text{ent}(f) = \infty\} \supset \{f \in C^0(X) \mid$ ent is contnuous at $f\} \supset \cap_{N=1}^{\infty} C_N^0(X)$. By Proposition 6, the later is a Baire residual set in $C^0(X)$. (5) and (6) are proved.

It is easy to prove that $C^0(X)$ is locally arc-connected and hence each arc connected component of $C^0(X)$ is open, if X is a manifold or a polyhedron. Because of that the spectral radius is constant on each arc connected component of $C^0(X)$,(7) follows Proposition 6.

References

[1] R.L. Adler, A.C. Konheim and M.H. McAndrew, *Topological entropy* , Trans. Amer. Math. Soc., 114 (1965), 309–319.

[2] Z.L. Zhou, *Some properties on topological entropy of self-mappings of polyhedron*, Scientia Sinica (series A), 27:2 (1984), 151–156.

[3] T.Y. Li and J.A. Yorke, *Period three implies chaos*, Amer. Math. Monthly, 82(1975), 985–992.

[4] Z.L. Zhou, *Chaos and total chaos*, Kexue Tongbao, 33: 18 (1988), 1494–1497.

[5] J. Palis, C.Pugh, M.Shub and D.Sullivan, *Genericity theorem in topological dynamics*, Springer Lecture Note, 468 (1975) , 241–250.

[6] K. Yano, *A remark on topological entropy of homeomorphisms*, Inventiones Math., 59 (1980), 215–220.

[7] C.R.F. Maunder, *Algebraic topology*, Cambridge University Press, 1980.

[8] Z.S. Zhang, *On the shift-invariant sets of endomorphisms*, Acta Mathematica Sinica (in Chinese), 27: 4(1984), 564–576.

[9] Z.L. Zhou, *Chaos and toplogical entropy*, Acta Mathematica Sinica (in Chinese), 31(1) (1988), 83–87.

Proceedings of the Asian Mathematical
Conference, 1990

The Structure of a Class of Archimedean Semigroups and Their Congruence Lattice

ZHU PINGYU
Department of Mathematics
Qinghai Normal University
Xining 810008, China

§1. The Structure of Archimedean Semigroups Containing a Primitive Idempotents

A semigroup S is called a (finite order) nilextension of K if K is an ideal of S and for any $x \in S$ there exists an n such that $x^n \in K(S^n \subseteq K)$. S is called an archimedean semigroup if for any $a, b \in S$ there exists an n such that $a^n \in SbS$. We know that a semigroup S is an archimedean semigroup containing a primitive idempotent if and only if S is a nilextension of a completely simple semigroup K[1] (called the kernel of S).

Definition 1. A semigroup T is called a (finite order) π-left (right) zero band if T is a (finite order) nilextension of a left (right) zero band I. The partial semigroup $Q = T - I$ is called the base of T.

It is clear that any π-left (right) zero band or any group is an archimedean semigroup containing a primitive idempotent. These three classes of semigroups are the materials used to construct any archimedean semigroup containing a primitive idempotent as follows.

Let T, T' be a pair of π-left and π-right zero bands having the same (isomorphic) base $Q = T - I = T' - \Lambda$, G be a group, P be a map from $\Lambda \times I$ to G, $(\lambda, i) \to p_{\lambda,i}$ and F be a partial homomorphism from $Q \times I$ to G, $(a, i) \to f_{a,i}$. Suppose that the following condition holds:

$$p_{\lambda,ai} f_{a,i} p_{\lambda a,i}^{-1} = p_{\lambda,aj} f_{a,j} p_{\lambda a,j}^{-1} \tag{1}$$

for any $a \in Q, i, j \in I$ and $\lambda \in \Lambda$. Let $K = I \times G \times \Lambda$ and $S = Q \cup K$. We define an operation $\circ$ in S by the following rule:

(1) Let $a, b \in Q$. If $ab \in Q$, $a \circ b = ab$. If $ab \notin Q, a \circ b = (i, g_{a,b}, \lambda)$, where $i(\lambda)$ is the product of a and b in $T(T')$ and $g_{a,b} = f_{a,bi} f_{b,i} p_{\lambda,i}^{-1}$.

(2) If $a \in Q$ and $x = (i, h, \lambda)$, then $a \circ x = (ai, f_{a,i}h, \lambda)$ and $x \circ a = (i, hp_{\lambda,ai} f_{a,i} p_{\lambda a,i}^{-1}, \lambda a)$.

(3) $(i, h, \lambda) \circ (j, k, \mu) = (i, hp_{\lambda,j}k, \mu)$.

Theorem 1.1. *$(S, \circ)$ is an archimedean semigroup containing a primitive idempotent. We denote it by $[T, G, T'; P, F]$ (called the product of a pair of π-left and π-right zero bands with a sandwich group).*

Conversely, let S be an archimedean semigroup containing a primitive idempotent and K be the kernel of S. Since K is a completely simple semigroup, then $K = M(I, G, \Lambda; P)$. Let $Q = S - K, T = Q \cup T$ and $T' = Q \cup \Lambda$. If $a \in Q$, then there exist a left transformation of $I(i \to ai)$ and a right transformation of $\Lambda(\lambda \to \lambda a)$ such that

$$a(i, h, \lambda) = (ai, h', \lambda), \quad (i, h, \lambda)a = (i, h'', \lambda a) \tag{2}$$

for any $(i, h, \lambda) \in K$. We define an operation $\circ$ in T as follows: For any $a, b \in Q$, if $ab \in Q$ (in S), $a \circ b = ab$. If $a \notin Q$, then $ab \in K$, and so $ab = (i_0, h, \lambda_0)$; set $a \circ b = i_0$. If $a \in Q, i, j \in I$, set $a \circ i = ai, i \circ a = i = i \circ j$. Similarly we define an operation $\circ$ in T' by the rule that $a \circ b = \lambda_0, \lambda \circ a = \lambda a$ and $a \circ \lambda = \lambda = \mu \circ \lambda$ (for any $a, b \in Q$ and $\lambda, \mu \in \Lambda$).

Lemma 1.2. *(1) $(T, \circ)$ and $(T', \circ)$ are a pair of π-left and π-right zero bands having the same base. (2) The Green relations $\mathcal{R}$ and $\mathcal{L}$ are congruences $T \simeq S/\mathcal{R}$ and $T' \simeq S/\mathcal{L}$.*

Lemma 1.3. *(1) The map $h \to h'$ defined in (2) above, denoted by $f_{a,i}$, is a left translation of G and map be regarded as an element of G.*

(2) The map $(a, i) \to f_{a,i}$ is a partial homomorphism $F : Q \times I \to G$.

(3) F satisfies the condition (1),in which P is the sandwich matrix of a completely simple semigroup K; thus we can construct a semigroup $[T, G, T'; P, F]$.

(4) $S \simeq [T, G, T'; P, F]$.

We summarize the results of this section in a theorem.

Theorem 1.4. *A semigroup S is an archimedean semigroup containing a primitive idempotent if and only if S is isomorphic to a product of a pair of π-left and π-right zero bands with a sandwich group.*

§2. The Congruence Lattice

Let L be a lattice and $a, b \in L$. We say that b covers a if $a < b$ and the interval $[a, b](= \{x \in L \mid a \leq x \leq b\}) = \{a, b\}$. We say that L is a semimodular lattice if, for any $a, b \in L$, a and b covering $a \wedge b$ implies that $a \vee b$ covers a and b. It is clear that any interval $[a, b]$ of a semimodular lattice L is a semimodular sublattice of L.

The congruence lattice of a semigroup S is denoted by $C(S)$.

Let $S = [T, G, T'; P, F]$ be an archimedean semigroup containing a primitive idempotent. In virtue of Lemma 1.2, $T \simeq S/\mathcal{R}$ is a quotient semigroup of S. Then $C(T)$ is isomorphic to the interval $[\mathcal{R}, S \times S]$ of $C(S)$. Thus we have

Theorem 2.1. *Let $S = [T, G, T'; P, F]$. If $C(S)$ is a semimodular lattice, then $C(T)$ and $C(T')$ are semimodular.*

Definition 2. Let $S = [T, G, T'; P, F]$ and K be the kernel of S. We say that S is a finite order archimedean semigroup if S is a finite order nilextension of K.

Remark. (1) If S is a finite archimedean semigroup, it is clear that S has a primitive idempotent. Since S is a finite order nilextension of K[2], then S is a finite order archimedean semigroup.

(2) If $S = [T, G, T'; P, F]$ is of finite order, then $T(T')$ is a finite order π-left (right) zero band

The following theorem is the main result of this section.

Theorem 2.2. *Let $S = [T, G, T'; P, F]$ be a finite order archimedean semigroup. If $C(T)$ and $C(T')$ are semimodular lattices, then $C(S)$, is semimodular.*

Next we consider the congruence lattice of a finite order π-left zero band.

Lemma 2.3. *Let T be a finite order π-left zero band and I be the kernel of T. If $|I| \leq 5$, then $C(T)$ is semimodular.*

Let $S = [T, G, T'; P, F]$ and $I(\Lambda)$ be the kernel of $T(T')$. We know that there exists a 1-1 map from $I(\Lambda)$ onto the set of regular $\mathcal{R}(\mathcal{L})$-classes of S. Thus we have

Theorem 2.4. *Let S be a finite order archimedean semigroup. If the number of regular $\mathcal{R}$-classes and the regular $\mathcal{L}$-classes of S do not exceed 5, then $C(S)$ is a semimodular lattice.*

References

[1] Bogdanovic, S., *Semigroups with a system of subsemigroups*, Novi Sad, 1985.

[2] Zhu Ping-Yu, *The structure of finite nilpotent semigroups*, J. of Math., (PRC) 9(3)(1989), 269–272.

[3] Zhu Ping-Yu, *On the congruence lattice of the nilextension of a semigroup*, Acta Math. Sinica, 33(4)(1990), 545–549.

[4] Howie, J. M., *An introduction to semigroup theory*, Academic Press, London, 1979.

Proceedings of the Asian Mathematical
Conference, 1990

On Stability of Noncompact Sets in a Class of Semidynamical Systems

ZHU SIMING
Department of Mathematics
Zhongshan University
Guangzhou 510275, China

§1. Preliminaries

Throughout this paper we assume X is a Hausdorff space. R, R^+, R^- are the set of real numbers, non-negative real numbers and non-positive real numbers, respectively. For a subset M of X, we denote by $\partial M, \overline{M}, M^\circ$ and $X\backslash M$ its boundary, closure, interior and supplementary respectively. I represents one of the following intervals: $(-\infty, 0], (-\alpha, 0]$ or $[-\alpha, 0]$ where $a \geq 0$.

Definition 1. Let $\pi : X \times R \to X$ be a continuous mapping. The couple (X, π) is called a semidynamical system (SDS) if π satisfies the following conditions:

(a) $\pi(x, 0) = x$ for each $x \in X$;

(b) $\pi(\pi(x, s), t) = \pi(x, s+t)$ for each $x \in X$, and $s, t \in R^+$.

To simplify the notations, we will write $\pi(x, t)$ as xt if no confusion is caused.

Similarly, we will denote by MK the set $\{xt \mid x \in M, t \in K\}$ where $M \subset X$ and $K \subset R$.

For SDSs, all the concepts on the positive aspects, such as positive trajectorypositive limit set, positive prolongational set, positive prolongational limit set, etc., coincide with those of dynamical systems (see [5]).

Let $X^* = X \cup \{\omega\}$ be the one point compactification space of X, where ω is an abstract element which is not in X. Let us consider the mapping $\pi^* : X^* \times R^+ \to X^*$, defined by

$$\pi^*(x, t) = \begin{cases} (x, t), & \text{if } x \in X \\ \omega, & \text{if } x = \omega \end{cases}$$

where $t \in R^+$.

It is known that if (X, π) is a dynamical system on a locally compact space X, then (X^*, π^*) is also a dynamical system defined on the Hausdorff space X^* (see [3], P.29). However, (X^*, π^*) is , in general, not a semidynamical system if (X^*, π^*) is only a SDS. We have the following theorem.

Theorem 1. *Let X be a locally compact space, then a SDS (X, π) can be extended to a SDS (X^*, π^*) iff π is continuous at infinity.*

Proof. Necessity, Let $x_i \to \infty, t_i \to t \geq 0$ where $t_i \geq 0$. x_i has no convergent subnets. We claim that it is not in every compact subset of X ultimately, which means that $x_i \to \omega$. By the assumption that (X^*, π^*) is SDS, we have $x_i t_i = \pi(x_i, t_i) = \pi^*(x_i, t_i) \to \omega$, so $x_i t_i$

is also not in every compact subset of X ultimately. We say that $x_j t_j \to \infty$. If this is not the case, then there exists a subnet $x_j t_j \to x \in X$. By the local compactness of X,$x_j t_j$ must be in a compact neighbourhood (neigh.) $\overline{U} \subset X$ of x ultimately, which contradicts the fact that $x_i t_i \to \omega$. Therefore π is continuous at infinity.

Sufficiency. To prove (X^*, π^*) is a SDS, it is sufficient to verify that π is continuous at (ω, t). Let $x_i \to \omega, t_i \to t \geq 0(t_i \geq 0)$. Then $x_i \to \omega$ implies $x_i \to \infty$ since X is locally compact. It follows from the assumption that $x_i t_i \to \infty$. Now holds $x_i t_i \to \infty$, which , together with the fact that $\pi^*(x_i, t_i) = \pi(x_i, t_i) = x_i t_i$, completes the proof of our theorem.

Note that for any dynamical system (X, π) on a locally compact space X, π is naturally continuous at infinity. By the above theorem one can see that (X^*, π^*) is a dynamical system.

Lemma 1. *Let (X, π) be a SDS, if π is a continuous at infinity, then the following-properties hold:*

(a) $F(A, S)$ is compact whenever $A \subset X, S \subset R^-$ are compact in the respective spaces,

(b) $F(x, T)$ is continuous for $x \in X$ and $t \in R^-$. Or more precisely, for any givenneigh. U of $F(X, T)$, there exist open neighs $V_\times \subset X$ of x and $V_t \subset R^-$ of t such that $F(V_\times, V_t) \subset U$.

Theorem 2. *Let (X, π) be a SDS. If there are no start points in X and π is continuous at infinity, then (X, π) can be globally backward extended. Furthermore if X is locally compact then the converse holds.*

Proof. By the assumption and Lemma above, $F(x, [-t, 0])$ is nonempty and compact and contains no start points, where $x \in X, t \in R^+$. The first part of this theorem follows from Property 2.9 of [5], (see P.32). The second part of this theorem is guaranteed by Theorem 2.7 and Theorem 2,6 of [8]. (see P. 178)

It is easy to see that $\pi^t : X \to X$ is surjective for any $t \in R^+$ is equivalent to that there exists a $t > 0$ such that π^t is surjective . If (X, π) has global backward extention then π^t is surjective for every $t \in R^+$. However, even if X is locally compact, the converse is not true, as is shown in the next example.

Example. Let $X = \{(x, y) \mid x \leq e^{-y}, (x, y) \in R^2\}$. Consider the SDS (X, π), defined by $\pi(p, t) = (x + t, -\ln(t + e^{-y}))(t \geq 0)$ if $p = p(x, y) \in M = \{(x, y) \in R^2 \mid x = e\}$, and

$$\pi(p, t) = \begin{cases} (x + t, y), & \text{as } 0 \leq t \leq e^{-y} - x \\ (x + t, -\ln(x + t)), & \text{as } t \geq e^{-y} - x \end{cases} \quad \text{if } p \in X \backslash M.$$

One can easily verify that for $t \geq 0$, π^t is surjective. Let $p \in M$, then $\sigma(t) = (x+t, -\ln(t+e^{-y}))$ $(-x < t \leq 0)$ is a maximal negative solution through p, which is not principal. Thus (X, π) cannot be backward extended. This example shows a mistake in Theorem 2.6 of [8].

Standing hypothesis. Throughout the rest of the text we will confine ourselves to the SDSs which satisfy the conditions in the first part of theorem 2. Moreprecisely, we assume that the SDS (X, π) considered contains no start points, and π is continuous at infinity.

We denote by xt the net $\{xt \mid t \in R^+\}$ X on R^+. we say $xt \to M$ if xt is in every neigh of M ultimately and $xt \to_f M$ if it is in every neigh. of M frequently.

Lemma 2. ([5], p.24) *Let M be a compact subset of X and let $x_i \to M$, by which we mean that x_i is in every neigh. of M ultimately, then any subnet of x_i has a contact point in M. Thus if $x_i \to M$, then there exist a subnet x_i of x_i and $x \in M$ such that $x_j \to x$.*

We denote by $(A_\omega^{-*}(M)), (A_\omega^-(M)), (A^-(M)), (A_s^-(M))$ the region of (negative weak) (negative weak) (negative) (negative strong) attraction of M, and $A_\omega^+(M), A^+(M), A_s^+(M)$ the region of positive weak attraction, positive attraction, positive strong attaction respectively.

Proposition 1 .

(a) $A_s^-(M) \subset A^-(M) \subset A_\omega^-(M) \subset A_\omega^{-*}(M)$;

(b) $A_\omega^{-*}(M)$ *is positively invariant, while* $A_\omega^-(M)$, $A^-(M)$ *and* $A_s^-(M)$ *are all negatively invariant.*

Let $x \in X$ and $L^-(x), D^-(x), J^-(x)$ represent the negative limit set, negative prolongation set and negative prolongational limit set, respectively. It is easy to verify that $D^-(x) = \{y \in X \mid x \in D^+(y)\}, J^-(x) = \{y \in X \mid x \in J^+(y)\}$ and $L^-(x) \subset J^-(x)$. Also we note that $D_-(x) = F(x) \cup J^-(x)$ and $\overline{F}(\overline{X}) = F(x) \cup L^-(x)$.

Let $A = \{x \in X \mid L^-(x) \cap M \neq \varphi\}, B = \{x \in X \mid \varphi \neq L^-(x) \subset M\}, C = \{x \in X \mid \varphi \neq J^-(x) \subset M\}$, then we have

Lemma 3. $A = A_\omega^{-*}(M), A^-(M) \subset B$, *and* $A_s^-(M) \subset C$ *if* M *is a compact subset of* X.

Proposition 2. *Let* $x \in X$. *Then* $D^-(x)$ *is weakly negatively invariant,* $J(x)$ *and* $L(x)$ *are both positively invariant and weakly negatively invariant.*

Proof. We will deal with the case of $J^-(x)$ only. Let $y \in J^-(X)$. According to the definition there exists a net $y_i \to y$, where $y_i t_i = x_i$ for some nets $x_i \to x$ and $t_i \to +\infty$. Given $t \geq 0$. Let $z_i = y_i t$, then $z_i \to yt, z^i \tau_i = x_i \to x$ where $\tau_i = t_i - t \to +\infty$, thus $yt \in J^-(x)$. The positive invariance of $J^-(x)$ is proved. We claim that $J^-(x)$ is weakly negatively invariant. In fact, if $\sigma_i(t)(t \leq 0)$ are negative solutions passing through $y_i \in X$, then by the standing hypothesis, $\sigma_i(-1)$ has a convergent subnet $\sigma_j(-1)$. Let $\sigma_j(-1) \to y_{-1} \in X$, then $y_{-1} \in J^-(x)$ since $\sigma_j(-1)\pi(1+t) = y_j t_j = x_j \to x$. Now $\sigma_j(-1+t)(t \leq 0)$ are the negative solutions passing throguh $\sigma_j(-1)$ and $\sigma_j(-1) \to y_{-1}$. By the same procedure as above we have a subset $\sigma_k(-2)$ of $\sigma_j(-2)$ and $y_{-2} \in X$ such that $\sigma_k(-2) \to y_{-2}$. Since $\sigma_k(-2)\pi(2+t_k) = y_k t_k = x_k \to x, \quad y_{-2} \in J^-(x)$. It is clear that $y_{-2}\pi 1 = y_{-1}$. Using this method we obtain a sequence y_{-n} satisfying $y_{-n} \in J^-(x), y_{-n+1}\pi 1 = y_{-n}$ for $n \geq 1$. Let $O^-(y) = \cup_{n=1}^\infty y_{-n}[n, 0]$. One can easily check that $O^-(y)$ is a negative trajectory of y and $O^-(y) \subset J^-(x)$, whcih completes the proof.

Note that for general SDS, $L^+(x)$ and $J^+(x)$ are not necessarily weakly negatively invariant. Their negative invariance can be guaranteed by the compactness in some sense of the phase space (see [5] and [6]).

§ 2. Negative Stability

Theorem 3. *A subset* $M \subset X$ *is negatively stable iff it is negatively invariant and* $M \subset A_s^-(M)$.

Remark. Theorem 3 is an analogue to theorem 5.4 in [7]. Note that we do not have the assumption that M is compact. Similarly, we can prove that M is positively stable iff M is positively invariant and $M \subset A_s^+(M)$.

Lemma 4. *Let* $x \in X$ *and* $M \subset X$, *then*

(a) $x \in A_\omega^+(M)$ *iff eigher* xt *is frequently in* M *as* $t \to +\infty$, *or* $L^+(x) \cap M \neq \varphi$.

(b) $x \in A_\omega^{-*}(m)$ *iff either* $F(x,-t)$ *has frequently nonempty intersection with* M *as* $t \to +\infty$ *or* $L^-(M) \cap M \neq \varphi$.

Theorem 4. *The subset* M *of* X *is both positively stable and negatively stable iff* M *is positively invariant and* $A_s^-(M) = M = A_s^+(M)$.

Remark. If (X,π) is a dynamical system, then $A_s^+(M)$ is naturally positively invariant. At this time it is unnecessary to assume that M is positively invariant in the sufficiency part of theorem 4.

Elady [7] has proved, under the assumption that X is a first countable locally compact space, $D^-(x)$ is connected if it is compact. We will give another sufficient condition for $D^-(x)$ to be connected.

Lemma 5. *Let* $x \in X$, *then* $D^-(x) = \cap\{\overline{F}(\overline{U}) \mid U$ *is an open neigh. of* $x\}$. *In particular,* $D^-(x)$ *is closed.*

Theorem 5. *Let* M *be a closed subset of* X *with compact boundary. Then for each* $x \in M, D^-(x) \subset M$ *if* M *is negatively stable. Moreover, if* M *is compact (and negatively stable), then* $D^-(x)$ *is connected.*

Lemma 6. *A subset* M *of* X *is negatively stable iff for each net* $y_i \in F(x_i,t_i)(t_i \leq 0), y_i \to M$ *whenever* $x_i \to M$.

Theorem 6. *Let* M *be a subset of* X. *Then* M *is negatively stable if each of its components is such. Moreover, if* M *is compact then the converse holds. The theorem above is an analogue to that in [7]. However we do not assume* X *to be locally compact.*

References

[1] J.L. Kelly, *General Topology*, Van Nostrand, N.J., 1955.

[2] K.M.Das and S.S Lakshmi, *Negative and Weak Negative Attractors in Semidynamical Systems* , Funkcial. Ekvac., 21(1978), 53–62.

[3] N.P. Bhatia and G.P. Szego, *Stability Theory of Dynamical Systems*, Springer-Verlag, 1970.

[4] N.P. Bhatia, *Attraction and Nonsaddle Sets in Dynamical Systems*, J. Diff. Eqs., 8 (1970), 229–249.

[5] N. P.Bhatia and O.Hajek, *Local Semidynamical Systems*, Lecture Notes in Mathematics 90, Springer-Verlag, New York/Berlin, 1969.

[6] P.N.Bajaj, *Stability of Noncompact Sets, "Nonlinear Systems and Applications"*, Proceedings of an International Conference on Nonlinear Systems and Applications, Acad. Press, New York/London, 1977, 381–388.

[7] S.Elaydi, *Semidynamical Systems with Nonunique Global Backwark Extentions II; the Negative Aspects*, Funkcial. Ekvac., 27 (1984), 85–100.

[8] S.Elaydi, *Semidynamical Systems with Nonunique Global Backward Extentions,* Funkcial. Ekvac., 26(1983), 173–187.

www.ingramcontent.com/pod-product-compliance
Lightning Source LLC
LaVergne TN
LVHW010538300926
849162LV00002B/174